The Mycota

Edited by
K. Esser

The Mycota

I *Growth, Differentiation and Sexuality*
1st edition ed. by J.G.H. Wessels and F. Meinhardt
2nd edition ed. by U. Kües and R. Fischer

II *Genetics and Biotechnology*
1st and 2nd edition ed. by U. Kück

III *Biochemistry and Molecular Biology*
1st and 2nd edition ed. by R. Brambl and G. Marzluf

IV *Environmental and Microbial Relationships*
1st edition ed. by D. Wicklow and B. Söderström
2nd edition ed. by C.P. Kubicek and I.S. Druzhinina

V *Plant Relationships*
1st edition ed. by G. Carroll and P. Tudzynski
2nd edition ed. by H.B. Deising

VI *Human and Animal Relationships*
1st edition ed. by D.H. Howard and J.D. Miller
2nd edition ed. by A.A. Brakhage and P.F. Zipfel

VII *Systematics and Evolution*
1st edition ed. by D.J. McLaughlin, E.G. McLaughlin, and P.A. Lemke†
2nd edition ed. by D. McLaughlin and J.W. Spatafora

VIII *Biology of the Fungal Cell*
1st and 2nd edition ed. by R.J. Howard and N.A.R. Gow

IX *Fungal Associations*
1st and 2nd edition ed. by B. Hock

X *Industrial Applications*
1st edition ed. by H.D. Osiewacz
2nd edition ed. by M. Hofrichter

XI *Agricultural Applications*
1st and 2nd edition ed. by F. Kempken

XII *Human Fungal Pathogens*
1st edition ed. by J.E. Domer and G.S. Kobayashi
2nd edition ed. by O. Kurzai

XIII *Fungal Genomics*
1st edition ed. by A.J.P. Brown
2nd edition ed. by M. Nowrousian

XIV *Evolution of Fungi and Fungal-Like Organisms*
Ed. by S. Pöggeler and J. Wöstemeyer

XV *Physiology and Genetics: Selected Basic and Applied Aspects*
Ed. by T. Anke and D. Weber

The Mycota

A Comprehensive Treatise
on Fungi as Experimental Systems
for Basic and Applied Research

Edited by K. Esser

IX *Fungal Associations* 2nd Edition

Volume Editor:
B. Hock

Springer

Series Editor

Professor Dr. Dr. h.c. mult. Karl Esser
Allgemeine Botanik
Ruhr-Universität
44780 Bochum, Germany

Tel.: +49 (234)32-22211
Fax.: +49 (234)32-14211
e-mail: Karl.Esser@rub.de

Volume Editor

Prof. Dr. Bertold Hock
Technische Universität München
Lehrstuhl für Bioanalytik
Alte Akademie 6
D-85354 Freising
Germany

Tel.: 08161 - 713396
Fax: 08161 - 232030
e-mail: hock@wzw.tum.de

ISBN 978-3-662-49589-6 978-3-642-30826-0 (eBook)
DOI 10.1007/978-3-642-30826-0
Springer Heidelberg New York Dordrecht London

Printed on acid-free paper

Springer is part of Springer Science+Business Media (www.springer.com)

Karl Esser
(born 1924) is retired Professor of General Botany and Director of the Botanical Garden at the Ruhr-Universität Bochum (Germany). His scientific work focused on basic research in classical and molecular genetics in relation to practical application. His studies were carried out mostly on fungi. Together with his collaborators he was the first to detect plasmids in higher fungi. This has led to the integration of fungal genetics in biotechnology. His scientific work was distinguished by many national and international honors, especially three honorary doctoral degrees.

Bertold Hock
(born 1939) is an emeritus Professor of Cell Biology at the Technische Universität München (Germany). He and his department became internationally recognized for their groundbreaking work in producing monoclonal and recombinant antibodies for ultrasensitive tests to detect pesticides and pollutants as well as specific antigens in mycorrhizal fungi. During his earlier career he worked together with his colleagues and the Institut für den Wissenschaftlichen Film (Göttingen) to produce several films and video discs on fungal development. He has served as a Visiting Professor at Washington State University (Pullman, Washington, USA), Program in Biophysics and Biochemistry, and at Jawaharlal Nehru University (New Delhi, India), Dept. of Microbiology.

Series Preface

Mycology, the study of fungi, originated as a sub discipline of botany and was a descriptive discipline, largely neglected as an experimental science until the early years of this century. A seminal paper by Blakeslee in 1904 provided evidence for self incompatibility, termed "heterothallism", and stimulated interest in studies related to the control of sexual reproduction in fungi by mating-type specificities. Soon to follow was the demonstration that sexually reproducing fungi exhibit Mendelian inheritance and that it was possible to conduct formal genetic analysis with fungi. The names Burgeff, Kniep and Lindegren are all associated with this early period of fungal genetics research.

These studies and the discovery of penicillin by Fleming, who shared a Nobel Prize in 1945, provided further impetus for experimental research with fungi. Thus began a period of interest in mutation induction and analysis of mutants for biochemical traits. Such fundamental research, conducted largely with Neurospora crassa, led to the one gene: one enzyme hypothesis and to a second Nobel Prize for fungal research awarded to Beadle and Tatum in 1958. Fundamental research in biochemical genetics was extended to other fungi, especially to Saccharomyces cerevisiae, and by the mid-1960s fungal systems were much favored for studies in eukaryotic molecular biology and were soon able to compete with bacterial systems in the molecular arena.

The experimental achievements in research on the genetics and molecular biology of fungi have benefited more generally studies in the related fields of fungal biochemistry, plant pathology, medical mycology, and systematics. Today, there is much interest in the genetic manipulation of fungi for applied research. This current interest in biotechnical genetics has been augmented by the development of DNA-mediated transformation systems in fungi and by an understanding of gene expression and regulation at the molecular level. Applied research initiatives involving fungi extend broadly to areas of interest not only to industry but to agricultural and environmental sciences as well.

It is this burgeoning interest in fungi as experimental systems for applied as well as basic research that has prompted publication of this series of books under the title The Mycota. This title knowingly relegates fungi into a separate realm, distinct from that of either plants, animals, or protozoa. For consistency throughout this Series of Volumes the names adopted for major groups of fungi (representative genera in parentheses) areas follows:

Pseudomycota

Division: Oomycota (Achlya, Phytophthora, Pythium)
Division: Hyphochytriomycota

Eumycota

Division:	Chytridiomycota (Allomyces)
Division:	Zygomycota (Mucor, Phycomyces, Blakeslea)
Division:	Dikaryomycota
Subdivision:	Ascomycotina
Class:	Saccharomycetes (Saccharomyces, Schizosaccharomyces)
Class:	Ascomycetes (Neurospora, Podospora, Aspergillus)
Subdivision:	Basidiomycotina
Class:	Heterobasidiomycetes (Ustilago, Tremella)
Class:	Homobasidiomycetes (Schizophyllum, Coprinus)

We have made the decision to exclude from The Mycota the slime molds which, although they have traditional and strong ties to mycology, truly represent nonfungal forms insofar as they ingest nutrients by phagocytosis, lack a cell wall during the assimilative phase, and clearly show affinities with certain protozoan taxa.

The Series throughout will address three basic questions: what are the fungi, what dothey do, and what is their relevance to human affairs? Such a focused and comprehensive treatment of the fungi is long overdue in the opinion of the editors.

A volume devoted to systematics would ordinarily have been the first to appear in this Series. However, the scope of such a volume, coupled with the need to give serious and sustained consideration to any reclassification of major fungal groups, has delayed early publication. We wish, however, to provide a preamble on the nature off ungi, to acquaint readers who are unfamiliar with fungi with certain characteristics that are representative of these organisms and which make them attractive subjects for experimentation.

The fungi represent a heterogeneous assemblage of eukaryotic microorganisms. Fungal metabolism is characteristically heterotrophic or assimilative for organic carbon and some nonelemental source of nitrogen. Fungal cells characteristically imbibe or absorb, rather thaningest, nutrients and they have rigid cellwalls. The vast majority of fungi are haploid organisms reproducing either sexually or asexually through spores. The spore forms and details on their method of production have been used to delineate most fungal taxa. Although there is amultitude of spore forms, fungal spores are basically only of two types: (i) asexual spores are formed following mitosis (mitospores) and culminate vegetative growth, and (ii) sexual spores are formed following meiosis (meiospores) and are borne in or upon specialized generative structures, the latter frequently clustered in a fruit body. The vegetative forms of fungi are either unicellular, yeasts are an example, or hyphal; the latter may be branched to form an extensive mycelium.

Regardless of these details, it is the accessibility of spores, especially the direct recovery of meiospores coupled with extended vegetative haploidy, that have made fungi especially attractive as objects for experimental research.

The ability of fungi, especially the saprobic fungi, to absorb and grow on rather simple and defined substrates and to convert these substances, not only into essential metabolites but into important secondary metabolites, is also noteworthy.

The metabolic capacities of fungi have attracted much interest in natural products chemistry and in the production of antibiotics and other bioactive compounds. Fungi, especially yeasts, are important in fermentation processes. Other fungi are important in the production of enzymes, citric acid and other organic compounds as well as in the fermentation of foods.

Fungi have invaded every conceivable ecological niche. Saprobic forms abound,especially in the decay of organic debris. Pathogenic forms exist with both plant and animal hosts. Fungi even grow on other fungi. They are found in aquatic as well as soil environments, and their spores may pollute the air. Some are edible; others are poisonous. Many are variously associated with plants as copartners in the formation of lichens and mycorrhizae, as symbiotic endophytes or as overt pathogens. Association with animal systems varies; examples include the predaceous fungi that trap nematodes, the micro fungi that grow in the anaerobic environment of the rumen, the many insect associated fungi and the medically important pathogens afflicting humans. Yes, fungi are ubiquitous and important.There are many fungi, conservative estimates are in the order of 100,000 species,and there are many ways to study them, from descriptive accounts of organisms found in nature to laboratory experimentation at the cellular and molecular level. All such studies expand our knowledge of fungi and of fungal processes and improve our ability to utilize and to control fungi for the benefit of humankind.

We have invited leading research specialists in the field of mycology to contributeto this Series. We are especially indebted and grateful for the initiative and leadership shown by the Volume Editors in selecting topics and assembling the experts. We have all been a bit ambitious in producing these Volumes on a timely basis and there in lies the possibility of mistakes and oversights in this first edition. We encourage the readership to draw our attention to any error, omission or inconsistency in this Series in order that improvements can be made in any subsequent edition.

Finally, we wish to acknowledge the willingness of Springer-Verlag to host this project, which is envisioned to require more than 5 years of effort and the publication of at least nine Volumes.

Bochum, Germany
Auburn, AL, USA
April 1994

Karl Esser
Paul A. Lemke
Series Editors

Addendum to the Series Preface

During the Fourth International Mycological Congress in Regensburg (1989) while relaxing in a beer garden with Paul Lemke (USA), Dr. Czeschlik (Springer-Verlag) discussed with us the possibility to publish a series about Fungi. We both were at first somewhat reserved, but after a comprehensive discussion this idea looked promising. We decided to name this new series *The Mycota.*

Then Paul Lemke and I created a program involving seven volumes covering a wide area of Mycology. The first volume was presented in 1994 at the Fifth International Mycological Congress in Vancouver (Canada). The other volumes followed step by step. After the early death of Paul Lemke (1995) I proceeded alone as Series Editor. However for Vols. X-XII I received support by Joan Bennett.

Since evidently the series was well accepted by the scientific community and since the broad area of Fungi was not completely covered, it was decided to proceed with eight more volumes. In addition, second editions of nine volumes were published and four more are in preparation.

I would like to thank Springer-Verlag, represented by Hanna G. Hensler-Fritton and Andrea Schlitzberger for their support and cooperation.

Bochum, Germany
June 2012

Karl Esser

Volume Preface to the Second Edition

Fungal associations play a prominent role in terrestrial ecosystems due to their impact on ecosystem health. Scientists as well as laymen have been fascinated for a long time by the fact that fungi can enter mutualistic symbioses with organisms of totally different origin, such as plants, algae and bacteria, resulting in new forms differing from the original partners.

This book focuses on mycorrhizas and lichens, but is also concerned with fungal–bacterial associations. Major progress has been achieved in this field during the last decade. It particularly concerns the analysis of molecular interactions, leading to a better understanding of crucial events at the level of genome, proteome and metabolome during colonization and subsequent steps. The communication by chemical signals plays a decisive role. Progress is also seen in the appreciation of ecological impacts caused by the interaction of these symbioses with their environment.

More than one decade after the appearance of the first edition of Volume IX of "The Mycota", a new edition of "Fungal Associations" was overdue. Again, internationally renowned scientists could be won as authors; many of whom formed their discipline and also contributed already to the first edition. I am most grateful to all authors and co-authors for their cooperation and willingness to update their former chapter or contribute a new chapter.

The size of the book has now been increased from 13 to 18 chapters. It guarantees a well-balanced overview of a rapidly expanding field. It is hoped that this second edition will encourage young scientists to enter a fascinating field, disseminate new ideas and serve as an attractive reference.

I would like to thank Springer Verlag, particularly Dr. Andrea Schlitzberger, for continued interest and active help. I am indebted to the series editor, Prof. Dr. Karl Esser, who accompanied the progress of the book with encouragement and on cordial terms.

Freising-Weihenstephan, Germany — Bertold Hock

Volume Preface to the First Edition

The vital role of fungal associations for all ecosystems was only recognized in the second half of the twentieth century. The growing realization that the relevance of fungal associations goes beyond classical mycology has greatly accelerated research in this field. The availability of new tools, provided especially by molecular biology, has triggered new approaches to the study of fungi as hosts as well as symbionts.

Due to be enormous amount of new work in the field of fungal associatious, it has been impossible to include all topics of interest in this Volume. Rather, it has been decided to concentrate in more depth on subjects such as mycorrhizae, lichens, as well as some new developments coming up more recently, e.g., *Geosiphon* and *Piriformospora* associations. Therefore a compromise had to be made, resulting in the omission of other important aspects such as fungal endophytes or symbioses with insects. Nevertheless, it is hoped that this Volume will contribute to a better understanding of fungal associations.

It has been a pleasure to edit this book, primarily due to the stimulating discussions with the series editor Prof. Karl Esser. I am indebted to Springer-Verlag for all the help and active cooperation during the preparation of this Volume. I am grateful to Stefanie Rauchalles who has eliminated many errors in the manuscripts and helped to attain consistency in the presentation of the chapters.

I hope that this book will help to answer questions concerning the complex fungal associations and provide guidance for future research.

Freising-Weihenstephan, April 2000 BERTOLD HOCK

Volume I Preface to the First Edition

Contents

List of Contributors

S. Abbà
(e-mail: simona.abba@unito.it, Tel.: +39 011 6705987, Fax: +39 011 6705962)
Dipartimento di Biologia Vegetale dell, Università di Torino and Istituto per la Protezione delle Piante del CNR, v. le Mattioli 25, 10125 Torino, Italy

A. Adya
Division of Biotechnology and Forensic Sciences, School of Contemporary Sciences, University of Abertay, Bell Street, Dundee, DD1 1HG, UK

A. Agrawal
Amity Institute of Microbial Technology, Amity University Uttar Pradesh, Noida, UP, 201303, India

M. Arora
Amity Institute of Microbial Technology, Amity University Uttar Pradesh, Noida, UP, 201303, India

U.S. Bagde
Applied Microbiology Laboratory, Department of Life Sciences, University of Mumbai, Mumbai, Maharashtra, 400098, India

M. Bakshi
(e-mail: mbakshi@amity.edu)
Amity Institute of Microbial Technology, Amity University Uttar Pradesh, Noida, UP, 201303, India

J.M. Barea
(e-mail: josemiguel.barea@eez.csic.es, Tel.: +34 958 181600 ext. 247, Fax: +34 958 129600)
Departamento de Microbiología del Suelo y Sistemas Simbióticos, Estación Experimental del Zaidín, CSIC, Prof. Albareda 1, 18008 Granada, Spain

J.W. Bennett
(e-mail: profmycogirl@yahoo.com, Tel.: +1 848-932-6223, Fax: +1 732-932-9441)
Department of Plant Biology and Pathology, Rutgers University, New Brunswick, NJ, 08901, USA

G. Berg
(e-mail: gabriele.berg@tugraz.at, Tel.: +43 316-873-8310, Fax: +43 316-873-8819)
Institute of Environmental Biotechnology, Graz University of Technology, Petersgasse 12, 8010 Graz, Austria

S. Bhardwaj
Amity Institute of Microbial Technology, Amity University Uttar Pradesh, Noida, UP, 201303, India

V.S. Bisaria
Department of Biochemical Engineering and Biotechnology, Indian Institute of Technology Delhi, New Delhi, India

I. Bodendiek
Faculty 2, Biology/Chemistry, Botany, University of Bremen, Leobener Str. 2, 28359 Bremen, Germany

P. Bonfante
(e-mail: paola.bonfante@unito.it, p.bonfante@ipp.cnr.it, Tel.: +39 011 670 5965, Fax: +39 011 670 5962)
Department of Life Science and Systems Biology, University of Torino, Viale P.A. Mattioli 25, 10125 Torino, Italy

M. Cardinale
(e-mail: m.cardinale@tugraz.at, Tel.: +43 316-873-8818, Fax: +43 316-873-8819)
Institute of Environmental Biotechnology, Graz University of Technology, Petersgasse 12, 8010 Graz, Austria

Y. Chen
Institute of Biotechnology, Zhejiang University, Hangzhou, 310058, P.R. China

C. Cruz
Department of Plant Biology, Faculty of Sciences, Campus Grande Block C-2, Piso-5, Sala 03, 1749-016 Lisbon, Portugal

Y. Dalpé
(e-mail: Yolande.Dalpe@AGR.GC.CA, Tel.: +1 613 759-1381, Fax: +1 613 759-1701)
Eastern Cereal and Oilseed Research Centre, Agriculture and Agri-Food Canada, Ottawa, Canada

A. Das
Amity Institute of Microbial Technology, Amity University Uttar Pradesh, Noida, UP, 201303, India

J.D.W. Dearnaley
(e-mail: john.dearnaley@usq.edu.au, Tel.: +61 7 4631 2804, Fax: +61 7 4631 1530)
Australian Centre for Sustainable Catchments and Faculty of Sciences, University of Southern Queensland, Toowoomba, 4350, Australia

J. Doidy
Pôle Interactions Plantes-Microorganismes, ERL CNRS 6300, UMR 1347 Agroécologie INRA/AgroSup/uB, BP 86510, 21065 cedex Dijon, France

M. Dua
School of Life Sciences, Jawaharlal Nehru University, New Delhi, 110067, India

E. Dumas-Gaudot
Pôle Interactions Plantes–Microorganismes, ERL CNRS 6300, UMR 1347 Agroécologie INRA/AgroSup/uB, BP 86510, 21065 cedex Dijon, France

N. Ferrol
Departamento de Microbiología del Suelo y Sistemas Simbióticos, Estación Experimental del Zaidín, Consejo Superior de Investigaciones Científicas, C. Profesor Albareda 1, 18008 Granada, Spain

J. Fontaine
Unité de Chimie Environnementale et Interactions sur le Vivant, Université du Littoral Côte d'Opale, Calais, France

P. Franken
(e-mail: franken@igzev.de, Tel.: +49 33701-78 215, Fax: +49 33701-55 391)
Leibniz Institute of Vegetable and Ornamental Crops, Theodor-Echtermeyer-Weg 1, 14979 Grossbeeren, Germany

A. Genre
Department of Life Science and Systems Biology, University of Torino, Viale P.A. Mattioli 25, 10125 Torino, Italy

V. Gianinazzi-Pearson
(e-mail: gianina@dijon.inra.fr, Tel.: +33 3 8069 3155, Fax: +33 3 8069 3753)
Pôle Interactions Plantes–Microorganismes, ERL CNRS 6300, UMR 1347 Agroécologie INRA/AgroSup/uB, BP 86510, 21065 cedex Dijon, France

S.K. Gosal
College of Basic Science and Humanities, Punjab Agriculture University, Ludhiana, Punjab, India

M. Grube
(e-mail: martin.grube@uni-graz.at, Tel.: +43 316-380-5655, Fax: +43 316-380-9883)
Institute of Plant Sciences, Karl-Franzens University, Holteigasse 6, 8010 Graz, Austria

R. Honegger
(e-mail: rohonegg@botinst.uzh.ch, Tel.: +41 44 634 82 48, Fax: +41 44 634 82 04)
Institute of Plant Biology, University of Zürich, Zollikerstrasse 107, 8008 Zürich, Switzerland

R. Hung
(e-mail: observationpost8@yahoo.com)
Department of Plant Biology and Pathology, Rutgers University, New Brunswick, NJ, 08901, USA

P. Jeffries
School of Biosciences, University of Kent, Canterbury, CT2 7NJ, Kent, UK

J.M. Johnson
Institut für Allgemeine Botanik und Pflanzenphysiologie, Friedrich-Schiller Universität Jena, Biologisch-Pharmazeutische Fakultät, Dornburger Str. 159, 07743 Jena, Germany

A.K. Johri
School of Life Sciences, Jawaharlal Nehru University, New Delhi, 110067, India

A.C. Kharkwal
Amity Institute of Microbial Technology, Amity University Uttar Pradesh, Noida, UP, 201303, India

N. Koele
(e-mail: koelen@landcareresearch.co.nz, Tel.: +64 3 321 9700, Fax: +64 3 321 9998)
Landcare Research, PO Box 40, Lincoln, 7640, New Zealand

A. Kohler
(e-mail: kohler@nancy.inra.fr)
'Ecogenomics of Interactions' Laboratory, UMR "Interactions Arbres/Micro-Organismes", INRA-Nancy, 54280 Champenoux, France

R.T. Koide
(e-mail: rxk13@psu.edu, Tel.: +1 814 863-0710, Fax: +1 814 863-6139)
The Pennsylvania State University, University Park, PA, 16802, USA

M. Kumar
School of Life Sciences, Jawaharlal Nehru University, New Delhi, 110067, India

S. Lee
(e-mail: slee@aesop.Rutgers.edu)
Department of Plant Biology and Pathology, Rutgers University, New Brunswick, NJ, 08901, USA

Y. Liu
Pôle Interactions Plantes-Microorganismes, ERL CNRS 6300, UMR 1347 Agroécologie INRA/AgroSup/uB, BP 86510, 21065 cedex Dijon, France

B. Lou
Institute of Biotechnology, Zhejiang University, Hangzhou, 310058, P.R. China

A. Lounès-Hadj Sahraoui
Unité de Chimie Environnementale et Interactions sur le Vivant, Université du Littoral Côte d'Opale, Calais, France

J. Ma
Institute of Biotechnology, Zhejiang University, Hangzhou, 310058, P.R. China

F. Martin
(e-mail: fmartin@nancy.inra.fr, Tel.: +33 383 39 40 80)
'Ecogenomics of Interactions' Laboratory, UMR, Interactions Arbres/Micro-Organismes, INRA-Nancy, 54280 Champenoux, France

E. Martino
(e-mail: elena.martino@unito.it, Tel.: +39 011 6705987, Fax: +39 011 6705962)
Dipartimento di Biologia Vegetale dell, Università di Torino and Istituto per la Protezione delle Piante del CNR, v. le Mattioli 25, 10125 Torino, Italy

F. Martos
School of Life Sciences, University of KwaZulu-Natal, Private Bag X01 Scottsville, Pietermaritzburg, 3209, South Africa

M.K. Meghvansi
Defence Research Laboratory, Defence Research Development Organization, Post Bag No. 2, Tezpur, 784001Assam, India

U. Nehls
(e-mail: nehls@uni-bremen.de, Tel.: +49 421 218 62901, Fax: +49 421 218 62914)
Faculty 2, Biology/Chemistry, Botany, University of Bremen, Leobener Str. 2, 28359 Bremen, Germany

F. Oberwinkler
(e-mail: franz.oberwinkler@uni-tuebingen.de, Tel.: +49 7071 66653)
Lehrstuhl Organismische Botanik, Universität Tübingen, Auf der Morgenstelle 1, 72076 Tübingen, Germany

R. Oelmüller
Institut für Allgemeine Botanik und Pflanzenphysiologie, Friedrich-Schiller Universität Jena, Biologisch-Pharmazeutische Fakultät, Dornburger Str. 159, 07743 Jena, Germany

S. Padhi
(e-mail: padhi@AESOP.Rutgers.edu)
Department of Plant Biology and Pathology, Rutgers University, New Brunswick, NJ, 08901, USA

S. Perotto
(e-mail: silvia.perotto@unito.it, Tel.: +39 011 6705987, Fax: +39 011 6705962)
Dipartimento di Biologia Vegetale dell, Università di Torino and Istituto per la Protezione delle Piante del CNR, v. le Mattioli 25, 10125 Torino, Italy

R.L. Peterson
(e-mail: lpeterso@uoguelph.ca, Tel.: +1 519-824-4210 ext. 56000, Fax: +1 519-837-1802)
Department of Molecular and Cellular Biology, University of Guelph, Guelph, Ontario, N1G 2W1, Canada

J.M. Plett
(e-mail: jonathan.plett@nancy.inra.fr)
'Ecogenomics of Interactions' Laboratory, UMR, Interactions Arbres/Micro-Organismes, INRA-Nancy, 54280 Champenoux, France

R. Prasad
Amity Institute of Microbial Technology, Amity University Uttar Pradesh, Noida, UP, 201303, India

K. Rastogi
Amity Institute of Microbial Technology, Amity University Uttar Pradesh, Noida, UP, 201303, India

G. Recorbet
Pôle Interactions Plantes-Microorganismes, ERL CNRS 6300, UMR 1347 Agroécologie INRA/AgroSup/uB, BP 86510, 21065 cedex Dijon, France

D. Redecker
Pôle Interactions Plantes-Microorganismes, ERL CNRS 6300, UMR 1347 Agroécologie INRA/AgroSup/uB, BP 86510, 21065 cedex Dijon, France

N. Requena
Karlsruhe Institute of Technology, Plant-Microbe Interactions Group, Hertzstrasse 16, 76187 Karlsruhe, Germany

M. Sancholle
Unité de Chimie Environnementale et Interactions sur le Vivant, Université du Littoral Côte d'Opale, Calais, France

A. Schüßler
(e-mail: arthur.schuessler@lmu.de, Tel.: +49 89 218074730, Fax: +49 89 218074702)
Genetics, Biocenter of the Ludwig-Maximilians-University Munich, Großhaderner Str. 2-4, 82152 Planegg-Martinsried, Germany

M.-A. Selosse
Centre d'Ecologie Fonctionnelle et Evolutive (CNRS, UMR 5157), 1919 Route de Mende, 34293 cedex 5 Montpellier, France

M. Sharma
Amity Institute of Microbial Technology, Amity University Uttar Pradesh, Noida, UP, 201303, India

I. Sherameti
Institut für Allgemeine Botanik und Pflanzenphysiologie, Friedrich-Schiller Universität Jena, Biologisch-Pharmazeutische Fakultät, Dornburger Str. 159, 07743 Jena, Germany

K.S. Sree
Amity Institute of Microbial Technology, Amity University Uttar Pradesh, Noida, UP, 201303, India

R.B. Srivastava
Defence Institute of Higher Altitude Research, Defence Research Development Organization, Leh-Ladakh, India

S. Talukdar
Amity Institute of Microbial Technology, Amity University Uttar Pradesh, Noida, UP, 201303, India

S. Tripathi
Amity Institute of Microbial Technology, Amity University Uttar Pradesh, Noida, UP, 201303, India

C.P. Upadhyaya
Department of Botany, Guru Ghasidas Central University, Koni Bilaspur, 495009, Chattisgarh, India

M. Vallino
(e-mail: marta.vallino@unito.it, Tel.: +39 011 6705987, Fax: +39 011 6705962)
Dipartimento di Biologia Vegetale dell, Università di Torino and Istituto per la Protezione delle Piante del CNR, v. le Mattioli 25, 10125 Torino, Italy

D. van Tuinen
Pôle Interactions Plantes–Microorganismes, ERL CNRS 6300, UMR 1347 Agroécologie INRA/AgroSup/uB, BP 86510, 21065 cedex Dijon, France

A. Varma
(e-mail: ajitvarma@aihmr.amity.edu, Tel.: +91 120 4392132, Fax: +91 120 2431268)
Amity Institute of Microbial Technology, Amity University Uttar Pradesh, Noida, UP, 201303, India

A. Waschke
Leibniz Institute of Vegetable and Ornamental Crops, Theodor-Echtermeyer-Weg 1, 14979 Grossbeeren, Germany

D. Wipf
Pôle Interactions Plantes-Microorganismes, ERL CNRS 6300, UMR 1347 Agroécologie INRA/AgroSup/uB, BP 86510, 21065 cedex Dijon, France

P.S. Won
Department of Molecular Biotechnology, Konkuk University, Seoul, 143701, South Korea

L. Zhang
Institute of Microbiology, Chinese Academy of Sciences, 8 Belertio Zhongguancun, Beijing, 100190, P.R. China

1 Exploring the Genome of Glomeromycotan Fungi

V. Gianinazzi-Pearson[1], D. van Tuinen[1], D. Wipf[1], E. Dumas-Gaudot[1], G. Recorbet[1], Y. Liu[1,3], J. Doidy[1], D. Redecker[1], N. Ferrol[2]

CONTENTS

I. Introduction

All fungi forming a mutualistic symbiosis with plant roots called arbuscular mycorrhiza were formerly grouped together in one order, the Glomales, placed in the Zygomycota (Morton 1993). Based on molecular analyses suggesting that arbuscular mycorrhizal fungi should be separated from other fungal taxa, they were transferred a decade ago to the **Glomeromycota,** a new phylum created specifically for them (Schüssler et al. 2001). Whilst members of this monophyletic group originated from the same common ancestor as the Ascomycota and Basidiomycota, they have **no obvious affinity to other major extant phylogenetic groups in the kingdom Fungi** (James et al. 2006) and they probably diverged from the other fungal lineages several hundred million years before plants colonized terrestrial habitats 400–500 million years ago (mya) (Heckman et al. 2001). Glomeromycotan fungi are complex but extremely successful organisms. They establish a compatible interaction with plants by either avoiding or suppressing plant defence reactions whilst redirecting host metabolic flow to their benefit without being detrimental to their host. The mechanisms by which this biotrophy is achieved are largely unknown but the Glomeromycota have accompanied land plants through evolution and survived across periods of important environmental change to become ecologically and agriculturally important symbionts which improve the overall fitness of very different plant taxa in terrestrial ecosystems world-wide (Smith and Read 2008). Substantial evidence has accumulated about how a rational use of the microsymbiont properties should significantly contribute to decreasing fertilizer and pesticide use in agriculture (Gianinazzi et al. 2010).

Although the Glomeromycota show considerable diversity between or within morphologically recognizable species (Rosendahl 2008), they share some singular biological traits

[1]Pôle Interactions Plantes-Microorganismes ERL CNRS 6300, UMR 1347 Agroécologie INRA/AgroSup/uB, BP 86510, Dijon cedex 21065, France; e-mail: gianina@dijon.inra.fr; tuinen@dijon.inra.fr; dwipf@dijon.inra.fr; dumas@dijon.inra.fr; recorbet@dijon.inra.fr; joan.doidy@voila.fr; Dirk.Redecker@dijon.inra.fr

[2]Departamento de Microbiología del Suelo y Sistemas Simbióticos, Estación Experimental del Zaidín, Consejo Superior de Investigaciones Científicas, C. Profesor Albareda 1, Granada 18008, Spain; e-mail: nuria.ferrol@eez.csic.es

[3]College of Life Science and Technology, Huazhong Agricultural, Wuhan 430070, China; e-mail: liuyihzau@hotmail.com

Fungal Associations, 2nd Edition
The Mycota IX
B. Hock (Ed.)

which limit experimental approaches that can be exploited to characterize their complexity. One particularity is their **reproduction through large asexual spores**, each of which is a single cell harbouring several hundreds or thousands of nuclei. The main mechanism for filling a spore with so many nuclei appears to be a massive influx of nuclei from subtending mycelium into the developing spore (Jany and Pawlowska 2010). **No sexual stage is known** for these fungi but, although they are assumed to only reproduce asexually, recent transcriptome analyses indicate that they do possess genetic information essential for sexual reproduction and meiosis (Tisserant et al. 2012). Absence of a sexual cycle raises questions about how Glomeromycota deal with deleterious mutations usually eliminated through meiosis and how they have adapted to new hosts or habitats during evolution. High polyploidy, with multiple gene copies in the same genome, has been speculated as a possible mechanism to buffer against mutational events (Pawlowska and Taylor 2004), whilst **nuclear exchange through hyphal anastomosis between different individuals of a same species** (Casana and Bonfante 1988; Giovannetti et al. 2001) provides the possibility for genetic flux and recombination events (Croll et al. 2009; Angelard and Sanders 2011). However, basic information concerning ploidy, karyosis, number of chromosomes or whether meiosis does occur in the Glomeromycota is still lacking, and evidence for genetic exchange, recombination or segregation is very limited.

Another particularity is that **all glomeromycotan fungi are obligate symbionts** which have so far proved to be incalcitrant to pure culture in the absence of a host root on which they depend as a carbon source. This introduces inherent **limitations in the application of standard techniques like genetic transformation or mutant generation/characterization**, and it greatly hinders advances in the knowledge about gene function in these organisms. As previously pointed out (Gianinazzi-Pearson et al. 2001), there is converging evidence that the Glomeromycota are an unusual group of fungi, and information about their genome structure, complexity and function is essential to understanding the processes regulating their symbiotic attributes, their reproductive biology and their apparent stability during coevolution in symbiosis with many different plant taxa. Despite the fact that the biology of glomeromycotan fungi makes them extremely difficult to manipulate experimentally, the advent of powerful molecular techniques has considerably furthered research during the past decade. The present chapter updates on the state of the art on Glomeromycota genomics within the past decade (Gianinazzi-Pearson et al. 2001), as well as on progress made through targeted and high-throughput sequencing programmes into an understanding of their basic biology.

II. Glomeromycota Genome Organization

A. Nuclear Genome Characteristics

Values for the DNA contents of glomeromycotan fungi vary depending on the analytical method and the genome of reference used (see Gianinazzi-Pearson et al. 2001). The **haploid genome size** was previously estimated to range from 128 to 1,065 Mb, depending on the fungal species studied, which further underlines the **high diversity within this phylum** (Hosny et al. 1998a). Important differences also exist within the same genus: the genome size of *Glomus* species was estimated by similar methods to be, for example, 177 Mb for *G. geosporum* and 375 Mb for *G. caledonium* (Hosny et al. 1998a) and 14.0–16.5 Mb for *G. intraradices* DAOM 197198 (syn. *G. irregulare*; Hijri and Sanders 2004). The apparently small genome of the latter (see below), together with its cultivability in vitro on root organ cultures, has made it an appropriate candidate for first sequencing attempts of a glomeromycotan genome (Lammers et al. 2004). To date, several genome sequencing programmes, based mainly on extensive whole-shotgun (WGS) sequencing, have generated altogether 345 Mbp that have been assembled in 163,968 contigs in a total of 52.5 Mb (Martin et al. 2008b). This corresponds to about fourfold the initially expected genome of *G. intraradices* DAOM 197198 and an estimated genome space of >150 Mb. Recent data provided by Sedzielewska et al. (2011) have confirmed that the genome of this fungus is in fact 10 times larger (155 Mb) than estimated by Hijri and Sanders (2004) and is within the range for the Glomeromycota. Correct genome assembly is hindered by the fact that critical genetic information, such as a genetic map, is not available for the Glomeromycota and that **multiple gene copies in the same genome** (Pawlowska and Taylor 2004) would lead to different assemblies with the same set of scaffold data. **High polymorphism**, which appears to be characteristic of glomeromycotan fungi (see Sanders and Croll 2010) and which has also been indicated from genome sequencing (Martin et al. 2008b), raises an additional question of functional genes versus pseudogenes, which are difficult to separate for most software packages used for whole genome assemblies. Recent information about functional genes is discussed below in Sect. IV.

The **very low GC content** (~30 %) observed from genome sequencing of *G. intraradices* DAOM 197198 is in agreement with the values of 30–35 % obtained by other methods for a range of Glomeromycota, and which are relatively low as compared to most other fungal taxa (Hosny et al. 1998b). The latter authors proposed that a mutational pressure from GC to AT may exist in the Glomeromycota and that, since these fungi proliferate under light-deprived conditions in soil and roots, their environment has not exerted any significant counter selection against AT-rich sequences so that GC contents have become low. The **relatively high proportion of methylated cytosine residues**, which are frequent in repeated sequences, is another particularity of the glomeromycotan genome (Hosny et al. 1998b). A number of repetitive non-coding DNA sequences have been characterized in several species in the Glomeromycota (Gollotte et al. 2006), and genome sequencing of *G. intraradices* DAOM 197198 has revealed that this fungus is also rich in small repeats. Interestingly, variable tandem repeats are considered to affect the rate of evolution of coding and regulatory sequences in other organisms (Gemayel et al. 2010).

B. Nuclear Ribosomal Genes

The **number of nuclear ribosomal gene copies** has been estimated to be 71–88 in representative glomeromycotan genomes (Gollotte et al. 2006), which is considerably less than in fungi with smaller genomes like yeast or *Cochliobolus heterostrophus* (see Gianinazzi-Pearson et al. 2001). The polymorphic characteristics of ribosomal genes have made them a choice target for phylogenetic and taxonomic studies over a wide range of eukaryotic organisms. The **three coding regions of ribosomal genes**, the small sub-unit (SSU) or 18S, the 5.8S unit and the large sub-unit (LSU) or 25S, are **separated by two internal transcribed spacers** (ITS1 and ITS2; Mitchell et al. 1995). The two non-coding regions (ITS), being under less functional pressure, are more variable and mutate more frequently (Sanders et al. 1995) than the three conserved coding regions. A comprehensive description of nuclear ribosomal genes in the Glomeromycota and their exploitation in phylogeny prior to 2001 is given by Gianinazzi-Pearson et al. (2001). **SSU, LSU and ITS regions are used for molecular phylogeny and taxonomy** of fungi in this phylum. The highly conserved SSU was the first region to be selected for phylogenetic analysis of the Glomeromycota, mainly because of the large number of eukaryotic sequences present in public databases (Simon et al. 1992). It enabled the initial dating back of this taxon to an ancestral *Glomus*-like fungus 353–462 mya (Simon et al. 1993). SSU sequences remain the most numerous available for glomeromycotan fungi and they are mainly at the origin of more recent taxonomic reorganizations (Schüssler et al. 2001; Schüssler and Walker 2010).

However, the SSU region does not allow species-level resolution in the Glomeromycota. Also, the first primer allowing identification of Glomeromycota in host roots, the SSU-based VANS1, is not well conserved across the phylum (Clapp et al. 1999). Consequently, other ribosomal regions, such as the ITS/5.8S region in combination with the SSU or the 5′ end of the LSU, have been exploited to generate more taxon-specific primers for monitoring fungal communities in roots (for example, see van Tuinen et al. 1998; Redecker 2000; Pivato et al. 2007). Comparison of the level of polymorphism between the different ribosomal regions of *G. mosseae* BEG12, *G. mosseae* BEG69, *G. coronatum* BEG22 and *G. intraradices*

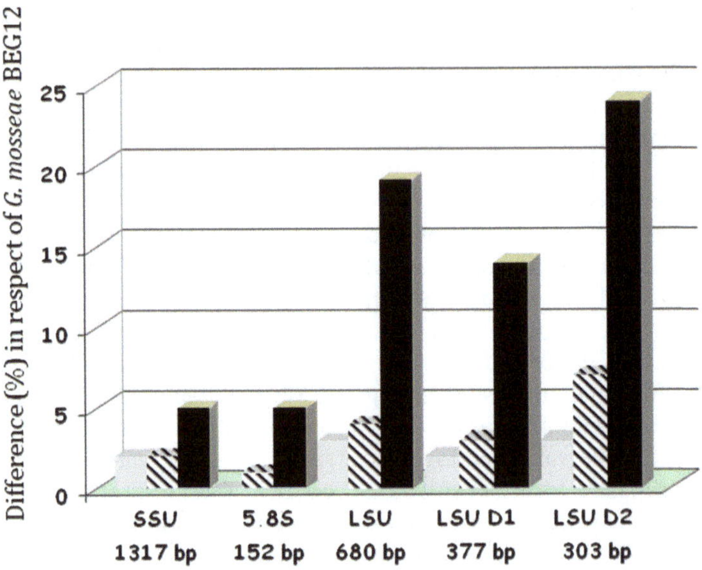

Fig. 1.1. Difference (expressed as a percentage) between sequences of *Glomus mosseae* (BEG12), *G. mosseae* BEG69 (*pale grey*), *G. coronatum* BEG22 (*hatched*) and *G. intraradices* AFTOL ID 48 (*black*) for the SSU, 5.8S, the 5′ end of the LSU (including the variable D1 and D2 domains) and the variable D1 and D2 domains alone

AFTOL ID48, clearly illustrates the **potentiality of the 5′ end of the LSU for diversity studies** (Fig. 1.1). Sequence differences between *G. mosseae* BEG12 and *G. coronatum* BEG22 or *G. intraradices* AFTOL ID 48 are greatest for the 5′ end of the LSU region and in particular for the variable D2 domain. In spite of this, a general caution applies to the use of nuclear ribosomal sequences for diversity studies in that they have a potentially high inter and intra-sporal heterogeneity which implies that several nuclear ribosomal variants may occur in the same isolate (Sanders et al. 1995; Sanders and Croll 2010).

C. The Mitochondrial Genome

The first **mitochondrially encoded genes** from the phylum Glomeromycota were identified from an amplified and sequenced region of the mitochondrial gene coding for the LSU of ribosomal RNA (Raab et al. 2005). In contrast to their nuclear-encoded counterparts, these genes do **not show variation within single spores and fungal isolates.** This is of particular importance as it has been proposed that the **nuclei in the coenocytic glomeromycotan mycelium are genetically heterogeneous,** resulting in a population of allelic variants with no fixed overall genotype (Kuhn et al. 2001). The first complete mitochondrial genome of a glomeromycotan fungus was sequenced from *G. intraradices* strain 494 by whole genome amplification and subsequent pyrosequencing (Lee and Young 2009). It has a size of 70.6 Kbp and a GC content of 37.2%, which is higher than the GC content of the nuclear genome. The genetic code used in the protein-coding genes is the standard code except that UGA is used for tryptophan, which is typical for many fungal mitochondrial genomes. The mitochondrial genome of *G. intraradices* contains a similar set of genes to that of other fungi: three subunits of ATP synthase (*atp6*, *atp8*, *atp9*), three of cytochrome c oxidase (*cox1*, *cox2*, *cox3*), seven of NADH dehydrogenase, and apocytochrome b (cob). Moreover, it contains the standard set of 26 tRNAs and ribosomal small subunit (*rns*) and large subunit (*rnl*) genes. Notably, all genes are encoded on the same strand. However, the *rps3* gene, which is found in the mitochondrial genome of other fungi, was apparently transferred to the nuclear genome in *G. intraradices*, and there is evidence for the presence of nuclear copies of other mitochondrial sequences (Lee and Young 2009). **The *G. intraradices* mitochondrial genome bears little resemblance to that of other fungi,** but this comes as little surprise since gene order in fungal mitochondrial genomes is not conserved among distantly related taxa.

A large part of the *G. intraradices* mitochondrial genome consists of introns and other non-coding sequences; only 24.6 % are coding sequences. A total of 26 introns were identified in strain 494, most of them belonging to type 1. As in other fungi, the mitochondrial genome of this glomeromycotan fungus is rich in **homing endonucleases,** enzymes that are often coded in introns and thought to function **as mobile elements** in the insertion of introns containing an endonuclease open reading frame (ORF) into intron-free alleles (Dalgaard et al. 1997). Homing endonucleases of the LAGLIDADG family were first reported in introns of the *rnl* gene (Raab et al. 2005). The complete mitochondrial genome sequence shows that four of these ORFs are within introns and one is attached to the *nad3* gene, whilst two GIY-YIG type homing endonuclease ORFs are in intergenic spacers. As mobile genetic elements, homing endonucleases are presumably transferred laterally. Comparison of different isolates and species of *Glomus* has provided evidence for the existence in glomeromycotan fungi of **horizontal transfer of *rnl* introns and the "homing cycle"** (Chevalier and Stoddard 2001), involving insertion, degeneration and loss (Thiéry et al. 2010). Due to this activity, introns containing homing endonuclease ORFs seem to show higher degrees of polymorphism. This polymorphism has been used to distinguish between isolates of *G. intraradices* in field studies (Börstler et al. 2010) and to track an inoculated isolate over several years in a field site (Sýkorov et al. 2011).

The involvement of fungal mitochondria in spore germination and early signal exchange with host root colonization (Besserer et al. 2006) highlights the need to study the coding sequences of mitochondrial genomes. On the other hand, non-coding regions are of direct importance as molecular markers of these fungi in ecological studies and in biotechnological applications. In addition, **mitochondrial markers to study inheritance of these organelles could provide important baseline data to help to elucidate the genetics** of the Glomeromycota and possible previously unrecognized sexual processes.

III. Nuclear Genome Evolution in the Glomeromycota

The **Glomeromycota are traditionally thought to be ancient asexuals.** This assumption is based on the fact that no morphological structures conclusively indicative of sexual reproduction have been observed in the whole phylum; a study reporting zygosporangia in *Gigaspora* (Tommerup and Sivasithamparam 1990) has never been confirmed. The lack of basic knowledge about their genetics leaves the evolutionary biology of the Glomeromycota open to a lot of speculation. The inability to culture these fungi separately from their host plants and to obtain stable transformants has contributed to this situation.

Considering the well-known **benefits of sexual reproduction** in avoiding the accumulation of deleterious mutations in the genome, which should inevitably lead to evolutionary meltdown processes such as Muller's ratchet (Muller 1932), it seems hard to understand how these fungi could persist as important actors of terrestrial ecology for more than 400 million years. Morphological stasis resulting in a relatively low number of described morphospecies and striking similarities of extant Glomeromycota with 400 million year old fossils have been cited as other lines of evidence for the lack of diversification caused by long term clonal propagation in conjunction with the asexual lifestyle (Sanders and Croll 2010). Ancient asexuals are often considered as "scandalous" exceptions of the rule as they challenge current theories of sex (Judson and Normark 1996). The bdelloid rotifers, a lineage of invertebrates solely consisting of parthenogenetically reproducing females for at least 40 million years, are often cited as a striking example where clonal organisms have evolved into more than 300 species in a similar way as sexual organisms (Fontaneto et al. 2007). Alternative **mechanisms to purge deleterious mutations from the genome** have therefore to be considered for Glomeromycota as they have been for bdelloid rotifers (Gladyshev and Arkhipova 2010).

When the nuclear rDNA genes were characterized as the first genetic components in the Glomeromycota, an **unusually high level of intra-organism polymorphism** was noted (see Gianinazzi-Pearson et al. 2001). In sexual organisms, this polymorphism is prevented by a mechanism known as "concerted evolution" that is thought to homogenize rDNA copies (Gandolfi et al. 2001). Intra-organismal polymorphism of nuclear rDNA or other genes has been demonstrated in many organisms, including fungi (O'Donnell 1992), but it seems to be exceptionally large in the Glomeromycota (Stockinger et al. 2009). Here, it was postulated that rDNA variation occurs between different genomes that are present in genetically different nuclei in the coenocytic mycelium (Kuhn et al. 2001; Hijri and Sanders 2005). **Heterokaryosis** was hypothesized to have arisen by hyphal anastomosis and accumulation of mutations (Bever and Wang 2005) and evoked as a possible mechanism to compensate for the absence of sexual processes in glomeromycotan species, although other authors have presented data for homokaryosis (Pawlowska and Taylor 2004). Different likely scenarios in this context were recently reviewed by Young (2008).

It is helpful to note in this context that numerous fungal lineages, though none as ancient as the Glomeromycota, were once thought to be asexual based on the absence of the expected morphological structures. In many cases, analyses of the population biology using molecular markers revealed evidence for recombination (Burt et al. 1996). Many possibilities may exist in fungi for cryptic sexuality or non meiotic recombination and low levels of recombination can be sufficient to prevent the accumulation of deleterious mutations. In contrast to other fungal groups there does not seem to be a uninucleate stage in the Glomeromycota, which would give the opportunity to reduce genetic variation to a single haploid genome but, as mentioned previously, a large number of nuclei are instead migrating into newly formed spores (Jany and Pawlowska 2010). However, some evidence for recombination in the genome of *G. intraradices* has been detected using molecular markers (Croll and Sanders 2009), whereas the genome of *Scutellospora castanea* was previously found to be predominantly clonal (Kuhn et al. 2001) and the life history of the cosmopolitan *G. etunicatum* was shown to be dominated by clonality with rare recombination, if at all (den Bakker et al. 2010).

A **parasexual mechanism for the exchange of genetic information** in fungi is through the formation of hyphal networks by **anastomoses** (hyphal bridges). This phenomenon has been well documented in the genus *Glomus* but the extent appears to differ between isolates. For example, it seems to be limited to within-isolate connections for *G. mosseae* originating from different geographic regions (Giovannetti et al. 2003), whilst genetically distinguishable isolates of *G. intraradices* from one field site in Switzerland formed anastomoses at a frequency dependent on their genetic relatedness (Croll et al. 2009). The *G. intraradices* isolates exchanged genetic markers, resulting in recombinant genotypes in offspring strains which showed altered symbiotic capabilities, indicating that this kind of genetic exchange may have some relevance for host plant fitness (Angelard et al. 2010).

However, basic parameters can differ substantially between members of the Glomeromycota. Their genome sizes vary greatly (see Sect. II.A), not much is known about genome structure outside the model species *G. intraradices*, and retrotransposons have been suggested to play an important role in the genome of at least one species (*S. castanea*; Gollotte et al. 2006). In addition, anastomosis formation has not been observed in the Gigasporaceae and other lineages, resulting in a completely different architecture of the mycelium and ruling out hyphal cross-bridges as a means to redistribute nuclei (Purin and Morton 2011). Also, in contrast to *G. intraradices*, *G. mosseae* was shown to have a rather uniform worldwide population structure, suggesting a different genetic disposition (Rosendahl et al. 2009). Due to these apparent differences across the Glomeromycota, it seems difficult to generalize questions of evolution or genetic exchange from any findings obtained with one or another fungus. Even though first steps towards understanding genome evolution have been taken in the model species *G. intraradices* (Sanders and Croll 2010), there remains much more to be explored for the rest of the phylum.

IV. The Symbiotic Genome of Glomeromycota

The **symbiotic genome** comprises those glomeromycotan **genes that are associated with development and functioning of the fungi in arbuscular mycorrhiza interactions**. Initial steps require genes permitting a switch from asymbiotic spore germination to pre-symbiotic stages of hyphal branching and appressorium formation at the root surface under the influence of host plant signals (Gianinazzi-Pearson et al. 2007). Once within roots, morphogenetic processes in the Glomeromycota lead to differentiation of characteristic intracellular haustoria-like structures, termed arbuscules, with the establishment of a symbiotic interface bordered by fungal and plant membranes, and thought to be the main site of nutrient and signal flow between the symbionts (Smith and Read 2008). Hyphae subsequently develop out from the mycorrhizal roots to form the extraradical mycelium (ERM). They provide extensive pathways for nutrient fluxes through the soil into mycorrhizal roots (Gianinazzi et al. 2010) which presumably rely on the regulation of fungal genes related to nutrient sensing, production of specific enzymes and resource partitioning between the fungal symbionts and host roots (Leake et al. 2004).

A. Transcriptome Features

In the absence of a full genome sequence, functional genomic studies of symbiotic traits in the Glomeromycota have so far primarily relied on **gene expression profiling.** Initial studies using targeted approaches, based on the assumption that a gene or gene product plays a role in developmental or metabolic processes, identified a certain number of genes encoding proteins with important nutritional and morphogenetic functions, such as phosphate transporters, H^+-ATPases, ammonium and amino acid transporters, carbohydrate metabolism, chitin synthases and β-tubulin (see Ferrol et al. 2004; Gianinazzi-Pearson et al. 2004). For example, as discussed below in the transportome section, the characterization of a high-affinity phosphate transporter in *G. versiforme* by heterologous screening of a cDNA library (Harrison and van Buuren 1995) provided a breakthrough in the understanding of fungal functioning in phosphate uptake by mycorrhizal plants.

Identification of a more comprehensive collection of fungal genes became possible during the last decade with the emergence of transcriptome technologies that allow analysis of the mRNA pool of a cell at any one moment. **Transcriptome studies of glomeromycotan fungi** initially focused on small collections of **expressed sequence tags (ESTs)** in cDNA libraries generated exclusively from activated spores (Stommel et al. 2001), germinated spores (Lammers et al. 2001; Lanfranco et al. 2002) or extraradical hyphae (Sawaki and Saito 2001; Jun et al. 2002). Clone sequencing revealed several interesting similarities to known genes which are consistent with postulated fungal activities in the symbiotic state. For example, evidence that **arginine** is probably the **preferred molecule for long-distance fungal transport of nitrogen to the host plant** was obtained by identifying a glutamine synthase gene from a *G. intraradices* cDNA library which is preferentially expressed in extraradical hyphae and a gene associated with arginine breakdown which is more highly expressed in the intraradical mycelium (Govindarajulu et al. 2005). Also, expression profiling of an acyl-CoA-dehydrogenase gene from the same library indicated mechanisms of lipid utilization in germinating spores and in extraradical mycelium, and expression analysis of genes coding for a malate synthase and an isocitrate lyase gene, containing motifs responsible for glyoxisomal targeting, reinforced the hypothesis that **substantial carbon fluxes within symbiotic hyphae involve the glyoxylate cycle** (Bago et al. 2002).

With the advent of techniques like **differential RNA display (DDRT)** and **suppression subtractive hybridization (SSH)**, snapshots of the glomeromycotan genome active in fungal structures associated with different stages in the mycorrhizal symbiosis became more easily accessible. Differential expression of numerous genes was revealed in a first SSH-based comparison of transcript profiles between germination hyphae and extraradical mycelium of *G. mosseae*, with the identification of a gene (*GmGin1*) probably involved in signalling during spore germination before symbiosis formation (Requena et al. 2002). Subsequent exploration, using DDRT, SSH and EST screening, of genetic determinants controlling the developmental switch from asymbiotic spore germination to presymbiotic hyphal branching of *Gigaspora rosea* and *Gig. gigantea*, stimulated by **a root exuded factor**, showed activation of genes encoding proteins involved in mitochondrial function, signal transduction, gene expression, DNA synthesis and cell cycle regulation (Tamasloukht et al. 2003, 2007). Induction of genes encoding mitochondrial proteins occurred before increases in respiratory activity, reorganization of the mitochondrial system and stimulation of fungal ramification, indicating that this **branching response is the result of a metabolic switch** (Tamasloukht et al. 2003). Strigolactones were later discovered to be a root exudate component which induces hyphal branching (Akiyama et al. 2005) and which provokes the responses of the mitochondrial apparatus in the fungus, leading to the conclusion that mitochondrial activation is a key event in the switch from asymbiotic to pre-symbiotic stages (Besserer et al. 2006, 2008).

Analysis of the fungal transcriptome during colonization of host plant tissues has been hampered by a low abundance of fungal transcripts (Maldonado-Mendoza et al. 2002). However, using SSH it was possible to detect transcriptome modifications in *G. mosseae* sporocarps triggered in synchrony with appressoria formation linked to recognition of a host root surface by fungal hyphae, and to show induction of genes with functions in signalling, transduction, general cell metabolism, defence or stress responses, or of unknown function during this early morphogenetic event (Breuninger and Requena 2004). Several of the identified genes code for proteins that have a potential role in **calcium-based signalling pathways**, indicating that **Ca^{2+}** could be involved as a **second messenger in the perception and transmission of a plant signal leading to appressorium formation**. Transcript profiling studies undertaken to understand molecular changes that accompany arbuscular mycorrhiza development have also led to the identification of a few fungal genes in symbiotic tissues. For example, DDRT analyses of mycorrhizal and non-mycorrhizal tomato gave a cDNA fragment with similarity to a phosphoglycerate kinase gene from *G. mosseae* that accumulates in higher amounts in colonized roots than in germinated spores (Harrier et al. 1998), and the same approach identified three differentially displayed cDNA fragments of *G. intraradices* in barley mycorrhiza which code for peptide sequences with similarities to proteins involved in gene regulation (Delp et al. 2000). A further six ESTs of *G. mosseae* up-regulated in mycorrhizal roots were identified in a SSH library of *M. truncatula*, two of which showed similarity to a thioredoxin homolog and to a peptidylprolyl *cis-trans* isomerase (Brechenmacher et al. 2004).

Studies aimed at understanding the molecular response of the ERM of Glomeromycota to various growth conditions, including nitrogen starvation (Capellazzo et al. 2008) and heavy metal stress (Waschke et al. 2006; Ouziad et al. 2005) and interaction with other microorganisms (Hildebrandt et al. 2006), have given access to more fungal genes. Among the ESTs of *G. intraradices* induced by heavy metals, for example, several stress-responsive genes were identified, particularly genes encoding enzymes involved in oxidative protection, such as CuZn-SOD, thioredoxins and glutathione-S-transferases, supporting the hypothesis that a primary strategy of the fungus to survive in heavy metal-polluted soils is to cope with the

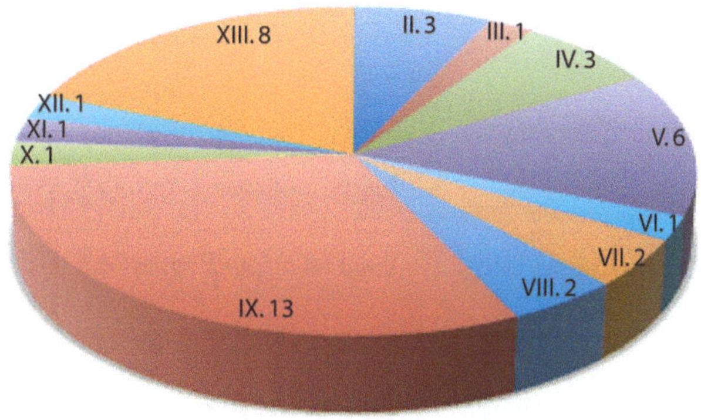

Fig. 1.2. Functional categories (roman numerals) and numbers of putative fungal genes from a *M. truncatula*/*G. intraradices* cDNA library: *II* cytoskeleton; *III* membrane transport, *IV* vesicular trafficking, secretion and protein sorting, *V* primary metabolism, *VI* secondary and hormone metabolism, *VII* chromatin and DNA metabolism, *VIII* gene expression and RNA metabolism, *IX* protein synthesis and degradation, *X* signal transduction and post-transcriptional modification, *XI* miscellaneous; *XII* defence and cell rescue, *XIII* no homology

heavy metal-induced oxidative stress (González-Guerrero et al. 2007; Benabdellah et al. 2009a).

Glomeromycotan genes have also been searched for amongst ESTs **in mycorrhizal root cDNA libraries.** In one approach, tblastx analysis of the Affymetrix GeneChip Medicago Genome Array identified 49 putative fungal genes, all present exclusively in *Medicago truncatula*/*G. intraradices* root cDNA libraries (Gomez et al. 2009). Further analysis of 10 of these genes, associated with the urea cycle, amino acid biosynthesis and cellular autophagy, showed they were expressed in laser-microdissected cortical cells containing arbuscules. These data confirm previous predictions by Govindarajulu et al. (2005) that the **urea cycle is active in the arbuscules** and provide the first molecular hint that **arbuscule turnover might involve autophagy.** In another analysis, a blastn search against 3034 EST contigs from a *M. truncatula*/*G. intraradices* cDNA library (Journet et al. 2002) was performed and 42 clones putatively corresponding to fungal genes were obtained. On the basis of their annotation, they were distributed in 12 different functional groups (van Tuinen et al. unpublished; Fig. 1.2); none of these fungal genes correspond to those reported by Gomez et al. (2009). The largest functional category, with 30% of the sequences, encodes proteins involved in protein synthesis and degradation; 14% are related to primary metabolism, suggesting a high metabolic activity of the mycorrhizal fungus during root colonization. The high number of orphan genes (19%), with unknown function or no homology to database sequences, could be an indication of the presence mycorrhiza-specific fungal genes.

Altogether, these different approaches have generated about 5200 ESTs from Glomeromycota that are publicly available in databases. In the absence of a full genome sequence, these ESTs have represented a valuable resource to identify candidate genes for targeted studies of, for example, glomeromycotan carbon metabolism (Lammers et al. 2001; Bago et al. 2002, 2003), nitrogen metabolism (Govindarajulu et al. 2005; Tian et al. 2010), sulfur metabolism (Allen and Shachar-Hill 2009), heavy metal homeostasis (Lanfranco et al. 2002; González-Guerrero et al. 2005, 2007) and redox homeostasis (Lanfranco et al. 2005; Benabdellah et al. 2009a, b; González-Guerrero et al. 2010a, b). Genes selected from these databases have also been exploited to show the impact of symbiosis-related plant genes on fungal activity during root interactions (Seddas et al. 2009; Kuznetsova et al. 2010).

Recently, the international *Glomus* consortium has expanded the EST repository of *G. intraradices* by generating cDNA libraries from different fungal structures that have been sequenced using high-throughput technologies. A robust set of non-redundant virtual transcripts (25,906, about 20 Mb) transcribed in quiescent and activated spores, ERM and

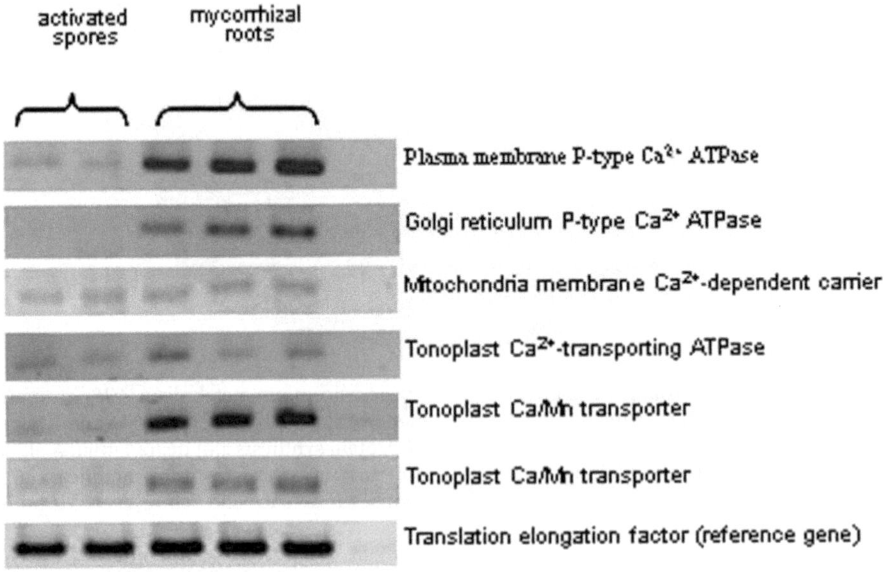

Fig. 1.3. Expression profiling of *G. intraradices* genes encoding proteins putatively involved in calcium homeostasis; amplification by RT-PCR of cDNA from activated spores and mycorrhizal roots of *M. truncatula*

symbiotic roots have been generated (Tisserant et al. 2012). This comprehensive *G. intraradices* transcriptome has provided evidence that **obligate biotrophy** in this fungus **cannot be explained by loss of metabolic complexity** and, as mentioned previously, **genetic information essential for sexual reproduction and meiosis is present** although a known sexual stage is absent from the glomeromycotan life-cycle. The virtual transcriptome of *G. intraradices* (based on >430,000 reads) has served to construct an oligoarray as a basis to pinpointing the peculiar biological traits of these organisms and to examine the functional responses of *G. intraradices* genes to symbiosis development. Gene expression profiling in ERM and symbiotic root tissues has, for example, shown that **pathways of amino acid and polyphosphate biosynthesis are highly expressed** in the fungus, which concords with **nutrient fluxes through mycelium during symbiotic interactions.** Up-regulation of genes encoding for membrane transporters, signal transduction pathways and small secreted proteins in intraradical mycelium and arbuscules, together with the lack of expression of hydrolytic enzymes acting on plant cell wall polysaccharides, are characteristics of *G. intraradices* that are shared with the ectomycorrhizal fungi *Laccaria bicolor* (Martin et al. 2008a) and *Tuber melanosporum* (Martin et al. 2010) and with the biotrophic pathogen *Blumeria graminis* (Spanu et al. 2010). In addition, data sets support biological traits identified in previous gene expression studies; for example, a number of *G. intraradices* genes encoding proteins putatively involved in Ca^{2+} homeostasis/signalling are up-regulated with mycorrhiza development (Fig. 1.3). Moreover, *G. intraradices* transcriptome features highlight the existence of *Glomus*-specific genes, including those coding for small secreted proteins, which are amongst the most highly up-regulated in the mycorrhizal association and which could be specific to the symbiotic state (Tisserant et al. 2012). The recent characterization of a secreted fungal effector (SP7) from *G. intraradices* points to a role of secreted proteins in managing the accommodation process of the fungus within plant roots (Kloppholz et al. 2011).

B. Proteome Insights

Much of the early studies of proteins in the Glomeromycota focussed on polypeptides or enzyme-active gene products, separated by gel electrophoresis, to investigate expressed functions in the fungal genome during the arbuscular mycorrhizal symbiosis or as polymorphic genetic markers in diagnostic taxonomy (see Gianinazzi-Pearson et al. 2001). However, with advances in bioinformatics and the development of mass spectrometry (MS; Rohrbough et al. 2007; Oeljeklaus et al. 2008), efficient **large-scale profiling of the fungal proteome** has become possible. Although initial proteomic studies conducted on the extra- and intra-radical stages of the glomeromycotan life cycle experienced difficulties in revealing the accumulation of fungal gene products (Bestel-Corre et al. 2002; Dumas-Gaudot et al. 2004), systematic nanoscale capillary liquid chromatography–tandem mass spectrometry (LC-MS/MS) has since proved successful in enlarging the coverage of the fungal proteins.

Proteomics performed on the ERM from in vitro mycorrhizal root organ cultures, using large-scale protein-profiling based on two-dimensional electrophoresis (2-DE) and MS-based identification analyses, gave the first extraradical glomeromycotan 2-DE reference map (438 spots from *G. intraradices* DAOM 181602), but only a limited number of fungal proteins could be identified (Dumas-Gaudot et al. 2004). A subsequent shotgun proteome analysis of in vitro-grown ERM, involving one-dimensional (1D)-PAGE nanoscale capillary liquid chromatography MS/MS (GeLC-MS/MS), led to the confident identification of 158 phenol-extracted fungal proteins, corresponding to 92 different (distinct or differentiable) isoforms after parsimony analysis, thus representing the **most comprehensive list of glomeromycotan proteins** so far identified (Recorbet et al. 2009). Over half (88) of the proteins retrieved from *G. intraradices* ERM were not previously indexed as related to mycorrhizal fungi in protein databases, and consequently represent **new protein candidates of the glomeromycotan life cycle.**

Biological process grouping of the GeLC-MS/MS-identified ERM proteins from *G. intraradices* indicates that the **majority of identified proteins have putative functions in sustaining energetic metabolism, protein synthesis, folding, transport and catabolism**, which suggests **an important protein turn-over and trafficking in the extra-radical phase** of the symbiotic fungus (Recorbet et al. 2009). Among these proteins are enzymes involved in dark CO_2 fixation, glycolysis/gluconeogenesis, pentose phosphate and glutamine biosynthesis-related pathways. Likewise, several proteins related to vesicular trafficking, including the GTP-binding protein Ypt1, the small GTPase SAR1 and three members of the Rab GTPase subfamily were identified, together with the signal-transducing proteins calcineurin, Rho1, Cpc2 and Bmh2, for which a role has been demonstrated in fungal morphogenesis. Bmh proteins are also involved in cell cycle regulation as being necessary for the initiation of DNA replication. Among proteins playing roles in the cell division cycle, the DNA damage checkpoint protein rad25, two isoforms of the AAA ATPase Cdc48 and a putative prohibitin were concomitantly identified in the ERM of *G. intraradices.* Current hypotheses on the mechanisms underlying the glomeromycotan cell cycle mostly refer to the switch from G0/G1 to S/M during root colonization and to DNA replication occurring during the production of mycelium from germinating spores (Bianciotto and Bonfante 1993; Bianciotto et al. 1995). Additionally, GeLC-MS/MS data have also pointed to two modules of enzymes related to cell redox homeostasis that accumulate in the ERM of glomeromycotan fungi, corresponding to the *trans*-sulfuration pathway/glutamyl cycle and the glutathione/thioredoxin system. Overall, this strategy has opened the way towards large-scale analyses of fungal genome responses and of metabolic adjustments to environmental cues, including nutrient supply, host recognition and stress-related stimuli.

Although arbuscules have been microdissected from mycorrrhizal roots for transcriptome analyses (Balestrini et al. 2007; Gomez et al. 2009; Kuznetsova et al. 2010), these fungal structures have not been isolated in sufficient amount and purity for proteomic analysis so that **intra-radical fungal gene products** have not yet been directly profiled on a large scale. To enlarge the coverage of intraradical fungal proteins, Recorbet et al. (2010) compared protein profiles of *G. intraradices* and *G. mosseae* in roots of *M. truncatula* using LC-MS/MS based on an enlarged pH gradient and two-dimensional gels coupled to quantitative analysis by Progenesis workstation. Over 2,000 protein spots were detected from mycorrhizal roots, of which confident identifications encompassed 21 fungal proteins. Homology-inferred functions were found to complement the working models so far proposed for functioning of the intra-radical mycelium with regards to carbon utilization, energy generation, redox homeostasis and protein turnover-related processes, thus representing **the largest set of *in planta*-expressed glomeromycotan proteins so far identified.** The proteins appear to belong to pathways active in other stages of the fungal life cycle, including the ERM (Recorbet et al. 2009, 2010).

Overall, this comparative 2-DE-based analysis provided evidence for the existence at the protein level of a conserved set of expressed genome functions associated with the mycorrhizal state of the symbiotic fungus. In relation to proteolytic processing, for example, a fungal subtilase was detected among the induced mycorrhiza-related proteins. Subtilisin-like proteases have been shown to be virulence factors in fungal pathogens of insects, nematodes and plants, including *Magnaporthe poae* (Bidochka and Khachatourians 1990; Tunlid et al. 1994; Sreedhar et al. 1999), and a serine proteinase was proposed to be a potential general feature of leaf infection by the mutalistic fungal plant endophyte *Acremonium typhinum* (Reddy et al. 1996). Some of the proteins identified in glomeromycotan symbionts may thus point to candidate genes required for mycorrhiza formation. In this context, it has been demonstrated that the **Glomeromycota and pathogenic fungi share some genetic components required for colonization of plant tissues** (Heupel et al. 2009; Tollot et al. 2009).

Because identification of a large majority of proteins from the Glomeromycota has been achieved by homology searches in databases for non-mycorrhizal fungi, they are unlikely to be specialized for the arbuscular mycorrhizal symbiosis. Nonetheless, high-throughput identification of protein orthologues in other biotrophic fungal species, even lacking complete genome coverage, remains a pertinent approach for deciphering some conserved key actors of biotrophy in glomeromycotan fungi. Regarding features and processes supposed to be unique to mycorrhizal fungi, which nowadays cannot be approached via homology searches in other species, broad insight into the genome and transcriptome of *G. intraradices* (see Sects. II.A and IV.A) will undoubtedly boost knowledge about the glomeromycotan proteome. Besides limited available genomic resources for Glomeromycota, global proteomic approaches also suffer from a restricted dynamic range resolution in that house-keeping proteins usually hinder the detection of low abundant polypeptides, which are regarded as important effectors in cell regulation pathways. A wide variety of fractionation tools are now available to cope with this issue in non-model organisms (Carpentier et al. 2008) and which could be adapted to proteomics of glomeromycotan fungi in mycorrhizal associations. Furthermore, although laser-capture micro-dissection (LCM)-based techniques require optimization of tissue preparation prior to LCM and GeLC-MS/MS, they represent a promising approach for broad spectrum profiling of glomeromycotan proteins and identification of corresponding genes active in mycorrhizal roots.

C. Transportome Genes

Current knowledge on the **transportome** in the Glomeromycota, that is, the complete repertoire of fungal genes encoding membrane transporters, ion exchangers and ion channels, is still in its infancy. In fact, only 14 transporter genes (for nine different substrates) have been characterized up to now in the glomeromycotan genome. This number is low compared to the number of putative transporters (around 500) that have recently been identified in the sequenced *G. intraradices* transcriptome (Tisserant et al. 2012; L. Casieri, personal communication).

1. Ion and Water Transporter Genes

In order to gain insight into mechanisms underlying the role of Glomeromycota in plant nutrient acquisition, studies have focussed on **fungal genes encoding transporters** potentially involved in the uptake from soil of solutes by the fungal hyphae, their transport through the mycelium and their transfer to the plant. Given the central role **phosphate (P_i) transport** plays in the arbuscular mycorrhizal symbiosis (Smith and Read 2008), the first fungal genes to be identified encoded proteins mediating P_i transport across the fungal membranes. The first gene characterized in detail encoded a P_i transporter of *G. versiforme* (*GvPT*; Harrison and van Buuren 1995), followed by homologous genes from *G. intraradices* (*GiPT*; Maldonado-Mendoza et al. 2001) and *G. mosseae* (*GmosPT*; Benedetto et al. 2005). *GvPT* codes for a high-affinity proton-coupled transporter and shares structural as well as sequence similarity with other plant and fungal high affinity phosphate transporters. The apparent K_m of GvPT, evaluated in a heterologous system, is in the micromolar range which is a value comparable to free P_i concentrations generally found in soil solution. *GvPT* and *GiPT* transcripts are in fact predominantly detected in the ERM, thus indicating a role in P_i acquisition from the soil. Moreover, *GiPT* expression appears to be regulated by phosphate as it responds to external P_i concentrations and to overall mycorrhiza P_i content (Maldonado-Mendoza et al. 2001). In contrast, *GmosPT* is highly expressed also in intraradical mycelium and in particular in arbuscules (Balestrini et al. 2007). This finding provides a new scenario for the plant–fungus nutrient exchanges suggesting that, at least when the plant is actively growing, the **fungus may regulate nutrient exchange at the symbiotic interface** (Balestrini et al. 2007).

Until now four different **genes encoding membrane transporters of nitrogen compounds** (Fig. 1.4) have been identified from Glomeromycota genomes: **one amino acid transporter** in *G. mosseae* (*GmosAAP1*; Cappellazzo et al. 2008) and two **ammonium transporters** (*GintAMT1*; Lopez-Pedrosa et al. 2006; *GintAMT2*; Pérez-Tienda et al. 2011) and one nitrate transporter (*GiNT*; Tian et al. 2010) in *G. intraradices*. GmosAAP1 is able to transport proline through a proton-coupled, pH- and energy-dependent process but it can bind non polar and hydrophobic amino acids, thus indicating a relatively specific substrate spectrum. *GmosAAP1* expression is detected in the ERM developing out from mycorrhizal roots, where transcript abundance could be increased by exposure to organic nitrogen, in particular when supplied at 2 mM concentrations (Cappellazzo et al. 2008). These findings suggest that the GmosAAP1 transporter plays a role in the first steps of amino acid acquisition by hyphae, allowing direct amino acid uptake from the soil and facilitating exploitation of soil nitrogen resources. *GintAMT1* and *GintAMT2* code for the ammonium transporter/methylamine permease/rhesus (AMT/Mep/Rh) protein family (TC#1. A.11) that mediate transport of NH_4^+ across biological membranes (Lopez-Pedrosa et al. 2006; Pérez-Tienda et al. 2011). Both genes functionally complement corresponding mutant yeast strains, and the apparent K_m of GintAMT1 has been evaluated in yeast to be in the micromolar range characteristic of a high-affinity and low capacity NH_4^+ transporter. *GintAMT1* and *GintAMT2* are differentially expressed during the fungal life cycle and in response to N, suggesting that the encoded proteins play different roles in the symbiosis. Whilst transcripts of both genes were detected in arbuscules and spores, *GintAMT1* was more highly expressed than *GintAMT2* in the ERM. Gene activity in the extraradical hyphae indicates an involvement in fungal nutrient uptake from the soil, but their expression also in arbuscules and spores suggests a role for these transporters in processes other than N uptake for nutrition, such as retrieval of NH_4^+ that leaks out during metabolism and, therefore, in fungal NH_4^+ retention. For the **nitrate transporter gene** *GiNT*, only a partial cDNA sequence encoding a putative high-affinity nitrate transporter has so far been obtained (Tian et al. 2010). Nitrate availability stimulates the expression of *GiNT* so that it may play a role in transporting NO_3^- into the ERM.

Heavy metal membrane transporter genes have also been identified in the glomeromycotan genome. In *G. intraradices*, a member of the cation diffusion facilitator (CDF) family *GinZnT1* was shown to decrease cytosolic Zn levels in yeast, although not to completely restore the phenotype of a *zrc1cot1* yeast mutant affected in two vacuolar CDF transporters (González-Guerrero et al. 2005). In yeast, CDF family members are involved in Zn, Fe and Co homeostasis by playing a role in metal efflux from the cytosol, either outside the cell or into intracellular organelles. The expression pattern of *GintZnT1* in response to short-time exposure to Zn and when *G. intraradices* is grown in the presence of high Zn concentrations, reveals a role for

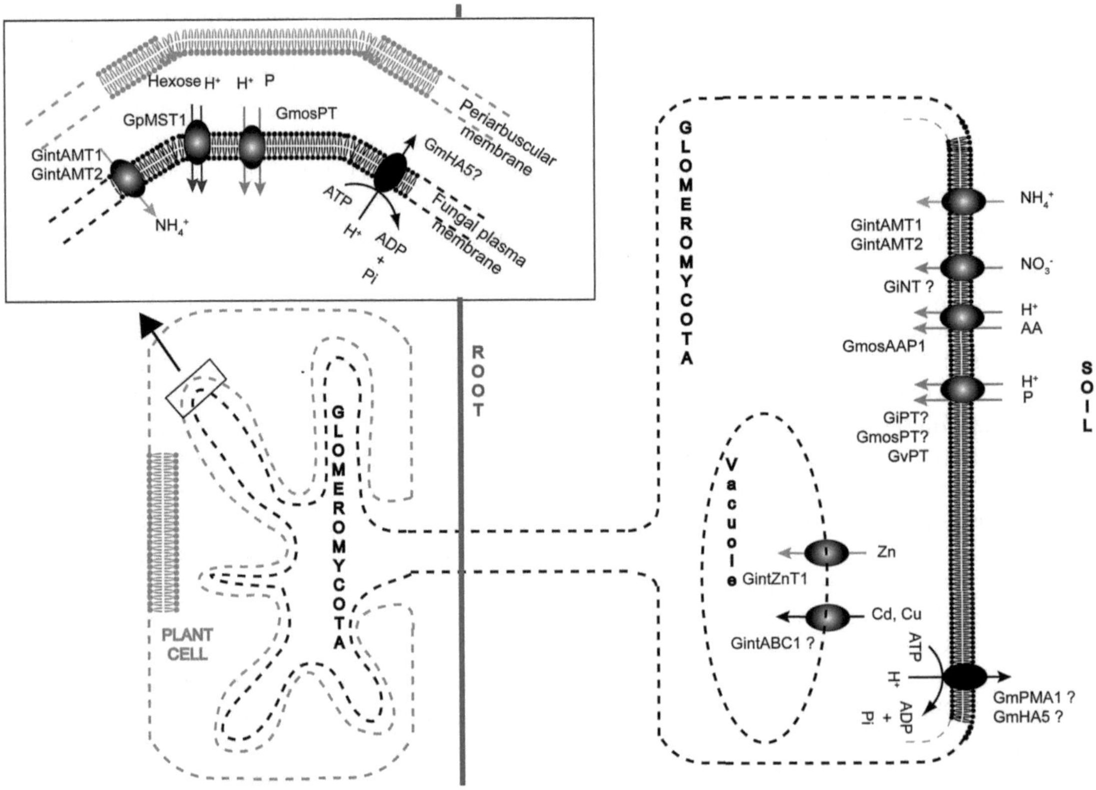

Fig. 1.4. Current knowledge of transporter genes from glomeromycotan fungi and their putative role in the arbuscular mycorrhiza: GpMST1 (*Geosiphon pyriformis* monosaccharide transporter 1; Schüssler et al. 2006, 2007), GmPMA1 (*Glomus mosseae* plasma membrane ATPase 1; Requena et al. 2003), GmHA5 (*G. mosseae* H^+-ATPase; Ferrol et al. 2000), GintABC1 (*G. intraradices* ABC transporter 1; González-Guerrero et al. 2010a), GintAMT1 (*G. intraradices* ammonium transporter 1; Lopez-Pedrosa et al. 2006), GintAMT2 (*G. intraradices* ammonium transporter 1; Pérez-Tienda et al., unpublished data), GiNT (*G. intraradices* nitrate transporter; Tian et al. 2010), GmosAAP1 (*G. mosseae* amino acid permease 1; Cappellazzo et al. 2008), GintZnT1 (*G. intraradices* zinc transporter 1; González-Guerrero et al. 2005), GiPT (*G. intraradices* phosphate transporter; Maldonado-Mendoza et al. 2001), GmosPT (*G. mosseae* phosphate transporter; Benedetto et al. 2005), GvPT (*G. versiforme* phosphate transporter; Harrison and van Buuren 1995) and GintAQP1 (*G. intraradices* aquaporin 1; Aroca et al. 2009). *Question marks* refer to transporters that have not been functionally characterized

GintZnT1 in Zn detoxification and in the protection of the mycorrhizal fungus against Zn stress. Recently, a gene member of the ATP-binding cassette (ABC) family of transporters that may also play a role in heavy metal detoxification has been reported in *G. intraradices* (*GintABC1*; González-Guerrero et al. 2010a). The fact that *GintABC1* is up-regulated by Cu and Cd, and that it has a high homology to the yeast Cd factor gene that transports *bis*-glutathione-Cd complexes across the vacuolar membranes (Li et al. 1996), suggests a putative role for GintABC1 in heavy metal tolerance by transporting the excess of metal into vacuoles in *G. intraradices*.

In addition to these ion transporters, **transmembrane water channels** or **aquaporins** can mediate the flux of small solutes including glycerol or ions, and in the fungal kingdom, five kinds of aquaporins have been described which are subdivided into orthodox aquaporins and aquaglyceroporins (Pettersson et al. 2005). Only one putative aquaporin gene has so far been described from the Glomeromycota (Aroca et al. 2009) although the beneficial effects of the symbiotic fungi include transport of water from soil to host plant roots (Smith and Read 1998). The function of the fungal gene *GintAQP1* (Fig. 1.4), identified in *G. intraradices*, has not yet been demonstrated. However, *GintAQP1* expression shows a variable response to different environmental stress and host plant, and it has been inferred that a certain compensatory mechanism exists between *GintAQP1* and host aquaporin gene expression pointing to communication mechanisms between mycelium and host root (Aroca et al. 2009).

2. Sugar Transporter Genes

Although in arbuscular mycorrhiza potential carbon compounds delivered to the fungal symbiont by the plant partner are soluble sugars, carboxylic acids and amino acid, **hexoses are considered to be the main source of carbon for the fungus** (Bago et al. 2000). In spite of this, only one gene encoding a **monosaccharide transporter** (*GpMST1*, Fig. 1.4) has been characterized from a glomeromycotan species (Schüssler et al. 2006, 2007). The gene was isolated from the unusual and culturable glomeromycotan fungus *Geosiphon pyriformis* which forms a unique symbiosis with cyanobacteria. The encoded protein, GpMST1, is a membrane domain transporter, which phylogenetically belongs to a new, not yet characterized, monosaccharide transporter (MST) clade. It functions as proton co-transporter (Fig. 1.4) with highest affinity for glucose, followed by mannose, galactose and fructose. The *GpMST1* gene has a very low GC content (32 %), typical of the Glomeromycota (see Sect. II. B) and contains six introns with unusual boundaries. *G. pyriformis* is not able to take up glucose, as is the case of "typical" arbuscular mycorrhizal Glomeromycota via the non-symbiotic plasma membrane (spores, germination hyphae, extraradical hyphae). It is supposed that GpMST1 may represent the type of MST that is responsible for the uptake of plant carbohydrates by "typical" glomeromycotan fungi at the symbiotic interface in mycorrhizal interactions. As such, it should enable the isolation and characterization of orthologues in mycorrhiza-forming Glomeromycota and so lead to a much better understanding of symbiotic carbon fluxes.

3. Plasma Membrane H^+-ATPase Encoding Genes

The activity of **proton-coupled membrane transporters** like phosphate or hexose transporters is dependent on electrochemical gradients which correspond to the H^+ gradient created by H^+-ATPases in the membranes. Plasma membrane H^+-ATPases are generally found in higher concentrations or activities in cell types that are specialized for intensive active transport, as is the case of the fungal and plant membranes at the symbiotic interface of arbuscular mycorrhiza (Gianinazzi-Pearson et al. 1991, 2000; Harrison 2005). In the Glomeromycota, two **H^+-ATPase genes**, *GmPMA1* (Requena et al. 2003) and *GmHA5* (Ferrol et al. 2000; Requena et al. 2003), have been isolated from the genome of *G. mosseae* (Fig. 1.4). *GmPMA1* is highly expressed during asymbiotic development, but its expression is not modified during symbiotic interactions with host roots, whereas *GmHA5* is induced upon plant recognition at the appressorium stage and up-regulated in the arbuscule (Balestrini et al. 2007). Both genes are highly expressed during intraradical development, but their expression is reduced in the ERM. Phosphate, the key nutrient transferred from fungus to plant in the symbiosis, induces *GmHA5* expression during asymbiotic growth, whereas sucrose has a negative effect. It has been suggested that different fungal H^+-ATPases isoforms might be recruited at different developmental stages of the mycorrhizal fungus, possibly in response to the different requirements during its life cycle (Requena et al. 2003).

With wider coverage of the transcriptome of *G. intraradices*, it is expected that many more glomeromycotan transporters will be characterized. Deciphering the function of individual proteins of the entire transportome, as well as understanding how transporters from intracellular organelles work together with those from the plasma membrane in different fungal structures and how they are regulated to ensure homeostasis, will give more insight into the symbiotic life style and physiology of these obligate biotrophs.

V. Concluding Remarks

Although obligate biotrophy, multinucleate nature, large genomes and asexuality of glomeromycotan fungi forming root symbiosis still remain challenges to understanding genome complexity, diversity and function in these organisms, the past decade has seen considerable advances in knowledge about their genomic make-up. Characterization of nuclear and mitochondrial genomes in the Glomeromycota has reinforced conclusions that they form a distinct, heterogeneous (Schüssler et al. 2001; James et al. 2006) and unusual (Gianinazzi-Pearson et al. 2001, 2004; Sanders and Croll 2010) ensemble within the fungal kingdom. However, much of the information generated by more recent research focuses on *G. intraradices* (syn. *G. irregulare*) as a model species and although symbiotic attributes of the genome may be broadly based across the Glomeromycota, basic parameters need to be more widely explored in order to draw conclusions for other members of the phylum. Also, whilst strategies like wide genome sequencing and comparative genomics have become amenable to research on glomeromycotan fungi, the lack of a stable transformation system remains a major drawback to manipulating their genome for functional studies of genes (Gianinazzi-Pearson et al. 2004; Helber and Requena 2008). The magnitude of intraspecific diversity within the genome still needs to be clearly defined for a large number of glomeromycotan species. This implies exploring target genes as well as rDNA in a larger spectrum of isolates and establishing more extensive sequence data sets to evaluate allele frequencies within individuals.

Acknowledgements Part of the work reported in this review article was supported by the Burgundy Regional Council and the Agence National de Recherche (project TRANSMUT, ANR-10-BLAN-1604-01), France, the Education Ministry of China (scholarship to Y.L.) and the Spanish Ministry of Science and Innovation (Project AGL2009-08868).

References

Akiyama K, Matsuzaki K, Hayashi H (2005) Plant sesquiterpenes induce hyphal branching in arbuscular mycorrhizal fungi. Nature 435:824–827

Allen JW, Shachar-Hill Y (2009) Sulfur transfer through an arbuscular mycorrhiza. Plant Physiol 149:549–560

Angelard C, Sanders IR (2011) Effect of segregation and genetic exchange on arbuscular mycorrhizal fungi in colonization of roots. New Phytol 189:652–657

Angelard C, Colard A, Niculita-Hirzel H, Croll D, Sanders IR (2010) Segregation in a mycorrhizal fungus alters rice growth and symbiosis-specific gene transcription. Curr Biol 20:1216–1221

Aroca R, Bago A, Sutka M, Paz JA, Cano C, Amodeo G, Ruiz-Lozano JM (2009) Expression analysis of the first arbuscular mycorrhizal fungi aquaporin described reveals concerted gene expression between salt-stressed and nonstressed mycelium. Mol Plant Microbe Interact 9:1169–1178

Bago B, Pfeffer PE, Shachar-Hill Y (2000) Carbon metabolism and transport in arbuscular mycorrhizas. Plant Physiol 124:949–950

Bago B, Zipfel W, Williams RM, Jun J, Arreola R, Lammers PJ, Pfeffer PE, Shachar-Hill Y (2002) Translocation and utilization of fungal storage lipid in the arbuscular mycorrhizal symbiosis. Plant Physiol 128:108–124

Bago B, Pfeffer PE, Abubaker J, Jun J, Allen JW, Brouillette J, Douds DD, Lammers PJ, Shachar-Hill Y (2003) Carbon export from arbuscular mycorrhizal roots involves the translocation of carbohydrate as well as lipid. Plant Physiol 131:1496–1507

Balestrini R, Gómez-Ariza J, Lanfranco L, Bonfante P (2007) Laser microdissection reveals that transcripts for five plant and one fungal phosphate transporter genes are contemporaneously present in arbusculated cells. Mol Plant Microbe Interact 9:1055–1062

Benabdellah K, Azcón-Aguilar C, Ferrol N (2009a) GintGRX1, the first characterized glomeromycotan glutaredoxin, is a multifunctional enzyme that responds to oxidative stress. Fungal Genet Biol 46:94–104

Benabdellah K, Valderas A, Azcón-Aguilar C, Ferrol N (2009b) PDX1, a protein involved in vitamin B6 biosynthesis, is transcriptionally up-regulated by oxidative stress in the arbuscular mycorrhizal fungus *Glomus intraradices*. New Phytol 184:682–693

Benedetto A, Magurno F, Bonfante P, Lanfranco L (2005) Expression profiles of a phosphate transporter gene (*GmosPT*) from the endomycorrhizal fungus *Glomus mosseae*. Mycorrhiza 15:620–627

Besserer A, Puech-Pages V, Kiefer P, Gomez-Roldan V, Jauneau A, Roy S, Portais JC, Roux C, Bécard G, Séjalon-Delmas N (2006) Strigolactones stimulate arbuscular mycorrhizal fungi by activating mitochondria. PLoS Biol 4:e226

Besserer A, Bécard G, Jauneau A, Roux C, Séjalon-Delmas N (2008) GR24, a synthetic analog of strigolactones, stimulates the mitosis and growth of the arbuscular mycorrhizal fungus *Gigaspora rosea* by boosting its energy metabolism. Plant Physiol 148:402–413

Bestel-Corre G, Dumas-Gaudot E, Poinsot V, Dieu M, van Tuinen D, Remacle J, Gianinazzi-Pearson V, Gianinazzi S (2002) Proteome analysis and identification of symbiosis-related proteins from *Medicago truncatula* Gaertn. by two-dimensional electrophoresis and mass spectrometry. Electrophoresis 23:122–137

Bever JD, Wang M (2005) Arbuscular mycorrhizal fungi – hyphal fusion and multigenomic structure. Nature 433:E3–E4

Bianciotto V, Bonfante P (1993) Evidence of DNA replication in an arbuscular mycorrhizal fungus in the absence of a host plant. Protoplasma 176:100–105

Bianciotto V, Barbiero G, Bonfante P (1995) Analysis of the cell cycle in an arbuscular mycorrhizal fungus by flow cytometry and bromodeoxyuridine labelling. Protoplasma 188:161–169

Bidochka MJ, Khachatourians GG (1990) Identification of *Beauveria bassiana* extracellular protease as a virulence factor in pathogenicity toward the migratory grasshopper, *Melanoplus sanguinipes*. J Invert Pathol 56:362–370

Börstler B, Thiery O, Sykorova Z, Berner A, Redecker D (2010) Diversity of mitochondrial large subunit rDNA haplotypes of *Glomus intraradices* in two agricultural field experiments and two semi-natural grasslands. Mol Ecol 19:1497–1511

Brechenmacher L, Weidmann S, van Tuinen D, Chatagnier O, Gianinazzi S, Franken P, Gianinazzi-Pearson V (2004) Expression profiling of up-regulated plant and fungal genes in early and late stages of *Medicago truncatula*-*Glomus mosseae* interactions. Mycorrhiza 14:253–262

Breuninger M, Requena N (2004) Recognition events in AM symbiosis: analysis of fungal gene expression at the early appressorium stage. Fungal Genet Biol 41:794–804

Burt A, Carter DA, Koenig GL, White TJ, Taylor JW (1996) Molecular markers reveal cryptic sex in the human pathogen *Coccidioides immitis*. Proc Natl Acad Sci USA 93:770–773

Cappellazzo G, Lanfranco L, Fitz M, Wipf D, Bonfante P (2008) Characterization of an amino acid permease from the endomycorrhizal fungus *Glomus mosseae*. Plant Physiol 147:429–437

Carpentier SC, Panis B, Vertommen A, Swennen R, Sergeant K, Renaut J, Laukens K, Witters E, Samyn B, Devreese B (2008) Proteome analysis of non-model organisms: a challenging but powerful approach. Mass Spectrom Rev 27:354–377

Casana MC, Bonfante P (1988) Intercellular hyphae and arbuscules of *Glomus fasciculatum* (Thaxter) Gerd & Trappe isolated after enzymatic digestion. Allionia 25:17–25

Chevalier BS, Stoddard BL (2001) Homing endonucleases: structural and functional insight into the catalysts of intron/intein mobility. Nucleic Acids Res 29:3757–3774

Clapp JP, Fitter AH, Young JPW (1999) Ribosomal small subunit sequence variation within spores of an arbuscular mycorrhizal fungus, *Scutellospora* sp. Mol Ecol 8:915–921

Croll D, Sanders IR (2009) Recombination in *Glomus intraradices*, a supposed ancient asexual arbuscular mycorrhizal fungus. BMC Evol Biol 9:13

Croll D, Giovannetti M, Kich AM, Sbrana C, Ehinger M, Lammers PJ, Sanders IR (2009) Nonself vegetative fusion and genetic exchange in the arbuscular mycorrhizal fungus *Glomus intraradices*. New Phytol 181:924–937

Dalgaard JZ, Klar AJ, Moser MJ, Holley WR, Chatterjee A, Mian IS (1997) Statistical modeling and analysis of the LAGLIDADG family of site-specific endonucleases and identification of an intein that encodes a site-specific endonuclease of the HNH family. Nucleic Acids Res 25:4626–4638

Delp G, Smith SE, Barker SJ (2000) Isolation by differential display of three partial cDNAs potentially coding for proteins from the VA mycorrhizal *Glomus intraradices*. Mycol Res 104:293–300

den Bakker HC, VanKuren NW, Morton JB, Pawlowska TE (2010) Clonality and recombination in the life history of an asexual arbuscular mycorrhizal fungus. Mol Biol Evol 27:2474–2486

Dumas-Gaudot E, Valot B, Bestel-Corre G, Recorbet G, St-Arnaud M, Dieu M, Raes M, Saravanan RS, Gianinazzi S (2004) Proteomics as a way to identify extra-radicular fungal proteins from *Glomus intraradices*-RiT-DNA carrot roots. FEMS Microbiol Ecol 48:401–411

Ferrol N, Barea JM, Azcón-Aguilar C (2000) The plasma membrane H^+-ATPase gene family in the arbuscular mycorrhizal fungus *Glomus mosseae*. Curr Genet 37:112–118

Ferrol N, Azcón-Aguilar C, Bago B, Franken P, Gollotte A, González-Guerrero M, Harrier LA, Lanfranco L, van Tuinen D, Gianinazzi-Pearson V (2004) Genomics of arbuscular mycorrhizal fungi. In: Arora DK, Khachatourians GG (eds) Applied mycology and biotechnology, vol 4, Fungal Genomics. Elsevier Science BV, St Louis, pp 379–403

Fontaneto D, Herniou EA, Boschetti C, Caprioli M, Melone G, Ricci C, Barraclough TG (2007) Independently evolving species in asexual bdelloid rotifers. PLoS Biol 5:914–921

Gandolfi A, Bonilauri P, Rossi V, Menozzi P (2001) Intraindividual and intraspecies variability of ITS1 sequences in the ancient asexual *Darwinula stevensoni* (Crustacea: Ostracoda). Heredity 87:449–455

Gemayel R, Vinces MD, Legendre M, Verstrepen KJ (2010) Variable tandem repeats accelerate evolution of coding and regulatory sequences. Annu Rev Genet 44:445–468

Gianinazzi S, Gollotte A, Binet M-N, van Tuinen D, Redecker D, Wipf D (2010) Agroecology: the key role of arbuscular mycorrhizas in ecosystem services. Mycorrhiza 20:519–530

Gianinazzi-Pearson V, Smith SE, Gianinazzi S, Smith FA (1991) Enzymatic studies on the metabolism of vesicular-arbuscular mycorrhizas. V. Is H^+-ATPase a component of ATP-hydrolysing enzyme activities in plant–fungus interfaces? New Phytol 117:61–74

Gianinazzi-Pearson V, Arnould C, Oufattole M, Arango M, Gianinazzi S (2000) Differential activation of H^+–ATPase genes by an arbuscular mycorrhizal fungus in root cells of transgenic tobacco. Planta 211:609–613

Gianinazzi-Pearson V, van Tuinen D, Dumas-Gaudot E, Dulieu H (2001) Exploring the genome of glomalean fungi. In: Hock B (ed) The mycota IX. Springer, Berlin, pp 3–17

Gianinazzi-Pearson V, Azcón-Aguilar C, Bécard G, Bonfante P, Ferrol N, Franken P, Gollotte A, Harrier LA, Lanfranco L, van Tuinen D (2004) Structural and functional genomics of symbiotic arbuscular mycorrhizal fungi. In: Tkacz JS, Lange L (eds) Advances in fungal biotechnology for industry, agriculture and medicine. Kluwer, New York, pp 405–424

Gianinazzi-Pearson V, Séjalon-Delmas N, Genre A, Jeandroz S, Bonfante P (2007) Plants and arbuscular mycorrhizal fungi: cues and communication in the early steps of symbiotic interactions. Adv Bot Res 46:181–219

Giovannetti M, Fortuna P, Citernesi AS, Morini S, Nuti MP (2001) The occurrence of anastomosis formation and nuclear exchange in intact arbuscular mycorrhizal networks. New Phytol 151: 717–724

Giovannetti M, Sbrana C, Strani P, Agnolucci M, Rinaudo V, Avio L (2003) Genetic diversity of isolates of *Glomus mosseae* from different geographic areas detected by vegetative compatibility testing and biochemical and molecular analysis. Appl Environ Microbiol 69:616–624

Gladyshev EA, Arkhipova IR (2010) Genome structure of bdelloid rotifers: shaped by asexuality or desiccation? J Hered 101:S85–S93

Gollotte A, L'Haridon F, Chatagnier O, Wettstein G, Arnould C, van Tuinen D, Gianinazzi-Pearson V (2006) Repetitive DNA sequences include retrotransposons in genomes of the Glomeromycota. Genetica 128:455–460

Gomez SK, Javot H, Deewatthanawong P, Torres-Jerez I, Tang Y, Blancaflor EB, Udvardi MK, Harrison MJ (2009) *Medicago truncatula* and *Glomus intraradices* gene expression in cortical cells harboring arbuscules in the arbuscular mycorrhizal symbiosis. BMC Plant Biol 9:10

González-Guerrero M, Azcón-Aguilar C, Mooney M, Valderas A, MacDiarmid CW, Eide DJ, Ferrol N (2005) Characterization of a *Glomus intraradices* gene encoding a putative Zn transporter of the cation diffusion facilitator family. Fungal Genet Biol 42:130–140

González-Guerrero M, Cano C, Azcón-Aguilar C, Ferrol N (2007) *GintMT1* encodes a functional metallothionein in *Glomus intraradices* that responds to oxidative stress. Mycorrhiza 17:327–335

González-Guerrero M, Benabdellah K, Valderas A, Azcón-Aguilar C, Ferrol N (2010a) *GintABC1* encodes a putative ABC transporter of the MRP subfamily induced by Cu, Cd and oxidative stress in *Glomus intraradices*. Mycorrhiza 20:137–146

González-Guerrero M, Oger E, Benabdellah K, Azcón-Aguilar C, Lanfranco L, Ferrol N (2010b) Characterization of a Cu, Zn superoxide dismutase gene in the arbuscular mycorrhizal fungus *Glomus intraradices*. Curr Genet 56:265–274

Govindarajulu M, Pfeffer PE, Jin HR, Abubaker J, Douds DD, Allen JW, Bucking H, Lammers PJ, Shachar-Hill Y (2005) Nitrogen transfer in the arbuscular mycorrhizal symbiosis. Nature 435:819–823

Harrier LA, Wright F, Hooker JE (1998) Isolation of the 3-phosphoglycerate kinase gene of the arbuscular mycorrhizal fungus *Glomus mosseae* (Nicol. and Gerd.) Gerdemann and Trappe. Curr Genet 34: 386–392

Harrison MJ (2005) Signaling in the arbuscular mycorrhizal symbiosis. Annu Rev Microbiol 59:19–42

Harrison MJ, van Buuren ML (1995) A phosphate transporter from the mycorrhizal fungus *Glomus versiforme*. Nature 378:626–629

Heckman DS, Geiser DM, Eidell BR, Stauffer RL, Kardos NL, Hedges SB (2001) Molecular evidence for the early colonization of land by fungi and plants. Science 293:1129–1133

Helber N, Requena N (2008) Expression of the fluorescence markers DsRed and GFP fused to a nuclear localization signal in the arbuscular mycorrhizal fungus *Glomus intraradices*. New Phytol 177:537–548

Heupel S, Roser B, Kuhn H, Lebrun MH, Villalba F, Requena N (2009) Erl1, a novel Era-like GTPase from *Magnaporthe oryzae*, is required for full root virulence and is conserved in the mutualistic symbiont *Glomus intraradices*. Mol Plant Microbe Interact 23:67–81

Hijri M, Sanders IR (2004) The arbuscular mycorrhizal fungus *Glomus intraradices* is haploid and has a small genome in the lower limit of eukaryotes. Fungal Genet Biol 41:253–261

Hijri M, Sanders IR (2005) Low gene copy number shows that arbuscular mycorrhizal fungi inherit genetically different nuclei. Nature 433:160–163

Hildebrandt U, Ouziad F, Marner FJ, Bothe H (2006) The bacterium *Paenibacillus validus* stimulates growth of the arbuscular mycorrhizal fungus *Glomus intraradices* up to the formation of fertile spores. FEMS Microbiol Lett 254:258–267

Hosny M, Gianinazzi-Pearson V, Dulieu H (1998a) Nuclear contents of eleven fungal species in Glomales. Genome 41:422–428

Hosny M, de Barros P, Gianinazzi-Pearson V, Dulieu H (1998b) Base composition of DNA from glomalean fungi: high amounts of methylated cytosine. Fungal Genet Biol 22:103–111

James TY, Kauff F, Schoch CL, Matheny PB, Hofstetter V, Cox C, Celio G, Gueidan C, Fraker E, Miadlikowska J et al (2006) Reconstructing the early evolution of the fungi using a six gene phylogeny. Nature 443:818–822

Jany JL, Pawlowska TE (2010) Multinucleate spores contribute to evolutionary longevity of asexual Glomeromycota. Am Nat 175:424–435

Journet EP, van Tuinen D, Gouzy J, Crespeau H, Carreau V, Farmer MJ, Niebel A, Schiex T, Jaillon O, Chatagnier O, Godiard L, Micheli F, Kahn D, Gianinazzi-Pearson V, Gamas P (2002) Exploring root symbiotic programs in the model legume *Medicago truncatula* using EST analysis. Nucleic Acids Res 30:5579–5592

Judson OP, Normark BB (1996) Ancient asexual scandals. Trends Ecol Evol 11:A41–A46

Jun J, Abubaker J, Rehrer C, Pfeffer PE, Shachar Hill Y, Lammers PJ (2002) Expression in an arbuscular mycorrhizal fungus of genes putatively involved in metabolism, transport, the cytoskeleton and the cell cycle. Plant Soil 244:141–148

Kloppholz S, Kuhn H, Requena N (2011) A secreted fungal effector of *Glomus intraradices* promotes symbiotic biotrophy. Curr Biol 21:1204–1209

Kuhn G, Hijri M, Sanders IR (2001) Evidence for the evolution of multiple genomes in arbuscular mycorrhizal fungi. Nature 414:745–748

Kuznetsova E, Seddas-Dozolme PMA, Arnould C, Tollot M, van Tuinen D, Borisov A, Gianinazzi S, Gianinazzi-Pearson V (2010) Symbiosis-related pea genes modulate fungal and plant gene expression during the arbuscule stage of mycorrhiza with *Glomus intraradices*. Mycorrhiza 20:427–443

Lammers PJ, Jun J, Abubaker J, Arreola R, Gopalan A, Bago B, Hernandez-Sebastia C, Allen JW, Douds DD, Pfeffer PE, Shachar-Hill Y (2001) The glyoxylate cycle in an arbuscular mycorrhizal fungus. Carbon flux and gene expression. Plant Physiol 127:1287–1298

Lammers P, Tuskan GA, DiFazio SP, Podila GK, Martin F (2004) Mycorrhizal symbionts of *Populus* to be sequenced by the United States Department of Energy's Joint Genome Institute. Mycorrhiza 14:63–64

Lanfranco L, Bolchi A, Cesale RE, Ottonello S, Bonfante P (2002) Differential expression of a metallothionein gene during the presymbiotic versus the symbiotic phase of an arbucular mycorrhizal fungus. Plant Physiol 130:58–67

Lanfranco L, Novero M, Bonfante P (2005) The mycorrhizal fungus *Gigaspora margarita* possesses a CuZn superoxide dismutase which is up-regulated during the symbiosis with legume hosts. Plant Physiol 137:1319–1330

Leake J, Johnson D, Donnelly D, Muckle G, Boddy L, Read D (2004) Networks of power and influence: the role of mycorrhizal mycelium in controlling plant communities and agroecosystem functioning. Can J Bot 82:1016–1045

Lee J, Young JPW (2009) The mitochondrial genome sequence of the arbuscular mycorrhizal fungus *Glomus intraradices* isolate 494 and implications for the phylogenetic placement of *Glomus*. New Phytol 183:200–211

Li ZS, Szczypka M, Lu YP, Thiele DJ, Rea PA (1996) The yeast cadmium factor protein (YCF1) is a vacuolar glutathione S-conjugate pump. J Biol Chem 271:6509–6517

Lopez-Pedrosa A, González-Guerrero M, Valderas A, Azcón-Aguilar C, Ferrol N (2006) *GintAMT1* encodes a functional high-affinity ammonium transporter that is expressed in the extraradical mycelium of *Glomus intraradices*. Fungal Genet Biol 43:102–110

Maldonado-Mendoza IE, Dewbre GR, Harrison MJ (2001) A phosphate transporter gene from the extraradical mycelium of an arbuscular mycorrhizal fungus *Glomus intraradices* is regulated in response to phosphate in the environment. Mol Plant Microbe Interact 14:1140–1148

Maldonado-Mendoza IE, Dewbre GR, Van Buuren ML, Versaw WK, Harrison MJ (2002) Methods to estimate the proportion of plant and fungal RNA in an arbuscular mycorrhiza. Mycorrhiza 12:67–74

Martin F, Aerts A, Ahren D, Brun A, Danchin EGJ, Duchaussoy F, Gibon J, Kohler A, Liindquist E, Pereda V et al (2008a) The genome of *Laccaria bicolor* provides insights into mycorrhizal symbiosis. Nature 452:88–92

Martin F, Gianinazzi-Pearson V, Hijri M, Lammers PJ, Requena N, Sanders IR, Shachar-Hill Y, Shapiro H, Tuskan GA, Young JPW (2008b) The long hard road to a completed *Glomus intraradices* genome. New Phytol 180:747–750

Martin F, Kohler A, Murat C, Balestrini R, Coutinho PM, Jaillon O, Montanini B, Morin E, Noel B, Percudani R et al (2010) Périgord Black Truffle genome uncovers evolutionary origins and mechanisms of symbiosis. Nature 464:1033–1038

Mitchell JI, Roberts PJ, Moss ST (1995) Sequence or structure? A short review on the application of nuclei acid sequence information to fungal taxonomy. Mycologist 9:67–75

Morton JB (1993) Problems and solutions for the integration of glomalean taxonomy, systematics biology, and the study of endomycorrhizal phenomena. Mycorrhiza 2:97–109

Muller HJ (1932) Some genetic aspects of sex. Am Nat 66:118–138

O'Donnell K (1992) Ribosomal DNA internal transcribed spacers are highly divergent in the phytopathogenic ascomycete *Fusarium sambucinum (Gibberella pulicaris)*. Curr Gen 22:213–220
Oeljeklaus S, Meyer HE, Warscheid B (2008) Advancements in plant proteomics using quantitative mass spectrometry. J Proteomics 72:545–554
Ouziad F, Hildebrandt U, Schmelzer E, Bothe H (2005) Differential gene expressions in arbuscular mycorrhizal-colonized tomato grown under heavy metal stress. J Plant Physiol 162:634–649
Pawlowska TE, Taylor JW (2004) Organization of genetic variation in individuals of arbuscular mycorrhizal fungi. Nature 427:733–737
Pérez-Tienda J, Snchez-Testillano P, Balestrini R, Fiorilli V, Azcón-Aguilar C, Ferrol N (2011) GintAMT2, a new member of the ammonium transporter family in the arbuscular mycorrhizal fungus *Glomus intraradices*. Fungal Genet Biol 48:1044–1055
Pettersson N, Filipsson C, Becit E, Brive L, Hohmann S (2005) Aquaporins in yeasts and filamentous fungi. Biol Cell 97:487–500
Pivato B, Mazurier S, Lemanceau P, Siblot S, Berta G, Mougel C, van Tuinen D (2007) *Medicago* species affect the community composition of arbuscular mycorrhizal fungi associated with roots. New Phytol 176:197–210
Purin S, Morton JB (2011) In situ analysis of anastomosis in representative genera of arbuscular mycorrhizal fungi. Mycorrhiza 21:505–514
Raab PA, Breenwald A, Redecker D (2005) Mitochondrial large ribosomal sub-unit sequences are homogenous within isolates of *Glomus* (arbuscular mycorrhizal fungi, Glomeromycota). Mycol Res 109:1315–1322
Recorbet G, Rogniaux H, Gianinazzi-Pearson V, Dumas-Gaudot E (2009) Fungal proteins in the extra-radical phase of mycorrhiza: a shotgun proteomic picture. New Phytol 181:248–260
Recorbet G, Valot B, Robert F, Gianinazzi-Pearson V, Dumas-Gaudot E (2010) Identification of *in planta*-expressed arbuscular mycorrhizal fungal proteins upon comparison of the root proteomes of *Medicago truncatula* colonised with two *Glomus* species. Fungal Genet Biol 47:608–618
Reddy PV, Lam CK, Belanger FC (1996) Mutualistic fungal endophytes express a proteinase that is homologous to proteases suspected to be important in fungal pathogenicity. Plant Physiol 111:1209–1218
Redecker D (2000) Specific PCR primers to identify arbuscular mycorrhizal fungi within colonized roots. Mycorrhiza 10:73–80
Requena N, Mann P, Hampp R, Franken P (2002) Early developmentally regulated genes in the arbuscular mycorrhizal fungus *Glomus mosseae*: identification of *GmGIN*1, a novel gene with homology to the C-terminus of metazoan hedgehog proteins. Plant Soil 244:129–139
Requena N, Breuninger M, Franken P, Ocon A (2003) Symbiotic status, phosphate, and sucrose regulate the expression of two plasma membrane H^+-ATPase genes from the mycorrhizal fungus *Glomus mosseae*. Plant Physiol 132:1540–1549
Rohrbough JG, Galgiani JN, Wysocki VH (2007) The application of proteomic techniques to fungal protein identification and quantification. Ann NY Acad Sci 1111:133–146
Rosendahl S (2008) Communities, populations and individuals of arbuscular mycorrhizal fungi. New Phytol 178:253–266
Rosendahl S, McGee P, Morton JB (2009) Lack of global population genetic differentiation in the arbuscular mycorrhizal fungus *Glomus mosseae* suggests a recent range expansion which may have coincided with the spread of agriculture. Mol Ecol 18:4316–4329
Sanders IR, Croll D (2010) Arbuscular mycorrhiza: a challenge to understand the genetics of the fungal partner. Annu Rev Genet 44:271–292
Sanders IR, Alt M, Groppe K, Boller T, Wiemken A (1995) Identification of ribosomal DNA polymorphisms among and within spores of the Glomales: application to studies on the genetic diversity of arbuscular mycorrhizal fungal communities. New Phytol 130:419–427
Sawaki H, Saito M (2001) Expressed genes in the extraradical hyphae of an arbuscular mycorrhizal fungus, *Glomus intraradices*, in the symbiotic phase. FEMS Microbiol Lett 195:109–113
Schüssler A, Walker C (2010) The Glomeromycota: a species list with new families and new genera. Royal Botanic Gardens Edinburgh, Kew; Botanische Staatssammlung, Munich; Oregon State University; www.amf-phylogeny.com
Schüssler A, Schwarzott D, Walker C (2001) A new fungal phylum, the Glomeromycota: phylogeny and evolution. Mycol Res 105:1413–1421
Schüssler A, Martin H, Cohen D, Fitz M, Wipf D (2006) Characterization of a carbohydrate transporter from symbiotic glomeromycotan fungi. Nature 444:933–936
Schüssler A, Martin H, Cohen D, Fitz M, Wipf D (2007) Arbuscular mycorrhiza – the first glomeromycotan sugar transporter is isolated and characterized. Plant Signal Behav 2:314–317
Seddas PMA, Arias CM, Arnould C, van Tuinen D, Godfroy O, Benhassou HA, Gouzy J, Morandi D, Dessaiint F, Gianinazzi-Pearson V (2009) Symbiosis-related plant genes modulate molecular responses in an arbuscular mycorrhizal fungus during early root interactions. Mol Plant Microbe Interact 22:341–351
Sedzielewska KA, Fuchs J, Temsch EM, Baronian K, Watzke R, Kunze G (2011) Estimation of the *Glomus intraradices* nuclear DNA content. New Phytol 192:794–797
Simon L, Lalonde M, Bruns TD (1992) Specific amplification of the 18S fungal ribosomal genes from vesicular-arbuscular endomycorrhizal fungi colonizing roots. Appl Environ Microbiol 58: 291–295
Simon L, Bousquet J, Lévesque RC, Lalonde M (1993) Origin and diversification of endomycorrhizal

fungi and coincidence with vascular land plants. Nature 363:67–69
Smith SE, Read DJ eds. (2008) Mycorrhizal symbiosis. Academic Press, London, pp 800
Spanu PD, Abbott JC, Amselem J, Burgis TA, Soanes DM, Stüber K, van Themaat EVL, Brown JKM et al (2010) Genome expansion and gene loss in powdery mildew fungi reveal tradeoffs in extreme parasitism. Science 330:1543–1546
Sreedhar L, Kobayashi DY, Bunting TE, Hillman BI, Belanger FC (1999) Fungal proteinase expression in the interaction of the plant pathogen *Magnaporthe poae* with its host. Gene 235:121–129
Stockinger H, Walker C, Schussler A (2009) '*Glomus intraradices* DAOM197198', a model fungus in arbuscular mycorrhiza research, is not *Glomus intraradices*. New Phytol 183:1176–1187
Stommel M, Mann P, Franken P (2001) Construction and analysis of an EST library using RNA from activated spores of the arbuscular mycorrhizal fungus *Gigaspora rosea*. Mycorrhiza 10:281–285
Sýkorová Z, Börstler B, Zvolenská S, Fehrer J, Gryndler M, Vosátka M, Redecker D (2011) Long-term tracing of *Rhizophagus irregularis* isolate BEG140 inoculated on *Phalaris arundinacea* in a coalmine spoil bank, using mitochondrial large subunit rDNA markers. Mycorrhiza. doi:10.1007/s00572-011-0375-1
Tamasloukht MB, Séjalon-Delmas N, Kluever A, Jauneau A, Roux C, Bécard G, Franken P (2003) Root factors induce mitochondrial-related gene expression and fungal respiration during development switch from asymbiosis to presymbiosis in the arbuscular mycorrhizal fungus *Gigaspora rosea*. Plant Physiol 131:1468–1478
Tamasloukht MB, Waschke A, Franken P (2007) Root exudates-stimulated RNA accumulation in the arbuscular mycorrhizal fungus *Gigaspora rosea*. Soil Biol Biochem 39:1824–1827
Thiéry O, Börstler B, Ineichen K, Redecker D (2010) Evolutionary dynamics of introns and homing endonuclease ORFs in a region of the large subunit of the mitochondrial rRNA in *Glomus* species (arbuscular mycorrhizal fungi, Glomeromycota). Mol Phylogen Evol 55:599–610
Tian C, Kasiborski B, Koul R, Lammers PJ, Bücking H, Shachar-Hill Y (2010) Regulation of the nitrogen transfer pathway in the arbuscular mycorrhizal symbiosis: gene characterization and the coordination of expression with nitrogen flux. Plant Physiol 153:1175–1187
Tisserant E, Kohler A, Dozolme-Seddas P, Balestrini R, Benabdellah K, Colard A, Croll D, Da Silva C, Gomez SK, Koul R, Ferrol N, Fiorilli V, Formey D, Franken Ph, Helber N, Hijri M, Lanfranco L, Lindquist E, Liu Y, Malbreil M, Morin E, Poulain J, Shapiro H, van Tuinen D, Waschke A, Azcón-Aguilar C, Bécard G, Bonfante P, Harrison MJ, Lammers P, Paszkowski U, Requena N, Rensing SA, Roux C, Sanders IR, Shachar-Hill Y, Tuskan G, Young JPW, Gianinazzi-Pearson V, Martin F (2012) The transcriptome of the arbuscular mycorrhizal fungus *Glomus intraradices* reveals functional tradeoffs in an obligate symbiont. New Phytol 193:755–769
Tollot M, Wong Sak Hoi J, van Tuinen D, Arnould C, Chatagnier O, Dumas B, Gianinazzi-Pearson V, Seddas PM (2009) An STE12 gene identified in the mycorrhizal fungus *Glomus intraradices* restores infectivity of a hemibiotrophic plant pathogen. New Phytol 181:693–707
Tommerup IC, Sivasithamparam K (1990) Zygospores and asexual spores of *Gigaspora decipiens*, an arbuscular mycorrhizal fungus. Mycol Res 94:897–900
Tunlid A, Rosen SE, Rask L (1994) Purification and characterization of an extracellular serine protease from the nematode-trapping fungus *Arthrobotrys oligospora*. Microbiology 140:1687–1695
van Tuinen D, Jacquot E, Zhao B, Gollotte A, Gianinazzi-Pearson V (1998) Characterization of root colonization profiles by a microcosm community of arbuscular mycorrhizal fungi using 25S rDNA-targeted nested PCR. Mol Ecol 7: 879–887
Waschke A, Sieh D, Tamasloukht M, Fischer K, Mann P, Franken P (2006) Identification of heavy metal-induced genes encoding glutathione S-transferases in the arbuscular mycorrhizal fungus *Glomus intraradices*. Mycorrhiza 17:1–10
Young JPW (2008) The genetic diversity of intraterrestrial aliens. New Phytol 178:465–468

2 Molecular Approaches to Arbuscular Mycorrhiza Functioning

P. Franken[1], A. Waschke[1], N. Requena[2]

CONTENTS

I. Introduction

Arbuscular mycorrhizal (AM) fungi are very particular in different aspects of their biology. Certain discrete features can be compared with those of other organisms, but the combination of so many relatively rare attributes makes them unique among the tree of life where they occupy their own distinct phylum, the Glomeromycota (Schüßler et al. 2001). These attributes, however, let AM fungi be very successful, because they seemed not to have evolved much since their appearance in the Ordovician together with the first bryophytes (Redecker et al. 2000). This is surprising taking into account that they are obligate symbionts not able to fulfil their life cycle without a plant and that they totally depend on an asexual mode of propagation. A further peculiarity is the absence of single cells. An AM fungus forms a coenocyticum, which is a tubular system from the spore up to the fine branches of the terminal structures, the arbuscules. Only during arbuscule senescence and spore formation is a separating membrane formed (Dickson and Smith 2001). These coenocytic hyphae are filled with organelles and nuclei. Each nucleus contains one probably haploid genome, but the genomes within a coenocyticum differ from each other (Hijri and Sanders 2004, 2005). Bacteria-like organisms living inside the cytoplasm of numerous AM fungi contribute even more to the 'supergenomes' of AM fungal coenocytica (Hosny et al. 1999; Bianciotto et al. 2000). Hyphae growing out from one spore can easily fuse and exchange their content, a process called anastomosis (Giovannetti et al. 1999). It is, however, unsettling that this occurs also between hyphae derived from the spores of different isolates (Croll et al. 2009). For all these characters it seems to be difficult to define the genetic unit which has to adapt to environmental conditions and evolves during time. The rules established for individuals cannot be applied, but the concepts of population biology might supply useful advice for AM fungal geneticists.

In the first instance, all these considerations do not affect the strategies and experimental approaches of molecular biologists. AM fungi have genes made of DNA which are transcribed to RNA and then translated to proteins. Many of these proteins are enzymes catalysing steps in biosynthetic pathways consuming and producing particular metabolites. These molecules can be identified and their amounts measured.

[1]Leibniz Institute of Vegetable and Ornamental Crops, Theodor-Echtermeyer-Weg 1, Grossbeeren 14979, Germany; e-mail: franken@igzev.de

[2]Karlsruhe Institute of Technology, Plant–Microbe Interactions Group, Hertzstrasse 16, Karlsruhe 76187, Germany

Fungal Associations, 2nd Edition
The Mycota IX
B. Hock (Ed.)

The results of such investigations provide new hypotheses and concepts about the mechanisms behind AM fungal development and their symbiotic functions. In the following, we will first summarise experimental systems and methodologies being used as molecular approaches for understanding AM fungal biology.

II. Experimental Systems

Because **AM fungi** are **obligate biotrophs**, they are not able to fulfil their life cycle without a plant and therefore cannot be axenically cultivated. However, the first steps of fungal development (asymbiotic spore germination, presymbiotic hyphal branching) can be analysed in vitro. This has been first carried out on **agar plates** amended with soil (Mosse 1959), but it turned out that pure buffered water agar was sufficient for analysing the influence of, e.g., inorganic and organic substrates (Siqueira et al. 1982), root exudates and flavonoids (Gianinazzi-Pearson et al. 1989) or bacteria (Tylka et al. 1991). A **liquid culture system** in chamber slides was introduced later to ease the harvest of high numbers of clean spores for different downstream methodologies (Tamasloukht et al. 2003). In order to analyse the influence of a living root on presymbiotic development, at first a **bicompartimental** system was established with a small Petri dish containing the fungus being placed in a larger plate with the root. This system showed the importance of root-derived CO_2 for hyphal growth (Bécard and Piché 1989). In double sandwich systems with a root covered by two membranes, physically separated spores germinated in close proximity to the host and showed typical branching patterns (Giovannetti et al. 1993). The next step in the fungal life cycle is the formation of hyphopodia on the epidermis of the roots upon physical contact. In order to enrich this developmental stage, Breuninger and Requena (2004) established a system in which early processes were synchronized and indeed enough hyphopodia could be harvested for the identification of genes expressed at this stage.

During symbiotic interactions with the plant, different structures can be distinguished, intraradical hyphae and arbuscules inside the roots and extraradical hyphae and spores (usually) outside. One important experimental system is the **root organ culture** originally developed by White (1943) and used by Mosse and Hepper (1975) for the first time for AM fungal research. An important step forward to permanent cultures was the **transformation of roots by *Agrobacterium rhizogenes*** (Tepfer 1984). Such hairy roots were for the first time used in 1987 (Mugnier and Mosse 1987), and Bécard and Fortin (1988) achieved a complete lifecycle in such a system. Further improvements was the introduction of **two compartments on a Petri plate** to separate the extraradical mycelium from the root (St-Arnaud et al. 1996) and the use of a liquid medium from which metabolically active hyphae could be easily harvested after approximately 2 weeks of growth (Sawaki and Saito 2001). Such root organ cultures can be applied for a lot of different analyses of AM fungal biology, but it has to be taken into account that the carbohydrates delivered to the fungus are not photosynthetically obtained in shoots, but artificially from the medium. Extraradical hyphae for molecular analyses have been also obtained from **pot cultures** by collecting them with forceps under a binocular after harvest of the roots (Harrison and van Buuren 1995), but this is a time-consuming procedure and the amounts obtained are low. Therefore a **compartment system** established by Redecker et al. (1995) was applied for RNA extraction from extraradical hyphae (Bütehorn et al. 1999). This consisted of a box with different compartments. One of these compartments was separated by a nylon mesh with a pore size too small for plant roots and therefore contained only extraradical hyphae of the AM fungus. Filled with glass beads, fungal material could be rapidly harvested without applying mechanical stress for a time span that would bias the results of molecular analyses. A further improvement was the **placement of little compartments in normal pot cultures** which allowed the in-growth of extraradical hyphae, but not roots (Requena et al. 1996). An improvement of the method is to fill these compartments with a mixture of soil and glass beads combining the advantage of rapid harvest and the interaction with soil particles possessing a nutrient exchange capacity (Neumann and George 2005).

In contrast to extraradical structures, hyphae colonising the roots are nearly impossible to obtain free of any plant tissue. A method developed by Saito (1995), where **intraradical hyphae** have been **purified by an enzyme cocktail** containing cellulases and pectinases, allowed the measurement of fungal enzymatic activities, but RNA was very much degraded and could not be further used (Lapopin and Franken, unpublished data). The ratio of fungal to plant material in a mycorrhiza turned out to be very low (Maldonado-Mendoza et al. 2002) and it was therefore very difficult to find fungal molecules among mixtures extracted from the symbiosis. Up to now, the most successful method in this direction is the isolation symbiotic material by **laser micro-dissection** (Balestrini et al. 2007). With this method, cells containing arbuscules which are highly enriched for fungal molecules as compared to root fragments can be harvested and also used for array hybridisation (Gomez et al. 2009).

III. Methodologies

Initial attempts to get insights into the **molecular basis of AM fungal biology** were directed towards the enzymes involved in phosphate metabolism, such as alkaline phosphatase (Gianinazzi-Pearson and Gianinazzi 1976). The first fungal **protein extracts** analysed by **polyacrylamide gel electrophoresis** were obtained from spores (Sanders et al. 1992). Changes in polypeptide patterns in germinating spores induced by the presence of root exudates were later detected by **2D gel electrophoresis** (Samra et al. 1996) and the identification of fungal proteins by **mass spectrometry** analysis was achieved for extraradical hyphae from root organ cultures (Dumas-Gaudot et al. 2004). Later, fungal proteins could be also detected inside roots (Recorbet et al. 2010). In order to obtain **sequence information** from AM fungi, spores and extraradical hyphae were used as pure fungal biological material for **DNA extractions** (Simon et al. 1992). DNA extracts were used for **PCR amplification** of fragments from ribosomal RNA and protein-encoding genes (Simon et al. 1992; Kaldorf et al. 1994) or for the construction of genomic libraries in plasmids and phages (Zézé et al. 1994; Franken and Gianinazzi-Pearson 1996). Activated spores and prior germination could both be used for RNA extractions and RT-PCR (Franken et al. 1997), as well as for a first collection of expressed sequence tags (EST; Stommel et al. 2001). At the same time, EST databases have been also established for extraradical hyphae (Sawaki and Saito 2001), for germinating spores (Lanfranco et al. 2002; Jun et al. 2002) and for hypopodia (Breuninger and Requena 2004).

Comparing sequence information from cDNA and genomic DNA resulted in the elucidation of gene structure in AM fungi (Requena et al. 1999). In order to get insights into particular functions which are associated with developmental stages, attempts were made to identify genes being up- or down-regulated by certain stimuli or at particular stages. This was first achieved by the means of a differential RNA display (Requena et al. 1999) and later also by subtractive hybridisation (Requena et al. 2002).

Another possibility to distinguish between the different stages of fungal development is the use of **plant mutants** where fungal development is blocked at a particular stage. In order to distinguish, e.g., between early- and late-stage development, a pea mutant was introduced as an experimental system for gene expression studies which could be colonized by AM fungi, but only truncated arbuscules were formed (Lapopin et al. 1999). However, no fungal gene could be identified which was particularly expressed at early stages, but a number of genes were shown to be dependent in their expression on arbuscule development (Kuznetsova et al. 2010).

Since most of the symbiotic structures of AM fungi are located inside the plant roots, which can hardly be separated, the localisation of gene products is important for understanding the function of genes. In situ hybridisation (Kaldorf et al. 1998) or in situ RT-PCR (Seddas et al. 2008) can be used to detect fungal RNA accumulation in intraradical structures, while antibodies have been used for the immunolocalisation of proteins (Peretto et al. 1995). Further functional characterisation of gene products was gained by heterologous expression of the genes in other organisms. Using a yeast strain defective in the respective function, it was in that way possible to analyse the characteristics of the fungal phosphate transporter (Harrison and van Buuren 1995). However, not only the enzymes present in every organism can be tested. Recently a transcriptional activator (Tollot et al. 2009) from an AM fungus was shown to complement the respective mutations in pathogenic fungi, restoring their ability to infect plant tissues. Also the regulation of AM fungal genes can be analysed in heterologous systems, as the *Glomus mosseae* 3-phosphoglycerate kinase gene promoter was inducible by different hexoses in yeast (Harrier and Paterson 2002).

Besides nucleic acids and proteins, molecular approaches need also to be targeted towards smaller molecules in order to understand the metabolism of AM fungi. Again direct measurements are hampered by the low availability of pure biological material. At first particular **lipids, sterols** and **fatty acids** were detected in mycorrhizal roots by **chromatographic methods** which were discussed to be specific for the fungal partner of the symbiosis (Nagy et al. 1980). Combining chromatography with mass spectroscopy extended the use of lipid analysis to the identification and quantification of AM fungi (Schmitz et al. 1991). Further fungal metabolites analysed with these techniques include **chitin** (Frey et al. 1994), **amino acids** (Jin et al. 2005) and **saccharides** (Hooker et al. 2007). An important step forward to the understanding of AM fungal metabolism was the introduction of **nuclear magnetic resonance (NMR) spectroscopy**. At first natural-abundance ^{13}C spectra were used to analyse the sugar composition in spores (Bécard et al. 1991), later metabolic routes in AM fungi could be followed by using ^{13}C-labelled substrates (Shachar-Hill et al. 1995).

In the following, we will show how the combination of experimental systems and molecular techniques were used to get insights into AM fungal development and interaction with the plant.

IV. Arbuscular Mycorrhizal Fungal Development

A. From Dormant Spores to Presymbiotic Hyphal Branching

Asexually formed **chlamydospores** are the dominant propagules of AM fungi and are able to survive in a dormant stage for years (Tommerup 1983). Physical stimuli as the combination of humidity and temperature activate these spores and result in RNA accumulation prior to germination in *Gigaspora rosea* (Franken et al. 1997). This transcript pattern shows no changes when the spores finally germinate (Franken et al. 2000). In contrast, *Glomus mosseae* **spores and sporocarps** contain the same transcripts during dormancy and activation. Changes in expression patterns cannot be observed before asymbiotic germination (Franken et al. 2000), which is also mirrored by changes in polypeptide patterns (Avio and Giovannetti 1998).

Comparing RNA accumulation in asymbiotic germ tubes and symbiotic hyphae resulted in the identification of the gene *GmGIN1* (Requena et al. 2002). It is only expressed at the early stages of fungal development and encodes a protein with two domains. The C-terminus is probably involved in sterol-mediated splicing and activates thereby the N-terminus of the protein. This N-terminus shows homology to small GTPases and is able to complement a mutation in a gene of *Magnaporthe oryzae* which is important for the full virulence of this pathogenic fungus (Heupel et al. 2010).

The switch from asymbiotic to **presymbiotic AM fungal development** in the vicinity of the plant root can be mimicked by treatment with root exudates or by particular compounds like flavonoids or strigolactones. The extensive branching of the germinating hyphae goes along with dramatic changes in polypeptide patterns (Samra et al. 1996). Interestingly the first molecular analysis down to a particular gene was conducted during the stimulation of presymbiotic development by a so-called mycorrhiza helper bacterium, a particular *Bacillus subtilis* strain (Requena et al. 1999).

Differential display and RT-PCR showed induction of the *G. mosseae* gene *GmFOX2* while the hyphae branched due to bacterial treatment. The encoded fatty acid oxidase enzyme could be involved in the modification of steroid-like signals or in the degradation of lipids for obtaining carbohydrates for further hyphal growth. This was substantiated by a later report showing that transcripts for an acyl-CoA dehydrogenase (β-oxidation of lipids) are increased 2 h after treatment with root exudates (Bücking et al. 2008). However, the first molecular signs of the developmental switch were immediate increases of NAD and ATP during the first 10 min after strigolactone application, indicating an activation of the energy metabolism (Besserer et al. 2008). Induction of gene expression, in contrast, could be identified only after 1 h of stimulation with purified root exudates (Tamasloukht et al. 2003, 2007). Most of the regulated genes identified in *Gigaspora rosea* 90 min after the stimulus were involved in mitochondrial functioning, indicating an increased respiration (Tamasloukht et al. 2003). One of the genes identified encoded a pyruvate carboxylase (Tamasloukht et al. 2003). The high induction of this gene could be necessary for the observed dark fixation of CO_2 and its transfer to oxalacetate (Bago et al. 1999). Addition of root exudates also enhanced the expression of a proton-pumping ATPase (Tamasloukht et al. 2007). The homologous gene *GmHA5* in *G. mosseae* was previously shown to be expressed upon plant recognition (Requena et al. 2003).

The assumption that proton gradients could play a role in the regulation of hyphal development was later supported by proton flux measurements showing pH gradient changes upon stimulus by the plant (Ramos et al. 2008). Root exudate also stimulated the expression of a gene encoding a glutamine-fructose-6-phosphate aminotransferase involved in chitin biosynthesis (Bücking et al. 2008) probably supporting the growth of the fungal cell wall. Interestingly, chitin biosynthesis seems not to be necessary during asymbiotic growth, based on the observation that chitin synthase genes are only expressed during symbiotic stages (Lanfranco et al. 1999; Ubalijoro et al. 2001).

The results obtained up to now indicate that many gene products are already present in small amounts in asymbiotic dormant, activated or germinating spores, but the expression of numerous genes is enhanced upon stimulation of presymbiotic hyphal branching. These genes encode mostly proteins necessary for signal exchange with the plant and for the supply of energy and structural components.

B. From Physical Contact to Production of New Spores

Upon physical contact the fungal hyphae differentiate to a swollen structure called the **appressorium** and more recently renamed to the **hyphopodium.** These structures are difficult to separate for functional analysis, but they could be at least enriched for identifying genes particularly induced at this stage of fungal development (Breuninger and Requena 2004). This study indicated Ca^{2+} signalling being important in the perception of plant signals by the fungus. This has been well documented on the plant side where perception of fungal diffusible molecules induces calcium oscillations in root epidermal cells (Navazio et al. 2007; Chabaud et al. 2011). Transcriptional analysis of two different H^{+}-ATPase genes revealed that one was mainly expressed during asymbiotic development, while the second was highly induced at appressorium formation (Requena et al. 2003).

Following hyphopodium formation, the fungus penetrates the root epidermal layer either between cells or directly through the cell apoplast, following the route formed by the **prepenetration apparatus** (Genre and Bonfante 2007). In a study comparing RNA accumulation patterns during different stages of fungal development, numerous genes expressed upon the development of hyphopodia on the root surface were identified (Seddas et al. 2009). Interestingly, this expression pattern could not be observed in plant mutants where hyphopodium formation is still possible but no further symbiotic colonisation of the root takes place. The transcriptional activator STE12 has been shown to be important for this step in plant-pathogenic fungi. The corresponding orthologous gene from *G. intraradices* is up-regulated during penetration of the root and was shown to complement the mutant in the pathogen *Colletotrichum lindemuthianum* which was unable to infect plant tissues (Tollot et al. 2009).

After penetration, the fungus colonises the root between and through the cortical cells (early stages) before finally developing **intracellular arbuscules** (late stages) in deeper cell layers of the cortex. At the same time the **extraradical hyphal network** is established in the surrounding soil. Genes being expressed only in root-internal hyphae – one of them exclusively at late stages – were found in a subtractive cDNA library, but they showed no homology to any known sequence (Brechenmacher et al. 2004). One gene important for this internal colonisation has been recently identified in the genome of *G. intraradices* (Kloppholz et al. 2011). *GintSP7* encodes a peptide which is translocated into the plant cell nucleus where it binds to a transcriptional activator of defence-related genes. Functional analysis showed that this fungal effector counteracts with the plant immune system and in this way allows the biotrophic colonisation of the root.

Another gene being up-regulated already at early stages of colonization by *Gigaspora margarita* encodes a superoxide dismutase (SOD), indicating oxidative stress upon colonization probably caused by particular defence reactions of the plant (Lanfranco et al. 2005). Interestingly, the transcripts were never detected in arbuscules, but to a high extent in vesicles of *Glomus intraradices* colonizing *M. truncatula* roots (Seddas et al. 2008). In pea, however, high amounts of *GintSOD* RNA were identified in laser-dissected arbuscules and decreased to a basic level at late stages of the symbiosis (Kuznetsova et al. 2010). A different accumulation was observed for transcripts putatively encoding a stearoyl-CoA desaturase involved in lipid metabolism. This gene was not expressed in the late-stage pea mutant (Kuznetsova et al. 2010), but its RNA could be detected already during appressoria development in *M. truncatula* (Seddas et al. 2008). It is currently not clear if the discrepancies between the different studies are due to the higher sensitivity of micro-dissection compared to in situ RT-PCR, or if gene expression patterns are highly dependent on the plant being colonised. Gene expression patterns in symbiotic extraradical hyphae have not been studied under pure developmental aspects such as spore production and will be discussed in the following paragraphs.

V. Arbuscular Mycorrhizal Fungal Functions

A. Mineral Nutrition of the Plant

1. Phosphorus

Nutrition of the plant by **phosphate** is probably the most often studied character of AM symbiosis. It is clear that AM fungi can bridge the depletion zone surrounding the roots with their extraradical mycelium, reaching a much greater soil volume than the plant can do. Whether they, however, are able to dissolve phosphate from organic resources chemically unavailable for the plant is a matter of debate (Smith and Read 2008). Hyphae are presumed to have a high and a low affinity system to take up orthophosphate (Thomson et al. 1990). However, up to now, only a gene for a high affinity **phosphate transporter (PT)** has been cloned and characterised in detail (Harrison and van Buuren 1995; Maldonado-Mendoza et al. 2001). This gene is mainly expressed in the extraradical mycelium according to its role in taking up phosphate from the soil during symbiosis, but *PT* RNA accumulation has been also shown in intraradical structures, indicating that the amount of phosphate released to the plant partner might be controlled by the fungus (Benedetto and Lanfranco 2005; Balestrini et al. 2007).

Expression of the gene in extraradical hyphae is induced by low phosphate concentrations, but the overall phosphate status of the mycorrhiza symbiosis seems also to play a role (Maldonado-Mendoza et al. 2001). Transport across the membrane is driven by a proton gradient which has to be built up by the activity of an H^+-ATPase.

Corresponding genes have been detected in the genome of *G. mosseae* (Ferrol et al. 2000). Thus, while one of them appears constitutively expressed (*PMA1*), a second one (*HA5*) is induced upon plant recognition and mainly expressed during symbiotic stages (Requena et al. 2003). This has been confirmed by laser micro-dissected arbuscules where *HA5* was shown to be induced (Balestrini et al. 2007). The putative orthologue of *HA5* in *G. rosea* is induced by root exudates (Tamasloukht et al. 2007), which corresponds to exudate-stimulated ATPase activity and uptake of ^{32}P in germinating spores (Lei et al. 1991).

Once taken up, phosphate must be transported along the hyphae towards the plant. It has been suggested that this transport occurs in the form of **polyphosphate (polyP) granules** located in vacuoles moved by cytoplasmic streaming (Cox and Tinker 1976; Cox et al. 1980). How polyP is synthesised in the extraradical hyphae of AM fungi (and also in other fungi) and how it is degraded in the arbuscule cells before inorganic phosphate is released into the interfacial apoplast is, however, unknown. The activity of **phosphatases** seems to be involved, because the inhibition of such enzymes resulted in much less efflux of phosphate from isolated intraradical hyphae (Kojima and Saito 2004). The alkaline phosphatase of AM fungi has been long thought to be involved in this step and its enzyme has been used as a marker for symbiotic functioning (Tisserant et al. 1993). The gene was cloned from two different fungi and expression analyses showed that RNA accumulation was higher in intra- than in extaradicular hyphae or germinating spores (Aono et al. 2004). However, Boddington and Dodd (1998) showed that the activity of the enzyme does not correlate with the efficiency of the symbiosis. This was confirmed by analysing the expression in a late pea mutant where much higher amounts of RNA were detected in roots containing only truncated arbuscules than in the corresponding wild type (Kuznetsova et al. 2010). All together, very little is known about the fungal molecular mechanisms behind the symbiotic transfer of phosphate in the AM symbiosis, although it is of central importance for the interaction.

2. Nitrogen

Nitrogen is present in soil mainly in organic form, but numerous soil micro-organisms convert this form to **ammonium** or **nitrate.** Different forms of organic and inorganic nitrogen can be taken up by extraradical hyphae (Hawkins et al. 2000) and by germinating spores (Gachomo et al. 2009). Genes for ammonium and nitrate transporters, as well as for an amino acid permease, have been cloned and their expression is induced in the extaradicular mycelium by low inorganic or organic N concentrations (Lopez-Pedrosa et al. 2006; Cappellazzo et al. 2008; Tian et al. 2010) or after N re-supply (Perez-Tienda et al. 2011). Nitrate has to be reduced inside the hyphae and the sequence with homology to a nitrate reductase gene was the first sequence from a protein-encoding gene to be identified in an AM fungus (Kaldorf et al. 1994). Its expression was, however, only analysed inside roots (Kaldorf et al. 1998). Ammonium is assimilated by glutamate and the resulting glutamine is converted to **arginine** (Bago et al. 2001; Breuninger et al. 2004; Govindarajulu et al. 2005). Most of the genes encoding the corresponding enzymes are identified and expression analysis shows that they are rapidly induced after supplying, e.g., nitrate to the extraradicular mycelium (Tian et al. 2010). Arginine has been shown to play a crucial role for the translocation of nitrogen to the intraradical mycelium (Cruz et al. 2007). There it is catabolised to obtain urea and finally ammonium which is released into the apoplast (Bago et al. 2001; Govindarajulu et al. 2005; Tian et al. 2010). The whole pathway was deduced from experiments with labelled substrates and from the analysis of metabolites and the RNA accumulation of known genes.

Regulation of N metabolism in the AM fungus can be, however, also accomplished on a different level. RNA accumulation of a gene encoding a glutamine synthetase was barely affected while the activity of the corresponding enzyme was differentially regulated by the nitrogen sources being applied to the mycelium (Breuninger et al. 2004).

In summary, much more is known about the molecular background of nitrogen metabolism in the fungi than is known for phosphorus. One has to be aware, however, that ammonium and nitrate are relatively mobile and depletion zones do therefore usually not exist as long as the soil is not very dry. Moreover, plants require ten times more nitrogen than phosphorus. Nevertheless, AM fungi at least partially contribute to the N nutrition of the plant, and accordingly an ammonium transporter induced in arbuscule-containing cells has been identified (Güther et al. 2009).

3. Other Mineral Nutrients

Beside phosphorus and nitrogen, AM fungi support also plant nutrition of many other minerals (Marschner and Dell 1994), but only the mechanisms behind the uptake of **sulfur** and its transfer to the host have been studied (Allen and Shachar-Hill 2009). The expression of several genes encoding, e.g., a high affinity sulfate permease were analysed and the route of sulfur was followed by labelling experiments. This indicated that the uptake of sulfate is repressed in the presence of sulfur-containing amino acids and that these amino acids or the assimilated sulfate can be later found in the protein pools of both the fungus and the plant. Beside this comprehensive study, only a Zn transporter gene has been analysed (González-Guerrero et al. 2005). The expression of this gene is induced upon treatment of extraradicular hyphae with low concentrations of **zinc**, but is repressed if high concentrations were applied. This pattern was, however, interpreted as a response to zinc toxicity and not as a contribution to the supply of this trace element to the fungus or the plant. Based on the positive effects of mycorrhizal symbiosis in the case of deficiency in many other mineral nutrients, more research is necessary to understand the mechanisms behind the function of AM fungi as biofertilisers of plants.

B. Carbohydrate Metabolism

As obligate biotrophs it has been thought that AM fungi are totally dependent on the carbohydrate supply by their plant partner, but several studies have shown that they are able to take up hexoses and acetate and show a significant level of CO_2 dark fixation (Bago et al. 1999). However, triacylglycerides (TAGs) were not produced (Bago et al. 1999), which are the main form of storage lipids in AM fungi (Beilby and Kidby 1980). TAGs are also not quantitatively synthesised in the extraradicular mycelium, but only within hyphae inside the root (Pfeffer et al. 1999). This deficiency seems to be due to the absence of fatty acid synthase activity (Trépanier et al. 2005) and this absence is discussed as a putative reason why AM fungi depend on symbiosis to fulfil their life cycle.

The **carbohydrate metabolism** of AM fungi as it occurs following the state of the art is described in detail by Smith and Read (2008). Briefly, hexoses are taken up by the root-internal hyphae and directly converted to trehalose and glycogen, enabling the maintenance of a constant sink for a steady transfer of C from the plant to the fungus. Finally TAGs are synthesised, transported together with glycogen to the extraradical hyphae and broken down again to hexoses for growth and energy metabolism. Apart from this general view on the C flow through AM fungi, the first gene cloned from an AM fungus involved in carbohydrate metabolism encoded a 3-phosphoglycerate kinase (Harrier et al. 1998). The enzyme presents a housekeeping function involved in glycolysis and gluconeogenesis. Both pathways are active during asymbiosis (Bago et al. 1999) and symbiosis (Pfeffer et al. 1999). The protein, however, shows a higher abundance in a mycorrhiza compared to germinating sporocarps (Harrier and Sawczak 2000).

In order to further study the regulation of the gene, the promoter was cloned and its regulation by different carbon sources was analysed in yeast (Harrier and Paterson 2002). This revealed an up-regulation by glycerol, glucose and raffinose. The induction by glycerol produced during the catabolism of lipids and trehalose could be important during asymbiotic growth, while glucose believed to be translocated from the plant to the fungus (Solaiman and Saito 1997) is probably the inducer in symbiotic intraradical hyphae. The role of raffinose is more difficult to understand, but it might be worthwhile to note that raffinose is used in some plant families as an alternative for carbohydrate translocation (Zimmermann and Ziegler 1975). Even more interesting, raffinose is highly effective in stimulating asymbiotic growth (Hildebrandt et al. 2006). Other genes were analysed in much less detail and only used to confirm hypotheses concerning the flux and use of carbohydrates in the different fungal structures (Bago et al. 2002, 2003).

Gene expression levels were also used to prove whether functional arbuscules are necessary for the carbohydrate metabolism of the fungus. This hypothesis could be confirmed, since all investigated genes showed a much lower RNA accumulation level in the late mutant, showing only truncated arbuscules compared to the wild type (Fig. 2.1). Finally, one study has to be mentioned where the expression of a gene for glucose-6-phosphate dehydrogenase was studied, an enzyme involved in the pentose phosphate pathway (Stewart et al. 2006). The expression of the gene was down-regulated when the root was treated with high phosphate concentrations, indicating a regulation of the C flux towards the fungus by the P status of the plant. An important step forward to elucidating the transfer system of carbohydrates was recently made by identification of a *G. intraradices* gene encoding the monosaccharide transporter MST2 (Helber et al. 2011). Further analysis showed that not only glucose, but also plant cell wall sugars such as xylose can serve as carbohydrate source for the fungus and that MST2 expression is crucial for mycorrhizal functioning.

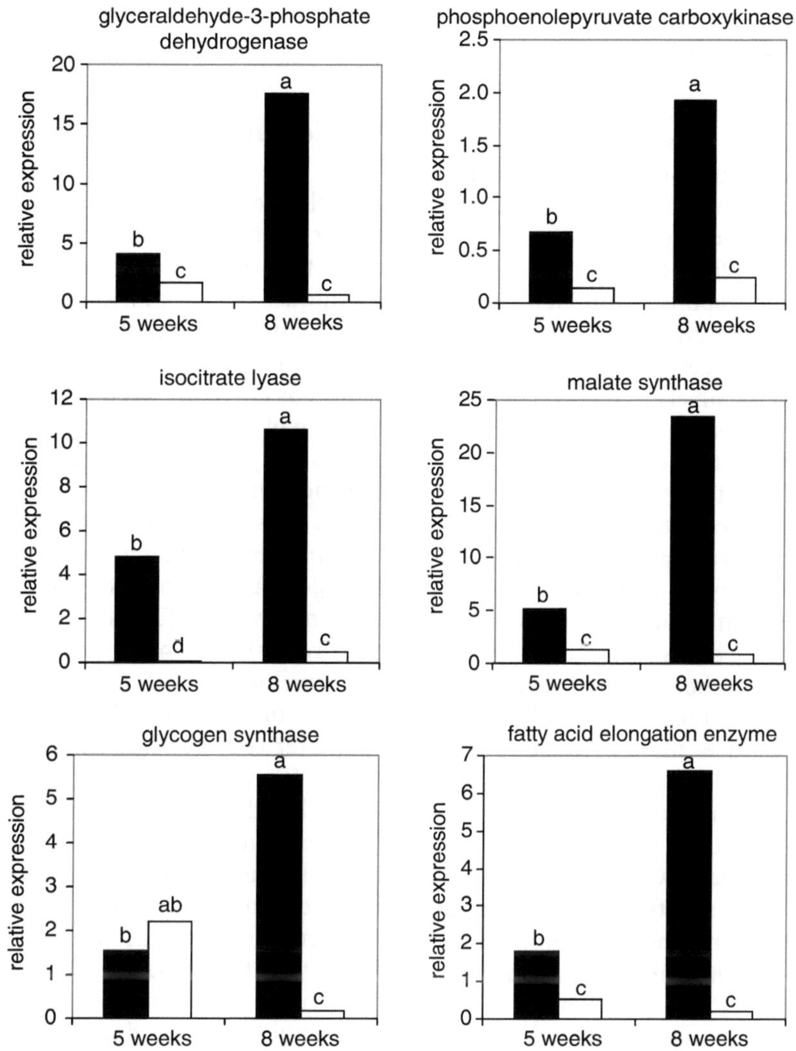

Fig. 2.1. RNA accumulation of genes involved in the carbohydrate metabolism. *Pisum sativum* cv. Finale (*black bars*) and the isogenic late mutant *PsSYM36* (*white bars*) were inoculated with *Glomus intraradices*. Five and eight weeks later, RNA was extracted from roots and used for quantitative RT-PCR with primers for genes encoding enzymes from the carbohydrate metabolism. Shown are the relative expression values calibrated by the gene for the translation elongation factor EF1-α. Significant different values are indicated by different *letters* above the columns. The factors pea genotype and time showed a significant interaction for all genes (two-way ANOVA; $P = 0.05$; $n = 3$)

C. Stress Response of Extraradical Hyphae

AM fungi are not only able to support the mineral nutrition of plants, they also increase the **tolerance for abiotic stress.** In addition to the comparison of mycorrhizal and non-mycorrhizal plants concerning their responses to, e.g., drought or heavy metal contamination, it is also important to analyse the molecular mechanisms on the fungal side being active, by the time extraradical hyphae are confronted with such detrimental conditions. The first gene identified as being involved in the response to heavy metals encoded a **metallothionein** in *G. rosea* (Stommel et al. 2001). A homologous gene in *G. margarita* showed constitutive expression throughout the life cycle and could be induced in symbiotic hyphae by copper, but not by cadmium (Lanfranco et al. 2002). Similarly, a metallothionein gene isolated from *G. intraradices* was able to restore heavy metal tolerance in a copper-sensitive yeast mutant (González-Guerrero et al. 2007). Expression analysis showed induction not only by copper, but also by paraquat, an agent inducing the generation of reactive oxygen species.

A number of other *G. intraradices* genes showing a similar response were identified and their function was confirmed in corresponding yeast mutants. Those genes encoded, e.g., a glutaredoxin (Benabdellah et al. 2009a), a pyridoxal 5′-phosphate synthase involved in vitamin B6 synthesis (Benabdellah et al. 2009b) and a superoxide dismutase (González-Guerrero et al. 2010a). Together with an ABC transporter not functionally characterized (González-Guerero et al. 2010b) they all responded to copper and paraquat. This indicates that the metal ion does not directly act as regulator, but that cellular responses to the heavy metal lead to ROS generation and that these ROS molecules act as signals to induce the expression of the corresponding genes. Recently, the molecular basis for an exclusion mechanism has been elucidated by cloning of a gene for an arsenite efflux pump (Gonzalez-Chavez et al. 2011). Together with the arsenate uptake by the high affinity phosphate transporter (see above), it could be the molecular basis for arsenate tolerance of mycorrhiza (Smith et al. 2010).

A classical response element to abiotic stress is **glutathione**. Hence it was not surprising to find a gene family encoding different isozymes of the glutathione S-transferase in a cDNA library enriched for *G. intraradices* heavy metal-induced genes (Waschke et al. 2006). One important molecule in the response of fungi to stress is **trehalose**. Measurements of trehalose content, RNA accumulation analysis of genes involved in the biosynthesis and degradation as well as complementation of a yeast mutant could show that trehalose metabolism plays a role in the response to the abiotic stressors heat shock and arsenate (Ocon et al. 2007). These analyses show that AM fungi possess the same machinery as many other fungi to respond to abiotic stress. It would be now interesting to compare the molecular responses among several AM fungal strains with different capacities to confer stress tolerance to plants.

VI. Concluding Remarks

This chapter describes how the analysis of particular genes (RNA accumulation and complementation assays) in combination with other techniques has given insights into the development and the metabolism of AM fungi. More information about developmentally regulated genes has currently been obtained by the hybridisation of arrays carrying probably most of the expressed genes identified in different cDNA libraries of *G. intraradices* (Tisserant et al. 2012). This information will be hopefully soon be complemented by information coming from a project aimed to obtain the whole genome sequence of this fungus (Martin et al. 2008). The greatest challenge for the future is, however, the development of a stabile transformation system for AM fungi. Transient expression of promoter reporter constructs has been already shown (Forbes et al. 1998; Helber and Requena 2008), but this is not sufficient to ascertain gene function during symbiosis. It is, however, questionable whether knocking out genes is possible in an organism supposedly containing several different genomes in a common cytoplasm. However, the RNAi technique successfully being used in plants might be also applied to AM fungi to at least down-regulate the expression of genes to an extent that the function can be studied.

Acknowledgements Astrid Waschke was supported by the DFG (German Research Foundation: SFB 395) and by the Ministries of Food, Agriculture, and Consumer Protection of the Federal Republic of Germany and of the states Brandenburg and Thüringen.

References

Allen JW, Shachar-Hill Y (2009) Sulfur transfer through an arbuscular mycorrhiza. Plant Physiol 149:549–560

Aono T, Maldonado-Mendoza IE, Dewbre GR, Harrison MJ, Saito M (2004) Expression of alkaline phosphatase genes in arbuscular mycorrhizas. New Phytol 162:525–534

Avio L, Giovannetti M (1998) The protein pattern of spores of arbuscular mycorrhizal fungi - comparison of species, isolates and physiological stages. Mycol Res 102:985–990

Bago B, Pfeffer PE, Douds DD, Brouillette J, Bécard G, Shachar-Hill Y (1999) Carbon metabolism in spores of the arbuscular mycorrhizal fungus *Glomus intraradices* as revealed by nuclear magnetic resonance spectroscopy. Plant Physiol 121:263–271

Bago B, Pfeffer P, Shachar-Hill Y (2001) Could the urea cycle be translocating nitrogen in the arbuscular mycorrhizal symbiosis? New Phytol 149:4–8

Bago B, Zipfel W, Williams RM, Jun J, Arreola R, Lammers PJ, Pfeffer PE, Shachar-Hill Y (2002) Translocation and utilization of fungal storage lipid in the arbuscular mycorrhizal symbiosis. Plant Physiol 128:108–124

Bago B, Pfeffer PE, Abubaker J, Jun J, Allen JW, Brouillette J, Douds DD, Lammers PJ, Shachar-Hill Y (2003) Carbon export from arbuscular mycorrhizal roots involves the translocation of carbohydrate as well as lipid. Plant Physiol 131:1496–1507

Balestrini R, Gomez-Ariza J, Lanfranco L, Bonfante P (2007) Laser microdissection reveals that transcripts for five plant and one fungal phosphate transporter genes are contemporaneously present in arbusculated cells. Mol Plant Microbe Interact 20:1055–1062

Bécard G, Fortin JA (1988) Early events of vesicular-arbuscular mycorrhiza formation on Ri T-DNA transformed roots. New Phytol 108:211–218

Bécard G, Piché Y (1989) Fungal growth stimulation by CO_2 and root exudates in vesicular-arbuscular mycorrhizal symbiosis. Appl Environ Microb 55:2320–2322

Bécard G, Doner LW, Rolin DB, Douds DD, Pfeffer PE (1991) Identification and quantification of trehalose in vesicular-arbuscular mycorrhizal fungi by *in vivo* ^{13}C NMR and HPLC analyses. New Phytol 118:547–552

Beilby JP, Kidby DK (1980) Biochemistry of ungerminated and germinated spores of the vesicular-arbuscular mycorrhizal fungus, *Glomus caledonius* - changes in neutral and polar lipids. J Lip Res 21:739–750

Benabdellah K, Merlos MA, Azcon-Aguilar C, Ferrol N (2009a) GintGRX1, the first characterized glomeromycotan glutaredoxin, is a multifunctional enzyme that responds to oxidative stress. Fungal Genet Biol 46:94–103

Benabdellah K, Azcon-Aguilar C, Valderas A, Speziga D, Fitzpatrick TB, Ferrol N (2009b) *GintPDX1* encodes a protein involved in vitamin B6 biosynthesis that is up-regulated by oxidative stress in the arbuscular mycorrhizal fungus *Glomus intraradices*. New Phytol 184:682–693

Benedetto A, Lanfranco L (2005) Expression profiles of a phosphate transporter gene (*GmosPT*) from the endomycorrhizal fungus *Glomus mosseae*. Mycorrhiza 15:620–627

Besserer A, Bécard G, Jauneau A, Roux C, Sejalon-Delmas N (2008) GR24, a synthetic analog of strigolactones, stimulates the mitosis and growth of the arbuscular mycorrhizal fungus *Gigaspora rosea* by boosting its energy metabolism. Plant Physiol 148:402–413

Bianciotto V, Lumini E, Lanfranco L, Minerdi D, Bonfante P, Perotto S (2000) Detection and identification of bacterial endosymbionts in arbuscular mycorrhizal fungi belonging to the family Gigasporaceae. Appl Environ Microb 66:4503–4509

Boddington CL, Dodd JC (1998) A comparison of the development and metabolic activity of mycorrhizas formed by arbuscular mycorrhizal fungi from different genera on two tropical forage legumes. Mycorrhiza 8:149–157

Brechenmacher L, Weidmann S, Van Tuinen D, Chatagnier O, Gianinazzi S, Franken P, Gianinazzi-Pearson V (2004) Expression profiling of up-regulated plant and fungal genes in early and late stages of *Medicago truncatula-Glomus mosseae* interactions. Mycorrhiza 14:253–262

Breuninger M, Requena N (2004) Recognition events in AM symbiosis: analysis of fungal gene expression at the early appressorium stage. Fung Genet Biol 41:794–804

Breuninger M, Trujillo CG, Serrano E, Fischer R, Requena N (2004) Different nitrogen sources modulate activity but not expression of glutamine synthetase in arbuscular mycorrhizal fungi. Fung Genet Biol 41:542–552

Bücking H, Abubaker J, Govindarajulu M, Tala M, Pfeffer PE, Nagahashi G, Lammers P, Shachar-Hill Y (2008) Root exudates stimulate the uptake and metabolism of organic carbon in germinating spores of *Glomus intraradices*. New Phytol 180:684–695

Bütehorn B, Gianinazzi-Pearson V, Franken P (1999) Quantification of beta-tubulin RNA expression during asymbiotic and symbiotic development of the arbuscular mycorrhizal fungus *Glomus mosseae*. Mycol Res 103:360–364

Cappellazzo G, Lanfranco L, Fitz M, Wipf D, Bonfante P (2008) Characterization of an amino acid permease from the endomycorrhizal fungus *Glomus mosseae*. Plant Physiol 147:429–437

Chabaud M, Genre A, Sieberer BJ, Faccio A, Fournier J, Novero M, Barker DG, Bonfante P (2011) Arbuscular mycorrhizal hyphopodia and germinated spore exudates trigger Ca^{2+} spiking in the legume and nonlegume root epidermis. New Phytol 189:347–355

Cox G, Tinker PB (1976) Translocation and transfer of nutrients in vesicular-arbuscular mycorrhizas. I. The arbuscule and phosphorus transfer: a quantitative ultrastructural study. New Phytol 7:371–378

Cox G, Moran KJ, Sanders F, Nockolds C, Tinker PB (1980) Translocation and transfer of nutrients in vesicular-arbuscular mycorrhizas. III. Polyphosphate granules and phosphorus translocation. New Phytol 84:649–659

Croll D, Giovannetti M, Koch AM, Sbrana C, Ehinger M, Lammers PJ, Sanders IR (2009) Nonself vegetative fusion and genetic exchange in the arbuscular mycorrhizal fungus *Glomus intraradices*. New Phytol 181:924–937

Cruz C, Egsgaard H, Trujillo C, Ambus P, Requena N, Martins-Loucao MA, Jakobsen I (2007) Enzymatic evidence for the key role of arginine in nitrogen translocation by arbuscular mycorrhizal fungi. Plant Physiol 144:782–792

Dickson S, Smith SE (2001) Cross walls in arbuscular trunk hyphae form after loss of metabolic activity. New Phytol 151:735–742

Dumas-Gaudot E, Valot B, Bestel-Corre G, Recorbet G, St-Arnaud M, Fontaine B, Dieu M, Raes M, Saravanan RS, Gianinazzi S (2004) Proteomics as a way to identify extra-radicular fungal proteins from *Glomus intraradices* – RiT-DNA carrot root mycorrhizas. FEMS Microbiol Ecol 48:401–411

Ferrol N, Barea JM, Azcon-Aguilar C (2000) The plasma membrane H^+-ATPase gene family in the arbuscular mycorrhizal fungus *Glomus mosseae*. Curr Genet 37:112–118

Forbes PJ, Millam S, Hooker JE, Harrier LA (1998) Transformation of the arbuscular mycorrhizal fungus *Gigaspora rosea* by particle bombardment. Mycol Res 102:497–501

Franken P, Gianinazzi-Pearson V (1996) Phage cloning of ribosomal RNA genes from the arbuscular mycorrhizal fungi *Glomus mosseae* (Nicol. & Gerd.) Gerd. & Trappe and *Scutellospora castanea* Walker. Mycorrhiza 6:167–173

Franken P, Lapopin L, Meyer-Gauen G, Gianinazzi-Pearson V (1997) RNA accumulation and genes expressed in spores of the arbuscular mycorrhizal fungus *Gigaspora rosea*. Mycologia 89:295–299

Franken P, Requena N, Bütehorn B, Krajinski F, Kuhn G, Lapopin L, Mann P, Rhody D, Stommel M (2000) Molecular analysis of the arbuscular mycorrhiza symbiosis. Arch Agron Soil Sci 45:271–286

Frey B, Vilarino A, Schuepp H, Arines J (1994) Chitin and ergosterol content of extraradical and intraradical mycelium of the vesicular-arbuscular mycorrhizal fungus *Glomus intraradices*. Soil Biol Biochem 26:711–717

Gachomo E, Allen JW, Pfeffer PE, Govindarajulu M, Douds DD, Jin HR, Nagahashi G, Lammers PJ, Shachar-Hill Y, Bucking H (2009) Germinating spores of *Glomus intraradices* can use internal and exogenous nitrogen sources for de novo biosynthesis of amino acids. New Phytol 184:399–411

Genre A, Bonfante P (2007) Check-in procedures for plant cell entry by biotrophic microbes. Mol Plant Microbe Interact 20:1023–1030

Gianinazzi-Pearson V, Gianinazzi S (1976) Enzymatic studies on the metabolism of vesicular-arbuscular mycorrhiza. Physiologie Végétale 14:833–841

Gianinazzi-Pearson V, Branzanti B, Gianinazzi S (1989) *In vitro* enhancement of spore germination and early hyphal growth of a vesicular-arbuscular mycorrhizal fungus by host root exudates and plant flavonoids. Symbiosis 7:243–255

Giovannetti M, Sbrana C, Avio L, Citernesi AS, Logi C (1993) Differential hyphal morphogenesis in arbuscular mycorrhizal fungi during pre-symbiotic phase. New Phytol 125:587–594

Giovannetti M, Azzolini D, Citernesi AS (1999) Anastomosis formation and nuclear and protoplasmic exchange in arbuscular mycorrhizal fungi. Appl Environ Microb 65:5571–5575

Gomez SK, Javot H, Deewatthanawong P, Torres-Jerez I, Tang YH, Blancaflor EB, Udvardi MK, Harrison MJ (2009) *Medicago truncatula* and *Glomus intraradices* gene expression in cortical cells harboring arbuscules in the arbuscular mycorrhizal symbiosis. BMC Plant Biol 9:10

Gonzalez-Chavez MDA, Ortega-Larrocea MD, Carrillo-Gonzalez R, Lopez-Meyer M, Xoconostle-Cazares B, Gomez SK, Harrison MJ, Figueroa-Lopez AM, Maldonado-Mendoza IE (2011) Arsenate induces the expression of fungal genes involved in as transport in arbuscular mycorrhiza. Fungal Biol 115:1197–1209

González-Guerero M, Benabdellah K, Valderas A, Azcon-Aguilar C, Ferrol N (2010) *GintABC1* encodes a putative ABC transporter of the MRP subfamily induced by Cu, Cd, and oxidative stress in *Glomus intraradices*. Mycorrhiza 20:137–146

González-Guerrero M, Azccón-Aguilar C, Mooney M, Valderas A, MacDiarmid CW, Eide DJ, Ferrol N (2005) Characterization of a *Glomus intraradices* gene encoding a putative Zn transporter of the cation diffusion facilitator family. Fung Genet Biol 42:130–140

González-Guerrero M, Cano C, Azcon-Aguilar C, Ferrol N (2007) *GintMT1* encodes a functional metal-lothionein in *Glomus intraradices* that responds to oxidative stress. Mycorrhiza 17:327–335

González-Guerrero M, Oger E, Benabdellah K, Azcon-Aguilar C, Lanfranco L, Ferrol N (2010) Characterization of a CuZn superoxide dismutase gene in the arbuscular mycorrhizal fungus *Glomus intraradices*. Curr Genet 56:265–274

Govindarajulu M, Pfeffer PE, Jin HR, Abubaker J, Douds DD, Allen JW, Bucking H, Lammers PJ, Shachar-Hill Y (2005) Nitrogen transfer in the arbuscular mycorrhizal symbiosis. Nature 435:819–823

Güther M, Neuhauser B, Balestrini R, Dynowski M, Ludewig U, Bonfante P (2009) A mycorrhizal-specific ammonium transporter from *Lotus japonicus* acquires nitrogen released by arbuscular mycorrhizal fungi. Plant Physiol 150:73–83

Harrier LA, Paterson LJ (2002) Inducibility studies with the arbuscular mycorrhizal fungus *Glomus mosseae* 3-phosphoglycerate kinase (PGK) gene promoter. Curr Genet 42:169–178

Harrier LA, Sawczak J (2000) Detection of the 3-phosphoglycerate kinase protein of *Glomus mosseae*. Mycorrhiza 10:81–86

Harrier LA, Wright F, Hooker JE (1998) Isolation of the 3-phosphoglycerate kinase gene of the arbuscular mycorrhizal fungus *Glomus mosseae* (Nicol. & Gerd.) Gerdemann & Trappe. Curr Genet 34:386–392

Harrison MJ, van Buuren ML (1995) A phosphate transporter from the mycorrhizal fungus *Glomus versiforme*. Nature 378:626–629

Hawkins HJ, Johansen A, George E (2000) Uptake and transport of organic and inorganic nitrogen by arbuscular mycorrhizal fungi. Plant Soil 226:275–285

Helber N, Requena N (2008) Expression of the fluorescence markers DsRed and GFP fused to a nuclear localization signal in the arbuscular mycorrhizal fungus *Glomus intraradices*. New Phytol 177:537–548

Helber N, Wippel K, Sauer N, Schaarschmidt S, Hause B, Requena N (2011) A versatile monosaccharide transporter that operates in the arbuscular mycorrhizal fungus *Glomus* sp is crucial for the symbiotic relationship with plants. Plant Cell 23:3812–3823

Heupel S, Roser B, Kuhn H, Lebrun MH, Villalba F, Requena N (2010) Erl1, a novel Era-like GTPase from *Magnaporthe oryzae*, is required for full root virulence and is conserved in the mutualistic symbiont *Glomus intraradices*. Mol Plant Microbe Interact 23:67–81

Hijri M, Sanders IR (2004) The arbuscular mycorrhizal fungus *Glomus intraradices* is haploid and has a small genome size in the lower limit of eukaryotes. Fung Genet Biol 41:253–261

Hijri M, Sanders IR (2005) Low gene copy number shows that arbuscular mycorrhizal fungi inherit genetically different nuclei. Nature 433:160–163

Hildebrandt U, Ouziad F, Marner FJ, Bothe H (2006) The bacterium *Paenibacillus validus* stimulates growth of the arbuscular mycorrhizal fungus *Glomus intraradices* up to the formation of fertile spores. FEMS Microbiol Let 254:258–267

Hooker JE, Piatti P, Cheshire MV, Watson CA (2007) Polysaccharides and monosaccharides in the hyphosphere of the arbuscular mycorrhizal fungi *Glomus* E3 and *Glomus tenue*. Soil Biol Biochem 39:680–683

Hosny M, Van Tuinen D, Jacquin F, Füller P, Zhao B, Gianinazzi-Pearson V, Franken P (1999) Arbuscular mycorrhizal fungi and bacteria: how to construct prokaryotic DNA-free genomic libraries from the Glomales. FEMS Microbiol Lett 170: 425–430

Jin H, Pfeffer PE, Douds DD, Piotrowski E, Lammers PJ, Shachar-Hill Y (2005) The uptake, metabolism, transport and transfer of nitrogen in an arbuscular mycorrhizal symbiosis. New Phytol 168:687–696

Jun J, Abubaker J, Rehrer C, Pfeffer PE, Shachar-Hill Y, Lammers PJ (2002) Expression in an arbuscular mycorrhizal fungus of genes putatively involved in metabolism, transport, the cytoskeleton and the cell cycle. Plant Soil 244:141–148

Kaldorf M, Zimmer W, Bothe H (1994) Genetic evidence for the occurence of assimilatory nitrate reductase in arbuscular mycorrhizal and other fungi. Mycorrhiza 5:23–28

Kaldorf M, Schmelzer E, Bothe H (1998) Expression of maize and fungal nitrate reductase genes in arbuscular mycorrhiza. Mol Plant Microbe Interact 11:439–448

Kloppholz S, Kuhn H, Requena N (2011) A secreted fungal effector of *Glomus intraradices* promotes symbiotic biotrophy. Curr Biol 21:1204–1209

Kojima T, Saito M (2004) Possible involvement of hyphal phosphatase in phosphate efflux from intraradical hyphae isolated from mycorrhizal roots colonized by *Gigaspora margarita*. Mycol Res 108:610–615

Kuznetsova E, Seddas-Dozolme PMA, Arnould C, Tollot M, Van Tuinen D, Borisov A, Gianinazzi S, Gianinazzi-Pearson V (2010) Symbiosis-related pea genes modulate fungal and plant gene expression during the arbuscule stage of mycorrhiza with *Glomus intraradices*. Mycorrhiza 20:427–443

Lanfranco L, Vallino M, Bonfante P (1999) Expression of chitin synthase genes in the arbuscular mycorrhizal fungus *Gigaspora margarita*. New Phytol 142:347–354

Lanfranco L, Bolchi A, Cesale RE, Ottonello S, Bonfante P (2002) Differential expression of a metallothionein gene during the presymbiotic versus the symbiotic phase of an arbucular mycorrhizal fungus. Plant Physiol 130:58–67

Lanfranco L, Novero M, Bonfante P (2005) The mycorrhizal fungus *Gigaspora margarita* possesses a CuZn superoxide dismutase that is up-regulated during symbiosis with legume hosts. Plant Physiol 137:1319–1330

Lapopin L, Gianinazzi-Pearson V, Franken P (1999) Comparative differential display analysis of arbuscular mycorrhiza in *Pisum sativum* and a mutant defective in late stage development. Plant Mol Biol 41:669–677

Lei J, Bécard G, Catford JG, Piché Y (1991) Root factor stimulate ^{32}P uptake and plasmalemma ATPase activity in the vesicular-arbuscular mycorrhizal fungus, *Gigaspora margarita*. New Phytol 118:289–294

Lopez-Pedrosa A, González-Guerrero M, Valderas A, Azcon-Aguilar C, Ferrol N (2006) *GintAMT1* encodes a functional high-affinity ammonium transporter that is expressed in the extraradical mycelium of *Glomus intraradices*. Fung Genet Biol 43:102–110

Maldonado-Mendoza IE, Dewbre GR, Harrison MJ (2001) A phosphate transporter gene from the extra-radical mycelium of an arbuscular mycorrhizal fungus *Glomus intraradices* is regulated in response to phosphate in the environment. Mol Plant Microbe Interact 14:1140–1148

Maldonado-Mendoza IE, Dewbre GR, van Buuren ML, Versaw WK, Harrison MJ (2002) Methods to estimate the proportion of plant and fungal RNA in an arbuscular mycorrhiza. Mycorrhiza 12:67–74

Marschner H, Dell B (1994) Nutrient uptake in mycorrhizal symbiosis. Plant Soil 159:89–102

Martin F, Gianinazzi-Pearson V, Hijri M, Lammers P, Requena N, Sanders IR, Shachar-Hill Y, Shapiro H, Tuskan GA, Young JPW (2008) The long hard road to a completed *Glomus intraradices* genome. New Phytol 180:747–750

Mosse B (1959) The regulated germination of resting spores and some observations on growth requirement of Endogones sp. causing VAM. Trans Br Mycol Soc 42:273–286

Mosse B, Hepper C (1975) Vesicular-arbuscular mycorrhizal infections in root organ-cultures. Physiol Plant Pathol 5:215–223

Mugnier J, Mosse B (1987) Vesicular-arbuscular mycorrhizal infection in transformed root-inducing T-DNA roots grown axenically. Phytopathol 77:1045–1050

Nagy S, Nordby HE, Nemec S (1980) Composition of lipids in roots of 6 citrus cultivars infected with the vesicular-arbuscular mycorrhizal fungus, *Glomus mosseae*. New Phytol 85:377–384

Navazio L, Moscatiello R, Genre A, Novero M, Baldan B, Bonfante P, Mariani P (2007) A diffusible signal from arbuscular mycorrhizal fungi elicits a transient cytosolic calcium elevation in host plant cells. Plant Physiol 144:673–681

Neumann E, George E (2005) Extraction of extraradical arbuscular mycorrhizal mycelium from compartments filled with soil and glass beads. Mycorrhiza 15:533–537

Ocon A, Hampp R, Requena N (2007) Trehalose turnover during abiotic stress in arbuscular mycorrhizal fungi. New Phytol 174:879–891

Peretto R, Bettini V, Favaron F, Alghisi P, Bonfante P (1995) Polygalacturonase activity and location in arbuscular mycorrhiza root of *Allium porrum* L. Mycorrhiza 5:157–163

Perez-Tienda J, Testillano PS, Balestrini R, Fiorilli V, Azcon-Aguilar C, Ferrol N (2011) GintAMT2, a new member of the ammonium transporter family in the arbuscular mycorrhizal fungus *Glomus intraradices*. Fungal Genet Biol 48: 1044–1055

Pfeffer PE, Douds DD, Bécard G, Shachar-Hill Y (1999) Carbon uptake and the metabolism and transport of lipids in an arbuscular mycorrhiza. Plant Physiol 120:587–598

Ramos AC, Facanha AR, Feijo JA (2008) Proton (H^+) flux signature for the presymbiotic development of the arbuscular mycorrhizal fungi. New Phytol 178:177–188

Recorbet G, Valot B, Robert F, Gianinazzi-Pearson V, Dumas-Gaudot E (2010) Identification of in planta-expressed arbuscular mycorrhizal fungal proteins upon comparison of the root proteomes of *Medicago truncatula* colonised with two *Glomus* species. Fungal Genet Biol 47:608–618

Redecker D, Thierfelder H, Werner D (1995) A new cultivation system for arbuscular mycorrhizal fungi on glass beads. Angew Bot 69:189–191

Redecker D, Kodner R, Graham LE (2000) Glomalean fungi from the Ordovician. Science 289:1920–1921

Requena N, Jeffries P, Barea JM (1996) Assessment of natural mycorrhizal potential in a desertified semiarid ecosystem. Appl Environ Microbiol 62:842–847

Requena N, Füller P, Franken P (1999) Molecular characterisation of *GmFOX2*, an evolutionary highly conserved gene from the mycorrhizal fungus *Glomus mosseae*, down-regulated during interaction with rhizobacteria. Mol Plant Microbe Interact 12:934–942

Requena N, Mann P, Hampp R, Franken P (2002) Early developmentally regulated genes in the arbuscular mycorrhizal fungus *Glomus mosseae*: Identification of *GmGIN1*, a novel gene with homology to the C-terminus of metazoan hedgehog proteins. Plant Soil 244:129–139

Requena N, Breuninger M, Franken P, Ocon A (2003) Symbiotic status, phosphate, and sucrose regulate the expression of two plasma membrane H^+-ATPase genes from the mycorrhizal fungus *Glomus mosseae*. Plant Physiol 132: 1540–1549

Saito M (1995) Enzyme activities of the internal hyphae and germinated spores of arbuscular mycorrhizal fungus, *Gigaspora margarita* Becker & Hall. New Phytol 129:425–431

Samra A, Dumas-Gaudot E, Gianinazzi-Pearson V, Gianinazzi S (1996) Soluble proteins and polypeptides profiles of spores of arbuscular mycorrhizal fungi. Interspecific variability and effects of host (myc+) and non-host (myc-) *Pisum sativum* root exudates. Agronomie 16:709–719

Sanders IR, Ravolanirina F, Gianinazzi-Pearson V, Gianinazzi S, Lemoine MC (1992) Detection of specific antigens in the vesicular-arbuscular mycorrhizal fungi *Gigaspora margarita* and *Acaulospora laevis* using polyclonal antibodies to soluble spore fractions. Mycol Res 96:477–480

Sawaki H, Saito M (2001) Expressed genes in the extraradical hyphae of an arbuscular mycorrhizal fungus, *Glomus intraradices*, in the symbiotic phase. FEMS Microbiol Lett 195:109–113

Schmitz O, Danneberg G, Hundeshagen B, Klingner A, Bothe H (1991) Quantification of vesicular-arbuscular mycorrhiza by biochemical parameters. J Plant Physiol 139:106–114

Schüßler A, Schwarzott D, Walker C (2001) A new fungal phylum, the Glomeromycota: phylogeny and evolution. Mycol Res 105:1413–1421

Seddas PMA, Arnould C, Tollot M, Arias CM, Gianinazzi-Pearson V (2008) Spatial monitoring of gene activity in extraradical and intraradical developmental stages of arbuscular mycorrhizal fungi by direct fluorescent in situ RT-PCR. Fung Genet Biol 45:1155–1165

Seddas PMA, Arias CM, Arnould C, Van Tuinen D, Godfroy O, Benhassou HA, Gouzy J, Morandi D,

Dessaint F, Gianinazzi-Pearson V (2009) Symbiosis-related plant renes modulate molecular responses in an arbuscular mycorrhizal fungus during early root interactions. Mol Plant Microbe Interact 22:341–351

Shachar-Hill Y, Pfeffer PE, Douds D, Osman SF, Doner LW, Ratcliffe RG (1995) Partitioning of intermediary carbon metabolism in vesicular-arbuscular mycorrhizal leek. Plant Physiol 108:7–15

Simon L, Lalonde M, Bruns TD (1992) Specific amplification of 18 S fungal ribosomal genes from vesicular-arbuscular endomycorrhizal fungi colonizing roots. Appl Environ Microb 58:291–295

Siqueira JO, Hubbell DH, Schenck NC (1982) Spore germination and germ tube growth of a vesicular-arbuscular mycorrhizal fungus *in vitro*. Mycologia 74:952–959

Smith SE, Read DJ (2008) Mycorrhizal symbiosis. Academic, London

Smith SE, Christophersen HM, Pope S, Smith FA (2010) Arsenic uptake and toxicity in plants: integrating mycorrhizal influences. Plant Soil 327:1–21

Solaiman MdZ, Saito M (1997) Use of sugars by intraradical hyphae of arbuscular mycorrhizal fungi revealed by radiorespirometry. New Phytol 136:533–538

St-Arnaud M, Hamel C, Vimard B, Caron M, Fortin JA (1996) Enhanced hyphal growth and spore production of the arbuscular mycorrhizal fungus *Glomus intraradices* in an in vitro system in the absence of host roots. Mycol Res 100:328–332

Stewart LI, Jabaji-Hare S, Driscoll BT (2006) Effects of external phosphate concentration on glucose-6-phosphate dehydrogenase gene expression in the arbuscular mycorrhizal fungus *Glomus intraradices*. Can J Microbiol 52:823–830

Stommel M, Mann P, Franken P (2001) Construction and analysis of an EST library using RNA from activated spores of the arbuscular mycorrhizal fungus *Gigaspora rosea*. Mycorrhiza 10:281–285

Tamasloukht M, Séjalon-Delmas N, Kluever A, Jauneau A, Roux C, Bécard G, Franken P (2003) Root factor induce mitochondrial-related-gene expression and fungal respiration during the developmental switch from asymbiosis to presymbiosis in the arbuscular mycorrhizal fungus *Gigaspora rosea*. Plant Physiol 131:1468–1478

Tamasloukht M, Waschke A, Franken P (2007) Root exudate-stimulated RNA accumulation in the arbuscular mycorrhizal fungus *Gigaspora rosea*. Soil Biol Biochem 39:1824–1827

Tepfer D (1984) Transformation of several species of higher-plants by *Agrobacterium rhizogenes* – sexual transmission of the transformed genotype and phenotype. Cell 37:959–967

Thomson BD, Clarkson DT, Brain P (1990) Kinetics of phosphorus uptake by the germ-tubes of the vesicular-arbuscular mycorrhizal fungus, *Gigaspora margarita*. New Phytol 116:647–653

Tian CJ, Kasiborski B, Koul R, Lammers PJ, Bucking H, Shachar-Hill Y (2010) Nitrogen transfer pathway in the arbuscular mycorrhizal symbiosis: gene characterization and the coordination of expression with nitrogen flux. Plant Physiol 153:1175–1187

Tisserant B, Gianinazzi-Pearson V, Gianinazzi S, Gollotte A (1993) *In planta* histochemical staining of fungal alkaline phosphatase activities for analysis of efficient arbuscular mycorrhizal infections. Mycol Res 97:245–250

Tisserant E, Kohler A, Dozolme-Seddas P, Balestrini R, Benabdellah K, Colard A, Croll D, Da Silva C, Gomez SK, Koul R, Ferrol N, Fiorilli V, Formey D, Franken P, Helber N, Hijri M, Lanfranco L, Lindquist E, Liu Y, Malbreil M, Morin E, Poulain J, Shapiro H, van Tuinen D, Waschke A, Azcón-Aguilar C, Bécard G, Bonfante P, Harrison MJ, Küster H, Lammers P, Paszkowski U, Requena N, Rensing SA, Roux C, Sanders IR, Shachar-Hill Y, Tuskan G, Young JPW, Gianinazzi-Pearson V, Martin F (2012) The transcriptome of the arbuscular mycorrhizal fungus *Glomus intraradices* (DAOM 197198) reveals functional tradeoffs in an obligate symbiont. New Phytol 193:755–769

Tollot M, Hoi JWS, Van Tuinen D, Arnould C, Chatagnier O, Dumas B, Gianinazzi-Pearson V, Seddas PMA (2009) An *STE12* gene identified in the mycorrhizal fungus *Glomus intraradices* restores infectivity of a hemibiotrophic plant pathogen. New Phytol 181:693–707

Tommerup IC (1983) Spore dormancy in vesicular-arbuscular mycorrhizal fungi. Trans Br Mycol Soc 81:37–45

Trépanier M, Bécard G, Moutoglis P, Willemot C, Gagne S, Avis TJ, Rioux JA (2005) Dependence of arbuscular-mycorrhizal fungi on their plant host for palmitic acid synthesis. Appl Environ Microb 71:5341–5347

Tylka GL, Hussey RS, Roncadori RW (1991) Axenic germination of vesicular-arbuscular mycorrhizal fungi: effects of selected *Streptomyces* species. Phytopathol 81:754–759

Ubalijoro E, Hamel C, McClung CR, Smith DL (2001) Detection of chitin synthase class I and II type sequences in six different arbuscular mycorrhizal fungi and gene expression in *Glomus intraradices*. Mycol Res 105:470–476

Waschke A, Sieh D, Tamasloukht M, Fischer K, Mann P, Franken P (2006) Identification of heavy metal-induced genes encoding glutathione S-transferases in the arbuscular mycorrhizal fungus *Glomus intraradices*. Mycorrhiza 17:1–10

White PR (1943) A handbook of plant tissue culture. Cattel, Lancaster

Zézé A, Dulieu H, Gianinazzi-Pearson V (1994) DNA cloning and screening of a partial genomic library from an arbuscular mycorrhizal fungus, *Scutellospora castanea*. Mycorrhiza 4:251–254

Zimmermann MH, Ziegler H (1975) List of sugars and sugar alcohols in sieve tube exudates. In: Zimmermann MH, Milburn JA (eds) Transport in plants 1: phloem transport. Springer, Berlin, pp 480–503

3 The Interface Between Plants and Mycorrhizal Fungi: Nutrient Exchange, Signaling and Cell Organization

A. Genre[1], P. Bonfante[1]

CONTENTS

I. Introduction

Mycorrhizal fungi are a heterogeneous group of soil fungi that are found in most terrestrial ecosystems where they form symbiotic associations called mycorrhizas with the roots of over 90 % of the plant families analyzed so far. This ecological success is the result of the major selective advantages that mycorrhizal interactions provide to both the plant and fungus. When we compare a mycorrhizal plant with one that is grown in the absence of its symbionts, the differences are striking: the mycorrhizal fungi influence plant growth and their capacity to absorb water and nutrients, and they protect them from pathogens (Smith and Read 2008). Beside improving plant health and overall fitness, mycorrhizas play a central role in nutrient cycles, in soil stability and—last but not least—in the survival and diffusion of mycorrhizal fungi. Most mycorrhizal fungi only produce propagules (sexual or asexual spores) when growing in association with their plant hosts. Some can grow as saprobes in pure culture (as in the case of most ectomycorrhizal fungi) while others, like the endomycorrhizal Glomeromycetes, are obligate biotrophs and do not even develop more than a few centimeters of mycelium in the absence of their hosts (Bonfante and Genre 2010). Structures related to arbuscular mycorrhizas (AM) have been repeatedly identified in fossils (Remy et al. 1994; Redecker et al. 2000; Dotzler et al. 2006); this evidence, coupled with the identification of symbiosis-specific genes throughout the plant kingdom, including the most basal clades, strongly supports the common idea that mycorrhizal symbiosis played a central role during the plants' conquest of dry land, around 450 million years ago. Mycorrhizas are therefore considered to have significantly contributed to shaping the Earth ecosystems as we know them (Brachmann and Parniske 2006; Smith and Read 2008). Today, their application in agricultural practices promises to be a fundamental tool to achieve sustainability in crop production and feed a constantly growing global population (Barrow et al. 2008; Fitter et al. 2011). These aspects of mycorrhizal interactions largely account for the increasing interest they raise in the scientific community as well as in agroindustrial companies and development agencies. Over the last 10 years, substantial advances have been made in our knowledge of many of the cellular and molecular mechanisms that underlie mycorrhizal associations; the advent of novel technologies, such as genome sequencing, high-throughput transcriptomics and in vivo confocal microscopy, has opened new ways to explore the hidden world of these fascinating subterranean

[1]Department of Life Science and Systems Biology, University of Torino, Viale P.A. Mattioli 25, 10125 Torino, Italy; e-mail: paola.bonfante@unito.it; p.bonfante@ipp.cnr.it

Fungal Associations, 2nd Edition
The Mycota IX
B. Hock (Ed.)

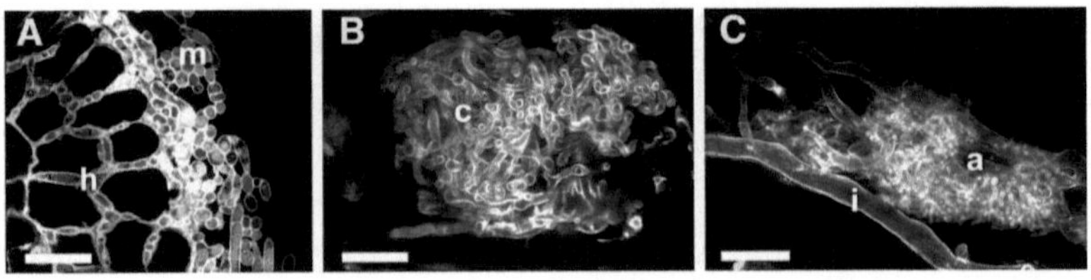

Fig. 3.1. Micrographs showing: (**A**) the fungal mantle (*m*) and Hartig net (*h*) in an ecetomycorrhiza of *Quercus* sp., (**B**) a hyphal coil (*c*) inside a root epidermal cell of *Calluna vulgaris*, and (**C**) an intercellular hypha (*i*) and an arbuscule (*a*) inside a root cortical cell of leek. In (**A**), (**B**), (**C**), *bars* 50, 16, 15 μm, respectively

symbioses. Such advancements were recently reviewed by Smith and Read (2008), Parniske (2008), Martin and Nehls (2009), Bonfante and Requena (2011).

Here we present a synopsis of the recent literature on the interactions between mycorrhizal fungi and their hosts, with particular focus on the intimate contact that develops between plant cells and fungal hyphae, in terms of molecular signaling, nutrient exchange and cell organization.

II. Root Colonization by Mycorrhizal Fungi

The impressive biodiversity of the plant and fungal taxa involved in mycorrhizal symbiosis is the first responsible for the profound anatomical diversity that characterizes such symbioses (Peterson et al. 2004); for this it is not surprising that anatomical features have dominated the taxonomy of mycorrhizas since the start of the last century (for references, see Smith and Read 2008). In spite of that, several aspects of plant–fungus interactions are common to all types of mycorrhiza and are here briefly summarized to provide the structural background for further discussion. Propagules of mycorrhizal fungi (sexual or asexual spores) can germinate independently of the host plant to develop a more or less extensive asymbiotic mycelium, depending on the fungal taxa. Hyphae are essential for survival, completion of the life cycle and mobilization of nutrients, and they can sometimes organize in highly structured bundles and rhizomorphs. In the presence of a host, mycorrhizal fungi switch to a presymbiotic and then symbiotic phase, where they adopt various colonization patterns and achieve different levels of cellular interactions with the host tissues (Fig. 3.1).

Ectomycorrhizal (EM) fungi do not penetrate the host cell wall and complete their colonization of the root in two major steps. After contacting the root surface, hyphae proliferate in tight aggregation to produce a tissue-like structure called the mantle (Bonfante and Genre 2010). Eventually, the mantle coats the whole root tip, which in the meantime has arrested its meristematic activity (Felten et al. 2009). Hyphae then progress into the apoplastic region around epidermal (in angiosperms) and cortical (in gymnosperms) cells, where they display a labyrinthine growth originating a network of intercellular hyphae termed the Hartig net (Fig. 3.2). The extensive surface that this structure develops between the plant cells is the most straightforward morphological evidence that the Hartig net is the site of nutrient exchange between the symbionts.

Endomycorrhizal fungi, by contrast, develop inside the lumen of plant cells and show a great variety of infection patterns. Among them, Glomeromycota, the makers of AM, are totally dependent on their hosts, and their asymbiotic mycelium cannot develop for more than a few days in their absence. As soon as they perceive the host root exudates, extraradical hyphae start branching over and over, in a rough strategy to improve the chances of eventually contacting the root surface. Epidermal contact rapidly leads to the differentiation of hyphopodia, the swollen and branched hyphae that tightly adhere to the plant cell wall. In most of the studied cases, the epidermal layer is crossed by simple, unbranched hyphae, which penetrate inside the plant cell lumen. Host cell integrity

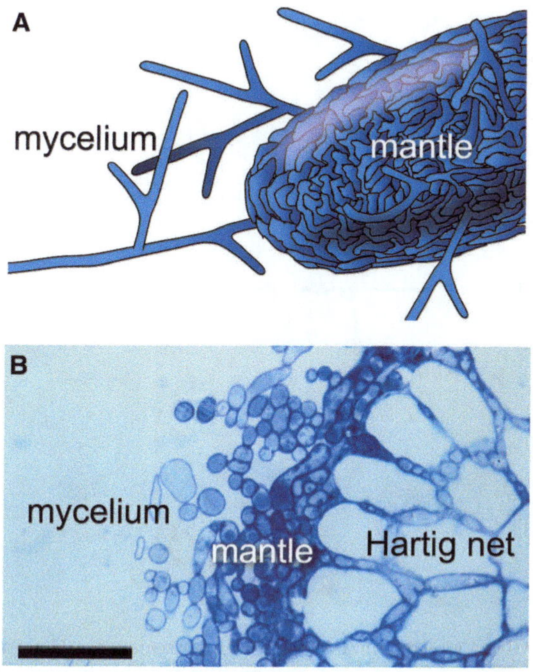

Fig. 3.2. The plant–fungus interface in ectomycorrhizas. (**A**) Scheme showing the connection of soil-borne mycelium with the mantle that envelopes a host root tip. (**B**) Micrograph showing a cross-section through a root tip from a *Quercus* sp./*Tuber melanosporum* mycorrhiza stained with trypan blue, highlighting the different hyphal organization in the mycelium, mantle and Hartig net. *Bar* 50 μm

is maintained by the invagination of its plasma membrane, which proliferates to engulf the developing hypha, physically separating the fungus from the plant cytoplasm. Hyphae can then traverse the outer cortical cells in a similar way, or sneak around them. In the cortical tissue hyphae often branch to spread the infection. Their final target are the inner cortical cells, where hyphae penetrate and ramify repeatedly to differentiate into arbuscules, the highly branched structures that give their name to this form of mycorrhizas and represent the main site of nutrient exchange (Fig. 3.3).

Ericoid mycorrhizal fungi also colonize the root cells of their host Ericaceae, but here each infection unit involves a single root cell: after the production of an ill-defined hyphopodium, the hypha penetrates into the epidermal cell and produces a terminal branched coil without moving to neighboring cells (Fig. 3.1; Perotto et al. 1995). Orchids also develop endomycorrhizas with Basidiomycetes, whose intracellular coils develop in the cortical cells of both protocorms and roots. In this case infection units extend to several cells (Peterson et al. 1998, 2004).

Other mycorrhizal morphologies include ectendomycorrhizas, the arbutoid and monotropoid mycorrhizas, all of which display distinct structural attributes and sometimes a mix of the previous examples (Smith and Read 2008).

In all cases, root colonization is accompanied by the proliferation of extraradical mycelium, which extends beyond the nutrient depletion zone that surrounds the root system, providing the structural basis for improving the fitness of mycorrhizal plants.

III. The Interface as a Unifying Feature of the Cellular Interactions in Mycorrhizas

All mycorrhizal interactions achieve full symbiotic functionality with the development of an extensive contact surface between the plant and fungal cells, where nutrients and signals are exchanged. In more detail, the exchange of molecules from the fungal to the plant cytoplasm and vice versa takes place through both partners' plasma membranes and cell walls (Figs. 3.2 and 3.3), defining a functional compartment known as the symbiotic interface (Scannerini and Bonfante 1983; Peterson and Massicotte 2004; Bücking et al. 2007; Bonfante et al. 2009).

Plant cell walls form a dynamic extracellular matrix that actively controls plant growth and development (Sattelmacher and Horst 2007). An analogous condition is found in fungi, where the wall determines hyphal growth, shape and responses (Durán and Nombela 2004). Among their multiple functions, cell walls are deeply involved in cell to cell signal exchange both within the same organism and between interacting organisms in heterologous communication. Not surprisingly, cell wall organization and composition within the interface compartment has attracted strong attention, both in ecto- and endomycorrhizas (Bonfante and Scannerini 1992; Tagu and Martin 1996; Martin et al. 1999). Important indications on cell wall remodeling in

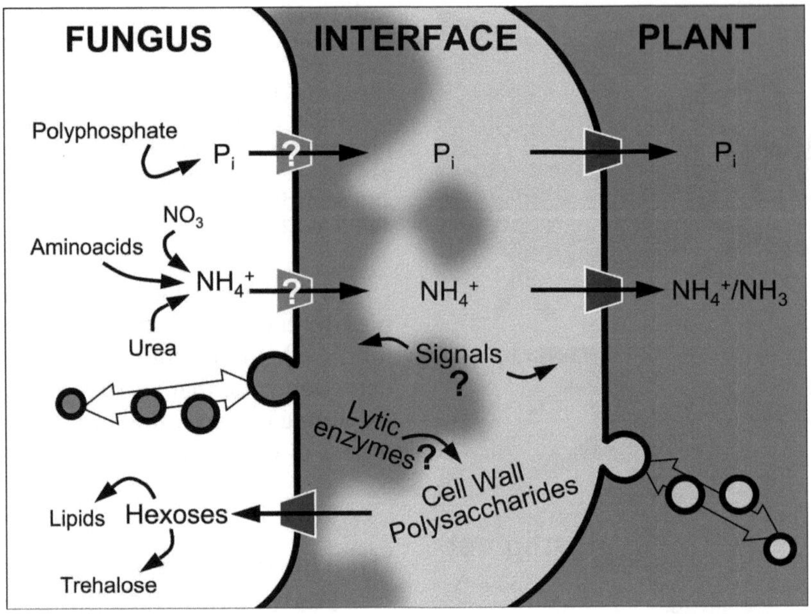

Fig. 3.3. Scheme of the symbiotic interface in arbuscular mycorrhizas, highlighting the major fluxes of nutrients, signaling molecules and membrane. *Question marks* label speculative pathways. Phosphate (P_i) derived from polyphosphate breakdown inside the hyphal vacuoles, reaches the interface space and is imported into the host cell via specific P transporters. Nitrogen is considered to reach the interface in the form of ammonium, which is then transported to the plant cytosol as ammonia by NH_4/NH_3 transporters. Hexose transporters are held responsible for the import of sugars from the interface to the fungal cytoplasm, where they feed fungal metabolism. Sugars have also been proposed to derive from the lysis of plant cell wall polysaccharides. The exchange of signal molecules through the interface is expected too, although direct evidence is still missing

EM have been obtained more recently, through transcriptomics analyses (Nehls 2003) and following the genomic sequencing of EM fungi. The genomes of *Laccaria bicolor* and *Tuber melanosporum* (Martin et al. 2008a, 2010) in fact revealed their mild capability in degrading the plant cell wall. Both EM fungi have a relatively small number of enzymes (e.g., pectin lyases and pectinases) that target plant cell wall components, compared with their pathogenic relatives. A few of these genes are however significantly upregulated in *Tuber* during symbiosis (Martin et al. 2010). These data strongly suggest that the colonization process requires a moderate weakening of the host plant wall, allowing the accommodation of the Hartig net in between root cells while limiting the entity of damage to avoid the elicitation of defense reactions.

On the same topic, but in the field of endomycorrhizas, high expectations still rely on the genome sequencing of *Glomus intraradices*, an AM fungus whose sequencing project was started in 2005 (Martin et al. 2008b) but has found major difficulties since. Much of the hassle is related to an unpredicted level of genome complexity, which is just one of the unique traits of Glomeromycetes (Bonfante and Genre 2010). However, preliminary data from the global transcriptomic profile of *G. intraradices* reveal a comparable scenario: cell wall-degrading enzymes are limited in number and their activity is finely regulated, as for other non-mutualistic biotrophs (McDowell 2011; Tisserant et al., 2012).

Among all symbiotic interfaces, the complex **intracellular interface of AM** has received large attention since its first descriptions. Over the last decade focus has mostly been put on the genetic control and the cellular mechanisms of AM interface biogenesis (Parniske 2008; Genre and Bonfante 2010). In these interactions, the interface compartment makes its first appearance in root epidermal cells, the first target of presymbiotic hyphae. As soon as a hyphopodium adheres to the epidermal cell wall, in fact, the plant cells start to assemble the secretory machinery that will build the intracellular compartment to host the fungus. Cytoplasm first aggregates at the contact site, then develops into a thick column predicting the subsequent route of the hypha in the cell lumen (Genre et al. 2005). All the elements of the secretory pathway are concentrated in the prepenetration apparatus (PPA): abundant endoplasmic reticulum, and many Golgi bodies and secretory vesicles (Genre et al. 2008). But a major role is played by nucleus, whose movements to and from the contact site accompany PPA development (Genre et al. 2005).

The result of this cytoplasmic bustle is the assembly of the novel membrane which engulfs the penetrating hypha as soon as this crosses the outer cell wall. The process of **interface biogenesis** is repeated in each penetrated cell throughout the root cortex, and broad PPAs have been observed prior to and during arbuscule development, which indicates that intracellular fungal accommodation relies on a conserved mechanism that is modulated depending on the symbiotic structure that is to be hosted by the colonized cell.

Forward genetics analyses have provided important evidence concerning the **genetic control of interface construction.** At least seven genes have been characterized whose mutation hampers interface development, directly affecting symbiosis establishment. The convergence between the signaling pathway that controls AM and the one that rules nodulation is well known and has been reviewed in detail in several occasions (Parniske 2008; Oldroyd et al. 2009).

We will only recall here that all of the seven genes controlling AM establishment constitute the so-called 'common SYM pathway'. Their gene products are in fact involved in the transduction of both fungal and rhizobial diffusible signals. Based on the nomenclature developed in *Lotus japonicus*, SYM genes include: *SYMRK*, a receptor-like kinase, directly or indirectly participating to the perception of rhizobial and AM fungal signals; two nucleoporins (*NUP 85* and *NUP133*); *CASTOR* and *POLLUX*, cation channels located in the nuclear envelope; *CCaMK*, a calcium- and calmodulin-dependent kinase localized in the nucleoplasm; and *CYCLOPS*, a nuclear protein of unknown function. The mutation of each of these SYM genes results to have a direct impact on interface biogenesis in AM. *Castor* mutants of *L. japonicus* show a lack of prepenetration responses in epidermal cells contacted by the hyphopodium, resulting in a hypersensitive-like response causing the plant cell to die when the fungus eventually breaches into its lumen (Novero et al. 2002). The phenotypes described for *dmi1-1*, *dmi2-2* and *dmi3-1* mutants (Ané et al. 2002), the *M. truncatula* orthologues of *CASTOR*, *SYMRK* and *CCaMK*, respectively, have been studied with more detail: the three of them lack PPA formation, which appears to be sufficient in this plant to limit fungal colonization to the root surface. By directly relating the success of the symbiotic interaction to interface biogenesis, such studies also highlight its strict dependency on the genotype of the plant, which therefore appears to have a strong hand in the process.

IV. Interface Composition

By involving the tight contact between fungal and plant walls, the chemical composition of mycorrhizal interfaces results from a combination of the two. Although in fact the hyphal and the host cell walls can normally be discriminated with ease, for example in transmission electron microscopy, the border is sometimes blurred and fragmented. In AM, in particular, the plant and fungal cell walls often show wide indentations, and the plant-derived components appear deeply deconstructed (Fig. 3.4). Whether this depends on the release of lytic enzymes by the fungus (Garcia-Garrido et al. 2000), or the localized self-digestion of wall macromolecules by the plant (as the evidence for focal exocytosis seems to suggest), plant and fungal walls, in these sites, clearly mix up in a complex and not fully understood way.

Molecules typical of the fungal wall, such as chitin and glucans, have been consistently located by affinity techniques in plant-contacting hyphae from a range of arbuscular, ericoid, orchid and ectomycorrhizal fungi (Bonfante 2001, and references within). A peculiar feature of AM fungi is the progressive thinning of their wall within the interface, concomitant to the process of root colonization. Arbuscule branches

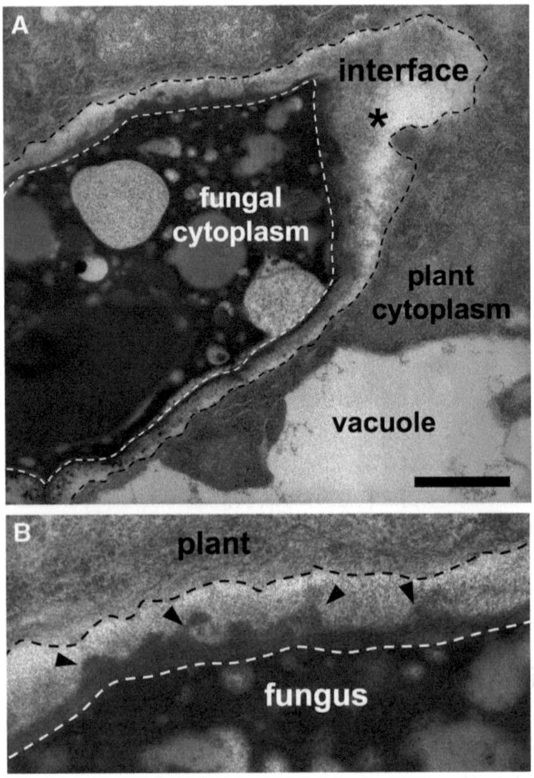

Fig. 3.4. Transmission electron micrographs showing the thin interface surrounding an arbuscule branch. *Dashed lines* mark the position of the fungal (*white*) and plant (*black*) membranes. Loose and deconstructed plant cell wall components are marked by the *asterisk* in (**A**). *Bar* 2 μm. (**B**) shows a 2.5× magnification of the same section, highlighting a wavy pattern in the border between plant and fungal wall components (*arrowheads*)

display the thinnest and most simplified wall, often mostly consisting of amorphous chitin chains. Comparable changes in wall morphology during colonization have not been observed in other endo- and ectomycorrhizal fungi, where the hyphal wall maintains a rather constant thickness during the saprotrophic/presymbiotic as during the symbiotic phase. However, biochemical and molecular analysis of EM fungi has highlighted the regulation of specific fungal cell wall proteins in mycorrhizal roots, including acidic polypeptides and hydrophobins (reviewed in Bonfante 2001; Balestrini et al. 2012).

Taken as a whole, these data indicate that **hyphal walls undergo drastic changes during symbiosis establishment** by mycorrhizal fungi, both in the presence or absence of macroscopic changes in wall morphology.

Moving to the **plant-derived component of the interface**, most studies have focused on endomycorrhizas, where intracellular hyphae are embedded by a plant-secreted matrix. Its composition has been deeply investigated by making use of enzymes or specific lectins and antibodies raised against a range of plant and fungus-derived molecules, which revealed that this compartment is a zone of high molecular complexity. Molecules typical of the plant primary wall, such as cellulose, glucans, polygalacturonans, hemicelluloses, hydroxyproline-rich proteins (HRGP) and arabinogalactan proteins have been repeatedly located in endomycorrhizal interfaces (Bonfante 2001). Remarkably, the distribution of different antigens (such as arabinogalactans and xyloglucans) is not homogeneous and—in analogy to the progressive thinning of AM fungal wall—significant differences in plant wall components can be found between the interface that surrounds large intracellular hyphae, fine arbuscule branches or collapsing arbuscules (Balestrini and Bonfante 2005). These findings suggest **that interface composition is developmentally regulated during the symbiosis establishment.**

Although most of the components of plant primary cell walls are present in the AM interface, they are not assembled into a structured wall. This may be the result of a weak lytic activity by the fungus, which hampers wall assembly mechanisms (Bonfante 2001) and may be related to nutrient provision (see below).

Gene expression analyses have confirmed the upregulation of HRGPs in AM roots and, remarkably, the concentration of HRGP transcripts in arbusculated cells (Balestrini and Bonfante 2005). Arabinogalactan proteins (AGPs) also occur in the interface of many plant/fungal combinations and are induced during mycorrhizal development (Bonfante 2001).

Also cell wall-localized enzymes, such as xyloglucan endo-transglycolases (XET), have been isolated from *Medicago truncatula* mycorrhizal roots (Maldonado-Mendoza et al. 2005). Local cell wall loosening related to fungal penetration, or xyloglucan modifications in the interface have been proposed as possible activities for these enzymes. XET are not the only plant cell-wall remodeling enzymes active in arbusculated cells. Expansins (Balestrini et al. 2005) have been localized by immuno-detection in colonized cells and expansin-related mRNAs were detected by in situ hybridization in the epidermal tissue during PPA assembly (Siciliano et al. 2007). The proposed function of these wall-remodeling enzymes is to achieve a diffuse cell wall loosening, leading to the observed expansion of arbusculated cells, although an involvement in maintaining the deconstructed status of the interface matrix cannot be ruled out (Balestrini and Bonfante 2005). A contribute to the fluidity of the interface materials may also come from the low interface pH (Guttenberg 2000), which can facilitate transglycosilation and enzymatic lysis of structural cell wall polysaccharides.

Altogether, the interface appears more and more as a highly dynamic compartment, not only in terms of the nutrient traffic, but also in its very structure.

V. Cell Walls as a Source of Signals and Nutrients

In **orchid mycorrhizas**, the nature of the interfacial material depends on the stage of the interaction. Hyphae at early stages of pelotons development are surrounded by a matrix that lacks pectins, cellulose and β-1,3 glucans. By contrast, the matrix around senescing pelotons and hyphae is rich in pectins, cellulose and β-1,3 glucans. This observation, made by Peterson et al. (1996), hinted at a possible role of cell wall polysaccharides as carbon sources for the symbiotic fungus. Orchid mycorrhizal fungi, in fact, are able to grow as saprobes in the soil; hence, active intracellular hyphae and pelotons can likely degrade the interface polysaccharides and uptake their derivatives. Similar hypotheses can be formulated for ericoid mycorrhizas (Perotto et al. 1995).

Whether or not this is the case for **AM interactions** remains to be understood. Indirect support to this hypothesis comes from the fact that no plant sugar transporter has been so far identified in the periarbuscular membrane (Bonfante and Genre 2010), which suggests that also in this interaction, fungus-bound carbon might reach the interface in the form of cell wall polysaccharides.

The degradation of cell wall polymers is known to be the source of **signaling oligosaccharides** that have a major role in eliciting plant defense reactions in pathogenic interactions (Humphrey et al. 2007). Unfortunately, our capability to investigate local signal exchange inside the interface space is extremely limited. Nevertheless, the stimulation of defense responses during mycorrhizal colonization is well known (Hause and Fester 2005) and could be explained with an at least partial activation of plant defense pathways by the oligosaccharides derived from the lysis of cell wall components in the interface.

On the same line, the active release of effector molecules by both ecto- and endomycorrhizal fungi cannot be ruled out, although it has not been described so far. Anyway, the fungal wall is acknowledged as a source of signaling molecules in pathogenic interactions, where chitin- and chitosan-related compounds play a major role in triggering plant defenses (Shibuya and Minami 2001), namely via the activation of membrane-bound receptor-like kinases (RLK). All mycorrhizal fungi display a large amount of chitin in their walls, but the involvement of **chitin-based molecules as signals** has been particularly investigated in AM, where the partial overlap with the signaling pathway activated by the chitin-like Nod factor suggested a chitinous nature for AM fungal signals too (Bucher et al. 2009). Indeed, lipo-chito-oligosaccharides with a very similar structure to Nod-factors have recently been characterized in exudates from *Glomus intraradices*-colonized roots (Maillet et al. 2011). When purified and applied to uncolonized roots, such molecules stimulate a few non-specific symbiotic responses, such as root branching, suggesting a role in pre-symbiotic signaling. Nevertheless, their isolation from colonized roots hints at a possible function in later stages, although their presence and possible role in infected cells remains to be investigated.

These observations (and speculations) propose a multifaceted role for both plant and fungal wall-related molecules in mycorrhizal interfaces. Cell wall function, in this unique apoplastic compartment, appears to go well beyond the structural aspects and rather combine these with nutrition and signaling.

VI. The Role of Membranes in Developing and Active Interfaces

On both sides of their combined cell wall domain, interfaces are enclosed by the plant and the fungal plasma membranes. While in the case of fungi, anyway, hyphal membrane maintains its usual position along the inner surface of the cell wall, the plant cell membrane, in endomycorrhizas, undergoes a dramatic proliferation to envelope the intracellular interface compartment. While the fully developed perifungal membrane is relatively easy to observe in electron micrographs, evidence concerning its biogenesis has so far been rather indirect.

A proliferation of Golgi bodies (Pumplin and Harrison 2009) and plastids (Strack and Fester 2006) has been described in young arbusculated cells, hinting at a burst in lipid biosynthesis and membrane dynamics. Abundant Golgi bodies, modified plastids, secretory vesicles and late endosomes (multivesicular bodies) are regularly observed in the PPA cytoplasmic aggregation (Genre et al., 2008). This strongly suggests that intense membrane dynamics are taking place in advance of fungal entry and intracellular growth. Recently (Genre et al., 2012), GFP labeling of SNARE and exocyst proteins has shed light on the dynamics of perifungal membrane biogenesis.

The **fundamental function of plant–fungus interfaces in mycorrhizas** is to allow **a two-way exchange of nutrients** (Balestrini and Bonfante 2005). Under this respect, the fungus can be considered as a sink for organic molecules derived from plant photosynthesis and as a source of mineral nutrients. The chemical nature of the molecules transferred and the mechanisms involved are still largely hypothetical, although several recent studies have identified a bunch of transporters and proposed a few molecules as the best candidates to cross the interface. In the case of AM, where the functional aspects of the interface have been more intensively investigated, several model plants have been studied, including *M. truncatula* (Liu et al. 2003; Küster et al. 2007; Gomez et al. 2009), rice (Guimil et al. 2005) and *L. japonicus* (Guether et al. 2009a). In all cases, **significant increases in the expression level** were observed for membrane transporters, transcription factors and proteins involved in cellular dynamics and cell wall synthesis. A mycorrhiza-specific plant phosphate transporter which is localized in the periarbuscular membrane is essential for active symbiosis in *M. truncatula* (Javot et al. 2007). Among the regulated *L. japonicus* genes, 47 putative transporters were identified, 28 of which may be important for nutrient acquisition. The strongest up-regulated gene of the array is a putative ammonium transporter (Guether et al. 2009b) whose transcripts have been quantified and localized in the arbusculated cells by using laser microdissection, although direct evidence of the protein localization in the periarbuscular membrane is still missing. In *Geosyphon pyriforme,* the only known glomeromycete living in symbiosis with cyanobacteria rather than plants, Schüssler and coworkers (2006) described a hexose importer, providing new momentum to the search of fungal sugar transporters expressed in arbuscules.

On the topic of nutrient fluxes through the interface, a special case is found in myco-heterotrophic plants. These photosyntetically inefficient (or sub-efficient) plants, in fact, partially or totally depend on mycorrhizal symbionts for their carbon supply, a functional feature that drastically differentiates these symbioses from the majority of mycorrhizas. This trophic condition is well known in young orchid protocorms, but extends to the adult phase for several orchids, monotropoids (Leake 1994), liverworts (Bidartondo et al. 2003) and achlorophyllous lycopodiacean gametophytes (Leake et al. 2008). The molecular mechanisms underlying this intriguing and ecologically important turnaround of the mycorrhizal paradigm, however, remain to be clarified (Selosse and Roy 2009).

VII. Conclusions

The use of the word **interface** in mycorrhizal symbiosis was proposed over 30 years ago (Scannerini and Bonfante 1983) based on merely morphological observations. Today the availability of new technologies and the success of molecular analyses and genome sequencing in both mycorrhizal plants and fungi have consistently extended the meaning of this term to the functional aspects of this specialized apoplastic compartment in nutrient exchange and, in general, in symbiosis functioning. Consequently, the very concept of interface became extremely popular to illustrate how mycorrhizas operate and to envisage their application in sustainable agricultural practices. On this line, we have so far regarded the plant–fungus exchange surfaces inside the root as one compartment. Nevertheless, thinking of the dynamic multitude of adjoined hyphae in the Hartig net, or the ephemeral three-dimensional complexity of arbuscules, the limitations of this simplified view become obvious. Understanding the dynamics of interface formation, functioning and senescence, or characterizing the composition and function of different interface subdomains (e.g. those that surround arbuscule branches versus simple transcellular hyphae)—these subjects are currently out of our reach and represent some of the most challenging objectives for future research in this field.

In parallel, the improvement of analytical technologies has also allowed the identification of the bioactive molecules released by both partners and keeping them reciprocally informed of their presence in the rhizosphere (Bonfante and Requena 2011). These investigations open new perspectives on the possible role of these molecules during the symbiotic phase: are signaling molecules also exchanged through the interface? The role of fungal effectors in pathogenic interactions is among the hot spots in molecular plant–microbe interactions, and the analogous search of effectors released by mycorrhizal fungi is opening a fascinating novel field of investigation.

Acknowledgements We are grateful to Raffaella Balestrini for kindly sharing her results and providing images of ectomycorrhizal roots and to Antonella Faccio for electron microscopy preparations. Contributions to this chapter have partially been funded by the Italian National project PRIN 2008, and by Regione Piemonte (CIPE-BioBITs).

References

Ané JM, Lévy J, Thoquet P, Kulikova O, de Billy F, Penmetsa V, Kim DJ, Debellé F, Rosenberg C, Cook DR, Bisseling T, Huguet T, Dénarié J (2002) Genetic and cytogenetic mapping of *DMI1*, *DMI2*, and *DMI3* genes of *Medicago truncatula* involved in Nod factor transduction, nodulation, and mycorrhization. Mol Plant Microbe Interact 15:110–118

Balestrini R, Bonfante P (2005) The interface compartment in arbuscular mycorrhizae: a special type of plant cell wall? Plant Biosyst 139:8–15

Balestrini R, Cosgrove DJ, Bonfante P (2005) Differential location of alpha-expansin proteins during the accommodation of root cells to an arbuscular mycorrhizal fungus. Planta 220:889–899

Balestrini R, Sillo F, Kohler A, Schneide G, Faccio A, Tisserant E, Martin F, Bonfante P (2012) Genome-wide analysis of cell wall-related genes in Tuber melanosporum. Current Genetics 58:165–177

Barrow JR, Lucero ME, Reyes-Vera I, Havstad KM (2008) Do symbiotic microbes have a role in plant evolution, performance and response to stress? Commun Integr Biol 1:69–73

Bidartondo MI, Bruns TD, Weiß M, Sergio C, Read DJ (2003) Specialized cheating of the ectomycorrhizal symbiosis by an epiparasitic liverwort. Proc R Soc Lond B 270:835

Bonfante P (2001) At the interface between mycorrhizal fungi and plants: the structural organization of cell wall, plasma membrane and cytoskeleton. In: Hock B (ed) The mycota IX. Fungal associations. Springer, Berlin, pp 45–91

Bonfante P, Genre A (2010) Mechanisms underlying beneficial plant–fungus interactions in mycorrhizal symbiosis. Nat Commun 1:48

Bonfante P, Requena N (2011) Dating in the dark: how roots respond to fungal signals to establish arbuscular mycorrhizal symbiosis. Curr Opin Plant Biol 14:1–7

Bonfante P, Scannerini S (1992) The cellular basis of plant–fungus interchanges in mycorrhizal associations. In: Allen M (ed) Functioning mycorrhizae. Chapman and Hall, New York, pp 65–101

Bonfante P, Balestrini R, Genre A, Lanfranco L (2009) Establishment and functioning of arbuscular mycorrhizas. In: Deising H (ed) The mycota V, 2nd edn, Plant relationships. Springer, Berlin, pp 45–91

Brachmann A, Parniske M (2006) The most widespread symbiosis on earth. PLoS Biol 4:e239

Bucher M, Wegmüller S, Drissner D (2009) Chasing the structures of small molecules in arbuscular mycorrhizal signaling. Curr Opin Plant Biol 12:500–507

Bücking H, Hans R, Heyser W (2007) The apoplast of ectomycorrhizal roots—site of nutrient uptake and nutrient exchange between the symbiotic partners. In: Sattelmacher B, Horst WJ (eds) The apoplast of higher plants: compartment of storage, transport and reactions. Springer, Berlin, pp 97–108

Dotzler N, Krings M, Taylor TN, Agerer R (2006) Germination shields in *Scutellospora* (Glomeromycota: Diversisporales, Gigasporaceae) from the 400 million-year-old Rhynie chert. Mycol Prog 5:178–184

Durán A, Nombela C (2004) Fungal cell wall biogenesis: building a dynamic interface with the environment. Microbiology 150:3099–3103

Felten J, Kohler A, Morin E, Bhalerao RP, Palme K, Martin F, Ditengou FA, Legue V (2009) The ectomycorrhizal fungus *Laccaria bicolor* stimulates

lateral root formation in poplar and *Arabidopsis* through auxin transport and signaling. Plant Physiol 151:1991–2005
Fitter AH, Helgason T, Hodge A (2011) Nutritional exchanges in the arbuscular mycorrhizal symbiosis: implications for sustainable agriculture. Fungal Biol Rev 25:68–72
Garcia-Garrido JM, Tribak M, Rejon-Palomares A, Ocampo JA, Garcia-Romera I (2000) Hydrolytic enzymes and ability of arbuscular mycorrhizal fungi to colonize roots. J Exp Bot 51:1443–1448
Genre A, Bonfante P (2010) The making of symbiotic cells in arbuscular mycorrhizal roots. In: Koltai H, Kapulnik Y (eds) Arbuscular mycorrhizas: physiology and function, 2nd edn. Springer, Dordrecht, pp 57–71
Genre A, Chabaud M, Timmers T, Bonfante P, Barker DG (2005) Arbuscular mycorrhizal fungi elicit a novel intracellular apparatus in *Medicago truncatula* root epidermal cells before infection. Plant Cell 17:3489–3499
Genre A, Chabaud M, Faccio A, Barker DG, Bonfante P (2008) Prepenetration apparatus assembly precedes and predicts the colonization patterns of arbuscular mycorrhizal fungi within the root cortex of both *Medicago truncatula* and *Daucus carota*. Plant Cell 20:1407–1420
Genre A, Ivanov S, Fendrych M, Faccio A, Zársky V, Bisseling T, Bonfante P (2012) Multiple exocytotic markers accumulate at the sites of perifungal membrane biogenesis in arbuscular mycorrhizas. Plant Cell Physiol 53:244–55
Gomez SK, Javot H, Deewatthanawong P, Torres-Jerez I, Tang YH, Blancaflor EB, Udvardi MK, Harrison MJ (2009) *Medicago truncatula* and *Glomus intraradices* gene expression in cortical cells harboring arbuscules in the arbuscular mycorrhizal symbiosis. BMC Plant Biol 9. doi:10.1186/1471-2229-9-10
Guether M, Balestrini R, Hannah M, He J, Udvardi MK, Bonfante P (2009a) Genome-wide reprogramming of regulatory networks transport cell wall synthesis and membrane biogenesis during arbuscular mycorrhizal symbiosis in *Lotus japonicus*. New Phytol 182:200–212
Guether M, Neuhauser B, Balestrini R, Dynowski M, Ludewig U, Bonfante P (2009b) A mycorrhizal specific ammonium transporter from *Lotus japonicus* acquires nitrogen. Plant Physiol 150:73–83
Guimil S, Chang H, Zhu T, Sesma A, Osbourn A, Roux C, Ioannidis V, Oakeley EJ, Docquier M, Descombes P, Briggs SP, Paszkowski U (2005) Comparative transcriptomics of rice reveals an ancient pattern of response to microbial colonization. Proc Natl Acad Sci USA 102:8066–8070
Guttenberg M (2000) Arbuscules of vesicular-arbuscular mycorrhizal fungi inhabit an acidic compartment within plant roots. Planta 211:299–304
Hause B, Fester T (2005) Molecular and cell biology of arbuscular mycorrhizal symbiosis. Planta 221: 184–196
Humphrey TV, Bonetta DT, Goring DR (2007) Sentinels at the wall: cell wall receptors and sensors. New Phytol 176:7–21
Javot H, Penmetsa RV, Terzaghi N, Cook DR, Harrison MJ (2007) A *Medicago truncatula* phosphate transporter indispensable for the arbuscular mycorrhizal symbiosis. Proc Natl Acad Sci USA 104:1720–1725
Küster H, Vieweg MF, Manthey K, Baier MC, Hohnjec N, Perlick AM (2007) Identification and expression regulation of symbiotically activated legume genes. Phytochemistry 68:8–18
Leake JR (1994) Transley review no. 69: the biology of myco-heterotrophic ('saprophytic') plants. New Phytol 127:171–216
Leake JR, Cameron DD, Beerling DJ (2008) Fungal fidelity in the myco- heterotroph-to-autotroph life cycle of Lycopodiaceae: a case of parental nurture? New Phytol 177:572–576
Liu J, Blaylock LA, Endre G, Cho J, Town CD, VandenBosch KA, Harrison MJ (2003) Transcript profiling coupled with spatial expression analyses reveals genes involved in distinct developmental stages of an arbuscular mycorrhizal symbiosis. Plant Cell 15:2106–2123
Maillet F, Poinsot V, Andre O, Puech-Pages V, Haouy A, Gueunier M, Giraudet D, Formey D, Martinez EA, Driguez H, Becard G, Dénarié J (2011) Fungal lipochitooligosaccharide symbiotic signals in arbuscular mycorrhiza. Nature 69:58–63
Maldonado-Mendoza IE, Dewbre GR, Blaycock L, Harrison MJ (2005) Expression of a xyloglucan endotransglucosylase/hydrolase gene Mt-XTH1 from Medicago truncatula is induced systemically in mycorrhizal roots. Gene 345:191–197
Martin F, Nehls U (2009) Harnessing ectomycorrhizal genomics for ecological insights. Curr Opin Plant Biol 12:508–515
Martin F, Laurent P, de Carvalho D, Voiblet C, Balestrini R, Bonfante P, Tagu D (1999) Cell wall proteins of the ectomycorrhizal basidiomycete *Pisolithus tinctorius*: identification, function, and expression in symbiosis. Fungal Genet Biol 27:161–174
Martin F, Aerts A, Ahren D, Brun A, Danchin EG, Duchaussoy F, Gibon J, Kohler A, Lindquist E, Pereda V, Salamov A, Shapiro HJ, Wuyts J, Blaudez D, Buee M, Brokstein P, Canback B, Cohen D, Courty PE, Coutinho PM, Delaruelle C, Detter JC, Deveau A, DiFazio S, Duplessis S, Fraissinet-Tachet L, Lucic E, Frey-Klett P, Fourrey C, Feussner I, Gay G, Grimwood J, Hoegger PJ, Jain P, Kilaru S, Labbe J, Lin YC, Legue V, Le Tacon F, Marmeisse R, Melayah D, Montanini B, Muratet M, Nehls U, Niculita-Hirzel H, Oudot-Le Secq MP, Peter M, Quesneville H,

Rajashekar B, Reich M, Rouhier N, Schmutz J, Yin T, Chalot M, Henrissat B, Kues U, Lucas S, Van de Peer Y, Podila GK, Polle A, Pukkila PJ, Richardson PM, Rouze P, Sanders IR, Stajich JE, Tunlid A, Tuskan G, Grigoriev IV (2008a) The genome of *Laccaria bicolor* provides insights into mycorrhizal symbiosis. Nature 452:88–92

Martin F, Gianinazzi-Pearson V, Hijri M, Lammers P, Requena N, Sanders IR, Shachar-Hill Y, Shapiro H, Tuskan GA, Young JPW (2008b) The long hard road to a completed *Glomus intraradices* genome. New Phytol 180:747–750

Martin F, Kohler A, Murat C, Balestrini R, Coutinho PM, Jaillon O, Montanini B, Morin E, Noel B, Percudani R, Porcel B, Rubini A, Amicucci A, Amselem J, Anthouard V, Arcioni S, Artiguenave F, Aury JM, Ballario P, Bolchi A, Brenna A, Brun A, Buée M, Cantarel B, Chevalier G, Couloux A, Da Silva C, Denoeud F, Duplessis S, Ghignone S, Hilselberger B, Iotti M, Marçais B, Mello A, Miranda M, Pacioni G, Quesneville H, Riccioni C, Ruotolo R, Splivallo R, Stocchi V, Tisserant E, Viscomi AR, Zambonelli A, Zampieri E, Henrissat B, Lebrun MH, Paolocci F, Bonfante P, Ottonello S, Wincker P (2010) Périgord black truffle genome uncovers evolutionary origins and mechanisms of symbiosis. Nature 464:1033–1038

McDowell JM (2011) Genomes of obligate plant pathogens reveal adaptations for obligate parasitism. Proc Natl Acad Sci USA 108:8921–8922

Nehls U (2003) Ectomycorrhizal development and function: transcriptome analysis. New Phytol 159:5–7

Novero M, Faccio A, Genre A, Stougaard J, Webb KJ, Mulder L, Parniske M, Bonfante P (2002) Dual requirement of the *LjSym4* gene for the mycorrhizal development in epidermal cells and cortical cells of *Lotus japonicus* roots. New Phytol 154: 741–749

Oldroyd GED, Harrison MJ, Paszkowski U (2009) Reprogramming plant cells for endosymbiosis. Science 324:753–754

Parniske M (2008) Arbuscular mycorrhiza: the mother of plant root endosymbioses. Nat Rev Microbiol 6:763–775

Perotto S, Peretto R, Faccio A, Schubert A, Varma A, Bonfante P (1995) Ericoid mycorrhizal fungi: cellular and molecular bases of their interactions with the host plant. Can J Bot 73:S557–S568

Peterson RL, Massicotte HB (2004) Exploring structural definitions of mycorrhizas, with emphasis on nutrient-exchange interfaces. Can J Bot 82: 1074–1088

Peterson RL, Bonfante P, Faccio A, Uetake Y (1996) The interface between fungal hyphae and orchid protocorm cells. Can J Bot 74:1861–1870

Peterson RL, Uetake Y, Zelmer C (1998) Fungal symbioses with orchid protocorms. Symbiosis 25: 29–55

Peterson RL, Massicotte HB, Melville LH, Phillips F (2004) Mycorrhizas: anatomy and cell biology. NRC Research Press, Ottawa

Pumplin N, Harrison MJ (2009) Live-cell imaging reveals periarbuscular membrane domains and organelle location in *Medicago truncatula* roots during arbuscular mycorrhizal symbiosis. Plant Physiol 151:809–819

Redecker D, Kodner R, Graham LE (2000) Glomalean fungi from the Ordovician. Science 289:1920–1921

Remy W, Taylor TN, Hass H, Kerp H (1994) Four hundred-million-year-old vesicular arbuscular mycorrhizae. Proc Natl Acad Sci USA 91: 11841–11843

Sattelmacher B, Horst WJ (2007) The apoplast of higher plants: compartment of storage, transport and reactions. Springer, Berlin Heidelberg New York

Scannerini S, Bonfante P (1983) Comparative ultra structural analysis of mycorrhizal associations. Can J Bot 61:917–943

Schuüssler A, Martin H, Cohen D, Fitz M, Wipf D (2006) Characterization of a carbohydrate transporter from symbiotic glomeromycotan fungi. Nature 444:933–936

Selosse MA, Roy M (2009) Green plants that feed on fungi: facts and questions about mixotrophy. Trends Plant Sci 14:64–70

Shibuya N, Minami E (2001) Oligosaccharide signalling for defense responses in plants. Physiol Mol Plant Pathol 59:223–233

Siciliano V, Genre A, Balestrini R, Cappellazzo G, DeWitt P, Bonfante P (2007) Transcriptome analysis of arbuscular mycorrhizal roots during development of the prepenetration apparatus. Plant Physiol 144:1455–1466

Smith SE, Read DJ (2008) Mycorrhizal symbiosis. Academic, New York

Strack D, Fester T (2006) Isoprenoid metabolism and plastid reorganization in arbuscular mycorrhizal roots. New Phytol 172:22–34

Tagu D, Martin F (1996) Molecular analysis of cell wall proteins expressed during the early steps of ectomycorrhiza development. New Phytol 133: 73–85

Tisserant E, Kohler A, Dozolme-Seddas P, Balestrini R, Benabdellah K, Colard A, Croll D, Da Silva C, Gomez SK, Koul R, Ferrol N, Fiorilli V, Formey D, Franken P, Helber N, Hijri M, Lanfranco L, Lindquist E, Liu Y, Malbreil M, Morin E, Poulain J, Shapiro H, van Tuinen D, Waschke A, Azcón-Aguilar C, Bécard G, Bonfante P, Harrison MJ, Küster H, Lammers P, Paszkowski U, Requena N, Rensing SA, Roux C, Sanders IR, Shachar-Hill Y, Tuskan G, Young JP, Gianinazzi-Pearson V, Martin F (2012) The transcriptome of the arbuscular mycorrhizal fungus Glomus intraradices (DAOM 197198) reveals functional tradeoffs in an obligate symbiont. New Phytol 193:755–69

4 Arbuscular Mycorrhiza: A Key Component of Sustainable Plant–Soil Ecosystems

P. Jeffries[1], J.M. Barea[2]

CONTENTS

I. Introduction

Sustainable plant–soil ecosystems occur when the utilization of mineral resources by plants is balanced by efficient biogeochemical cycling, such that nutrients are not rapidly exhausted and plant communities can exist in a stable form for prolonged periods. **Sustainable ecosystems** are the target of ecologically sound management strategies in agriculture as well as being the normal situation in natural, undisturbed ecosystems. The concept of sustainability in agriculture aims to conserve the productive capacity of the soil, minimizing energy and resource use and optimizing the rate of turnover and recycling of matter and nutrients. Sustainability demands the efficient utilization of nutrients by plants and this process is facilitated through mycorrhizal associations. The critical role that **mycorrhizal symbioses** play in plant nutrition (Smith and Read 2008) is now widely accepted. Almost all plants form mycorrhizal associations and a variety of types are formed depending on the plant taxa involved. However, it is the **arbuscular mycorrhizal (AM) relationship** that is most common and over 80 % of plant species are capable of forming these structures when associated with arbuscular mycorrhizal fungi (AMF). The AMF are ubiquitous soil-borne microbial fungi, whose origin and divergence have been dated back over 450 million years (Redecker et al. 2000). The AM symbiosis has an ancient origin and the rhizomes of the first primitive terrestrial plants contained fungal structures almost identical to their modern counterparts (e.g. Remy et al. 1994; Kenrick 2003; Honrubia 2009). As a consequence of this co-evolution, the benefits of the relationship became an integral component of plant ecology in both natural and agricultural ecosystems (Brundrett 2002). In this chapter we consider the concepts implicit in sustainable ecosystems and describe how the AM symbiosis is crucial in maintaining sustainability. We also discuss how the biodiversity of AMF populations can influence plant community dynamics in both natural and agricultural environments, and we give examples where manipulation of the AM symbiosis can help restore sustainability to disturbed environments. Adverse conditions of

[1]School of Biosciences, University of Kent, Canterbury, CT2 7NJ KentUK
[2]Departamento de Microbiología del Suelo y Sistemas Simbióticos, Estación Experimental del Zaidín, CSIC, Prof. Albareda 1, 18008 GranadaSpain; e-mail: josemiguel.barea@eez.csic.es

Fungal Associations, 2nd Edition
The Mycota IX
B. Hock (Ed.)

differing origin affect the stability of both natural and agricultural systems, and thus plants must be able to cope with these stresses. Thus, AM formation can be considered as an adaptive strategy which provides the plant with an increased ability for nutrient capture and cycling in soils with low nutrient availability and an increased tolerance to environmental stresses. In sustainable agriculture the AM symbiosis plays a key role in helping the plant not only to survive but also to be productive under adversity (Jeffries et al. 2003).

II. Sustainable Plant–Soil Ecosystems

The concept of sustainability is often used in relation to environmental changes that have been induced as a result of the activities of mankind. This concept has not changed over the past 10 years since the first edition of this book. Indeed, sustainability may still be defined as **'the successful management of resources to satisfy changing human needs while maintaining or enhancing the quality of the environment and conserving resources'** (Bohlool et al. 1992). In this context, sustainability also takes into account the protection of natural resources, such as soil water, non-renewable energy resources and environmental quality (Ladha 1992). Maintaining the quality and sustainability of soil resources is a key issue, not only for optimizing the stability and productivity of natural ecosystems, but also to prevent erosion and minimize negative environmental stresses (Buscot 2005; Chaudhary et al. 2009). A sustainable ecosystem approach involves the rational use of natural resources rather than their exploitation. It is based on the use of renewable inputs and in the optimization of resource utilization to reach a balanced environmental relationship (Lal 1989). To achieve sustainability it is necessary to prevent both environmental pollution and the depletion of agricultural and forestry resources, whilst preserving the structure and diversity of natural plant communities (Klironomos 2002). A sustainable approach also aims to provide vegetation cover, thus reducing the susceptibility of soils to erosion, and also to lower the energy-based inputs (Lal 1989; Bethlenfalvay and Linderman 1992; Peoples and Craswell 1992; Wolfe et al. 2005). These concepts apply to the preservation of natural plant communities but the term sustainable has been increasingly applied to agriculture as attempts are made to provide long-term sustained yields through the use of ecologically sound management technologies such as crop diversification, organic soil management and biological pest and disease control (Altieri 1994, 2004).

Sustainable agriculture can be envisaged as a key component of current and future trends in plant productivity for both developing and developed countries. In the 1990 Food, Agriculture, Conservation and Trade Act (GAO/RCED 1992), the United States Congress discussed sustainable agriculture as 'an integrated system of plant and animal production practices having a site-specific application that will, over the long term: (a) satisfy human food and fibre needs, (b) enhance environmental quality and the natural resource base upon which the agricultural economy depends, (c) make the most efficient use of non-renewable resources and on-farm resources and integrate, where appropriate, natural biological cycles and controls, (d) sustain the economic viability of farm operations, and (e) enhance the quality of life for farmers and society as a whole'.

Sustainable agriculture can thus be considered as the maintenance of soil fertility and structure over a long period of time such that the economic yields from crop plants can be achieved through the minimum inputs of fertilizer necessary to reach such yields. It must achieve economic and sustained production and yet also preserve the resource base. However, true sustainability in which outputs are balanced by inputs may only be achieved in stable natural ecosystems. The term 'sustainable agriculture' is a contradiction as agriculture inevitably involves the artificial manipulation of plant ecosystems for production of food. Inputs are thus no longer balanced to outputs, and natural relationships are disturbed, diversity is decreased and the sustainability is altered. In agricultural development, priority is often given to increasing yields and this can result in a dependence on the high inputs of artificial readily soluble fertilizers that are necessary to sustain them. It is difficult to develop any form of agriculture that could be truly sustainable. Instead, a modification of existing strategies is necessary in sustainable agriculture such that fertilizer inputs are reduced but not eliminated and that maximum use is made of the soil microbiota in efficient nutrient capture and in cycling nutrients to the plant root system. Soil microbes are thus critical for essential ecosystem functions such as the biogeochemical cycling of nutrients and matter and the maintenance of plant health and soil quality (Barea et al. 2005a; Avis et al. 2008; Chaudhary et al. 2009; Richardson et al. 2009). In this context, mycorrhizal relationships are particularly important and will be discussed in the next section.

III. Importance of AM in Soil Fertility

This topic has been reviewed in depth (e.g. see Bethlenfalvay and Linderman 1992; Jeffries and Barea 1994) but more recent publications have highlighted the significant role that AMF play either in sustainable agriculture or in natural plant communities (e.g. van der Heijden 2002; Jeffries et al. 2003; Barea et al. 2005a; Richardson et al. 2009; Gianinazzi et al. 2010) and thus contribute to the maintenance of soil fertility. These studies have reinforced the view that the **soil microbiota is a key component of sustainable systems** when used as a natural resource tool (Barea et al. 2005a; Avis et al. 2008; Mallik and Williams 2008; Chaudhary et al. 2009; Richardson et al. 2009). The diversity of the microbiota is a major issue (Kennedy and Smith 1995) and molecular tools now offer the means to study this topic in depth (Robinson-Boyer et al. 2009). Micro-organisms conduct activities which are crucial to the establishment, development, nutrition and health of plants (Barea et al. 2005a, b; Richardson et al. 2009; Dessaux et al. 2010). Microbial populations in soil actively develop around plant roots, within the rhizosphere, where they are stimulated by root exudates, plant residues and other organic substrates supplied by the plant. The beneficial activities of the rhizosphere microbiota include the increased availability of plant nutrients, improvement of nutrient uptake and protection against root pathogens (Kloepper 1992; Linderman 1992; Glick 1995; Barea et al. 1997, 2002b, 2005b; Faure et al. 2009; Hartmann et al. 2009; Jones et al. 2009; Lambers et al. 2009; Richardson et al. 2009; Dessaux et al. 2010).

Mycorrhizal associations are arguably the most significant agents of **nutrient cycling** between plant and soil. The mycorrhizal relationship improves the nutritional status of the plant, and thereby plant health, which then enables it to cope more effectively with cultural or environmental stress, either biotic (e.g. pathogen attack) or abiotic (e.g. drought, salinity, heavy metals, organic pollutants) as recently reviewed (Jeffries et al. 2003; Rillig and Mummey 2006; Turnau et al. 2006; Pozo and Azcón-Aguilar 2007; Barea et al. 2008; Finlay 2008; Ruíz-Lozano et al. 2008; Azcón and Barea 2010). Through the agency of an inter-linked and extensive soil mycelium, the mycorrhizal fungi are critical agents that affect the distribution and movement of nutrients within the soil ecosystem (Jakobsen 2004; Richardson et al. 2009). The major flux is the **transfer of carbon from plant to fungus** (and thereby to the soil) and the reciprocal **movement of phosphate and ammonium from fungus to plant.** Plant P availability is the most limiting factor for crop yield in 30–40% of the world's arable soils (Vance et al. 2003; Barea et al. 2008). The role of AMF in phosphate exchange is well documented and it is recognised that the majority of P taken up by plants comes via the fungal partner, but more emphasis is now being placed on N and C movements. Some 5–10% of photosynthetic carbon in plants resides within the mycorrhizal component (Bryla and Eissenstat 2005) and turnover of carbon in the soil mycelium is rapid. Fitter et al. (2011) estimate that AMF account for 5–10% of total soil biomass—an incredible statistic! Smith et al. (2009) recently discussed the carbon economy issues in the AM symbiosis in the context of nutrient trade and ecological sustainability. The acquisition of N via the fungal partner can also be substantive (Leigh et al. 2009), despite the high N demand by the fungi themselves, and the global pool of N in AMF mycelia is at least as big as that in fine roots (Hodge and Fitter 2010). Whatever the case, the

hyphal network within the soil is a vital component of the soil ecosystem and is the functional organ for the uptake and translocation of these key nutrients to and from mycorrhizal stuctures. Many reviews (e.g. Barea et al. 2005a; Smith and Read 2008; Azcón and Barea 2010) also describe the well established role of the extraradical mycelium in the uptake of water and other mineral nutrients, and the enhanced role of AMF in drought conditions is discussed later. It is clear that the network is essential for the continued cycling of nutrients within the plant community, and once it is lost nutrient sequestration or leaching will occur at a faster rate than in its presence. Anastomosis of hyphae between mycelia of the same species is an important factor in establishing a functional network that interlinks host plants and is capable of long-distance transport and retention of nutrients in readily available plant pools (Mikkelsen et al. 2008). The hyphal network is thus vital for the maintenance of sustainable plant yields, both in a natural ecosystem and in low-input agricultural situations where nutrients are limiting.

It is clear that P, C, N and other mineral nutrients can be transported from remote sources in the soil by hyphae of AMF (e.g. Cooper and Tinker 1978; Zhu and Miller 2003; Barea et al. 2005a, 2008). The fungi take up P from the same pool of soluble ions as do roots and thus act as an extension to the root system. There is evidence that phosphatase activity is higher in the rhizosphere soil around arbuscular mycorrhiza than around that of non-mycorrhizal roots (Dodd et al. 1987; Tawaraya et al. 2006; Barea et al. 2008) but there is no clear evidence that this is a fungus-mediated phenomenon that allows alternative P sources to be accessed. Many studies have also demonstrated that inter-plant bridges formed by AMF can provide channels for direct nutrient transfer between the AM of different plants (Azcón and Barea 2010). This nutrient transfer may be sufficient to sustain significant enhancement of both growth and nutrient composition of receiver plants, in some cases within 6 weeks of commencement of experimentation (Francis et al. 1986). Lekberg et al. (2010) used a monoxenic laboratory system to show how four times as much carbon was transferred to fungal structures within C-limited receiver plants than to C-rich donor plants, whereas C-rich plants received 10 times more P than their C-limited equivalents.

In order to function effectively, the mycelium must be allowed to spread into the soil and remain intact. Destruction of the hyphae can occur through soil disturbance. For example, tillage or similar agricultural practices which disrupt the mycelial network have serious effects on its capacity to translocate nutrients over any significant distance (see later). Grazing by small invertebrates can also destroy parts of the mycelial network. As much of the soil biomass may consist of mycorrhizal hyphae, they are probably major components of soil food chains (Fitter et al. 2011). Mycorrhizal hyphae are involved in the formation of stable soil aggregates, a process crucial for soil conservation and for good soil tilth (Bethlenfalvay and Schüepp 1994; Rillig and Mummey 2006; Wright et al. 2007; Kohler et al. 2010). This effect is critical in the development of sustainable ecosystems especially in eroding soils. The soil mycelium first develops as a skeletal structure which holds soil particles by simple entanglement. Later, the roots and hyphae provide conditions necessary to form microaggregates by means of physicochemical mechanisms which involve the incorporation of organic debris (Tisdall 1991). Binding agents, mostly of microbial origin, participate in the process of cementation and stabilization of microaggregates. Hyphae of AMF produce copious amounts of a recalcitrant glycoprotein termed glomalin, the presence of which is closely correlated with soil aggregate stability (Rillig 2004; Wright et al. 2007; Curaqueo et al. 2011).

IV. Biodiversity of AMF and Relevance to Sustainability

As discussed earlier, AM symbioses enhance the relative abilities of plants to compete for limiting nutrients. Integration of individual plants into the collective nutrient-gathering capacity of the community assures survival, and the role of the soil mycelium is again crucial. Research on the impact of AMF on plant community composition and functioning has concluded that **the diversity and activity of mycorrhizal fungi is a key mechanism for linking biodiversity and ecosystem functioning** (Read 1989; Hart and Klironomos 2002; Kennedy et al. 2007; Martínez-García and Pugnaire 2009). Conversely, diversity and structure of plant cover can influence the diversity of AM fungal populations (Bever et al. 2002; Read 2002; Wolfe et al. 2005).

A pioneering experiment on the impact of AM fungal diversity on plant community diversity by Grime et al. (1987) showed that the presence of undefined mixtures of AM fungi increased the floristic diversity in a microcosm trial. The first demonstration that the diversity and identity of AM fungi, and not merely their presence, was a determinant of plant diversity and/or ecosystem productivity came from the field plots and mesocosms studies of van der Heijden et al. (1998b). This influential publication prompted a number studies to show that AM fungal diversity could affect the species composition and functioning of plant communities.

Several mechanisms/factors have been proposed to be responsible for the ecological interactions between plant and fungal communities, and among these are: (1) the functional specificity of the different plant–fungus combinations (van der Heijden et al. 1998a, b, 2006; Klironomos 2002), (2) the mycorrhizal dependency/responsiveness of the plant species involved, (3) the dominant versus subordinate character of these species in the community (O'Connor et al. 2002; van der Heijden 2002; van der Heijden et al. 2006), (4) the 'niche differentiation in P use' (Reynolds et al. 2006; Vogelsang et al. 2006) and (5) the so-called 'sampling effect' (Vogelsang et al. 2006).

Key concepts can be drawn from consolidated knowledge of these aspects of AMF ecology. For example, despite the fact that specificity *sensu stricto* does not exist in AM associations, in which many plants in the community can be colonized simultaneously by several species of AM fungi, it is accepted that different AM fungal taxa induce more positive responses in some plant species than in others (Sanders 2002). It also seems that not every fungus can colonize every plant in the community (Barea et al. 2008). It is also noteworthy that not all plants form AM associations and that not all plants that form AM associations obtain nutritional benefits from AM fungi under all growth conditions (O'Connor et al. 2002). The differential effects on plant growth and development of specific plant–fungus associations, known as their 'functional compatibility/specificity' (Gianinazzi-Pearson et al. 1988), has a critical significance in AM ecology/functioning.

The results of van der Heijden et al. (1998b) were re-analysed by Read (1989) and Hart and Klironomos (2002) in discussing the specificity relationships in the AM symbiosis. These review articles highlight the role of AM fungi in increasing productivity and/or diversity of plant communities and support the hypothesis that the differential effects of specific plant–fungus combinations (functional compatibility) explain the original observations. In particular, the ability of specific fungi to supply plants with P was found to be the basis of the differential effects of functional compatibility (van der Heijden et al. 1998b).

However, more recent work investigating the influence of AM fungi on the diversity and/or productivity of plant communities has presented some conflicting results. Van der Heijden (2002) re-analysed publications on this topic and concluded that the **main factors influencing the ecological interactions between plants and AMF** were the **degree of mycorrhizal dependency (MD) or mycorrhizal responsiveness (MR) of the plant species involved and their dominant or subordinate character in the plant community**. The terms MD and MR are often used synonymously in an ecological context. However, Janos (2007) distinguishes between these two terms and considers that MR is represented by the difference in growth between plants with and without AM colonization, at any level of P availability, being also a measure of AM fungal effectiveness, and MD is defined by the lowest level of P availability at which plants can grow without AM colonization. All in all, MD is a measure of the extent to which a plant benefits from association with AMF, which is in turn is related to the ability of AMF to supply P to the plant (van der Heijden et al. 1998b). Hence, van der Heijden (2002) concluded that the differential responses of a plant species to different AM fungal species are higher when the target plant species has a higher MD. The presence of AMF promotes plant diversity when most of the subordinate plant species in the community have a high MD so that their growth is stimulated specifically (Grime et al. 1987; van der Heijden et al. 1998b). Conversely, AM formation reduces the diversity of plant communities when most of the plant species have a negative or low MD (Hartnett and Wilson 1999). Moreover, when dominant plant species in the community have a high degree of mycorrhizal responsiveness, and thereby derive greater benefit from AM fungi, plant diversity is not promoted, irrespective of the level of MD of the subordinate species (O'Connor et al. 2002). By labelling the dominant but low mycorrhiza-dependent plant species with ^{14}C, Grime et al. (1987) also found a belowground transport of C via the AM mycelium to the subordinate highly mycorrhiza-dependent species. This C and P redistribution facilitates the establishment and growth of subordinate species to increase plant diversity in the community.

With regard to the impact of AM fungal diversity on community productivity, van der Heijden (2002) concluded that AMF enhance community productivity when the communities are dominated by mycorrhiza-dependent plant species, able to reap the benefit from P supply from AMF. When the community is dominated by plant species with a low MD, van der Heijden (2002) concluded that there was no increase in productivity, because P supply by AMF is a less influential process. More recently van der Heijden et al. (2006) showed that mixtures of mycorrhiza-dependent and non-dependent plant species do not result in an increase in the productivity of a grassland community as a whole.

In the context of the mechanisms underlying the relationship between AM fungal diversity and plant community diversity and/or productivity it is important to consider whether the benefits generated by individual fungal isolates are greater than those produced by a mixed inoculum. Vogelsang et al. (2006) found that AM fungal identity benefited plant diversity and productivity more than the diversity *per se*. These results conflict with those of van der Heijden et al. (1998b), who correlated AM fungal diversity *per se* with plant diversity and productivity. However, these conflicting reports might be explained by the confounding effects of plant species dominance or mycorrhiza-dependency (van der Heijden et al. 2006). Since different plant–AMF combinations access different P sources, Vogelsang et al. (2006) also investigated whether competition for limited P resources could favour AM fungus-mediated P-niche partitioning. These authors manipulated the P sources in mesocosm experiments and found little support for such AM fungally facilitated complementarity in P use by the community.

The influence of AMF on the diversity of the plant communities has been discussed above, but the converse should also be considered, i.e. fungal biodiversity is influenced by the aboveground plant community. Community studies of AMF have taken a range of approaches, either using indirect methods such as isolation of spores from soil or from trap cultures, or direct methods via isolation of fungal DNA from roots. The AMF were formerly included in the order Glomales, Zygomycota (Redecker et al. 2000), but they now comprise a unique phylum, the Glomeromycota (Schüßler et al. 2001) which has been widely accepted (Rosendahl 2008; Helgason and Fitter 2009; Gamper et al. 2010). Molecular analyses have shown that there is considerable genetic diversity within morphologically recognizable species within the Glomeromycota, and a recent authoritative review suggests that there is currently no formalized operational species concept in this group (Rosendahl 2008). Many recent analyses cluster sequence data into 'species groups' or 'phylogenetic clusters'.

Plant invasions have been shown to impact on belowground fungal diversity Kilvin and Hawkes (2011), although Dumbrell et al. (2010a, b) have suggested that soil factors such as pH and stochastic factors are more important than the aboveground composition of the plant community. In semi-natural grassland communities, Schnoor et al. (2011) showed that soil disturbance had a stronger influence on AMF communities than host plants, whilst Schalmuk and Cabello (2010) suggested that the higher levels of Glomeraceae found in no-tillage systems could be related partially to the lack of disruption of hyphae and the composition of soil propagules in these systems. Nevertheless, belowground biological systems need the same considerations as aboveground systems because AM are so essential in sustainable land management practices. Different genera and families of

AMF have unique patterns of mycelial development (Boddington and Dodd 1998; Voets et al. 2006) and also different mechanisms for control of P transfer to the host. It is important to know: what species are present in the environment as well as monitoring their growth inside the roots; how the community structure is changing during the restoration period; which species are significant and at what time.

The **maintenance of fungal biodiversity** is important since, as discussed above, differential benefits can be conferred to the plant hosts by different AMF depending on host plant or environmental conditions (Sieverding and Howeler 1985; van der Heijden et al. 1998a, b). It is also clear that different isolates of the same species differ in their effectiveness in conferring benefits to similar host plants grown under standard conditions (Bethlenfalvay et al. 1989; Herrera-Peraza et al. 2011). In a natural ecosystem these mixed populations co-exist, with certain fungi becoming dominant in particular patches and subsequently being replaced as environmental conditions change. Once this equilibrium is disturbed, for example by crop monoculture or by the use of fungicides, the population dynamics are disrupted and a bias can develop towards a few or even one dominant fungus (Johnson et al. 1992). In some environments tillage and fertilizer use have led to fewer species of AMF being found in the soil (Schenck and Kinloch 1980; Hetrick and Bloom 1983), while more unusually in others, agricultural use may lead to greater diversity (Abbott and Robson 1977). Soil type, land use and agronomic management can all affect the composition of AMF populations (Brito et al. 2008; Oehl et al. 2009, 2010; Curaqueo et al. 2011) When non-mycotrophic crops such as canola are grown in monoculture, there is evidence for a general decline in mycorrhizal propagules (Gavito and Miller 1998). The longer the soil is cropped with a non-host species, the more pronounced will be the loss of mycorrhizal infectivity and diversity within the soil.

Where sustainable agriculture is desired it is necessary to maintain a diversity of mycorrhizal fungi as mixed crops will benefit more from a mixed population of symbionts. The diversity of AMF and spatial turnover of populations is often greater in organic agriculture than in conventional farm management (van der Gast et al. 2011). In non-agricultural situations, plant diversity has been related to AM diversity and it follows that maintenance of a sustainable mixed plant population depends on the maintenance of a diverse AMF population and vice versa.

It is clear that there is no single mycorrhizal effect on plant communities (Allen 1991). As a consequence, it is necessary to have information about the individual fungi that are functional within an ecosystem in order to be able to predict their behaviour. For this purpose it is necessary to identify and discriminate individual AMF within the roots. This used to be a daunting prospect, but **the molecular tools** are now available for a challenging dissection of mycorrhizal population dynamics (Robinson-Boyer et al. 2009). Diversity studies of AM fungal communities are being based the sequence analysis of the small-subunit (18S) ribosomal DNA of the spores and/or the mycelia from AM fungi (Hempel et al. 2007; Morton 2009; Öpik et al. 2008, 2010; Oehl et al. 2009; Gamper et al. 2010). Although molecular techniques have gained prominence, identification approaches based on morphological criteria are still important and complement molecular methods (Barea et al. 2007; Morton 2009; Oehl et al. 2009). However, alternative molecular tools now exist to quantitatively analyse the effect of environment, management or inoculation of soils on more diverse AMF communities. Q-PCR can be used for simultaneous specific and quantitative investigations of particular taxa of AMF in roots and soils colonized by several taxa (Gamper et al. 2008) providing the constraints outlined above are taken into consideration. In addition, new techniques of high-throughput sequencing (e.g. pyrosequencing) are rapidly developing (Roesch et al. 2007) and are now being used for AMF (Lumini et al. 2010). For bacterial community analysis, this approach is already replacing community fingerprinting methods and clone libraries (Jonasson et al. 2002). Microsatellite (simple sequence repeats) analysis is also starting to impact as a tool to discriminate individual strains within species complexes (Croll et al.

2008; Mathimaran et al. 2008a, b). These results will be key to understanding how ecosystem management affects potential sustainability via indirect effects on AMF populations belowground. For molecular identification, the PCR-amplified rDNA fragments (amplicons) are usually subjected to cloning, fingerprinting and sequencing. Both universal (NS1, NS4, NS31, NS41) and glomalean-specific (AM1, AML1, AML2, Glo1) primers have been used for either a second classical or a nested PCR amplification. The ITS rDNA regions (Krueger et al. 2009) have been also used to characterize new AMF species (Palenzuela et al. 2010, 2011). Temporal temperature gradient electrophoresis (TGGE), terminal restriction fragment length polymorphism (T-RFLP) or single-strand conformation polymorphism (SSCP) electrophoresis fingerprinting approaches have also been used, but sequencing approach are now so simple and cheap that they are often preferred. After sequencing, the phylotypes (operative taxonomic units; OTUs) are defined and compared with International databases to establish their phylogenetic relationships (Barea et al. 2011). Recent studies (Hempel et al. 2007; Öpik et al. 2008, 2010; Sánchez-Castro et al. 2008; Alguacil et al. 2009a, b, 2011) have revealed a hidden diversity, with many AM fungal sequences detected in plant roots that cannot be related to known AM taxa present in the surrounding soil as spores. These molecular studies clearly show that populations of spores in soil do not reflect completely the fungi present in roots and vice versa.

Consideration of AMF ecology is important in sustainable strategies of agriculture or restoration of natural ecosystems. The beneficial effects of AMF on plant growth have led to their development as **bioinoculants for forestry, agriculture and horticulture** (Ijdo et al. 2011). Once the importance of the AM symbiosis is recognized, a decision must be made as to whether the native population of AMF suffices as the starting material from which to develop a sustainable system. If not, it will be necessary to augment the native species with inoculum of exotic or indigenous isolates. If possible it is best to use **indigenous isolates** that are already adapted to the prevailing conditions at the fieldsite. In a study involving reciprocal transplanting of AMF and hosts across the Great Basin, United States (Weinbaum et al. 1996), AMF had higher survival on the site from which they were collected and with the host plant population with which they were originally associated. In contrast, although transplanted AMF survived and spread for at least three growing seasons, the populations declined significantly at the exotic site and with exotic hosts. More recently, Pelligrino et al. (2011) also concluded that the use of native AMF, produced on farm using mycotrophic plants, was as effective as, or more effective than, the use of exotic AM fungal isolates and the concept of using native inoculants rather than exotic ones is now widely accepted. Evidence for the functional divergence of strains within AMF species (Antunes et al. 2010) was shown by isolates from contrasting climates which consistently and differentially altered the growth of two grass species of differing thermal optima.

In cases where the native population is adequate but of low infectivity, it may be possible to rapidly increase inoculum levels using appropriate management practices (see Barea and Jeffries 1995). Unfortunately, there is no easy way to assess the suitability of the indigenous population. Principal component analysis of a number of factors was used by Hamel et al. (1997) to show that indigenous populations of AMF and soil aggregate stability were the major determinants of the response of leek (*Allium porrum* L.) to inoculation with *Glomus intraradices* or *G. versiforme*. Alternatively, young, fresh roots of native plants growing in the soil may be collected and stained to determine AM infection, but this may not always be possible. Differences in the proportion of roots colonized by particular species of AMF can undergo considerable seasonal fluctuations, particularly in temperate climates (Jakobsen and Nielsen 1983; Dodd and Jeffries 1986), and an absence of infection at a particular time may not mean that the soil does not contain propagules of AMF. One of the best **indicators of infectiveness** occurs when highly mycotrophic trap plants are grown in the test soil. If they do not become infected, or become infected only slowly, the soil will require inoculation.

Because of these difficulties in approach, some successful examples of assessing the need for inoculation will be described.

Introduction of exotic inoculum may only be necessary if a native population is either absent or has a low inoculum potential, or if the native fungi are ineffective for the crop species that are being planted. For example, when the top layer of a soil profile is degraded or eroded most AM propagules are lost. Cuenca et al. (1998a) reported that the loss of the topsoil organic layer from previously undisturbed communities caused a sharp decrease in the numbers and diversity of AMF spores recovered from the underlying substrate. Soil disturbance is also responsible for the destruction of the AM hyphal network, thereby affecting inoculum potential. Thus, consideration of the AM symbiosis is usually recommended in the establishment of plants in degraded soils (Skujins and Allen 1986; Stahl et al. 1988; Jasper et al. 1989; Cuenca et al. 1998b; Requena et al. 2001; Allen 2007; Barea et al. 2011). As most of the spores and mycelia of AMF are found in the upper few centimetres of soil, subsoils excavated during mining operations usually lack propagules of AMF. The role of AM in the restoration of surface-mined lands has been reviewed (Miller and Jastrow 1992; Turnau et al. 2006).

V. Factors Affecting the Function of AM Symbiosis

The populations of indigenous AMF in agricultural soils can be manipulated by a variety of **management strategies** including crop rotation, pre-cropping or intercropping with mycorrhizally reponsive plants, the selective use of biocides, through the choice of appropriate inorganic P and N sources or through the use of organic fertilizers (Brito et al. 2008). Soil fumigants obviously severely reduce inoculum potential of AMF populations and, despite the decrease in soil fumigation worldwide, there are still reports of an absence of AMF in soils with a history of these treatments (e.g. Cavagnaro and Martin 2011).

The application of organic fertilizers is a common practice adopted in sustainable agriculture, and organic matter tends to increase the amount of mycelium of AMF in the soil (Joner and Jakobsen 1995; Brito et al. 2008). Other experiments suggest that dairy manure treatment does not increase mycorrhizal colonization of corn or wheat in the field, nor does it increase sporulation, despite a positive effect in pot experiments Tarkalson et al. (1998). del Val et al. (1999) selected ecotypes of AMF from soils that had received long-term applications of metal-contaminated sewage sludge. Differential tolerances to heavy metals were observed across the isolates obtained, and the growth of the external mycelium was particularly affected. Selection of appropriate isolates will be necessary in order to guarantee the effectiveness of AM symbioses (Baar 2008; Cuenca et al. 2008; Barea et al. 2011).

Disturbance of soil can reduce soil fertility due to destruction of the soil mycelium of AMF. Shoot P and N concentrations were shown to be much lower in maize plants grown in disturbed soil (Evans and Miller 1990) indicating that the effects were a result of the destruction of a pre-existing mycelial network. Ploughing of the soil reduces the abundance of soil mycelium and results from reduced tillage experiments have resulted in higher plant biomass yields, and this has been attributed to an AM effect (Krauss et al. 2010; Maiti et al. 2011), leading Fitter et al. (2011) to suggest that reduced tillage can offer improved ecosystem services without affecting crop yield. Nevertheless, some reports suggest that soil disturbance may not affect some AMF so negatively, and inoculation of soils with a fungus insensitive to disturbance can compensate for the loss of contribution of a sensitive one (Duan et al. 2011). Grazing pressure from herbivorous animals can also result in disturbance of the mycelium and a reduction in mycorrhizal benefits—natural ecosystems and organic agriculture will be more affected than intensive agriculture where pesticides reduce invertebrate populations.

Sustainable systems seek to maximize the beneficial effects of the natural soil microbiota. The AMF, as key components of soil microbiota (Gianinazzi et al. 2010), are immersed in a framework of interactions with other soil micro-organisms that are fundamental to rhizosphere functioning. Some of these interactions have beneficial consequences for plant health and productivity (Artursson et al. 2005; Barea et al. 2002a, 2005b). Particularly relevant to P and N cycling in ecosystems are interactions between AMF and phosphate-solubilizing micro-organisms or N_2-fixing rhizobial bacteria (Barea et al. 2005a, 2007; Azcón and Barea 2010). Ever since Asai (1944) concluded that nodulation by rhizobial bacteria appeared to be dependent on mycorrhiza formation by the common host legume, there have been many studies of this **tri-partite relationship**. Subsequently, both the widespread presence of the AM symbiosis in nodulated legumes and the impact of AM fungi in improving nodulation and N_2-fixation have been recognized (Barea and Azcón-Aguilar 1983; Hayman 1986; Mosse 1986; Zaidi et al. 2003; Barea et al. 2005c; Siviero et al. 2008; Azcón and Barea 2010). Despite the positive effect of AM fungi on nodule formation and function, some reports on mycorrhiza–legume interaction are contradictory. For example, Franzini et al. (2010) found in an experiment that the tested AM fungi did not improve nodule formation and function.

Other micro-organisms are also known to have synergistic effects on mycorrhiza establishment, in particular, '**mycorrhiza helper bacteria**' (Garbaye 1994), are known to stimulate mycelial growth and/or improve mycorrhizal formation (Barea et al. 2005b). Likewise, the establishment of AMF in the root cortex can affect rhizosphere microbial populations, due in part to AM formation changing many key aspects of plant physiology, such as the mineral nutrient composition of plant tissues, hormonal balance and the patterns of C allocation. As a consequence, the AM symbiotic status usually modifies root exudation and the chemical characteristics of root exudates (Tawaraya et al. 2006). In addition, the development of an external AM mycelium introduces physical modifications to the soil environment surrounding roots (Barea et al. 2005b). The microbial populations change both quantitatively and qualitatively in the rhizosphere of AM plants, the so-called ***mycorrhizosphere*** (*sensu lato*), resulting in features that differ from those of a non-mycorrhizal plant (Johansson et al. 2004; Barea et al. 2005b). The use of molecular techniques has demonstrated specificity in the bacterial populations associated with AM roots (Offre et al. 2007), AM mycelium (Toljander et al. 2005; Rillig and Mummey 2006; Rillig et al. 2006) and AMF spores (Roesti et al. 2005). In addition to their effects on natural bacterial populations, AM formation also affects the establishment in the rhizosphere of inocula of plant growth promoting rhizobacteria (PGPR; Barea et al. 2005b).

VI. Examples Promoting AM Symbiosis

Having established that AMF have an important role in the maintenance of soil structure and fertility in sustainable practices, it is essential to consider their behaviour in strategies designed to establish sustainable practices in agriculture or restoration of natural ecosystems. An increasing demand for low-input agriculture has resulted in greater interest in the manipulation and use of beneficial soil micro-organisms because of their positive impacts on plant growth and health and on soil quality and fertility. The strategic management of beneficial soil microbes can reduce the use of chemicals and energy in agriculture leading to a more economical and sustainable production, while minimizing environmental degradation. These biological interventions are becoming more attractive as the use of chemicals for fumigation and disease control is progressively discouraged and fertilizers have become more and more expensive (Atkinson 2009). Agro-biotechnological approaches include the **use of microbial inoculants**. Legumes, in particular, are potential beneficiaries and the target microbes are obviously the AM fungi and rhizobia, although PGPR can also be included and tested under pot/field conditions (Barea et al. 2002c; Wani et al. 2007; Zaidi and Khan 2007; Azcón and Barea 2010).

While the technology for the production of inexpensive rhizobial and free-living PGPR is commercially available, constraints on the **production of inocula** and the **development of inoculation techniques** have limited the use of AMF inoculants. The difficulty in culturing obligate symbionts such as AMF in the absence of their host plant is a major obstacle (Baar 2008). Despite these problems, several companies worldwide are producing AM inoculum products which are now commercially available (Gianinazzi and Vosátka 2004; Vosátka et al. 2008; Ijdo et al. 2011). Selection of the appropriate AM fungi is a key step (Estaún et al. 2002), and specific procedures are required to multiply AMF and to produce high quality inocula (von Alten et al. 2002). Recent developments in AM-inoculum production systems range from nursery plots (Cuenca et al. 2008; Koltai et al. 2008) to *in vitro* monoxenic root organ cultures (Bago and Cano 2005; Ijdo et al. 2011). The resulting materials (spores, hyphae, root fragments, etc.) are added to different carriers, resulting in a wide range of formulations, including encapsulation. These can be applied on agronomical scales using application methods (Cuenca et al. 2008; Vosátka et al. 2008) such as hydroseeding (Estaún et al. 2007). The response in plants to inoculation with a particular strain can depend significantly on the soil properties where it is introduced (Herrera-Peraza et al. 2011). There is controversy whether 'generic products', containing a mixture of isolates AMF, potentially suitable for a range of applications, are more appropriate for the market than those with precise formulations and AMF specifically tuned to particular end-users (Smith and Read 2008).

Large-scale inoculation in intensive farming systems is not practical, and management of indigenous populations is preferred (Brito et al. 2008). However, at a relatively small-scale such as nursery production, AM inoculation is feasible and advantageous. Inoculation of seedlings is used to establish selected fungi in roots before potting on or planting out into the field. Inoculation is appropriate where transplanting is part of the normal production system, as in many horticultural and plantation crops. Farmers can also produce 'home-grown' inoculum of highly colonized roots and soil to be applied to plots immediately before planting a crop which could make a valuable contribution to food production in many relatively small, low input systems. Management strategies for larger scale inoculum build-up include the use of pastures, sequential cropping or intercropping. Biodynamic and organic farm management results in higher per cent colonization of roots of pasture and annual crops than conventional management.

Agricultural soils are usually threatened by a great array of **stress factors of either biotic or abiotic origin**, which negatively affect crop productivity (Buscot 2005). The presence of soil-borne plant pathogens is the most common biotic stress, while abiotic stresses include drought/salinity, persistent organic pollutants (fungicides, herbicides, hydrocarbons, etc.), heavy metals and radionucleides. These abiotic factors negatively affect not only plant growth but also AMF populations, although AMF strains adapted to stress may be found in these situations. As a consequence, these adapted strains may be used to 'tailor' a mycorrhizosphere to help plants to offset the negative impact of stress situations (Barea et al. 2005b). In addition, the influence of climatic change on AM formation and function has also been studied, and the temperature component of climatic change seems more influential than its CO_2 component (Gavito et al. 2003). Populations of AMF were affected by long-term experimental climatic manipulation (Staddon et al. 2003; Vargas et al. 2010), either indirectly as a consequence of the impact of the climatic manipulation on plant communities, or directly by affecting the AM mycelium particularly in drier conditions.

Here we will restrict our review to how AMF can help plants under the following environmental stresses: (1) soil-borne pathogen attack, (2) aggressive agricultural weeds, (3) drought/salinity stress, (4) the presence of toxic pollutants and (5) desertification.

A. Soil-Borne Pathogen Attack

The establishment of AM in plant roots reduces damage caused by **soil-borne plant pathogens.**

This AM effect is not exerted with the same effectiveness by all AM fungi, and is not applicable to all pathogens, and is not expressed in all substrata or in all environmental conditions. Prior colonization by selected AMF can protect plants against pathogenic fungi, such as *Phytophthora*, *Gaeumannomyces*, *Fusarium*, *Thielaviopsis*, *Pythium*, *Rhizoctonia*, *Sclerotium*, *Verticillium* and *Aphanomyces*, or nematodes such as *Rotylenchus*, *Pratylenchus* and *Meloidogyne* (Whipps 2004; Barea et al. 2005b; Pozo and Azcón-Aguilar 2007; Pozo et al. 2009). Different mechanisms may explain the contribution of AM fungi in biological control of plant pathogens (Azcón-Aguilar et al. 2002; Elmer 2002; Barea et al. 2005b). One simple mechanism is by compensating for the damage caused by pathogens via the growth promotion effects. Alternatively, control may be exerted via microbial changes that are produced as the mycorrhizosphere develops. There is strong evidence that the shifts and resulting microbial equilibrium positively influence the growth and health of the plants thus contributing to biological control of root pathogens. **Activation of plant defence mechanisms that induce a systemic resistance reaction of the plant** is also possible (Pozo and Azcón-Aguilar 2007; Pozo et al. 2009) and this might also affect foliar pathogens as well as root pathogens. Although the molecular basis of the mechanisms involved in inducing systematic resistance has become clear, the feasibility of this mechanism and its impact on biological control needs further research.

Vestberg et al. (2004) conducted a comprehensive experiment that merits a more in-depth discussion. Seven nursery experiments were carried out to test different conditions and/or inoculation patterns on the effect of five diverse rhizosphere micro-organisms involved in the biological control of strawberry diseases (crown rot caused by *Phytophthora cactorum* and red stele caused by *P. fragariae*). The micro-organisms tested were registered strains of the AMF *Glomus mosseae*, the biocontrol bacteria *Bacillus subtilis* and *Pseudomonas fluorescens*, and the biocontrol fungi *Trichoderma harzianum* and *Gliocladium catenulatum*. Inocula from these microbes were applied singly or in dual mixtures. In most experiments, all the inoculated micro-organisms except *T. harzianum* and *G. mosseae* established in the rhizosphere. The growth-promoting effects were not consistent and dual inoculation did not increase growth to any greater extent than single inoculation. In some treatments, a decrease in crown rot shoot symptoms was found, with the mixture *T. harzianum* + *G. catenulatum* being most effective. The general conclusion was that: 'the great variation between experiments indicates that more studies are needed for optimization of the whole plant–substrate–micro-organism system'.

In summary, the use of mycorrhizosphere interactions to increase root resistance or tolerance to pathogen attack has biotechnological promise. However, effectiveness varies with microbial combinations, substrata and environmental conditions, thus more research is needed for the successful application of microbial consortia in sustainable agricultural practices.

B. Aggressive Agricultural Weeds

As many agricultural weeds are non-mycorrhizal or originate from ruderal environments where AMF are rare or absent, Rinaudo et al. (2010) hypothesized that AMF may suppress weed growth, a mycorrhizal attribute which has hardly been considered. Accordingly, these authors investigated the **impact of AMF on weed growth** in experimental microcosms where a mycotrophic crop (sunflower) was grown together with six widespread weed species. The presence of AMF reduced total weed biomass while sunflower benefitted from AMF via enhanced phosphorus nutrition. The results indicate that the stimulation of AMF in agro-ecosystems may suppress some aggressive weeds and suggest a possible applicability of the AM symbiosis in weed control, an agricultural practice in the context of sustainability issues. Looking for increase the knowledge on the impact of AMF on weed control, López-Ráez et al. (2011a) discussed the potential use of some rhizosphere signal molecules as new biological control strategies against weeds. Particularly, they focus on the role of strigolactones, a new class of plant hormones emerging as important signal molecules for some rhizosphere processes (Gómez-Roldán et al. 2008; Umehara et al. 2008; Faure et al. 2009). Strigolactones are exuded into the soil, where they are known to act both as host detection signals for AM formation (Akiyama et al. 2005; Parniske 2008) and as germination stimulants for root parasitic plant seeds (Bouwmeester et al. 2007). López-Ráez et al. (2011b) found that strigolactone production is significantly reduced upon AM symbiosis establishment. Thus they suggested the potential of the AM symbiosis for controlling root parasitic weeds.

The information from the studies by Rinaudo et al. (2010) and López-Ráez et al. (2011a, b) therefore suggest a potential role of AM symbiosis as a new control method against aggressive weeds, an AMF attribute that deserves further attention.

C. Salinity/Drought Conditions

Drought and salinity, together with **extreme temperatures**, are the most common environmental stress factors experienced by crop plants (Augé 2001; Ruíz-Lozano 2003). All share a common osmotic component since they cause a dehydration of plant tissues. **Osmotic stresses** are known to have a major adverse effect on survival, normal development and productivity of crop plants. Ecophysiological approaches have demonstrated that the AM symbiosis often results in altered rates of water movement into, through and out of the host plants, with beneficial consequences for tissue hydration and plant physiology (Ruíz-Lozano 2003). Consequently, the AM symbiosis can protect crop plants against the detrimental effects of water deficit and that the AM contribution to plant drought tolerance results from a combination of physical, nutritional and cellular effects.

There are several mechanisms by which the AM symbiosis can alleviate plant drought stress, including: (1) direct uptake of water by the fungal hyphae from soil areas inaccessible to plant roots and the water transfer to the host plant, (2) a better osmotic status of AM plants which allows these plants to maintain a favourable gradient of water flow from soil into their roots, (3) the enhancement of plant gas exchange by the AM symbiosis, which maintains the correct stomatal opening and CO_2 assimilation, (4) an increase in the soil water retention properties through the formation of stable soil aggregates, (5) the stimulation of assimilative activities essential for plant growth such as nitrate reductase activity, which is strongly inhibited in plants by water deficit, (6) the protection of the host plant against the oxidative damage generated by drought and (7) the activation of aquaporin expression. This last mechanism has received significant attention (Augé 2001; Ruíz-Lozano 2003; Ruíz-Lozano and Aroca 2008; Ruíz-Lozano et al. 2008).

The **aquaporins** are proteinic channels which facilitate a passive water flow through membranes along a gradient of water potential. They are responsible for the cytosolic osmoregulation, transport of water, transport of small molecules (NH_4, urea, glycerol, CO_2) and transport of K^+ (osmotic adjustment). Modulation of particular aquaporins by AM symbiosis results in a better regulation of plant water status and contributes to the overall plant resistance to water stress. Aquaporins may not only be involved in regulation of plant water status but also in the symbiotic exchange processes between the fungus and the plant. Recent experiments have corroborated a positive effect of the interactions between AM fungi and nodulating rhizobia in legumes under drought conditions (Goicoechea et al. 2000; Ruiz-Lozano et al. 2001; Echeverria et al. 2008). Porcel et al. (2006) concluded that AM-inoculated soybean plants respond to drought stress by down-regulating the expression of aquaporin genes. This could be a mechanism to decrease membrane water permeability to allow cellular water conservation. Aquaporin gene expression in mycorrhizal *Phaseolus vulgaris* depended on the particular stress conditions: drought, cold and salinity (Aroca et al. 2007), and H_2O_2 was apparently involved as a signalling molecule (Benabdellah et al. 2009).

Alleviation of the **oxidative damage** also seems to be involved in the drought-protective activity of AM (Porcel et al. 2003; Porcel and Ruíz-Lozano 2004). Indigenous drought-tolerant AM fungi have been shown to also improve nutrient acquisition, gas exchange, nitrate reductase (Caravaca et al. 2004a, b, c), water transport and root development (Marulanda et al. 2006) in the shrub legume *Retama sphaerocarpa* under dry conditions.

D. Bioremediation of Polluted Soils

The impact of AM inoculation on plants used for the bioremediation of soils containing pollutants has been investigated, mostly for **heavy metals (HM)** and **polycyclic aromatic hydrocarbons (PAH)**—here we will concentrate on the former. Mycorrhizal fungi can become adapted to grow under such stress situations, thus selection of these strains provides a source of inoculants to enhance bioremediation. Most phytoremediation studies concern the heavy metals Zn, Cu, Cd, Pb or N (Leyval et al. 2002; Turnau et al. 2006; Jasper 2007; Maki et al. 2008; Teng et al. 2008; Azcón et al. 2009), but bioremediation of other metals, like arsenic, has also been attempted (Dong et al. 2008). The mycorrhizal fungi apparently sequester HM on their soil mycelium, thus AM plants translocate less HM to their shoots than the corresponding non-AM controls as shown for example in herbaceous (Díaz et al. 1996; Redon et al. 2009) or tree legumes (Lin et al. 2007). Rhizobacteria and AM fungi have been found to interact synergistically to benefit phytoremediation and the selection of combinations of HM-adapted microbial components is necessary to optimise remediation (Biró et al. 1998). An example is given by Vivas et al. (2003a, b, c, d, e, 2005a, b, 2006a, b, c) for *Trifolium* planted in an agricultural soil from the Nagyhörcsök Experimental Station (Hungary), that was contaminated in 1991 with suspensions of 13 microelement salts. Autochthonous bacteria and AMF were isolated from this soil and used as co-inoculants to successfully increase biomass, N and P content and rhizobial nodulation. Although inoculated plants had lower HM concentrations, the total HM content in plant shoots was due to the increased biomass, which suggested this strategy could be used for both phytostabilization and phytoextraction. Inoculated plants showed higher dehydrogenase, phosphatase and gluconase activities and increased auxin levels in the mycorrhizosphere, indicating enhanced microbial activities which were related to improved plant development (Vivas et al. 2005a, b; 2006a, b, c; Azcón et al. 2009). The molecular mechanisms involved in HM tolerance in AM inoculated plants have been recently discussed (González-Guerrero et al. 2009).

E. AM and the Restoration of Desertified Ecosystems

One of the most serious world problems affecting marginal agricultural lands is **desertification**. It is a complex and dynamic process which is claiming several hundred million hectares annually (Francis and Thornes 1990; Morgan et al. 1990). Human activities can cause or accelerate desertification and the loss of most plant species and their corresponding symbionts. Revegetation of desertified ecosystems is problematical but management practices have been developed and recently reviewed (Barea et al. 2011). These range from the pioneering study of Herrera et al. (1993) to the present day where the reintroduction of native species has been accompanied by the introduction of appropriate symbionts, including mycorrhizal fungi, in order to aid the restoration of soil fertility. The use of fertilizer inputs to accelerate this process is not a feasible strategy, as fertilizer application encourages the growth of ruderal weeds which are not adapted to the local edaphic conditions. These weeds will out-compete and suppress the pioneer plants such that the resulting plant community will not survive in the absence of continued addition of fertilizer. Instead, the growth of highly mycotrophic pioneer species has been successfully encouraged under conditions of low nutrient availability (see Barea et al. 2011). As a result of desertification processes, disturbance of natural plant communities is often accompanied, or preceded by, a loss of physical, chemical and biological soil properties, such as soil structure, plant nutrient availability, organic matter content, microbial activity, etc. (Jeffries and Barea 2001) which are fundamental for soil quality and soil structure (Miller and Jastrow 2000; Buscot 2005) and thus limit re-establishment of the natural plant cover. In particular, **desertification causes disturbance of plant–microbe symbioses** and thus the recovery of populations of AM fungi and rhizobial bacteria is to the integral restoration of a degraded area (Jeffries and Barea 2001). For example, Galleguillos et al. (2000) showed positive benefits to the shrub legume *Medicago arborea* from co-inoculation with AM fungi, *R. meliloti* strains and PGPR confirming that inoculation with mixtures of microbial symbionts could aid the recovery of desertified ecosystems in semi-arid areas.

Requena et al. (2001) carried out a long-term field experiment to ascertain the impact of inoculation with indigenous microsymbionts as part of a reclamation strategy in a desertified semi-arid Mediterranean ecosystem in southeast Spain. The existing natural vegetation was a degraded shrubland where *Anthyllis cytisoides* was the dominant species (Requena et al. 1997). *Anthyllis* seedlings inoculated with an indigenous rhizobial and AM fungal inocula, were transplanted to field plots for a 5-year trial. The tailored mycorrhizosphere not only enhanced establishment of the target legume but also increased soil fertility and quality. This included enhanced seedling survival rates, growth, P-acquisition, N-fixation and N-transfer from N-fixing to associated non-fixing species in the natural succession. The improvement in the physico-chemical properties in the soil around the *Anthyllis* plants was shown by the increased levels of N, organic matter and number of hydrostable soil aggregates. The role of the AMF, in co-operation with other microbes, in the formation of water-stable soil aggregates (Rillig and Mummey 2006) is relevant here. **Glomalin-related proteins** produced by the external hyphae of AM fungi were involved (Miller and Jastrow 2000; Bedini et al. 2009).

The increase in N content in the rhizosphere of the legume can be accounted for by the supply of N-richer root exudates due to an improvement in nodulation and N-fixing capacity resulting from inoculation with both microsymbionts. The dually inoculated shrub legumes were a source of AM fungal inoculum for the surrounding area and in improving N nutrition for non-N-fixing vegetation.

In follow-up studies, *Retama sphaerocarpa*, was selected as a target species for revegetation programs in desertified areas. Inoculation with native AMF improved plant establishment (Caravaca et al. 2003b) and enzymatic activities related with C, N and P cycling and soil aggregation (Alguacil et al. 2005). Furthermore, the application of composted urban residues increased the beneficial response (Caravaca et al. 2003a, c). Similar complementary effects have been found for other organic amendments such as sewage sludge (Alguacil et al. 2004; Caravaca et al. 2005a, b) or composted dry olive residues (Caravaca et al. 2006). The synergistic benefits of AM inoculation in combination with organic amendments has been further demonstrated for *Dorycium pentaphyllum*, using *Aspergillus niger*-treated organic amendments and rock phosphate additions, sugar beet residues (Caravaca et al. 2004a) or dry olive cake residues (Medina et al. 2004; Alguacil et al. 2008).

Apart from these field experiments carried out in Mediterranean ecosystems, Bhatia et al. (1998) demonstrated **that dual inoculation with rhizobia and AM fungi** helped the establishment and biomass production of a woody legume (*Prosopis* sp.) in wasteland in India. More recently, Siviero et al. (2008) and Schiavo et al. (2009) demonstrated the successful effects of dual inoculation with AM fungi and rhizobia on the performance of representative tree legumes in Brazil. These and other associated data thus suggest that the management of appropriate microsymbionts can help legumes to promote the stabilization of a self-sustaining ecosystem. The mycorrhizal shrub/tree legumes, act as 'fertility islands' (Caravaca et al. 2005a), which serve as sources of symbiont inocula for the surrounding area and improve N nutrition for the non N_2-fixing vegetation in stressed ecosystems.

VII. Conclusions

Mycorrhizal dynamics are a key component of ecosystem sustainability. Understanding and managing this symbiosis is important for strategies for maintaining or recovering sustainable plant communities, either in natural systems or in low input agriculture. Plant ecosystem functioning is governed largely by soil microbial dynamics and this may fluctuate or falter in non-sustainable systems. Ecosystem sustainability is significantly impaired by a loss of soil biodiversity, and it is known that diversity of AMF is strikingly low in arable sites compared with natural woodland (Helgason et al. 1998). Mycorrhizal fungi are an essential component of the soil microbiota in developing sustainable agricultural practices, because they enhance plant growth and nutrient uptake while at the same time stabilizing soil aggregates, making the soil less susceptible to erosion (Schreiner and Bethlenfalvay 1995). The loss of mycorrhizal diversity within the soil will have important effects on sustainability and plant community dynamics. In this review we have highlighted some of the evidence that has confirmed this prediction, and have discussed the ways in which AMF function can be affected by external factors. Mycorrhizal technology should be a component in sustainable strategies in the future, whereby application of AMF can reduce fertilizer inputs yet promote healthy plant growth. We maintain our earlier assertion (Jeffries and Barea 2001) that arbuscular mycorrhizal should be considered as an essential natural resource for ensuring growth and health of plants and fully deserve their title as 'biological fertilizers' and 'bioprotectors'.

Acknowledgements We thank our numerous colleagues over a series of EU COST actions for their collaborations and sustainable partnerships that have contributed to our current knowledge of mycorrhizal ecology. One of us (J.M.B.) is also grateful for support from the Spanish National Research Programme (R&D +i)-European Union (Feder) CGL2009-08825/BOS and from the Andalucian (Spain) Government, PAIDI (R&D+i) Programme: P07-CVI-02952.

References

Abbott LK, Robson AD (1977) The distribution and abundance of vesicular-arbuscular endophytes in some Western Australian soils. Austral J Bot 25:515–522

Akiyama K, Matsuzaki K, Hayashi H (2005) Plant sesquiterpenes induce hyphal branching in arbuscular mycorrhizal fungi. Nature 435:824–827

Alguacil MM, Caravaca F, Díaz G, Marín P, Roldán A (2004) Establishment of *retama sphaerocarpa* L. seedlings on a degraded semiarid soil as influenced by mycorrhizal inoculation and sewage-sludge amendment. J Plant Nutr Soil Sci 167:637–644

Alguacil MM, Caravaca E, Roldán A (2005) Changes in rhizosphere microbial activity mediated by native or allochthonous AM fungi in the reafforestation of a Mediterranean degraded environment. Biol Fert Soils 41:59–68

Alguacil MM, Lumini E, Roldán A, Salinas-García JR, Bonfante P, Bianciotto V (2008) The impact of tillage practices on arbuscular mycorrhizal fungal diversity in subtropical crops. Ecol Appl 18:527–536

Alguacil MM, Roldán A, Torres MP (2009a) Assessing the diversity of AM fungi in arid gypsophilous plant communities. Environ Microbiol 11:2649–2659

Alguacil MM, Roldán A, Torres MP (2009b) Complexity of semiarid gypsophilous shrub communities mediates the AMF biodiversity at the plant species level. Microbiol Ecol 57:718–727

Alguacil MM, Torres MP, Torrecillas E, Díaz G, Roldán A (2011) Plant type differently promote the arbuscular mycorrhizal fungi biodiversity in the rhizosphere after revegetation of a degraded, semiarid land. Soil Biol Biochem 43:167–173

Allen MF (1991) The ecology of mycorrhizae. Cambridge University Press, Cambridge

Allen MF (2007) Mycorrhizal fungi: highways for water and nutrients in arid soils. Vadose Zone J 6:291–297

Altieri MA (1994) Sustainable agriculture. Encylop Agric Sci 4:239–247

Altieri MA (2004) Linking ecologists and traditional farmers in the search for sustainable agriculture. Front Ecol Environ 2:35–42

Antunes PM, Koch AM, Morton JB, Rillig MC, Klironomos JN (2010) Evidence for functional divergence in arbuscular mycorrhizal fungi from contrasting climatic origins. New Phytol 189:507–514

Aroca R, Porcel R, Ruíz-Lozano JM (2007) How does arbuscular mycorrhizal symbiosis regulate root hydraulic properties and plasma membrane aquaporins in *Phaseolus vulgaris* under drought, cold or salinity stresses? New Phytol 173:808–816

Artursson V, Finlay RD, Jansson JK (2005) Combined bromodeoxyuridine immunocapture and terminal-restriction fragment length polymorphism analysis highlights differences in the active soil bacterial metagenome due to *Glomus mosseae* inoculation or plant species. Environ Microbiol 7:1952–1966

Asai T (1944) Die bedeutung der mikorrhiza für das pflanzenleben. Jpn J Bot 12:359–408

Atkinson D (2009) Soil microbial resources and agricultural policies. In: Azcón-Aguilar C, Barea J, Gianinazzi S, Gianinazzi-Pearson V (eds) Mycorrhizas functional processes and ecological impact. Springer, Berlin, pp 33–45

Augé RM (2001) Water relations, drought and vesicular-arbuscular mycorrhizal symbiosis. Mycorrhiza 11:3–42

Avis TJ, Gravel V, Antoun H, Tweddell RJ (2008) Multifaceted beneficial effects of rhizosphere microorganisms on plant health and productivity. Soil Biol Biochem 40:1733–1740

Azcón R, Barea JM (2010) Mycorrhizosphere interactions for legume improvement. In: Khan MS, Zaidi A, Musarrat J (eds) Microbes for legume improvement. Springer, Vienna, pp 237–271

Azcón R, Medina A, Roldán A, Biró B, Vivas A (2009) Significance of treated agrowaste residue and autochthonous inoculates (arbuscular mycorrhizal fungi and *Bacillus cereus*) on bacterial community structure and phytoextraction to remediate heavy metals contaminated soils. Chemosphere 75:327–334

Azcón-Aguilar C, Jaízme-Vega MC, Calvet C (2002) The contribution of arbuscular mycorrhizal fungi to the control of soil-borne plant pathogens. In: Gianinazzi S, Schüepp H, Barea JM, Haselwandter K (eds) Mycorrhiza technology in agriculture: from genes to bioproducts. Birkhäuser, Basel, pp 187–197

Baar J (2008) From production to application of arbuscular mycorrhizal fungi in agricultural systems: requirements and needs. In: Varma A (ed) Mycorrhiza: state of the art, genetics and molecular biology, eco-function, biotechnology, eco-physiology, structure and systematics, 3rd edn. Springer, Berlin, pp 361–373

Bago B, Cano C (2005) Breaking myths on arbuscular mycorrhizas in vitro biology. In: Declerck S, Strullu FG, Fortin JA (eds) In vitro culture of mycorrhizas, vol 4, Soil biology. Springer, Berlin, pp 111–138

Barea JM, Azcón-Aguilar C (1983) Mycorrhizas and their significance in nodulating nitrogen-fixing plants. In: Brady N (ed) Advances in agronomy, vol 36. Academic, New York, pp 1–54

Barea JM, Jeffries P (1995) Arbuscular mycorrhizas in sustainable plant–soil systems. In: Hock B, Varma A (eds) Mycorrhizae: function, molecular biology and biotechnology. Springer, Berlin, pp 521–560

Barea JM, Azcón-Aguilar C, Azcón R (1997) Interactions between mycorrhizal fungi and rhizosphere micro-organisms within the context of sustainable soil–plant systems. In: Gange AC, Brown VK (eds) Multitrophic interactions in terrestrial systems. Blackwell, Cambridge, pp 65–77

Barea JM, Azcón R, Azcón-Aguilar C (2002a) Mycorrhizosphere interactions to improve plant fitness and soil quality. Anton Leeuw Int J Gen Mol Microbiol 81:343–351

Barea JM, Palenzuela J, Azcón R, Ferrol N, Azcón-Aguilar C (2002b) Micorrizas y restauración de la cubierta vegetal en ambientes mediterráneos. In: Barea-Azcón JM, Ballesteros E, Luzón JM, Moleón M, Tierno JM, Travesi R (eds) Biodiversidad y conservación de fauna y flora en ambientes mediterráneos. Granada, Sociedad Granatense de Historia Natural, pp 83–105

Barea JM, Toro M, Orozco MO, Campos E, Azcón R (2002c) The application of isotopic (P^{32} and N^{15}) dilution techniques to evaluate the interactive effect of phosphate-solubilizing rhizobacteria, mycorrhizal fungi and *Rhizobium* to improve the agronomic efficiency of rock phosphate for legume crops. Nutr Cycl Agroecosyst 63:35–42

Barea JM, Azcón R, Azcón-Aguilar C (2005a) Interactions between mycorrhizal fungi and bacteria to improve plant nutrient cycling and soil structure. In: Buscot F, Varma A (eds) Microorganisms in soils: roles in genesis and functions. Springer, Berlin, pp 195–212

Barea JM, Pozo MJ, Azcón R, Azcón-Aguilar C (2005b) Microbial co-operation in the rhizosphere. J Exp Bot 56:1761–1778

Barea JM, Werner D, Azcón-Aguilar C, Azcón R (2005c) Interactions of arbuscular mycorrhiza and nitrogen fixing symbiosis in sustainable agriculture. In: Werner D, Newton WE (eds) Nitrogen fixation in agriculture, forestry, ecology, and the environment. Kluwer, Rotterdam, pp 199–222

Barea JM, Palenzuela J, Cornejo P, Sánchez I, Navarro C, Quiñones PB, Azcón R, Ferrol N, Azcón-Aguilar C (2007) Significado, diversidad e impacto de los hongos de las micorrizas arbusculares en ambientes mediterráneos. In: Barea-Azcón JM, Moleón M, Travesí R, Ballesteros E, Luzón JM, Tierno JM (eds) Biodiversidad y conservación de fauna y flora en ambientes mediterráneos. Sociedad Granatense de Historia Natural, Granada, pp 155–185

Barea JM, Ferrol N, Azcón-Aguilar C, Azcón R (2008) Mycorrhizal symbioses. In: White PJ, Hammond JP (eds) The ecophysiology of plant–phosphorus interactions, vol 7, Plant ecophysiology series. Springer, Dordrecht, pp 143–163

Barea JM, Palenzuela J, Cornejo P, Sánchez-Castro I, Navarro-Fernández C, Lopéz-García A, Estrada B, Azcón R, Ferrol N, Azcón-Aguilar C (2011) Ecological and functional roles of mycorrhizas in semi-arid ecosystems of Southeast Spain. J Arid Environ 75:1292–1301

Bedini S, Pellegrino E, Avio L, Pellegrini S, Bazzoffi P, Argese E, Giovannetti M (2009) Changes in soil aggregation and glomalin-related soil protein content as affected by the arbuscular mycorrhizal fungal species *Glomus mosseae* and *Glomus intraradices*. Soil Biol Biochem 41:1491–1496

Benabdellah K, Azcon-Aguilar C, Valderas A, Speziga D, Fitzpatrick TB, Ferrol N (2009) GintPDX1 Encodes a protein involved in vitamin B6 biosynthesis that is up-regulated by oxidative stress in the arbuscular mycorrhizal fungus *Glomus intraradices*. New Phytol 184:682–693

Bethlenfalvay GJ, Linderman RG (1992) Mycorrhizae in sustainable agriculture. ASA, Madison

Bethlenfalvay GJ, Schüepp H (1994) Arbuscular mycorrhizas and agrosystem stability. In: Gianinazzi S, Schüepp H (eds) Impact of arbuscular mycorrhizas on sustainable agriculture and natural ecosystems. Birkhäuser, Basel, pp 117–131

Bethlenfalvay GJ, Brown MS, Franson RL, Mihara KL (1989) The *Glycine-Glomus-Bradyrhizobium* symbiosis IX. Nutritional, morphological and physiological responses of nodulated soybean to geographic isolates of the mycorrhizal fungus *Glomus mosseae*. Physiol Plant 76:226–232

Bever JD, Pringle A, Schultz PA (2002) Dynamics within the plant—arbuscular mycorrhizal fungal mutualism: testing the nature of community feedback. In: van der Heijden MGA, Sanders IR (eds) Mycorrhizal ecology. Springer, Berlin, pp 267–292

Bhatia NP, Adholeya A, Sharma A (1998) Biomass production and changes in soil productivity during longterm cultivation of *Prosopis juliflora* (Swartz) DC inoculated with VA mycorrhiza and *Rhizobium* spp. in a semi-arid wasteland. Biol Fert Soils 26:208–214

Biró B, Köves-Péchy K, Vörös I, Kádár I (1998) Toxicity of some field applied heavy metal salts to the rhizobial and fungal microsymbionts of alfalfa and red clover. Agrokem Talajtan 47:265–277

Boddington CL, Dodd JC (1998) A comparison of the development and metabolic activity of mycorrhizas formed by arbuscular mycorrhizal fungi from different genera on two tropical forage legumes. Mycorrhiza 8:149–157

Bohlool BB, Ladha JK, Garrity DP, George T (1992) Biological nitrogen fixation for sustainable agriculture: a perspective. Plant Soil 141:1–11

Bouwmeester HJ, Roux C, López-Ráez JA, Becard G (2007) Rhizosphere communication of plants, parasitic plants and AM fungi. Trends Plant Sci 12: 224–230

Brito I, Goss MJ, de Carvalho M, van Tuinen D, Antunes PM (2008) Agronomic management of indigenous mycorrhizas. In: Varma A (ed) Mycorrhiza: state of the art, genetics and molecular biology, eco-function, biotechnology, eco-physiology, structure and systematics, 3rd edn. Springer, Berlin, pp 375–402

Brundrett MC (2002) Coevolution of roots and mycorrhizas of land plants. New Phytol 154:275–304

Bryla DR, Eissenstat DM (2005) Respiratory costs of mycorrhizal associations. In: Lambers H, Ribas Carbo M (eds) Plant respiration. Springer, Dordrecht, pp 207–224

Buscot F (2005) What are soils? In: Buscot F, Varma S (eds) Microorganisms in soils: roles in genesis and functions. Springer, Berlin, pp 3–18

Caravaca F, Barea JM, Palenzuela J, Figueroa D, Alguacil MM, Roldán A (2003a) Establishment of shrubs species in a degraded semiarid site after inoculation with native or allochthonous arbuscular mycorrhizal fungi. Appl Soil Ecol 22:103–111

Caravaca F, Alguacil MM, Figueroa D, Barea JM, Roldán A (2003b) Re-establishment of *Retama sphaerocarpa* as a target species for reclamation of soil physical and biological properties in a semi-arid Mediterranean area. For Ecol Manage 182:49–58

Caravaca F, Figueroa D, Alguacil MM, Roldán A (2003c) Application of composted urban residue enhanced the performance of afforested shrub species in a degraded semiarid land. Bioresour Technol 90:65–70

Caravaca F, Alguacil MM, Azcón R, Díaz G, Roldán A (2004a) Comparing the effectiveness of mycorrhizal inoculation and amendment with sugar beet, rock phosphate and *Aspergillus niger* to enhance field performance of the leguminous shrub *Dorycnium pentaphyllum* L. Appl Soil Ecol 25:169–180

Caravaca F, Alguacil MM, Vassileva M, Díaz G, Roldán A (2004b) AM fungi inoculation and addition of microbially-treated dry olive cake-enhanced afforestation of a desertified Mediterranean site. Land Degrad Dev 15:153–161

Caravaca F, Figueroa D, Barea JM, Azcon-Aguilar C, Roldan A (2004c) Effect of mycorrhizal inoculation on nutrient acquisition, gas exchange, and nitrate reductase activity of two Mediterranean-autochthonous shrub species under drought stress. J Plant Nutr 27:57–74

Caravaca F, Alguacil MM, Barea JM, Roldán A (2005a) Survival of inocula and native AM fungi species associated with shrubs in a degraded Mediterranean ecosystem. Soil Biol Biochem 37:227–233

Caravaca F, Alguacil MM, Díaz G, Marín P, Roldán A (2005b) Nutrient acquisition and nitrate reductase activity of mycorrhizal *Retama sphaerocarpa* L. seedlings afforested in an amended semiarid soil under two water regimes. Soil Use Manage 21:10–16

Caravaca F, Alguacil MM, Azcón R, Roldán A (2006) Formation of stable aggregates in rhizosphere soil of *Juniperus oxycedrus*: effects of AM fungi and organic amendments. Appl Soil Ecol 33:30–38

Cavagnaro TR, Martin AW (2011) Arbuscular mycorrhizas in southeastern Australia processing tomato farm soils. Plant Soil 340:327–336

Chaudhary VB, Bowker MA, O'Dell TE, Grace JB, Redman AE, Rillig MC, Johnson NC (2009) Untangling the biological contributions to soil stability in semiarid shrublands. Ecol Appl 19:110–122

Cooper KM, Tinker PB (1978) Translocation and transfer of nutrients in vesicular-arbuscular mycorrhizas. II uptake and translocation of phosphorus, zinc and sulphur. New Phytol 81:43–52

Croll D, Wille L, Gamper HA, Mathimaran N, Lammers PJ, Corradi N, Sanders IR (2008) Genetic diversity and host plant preferences revealed by simple sequence repeat and mitochondrial markers in a population of the arbuscular mycorrhizal fungus *Glomus intraradices*. New Phytol 178:672–687

Cuenca G, De Andrade Z, Escalante G (1998a) Diversity of glomalean spores from natural communities and revegetated communities growing on nutrient-poor tropical soils. Soil Biol Biochem 30:711–719

Cuenca G, De Andrade Z, Escalante G (1998b) Arbuscular mycorrhizas in the rehabilitation of fragile degraded tropical lands. Biol Fertil Soils 26:107–111

Cuenca G, Cáceres A, González MG (2008) AM inoculation in tropical agriculture: field results. In: Varma A (ed) Mycorrhiza: state of the art, genetics and molecular biology, eco-function, biotechnology, eco-physiology, structure and systematics, 3rd edn. Springer, Berlin, pp 403–417

Curaqueo G, Barea JM, Acevedo E, Rubio R, Cornejo P, Borie F (2011) Effects of different tillage system on arbuscular mycorrhizal fungal propagules and physical properties in a Mediterranean agroecosystem in central Chile. Soil Till Res 113:11–18

del Val C, Barea JM, Azcón-Aguilar C (1999) Assessing the tolerance to heavy metals of arbuscular mycorrhizal fungi isolated from sewage sludge contaminated soils. Appl Soil Ecol 11:261–269

Dessaux Y, Hinsinger P, Lemanceau P (2010) Rhizosphere: achievements and challenges. Springer, Berlin

Díaz G, Azcón-Aguilar C, Honrubia M (1996) Influence of arbuscular mycorrhizae on heavy metal (Zn and Pb) uptake and growth of *Lygeum spartium* and *Anthyllis cytisoides*. Plant Soil 180:241–249

Dodd JC, Jeffries P (1986) Early development of vesicular–arbuscular mycorrhizas in autumn-sown cereals. Soil Biol Biochem 18:149–154

Dodd JC, Burton CC, Burns RG, Jeffries P (1987) Phosphatase activity associated with the roots and rhizosphere of plants infected with vesicular–arbuscular mycorrhizal fungi. New Phytol 107:163–172

Dong Y, Zhu YG, Smith FA, Wang YS, Chen BD (2008) Arbuscular mycorrhiza enhanced arsenic resistance of both white clover (*Trifolium repens* Linn.) and ryegrass (*Lolium perenne* L.) plants in an arsenic-contaminated soil. Environ Pollut 155:174–181

Duan T, Facelli E, Smith SE, Smith FA, Nan Z (2011) Differential effects of soil disturbance and plant residue retention on function of arbuscular mycorrhizal (AM) symbiosis are not reflected in colonization of roots or hyphal development in soil. Soil Biol Biochem 43:571–578

Dumbrell AJ, Nelson M, Helgason T, Dytham C, Fitter AH (2010a) Relative roles of niche and neutral processes in structuring a soil microbial community. ISME J 4:337–345

Dumbrell AJ, Nelson M, Helgason T, Dytham C, Fitter AH (2010b) Idiosyncrasy and overdominance in the structure of natural communities of arbuscular mycorrhizal fungi: is there a role for stochastic processes? J Ecol 98:419–428

Echeverria M, Scambato AA, Sannazzaro AI, Maiale S, Ruíz OA, Menéndez AB (2008) Phenotypic plasticity with respect to salt stress response by *Lotus glaber*: the role of its AM fungal and rhizobial symbionts. Mycorrhiza 18:317–329

Elmer WH (2002) Influence of formononetin and NaCl on mycorrhizal colonization and fusarium crown and root rot of asparagus. Plant Dis 86:1318–1324

Estaún V, Camprubí A, Joner EJ (2002) Selecting arbuscular mycorrhizal fungi for field application. In: Gianinazzi S, Schüepp H, Barea JM, Haselwandter K (eds) Mycorrhiza technology in agriculture, from genes to bioproducts. Birkhauser, Basel, pp 249–259

Estaún V, Vicente S, Calvet C, Camprubí A, Busquets M (2007) Integration of arbuscular mycorrhiza inoculation in hydroseeding technology. Effects on plant growth and inter-species competition. Land Degrad Dev 18:621–630

Evans DG, Miller MH (1990) The role of the external mycelial network in the effect of soil disturbance upon vesicular arbuscular mycorrhizal colonization of maize. New Phytol 114:65–71

Faure D, Vereecke D, Leveau JHJ (2009) Molecular communication in the rhizosphere. Plant Soil 321:279–303

Finlay RD (2008) Ecological aspects of mycorrhizal symbiosis: with special emphasis on the functional diversity of interactions involving the extraradical mycelium. J Exp Bot 59:1115–1126

Fitter AH, Helgason T, Hodge A (2011) Nutritional exchanges in the arbuscular mycorrhizal symbiosis: implications for sustainable agriculture. Fungal Biol Rev 25:68–72

Francis CF, Thornes JB (1990) Matorral: erosion and reclamation. In: Albaladejo J, Stocking MA, Díaz E (eds) Soil degradation and rehabilitation in Mediterranean environmental conditions. CSIC, Murcia, pp 87–115

Francis R, Finlay RD, Read DJ (1986) Vesicular–arbuscular mycorrhiza in natural vegetation systems. IV transfer of nutrients in inter- and intra-specific combinations of host plants. New Phytol 102:103–111

Franzini VI, Azcón R, Latanze-Mendes F, Aroca R (2010) Interaction between *Glomus* species and *Rhizobium* strains affect the nutritional physiology of drought stressed legume hosts. J Plant Physiol 167:614–619

Galleguillos C, Aguirre C, Barea JM, Azcón R (2000) Growth promoting effect of two *Sinorhizobium meliloti* strains (a wild type and its genetically modified derivative) on a non-legume plant species in specific interaction with two arbuscular mycorrhizal fungi. Plant Sci 159:57–63

Gamper HA, Young JPW, Jones DL, Hodge A (2008) Real-time PCR and microscopy: are the two methods measuring the same unit of arbuscular mycorrhizal fungal abundance? Fung Gene Biol 45:581–596

Gamper HA, van der Heijden MGA, Kowalchuk GA (2010) Molecular trait indicators: moving beyond phylogeny in arbuscular mycorrhizal ecology. New Phytol 185:67–82

GAO/RCED (1992) GAO/RCED 92-233 Sustainable agriculture. United States Congress Food, Agriculture Conservation, and Trade Act of 1990, P. L. no 101-624, Washington

Garbaye J (1994) Helper bacteria, a new dimension to the mycorrhizal symbiosis. New Phytol 128:197–210
Gavito ME, Miller MH (1998) Changes in mycorrhiza development in maize induced by crop management practices. Plant Soil 198:185–192
Gavito ME, Schweiger P, Jakobsen I (2003) P uptake by arbuscular mycorrhizal hyphae: effect of soil temperature and atmospheric CO_2 enrichment. Global Change Biol 9:106–116
Gianinazzi S, Vosátka M (2004) Inoculum of arbuscular mycorrhizal fungi for production systems: science meets business. Can J Bot 82:1264–1271
Gianinazzi G, Gollotte A, Binet M-N, Van Tuinen D, ReD WD (2010) Agroecology: the key role of arbuscular mycorrhizas in ecosystem services. Mycorrhiza 20:519–530
Gianinazzi-Pearson V, Gianinazzi S, Scannerini S, Smith DC, Bonfante-Fasolo P (1988) Morphological integration and functional compatibility between symbionts in vesicular–arbuscular endomycorrhizal associations. Cell to cell signals in plant, animal and microbial symbiosis: NATO ASI, Ser H. Cell Biol 1988:73–84
Glick BR (1995) The enhancement of plant growth by free-living bacteria. Can J Microbiol 41:109–117
Goicoechea N, Antolín MC, Sánchez-Díaz M (2000) The role of plant size and nutrient concentrations in associations between *Medicago*, and *Rhizobium* and/or *Glomus*. Biol Plant 43:221–226
Gómez-Roldán V, Fermas S, Brewer PB, Puech-Pages V, Dun EA, Pillot JP, Letisse F, Matusova R, Danoun S, Portais JC, Bouwmeester H, Becard G, Beveridge CA, Rameau C, Rochange SF (2008) Strigolactone inhibition of shoot branching. Nature 455:189–122
González-Guerrero M, Benabdellah K, Ferrol N, Azcón-Aguilar C (2009) Mechanisms underlying heavy metal tolerance in arbuscular mycorrhizas. In: Azcón-Aguilar C, Barea JM, Gianinazzi S, Gianinazzi-Pearson V (eds) Mycorrhizas functional processes and ecological impact. Springer, Berlin, pp 107–122
Grime JP, Mackey JML, Hillier SH, Read DJ (1987) Floristic diversity in a model system using experimental microcosms. Nature 328:420–422
Hamel C, Dalpé Y, Furlan V, Parent S (1997) Indigenous populations of arbuscular mycorrhizal fungi and soil aggregate stability are major determinants of leek (*Allium porrum* L.) response to inoculation with *Glomus intraradices* Schenck & Smith or *Glomus versiforme* (Karsten) Berch. Mycorrhiza 7:187–196
Hart M, Klironomos JN (2002) Diversity of arbuscular mycorrhizal fungi and ecosystem functioning. In: van der Heijden MGA, Sanders IR (eds) Mycorrhizal ecology. Springer, Berlin, pp 225–242
Hartmann A, Schmid M, van Tuinen D, Berg G (2009) Plant-driven selection of microbes. Plant Soil 321:235–257
Hartnett DC, Wilson GWT (1999) Mycorrhizae influence plant community structure and diversity in tallgrass prairie. Ecology 80:1187–1195
Hayman DS (1986) Mycorrhizae of nitrogen-fixing legumes. MIRCEN J 2:121–145
Helgason T, Fitter AH (2009) Natural selection and the evolutionary ecology of the arbuscular mycorrhizal fungi (phylum Glomeromycota). J Exp Bot 60:2465–2480
Helgason T, Daniell TJ, Husband R, Fitter AH, Young JPW (1998) Ploughing up the wood-wide web? Nature 394:431
Hempel S, Renker C, Buscot F (2007) Differences in the species composition of arbuscular mycorrhizal fungi in spore, root and soil communities in a grassland ecosystem. Environ Microbiol 9:1930–1938
Herrera MA, Salamanca CP, Barea JM (1993) Inoculation of wooody legumes with selected arbuscular mycorrhizal fungi and rhizobia to recover desertified Mediterranean ecosystems. Appl Environ Microbiol 59:129–133
Herrera-Peraza RA, Hamel C, Fernández F, Ferrer RL, Furrazola E (2011) Soil-strain compatability: the key to effective use of arbuscular mycorrhizal inoculants? Mycorrhiza 21:183–193
Hetrick BAD, Bloom J (1983) Vesicular-arbuscular mycorrhizal fungi associated with native tall grass prairie and cultivated winter wheat. Can J Bot 61:2140–2146
Hodge A, Fitter AH (2010) Substantial nitrogen acquisition by arbuscular mycorrhizal fungi from organic material has implications for N cycling. Proc Natl Acad Sci USA 107:13754–13759
Honrubia M (2009) The mycorrhizae: a plant–fungus relation that has existed for more than 400 million years. An Jard Bot Madrid 66:133–144
Ijdo M, Cranenbrouck DS (2011) Methods for large-scale production of AM fungi: past, present, and future. Mycorrhiza 21:1–16
Jakobsen I (2004) Hyphal fusion to plant species connections—giant mycelia and community nutrient flow. New Phytol 164:4–7
Jakobsen I, Nielsen NE (1983) Vesicular-arbuscular mycorrhiza in field grown crops I mycorrhizal infection in cereals and peas at various times and soil depths. New Phytol 93:401–413
Janos DP (2007) Plant responsiveness to mycorrhizas differs from dependence upon mycorrhizas. Mycorrhiza 17:75–91
Jasper DA (2007) Beneficial soil microorganisms of the Jarrah forest and their recovery in bauxite. Southwest Austral Restor Ecol 15:S74–S84
Jasper DA, Abbott LK, Robson AD (1989) Soil disturbance reduces the infectivity of external hyphae of vesicular-arbuscular mycorrhizal fungi. New Phytol 112:93–99
Jeffries P, Barea JM (1994) Biogeochemical cycling and arbuscular mycorrhizas in sustainability of plant–soil systems. In: Gianinazzi S, Schuepp H (eds)

Impact of arbuscular mycorrhizas on sustainable agriculture and natural ecosystems. Birkhauser, Basel, pp 101–115

Jeffries P, Barea JM (2001) Arbuscular mycorrhiza—a key component of sustainable plant-soil ecosystems. In: Hock B (ed) The mycota, vol IX, Fungal associations. Springer, Berlin, pp 95–113

Jeffries P, Gianinazzi S, Perotto S, Turnau K, Barea JM (2003) The contribution of arbuscular mycorrhizal fungi in sustainable maintenance of plant health and soil fertility. Biol Fertil Soils 37:1–16

Johansson JF, Paul LR, Finlay RD (2004) Microbial interactions in the mycorrhizosphere and their significance for sustainable agriculture. FEMS Microbiol Ecol 48:1–13

Johnson CR, Copeland PJ, Crosskston RK, Pfleger FL (1992) Mycorrhizae: a possible explanation for yield decline associated with continuous cropping of corn and soybean. Agron J 84:387–390

Jonasson J, Olofsson M, Monstein HJ (2002) Classification, identification and subtyping of bacteria based on pyrosequencing and signature matching of 16S rDNA fragments. APMIS 110:263–272

Joner EJ, Jakobsen I (1995) Growth and extracellular phosphatase activity of arbuscular mycorrhizal hyphae as influenced by soil organic matter. Soil Biol Biochem 27:1153–1159

Jones DL, Nguyen C, Finlay RD (2009) Carbon flow in the rhizosphere: carbon trading at the soil-root interface. Plant Soil 321:5–33

Kennedy AC, Smith KL (1995) Soil microbial diversity and the sustainability of agriculture soils. Plant Soil 170:75–86

Kennedy PG, Hortal S, Bergemann SE, Bruns TD (2007) Competitive interactions among three ectomycorrhizal fungi and their relation to host plant performance. J Ecol 95:1338–1345

Kenrick P (2003) Fishing for the first plants. Nature 425:248–249

Kilvin SN, Hawkes CV (2011) Differentiating between effects of invasion and diversity: impacts of aboveground plan communities on belowground fungal communities. New Phytol 189:526–535

Klironomos JN (2002) Another form of bias in conservation research. Science 29:749–749

Kloepper JW (1992) Plant growth-promoting rhizobacteria as biological control agents. In: Blaine F, Metting J Jr (eds) Soil microbial ecology. Applications in agriculture forestry and environmental management. Dekker, New York, pp 255–274

Kohler J, Caravaca F, Roldan A (2010) An AM fungus and a PGPR intensify the adverse effects of salinity on the stability of rhizosphere soil aggregates of *Lactuca sativa*. Soil Biol Biochem 42:429–434

Koltai H, Meir D, Shlomo E, Resnick N, Ziv O, Wininger S, Ben-Dor B, Kapulnik Y (2008) Exploiting arbuscular mycorrhizal technology in different cropping systems under greenhouse conditions in semi-arid regions. In: Kubota C, Kacira M (eds) Acta Horticulturae proceedings of the international workshop on greenhouse environmental control and crop production in semi-arid regions. ISHS, Tucson, pp 223–228

Krauss M, Berner A, Burger D, Wiemken A, Niggli U, Mäder P (2010) Reduced tillage in temperate organic framing: implications for crop management and forage production. Soil Use Manage 26:12–20

Krueger M, Stockinger H, Krueger C, Schüßler A (2009) DNA-based species level detection of Glomeromycota: one PCR primer set for all arbuscular mycorrhizal fungi. New Phytol 183:212–223

Ladha JK (1992) Extended versions of papers presented in the symposium role of biological nitrogen-fixation in sustainable agriculture at the 13th international-congress of soil science—Kyoto, Japan, 1990—preface. Plant Soil 141:R7–R7

Lal R (1989) Conservation tillage for sustainable agriculture: tropics versus temperate environments. Adv Agron 42:85–185

Lambers H, Mougel C, Jaillard B, Hinsinger P (2009) Plant-microbe-soil interactions in the rhizosphere: an evolutionary perspective. Plant Soil 321:83–115

Leigh J, Hodge A, Fitter AH (2009) Arbuscular mycorrhizal fungi can transfer substantial amounts of nitrogen to their host plant from organic material. New Phytol 181:199–207

Lekberg Y, Hammer EC, Olsson PA (2010) Plants as resource islands and storage units—adopting the mycocentric view of arbuscular mycorrhizal networks. FEMS Microbiol Ecol 74:336–345

Leyval C, Joner EJ, del Val C, Haselwandter K (2002) Potential of arbuscular mycorrhizal fungi for bioremediation. In: Gianinazzi S, Schüepp H, Barea JM, Haselwandter K (eds) Mycorrhizal technology in agriculture. Birkhäuser, Basel, pp 175–186

Lin AJ, Zhang XH, Wong MH, Ye ZH, Lou LQ, Wang YS, Zhu YG (2007) Increase of multi-metal tolerance of three leguminous plants by arbuscular mycorrhizal fungi colonization. Environ Geochem Health 29:473–481

Linderman RG (1992) Vesicular-arbuscular mycorrhizal and soil microbial interactions. In: Bethlenfalvay GJ, Linderman RG (eds) Mycorrhizae in sustainable agriculture. ASA, Madison, pp 45–70

López-Ráez JA, Bouwmeester H, Pozo MJ (2011a) Communication in the rhizosphere, a target for pest management. In: Lichtfouse E (ed) Sustainable agriculture reviews, vol 8. Springer, Dordrecht, pp 109–133

López-Ráez JA, Charnikhova T, Fernández I, Bouwmeester H, Pozo MJ (2011b) Arbuscular mycorrhizal symbiosis decreases strigolactone production in tomato. J Plant Physiol 168:294–297

Lumini E, Orgiazzi A, Borriello R, Bonfante P, Bianciotto V (2010) Disclosing arbuscular mycorrhizal fungal biodiversity using a pyrosequencing approach. Environ Microbiol 12:2165–2179

Maiti D, Variar M, Singh RK (2011) Optimising tillage schedule for maintaining activity of the arbuscular mycorrhizal fungal population in a rainfed upland rice (*Oryza sativa* L.) agro-ecosystem. Mycorrhiza 21:167–171

Maki T, Nomachi M, Yoshida S, Ezawa T (2008) Plant symbiotic microorganisms in acid sulfate soil: significance in the growth of pioneer plants. Plant Soil 310:55–65

Mallik MAB, Williams RD (2008) Plant growth promoting rhizobacteria and mycorrhizal fungi in sustainable agriculture and forestry. In: Zeng RS, Mallik AU, Luo SM (eds) Allelopathy in sustainable agriculture and forestry. Springer, Berlin, pp 321–345

Martínez-García LB, Pugnaire FI (2009) Interacciones entre las comunidades de hongos formadores de micorrizas arbusculares y de plantas. Algunos ejemplos en los ecosistemas semiáridos. Ecosistemas 18:44–54

Marulanda A, Barea JM, Azcón R (2006) An indigenous drought-tolerant strain of *Glomus intraradices* associated with a native bacterium improves water transport and root development in *Retama sphaerocarpa*. Microb Ecol 52:670–678

Mathimaran N, Falquet L, Ineichen K, Picard C, Redecker D, Boller T, Wiemken A (2008a) Microsatellites for disentangling underground networks: strain-specific identification of *Glomus intraradices*, an arbuscular mycorrhizal fungus. Fungal Genet Biol 45:812–817

Mathimaran N, Falquet L, Ineichen K, Picard C, Redecker D, Wiemken A, Boller T (2008b) Unexpected vagaries of microsatellite loci in *Glomus intraradices*: length polymorphisms are rarely caused by variation in repeat number only. New Phytol 180:568–570

Medina A, Vassileva M, Caravaca F, Roldán A, Azcón R (2004) Improvement of soil characteristics and growth of *Dorycnium pentaphyllum* by amendment with agrowastes and inoculation with AM fungi and/or the yeast *Yarrowia lipolytica*. Chemosphere 56:449–450

Mikkelsen BL, Rosendahl S, Jakobsen I (2008) Underground resource allocation between individual networks of mycorrhizal fungi. New Phytol 180:890–898

Miller RM, Jastrow JD (1992) The application of VA mycorrhizae to ecosystem restoration and reclamation. In: Allen MF (ed) Mycorrhizal functioning an integrative plant fungal process. Chapman and Hall, New York, pp 438–467

Miller RM, Jastrow JD (2000) Mycorrhizal fungi influence soil structure. In: Kapulnik Y, Douds DD (eds) Arbuscular mycorrhizas: physiology and function. Kluwer, Dordrecht, pp 3–18

Morgan RPC, Rickson RJ, Wright W (1990) Regeneration of degraded soils. In: Albaladejo J, Stocking MA, Díaz E (eds) Soil degradation and rehabilitation in Mediterranean environmental conditions. CSIC, Murcia, pp 69–85

Morton JB (2009) Reconciliation of conflicting phenotypic and rRNA gene phylogenies of fungi in glomeromycota based on underlying patterns and processes. In: Azcón-Aguilar C, Barea JM, Gianinazzi S, Gianinazzi-Pearson V (eds) Mycorrhizas functional processes and ecological impact. Springer, Berlin, pp 137–154

Mosse B (1986) Mycorrhiza in a sustainable agriculture. Biol Agric Hortic 3:191–209

O'Connor PJ, Smith SE, Smith EA (2002) Arbuscular mycorrhizas influence plant diversity and community structure in a semiarid herbland. New Phytol 154:209–218

Oehl F, Sieverding E, Ineichen K, Mader P, Wiemken A, Boller T (2009) Distinct sporulation dynamics of arbuscular mycorrhizal fungal communities from different agroecosystems in long-term microcosms. Agric Ecosyst Environ 134:257–268

Oehl F, Laczko E, Bogenrieder A, Stahr K, Bosch R, van der Heijden M, Sieverding E (2010) Soil type and land use intensity determine the composition of arbuscular mycorrhizal fungal communities. Soil Biol Biochem 42:724–738

Offre P, Pivato B, Siblot S, Gamalero E, Corberand T, Lemanceau P, Mougel C (2007) Identification of bacterial groups preferentially associated with mycorrhizal roots of *Medicago truncatula*. Appl Environ Microbiol 73:913–921

Öpik M, Saks Ü, Kennedy J, Daniell T (2008) Global diversity patterns of arbuscular mycorrhizal fungi-community composition and links with functionality. In: Varma A (ed) Mycorrhiza: state of the art, genetics and molecular biology, eco-function, biotechnology, eco-physiology, structure and systematics, 3rd edn. Springer, Berlin, pp 89–111

Öpik M, Vanatoa A, Vanatoa E, Moora M, Davison J, Kalwij JM, Reier U, Zobel M (2010) The online database MaarjAM reveals global and ecosystemic distribution patterns in arbuscular mycorrhizal fungi (Glomeromycota). New Phytol 188:223–241

Palenzuela J, Barea JM, Ferrol N, Azcón-Aguilar C, Oehl F (2010) *Entrophospora nevadensis*, a new arbuscular mycorrhizal fungus from Sierra Nevada National Park (southeastern Spain). Mycologia 102:624–632

Palenzuela J, Barea JM, Ferrol N, Oehl F (2011) *Ambispora granatensis*, a new arbuscular mycorrhizal fungus, associated with *Asparagus officinalis* in Andalucia (Spain). Mycologia 103:333–340

Parniske M (2008) Arbuscular mycorrhiza: the mother of plant root endosymbioses. Nat Rev Microbiol 6:763–775

Pelligrino E, Bedini S, Avio L, Bonari E, Giovannetti M (2011) Field inoculation effectiveness of native and exotic arbuscular mycorrhizal fungi in a Mediterranean agricultural soil. Soil Biol Biochem 43:367–376

Peoples MB, Craswell ET (1992) Biological nitrogen fixation: investments, expectations and actual contributions to agriculture. Plant Soil 141:13–39

Porcel R, Ruíz-Lozano JM (2004) Arbuscular mycorrhizal influence on leaf water potential, solute accumulation, and oxidative stress in soybean plants subjected to drought stress. J Exp Bot 55:1743–1750

Porcel R, Barea JM, Ruíz-Lozano JM (2003) Antioxidant activities in mycorrhizal soybean plants under drought stress and their possible relationship to the process of nodule senescence. New Phytol 157:135–143

Porcel R, Aroca R, Azcón R, Ruíz-Lozano JM (2006) PIP aquaporin gene expression in arbuscular mycorrhizal *Glycine max* and *Lactuca sativa* plants in relation to drought stress tolerance. Plant Mol Biol 60:389–404

Pozo MJ, Azcón-Aguilar C (2007) Unraveling mycorrhiza-induced resistance. Curr Opin Plant Biol 10:393–398

Pozo MJ, Verhage A, García-Andrade J, García JM, Azcón-Aguilar C (2009) Priming plant defence against pathogens by arbuscular mycorrhizal fungi. In: Azcón-Aguilar C, Barea JM, Gianinazzi S, Gianinazzi-Pearson V (eds) Mycorrhizas functional processes and ecological impact. Springer, Berlin, pp 123–135

Read D (1989) Mycorrhizas and nutrient cycling in sand dune ecosystems. Proc R Soc Edinb 86B:89–110

Read D (2002) Towards ecological relevance-progress and pitfalls in the path towards an understanding of mycorrhizal functions in nature. In: van der Heijden MGA, Sanders IR (eds) Mycorrhizal ecology. Springer, Berlin, pp 3–24

Redecker D, Kodner R, Graham LE (2000) Glomalean fungi from the Ordovician. Science 289:920–1921

Redon PO, Beguiristain T, Leyval C (2009) Differential effects of AM fungal isolates on *Medicago truncatula* growth and metal uptake in a multimetallic (Cd, Zn, Pb) contaminated agricultural soil. Mycorrhiza 19:187–195

Remy W, Taylor TN, Haas H, Kerp H (1994) Four hundred-million-year-old vesicular-arbuscular mycorrhizae. Proc Natl Acad Sci USA 91:11841–11843

Requena N, Jimenez I, Toro M, Barea JM (1997) Interactions between plant-growth-promoting rhizobacteria (PGPR), arbuscular mycorrhizal fungi and *Rhizobium* spp. In the rhizosphere of *Anthyllis cytisoides*, a model legume for revegetation in Mediterranean semi-arid ecosystems. New Phytol 136:667–677

Requena N, Pérez-Solis E, Azcón-Aguilar C, Jeffries P, Barea JM (2001) Management of indigenous plant–microbe symbioses aids restoration of desertified ecosystems. Appl Environ Microbiol 67:495–498

Reynolds HL, Vogelsang KM, Hartley AE, Bever JD, Schultz PA (2006) Variable responses of old-field perennials to arbuscular mycorrhizal fungi and phosphorus source. Oecologia 147:348–358

Richardson AE, Barea JM, McNeill AM, Prigent-Combaret C (2009) Acquisition of phosphorus and nitrogen in the rhizosphere and plant growth promotion by microorganisms. Plant Soil 321:305–339

Rillig MC (2004) Arbuscular mycorrhizae, glomalin, and soil aggregation. Can J Soil Sci 84:355–363

Rillig MC, Mummey DL (2006) Mycorrhizas and soil structure. New Phytol 171:41–53

Rillig MC, Mummey DL, Ramsey PW, Klironomos JN, Gannon JE (2006) Phylogeny of arbuscular mycorrhizal fungi predicts community composition of symbiosis-associated bacteria. FEMS Microbiol Ecol 57:389–395

Rinaudo V, Barberi P, Giovannetti M, van der Heijden MGA (2010) Mycorrhizal fungi suppress aggressive agricultural weeds. Plant Soil 333:7–20

Robinson-Boyer L, Grzyb I, Jeffries P (2009) Shifting the balance from qualitative to quantitative analysis of arbuscular mycorrhizal communities in field soils. Fungal Ecol 2:1–9

Roesch LF, Fulthorpe RR, Riva A, Casella G, Hadwin AKM, Kent AD, Daroub SH, Camargo FAO, Farmerie WG, Triplett EW (2007) Pyrosequencing enumerates and contrasts soil microbial diversity. ISME J 1:283–290

Roesti D, Ineichen K, Braissant O, Redecker D, Wiemken A, Aragno M (2005) Bacteria associated with spores of the arbuscular mycorrhizal fungi *Glomus geosporum* and *Glomus constrictum*. Appl Environ Microbiol 71:6673–6679

Rosendahl S (2008) Communities, populations and individuals of arbuscular mycorrhizal fungi. New Phytol 178:253–266

Ruíz-Lozano JM (2003) Arbuscular mycorrhizal symbiosis and alleviation of osmotic stress. New perspectives for molecular studies. Mycorrhiza 13:309–317

Ruíz-Lozano JM, Aroca R (2008) Last insights into the role of aquaporins in the alleviation of osmotic stress by arbuscular mycorrhizal symbiosis. In: Van Dijk T (ed) Microbial ecology research trends. Nova Science, New York, pp 139–154

Ruiz-Lozano JM, Collados C, Barea JM, Azcon R (2001) Arbuscular mycorrhizal symbiosis can alleviate drought-induced nodule senescence in soybean plants. New Phytol 151:493–502

Ruíz-Lozano JM, Porcel R, Aroca R (2008) Evaluation of the possible participation of drought-induced genes in the enhanced tolerance of arbuscular mycorrhizal plants to water deficit. In: Varma A (ed) Mycorrhiza: state of the art, genetics and molecular biology, eco-function, biotechnology, eco-physiology, structure and systematics. Springer, Berlin, pp 185–207

Sánchez-Castro I, Barea JM, Ferrol N (2008) Analyzing the community composition of arbuscular mycorrhizal fungi colonizing the roots of representative shrubland species in a Mediterranean ecosystems (Granada, Spain). Book of abstracts plant–microbial interactions, 2–6 July, Krakow, pp 38–39

Sanders IR (2002) Ecology and evolution of multigenomic arbuscular mycorrhizal fungi. Am Nat 160: S128–S141

Schalmuk S, Cabello M (2010) Arbuscular mycorrhizal fungal propagules from tillage and non-tillage systems: possible effects on Glomeromycota diversity. Mycologia 102:261–268

Schenck NC, Kinloch RA (1980) Incidence of mycorrhizal fungi on six field crops in monoculture on a newly cleared woodland site. Mycologia 72:445–456

Schiavo JA, Busato JG, Martins MA, Canellas LP (2009) Recovery of degraded areas revegeted with *Acacia mangium* and Eucalyptus with special reference to organic matter humification. Sci Agric 66:353–360

Schnoor TK, Lekberg Y, Rosendahl S, Olsson PA (2011) Mechanical soil disturbance as a determinant of arbuscular mycorrhizal fungal communities in semi-natural grassland. Mycorrhiza 21:211–220

Schreiner RP, Bethlenfalvay GJ (1995) Mycorrhizal interactions in sustainable agriculture. Crit Rev Biotechnol 15:271–285

Schüßler A, Schwarzott D, Walker C (2001) A new fungal phylum, the Glomeromycota, phylogeny and evolution. Mycol Res 105:1413–1421

Sieverding E, Howeler RH (1985) Influence of species of VA mycorrhizal fungi on cassava yield response to phosphorus fertilization. Plant Soil 88:213–221

Siviero MA, Motta AM, Lima DDS, Birolli RR, Huh SY, Santinoni IA, Murate LS, de Castro CMA, Miyauchi MYH, Zangaro W, Nogueira MA, Andrade G (2008) Interaction among N-fixing bacteria and AM fungi in Amazonian legume tree (*Schizolobium amazonicum*) in field conditions. Appl Soil Ecol 39:144–152

Skujins J, Allen MF (1986) Use of mycorrhizae for land rehabilitation. Mircen J 2:161–176

Smith SE, Read DJ (2008) Mycorrhizal symbiosis, 3rd edn. Elsevier, New York

Smith FA, Grace EJ, Smith SE (2009) More than a carbon economy: nutrient trade and ecological sustainability in facultative arbuscular mycorrhizal symbioses. New Phytol 182:347–358

Staddon PL, Thompson K, Jakobsen I, Grime JP, Askew AP, Fitter AH (2003) Mycorrhizal fungal abundance is affected by long-term climatic manipulations in the field. Global Change Biol 9:186–194

Stahl PD, Williams SE, Christensen M (1988) Efficacy of native vesicular-arbuscular mycorrhizal fungi after severe soil disturbance. New Phytol 110:347–354

Tarkalson DD, Von Jolley D, Robbins CW, Terry RE (1998) Mycorrhizal colonization and nutrition of wheat and sweet corn grown in manure-treated and untreated topsoil and subsoil. J Plant Nutr 21:1985–1999

Tawaraya K, Naito M, Wagatsuma T (2006) Solubilization of insoluble inorganic phosphate by hyphal exudates of arbuscular mycorrhizal fungi. J Plant Nutr 29:657–665

Teng Y, Luo Y-M, Gao J, Li Z-G (2008) Combined remediation effects of arbuscular mycorrhizal fungi-legumes-*Rhizobium* symbiosis on PCBs contaminated soils. Huanjing Kexue 29:2925–2930

Tisdall JM (1991) Fungal hyphae and structural stability of soil. Aust J Soil Res 29:729–743

Toljander JF, Artursson V, Paul LR, Jansson JK, Finlay RD (2005) Attachment of different soil bacteria to arbuscular mycorrhizal fungal extraradical hyphae is determined by hyphal vitality and fungal species. FEMS Microbiol Lett 254:34–40

Turnau K, Jurkiewicz A, Língua G, Barea JM, Gianinazzi-Pearson V (2006) Role of arbuscular mycorrhiza and associated microorganisms in phytoremediation of heavy metal-polluted sites. In: Prasad MNV, Sajwan KS, Naidu R (eds) Trace elements in the environment biogeochemistry, biotechnology and bioremediation. CRC, Boca Raton, pp 235–252

Umehara M, Hanada A, Yoshida S, Akiyama K, Arite T, Takeda-Kamiya N, Magome H, Kamiya Y, Shirasu K, Yoneyama K, Kyozuka J, Yamaguchi S (2008) Inhibition of shoot branching by new terpenoid plant hormones. Nature 455:195–U129

van der Gast CJ, Gosling P, Tiwari B, Bending GD (2011) Spacial scaling of arbuscular mycorrhizal fungal diversity is affected by farming practice. Environ Microbiol 13:241–249

van der Heijden MGA (2002) Arbuscular mycorrhizal fungi as a determinant of plant diversity: in search for underlying mechanisms and general principles. In: van der Heijden MGA, Sanders IR (eds) Mycorrhizal ecology. Springer, Berlin, pp 243–265

van der Heijden MGA, Boller T, Wiemken A, Sanders IR (1998a) Different arbuscular mycorrhizal fungal species are potential determinants of plant community structure. Ecology 79:2082–2091

van der Heijden MGA, Klironomos JN, Ursic M, Moutoglis P, Streitwolf-Engel R, Boller T, Wiemken A, Sanders ER (1998b) Mycorrhizal fungal diversity determines plant biodiversity, ecosystem variability and productivity. Nature 396:69–72

van der Heijden MGA, Streitwolf-Engel R, Riedl R, Siegrist S, Neudecker A, Ineichen K, Boller T, Wiemken A, Sanders IR (2006) The mycorrhizal contribution to plant productivity, plant nutrition and soil structure in experimental grassland. New Phytol 172:739–752

Vance CP, Stone CU, Allan DL (2003) Phosphorus acquisition and use: critical adaptations by plants for securing a non-renewable resource. New Phytol 157:423–447

Vargas R, Baldocchi DD, Querejeta JI, Curtis PS, Hasselquist NJ, Janssens IA, Allen MF, Montagnani L (2010) Ecosystem CO_2 fluxes of arbuscular and ectomycorrhizal dominated vegetation types are differentially influenced by precipitation and temperature. New Phytol 185:226–236

Vestberg A, Kukkonen S, Saari K, Parikka P, Huttunen J, Tainio L, Devos D, Weekers F, Kevers C, Thonart P, Lemoine MC, Cordier C, Alabouvette C, Gianinazzi S (2004) Microbial inoculation for improving the growth and health of micropropagated strawberry. Appl Soil Ecol 27:243–258

Vivas A, Azcón R, Biró B, Barea JM, Ruíz-Lozano JM (2003a) Influence of bacterial strains isolated from lead-polluted soil and their interactions with arbuscular mycorrhizae on the growth of *Trifolium pratense* L. under lead toxicity. Can J Microb 49:577–588

Vivas A, Biró B, Campos E, Barea JM, Azcón R (2003b) Symbiotic efficiency of autochthonous arbuscular mycorrhizal fungus (*G. Mosseae*) and *Brevibacillus* sp. Isolated from cadmium polluted soil under increasing cadmium levels. Environ Pollut 126:179–189

Vivas A, Marulanda A, Gómez M, Barea JM, Azcón R (2003c) Physiological characteristics (SDH and ALP activities) of arbuscular mycorrhizal colonization as affected by *Bacillus thuringiensis* inoculation under two phosphorus levels. Soil Biol Biochem 35:987–996

Vivas A, Marulanda A, Ruíz-Lozano JM, Barea JM, Azcón R (2003d) Influence of a *Bacillus* sp. on physiological activities of two arbuscular mycorrhizal fungi and on plant responses to PEG-induced drought stress. Mycorrhiza 13:249–256

Vivas A, Vörös A, Biró B, Barea JM, Ruíz-Lozano JM, Azcón R (2003e) Beneficial effects of indigenous Cd-tolerant and Cd-sensitive *Glomus mosseae* associated with a Cd-adapted strain of *Brevibacillus* sp. In improving plant tolerance to Cd contamination. Appl Soil Ecol 24:177–186

Vivas A, Barea JM, Azcón R (2005a) *Brevibacillus brevis* isolated from cadmium- or zinc-contaminated soils improves *in vitro* spore germination and growth of *Glomus mosseae* under high Cd or Zn concentrations. Microb Ecol 49:416–424

Vivas A, Barea JM, Azcón R (2005b) Interactive effect of *Brevibacillus brevis* and *Glomus mosseae*, both isolated from Cd contaminated soil, on plant growth, physiological mycorrhizal fungal characteristics and soil enzymatic activities in Cd polluted soil. Environ Pollut 134:257–266

Vivas A, Barea JM, Biro B, Azcon R (2006a) Effectiveness of autochthonous bacterium and mycorrhizal fungus on *Trifolium* growth, symbiotic development and soil enzymatic activities in Zn contaminated soil. J Appl Microbiol 100:587–598

Vivas A, Biro B, Nemeth T, Barea JM, Azcón R (2006b) Nickel-tolerant *Brevibacillus brevis* and arbuscular mycorrhizal fungus can reduce metal acquisition and nickel toxicity effects in plant growing in nickel supplemented soil. Soil Biol Biochem 38:2694–2704

Vivas A, Biró B, Ruíz-Lozano JM, Barea JM, Azcón R (2006c) Two bacterial strains isolated from a Zn-polluted soil enhance plant growth and mycorrhizal efficiency under Zn-toxicity. Chemosphere 62:1523–1533

Voets L, de la Providencia IE, Declerck S (2006) Glomeraceae and Gigasporaceae differ in their ability to form hyphal networks. New Phytol 172:185–188

Vogelsang KM, Reynolds HL, Bever JD (2006) Mycorrhizal fungal identity and richness determine the diversity and productivity of a tallgrass prairie system. New Phytol 172:554–562

von Alten H, Blal B, Dodd JC, Feldman F, Vosátka M (2002) Quality control of arbuscular mycorrhizal fungi inoculum in Europe. In: Gianinazzi S, Schüepp H, Barea JM, Haselwandter K (eds) Mycorrhiza technology in agriculture, from genes to bioproducts. Birkhauser, Basel, pp 281–296

Vosátka M, Albrechtová J, Patten R (2008) The international marked development for mycorrhizal technology. In: Varma A (ed) Mycorrhiza: state of the art, genetics and molecular biology, eco-function, biotechnology, eco-physiology, structure and systematics, 3rd edn. Springer, Berlin, pp 419–438

Wani PA, Khan MS, Zaidi A (2007) Synergistic effects of the inoculation with nitrogen-fixing and phosphate-solubilizing rhizobacteria on the performance of field-grown chickpea. J Plant Nutr Soil Sci 170:283–287

Weinbaum BS, Allen MF, Allen EB (1996) Survival of arbuscular mycorrhizal fungi following reciprocal transplanting across the Great Basin, USA. Ecol Appl 6:1365–1372

Whipps JM (2004) Prospects and limitations for mycorrhizas in biocontrol of root pathogens. Can J Bot 82:1198–1227

Wolfe BE, Husband BC, Klironomos JN (2005) Effects of a belowground mutualism on an aboveground mutualism. Ecol Lett 8:218–223

Wright SF, Green VS, Cavigelli MA (2007) Glomalin in aggregate size classes from three different farming systems. Soil Till Res 94:546–549

Zaidi A, Khan MS (2007) Stimulatory effects of dual inoculation with phosphate solubilising microorganisms and arbuscular mycorrhizal fungus on chickpea. Aust J Exp Agric 47:1016–1022

Zaidi A, Khan MS, Amil M (2003) Interactive effect of rhizotrophic microorganisms on yield and nutrient uptake of chickpea (*Cicer arietinum* L.). Eur J Agron 19:15–21

Zhu YG, Miller RM (2003) Carbon cycling by arbuscular mycorrhizal fungi in soil–plant systems. Trends Plant Sci 8:407–409

5 The *Geosiphon–Nostoc* Endosymbiosis and Its Role as a Model for Arbuscular Mycorrhiza Research

A. Schüßler[1]

CONTENTS

I. Introduction

It was suggested in the previous edition of this volume that the *Geosiphon pyriformis–Nostoc punctiforme* symbiosis can serve as an important model symbiosis for the arbuscular mycorrhiza (AM). Due to its close phylogenetic relationship with the AM fungi (AMF; *Glomeromycota*), *G. pyriformis* attracted interest from the field of AM research and, in fact, eventually became one of the best studied glomeromycotan fungi. The AM symbiosis is formed by ~80 % of vascular plants and is one of the most important biotic factors in land ecosystems (Smith and Read 2008). Many lower plants also form associations with AMF (Read et al. 2000; Schüßler 2000) and, although not involving roots, these also are known as AM because of the evolutionary and functional homology. For both, the AM and the *Geosiphon* symbiosis, the association is facultative for the photoautotroph (*N. punctiforme* can be cultivated without the fungus) and obligatory for the fungus (*G. pyriformis* is an obligatory symbiont). The *Geosiphon* fungus might not be restricted to partnership with the cyanobacterium, but may also be symbiotic with plants. If so, an ecological network could exist among AMF, vascular plants, bryophytes and cyanobacteria, with extensive sharing and exchange among the symbiotic partners.

The *Geosiphon* symbiosis has indeed demonstrated its potential for the study of fundamental mechanisms and evolutionary questions relating to AM (Schüßler 2002; Schüßler et al. 2007, 2008; Schüßler and Walker 2011). However, although influencing most fields of plant sciences and terrestrial ecology, AM research is much under-represented in Germany. Nevertheless, the limited new discoveries of the last decade and the great potential of *G. pyriformis* research are reviewed herein, in relation to its life history, evolution, ultrastructure, nutrient transport and physiological activity.

Geosiphon pyriformis (*Glomeromycota*) is a coenocytic (non-septate) soil fungus and, to

[1]Genetics, Biocenter of the Ludwig-Maximilians-University Munich, Großhaderner Str. 2-4, 82152 Planegg-Martinsried-Germany; e-mail: arthur.schuessler@lmu.de

Fungal Associations, 2nd Edition
The Mycota IX
B. Hock (Ed.)

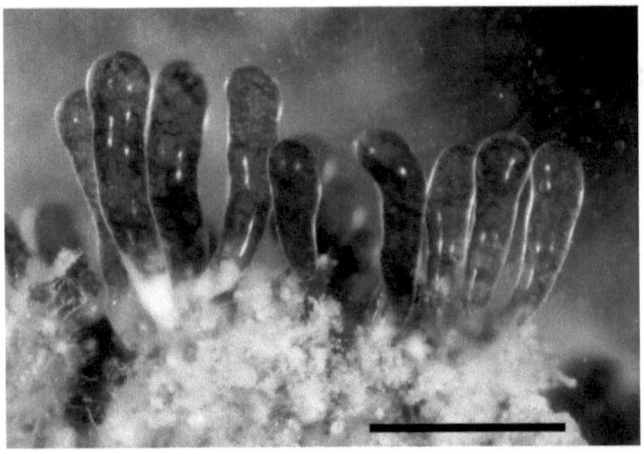

Fig. 5.1. *Geosiphon pyriformis* bladders, harvested from a laboratory culture on sterilized natural substrate still attached to the base of the bladders. *Bar* 1 mm

date, the only fungus known to live in endocytobiotic association with cyanobacteria. The symbiotic nature of this association was first recognised by von Wettstein (1915), who described it as a symbiosis between a heterotrophic siphonal chlorophyceaen alga and *Nostoc*. The fungal nature of the macrosymbiont was then recognised by Knapp (1933). *G. pyriformis* lives together with the cyanobacteria at the surface of humid, nutrient poor soil, where its hyphae come into contact with free-living *Nostoc* cells. The cyanobacteria may be incorporated by the fungus at the hyphal tip, which then swells and forms a unicellular 'bladder' up to 2 mm (occasionally more) in length and appearing on the soil surface (Fig. 5.1). Inside such a bladder the cyanobacteria proliferate within one large compartment, the symbiosome.

The species name, *G. pyrifome*, in the past often was used for both the fungus and the symbiosis, partly because this association historically was regarded as a 'phycomycetous lichen'. However, endosymbiotic associations are excluded from several modern lichen definitions (e.g., Hawksworth and Honegger 1994). Moreover, *Geosiphon* phylogenetically belongs in a group of fungi that are not known to form lichens and the species name should be used for the fungus only (Schüßler 2002), to reflect the functional and evolutionary links with the AM, rather than the lichen-forming fungi. The association between the fungus and cyanobacteria therefore is referred to as *Geosiphon* symbiosis or *Geosiphon–Nostoc* symbiosis and the orthographically correct (according to the gender in ancient Latin) species name of the fungus is *G. pyriformis* (Schüßler 2002).

Due to the physiological activity of the cyanobacterial endosymbiont the consortium is C- and N-autotrophic. In the past, some authors have considered *G. pyriformis* to be a 'primitive' endocytobiotic system, because the photobiont can be experimentally separated and cultivated without the fungal partner. Indeed, this symbiotic system provided historical insights into the evolutionary steps which finally gave rise to the organelles within the eukaryotic cells (Schnepf 1964).

The *Geosiphon* symbiosis is often referred to as being rare. Only six reports of this symbiosis having been found in nature are known. However these range from eastern Germany to Austria (Fig. 5.2) and thus the symbiosis appears to be geographically rather widespread but, due to its small size, rarely found. Presently, locations around the small village Bieber in the Spessart Mountains (Germany) are the only known stable natural habitats world-wide (Mollenhauer 1992; Schüßler and Wolf 2005).

II. Phylogeny of the Symbiotic Fungus and the Cyanobacterium

Knapp (1933) recognised *G. pyriformis* (as *G. pyriforme*) as a phycomycete (aseptate fungus) and Mollenhauer (1992), based on suggestions by Walter Gams, first proposed that it might be related to *Glomus*-like fungi. Because *Glomus* spp. form ecologically and economically important AM with plants, the verification of this suggestion made it conceivable that *Geosiphon* might also be capable of associating with plants to form AM.

A. Phylogeny of the Fungus *Geosiphon pyriformis*

In the twentieth century, the taxonomy of AMF was based mainly on the characteristics of the spore structure and development. Thus, Schüßler et al. (1994) compared morphological and

Fig. 5.2. Habitats reported for the *Geosiphon* symbiosis, indicated by *filled circles*. Since the 1990s, the only stable habitat known is close to the small village Bieber in the Spessart Mountains. From Schüßler and Wolf (2005)

ultrastructural criteria of *Geosiphon* spores with those of some AMF. This investigation indeed revealed similarities between *G. pyriformis* and an AMF named *Glomus versiforme* BEG47 (recently reclassified as *Diversispora epigaea*; Schüßler and Walker 2010; Schüßler et al. 2011). Final evidence showing that *Geosiphon* is closely related to AMF (at that time defined at the ordinal level, as *Glomerales*, and placed in the *Zygomycetes*; Morton and Benny 1990) was provided by Gehrig et al. (1996), based on analyses of the small subunit ribosomal RNA (SSU rRNA) genes of *G. pyriformis* and *D. epigaea* BEG47. The phylogenetic trees clearly showed that the *Glomeromycota*, together with *Geosiphon*, form a distinct branch not closely related with any other group of the 'zygomycetes' sequenced so far. Further analyses of SSU rRNA genes eventually led to the erection of the *Glomeromycota*, a widely accepted fungal phylum (Schüßler et al. 2001b) comprising the 'arbuscular mycorrhizal (AM) and related fungi'.

It also became obvious that species placed in *Glomus*, such as *Diversispora epigaea* BEG47 noted above, are separated by large phylogenetic distances and that *Glomus* is non-monophyletic (Schwarzott et al. 2001), which later was confirmed by multi-gene analyses (James et al. 2006). Studies from the last decade thus unequivocally showed that *G. pyriformis* indeed is a member of the *Glomeromycota*. Previous sequence analyses (Sawaki et al. 1999; Schüßler 1999; Schüßler et al. 2001a; Redecker et al. 2000b) showed it to be closely related to an AMF forming two different spore morphs, at the time named *Acaulospora gerdemannii* (synanamorph *Glomus leptotichum*). Nowadays, the clade containing these lineages is defined as the order *Archaeosporales* and represents one of the basal main phylogenetic lineages in the *Glomeromycota* (Schüßler et al. 2001b). In this order, the *Geosiphonaceae* clusters as a sister to the *Ambisporaceae*, thus appearing to be more derived than the *Archaeosporaceae*, which branched earlier. The probable most ancestral order in the AMF is the *Paraglomerales* (Krüger et al. 2012). In fact, studies on the *Geosiphon* symbiosis eventually led to the phylogenetically based, revised classification of the *Glomeromycota* (Schüßler and Walker 2010).

B. Phylogeny of the Cyanobacterium *Nostoc punctiformis*

The endosymbiont (endocytobiont) in the *Geosiphon* symbiosis is ***N. punctiforme*** (Figs. 5.3, 5.4), which belongs to a clade of cyanobacteria containing many symbiosis-forming members. In laboratory cultures (Schüßler and Wolf 2005) a strain was used that originally was isolated

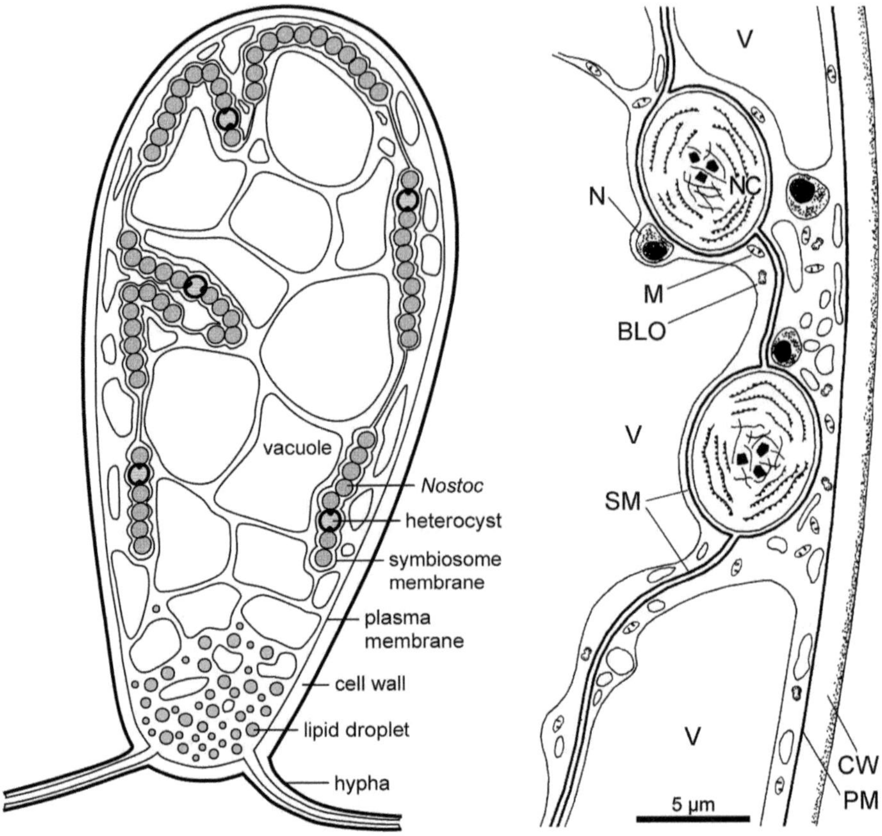

Fig. 5.3. Schematic drawings of the *Geosiphon pyriformis* symbiotic bladder compartmentation. *Left* Overview (only approx. in scale; bladder about 1 mm in length). *Right* Detail (in scale, reproduced from electron microscopy images). *BLO Mollicutes*-related endobacteria; *CW* cell wall; *M* mitochondrium; *N* nucleus; *NC Nostoc* cell; *PM* plasma membrane; *SM* symbiosome membrane; *V* vacuole

from the *Geosiphon* symbiosis (Mollenhauer 1992). However, various other strains of *N. punctiforme* from other symbiotic systems (e.g., *Anthoceros, Blasia, Gunnera*) are also capable of forming symbiosis with *G. pyriformis*. In the field, *G. pyriformis* was usually found together with *Anthoceros*, and the cyanobionts of *G. pyriformis* associate in symbioses with *Anthoceros* and *Blasia* (Mollenhauer 1992).

C. Phylogeny of the *Mollicutes*-Related 'BLO' Endobacteria

Geosiphon pyriformis harbours another prokaryotic endosymbiont, the so-called bacteria-like organisms (BLOs; Figs. 5.4B, 5.5A), which are not enclosed by a host membrane (Schüßler et al. 1996). These endobacteria show the typical ultrastructure of those found in most of the AMF so far investigated for their presence. The BLO-endobacteria occur in diverse branches of the *Glomeromycota* and were considered as Gram-positive widespread symbionts of the AMF (Schüßler et al. 1994), but little else was known about them.

Another type of AM fungal endobacterium that is found in the *Gigasporaceae* has been better investigated. These bacteria are enclosed by a host membrane, Gram-negative and probably restricted to the family *Gigasporaceae*. At least from one gigasporacean species, *Gigaspora margarita*, a culture is known to also host the BLO-type of endobacteria (Kuga et al. 2008; Long et al. 2009). The BLOs are the widespread, ancestral and typical endobacteria in AMF.

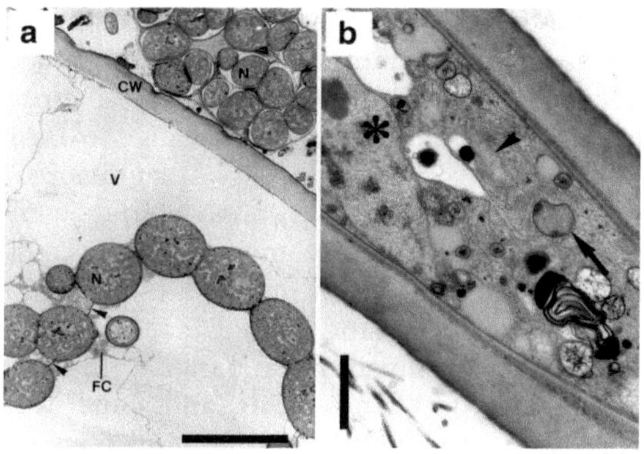

Fig. 5.4. Electron micrographs of *Geosiphon pyriformis* and *Nostoc punctiforme*, glutaraldehyde/OsO_4 fixation. From Schüßler et al. (1994). (**a**) Part of a symbiotic *G. pyriformis* bladder in cross-section. A vegetative colony of free-living *N. punctiforme* (*upper right*) is attached to the outside of the bladder. Endosymbiotic *Nostoc* cells are much larger than the free-living ones. Parts of the symbiosome where no endosymbiotic cyanobacteria are located are marked by *arrowheads*. *CW* Cell wall; *FC* fungal cytoplasm; *N Nostoc* cell; *V* vacuole. *Bar* 10 μm. (**b**) Tangentially sectioned, relatively thick *G. pyriformis* hypha. Nucleus (*asterisk*), BLO endobacterium (*arrow*) and a mitochondrium (*arrowhead*) are marked. Dark deposits, probably polyphosphate precipitates, can be seen within small vacuoles. *Bar* 1 μm

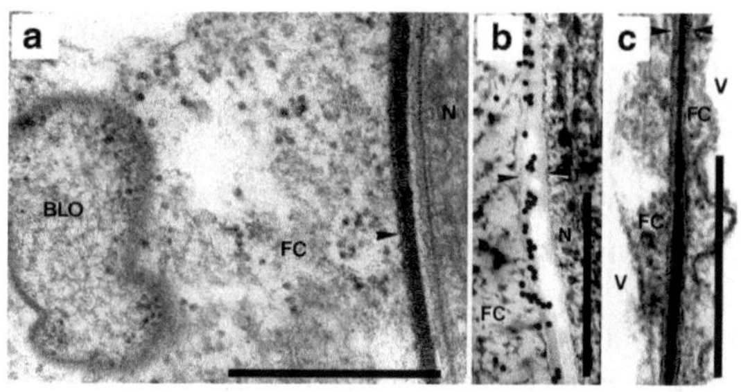

Fig. 5.5. Electron micrographs of the symbiotic interface in the *Geosiphon–Nostoc* symbiosis. *BLO Mollicutes*-related endobacterium; *FC* fungal cytoplasm; *N Nostoc* cell; *V* vacuole. *Bars* 0.5 μm. From Schüßler et al. (1996). (**a**) Conventional fixation (glutaraldehyde/OsO_4), the symbiosome membrane is marked by an *arrowhead*. (**b**) Sample after freeze substitution, without OsO_4. The thin layer attached to the symbiosome membrane is labelled by WGA-gold. The plasma membranes of both symbiosis partners, enclosing the symbiosome space, are marked by *arrowheads*. (**c**) Conventional fixation (glutaraldehyde/OsO_4), part of the symbiosome where no cyanobacteria are located (compare also Fig. 5.4a). The fungal symbiosome membrane, enclosing the osmiophilic layer, is marked by *arrowheads*

New findings regarding their phylogeny and occurrence in diverse AMF lineages (Naumann et al. 2010) showed that the BLOs are related to the cell wall-lacking *Mollicutes*. This was surprising as they possess a cell wall similar to that of the Gram-positive bacteria.

We nowadays know that the BLO endobacteria represent a monophyletic lineage of bacteria that are laterally transferred and probably diversified within the AMF for more than 450 million years (MY). Their phylogeny and biotrophic lifestyle is shared with members of the related genus Mycoplasma, despite the obvious difference of possessing a murein sacculus.

D. Implications for the Interpretation of Arbuscular Mycorrhiza Evolution

Most vascular plant species form AM (Smith and Read 2008), including gametophytes and sporophytes of many ferns (Peterson et al. 1981) and *lycopods* (Schmid and Oberwinkler 1993). Also, except for mosses, all groups of bryophytes contain species with AM associations (Stahl 1949; Ligrone 1988; Ligrone and Lopes 1989), indicating an early origin of the AM.

From fossil cryptospore assemblages sharing characters with those of extant liverworts (found in what was eastern Gondwana; Rubinstein et al. 2010), it is estimated that land plants are more than 470 MY old (Early Middle Ordovician). The diversity of these assemblages implies an earlier, perhaps even Cambrian, origin of embryophytes. Early vascular plants already existed ~420 MY ago (Middle Silurian; Cai et al. 1996). A recent molecular clock study (Smith et al. 2010) suggested an origin of land plants around ~477 MY, but this dating in fact refers to the split between bryophytes and the remaining lineages, not to the (presumably earlier) origin of the land plant lineage itself. Therefore, a minimum age of 420 MY for the liverwort–vascular plant divergence must be assumed and bryophyte-like land plants already were present more than 470 MY ago. The AMF also have an ancient fossil record history of >460 MY. Many of the oldest and best preserved fossils of AMF in association with plants are known from the Rhynie chert, radiometrically dated to the early Devonian (~400 MY ago; e.g., Remy et al. 1994; Dotzler et al. 2009). The oldest known fossils of AMF spores and hyphae are from ~460 MY old Ordovician dolomite rock of Wisconsin (Redecker et al. 2000a) and it was concluded and appears likely that terrestrial AMF already existed at a time when the land flora most likely consisted only of bryophyte-like 'lower' plants, 460 MY ago.

Altogether, these data provide support for the hypothesis of Pirozynsky and Malloch (1975) that symbioses with fungi played a crucial role in the colonisation of the land by plants, with a partnership between two aquatic types of organism, algae and 'phycomycetous' fungi, as the initial step of land plant evolution. A mycotrophic lifestyle could have been essential for an efficient supply of plants with water and nutrients from the soil (Malloch et al. 1980; Simon et al. 1993; Marschner and Dell 1994). However, molecular clock estimates usually date the origin of the AM fungal lineage earlier than that of land plants. If this holds true, it indicates that there were other types of associations formed by AMF before land plants existed, whether saprobically, parasitically, or already mutualistically. *Geosiphon pyriformis*, representing a symbiotic association between a glomeromycotan fungus and a photoautotrophic prokaryote, may reflect such an ancestral partnership and thus, indirectly but substantially, it supports the view of Pirozynski and Malloch (1975). It is plausible to assume that, in the beginning of terrestrial plant life, other associations also existed between glomeromycotan fungi and photoautotrophic organisms (like cyanobacteria). The present knowledge regarding AMF and AM symbiosis evolution was recently discussed and reviewed in Schüßler and Walker (2011).

In summary, glomeromycotan fungi may have adapted to symbiotic life more than 500 MY ago, and the *Geosiphon* symbiosis perhaps reflects such an early stage of land-colonising AMF symbioses. Without fossil support this is speculative, but *G. pyriformis* clearly confirms the ability of glomeromycotan fungi to form symbioses with prokaryotic photoautotrophic organisms. Therefore, cyanobacterial symbioses formed by glomeromycotan fungi could have been an ecologically important step for the colonisation of the land habitat.

Arbuscular mycorrhizal fungi form symbioses with most land plants, and some AMF (e.g., *Claroideoglomus claroideum*) can be symbiotic with such widely divergent photoautotrophs as hornworts and vascular plants (Schüßler 2000), and even cyanobacteria in the case of *Geosiphon*. There must be some very fundamental mechanisms of plant–microorganism interactions present amongst the different AM (-like) associations. When conducting eco-physiological studies involving plants, it is important to consider that in nature the mycorrhizal fungal partners are the main facilitators of nutrient uptake, rather than the plant roots alone. If, as is thought, mechanisms of nutrient acquisition by land plants co-evolved since their origin with the AMF, ecologically and economically important questions might be answered by using the *Geosiphon* symbiosis as a model.

Against this background, the interesting question arises as to whether *G. pyriformis* itself can act as fungal partner to form AM. This question remains unanswered. The molecular probes to screen for the occurrence of *Geosiphon* in the field (soil and plant roots) were developed (unpublished data), but lack of funds prevents their application. If *Geosiphon* indeed forms AM with plants, a complex network of biotic interactions would exist in the natural habitat (Schüßler and Wolf 2005). Within such a network, symbiotic *Nostoc* could be exchanged between *Geosiphon* and bryophytes, and *Geosiphon* could simultaneously form endocytosymbiosis with *Nostoc* and AM with plants, possibly thus delivering N_2 fixed by the cyanobacteria to the plants.

III. Initiation and Development of the Symbiotic Association

A. Development of the *Geosiphon* Symbiosis

Development of the *Geosiphon–Nostoc* association was studied microscopically (Mollenhauer et al. 1996). Initially the cells of the cyanobacterium *N. punctiforme* live freely with the future fungal partner at or close to the soil surface. There, the partners come into contact, but a successful interaction of the fungus with *Nostoc* to form the symbiosis depends on the appropriate developmental stage of the cyanobacterium (see Wolf and Schüßler 2005). The life cycle of *Nostoc* starts from akinetes (spore-like resting stages) leading to vegetative colonies. These colonies release motile trichomes (hormogonia) which are positively phototactic in dim light and negatively in strong light. As a consequence, the hormogonia often congregate just below the soil surface where they spread and meet their symbiotic partners. They eventually undergo a transformation into an aseriate stage called primordium. This stage differentiates into so called vegetative cells, which then form gelatinous colonies ('thalli'). Only the very early primordial stage of *Nostoc* can interact with the future fungal partner to give rise to the symbiotic consortium.

The life cycle of the fungal partner of the association starts from resting spores formed in the upper soil layer. These contain several different storage compounds (Schüßler et al. 1994). The spores germinate by the outgrowth of a hypha (sometimes more than one), which branches to form a small mycelium of a few centimetres (max. 2–3 cm) inside the soil. As they grow, they come into contact with the *Nostoc* primordia, which they may incorporate if in a compatible developmental stage. When contact is made with such a compatible early *Nostoc* primordium, a portion of fungal cytoplasm bulges out just below the apex of the hypha. This bulging process is repeated several times so that finally the hyphal tip forms an irregularly shaped mantle surrounding a part of a *Nostoc* primordium, usually about 5–15 cells, incorporating it into the fungal structure, after which the fungal bladder develops from this *Nostoc*-containing structure.

Each single incorporation event results in the formation of a pear-shaped above-ground bladder of up to 2 mm length (Knapp 1933). Each bladder represents a polyenergid cell, coenocytic with the fungal mycelium, in which the symbiotic *Nostoc* cells divide and become physiologically active. Laboratory cultivation experiments have shown that, as for AM, phosphate limitation (1–2 μM) in the nutrient solution triggers the stable establishment of the symbiosis. N-limitation seems not to be a crucial factor. The same situation is found in the natural habitat, so P-limitation seems to be a driving factor for establishing this symbiosis.

Within the first hours after incorporation into the fungal cytoplasm, the *Nostoc* filaments become deformed and some cells may die during this process. The photosynthetic pigments of the cells bleach considerably (Mollenhauer et al. 1996; Schüßler and Wolf 2005). These alterations and significant changes in ultrastructure (unpublished data) suggest that, during the initial state of endocytotic life, the incorporated cyanobacteria suffer severe stress. Within 2–3 days, the enclosed *Nostoc* cells recover and begin to multiply and grow to become as much as six times the volume of free-living cells (Schüßler et al. 1996). In the standard growth medium used to culture the *Geosiphon* symbiosis, under phosphate limitation, the endosymbiotic cyanobacteria divide much faster and form a much higher biomass compared to the free-living ones (unpublished data). In the symbiosis, the *Nostoc* cells arrange in filaments in which heterocysts are formed with the same frequency as in the filaments outside the bladders (when cultured under nitrogen limitation). Individual, mature *Geosiphon* bladders can then reach more than 2 mm in length, possess a high turgor pressure and may reach ages of up to 6 months in laboratory cultures. Details of this development are also documented in a scientific film available in English and German (Mollenhauer and Mollenhauer 1997).

B. Specificity of Partner Recognition

The initial reaction between the partners leading to the symbiosis establishment is specific. Cells of particular strains of *N. punctiforme* can be incorporated by *Geosiphon*, resulting in the formation of functional symbioses. For other strains, although cells are incorporated, the formation of symbiotic bladders is blocked in an early stage of development. Yet other *N. punctiforme* strains are not incorporated at all by the fungus. Further evidence for a specific recognition process is the fact that, among the various developmental stages of *Nostoc*, only the early primordia are incorporated. Not only is the physiological activity of the primordia different from the other stages of the *Nostoc* life cycle (Bilger et al. 1994) but also the composition of the gelatinous envelope. When differentiating into primordia, a slime, containing mannose, is produced by the cells, whereas other sugars within the extracellular glycoconjugates could be detected only in earlier or later stages of the life cycle. It is important to note that heterocysts (specialised N_2-fixing cells), differentiating at regular spacing along the filaments of the *Nostoc* primordia when growing under nitrogen limitation, are never enclosed by the fungal cytoplasm. They always remain outside the fungal hypha during the incorporation process (Mollenhauer et al. 1996) and are not surrounded by a newly appearing mannose-containing glycoconjugate (Schüßler et al. 1997), indicating a specific recognition of the early primordial surface by the fungus.

It is not known what triggers the recognition process and the morphological changes during the symbiosis establishment. Microscopical studies give no hints for any chemotactic or otherwise directed growth towards the respective symbiosis partner, but the symbiosis-compatible *Nostoc* stage can be synchronised (Wolf and Schüßler 2005) and exists only for approx. 3–12 h during the life cycle. Some unpublished data indicate that the specific partner recognition may be based on the glycoconjugate composition of the cyanobacterial envelope, perhaps by a lectin-mediated process.

IV. Structure and Compartmentation of the *Geosiphon* Symbiosis

A. *Geosiphon* Bladders and Spores

The *Geosiphon* bladder is effectively a multikaryotic cell, coenocytic with the fungal mycelium in the soil. It shows a strong polarity and has a photosynthetically active region in the apical part exposed to light and air, and a smaller storage region in the basal part embedded in the soil surface, containing many lipid droplets that tend to give it a milky white appearance. The centre of the bladder is highly vacuolated. Schematic drawings of the compartmentation of *Geosiphon* are shown in Fig. 5.3. Ultrastructural observations (Figs. 5.4, 5.5) show the *G. pyriformis* symbiosis as an endosymbiotic system with very close contact between the partners. In fact, it is composed of three organisms: (1) the fungus, supplying the consortium with inorganic nutrients like phosphate, trace elements and water, (2) the cyanobacteria, supplying the consortium with carbohydrates by photosynthesis and, at least under some conditions, nitrogen compounds by N_2 fixation, (3) the 'bacteria-like organisms' (BLOs), which are *Mollicutes*-related endobacteria, with yet unknown function.

Within the bladders, the cyanobacteria are located peripherally in a single compartment, the symbiosome, which is cup-shaped (often with invaginations, Fig. 5.3). The *Nostoc* cells divide and are physiologically active as endosymbionts in this compartment. They are much larger than free-living vegetative cells (Fig. 5.4A), reaching an approximately sixfold increased volume, probably caused by the high osmotic pressure inside the bladders. The iso-osmolar concentration of sorbitol was measured with oil-filled microcapillaries and determined to be 220–230 mM, corresponding to a turgor pressure (P) of about 0.6 MPa (= 6 bar; Schüßler et al. 1995). In many symbioses with plants cyanobacteria are known to increase in size (Grilli Caiola 1992; Johansson and Bergman 1992), probably as a reaction to the higher osmotic pressure of the surrounding medium. High NaCl concentrations are also known to cause an increase in the volume of cyanobacteria (Erdmann and Schiewer 1984). Despite the increase in size, the *Nostoc* cells inside the *Geosiphon* bladder have an almost normal ultrastructural appearance (Fig. 5.4A). They contain a high number of thylakoids and carboxysomes; one alteration is that the outer membrane is hardly recognizable electron microscopically. Heterocysts are formed with the same frequency as in free-living colonies, but their cell wall is thinner in the symbiosis, possibly indicating a lower O_2 concentration in the surrounding environment. The lack of increased heterocyst frequency indicates that in the *Geosiphon* symbiosis the major role of the cyanobacteria is photosynthesis. In contrast, in symbioses with plants, e.g., bryophytes or in the endosymbiosis with *Gunnera*, the main role usually is N_2 fixation, reflected by a great increase in the relative number of heterocysts.

Within the cytoplasm of the fungus, glycogen granules exist as storage compounds. Many lipid droplets are found in the basal part of the bladder. No dictyosomes could be found; microtubules can rarely be observed. Fixation of the bladders during preparation for electron microscopy was often inadequate, probably due to the low cell wall permeability, but could be improved by using microwave acceleration.

The ultrastructure of the asexually formed *G. pyriformis* spores was described by Schüßler et al. (1994). Preparation of the spores for electron microscopy is even more difficult than for the bladders. This problem, caused by the thick spore wall being only slowly permeable to fixatives, also exists with other glomeromycotan species (Maia et al. 1993). Two main storage compounds occur inside the spores: lipid droplets of different size, and 'structured granules' that occupy about 25 % of the spore volume. The latter are thought to be storage compounds and are discussed below with respect to element analysis. They show paracrystalline inclusions, as also found in spores of some AMF. Small vacuoles are found in germinating spores and hyphae, which often contain dark deposits (Fig. 5.4B). These are similar to the deposits in AMF and probably are polyphosphate precipitate granules. In general, the ultrastructural appearance of *G. pyriformis* is similar to that of AMF.

B. The Symbiotic Interface Between the Partners

The ultrastructure of the *Geosiphon* symbiosis was first studied by Schnepf (1964) and was the crucial investigation leading to the theory of the compartmentation of the eukaryotic cell. Later ultrastructural studies (Schüßler et al. 1996) showed that inside the fungal bladder the symbiotic *Nostoc* cells are located in a single peripheral, irregularly cup-shaped compartment, the symbiosome. The space between the symbiosome membrane and the wall of the enclosed *Nostoc* cells is only 30–40 nm thick and contains a layer of electron microscopically opaque and amorphous-appearing material (Fig. 5.5) which was originally assumed to be slime produced by the endosymbiont (Schnepf 1964). Confocal laser scanning microscopical (CLSM) studies by means of affinity techniques with six fluorescence-labelled lectins with different carbohydrate-specificity revealed that this amorphous layer inside the symbiosome contains chitin. Labelling with WGA-gold conjugates (Fig. 5.5B) confirmed these results (Schüßler et al. 1996). Thus, the electron opaque layer within the symbiosome contains chitin and represents a 'rudimentary' fungal cell wall, showing that the symbiosome membrane surrounding the *Nostoc* cells is homologous to a fungal plasma membrane.

Clear similarities exist between the fungal cell wall material present in the symbiosome space of the *Geosiphon* symbiosis and the thin arbuscular cell wall bordering the symbiotic AM fungus from the colonised plant cell in the AM: both are electron-dense after OsO_4 fixation, about 30–40 nm thick and show the same amorphous structure and appearance. Considering also the phylogenetic position of *G. pyriformis* and the known or proposed nutrient flows between the symbiotic partners, it was suggested that the symbiotic interface in the AM and the *Geosiphon* symbiosis are homologous (Schüßler et al. 1996). The main difference between the symbioses is the relation of macro- and microbiont. In the *Geosiphon* symbiosis the photoautotrophic partner (cyanobacterium) is the microsymbiont, whereas in the AM it is the macrosymbiont (plant; Fig. 5.6).

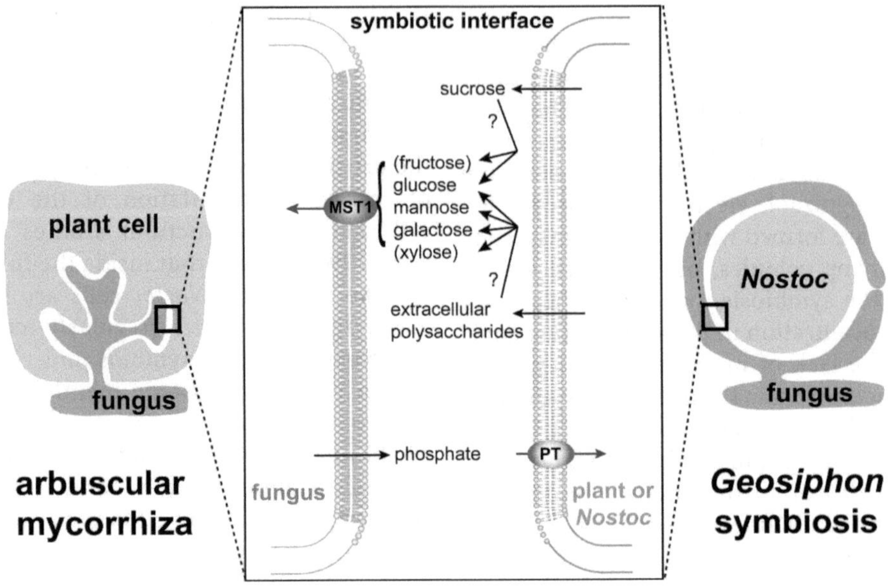

Fig. 5.6. Interfaces of the arbuscular mycorrhiza and *Geosiphon* symbiosis and indicated role of the GpMST1 type of sugar transporters at the symbiotic interface. From Schüßler et al. (2006)

V. Metabolism and Nutrient Acquisition

A. Photosynthetic Carbon Acquisition and Atmospheric Nitrogen Fixation

The *Geosiphon* symbiosis fixes carbon photosynthetically, as shown by ^{14}C-tracer experiments (Kluge et al. 1991). In light, ^{14}C is trapped mainly by phosphate esters, polyglucanes, free sugars (among them trehalose and raffinose), some amino acids and organic acids. The bladders also perform dark CO_2 fixation, at rates much lower than in light. As expected, the labelled products of dark CO_2 fixation were organic acids (malic and fumaric acid) and some amino acids. High photosynthetic activity of the endosymbiotic *Nostoc* cells was also shown by photosystem II chlorophyll–fluorescence kinetics (Bilger et al. 1994), in which the symbiotic *Nostoc* cells achieved much higher steady-state quantum yields and electron transport rates than free-living *Nostoc*. The reason for this different behaviour is unknown, but possibly CO_2 concentration, and thus the availability of the major photosynthetic substrate for the photobionts, is higher inside than outside the bladder.

The capability of N_2 fixation is indicated by the occurrence of heterocysts, and nitrogenase activity was shown for the bladders (Kluge et al. 1992). Nevertheless, the major role of the endosymbiotic *Nostoc* seems to be photosynthesis. However, matter exchange between the partners is still poorly investigated, and it is even possible that the second bacterial endosymbiont (BLO), which is typical for most glomeromycotan fungi, contributes to N_2 fixation.

B. Uptake of Nutrients from the Outside

For the endosymbiotic *Nostoc* cyanobacteria all inorganic nutrients except N have to be provided by the fungus, as the cyanobacteria live intracellularly.

As shown by electrophysiological experiments (unpublished data), inorganic ions (nitrate, chloride) and small organic molecules (e.g., glycin, cystein) lead to rapid, transient depolarisation of the plasmamembrane potential of the *G. pyriformis* bladders, indicating that these substances are actively taken up from the outside.

In contrast, there were no changes in membrane potential if hexoses (e.g., glucose) and larger amino acids were applied. Metabolism of radioactively labelled hexoses by the bladders also could not be detected after the usual incubation times. Low cell wall permeability was discussed as the likely reason for the lack of uptake of monosaccharides. This theory is supported by observations showing that the presence of solutes with large molecule radii leads to irreversible cytorrhysis, i.e., collapse of the whole bladder including the cell wall, whereas in presence of small solutes plasmolysis occurred (or cytorrhysis was quickly reversed). This different behaviour must be caused by a selective permeability of the bladder wall.

By systematically using solutes with known molecular radii for cytorrhysis experiments, it was shown that the limiting pore radius of the *G. pyriformis* bladder wall is approx. 0.5 nm, which, compared with other cell wall types, is very small (Schüßler et al. 1995). Such a pore size is too small for an efficient permeation by, e.g., hexose molecules from the outside, but it allows permeation of inorganic hydrated ions like phosphate. Provided that such a small pore size holds true also for the hyphal cell wall, the fungus would not be capable of saprobic acquisition of organic molecules such as glucose, sucrose, larger amino acids, etc. However, cell wall permeability is a complex topic and, e.g., the thin hyphae formed by AMF, known as 'branching absorbing structures', might possess different cell wall permeability.

Because AMF obtain up to 20 % of the plant-fixed CO_2, putatively as monosaccharides, the study of a glomeromycotan sugar transporter that could play a role in the C-transfer from plants to AMF was an important goal. Until recently only one such glomeromycotan monosaccharide transporter was characterised, and this was from the *Geosiphon* symbiosis GpMST1 (Schüßler et al. 2006, 2007). This putatively symbiosome-membrane located transporter was demonstrated also to transport sugars potentially deriving from plant cell-wall material (Fig. 5.6). Only recently, the second MST was characterised from the AMF *Rhizophagus irregularis* (Helber et al. 2011).

An interesting ecological aspect is that preliminary studies show that the intracellular cyanobacteria are protected against heavy metals, which accumulate in the fungus (Scheloske et al. 2001). Therefore, as also known for the AM, the photobiont seems to be protected against abiotic stress factors in the *Geosiphon* symbiosis.

C. Element Composition and Distribution Within the *Geosiphon* Symbiosis

It is not yet known why AMF cannot be cultured axenically and there is little information available about their trace element requirements and general element composition. For this and other reasons, considering the fact that these fungi supply the majority of land plants with inorganic elements, studies on the element composition and transport processes are interesting topics. We have used proton-induced X-ray emission (PIXE) measurements on *G. pyriformis* to obtain first indications on the macro- and microelement composition of the spores and symbiotic bladders. The element content of some subcellular compartments could be quantitatively measured and, by a differential approach, calculated. PIXE, combined with scanning transmission ion microscopy (STIM) allowed elemental concentrations to be quantified with a lateral resolution in the 1 μm range and with high accuracy and precision (Maetz et al. 1999a).

Results of the PIXE studies on the *G. pyriformis* bladders (Maetz et al. 1999b) showed that the fungal partner of the symbiosis, grown on a poor nutrient solution (e.g., containing 1 μM phosphate), accumulates high amounts of Cl (about 2.5 %) and K (about 8 %), which both seem to play a major role in osmoregulation of the fungus (all values given here are related to dry weight, ppm = μ g/g DW). The symbiosome (including the cyanobacteria) contains only small amounts of these elements. This is in line with presumed high concentration of monovalent ions in the fungal vacuoles. P also is accumulated by the fungus in high concentrations (about 2 %), but not in the symbiosome, and thus is probably stored as polyphosphate in the vacuoles, as for AM (and many other) fungi. The macroelements Mg, S and Ca and the microelements Fe, Mn, Cu and Zn occur in concentrations comparable to those found in plants. Se concentration is below 1 ppm. Mo is present within the symbiosome in very low amounts, compared to the rest of the bladder, although Mo is a constituent of nitrogenase, required for N_2-fixation of the cyanobacteria. Reasons for this might be that other Mo-enzymes (e.g., nitrate reductase, sulfite oxidase) occur in relevant amounts in the fungal cytoplasm or that Mo is located in the fungal vacuole. Mn and Ni, on the contrary, are present in the symbiosome in much higher amounts than in the rest of the bladder. Much of the Mn (approx. 50 ppm, which is comparable to values found in plant leaves) probably is contained in the water-cleaving Mn protein of photosystem II. Some may be from other enzymes, e.g., Mn-superoxide dismutase (SOD). A likely candidates for enzymes containing the approx. 50 ppm Ni is cyanobacterial (or secreted fungal) urease; other Ni-containing bacterial enzymes are Ni-SOD and NiFe-hydrogenases.

Unpublished results on the element composition of the *Geosiphon* spores show that the structured granules (SGs), which are 4–6 μm in diameter, located each within a vesicle, together occupy about 25 % of the spore volume and contain most of the total P, K and S. The S concentration of the spore cell wall is ~0.25 %, probably because of a high protein content, similar to that of an AMF species (Bonfante and Grippiolo 1984). Compared with the bladders, Cl and K are concentrated within the spores in much lower amounts.

VI. Conclusions

The *Geosiphon* symbiosis is the only known fungal endocyanosis, and the fungal partner doubtless belongs in the *Glomeromycota.* Because of its size and structure the *Geosiphon* symbiosis has considerable advantages for the study of particular physiological features of AMF, e.g., by electrophysiological measurements or microscopic observations. Moreover, the bladders can easily be microinjected. A further experimental advantage is that only one eukaryote is present in the symbiotic stage. This means that this organism is well suited for molecular biological studies dealing with the expression of symbiotically expressed fungal genes. Its use would greatly facilitate, e.g., the sequencing of the fungal, symbiotically expressed transcriptome. In contrast to the AM symbiosis, where in the symbiotic stage the eukaryotic plant is always present, from the *Geosiphon* symbiosis the fungal mRNA can be easily and specifically isolated, because the prokaryotic mRNA of *Nostoc* lacks stable poly (A) tails. cDNA libraries of the symbiosis have been produced, including those from subtractive hybridisation attempts, and differentially expressed genes could be identified. It should be possible to identify, e.g., specific fungal genes related to the acquisition and metabolism of inorganic nutrients such as P, N and S.

Altogether, at the molecular level, the *Geosiphon* symbiosis appears to be a very promising model symbiosis and could lead to major advances in AM research, but we have already lost many years of possible progress, and unfortunately the laboratory cultures of the symbiosis are no longer available. Currently, research on the symbiosis is limited to organism-independent (e.g., based on cDNA libraries) and cross-financed minor topics, as research projects received no funding. A well funded research project with this symbiosis would likely produce a wealth of evidence to reveal fundamental information about the mechanisms and processes involved in the arbuscular mycorrhizal symbiosis.

References

Bilger W, Büdel B, Mollenhauer R, Mollenhauer D (1994) Photosynthetic activity of two developmental stages of a *Nostoc* strain (cyanobacteria) isolated from *Geosiphon pyriforme* (Mycota). J Phycol 30:225–230

Bonfante P, Grippiolo R (1984) Cytochemical and biochemical observations on the cell wall of the spore of *Glomus epigaeum*. Protoplasma 123:140–151

Cai CY, Ouyang S, Wang Y, Fang ZJ, Rong JY, Geng LY, Li XX (1996) An early Silurian vascular plant. Nature 379:592

Dotzler N, Krings M, Hass H, Walker C, Taylor TN, Agerer R (2009) Acaulosporoid glomeromycotan spores with germination shields from 400-million-year-old Rhynie chert. Mycol Prog 8:9–18

Erdmann N, Schiewer U (1984) Cell size changes as indicator of salt resistance of blue green algae. Arch Hydrobiol Suppl (Algol Stud) 67: 431–439

Gehrig H, Schüßler A, Kluge M (1996) *Geosiphon pyriforme*, a fungus forming endocytobiosis with *Nostoc* (cyanobacteria), is an ancestral member of the *Glomales*: evidence by SSU rRNA analysis. J Mol Evol 43:71–81

Grilli Caiola M (1992) Cyanobacteria in symbiosis with bryophytes and tracheophytes. In: Reisser W (ed) Algae and symbiosis: plants, animals, fungi, viruses, interactions explored. Biopress, Bristol, pp 231–253

Hawksworth DL, Honegger R (1994) The lichen thallus: a symbiotic phenotype of nutritionally specialized fungi and its response to gall producers. In: Williams MAJ (ed) Plant galls. Organisms, interactions, populations, Systematics association, special vol 49. Clarendon, Oxford, pp 77–98

Helber N, Wippel K, Sauer N, Schaarschmidt S, Hause B, Requena N (2011) A Versatile monosaccharide transporter that operates in the arbuscular mycorrhizal fungus *Glomus* sp is crucial for the symbiotic relationship with plants. Plant Cell 23:3812–3823

James TY, Kauff F, Schoch C, Matheny B, Hofstetter V, Cox C, Celio G, Guiedan C, Fraker E, Miadlikowska J, Lumbsch HT, Rauhut A, Reeb V, Arnold AE, Amtoft A, Stajich JE, Hosaka K, Sung G-H, Johnson D, O'Rourke B, Crockett M, Binder M, Curtis JM, Slot JC, Wang Z, Wilson AW, Schüßler A, Longcore JE, O'Donnell K, Mozley-Standridge S, Porter D, Letcher PM, Powell MJ, Taylor JW, White MM, Griffith GW, Davies DR, Humber RA, Morton JB, Sugiyama J, Rossman AY, Rogers JD, Pfister DH, Hewitt D, Hansen K, Hambleton S, Shoemaker RA, Kohlmeyer J, Volkmann-Kohlmeyer B, Spotts RA, Serdani M, Crous PW, Hughes KW, Matsuura K, Langer E, Langer G, Untereiner WA, Lücking R, Büdel B, Geiser DM, Aptroot A, Diederich P, Schmitt I, Schultz M, Yahr R, Hibbett DS, Lutzoni F, McLaughlin DJ, Spatafora JW, Vilgalys R (2006) Reconstructing the early evolution of the *Fungi* using a six-gene phylogeny. Nature 443:818–822

Johansson C, Bergman B (1992) Early events during the establishment of the *Gunnera*/*Nostoc* symbiosis. Planta 188:403–413

Kluge M, Mollenhauer D, Mollenhauer R (1991) Photosynthetic carbon assimilation in *Geosiphon pyriforme* (Kützing) F.v. Wettstein, an endosymbiotic association of fungus and cyanobacterium. Planta 185:311–315

Kluge M, Mollenhauer D, Mollenhauer R, Kape R (1992) *Geosiphon pyriforme*, an endosymbiotic consortium of a fungus and a cyanobacterium (*Nostoc*), fixes nitrogen. Bot Acta 105:343–344

Knapp E (1933) Über *Geosiphon pyriforme* Fr.v. Wettst., eine intrazelluläre Pilz-Algen-Symbiose. Ber Dtsch Bot Ges 51:210–217

Krüger M, Krüger C, Walker C, Stockinger H, Schüßler A (2012) Phylogenetic reference data for systematics and phylotaxonomy of arbuscular mycorrhizal fungi from phylum to species-level. New Phytol 193:970–984

Kuga Y, Saito K, Nayuki K, Peterson RL, Saito M (2008) Ultrastructure of rapidly frozen and freeze-substituted germ tubes of an arbuscular mycorrhizal fungus and localization of polyphosphate. New Phytol 178:189–200

Ligrone R (1988) Ultrastructure of a fungal endophyte in *Phaeoceros laevis* (L.) Prosk. (*Anthocerophyta*). Bot Gaz 149:92–100

Ligrone R, Lopes C (1989) Cytology and development of a mycorrhiza-like infection in the gametophyte of *Conocephalum conicum* (L.) Dum. (*Marchantiales*, *Hepatophyta*). New Phytol 111:423–433

Long L, Yao Q, Ai Y, Deng M, Zhu H (2009) Detection of a novel bacterium associated with spores of the arbuscular mycorrhizal fungus *Gigaspora margarita*. Can J Microbiol 55:771–775

Maetz M, Przybylowicz WJ, Mesjasz-Przybylowicz J, Schüßler A, Traxel K (1999a) Low dose nuclear microscopy as a necessity for accurate quantitative microanalysis of biological samples. Nucl Instrum Method B 158:292–298

Maetz M, Schüßler A, Wallianos A, Traxel K (1999b) Subcellular trace element distribution in *Geosiphon pyriforme*. Nucl Instrum Method B 150:200–207

Maia LC, Kimbrough JW, Erdos G (1993) Problems with fixation and embedding of arbuscular mycorrhizal fungi (*Glomales*). Mycologia 85:323–330

Malloch DW, Pirozynski KA, Raven PH (1980) Ecological and evolutionary significance of mycorrhizal symbiosis in vascular plants (a review). Proc Natl Acad Sci USA 77:2113–2118

Marschner H, Dell B (1994) Nutrient uptake in mycorrhizal symbiosis. Plant Soil 159:89–102

Mollenhauer D (1992) *Geosiphon pyriforme*. In: Reisser W (ed) Algae and symbiosis: plants, animals, fungi, viruses, interactions explored. Biopress, Bristol, pp 339–351

Mollenhauer D, Mollenhauer R (1997) Endosymbiosis between *Nostoc* and *Geosiphon pyriforme*. Institut für den Wissenschaftlichen Film, Göttingen, Film C1955

Mollenhauer D, Mollenhauer R, Kluge M (1996) Studies on initiation and development of the partner association in *Geosiphon pyriforme* (Kütz.) v.

Wettstein, a unique endocytobiotic system of a fungus (*Glomales*) and the cyanobacterium *Nostoc punctiforme* (kütz.) Hariot. Protoplasma 193:3–9

Morton JB, Benny GL (1990) Revised classification of arbuscular mycorrhizal fungi (Zygomycetes): a new order, *Glomales*, two new suborders, *Glomineae* and *Gigasporineae*, and two new families, *Acaulosporaceae* and *Gigasporaceae*, with an emendation of *Glomaceae*. Mycotaxon 37:471–491

Naumann M, Schüßler A, Bonfante P (2010) The obligate endobacteria of arbuscular mycorrhizal fungi are ancient heritable components related to the *Mollicutes*. ISME J 4:862–871

Peterson RL, Howarth MJ, Whittier DP (1981) Interactions between a fungal endophyte and gametophyte cells in *Psilotum nudum*. Can J Bot 59:711–720

Pirozynski KA, Malloch DW (1975) The origin of land plants: a matter of mycotrophism. Biosystems 6:153–164

Read DJ, Duckett JG, Francis R, Ligrone R, Russell A (2000) Symbiotic fungal associations in lower land plants. Philos Trans R Soc B 355:815–831

Redecker D, Kodner R, Graham LE (2000a) Glomalean fungi from the Ordovician. Science 289:1920–1921

Redecker D, Morton JB, Bruns TD (2000b) Ancestral lineages of arbuscular mycorrhizal fungi (Glomales). Mol Phylogenet Evol 14:276–284

Remy W, Taylor TN, Hass H, Kerp H (1994) Four hundred-million-year-old vesicular arbuscular mycorrhizae. Proc Natl Acad Sci USA 91:11841–11843

Rubinstein CV, Gerrienne P, De La Puente GS, Astini RA, Steemans P (2010) Early middle Ordovician evidence for land plants in Argentina (Eastern Gondwana). New Phytol 188:365–369

Sawaki H, Sugawara K, Saito M (1999) Phylogenetic position of an arbuscular mycorrhizal fungus, *Acaulospora gerdemannii*, and its synanamorph *Glomus leptotichum*, based upon 18S rRNA gene sequence. Mycoscience 39:477–480

Scheloske S, Maetz M, Schüßler A (2001) Heavy metal uptake of *Geosiphon pyriforme*. Nucl Instrum Method B 181:659–663

Schmid E, Oberwinkler F (1993) Mycorrhiza-like interactions between the achlorophyllous gametophyte of *Lycopodium clavatum* L. and its fungal endophyte studied by light and electron microscopy. New Phytol 124:69–81

Schnepf E (1964) Zur Feinstruktur von *Geosiphon pyriforme*. Arch Mikrobiol 49:112–131

Schüßler A (1999) Glomales SSU rRNA gene diversity. New Phytol 144:205–207

Schüßler A (2000) *Glomus claroideum* forms an arbuscular mycorrhiza-like symbiosis with the hornwort *Anthoceros punctatus*. Mycorrhiza 10:15–20

Schüßler A (2002) Molecular phylogeny, taxonomy, and evolution of *Geosiphon pyriformis* and arbuscular mycorrhizal fungi. Plant Soil 244:75–83

Schüßler A, Walker C (2010) The *Glomeromycota*: a species list with new families and new genera. Arthur Schüßler, Christopher Walker, Gloucester. Published in libraries at the Royal Botanic Garden Edinburgh, The Royal Botanic Garden Kew, Botanische Staatssammlung Munich, and Oregon State University; freely available online at www.amf-phylogeny.com

Schüßler A, Walker C (2011) Evolution of the 'plant-symbiotic' fungal phylum, *Glomeromycota*. In: Pöggeler S, Wöstemeyer J (eds) The Mycota XIV. Evolution of fungi and fungal-like organisms. Springer, Berlin, pp 163–185

Schüßler A, Wolf E (2005) *Geosiphon pyriformis*—a glomeromycotan soil fungus forming endosymbiosis with cyanobacteria. In: Declerck S, Strullu D-G, Fortin JA (eds) In vitro culture of mycorrhizas, vol 4, Soil Biology. Springer, Berlin, pp 271–289

Schüßler A, Mollenhauer D, Schnepf E, Kluge M (1994) *Geosiphon pyriforme*, an endosymbiotic association of fungus and cyanobacteria: the spore structure resembles that of arbuscular mycorrhizal (AM) fungi. Bot Acta 107:36–45

Schüßler A, Schnepf E, Mollenhauer D, Kluge M (1995) The fungal bladders of the endocyanosis *Geosiphon pyriforme*, a *Glomus*-related fungus: cell wall permeability indicates a limiting pore radius of only 0.5 nm. Protoplasma 185:131–139

Schüßler A, Bonfante P, Schnepf E, Mollenhauer D, Kluge M (1996) Characterization of the *Geosiphon pyriforme* symbiosome by affinity techniques: confocal laser scanning microscopy (CLSM) and electron microscopy. Protoplasma 190:53–67

Schüßler A, Meyer T, Gehrig H, Kluge M (1997) Variations of lectin binding sites in extracellular glycoconjugates during the life cycle of *Nostoc punctiforme*, a potentially endosymbiotic cyanobacterium. Eur J Phycol 32:233–239

Schüßler A, Gehrig H, Schwarzott D, Walker C (2001a) Analysis of partial *Glomales* SSU rRNA genes: implications for primer design and phylogeny. Mycol Res 105:5–15

Schüßler A, Schwarzott D, Walker C (2001b) A new fungal phylum, the *Glomeromycota*: phylogeny and evolution. Mycol Res 105:1413–1421

Schüßler A, Martin H, Cohen D, Fitz M, Wipf D (2006) Characterization of a carbohydrate transporter from symbiotic glomeromycotan fungi. Nature 444:933–936

Schüßler A, Martin H, Cohen D, Fitz M, Wipf D (2007) Arbuscular mycorrhiza – studies on the *Geosiphon* symbiosis lead to the characterization of the first glomeromycotan sugar transporter. Plant Signal Behav 2:431–434

Schüßler A, Martin H, Cohen D, Wipf D (2008) The *Geosiphon–Nostoc* symbiosis as a tool to characterize symbiotic nutrient transporters in the arbuscular mycorrhiza symbiosis. In: Lorito M, Woo S,

Scala F (eds) Biology of molecular plant–microbe interactions, vol 6. International Society for Molecular Plant–Microbe Interactions, St. Paul, pp 1–6, paper 24

Schüßler A, Krüger M, Walker C (2011) Revealing natural relationships among arbuscular mycorrhizal fungi: culture line BEG47 represents *Diversispora epigaea*, not *Glomus versiforme*. PLoS One 6: e23333

Schwarzott D, Walker C, Schüßler A (2001) *Glomus*, the largest genus of the arbuscular mycorrhizal fungi (Glomales), is non-monophyletic. Mol Phylogenet Evol 21:190–197

Simon L, Bousquet J, Lévesque RC, Lalonde M (1993) Origin and diversification of endomycorrhizal fungi and coincidence with vascular land plants. Nature 363:67–69

Smith SE, Read DJ (2008) Mycorrhizal symbiosis, 3rd edn. Academic, New York

Smith SA, Beaulieu JM, Donoghue MJ (2010) An uncorrelated relaxed-clock analysis suggests an earlier origin for flowering plants. Proc Nat Acad Sci USA 107:5897–5902

Stahl M (1949) Die Mycorrhiza der Lebermoose mit besonderer Berücksichtigung der thallösen Formen. Planta 37:103–148

von Wettstein F (1915) *Geosiphon* Fr. v. Wettst., eine neue, interessante siphonee. Österr Bot Z 65:145–156

Wolf E, Schüßler A (2005) Phycobiliprotein fluorescence of *Nostoc punctiforme* changes during the life cycle and chromatic adaptation: characterization by spectral confocal laser scanning microscopy and spectral unmixing. Plant Cell Environ 28:480–491

6 De-Constructing a Mutualist: How the Molecular Blueprints of Model Symbiotic Fungi Are Changing Our Understanding of Mutualism

J.M. Plett[1], A. Kohler[1], F. Martin[1]

CONTENTS

I. No Man is an Island: Plant Life Depends on Mutualistic Relationships

Since their first emergence on land, plant life has had to overcome harsh and unforgiving environmental factors to survive. Unlike animal and bacterial life, plants did not have the luxury of movement to avoid stressful conditions. This applied a variety of selective pressures on plants to develop both endogenously encoded coping mechanisms as well as mutualistically favorable relationships with other organisms. One of the most fundamental factors that plants must overcome in order to survive is the ability to acquire the limited nutrients and water from the soil. In temperate forests, approximately 90 % of available nutrients are locked into organic polymers (Wild 1988), while in arctic environments this can rise higher than 99 % (Kielland 1990). The majority of plants favor the use of inorganic nutrient supplies (Cole 1981; Mengel and Kirby 1987; Brady 1990). As the pools of inorganic nutrients are not sufficient to support the plant life found in these environments, another mechanism must exist to supply the nutrient needs in these ecosystems. In general, it is hypothesized that most plants rely on mycorrhizal fungal symbiosis (Tibbett et al. 1998). Although this group of fungi are typically poor in the enzymes necessary to utilize humus-bound nutrients (Lundeberg 1970; Tibbett et al. 1999; Buée et al. 2005; Courty et al. 2006, 2007; Martin et al. 2008, 2010; Eastwood et al. 2011), they do encode genes for excreted proteins involved in habitat engineering and in nutrient acquisition necessary to meet the needs of the plant throughout the different seasons. Through such associations plants cover a vast portion of the earth from the equator to 72°N in the central Siberian plateau and from below sea level to 5,200 m in altitude in the Andes mountains. The ecological importance of this interaction and the complexity of unraveling the interplay between the two partners in a mutualistic symbiosis make the study of mycorrhizal fungi and their host plants intellectually, environmentally and economically interesting.

[1]'Ecogenomics of Interactions' Laboratory, UMR "Interactions Arbres/Micro-Organismes", INRA-Nancy, 54280 Champenoux, France; e-mail: fmartin@nancy.inra.fr

Fungal Associations, 2nd Edition
The Mycota IX
B. Hock (Ed.)

One class of mycorrhizal fungi, typically found in forest environments, are the ectomycorrhizal (ECM) fungi. The ECM fungi are not a phylogenetically distinct group, but an assemblage of very different fungal species, mainly Basidiomycota, that have independently developed a mutualistic symbiotic lifestyle. Unlike the much older arbuscular (AM) mycorrhizal fungi that are found ubiquitously in most terrestrial environments, ECM fungi are not obligate biotrophs and can survive independently on soil substrates as long as there is a reservoir of simple sugars present. Without a host, however, these fungi are severely limited in their speed of growth and reproduction as they are not efficient carbon scavengers. Symbiosis is established between an ECM fungi and a favorable host when running, branching hyphae of ECM fungi encounter host root tips whereupon the hyphae form a sheath around the root (mantle) followed by an inward growth between root cells into the plant apoplastic space (forming the Hartig net; Horan et al. 1988; Selosse et al. 2006). ECM fungal hyphae normally never penetrate plant cells but remain in the apoplast where, through a suite of secreted proteins, they establish a symbiotic interface with plant cells across which nutrients are exchanged. The Hartig net can range in depth from just around the root epidermal cells (in hardwood species) to several cell layers deep within the root (in conifer hosts). Given the high density of ECM fungal spores, propagules and hyphae within the soils of forest ecosystems, most lateral roots of trees and shrubs are colonized within days after their emergence in the upper soil profiles (Taylor et al. 2000; Ruess et al. 2003; Adams et al. 2006).

The mutualistic aspect of the interaction between ECM fungi and plant roots requires that there is a bi-directional exchange of nutrients. The hyphal ECM network within the bulk soil collects and supplies approximately 70 % of a tree or shrubs needs for nitrogen and phosphorus (Horan et al. 1988; Brandes et al. 1998; Selosse et al. 2006) while in return the plant provides approximately 20 % of the carbon it assimilates through photosynthesis in the form of sugar (Högberg et al. 2001; Leake et al. 2004). Deep sequencing has been used to show the concurrent expression of mycorrhizal transporter genes in *L. bicolor* and aspen (Larsen et al. 2011). Using this technique the authors were able to demonstrate that the expression of amino acid transporters is enriched in the plant partner while sugar transporters were regulated in the fungal partner, leading Larsen and colleagues to hypothesize that amino acids are the major metabolite shared during interactions with ECM fungi. Similarly, a large number of amino acid transporters showed increased expression in *T. melanosporum* during symbiosis (Martin et al. 2010). This mutualistic association allows host trees to grow efficiently in low-nutrient and marginal environments (Read and Perez-Moreno 2003).

The underlying mechanisms that shape the mycorrhizal root are based on physical changes as well as on signaling between the two partners. Mechanistically, hyphae divide while concurrently losing septation and apical coherence. In the host, connections between plant cells are lost and cells in contact with fungal hyphae change shape to more ovoid, regular structures (Kottke and Oberwinkler 1987). The orchestration of these elementary processes depends on general growth factors (e.g., auxins, ethylene) that promote cell divisions and regulate tissue size (Rupp and Mudge 1985; Rupp et al. 1989; Jambois et al. 2005; Felten et al. 2009; Splivallo et al. 2009) and on proteins that control the orientation of cell divisions and on cell rearrangements. To allow for communication between the two organisms in symbiotic tissues, fungal and root cells are thought to have developed a suite of molecules that sense changes to their immediate environment (e.g., cell-anchor receptors, hydrophobins; Correa et al. 1996) as well as a vast army of small secreted proteins that act to divert defenses and, on the part of the fungus, to control host cell function (Martin et al. 2007, 2008, 2010; Plett et al. 2011). A number of these signaling pathways and developmental gene networks are common among ECM fungi, suggesting that there is a common mutualistic 'toolbox'. The more recent discovery, however, of lineage-specific mycorrhizal effector proteins has added an intriguing dimension to our understanding of how fungi have also developed unique signaling pathways to gain the upper hand over their plant host and potentially their competition within the rhizosphere. Although our understanding of the physiological mechanisms and signaling pathways at play during mutualistic development and functioning need

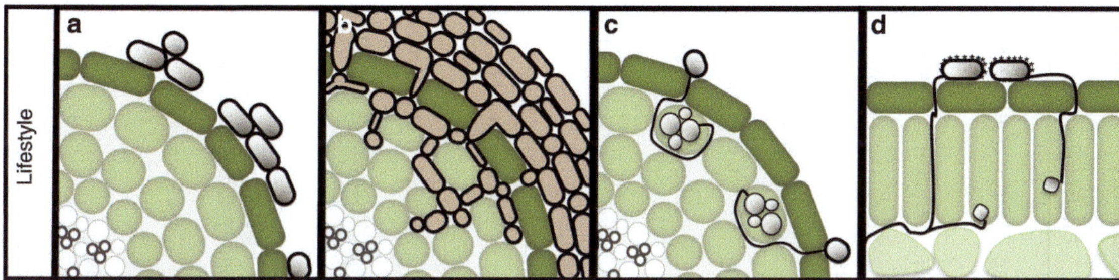

	a	b	c	d
Complete	*Auricularia delicata* (WR) *Bjerkandera adusta* (WR) *Coniophora puteana* (BR) *Dacryopinax* sp. (BR) *Dichomitus squalens* (WR) *Fomitiporia mediterranea* (WR) *Fomitopsis pinicola* (BR) *Ganoderma lucidum* (WR) *Gloeophyllum trabeum* (BR) *Gymnopus luxurians* (Litter) *Hypholoma sublatericium* (WR) *Phlebia brevispora* (WR) *Postia placenta* (*Punctularia strigosozonata* (WR) *Serpula lacrymans* (BR) *Stereum hirsutum* (WR) *Trametes versicolor* (WR) *Wolfiporia cocos* (BR)	*Hebeloma cylindrosporum* *Laccaria bicolor* *Oidiodendron maius* *Paxillus involutus* *Piloderma croceum* *Pisolithus microcarpus* *Pisolithus tinctorius* *Tuber melanosporum*	*Piriformospora indica*	*Melampsora larici-populina*
In progress	*Armillaria mellea* (WR, Path) *Plicaturopsis crispa* (WR) *Hydnomerulius pinastri* (BR) *Sphaerobolus stellatus* (Litt.)	*Amanita muscaria* *Boletus edulis* *Cenococcum geophilum* *Laccaria amethystina* *Lactarius quietus* *Meliniomyces bicolor* *Paxillus rubicundulus* *Rhizoscyphus ericeae* *Scleroderma citrinum* *Sebacina vermifera* *Suillus luteus* *Terfezia boudieri* *Thelephora terrestris* *Tricholoma matsutake* *Tuber aestivum*, *T. magnatum* *Tulasnella calospora*	*Glomus intraradices*	

Fig. 6.1. Selected sequencing projects for fungi interacting with tree systems within the saprotrophy to pathogenesis lifestyle continuum: (**A**) saprotrophs, (**B**) ectomycorrhizal fungi, (**C**) symbiotic endophytes and endomycorrhizal fungi, (**D**) pathogenic fungi. *WR* white rots, *BR* brown rots, *Litt* litter decayers, *Path* pathogens

further refinement, the past few years have brought exciting discoveries in this area that we will elaborate upon in this chapter.

Since the first published genome of an ECM fungus (Martin et al. 2008), the number of ECM genome sequencing programs completed, or in progress, are growing rapidly (Martin et al. 2010; 2011; Fig. 6.1) as are the molecular tools necessary to manipulate these fungi to better study the role of individual genes or gene products (e.g., RNA silencing; Kemppainen et al. 2009). These projects, along with the sequencing projects of ECM plant hosts such as *Populus* spp. (Tuskan et al. 2006) or *Eucalyptus* spp. (http://eucalyptusdb.bi.up.ac.za/) allow for microarray analysis of signaling events in both partners during the establishment of symbiosis and the regulation of these transcriptomes in response to developmental and environmental cues. More recently deep sequencing methods like 454 pyrosequencing or Illumina allow not only whole-transcriptome sequencing, but also allow for the detection of new transcripts, to better define the exon/intron boundaries of genes and to detect alternative splicing events (Larsen et al. 2011; Tisserant et al. 2011). Using these techniques it appears that ontogenic and metabolic programs that lead to the development of symbiosis are driven not only by the differential expression of transcription factors (Montanini et al. 2011) and transduction pathways conserved with other fungi (i.e., transcription factors also found in saprotrophic fungi), but also by the expression of novel symbiosis-specific gene networks (Martin et al. 2008, 2010; Plett et al. 2011). Table 6.1 summarizes recent ECM transcriptome studies.

The aim of this chapter is to highlight the most recent work into the understanding of signaling within the rhizosphere during the establishment of ECM symbiosis and to illustrate the way functional genomics is altering our

Table 6.1. Current published works investigating the transcriptional control of the colonization of plant tissues by ectomycorrhizal fungi

Organisms	Method	Experiment/Major findings	References
Whole transcriptome analyses			
Laccaria/Pseudotsuga *Laccaria/Populus*	*Laccaria* whole-genome oligoarray	Comparisons of gene expression in ECM root tips (mature mycorrhizae), free-living mycelium and fruiting bodies. ECM synthesized in pots	Martin et al. (2008)
Laccaria/Populus	*Populus* whole-genome array	Gene expression analysis of the early phase of the interaction between *Poplar* roots and *Laccaria*	Felten et al. (2009)
Paxillus/Populus	*Populus* whole-genome array	Comparative transcriptome analysis indicated EM-related genes whose transcript abundances were independent of salt stress and a set of salt stress-related genes that were common to EM non-salt-stressed and non-EM salt-stressed plants	Luo et al. (2009)
Laccaria/Populus	RNA-Seq	Using deep RNA sequencing for the structural annotation of the *Laccaria bicolor* mycorrhizal transcriptome	Larsen et al. (2010)
Tuber	*Tuber* whole-genome oligoarray	Comparison of gene expression in ECM root tips, free-living mycelium and fruiting bodies. ECM synthesized in pots.	Martin et al. (2010)
Laccaria/Populus	*Populus* whole-genome array	Identification of quantitative trait loci affecting ectomycorrhizal symbiosis in an interspecific F1 *Poplar* cross and differential expression of genes in ectomycorrhizas of the two parents: *P. deltoides* and *P. trichocarpa*	Labbe et al. (2010)
Laccaria/Populus	RNA-Seq	Next generation short-read transcriptomic sequencing data from fully formed ectomycorrhizae between *Laccaria bicolor* and aspen (*Populus tremuloides*) roots. The transcriptomic data was used to map expressed gene to specific metabolic pathways	Larsen et al. (2011)
Populus/MISSP7 peptide from *Laccaria*	*Populus* whole-genome array	MISSP7 is secreted by the fungus, imported into the plant cell, and targeted to the plant nucleus where it alters the transcriptome of the plant cell	Plett et al. (2011)
Tuber	*Tuber* whole-genome array	The Perigord black truffle responds to cold temperature with an extensive reprogramming of its transcriptional activity	Zampieri et al. (2011)
Tuber	RNA-Seq	Comparison of gene expression in ECM root tips, free-living mycelium and fruiting bodies. ECM synthesized in pots. Identification of novel transcripts, antisense transcripts, new exons, untranslated regions	Tisserant et al. (2011)
Other gene expression profiling			
Laccaria/Pinus	Microarray 2109 ESTs	*Pinus sylvestris* roots challenged with the ectomycorrhizal fungus *Laccaria bicolor*	Heller et al. (2008)
Laccaria/Pinus	Microarray 2109 ESTs	Comparative analysis of transcript abundance in *Pinus sylvestris* after challenge with a saprotrophic, pathogenic or mutualistic fungus	Adomas et al. (2008)
Hydnangium/Eucalyptus	Subtractive cDNA library	The fungus was cultured in the presence of *Eucalyptus grandis* roots, but with no contact. Genes that code for proteins related to carbohydrate, amino acid, and energy metabolisms, transcription, and protein synthesis, cellular communication, signal transduction, stress response, transposons, and proteins related to the biogenesis of cell components were identified among the 131 differentially expressed sequence tags	Da Silva et al. (2010)

thinking about the mechanisms used by mutualistic organisms during the establishment of a relationship between an ECM fungus, their hosts and the organisms that surround them.

II. The Rhizosphere: Communication Amongst the Teaming Masses

The rhizosphere, defined as the compartment of soil most closely associated with the roots of plants, teams with an unbelievable diversity of bacteria, fungi, nematodes and invertebrate species (Buée et al. 2009). Within this niche, mycorrhizal fungi must seek and perceive the presence of host plants while filtering the signals from other organisms, both beneficial and detrimental, around them. In this section we will address how ECM fungi participate both passively and pro-actively in the communication within the rhizosphere.

A. A Trail of Breadcrumbs: ECM Fungi React to a Concentration Gradient of Different Compounds to Find a Receptive Host

It is likely that prospecting hyphae originating from spore germination, or from nascent fungal colonies or colonized roots, sense the presence of a root by perceiving modifications in nutrient components of the rhizosphere compared to bulk soil (the soil not in intimate association with plant roots). It is along a gradient, both positive and negative, of different compounds that these hyphae will grow until they meet a lateral root. For example, plant roots create a depleted nitrogen and phosphorus zone around their roots through absorption of this compound. Conversely, roots excrete sugars and other carbohydrates which locally enrich the rhizosphere. This N/P depletion and C enrichment could serve as trophic signals by soil-borne fungi to denote the presence of a root system (Pérez-García et al. 2001). It is likely that the large majority of these trophic changes are not meant to specifically signal ECM fungi as they could be used by saprotrophic or pathogenic fungi to recognize the presence of a plant, but rather that ECM fungi have taken 'advantage' of these concentration gradients to find host plants. Compounds from two groups found within the root exudate, the flavonoids and the strigolactones, have received special attention in this regard due to their affect on mycorrhizal fungi physiology.

Over 4,000 different flavones have been isolated from root exudates, each differently functionalized. It is thought that a number of these flavonoids have a role as chemical attractants or repellants to mycorrhizal fungi. For example, a range of different flavonoids induces germination in the spores of arbuscular and ectomycorrhizal fungi (Fries et al. 1989; Kikuchi et al. 2007). Further, flavonoids from *Eucalyptus* root exudates stimulate the growth and branching of the ECM fungus *Pisolithus microcarpus*, an effect likely to ameliorate the possibility that a hyphal tip will encounter a root (Köttke and Oberwinkler 1987; Horan et al. 1988; Lagrange et al. 2001; Martin 2007). More interestingly is the fact that the flavonoid profile of root exudates changes based on the physiological status of the root system (Morandi et al. 1984; Larose et al. 2002). This phenomenon, best studied in root systems colonized by arbuscular fungi, demonstrates that there is a distinct shift in the flavonoid profiles after the initial colonization of a root by the fungus, a shift that is concurrent with reduced colonization of newly emerged roots (Vierheilig et al. 2000a, b; Catford et al. 2003, 2006; Vierheilig 2004; Meixner et al. 2005; Meixner et al. 2007). While arbuscular mycorrhizal fungi are distinctly different from ectomycorrhizal fungi in both lifestyle and growth habits, these data would suggest that there may be a link between the shift in flavonoid profiles and the production of a signaling agent within the root exudate that acts as a chemoattractant for mycorrhizal fungi and would warrant further study to see whether similar instances occur in root systems colonized by ectomycorrhizal fungi. Energetically, the loss of a chemoattractant in the exudates of roots already colonized by a mutualistic fungus would be mutually beneficial as it would be expensive in time and resources both for the plant to produce the secondary metabolites when it was already benefiting from a mutualistic exchange as well as for the free-living mycelium of a fungus to search out a non-receptive host root system as it was already colonized.

It is likely that flavonoids do not act alone, but rather with other root exudate compounds. Strigolactones, a class of sesquiterpene lactones found in the root exudates of many plants, are another important signal in the rhizosphere. Originally identified in *Lotus japonicus*, strigolactones were initially linked to the stimulation of seed germination of parasitic weeds such as *Striga* and *Orobanche* (Akiyama et al. 2005). Strigolactones are produced by a wide range of mono- and dicotyledonous plants and the concentration at which they are produced by roots coincides with the host specificity of AM fungi (Akiyama et al. 2005). Currently, as with the study of the effect of flavonoids on the growth of fungi, the majority of studies considering the effect of these compounds have looked at AM fungi, but they are also likely to affect ECM fungi. For example, as found with the flavonoids, these compounds have proven to be very important in the control of hyphal growth and branching in the proximity to the root (Bouwmeester et al. 2003; Akiyama et al. 2005; Besserer et al. 2008; Bonfante and Genre 2010). As with flavonoids, strigolactones decrease rapidly in the root exudates of plants that are not receptive to colonization, although strigolactone production seems to be linked more closely to nutrient status of the root (Akiyama and Hayashi 2008). This finding further reinforces the notion that the root secretome changes such that mutualistic symbiotic fungi are no longer attracted to a previously colonized root. Beyond hyphal branching, application of the synthetic strigolactone GR24 at very low concentrations (10^{-8} M) greatly increases both the number of mitochondria and nuclei within fungal hyphae of *Glomus intraradices*, *G. irregulare* and *G. claroideum*. This is thought to correlate with an increase in metabolism as well as the growth potential for the fungus as it prospects throughout the soil in search of a host (Besserer et al. 2006, 2008). It remains to be determined whether ECM fungi are responsive to strigolactones; but, should there be a conservation of strigolactone responsiveness between the two classes of mycorrhizal fungi, this would reinforce the idea that the earliest land plants already used this class of molecules to communicate with their symbiotic partners (Brachmann and Parniske 2006).

The mechanism by which root exudates are sensed by the ECM fungus, and what signaling cascades are activated in response to these varying compounds, is an area that is not well understood but which deserves a re-examination given the current data available for different model ECM fungi (e.g., *Laccaria bicolor*, *Tuber melanosporum*). Early work demonstrated that upon receipt of different signals from a plant host, *L. bicolor* genes involved in transduction pathways were among the most abundant of the mRNAs in ectomycorrhizal tissues (Voiblet et al. 2001). The regulated genes correspond to the different sub-units of G-proteins (e.g., a fungal G-protein α subunit), response regulators (e.g., Ras protein kinases) and calcium-binding proteins (e.g., calmodulin; Table 6.2). A Ras cDNA has also been described in *L. bicolor* and may be involved in the control of cell growth and proliferation of hyphae (Sundaram et al. 2001). Similarly, *T. melanosporum* genome encodes the signaling genes documented in other filamentous fungi that are involved in pathways controlling stress response, filamentous growth, virulence and mating. During the interaction with the host plant most signaling transcripts are strongly expressed, although very few are specific to mycorrhizal root tips (Martin et al. 2010; Table 6.2).

It is interesting to note that very similar signaling cascades to those found in ECM fungi during their interaction with host tissues are also used by pathogenic fungi in the early interaction with plant hosts. For example, filamentous pathogenic fungi contain two classes of response regulators that are important during host interactions—the first class consists of response regulators that bear homology to the yeast Ssk1 response regulator (Calera and Calderone 1999) while the second class is populated by proteins similar to the yeast Skn7 protein that contains a heat-shock factor-type DNA binding domain at the N-terminus and a response regulator at the C-terminus of the protein (Grant 2001). The former class is responsible for response to various different

Table 6.2. A comparison of signaling genes differentially regulated by the colonization process in *Laccaria bicolor* (Martin et al. 2008) and in *Tuber melanosporum* (Martin et al. 2010)

Sequence ID	Predicted function	*Laccaria bicolor*	*Tuber melanosporum*
LACB00S00018030	G-Protein, alpha subunit	300	
LACB00S00005082	G-Protein, alpha subunit	10	
LACB00S00005175	G-Protein coupled receptor	5	
LACB00S00012836	G-Protein, alpha subunit	5	
LACB00S00015797	Mitogen-activated protein kinase	4	
LACB00S00007826	G-Protein, alpha subunit	3	
LACB00S00002342	Serine/threonine kinase	2	
LACB00S00017586	G-Protein, beta subunit	2	
LACB00S00005178	Hypothetical signalling protein	2	
LACB00S00005987	cAMP phosphodiesterase	2	
LACB00S00003242	Protein kinase C	2	
LACB00S00008843	Cyclic nucleotide phosphodiesterase	2	
LACB00S00008315	RAS-like protein	2	
LACB00S00002188	G-Protein, gamma subunit	2	
LACB00S00017200	G-Protein, gamma subunit	−2	
LACB00S00008666	Mitogen-activated protein kinase	−2	
LACB00S00008500	cAMP-dependent protein kinase	−2	
LACB00S00004606	STE3 pheremone receptor	−2	
LACB00S00007300	G-Protein, alpha subunit	−2	
LACB00S00002858	Mitogen-activated protein kinase kinase kinase	−2	
LACB00S00000385	Mitogen-activated protein kinase kinase	−2	
LACB00S00001443	Hypothetical signalling protein	−2	
LACB00S00009375	Hypothetical signalling protein	−2	
LACB00S00017159	STE3 pheremone receptor	−2	
LACB00S00004341	cAMP-dependent protein kinase	−2	
LACB00S00001920	Mitogen-activated protein kinase	−2	
LACB00S00000593	PXA domain	−3	
LACB00S00017157	STE3 pheremone receptor	−5	
LACB00S00010542	Hypothetical signalling protein	−5	
LACB00S00017155	STE3 pheremone receptor	−6	
LACB00S00011898	G-Protein, alpha subunit	−7	
LACB00S00016542	G-Protein, alpha subunit	−8	
LACB00S00003696	G-Protein, alpha subunit	−11	
LACB00S00018994	G-Protein, alpha subunit	−100	
GSTUMT00012481001	GPCR-like protein		94
GSTUMT00012510001	STE3, pheromone A receptor		13
GSTUMT00003004001	Mitogen-activated protein kinase kinase		4
GSTUMT00010364001	GPCR-like protein		4
GSTUMT00000073001	Ras-related GTPase		3
GSTUMT00000085001	AMP-dependent protein kinase catalytic subunit		2
GSTUMT00003009001	PDEaseJI, cAMP phosphodiesterases class-ll		2
GSTUMT00012376001	GPCR-like protein		-3

stresses imposed during the colonization of plant hosts such as osmotic stress (Calera et al. 2000; Chauhan et al. 2003). Deletion of the genes in this class also reduce the adhesion of fungal cells to host tissues through down-regulation of adhesins (Chauhan et al. 2003). Additionally the Skn7-like response regulators mediate different stress responses and are also thought to be important in hyphal differentiation and virulence (Singh et al. 2004; Fassler and West 2011). These data would suggest that these signaling proteins are fundamental for the proper interaction between two organisms, a supposition that is supported by experiments which have shown that a disruption to any one of these transduction pathways in pathogenic systems often leads to a loss in pathogenicity (Mitchell and Dean 1995; Regenfelder et al. 1997). It remains to be seen whether disruption of these pathways would likewise disrupt the homing ability of ECM fungal hyphae.

B. Mycorrhizal Fungi Signals Alter Root Architecture and Function

While the growth and development of ECM fungi is influenced by the host plant, ECM fungi also wield a level of control over the functioning of their host plant through hormone signals and effector-like proteins produced and secreted during colonization. Indole-3-acetic acid is a naturally occurring auxin which is produced by many bacteria (e.g., *Sinorhizobium meliloti*, *Pseudomonas savastanoi*, *Erwinia herbicola*; Hamill 1993) and fungi, including ectomycorrhizal fungi (Gogala 1991; Beyrle 1995). Many studies indicate that changes in auxin balance are a prerequisite for mycorrhiza organogenesis (Rupp and Mudge 1985; Rupp et al. 1989; Gogala 1991; Gay et al. 1994; Gea et al. 1994; Beyrle 1995; Karabaghli-Degron et al. 1998; Kaska et al. 1999) and that auxins serve as very potent morphogenetic signals towards root systems. At low concentrations, they increase root growth and stimulate the formation of new meristems and lateral roots which then act as new targets for colonization by ectomycorrhizal hyphae. It has been demonstrated that when auxin production is increased in fungal hyphae, mycorrhizal root tips increased three to five times as compared to contacts between tree roots and wild-type mycelium (Gay et al. 1994). These morphogenic effects are dependent on the maintenance of a specific concentration gradient of auxin within the host root as auxin transport inhibitors disrupt mycorrhizal root formation (Karabaghli-Degron et al. 1998; Kaska et al. 1999). These hormonal gradients are probably regulated by pools of conjugated auxins as the active auxins are known to be very unstable. Recent lines of evidence have also suggested that unknown fungal effectors may also act on auxin homeostasis within the plant such that lateral roots are initiated (Felten et al. 2009; Splivallo et al. 2009). This supposition came about as host roots in contact with *L. bicolor* resulted in alterations to the expression of genes involved in the polar transport of auxin (*PtaPIN* and *PtaAUX* genes), auxin conjugation (*PtaGH3*) and auxin signaling (*PtaIAA*). These differences resulted in an accumulation of auxin within root tips and lateral root induction, an effect that can be repressed by pharmacological inhibitors of auxin signaling or in mutants defective in auxin transport and signaling (Felten et al. 2009). Similarly, in the contact between the ECM fungus *Tuber borchii* and *A. thaliana*, changes to root architecture induced by the presence of the fungus required an intact auxin-signaling network (Splivallo et al. 2009). The current working model in ECM host plants is that a/some diffusible effector(s) from ECM fungi, be they auxin or another effector-like compound, activate PIN auxin transporters such that basipetal auxin transport is enhanced. This, in turn, primes the creation of an increased number of lateral root initials in the expansion zone that later give rise to mature lateral roots (Felten et al. 2009). Therefore, fungal effectors control auxin signaling within plant cells and whole tissues leading to altered root architecture which may favor colonization. Phytopathogenic fungi and bacteria also play on the auxin signaling network of their plant hosts to improve colonization. The obligate biotroph *Plasmodiophora brassicae*, for instance, infects both roots and hypocotyls of the Brassicaceae causing swelling of root tissues and enlargement of root cells (Ingram and Tommerup 1972). Auxin and amide conjugates of auxin accumulate at higher concentrations in the infected plant tissues when compared to non-infected roots, although the origin of the auxin is in question (Ludwig-Müller et al. 1993, 1996). The plant pathogenic bacterium *Pseudomonas syringae* produces an effector protein that it injects into the plant cell through a type III secretion system which promotes host auxin biosynthesis (Kunkel et al. 1993; Chen et al. 2007). This increase in auxin is associated with increased pathogenicity of *P. syringae* (Chen et al. 2007), an effect that is likely due to the antagonistic role of the auxin-signaling pathway in repressing the salicylic acid defense system (Wang et al. 2007). Thus both mutualistic and pathogenic organisms manipulate the host auxin pathway to increase colonization success. It will be interesting to see whether ECM fungi also use a range of effectors to promote or repress other plant hormone signaling transduction in order to benefit biotrophic colonization (e.g., promote jasmonic acid signaling versus salicylic acid-induced signaling).

One lineage of ECM fungi are also able to produce abundant indolic compounds that, while they are not auxin or auxin-like compounds, affect plant-based auxin signaling. Excreted by *Pisolithus* hyphae, this molecule—hypaphorine—is a tryptophane betaine and is highly accumulated in hyphal cells in contact with host roots. Hypaphorine is able to stimulate the expression of auxin-regulated genes in *Eucalyptus* and it was hypothesized that hypaphorine might act as an auxin-like compound (Nehls et al. 1998). Hypaphorine is likely not an auxin mimic but, more surprisingly, counteracts auxin action on tap root elongation or root hair elongation (Ditengou and Lapeyrie 2000; Ditengou et al. 2000). It is probable therefore that hypaphorine acts in the plant auxin pathway by competing with auxins for auxin-binding proteins and receptors (Kawano et al. 2001). To date this compound has not been found in other fungi, either mutualistic or pathogenic, and thus its role in the establishment of symbiosis might be limited to the *Pisolithus* genus.

Hypaphorines are not the only strain- or genus-specific signaling agents encoded by ECM fungi used during the establishment of host colonization. Upon the release of the genome of *L. bicolor*, it was found that over 12 small secreted proteins (SSPs) with no known homology were up-regulated during the late stages of colonization (Martin et al. 2008). The size of these proteins, their predicted secretion signal and their expression in symbiotic tissues are reminiscent of pathogenic 'effector' proteins, small proteins that are secreted and utilized by pathogenic organisms to control host cell function. It was questioned at the time whether or not the SSPs up-regulated during mutualistic exchanges between *L. bicolor* and its host (dubbed the *m*ycorrhiza *i*nduced small *s*ecreted *p*roteins; MiSSPs) may encode effector-like proteins to control host plant function to favor colonization. Recently the first of these MiSSPs, MiSSP7, was characterized and the results obtained furthered the idea that mycorrhizal fungi encode effector-like proteins (Plett et al. 2011). *MiSSP7* was identified as the most highly regulated MiSSP in *L. bicolor* during establishment and maintenance of mycorrhizal root tips (Martin et al. 2008). The expression of *MiSSP7* is induced by root exudates (both host and non-host) before physical contact between the two organisms and continues at a high level through to a mature mycorrhizal root tip. Current research into isolating the compounds within these exudates that induces MiSSP7 production would suggest that it is not due to external carbon sources but rather that flavonoids within the rhizosphere are responsible, at least in part, for the induction of this gene (Plett and Martin 2011). MiSSP7 was determined to be one of the essential signals for controlling the in-growth of fungal hyphae into the apoplastic space of the root as RNAi mutant lines of *L. bicolor* with lowered production of MiSSP7 were not able to establish a functional mycorrhizal root tip. These mutant lines of *L. bicolor* with reduced levels of MiSSP7 were only able to colonize root tissues upon either the external application of MiSSP7 protein to roots in contact with the mutant fungus or by heterologous expression of MiSSP7 in the plant host roots (Plett et al. 2011). The site of action of MiSSP7 is in the nucleus of the plant cell where it affects the expression of a number of different genes involved in plant defense, cell wall remodeling and signaling pathways. MiSSP7 enters host plant cells through the binding of an RXLR-like region to plant phospholipids, followed by endocytosis—a mechanism of entry also used by numerous pathogenic effector proteins (Fig. 6.2; Dou et al. 2008; Kale et al. 2010). These data indicate that MiSSP7 is a genuine symbiotic effector

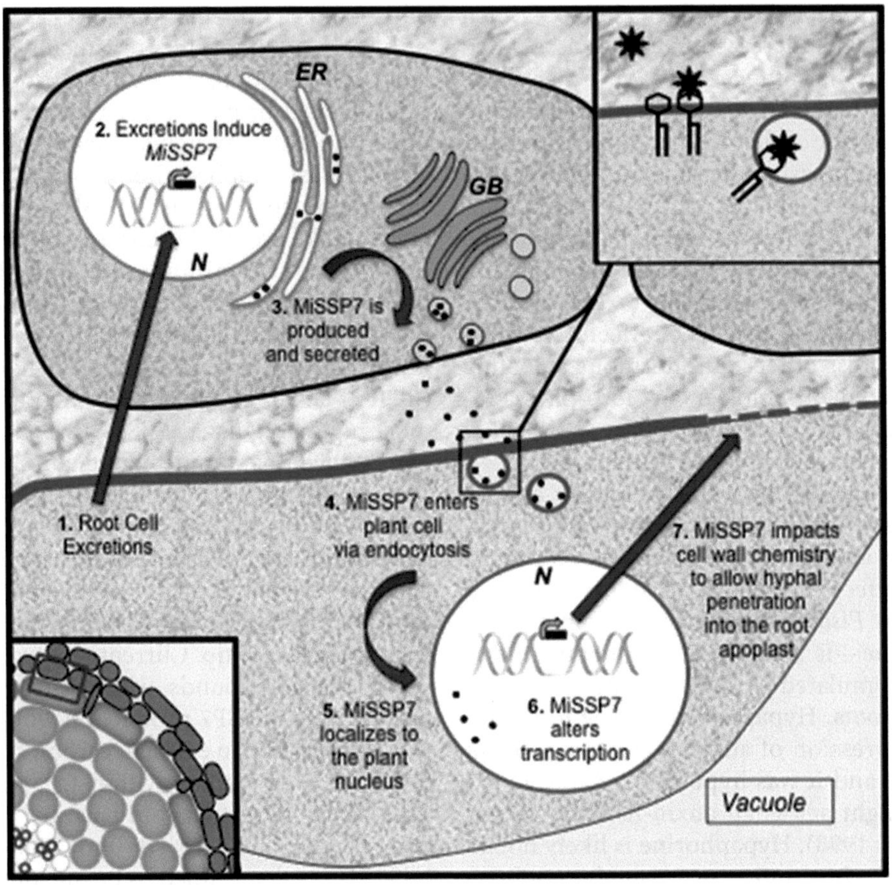

Fig. 6.2. Graphical representation of a transverse cross-section of a poplar root (*inset*) undergoing colonization by the ectomycorrhizal fungus *Laccaria bicolor*. The main figure represents the interface between plant epidermal cells and fungal cells. During the establishment of symbiosis we demonstrate that diffusible root secretions (*1*) induce the expression (*2*) and production of the mycorrhizal induced small secreted protein 7 (MiSSP7). This small protein, 68 amino acids in length, is secreted from the hyphal cell into the symbiotic interface (*gray*) area between cells (*3*). Our results show that MiSSP7 is taken up into the plant cell via endocytosis following binding to cell surface phosphatidyinositol-3-phosphate (*4*) and concentrates in the plant nucleus (*5*) where it alters transcription of genes associated with plant cell wall construction and direction of root architecture (*6*). We hypothesize that this, in turn, alters the plant cell wall chemistry such that fungal hyphae can penetrate within the apoplastic space of the root and establish the complex network of fungal and plant tissues named the Hartig net (*7*). It is within this network that the mutualistic exchange of nutrients between partners occurs. *N* nucleus, *ER* endoplasmic reticulum, *GB* golgi bodies (Taken from Plett et al. (2011))

protein, much like those employed by pathogens. It is also interesting to note that effector proteins that prepare the root for symbiosis even before physical contact between the two partners have also been found in symbiotic bacteria (Kambara et al. 2009) and in the much more ancient AM fungi (Kloppholz et al. 2011). SP7 is the first identified effector protein from the AM fungus *G. intraradices* and shares some characteristics with MiSSP7. SP7 is produced upon receipt of plant-based signals and enters the host plant cell. SP7 also appears to enter the plant cell through interactions with membrane phospholipids, although this is thought to be mediated by a series of imperfect, hydrophilic repeats on the amino end of the protein rather than by an RXLR-like motif as found in pathogens and MiSSP7. In the host plant cell, SP7 localizes to the plant nucleus where it interacts with ethylene response factor 19 (ERF19) to repress plant defense signaling (Kloppholz et al. 2011). When SP7 was heterologously expressed in the hemibiotrophic fungus *Magnaporthe oryzae*, it acted to prolong the biotrophic phase and suppress the HR response within plant cells. Therefore the two proteins, MiSSP7 and SP7, appear to closely mirror pathogenic effectors in their role and localization within plant tissues.

The role of SP7 in subjugating the ethylene signaling pathway is interesting in light of recent work by Labbé et al. (2010), Splivallo et al. (2009) and Camehl et al. (2010) in diverse symbiotic fungi which have also suggested that the ethylene pathway is a key determinant in whether or not a fungus is able to colonize plant tissues. In their work, Labbe and coworkers (2010) analyzed the F1 pedigree of a *P. deltoides* × *P. trichocarpa* cross for quantitative trait loci associated with mycorrhization of plant roots by *L. bicolor*. The two parental strains of this cross show contrasting levels of colonization by *L. bicolor*, with *P. trichocarpa* easily colonized, and with *P. deltoides* colonized at a much lower level. By comparing colonization rates and whole genome expression analysis this group was able to map a number of significant QTLs, one of which contained an ethylene responsive ethylene binding protein (EREBP4) which is highly up-regulated during root colonization in *P. trichocarpa* and not in *P. deltoides*. Considering the role of ethylene in communicating stressful situations between plant organs, and the proposed role for ethylene in limiting the development of fungal hyphae within plant tissues (Camehl et al. 2010), this is a very interesting result as it would have been anticipated that ethylene signaling would be higher in *P. deltoides* given its recalcitrance to fungal colonization. Splivalo and colleagues (2009), meanwhile, found that an intact ethylene network was required for auxin to aid fungal colonization of roots. Therefore ethylene is likely to have a much more complicated role during fungal colonization of roots than simply limiting fungal in-growth. The occurrence, however, of a convergent strategy by evolutionarily different fungi to target the same signaling pathway (i.e., ethylene) to promote (or at least not inhibit) colonization is consistent with the proposal of conserved 'hub' proteins or pathways that are targeted by effector proteins of pathogenic organisms to colonize plant tissues (Mukhtar et al. 2011). In this study, 3,148 protein interactions were determined in the colonization of *Arabidopsis thaliana* leaf tissues by two pathogens separated by two billion years of evolutionary time. Within this interactome, there were 15 protein hubs with degrees greater than 50, 7 of which were targeted by the effectors of **both** pathogens. It was concluded that the convergence of pathogenic effectors on these same protein hubs was not by chance alone but more likely by the co-evolution of the host:pathogen interaction and, more specifically, by the co-evolution of the effectors with the plant immune system. Thus, over time, effectors in both organisms developed to exploit the same 'weaknesses' in the immune response to allow development of the pathogen. Given that ECM, AM fungi and symbiotic endophytes also encode effector proteins and that these fungi appear to affect similar host systems (e.g., ethylene signal transduction) it may be anticipated that mutualistic fungi also target conserved host signaling hubs. We shall have to await further interaction studies in mutualistic systems before this hypothesis can be substantiated.

C. Mycorrhizal Fungi Influence the Developmental Programs of Soil Bacteria

Like their effect on plants, mycorrhizal fungi excrete different metabolites, hormones and proteins into the soil that act as rhizospheric signals which they use to affect the organisms within their environment. It would appear, based on current research, that these signals may play a role in structuring rhizospheric bacterial populations.

Bacteria from many different clades associate both loosely and tightly with ectomycorrhizal root tips and are a third component of mycorrhizal associations. Research concerning these bacterial populations has been gaining popularity since Garbaye (1994) described these 'helper bacteria' such that we now have a more in-depth understanding of fungal–bacterial interactions, interactions that are more widespread than originally expected (Frey-Klett et al. 2007). Given the fact that these helper bacteria colonize the surface of fungal hyphae and, in some cases, reside within the fungal hyphae as endobacteria, an understanding of how mycorrhizal fungi affect these organisms is important in understanding the plant–fungus–soil interface. Interestingly, the bacterial communities that associate with a certain ECM fungi or ECM colonized root tip tend to be similar, suggesting that the ECM fungus may play a role in 'cultivating' or attracting these specific genera (Nurmiaho-Lassila et al. 1997; Frey-Klett et al. 2005). Using microscopic techniques, purification, isolation and large-scale sequencing, these bacterial populations are being described, with the dominant genera being the *Pseudomonas*, *Rhodococcus*, *Streptomycetes*, *Burkholderia* and *Bacillus* (Abdel-Fattah and Mohamedin 2000; Ochsenreiter et al. 2003; De Boer et al. 2005; Schrey et al. 2005; Bomberg and Timonen 2007; Schrey and Tarkka 2008). Bacteria from these genera promote the formation of a mutualistic relationship between the fungus and plant roots by complementing the functions of mycorrhizal root tips, including nutrient acquisition and biological control of host plants (Artursson et al. 2006; Frey-Klett et al. 2007; Tarkka and Frey-Klett 2008). During the interaction between *L. bicolor* and the roots of Douglas fir, *Pseudomonas fluorescens* is attracted through the soil by a gradient of different fungal extracts (e.g., trehalose). The most studied *P. fluorescens* strain in this relationship is BBc6R8, a strain known to improve the growth and survival of *L. bicolor* hyphae (Brulé et al. 2001; Deveau et al. 2007). It is likely that the positive effect of *P. fluorescens* BBc6R8 on the growth of fungal hyphae is due in part to the production and secretion of thiamine, an essential cofactor of several central enzymes of carbon metabolism (Tarkka et al. 2009), a compound that has been previously associated with growth promotion of the yeasts *Debaryomyces vanrijiae* (Rikhvanov et al. 1999) and of *Saccharomyces cerevisiae* (Chaucheyras-Durand and Fonty 2001). The role played by streptomycete bacteria in the rhizosphere is also intriguing as they can have beneficial effects on ECM plant host species. Within the context of mutualistic symbiosis, and with mycorrhizal fungi, *Streptomycetes* promote the formation of symbioses between fungi and plant hosts by suppressing the growth of pathogenic bacteria while increasing the growth rates of different ECM fungi (Schrey et al. 2005). As such, the concurrent innoculation of sorghum with mycorrhizal fungi and *Streptomyces coelicolor* increases the root colonization of the sorghum (Abdel-Fattah and Mohamedin 2000) while the mycorrhization rate is significantly increased between *Picea abies* and the ECM fungus *Amanita muscaria* when either *Streptomyces* sp. AcH505 or *S. setonii* are also present (Maier 2003). But if these genera of bacteria always associate with the same ECM fungi, how do the fungi affect the bacterial populations around them? One possible mechanism of control could be through the production of proteins that disrupt or modify quorum sensing between certain bacterial species.

Quorum sensing is a mechanism by which bacteria can sense their population density and thereby their growth rate as a population. Quorum sensing molecules may also coordinate bacterial virulence factors when colonizing host tissues. First identified in *Vibrio fischeri*, an aquatic bacterium, quorum sensing is often mediated by small molecules called autoinducers (AIs) that act to transduce messages between bacteria. While there are a number of different molecules that act as AIs (e.g., diketopiperazine, furanosyl borate diester or γ-butyrolactone), the most common class of molecules that act as AIs are *N*-acyl homoserine lactones (AHLs). AHLs are typically synthesized at a basal level by each bacterium which, upon reaching a certain critical level in the medium surrounding the bacteria, induce the expression of different suites of genes which restrict growth of the colony or to serve a number of other purposes (Walker et al. 2003). In the last decade, however, it was found that quorum sensing molecules were made by more organisms than just bacteria. Root exudates from *Pisum sativum* contain a number of active compounds that can mimick AHLs, although their chemical nature has not been elucidated (Teplitski et al. 2000; Knee et al. 2001; Fray 2002; Uroz et al. 2003; Rasmussen et al. 2005). Conversely, a number of other plant species have been reported to excrete compounds that act as AHL blockers (Walker et al. 2003). More recently it was found that ECM fungi also produce compounds that interfere with bacterial quorum sensing systems (Uroz and Heinonsalo 2008). These compounds exhibited AHL lactonase activity, enzymes that convert AHLs to the inactive *N*-acyl homoserine (AHS) derivative. The discovery of a mechanism by which ECM fungi can disrupt AHL signaling through inactivation of AHLs to AHSs, or the inhibition of their perception, could well be a strategy developed by the fungi to interfere with the growth of populations of bacteria with deleterious functions. Through this mechanism the fungus could control the surrounding bacterial community and may be a pathway that would partially explain the reason why only certain bacteria are present in or on mycorrhizal root tips (Nurmiaho-Lassila et al.. 1997; Frey-Klett et al. 2005).

III. The Porous Barrier: Construction of the Symbiotic Interface

After an ECM fungus has followed rhizospheric signals in search of a plant host, after the fungus has released signaling compounds to restructure both the host tissue and the bacterial communities surrounding it, the fungus must attach to the plant tissue and form an interface between itself and the host that will both protect the fungal hyphae from plant based defenses but will also allow for the continued exchange of nutrients as well as signaling molecules. A wide range of proteins and compounds restructure both the plant and fungal cell wall to form the symbiotic interface, most of which remain to be properly characterized. We will discuss here three classes of these compounds: (1) cell wall remodeling enzymes, (2) secreted symbiosis-regulated acidic polypeptides (SRAPs) and (3) the hydrophobins, the latter two of which have a putative role in the adhesion of fungal hyphae to plant tissues and in 'hiding' and protecting the fungal hyphae from plant defenses.

To form the Hartig net, the fungal hyphae of ECM fungi must alter the chemistry of the plant cell walls and the connections that hold cells together in a cohesive mass. In part, the fungus may use effectors to alter the chemistry of plant cell walls during root morphogenesis in such a way that the fungal hyphae have an easier time to enter into the plant apoplastic space (e.g., through the release of auxin which may alter proton levels; Gay et al. 1994). As discussed above, the effector protein MiSSP7 from *L. bicolor* induces the transcription of a number of host plant enzymes (e.g., pectin methylesterases) associated with cell wall remodeling. Other effectors with similar roles have yet to be characterized in *L. bicolor* or in any other ECM fungus. In the absence of effector-like proteins, however, ECM fungi may also use enzymes to facilitate the loosening of the contact between cell walls. From one aspect, ECM fungi are hampered in this process by a large reduction in the number of carbohydrate-active enzymes used by saprotrophs and pathogens to degrade plant cell wall oligosaccharides and polysaccharides. The genomes of both *L. bicolor* and *T. melanosporum* show reduction in almost all glycosyl hydrolase, hemicellulose- and pectin-degrading families, culminating in the complete absence of several key families (Martin et al. 2008, 2010). For instance, in *L. bicolor*, there is only one glucanase candidate with a fungal cellulose-binding module found in the genome (GH5) while there are no cellulases from families GH6 and GH7. While genes coding cell wall-active enzymes are regulated in both *L. bicolor* and *T. melanosporum* during contact with host root tissues, it is interesting to note that only two of these genes are highly upregulated in both *L. bicolor* and *T. melanosporum* ectomycorrhizas (the CBM1-GH5 glucanase and a GH28 pectinase). This specific expression suggested at the time that these two proteins may play a key role during the establishment of symbiosis (Martin et al. 2010). ECM fungi may also use peptidases and proteases to manipulate host cell walls. The complement of secreted peptidases in the *T. melanosporum* genome, for instance, are similar to a range of other sequenced fungi (saprotrophic, mutualistic, pathogenic). Further, a number of proteases are regulated in ectomycorrhizal root tips of both *T. melanosporum* and *L. bicolor* (Martin et al. 2008, 2010; Table 6.3). Phytopathogenic fungi also use extracellular proteases to aid in

Table 6.3. A comparison of proteases regulated by the colonization of host roots in *Laccaria bicolor* (Martin et al. 2008) and in *Tuber melanosporum* (Martin et al. 2010)

Protein ID	MEROPS family	Holotype	*Laccaria bicolor*	*Tuber melanosporum*
303920	S08A	Subtilisin	38	
294888	S08A	Subtilisin	20	
328926	S01A	Polyserase	12	
302848	A01A	Pepsin	10	
293118	S08A	Subtilisin	10	
294125	S08A	Subtilisin	7	
322095	C01A	Papain	6	
318140	A01A	Pepsin A	5	
321341	A01A	Aspartic peptidase	4	
191892	A01A	Aspartic peptidase	3	
304862	S08A	Subtilisin	2	
310796	S08A	Subtilisin	−2	
294979	S08A	Subtilisin	−2	
252730	S10	Carboxypeptidase	−100	
118001	S0X9	Prolyl oligopeptidase		9795
1164001	M28A	Aminopeptidase		164
122001	S09B	Dipeptidyl peptidase		151
8292001	M28A	Aminopeptidase		21
9607001	S08A	Subtilisin		14
11308002	I51	Inhibitor		3
3552001	S09X	Peptidase		2
4908001	S28	Carboxypeptidase		−2
5184001	A01A	Pepsin A		−2
1312001	S08A	Cerevisin(*Saccharomyces cerevisiae*)		−3
118001	S09B	Dipeptidyl peptidase		−8
11308003	S09X	Prolyl oligopeptidase		−11
11308001	S09X	Prolyl oligopeptidase		−20

host colonization and plant cell wall de- or reconstruction. Two examples of this include the expression of an aspartyl protease from *Botrytis cinerea* that hydrolyzes the cell walls of apple (Urbanek and Kaczmarek 1985) while *Fusarium culmorum* produces an acid protease that has the ability to hydrolyze a number of vegetable proteins (Urbanek and Yirdaw 1984). These results suggest that fungi living on plant tissues utilize a series of similar enzymes to modify host cell walls, but that the loss of certain plant cell wall-degrading enzymes and the regulation of very few of these genes in ECM fungi may represent an adaptation to the mutualistic lifestyle. The different contingent of cell wall-active enzymes may also be a manner by which the host plant may differentiate between mutualistic fungi and more predatory fungi.

The formation of mycorrhizal root tips never occurs in a static environment. Environmental conditions are constantly changing as are the presence of other organisms in the rhizosphere. Similarly, the root and the fungal hyphae are constantly growing. Therefore, the fungal hyphae must excrete a mucilage of different compounds to firmly attach itself to the root in order to commence colonization. SRAPs are thought to be involved during this stage of symbiosis as different classes of SRAPs are upregulated early in the interaction between ECM fungi and their plant hosts (Martin et al. 1999, 2008). While a large number of different SRAP classes have been identified in fungi (Hilbert et al. 1991; Burgess et al. 1995), one class in particular, the 32-kDa class of SRAPs of *P. tinctorius* (*PtSRAP32*), have received attention in their role as potential adhesins. The evidence that the *PtSRAP32* family is involved in adhesion is threefold: (1) timing of expression, (2) an encoded cell afhesion RGD motif and (3) its location on the outer hyphal cell wall. Expression of one *PtSRAP32* (*PtSRAP32-1*) is especially high at

3 days post-contact between fungal hyphae and *Eucalyptus globus* roots, a time period when hyphae are aggregating around plant roots. The *PtSRAP32* class is composed of at least six isoforms (Laurent 1995; Laurent et al. 1999) that range in pI from 4.5 to 5.5 and are secreted to the plant/hyphal interface during colonization of the plant root (De Carvalho 1994). RGD motifs, such as the motif found in SRAPs, are thought to be involved in adhesion. For example, fibronectin RGD containing proteins act as a cell attachment site and is the recognition sequence for integrin receptors (Pierchbacher and Ruoslahti 1984; Ruoslahti 1996). An RGD-containing protein from the fungus *Lentinus edodes* has also been shown to induce flocculation in yeast (Kondoh et al. 1995) and aggregation of hyphae in *Schizophyllum commune* (Yasuda and Shishido 1997) giving further evidence for the role of RGD proteins in cell to cell adhesion. Pathogens also use a wide variety of RGD-containing proteins in interactions with plant hosts. *Candida albicans* utilizes secreted RGD proteins to attach to human cell lines (Bendel and Hostetter 1993) as do *Pneumocystis carinii* (Narasimhan et al. 1994) and *Aspergillus fumigatus* (Gil et al. 1996; Penalver et al. 1996). In plant pathogens, the rust fungus *Uromyces vignae* uses RGD containing proteins (RGDS, RGDSP) to control and repress the defense reaction by the plant host. This suppression is correlated to increased pathogenicity of the fungus and is linked to the ability of these proteins to disrupt the connection between the plasma membrane and the plant cell wall—a connection necessary to the proper maintenance of plant cell wall-associated defenses (Hostetter 2000). While *PtSRAP32*s have yet to been shown as necessary for fungus/plant cell adhesion or for the control of cellular defenses during symbiosis, their similarity to pathogenic RGD motif proteins would suggest a critical role during fungus/host interactions.

ECM fungi are specialists in avoiding the plant-encoded defense system as, without this 'stealth' factor, mutualism would be more difficult to establish (Spanu and Bonfante-Fasolo 1988; Spanu et al. 1989; Albrecht et al. 1994). The ability to avoid plant defenses may be controlled by differential gene expression of fungal elicitors or through the secretion of proteins that coat the hyphal surface to mask the antigenicity of the fungal cell wall. Hydrophobins are thought to be important in this respect in both mutualistic mycorrhizal and lichen symbioses (Duplessis et al. 2001; Tagu et al. 1996, 1998, 2001; Trembley et al. 2002a, b; Scherrer et al. 2002; Scherrer and Honegger 2003; Rajashekar et al. 2007). Two hydrophobins from the ECM fungus *Pisolithus microcarpus* (formerly *P. tinctorius*), *hydPt-1* and *hydPt-2*, are up-regulated during the early stages of plant colonization (Tagu et al. 1996; Voiblet et al. 2001; Duplessis et al. 2005). Similarly, increased accumulation of hydrophobin transcripts was observed in *Paxillus involutus*/*Betula pendula* ectomycorrhiza (Le Quéré et al. 2006). Hydrophobins are small secreted, moderately hydrophobic, self-assembling polypeptides with a conserved distribution of eight cystein residues that are crucial for proper protein folding (Sunde et al. 2008) that are involved in fungal adhesion as well as in protecting the fungus from plant defenses (Wessels et al. 1991; Kershaw and Talbot 1998; Duplessis et al. 2001; Wösten 2001; Linder et al. 2005). Their roles are related to their amphipathic structure: they accumulate at the surface of hyphae with their hydrophobic domains directed outward, aiding the aggregation of hyphae or adhesion to hydrophobic host surfaces. In protecting hyphae from the plant defense system, hydrophobins are thought to bind to fungal β-1,3-D glucans and chitin, known elicitors of the plant defense system (Sharp et al. 1984a, b; Roby et al. 1987; Templeton et al. 1994). Due to their incredible stability, hydrophobin layers have also been proposed to protect the fungal hyphae against plant proteases and reactive oxygen species (Wösten and Wessels 1997). The small size of these proteins, their induction by the formation of a mycorrhizal root tip, their secretion and their proposed role in mediating host:fungus interactions would suggest that these proteins form a subclass of MiSSPs whose role is principally in the root apoplastic space, making them a very interesting candidate for research into understanding how the establishment of a mutualistic symbiosis occurs.

The genome of *L. bicolor* S238N-H82 encodes 12 hydrophobin genes (Martin et al. 2008). Like all other basidiomycete fungi sequenced to date, these identified hydrophobins carried the signature traits of class I hydrophobin genes with no chimeric class I/class II hydrophobins as was recently found in a number of *Aspergillus* hydrophobins (Jensen et al. 2010). A number of these hydrophobins are expressed during the formation of ectomycorrhizal root tips, although their individual roles have yet to be elucidated (Plett et al. 2012). Intriguingly, the variety of hydrophobins expressed depends upon the plant host being colonized; if the host is easily colonized, far fewer hydrophobins are expressed as opposed to a host that is not easily colonized. As hydrophobins are regulated by changes in the external environment (e.g., during host colonization; Wessels 1996) or are produced to hide antigenic hyphal compounds from the plant immune system (Aimanianda et al. 2009), or to act as a base for enzymes that loosen connections between plant tissues (Corvis et al. 2005, 2006, 2007; Qin et al. 2007; Zhao et al. 2007; Wang et al. 2010) it is possible that the large variety of hydrophobins encoded by *L. bicolor* strains are a necessary adaptation to aid in the colonization of different hosts. While it would be attractive to broadly conclude that hydrophobin expression varies inversely with the ease of host colonization, it is likely that the story is far more complicated. While it has been found that hydrophobin expression correlates to the pathogenesis of animal pathogens, in plant pathogens this same link has not be absolutely correlated to the ability of different fungal isolates to colonize plant tissues (Parta et al. 1994; Thau et al. 1994; Brasier et al. 1995; Bowden et al. 1996).

There are a number of interesting differences in *L. bicolor* hydrophobins as compared to classical hydrophobin genes that has led to questions concerning their role during fungal morphogenesis and their use during the interaction with plant hosts. A first defining characteristic of hydrophobins encoded by *L. bicolor* is the level of hydrophobicity within the core of the consensus sequence. The degree of hydrophobicity of the region between cystein doublets, as defined by hydropathy plots, is much greater in the hydrophobins of *L. bicolor* when compared to other class I (e.g., SC3 from *Schizophyllum commune*) or even to class II hydrophobins (e.g., HFBI from *Trichoderma reesei*; Kim et al. 2005; Jensen et al. 2010). A second characteristic observed in a subset of the hydrophobins encoded by *L. bicolor* (*LbH3*, *LbH11*, *LbH14*) is the presence, or absence, of a cystein residue outside of, or within, the normal hydrophobin consensus sequence. Additional cystein residues have been identified outside the canonical hydrophobin domains in a number of hydrophobins in *Aspergillus* species (Jensen et al. 2010) and may suggest that the mature protein has an altered function, although this has not been proven. For example, in the rice pathogen *Magnaporthe grisea*, mutations in any of the eight cysteins of hydrophobin MPG1 caused a defect in protein secretion (Kershaw et al. 2005). Therefore, an avenue of future research would be to characterize the role of these hydrophobins which have novel characteristics during the colonization of plant tissues. As our current models of how hydrophobins act in mutualistic systems have been largely informed by comparisons to the roles of hydrophobins in the interaction between pathogenic fungi and their plant hosts (Talbot et al. 1993; Zhang et al. 1994; Kazmierczak et al. 1996; Talbot et al. 1996; Holder and Keyhani 2005) we may still be missing a wealth of information regarding the role of novel hydrophobins encoded by mutualistic fungi like *L. bicolor*.

IV. The Mutualistic Blueprint: Building Predictive Models to Maintain Future Sustainability of our Forests

Similar to pathogens in their methods of colonization while also similar to saprotrophs in their ability to mine soils for nutrients, mutualistic ectomycorrhizal fungi occupy an interesting niche between the two lifestyles (Fig. 6.1). Given their lack of a negative impact upon our daily lives, as is the case with pathogenic organisms, ECM fungi have received far less attention in recent decades as compared to the former

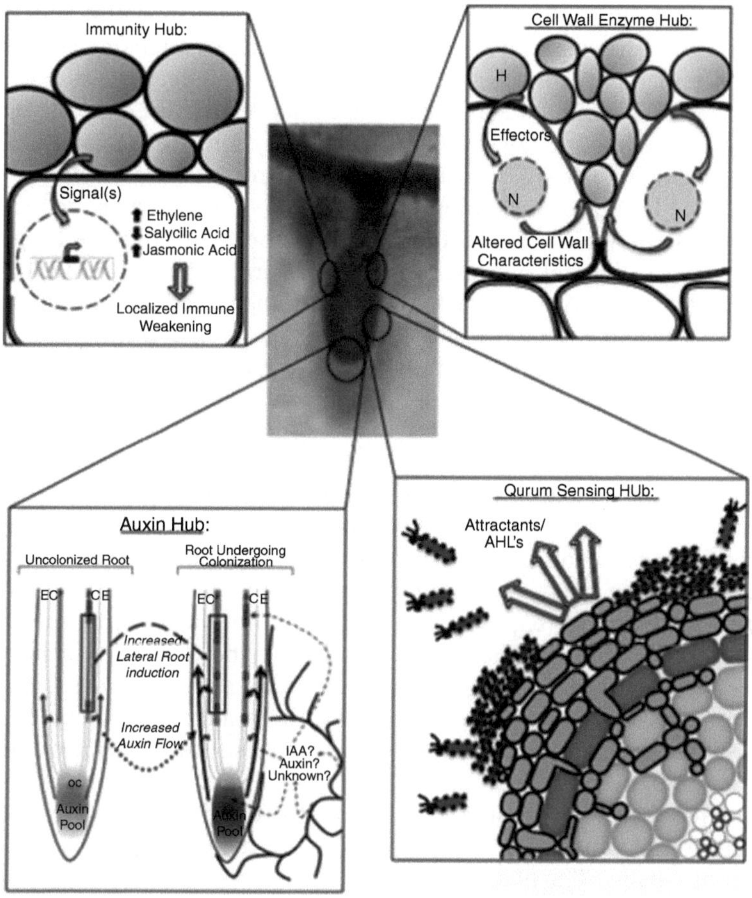

Fig. 6.3. Modelling the different signaling hubs affected by the mycorrhization process. Effectors and signaling molecules from ectomycorrhizal fungi affect defense responses in plant tissues, re-route auxin flows and maximas within lateral roots and alter the connections and chemistry of plant cell walls. Concurrently, ectomycorrhizal fungi likely cultivate a certain set of 'helper' bacteria to the hyphal surface using a combination of attractants and quorum-sensing mimics

two classes of fungi, a lack that is unfortunate in view of their essential role in nutrient cycling in forest environments. As the mutualistic interaction requires a finely tuned, and likely precarious, dialog between two partners in order to function properly it is critical that we begin to focus on a better understanding how to foster these relationships, because factors such as global climate change and increasing anthropomorphic infringement on natural environments could easily lead to the disruption of the continued success of these interactions with untold damage to the sustainability of forest habitats and the ecosystems they support. Thankfully, in recent years, increased effort has been put into establishing genomic and molecular resources such that we can better understand this complex biological system. Given their similarities to pathogens and saprotrophs, it should also be possible to use models built in these systems to inform the theoretical models built concerning ECM:host communication and colonization.

One of the most global lessons from pathogenic systems that will likely be applicable to mutualistic systems is the existence of conserved 'hubs' within protein networks of the host plant and rhizosperic community upon which fungal effectors may act to achieve host colonization while avoiding/manipulating the plant immune system (Mukhtar et al. 2011). As covered in this chapter, ECM fungi likely first target the plant auxin system, where they use a variety of methods to manipulate the auxin gradient within the root such that more lateral roots are produced and thus providing more surface area to colonize (Fig. 6.3). Fungal effectors then attack the second hub, that of defense-related hormone signaling (e.g., ethylene, salicylic acid), to create localized weakening of the plant immune system which would then allow for the beginning of hyphal penetration into the root apoplastic space. A third wave of effectors, like MiSSP7, would then harness the power of plant encoded cell wall-active enzymes to modify cell:cell attachments and plant cell wall rigidity, allowing further hyphal penetration within the root tissues. Once within the apoplastic space, and upon establishment of nutrient transfer, the hyphae must continually protect themselves from detection by the plant immune system, likely using masking proteins (e.g., hydrophobins) and decoys (e.g., secreted effector proteins) as diversionary tactics. Concurrently to these actions, ECM fungi also appear to cultivate 'helper' bacteria to the rhizosphere that aid the fungus by providing extra nutrients (e.g., thiamine), by fending off pathogenic fungi and bacteria, etc. In this step ECM fungi play upon bacterial sensory interaction hubs such as quorum sensing. The plant, meanwhile, also takes advantage of conserved signaling hubs in ECM fungi to attract this class of organism within soil profiles and likely to control the aggressiveness of the fungi during colonization.

The precision and tuning of this process is its biggest success, as demonstrated by nearly all land plants being colonized by mycorrhizal fungi, but it is also its biggest weakness. Interruption at any one of these hubs is likely to greatly change the nature of the interaction between the organisms, if not completely abolish the mutualism. This is why excessive 'mining' type management of vast stands of boreal forests and the forecasted effect of climate change are such a threat to these systems. Effort should be dedicated to comparative genomics between the three different forest fungal lifestyles (saprotroph, mutualistic, pathogenic) to identify common and divergent molecular 'toolboxes' related to their lifestyles. Within these toolboxes we should delve deeper into the complement of nutrient transporters used during mutualism; we need to understand the roles of effector proteins and hormones used to restructure the plant host cell (both physically and transcriptomically) to promote mutualism; and we should also research the different signal transduction systems both within the plant and the fungus that are activated during the symbiotic interaction. Given the predicted common signaling hubs maintained between evolutionarily disparate organisms, future effort must also be placed into understanding the putative targets within these interactomes of both host and colonizer. We also must step outside of model systems and look at the impact of new climate parameters upon small-scale ecosystems and, hopefully, on population dynamics within whole ecosystems. By using candidate markers identified within the research areas listed above, we will have the tools necessary to evaluate ECM functioning in situ. By studying these systems, and their adaptive plasticity under different parameters, we will know better how ECM fungi will fare under new climactic conditions, and we will know how to better manage the microbiome of forests.

References

Abdel-Fattah GM, Mohamedin AH (2000) Interactions between a vesicular–arbuscular mycorrhizal fungus (*Glomus intraradices*) and *Streptomyces coelicolor* and their effects on sorghum plants grown in soil amended with chitin of brawn scales. Biol Fertil Soils 32:401–409

Adams F, Reddell P, Webb MJ, Shipton WA (2006) Arbuscular mycorrhizas and ectomycorrhias on *Eucalyptus grandis* (*Myrtaceae*) trees and seedlings in native forests of tropical north-eastern Australia. Aust J Bot 54:271–281

Adomas A, Heler G, Olson A, Osborne J, Karlsson M, Nahalkova J, Van Zyl L, Sederoff R, Stenlid J, Finlay R, Asiegbu FO (2008) Comparative analysis of transcript abundance in *Pinus sylvestris* after challenge with a saprotrophic, pathogenic or mutualistic fungus. Tree Physiol 28:885–897

Aimanianda V, Bayry J, Bozza S, Kniemeyer O, Perruccio K, Elluru SR, Clavaud C, Paris S, Brakhage AA, Kaveri SV, Romani L, Latge J-P (2009) Surface hydrophobin prevents immune recognition of airborne fungal spores. Nature 460:1117–1121

Akiyama K, Hayashi H (2008) Plastid-derived strigolactones show the way to roots for symbionts and parasites. New Phytol 178:695–698

Akiyama K, Matsuzaki K, Hayashi H (2005) Plant sesquiterpenes induce hyphal branching in arbuscular mycorrhizal fungi. Nature 435:824–827

Albrecht C, Asselin A, Piché Y, Lapeyrie F (1994) Chitinase activities are induced in *Eucalyptus globulus* roots by ectomycorrhizal or pathogenic fungi, during early colonization. Physiol Plant 91:104–110

Artursson V, Finlay RD, Jansson JK (2006) Interactions between arbuscular mycorrhizal fungi and bacteria and their potential for stimulating plant growth. Environ Microbiol 8:1–10

Bendel CM, Hostetter MK (1993) Distinct mechanisms of epithelial adhesion for *Candida albicans* and *Candida tropicalis*. J Clin Invest 92:1840–1849

Besserer A, Puech-Pages V, Kiefer P, Gomez-Roldan V, Jauneau A, Roy S, Portais JC, Roux C, Becard G, Sejalon-Delmas N (2006) Strigoactones stimulate arbuscular mycorrhizal fungi by activating mitochondria. PLoS Biol 4:e226

Besserer A, Becard G, Jauneau A, Roux C, Sejalon-Delmas N (2008) GR24, a synthetic analog of strigolactones, stimulates the mitosis and growth of the arbuscular mycorrhizal fungus *Gigaspora rosea* by boosting its energy metabolism. Plant Physiol 148:402–413

Beyrle H (1995) The role of phytohormones in the function and biology of mycorrhizas. In: Varma AK, Hock B (eds) Mycorrhiza: structure, molecular biology and function. Springer, Berlin, p 365

Bomberg M, Timonen S (2007) Distribution of Cren- and Euryarchaeota in Scots pine mycorrhizospheres and boreal forest humus. Microb Ecol 54:406–416

Bonfante P, Genre A (2010) Mechanisms underlying beneficial plant-fungus interactions in mycorrhizal symbiosis. Nat Commun 1:48

Bouwmeester HJ, Matusova R, Zhongkui S, Beale MH (2003) Secondary metabolite signalling in host-parasitic plant interactions. Curr Opin Plant Biol 6:358–364

Bowden CG, Smalley E, Guries RP, Hubbes M, Temple B, Horgen PA (1996) Lack of association between cerato-ulmin production and virulence in *Ophiostoma novo-ulmi*. Mol Plant Microbe Interact 9: 556–564

Brachmann A, Parniske M (2006) The most widespread symbiosis on Earth. PLoS Biol 4:1111–1112

Brady NC (1990) The nature and properties of soils, 10th edn. Macmillan, New York

Brandes B, Godbold DL, Kuhn AJ, Jentschke G (1998) Nitrogen and phosphorus acquisition by the mycelium of the ectomycorrhizal fungus *Paxillus involutus* and its effect on host nutrition. New Phytol 140:735–743

Brasier CM, Kirk SA, Tegli S (1995) Naturally occurring noncerato-ulmin producing mutants of *Ophiostoma novo-ulmi* are pathogenic but lack aerial mycelium. Mycol Res 99:436–440

Brulé C, Frey-Klett P, Pierrat JC, Courier S, Gerard F, Lemoine MC, Rousselet JL, Sommer G, Garbaye J (2001) Survival in the soil of the ectomycorrhizal fungus *Laccaria bicolor* and the effect of a mycorrhiza helper *Pseudomonas fluorescens*. Soil Biol Biochem 33:1683–1694

Brundrett MC (2002) Coevolution of roots and mycorrhizas of land plants. New Phytol 154:275–304

Buée M, Vairelles D, Garbaye J (2005) Year-round monitoring of diversity and potential metabolic activity of the ectomycorrhizal community in a beech (*Fagus silvatica*) forest subjected to two thinning regimes. Mycorrhiza 15:235–245

Buée M, De Boer W, Martin F, van Overbeek L, Jurkevitch E (2009) The rhizosphere zoo: an overview of plant-associated communities of microorganisms, including phages, bacteria, archaea and fungi, and of some of their structuring factors. Plant Soil 321:189–212

Burgess T, Laurent P, Dell B, Malajczuk N, Martin F (1995) Effect of the fungal isolate aggressivity on the biosynthesis of symbiosisrelated polypeptides in differentiating eucalypt ectomycorrhiza. Planta 195:408–417

Camehl I, Sherameti I, Venus Y, Bethke G, Varma A, Lee J, Oelmuller R (2010) Ethylene signalling and ethylene-targeted transcription factors are required to balance beneficial and nonbeneficial traits in the symbiosis between the endophytic fungus Piriformospora indica and Arabidopsis thaliana. New Phytol 185:1062–1073

Calera JA, Calderone RA (1999) Identification of a putative response regulator, two-component phosphorelay gene (CaSSK1) from *Candida albicans*. Yeast 15:1243–1254

Calera JA, Zhao XJ, Calderone R (2000) Defective hyphal development and avirulence caused by a deletion of the *SSK1* response regulator gene in *Candida albicans*. Infect Immun 68:518–525

Catford JG, Staehelin C, Lerat S, Piché Y, Vierheilig H (2003) Suppression of arbuscular mycorrhizal colonization and nodulation in split-root systems of alfalfa after pre-inoculation and treatment with Nod factors. J Exp Bot 54:1481–1487

Catford JG, Staehelin C, Larose G, Piché Y, Vierheilig H (2006) Systemically suppressed isoflavonoids and their stimulating effects on nodulation and mycorrhization in alfalfa split-root systems. Plant Soil 285:257–266

Chapin FS III, Moilanen L, Kielland K (1993) Preferential use of organic nitrogen by non-mychorrhizal Arctic sedge. Nature 361:150–153

Chauhan N, Inglis D, Roman E, Pla J, Li D, Calera JA, Calderone R (2003) *Candida albicans* response regulator gene *SSK1* regulates a subset of genes whose functions are associated with cell wall biosynthesis and adaptation to oxidative stress. Eukaryot Cell 2:1018–1024

Chaucheyras-Durand F, Fonty G (2001) Establishment of cellulolytic bacteria and development of fermentative activities in the rumen of gnotobioti-

cally-reared lambs receiving the microbial additive Saccharomyces cerevisiae CNCM I-1077. Reprod Nutr Dev 41:57–68
Chen Z, Agnew JL, Cohen JD, He P, Shan L, Sheen J, Kunkel BN (2007) *Pseudomonas syringae* type III effector AvrRpt2 alters *Arabidopsis thaliana* auxin physiology. Proc Natl Acad Sci USA 104:20131–20136
Cole DW (1981) Nitrogen uptake and translocation by forest ecosystems. Ecol Bull 33:219–232
Correa A, Staples RC, Hoch HC (1996) Inhibition of thigmostimulated cell differentiation with RDG-peptides in *Uromyces* germlings. Protoplasma 194:91–102
Corvis Y, Walcarius A, Rink R, Mrabet NT, Rogalska E (2005) Preparing catalytic surfaces for sensing applications by immobilizing enzymes via hydrophobin layers. Anal Chem 77:1622–1630
Corvis Y, Brezesinski G, Rink R, Walcarius A, Van der Heyden A, Mutelet F, Rogalska E (2006) Analytical investigation of the interactions between SC3 hydrophobin and lipid layers: elaborating of nanostructured matrixes for immobilizing redox systems. Anal Chem 78:4850–4864
Corvis Y, Trzcinska K, Rink R, Bilkova P, Gorecka E, Bilewicz R, Rogalska E (2007) Electron donor-acceptor fullerene derivative retained on electrodes using SC3 hydrophobin. J Phys Chem C 111: 1176–1179
Courty P-E, Pouysegur R, Buee M, Garbaye J (2006) Laccase and phosphatase activities of the dominant ectomycorrhizal types in a lowland oak forest. Soil Biol Biochem 38:1219–1222
Courty P-E, Breda N, Garbaye J (2007) Relation between oak tree phenology and the secretion of organic matter degrading enzymes by *Lactarius quietus* ectomycorrhizas before and during bud break. Soil Biol Biochem 39:1655–1663
da Silva CI, de Queiroz MV, Costa MD, Kasuva MCM, de Araujo EF (2010) Identification of differentially expressed genes of the fungus *Hydnangium* sp. during the pre-symbiotic phase of the ectomycorrhizal association with *Eucalyptus grandis*. Mycorrhiza 20:531–540
De Boer W, Folman LB, Summerbell RC, Boddy L (2005) Living in a fungal world: impact of fungi on soil bacterial niche development. FEMS Microbiol Rev 29:795–811
De Carvalho D (1994) Contribution à l'étude des protéines régulées par la symbiose ectomycorhizienne. PhD thesis, Ecole Nationale de Génie Rurale et Forestier, Nancy
Deveau A, Palin B, Delaruelle C, Peter M, Kohler A, Pierrat JC, Sarniguet A, Garbaye J, Martin F, Frey-Klett P (2007) The mycorrhiza helper *Pseudomonas fluorescens* BBc6R8 has a specific priming effect on the growth, morphology and gene expression of the ectomycorrhizal fungus *Laccaria bicolor* S238N. New Phytol 175:743–755
Ditengou FA, Lapeyrie F (2000) Hypaphorine from the ectomycorrhizal fungus *Pisolithus tinctorius* counteracts activities of indole-3-acetic acid and ethylene but not synthetic auxins in eucalypt seedlings. Mol Plant Microbe Interact 13:151–158
Ditengou FA, Béguiristain T, Lapeyrie F (2000) Root hair elongation is inhibited by hypaphorine, the indole alkaloid from the ectomycorrhizal fungus *Pisolithus tinctorius*, and restored by IAA. Planta 211:722–728
Dou D, Kale SD, Wang X, Jiang RHY, Bruce NA, Arredondo FD, Zhang X, Tyler BM (2008) RXLR-mediated entry of phytophthora sojae effector Avr1b into soybean cells does not require pathogen-encoded machinery. Plant Cell 20:1930–1947
Duplessis S, Sorin C, Voiblet C, Martin F, Tagu D (2001) Cloning and expression analysis of a new hydrophobin cDNA from the ectomycorrhizal basidiomycete *Pisolithus*. Curr Genet 39:335–339
Duplessis S, Courty PE, Tagu D, Martin F (2005) Transcript patterns associated with ectomycorrhiza development in *Eucalyptus globulus* and *Pisolithus microcarpus*. New Phytol 165:599–611
Eastwood DC, Floudas D, Binder M, Majcherczyk A, Schneider P et al (2011) The plant cell wall-decomposing machinery underlies the functional diversity of forest fungi. Science 333: 762–765
Fassler JS, West AH (2011) Fungal Skn7 stress responses and their relationship to virulence. Eukaryot Cell 10:156–167
Felten J, Kohler A, Morin E, Bhalerao RP, Palme K, Martin F, Ditengou FA, Legue V (2009) The ectomycorrhizal fungus *Laccaria bicolor* stimulates lateral root formation in poplar and *Arabidopsis* through auxin transport and signaling. Plant Physiol 151:1991–2005
Fray RG (2002) Altering plant-microbe interaction through artificially manipulating bacterial quorum sensing. Ann Bot 89:245–253
Freeling M (2009) Bias in plant gene content following different sorts of duplication: tandem, whole-genome, segmental, or by transposition. Annu Rev Plant Biol 60:433–533
Frey-Klett P, Chavatte M, Clausse M-L, Courrier S, Le Roux C, Raaijmakers J, Martinotti MG, Pierrat J-C, Garbaye J (2005) Ectomycorrhizal symbiosis affects functional diversity of rhizosphere fluorescent pseudomonads. New Phytol 165:317–328
Frey-Klett P, Garbaye J, Tarkka MT (2007) The mycorrhiza helper bacteria revisited. New Phytol 176: 22–36
Fries N, Serck-Hanssen K, Dimberg LH, Theander O (1989) In vitro enhancement of spore germination and early hyphal growth of a vesicular-arbuscular mycorrhizal fungus by host root exudates and plant flavonoids. Symbiosis 7:243–255
Garbaye J (1994) Helper bacteria: a new dimension to the mycorrhizal symbiosis. New Phytol 128:197–210

Gay G, Normand L, Marmeisse R, Sotta B, Debaud JC (1994) Auxin overproducer mutants of *Hebeloma cylindrosporum* Romagnési have increased mycorrhizal activity. New Phytol 128:645

Gea L, Normand L, Vian B, Gay G (1994) Structural aspects of ectomycorrhiza of *Pinus pinaster* (Ait.) Sol. formed by an IAA-overproducer mutant of *Hebeloma cylindrosporum* Romagnési. New Phytol 128:659

Gil ML, Penalver MC, Lopez-Ribot JL, O'Connor JE, Martinez JP (1996) Binding of extracellular matrix proteins to *Aspergillus fumigatus* conidia. Infect Immun 64:5239–5247

Gogala N (1991) Regulation of mycorrhizal infection by hormonal factors produced by hosts and fungi. Cell Mol Life Sci 47:331–340

Grant CM (2001) Role of the glutathione/glutaredoxin and thioredoxin systems in yeast growth and response to stress conditions. Mol Microbiol 39:533–541

Hamill JD (1993) Alterations in auxin and cytokinin metabolism of higher plants due to expression of specific genes from pathogenic bacteria: a review. Aust J Plant Physiol 20:405

Heller G, Adomas A, Li G, Osborne J, van Zyl L, Sederoff R, Finlay RD, Stenlid J, Asiegbu FO (2008) Transcriptional analysis of *Pinus sylvestris* roots challenged with the ectomycorrhizal fungus *Laccaria bicolor*. BMC Plant Biol 8:19

Hilbert JL, Costa G, Martin F (1991) Ectomycorrhizal synthesis and polypeptide changes during the early stage of eucalypt mycorrhiza development. Plant Physiol 97:977–984

Högberg P, Hordgre A, Buchmann N, Taylor AFS, Ekblad A, Högberg MN, Nyberg G, Ottosson-Löfvenius M, Read DJ (2001) Largescale forest girdling shows that current photosynthesis drives soil respiration. Nature 411:789–792

Holder DJ, Keyhani NO (2005) Adhesion of the entomopathogenic fungus *Beauveria* (*Cordyceps*) *bassiana* to substrata. Appl Environ Microbiol 71: 5260–5266

Horan DP, Chilvers GA, Lapeyrie F (1988) Time sequence of the infection process in eucalypt ectomycorrhizas. New Phytol 109:451–458

Hostetter M (2000) RGD-mediated adhesion in fungal pathogens of humans, plants and insects. Curr Opin Microbiol 3:344–348

Ingram DS, Tommerup IC (1972) The life history of *Plasmodiophora brassicae* Woron. Proc R Soc Lond Ser B 180:103–112

Jambois A, Daupin A, Kawano T, Ditengou F, Bouteau F, Legué V, Lapeyrie F (2005) Competitive anagonism between IAA and indole alkaloid hypaphorine must contribute to regulate ontogenesis. Physiol Plant 123:120–129

Jensen BG, Andersen MR, Pedersen MH, Frisvad JC, Sondergaard I (2010) Hydrophobins from *Aspergillus* species cannot be clearly divided into two classes. BMC Res Notes 3:344

Kale SD, Gu B, Capelluto DGS, Feldman E, Rumore A, Arredondo FD, Hanlon R, Fudal I, Rouxel T, Lawrence CB, Shan W, Tyler BM (2010) External lipid PI3P mediates entry of eukaryotic pathogen effectors into plant and animal host cells. Cell 142:284–295

Kambara K, Ardissone S, Kobayashi H, Saad MM, Schumpp O, Broughton WJ, Deakin WJ (2009) Rhizobia utilize pathogen-like effector proteins during symbiosis. Mol Microbiol 71:92–106

Karabaghli-Degron C, Sotta B, Bonnet M, Gay G, Le Tacon F (1998) The auxin transport inhibitor 2,3,5-triiodobenzoic acid (TIBA) inhibits the stimulation of in vitro lateral root formation and the colonization of the tap-root cortex of Norway spruce (*Picea abies*) seedlings by the ectomycorrhizal fungus *Laccaria bicolor*. New Phytol 140:723

Kaska DD, Myllylä R, Cooper JB (1999) Auxin transport inhibitors act through ethylene to regulate dichotomous branching of lateral root meristems in pine. New Phytol 142:49

Kawano T, Kawano N, Hosoya H, Lapeyrie F (2001) Fungal auxin antagonist hyphphorine competitively inhibits indole-3-acetic acid-dependent superoxide generation by horseradish peroxidase. Biochem Biophys Res Commun 288:546–551

Kazmierczak P, Pfeiffer P, Zhang L, Van Alfen NK (1996) Transcriptional repression of specific host genes by the mycovirus *Cryphonectria*-hypovirus-1. J Virol 70:1137–1142

Kemppainen M, Duplessis S, Martin F, Pardo AG (2009) RNA silencing in the model mycorrhizal fungus *Laccaria bicolor*: gene knock-down of nitrate reductase results in inhibition of symbiosis with *Populus*. Environ Microbiol 11:1878–1896

Kershaw MJ, Talbot NJ (1998) Hydrophobins and repellents: proteins with fundamental roles in fungal morphogenesis. Fungal Genet Biol 23:18–33

Kershaw MJ, Thornton CR, Wakley GE, Talbot NJ (2005) Four conserved intramolecular disulphide linkages are required for secretion and cell wall localization of a hydrophobin during fungal morphogenesis. Mol Microbiol 56:117–125

Kielland K (1990) Processes controlling nitrogen release and turnover in arctic tundra. PhD thesis, University of Alaska, Alaska

Kikuchi K, Matsushita N, Suzuki K, Hogetsu T (2007) Flavonoids induce germination of basidiospores of the ectomycorrhizal fungus *Suillus bovinus*. Mycorrhiza 17:563–570

Kim S, Ahn I-P, Rho H-S, Lee Y-H (2005) *MPH1*, a *Magnaporthe grisea* hydrophobin gene, is required for fungal development and plant colonization. Mol Microbiol 57:1224–1237

Kloppholz S, Kuhn H, Requena N (2011) A secreted fungal effector of *Glomus intraradices* promotes symbiotic biotrophy. Curr Biol 21:1204–1209

Knee EM, Gong FC, Gao M, Teplitski M, Jones AR, Foxworthy A, Mort AJ, Bauer WD (2001) Root mucilage from pea and its utilization by rhizo-

sphere bacteria as a sole carbon source. Mol Plant Microbe Interact 14:775–784
Kondoh O, Muto A, Kajiwara S, Takagi J, Saito Y, Shishido K (1995) Fruiting body-specific cDNA, mfbAc, from the mushroom *Lentinus edodes* encodes a high-molecular-weight cell-adhesion protein containing an Arg-Gly-Asp motif. Gene 154:31–37
Köttke I, Oberwinkler F (1986) Root-Fungus interactions observed on initial stages of mantle formation and Hartig net establishment in mycorrhizas of *Amanita muscaria* on *Picea abies* in pure culture. Can J Bot 64:2348–2354
Kottke I, Oberwinkler F (1987) Cellular structure and function of the Hartig net: coenocytic and transfer cell-like organization. Nord J Bot 7:85–95
Kroehler CJ, Linkins AE (1991) The absorption of inorganic phosphate from ^{32}P labeled inositol hexaphosphate by *Eriophorum vainatum*. Oecologia 85:424–428
Kunkel BN, Bent AF, Dahlbeck D, Innes RW, Staskawicz BJ (1993) RPS2, an *Arabidopsis* disease resistance locus specifying recognition of *Pseudomonas syringae* strains expressing the avirulence gene avrRpt2. Plant Cell 5:865–875
Labbé J, Jorge V, Kohler A, Vion P, Marcais B, Bastien C, Tuskan GA, Martin F, Le Tacon F (2010) Identification of quantitative trait loci affecting ectomycorrhizal symbiosis in an interspecific F1 poplar cross and differential expression of genes in ectomycorrhizas of the two parents: *Populus deltoides* and *Populus trichocarpa*. Tree Genet Genomics 7:617–627
Lagrange H, Jay-Allemand C, Lapeyrie F (2001) Rutin, the phenolglycoside from eucalyptus root exudates, stimulates *Pisolithus* hyphal growth at picomolar concentrations. New Phytol 149:349–355
Larose G, Chenevert R, Moutoglis P, Gagne S, Piché Y, Vierheilig H (2002) Flavonoid levels in roots of *Medicago sativa* are modulated by the developmental stage of the symbiosis and the root colonizing arbuscular mycorrhizal fungus. J Plant Physiol 159:1329–1339
Larsen PE, Trivedi G, Sreedasyam A, Lu V, Podila GK, Collart FR (2010) Using deep RNA sequencing for the structural annotation of the *Laccaria bicolor* mycorrhizal transcriptome. PLoS One 5:e9780
Larsen PE, Sreedasyam A, Trivedi G, Podila GK, Cseke LJ, Collart FR (2011) Using next generation transcriptome sequencing to predict an ectomycorrhizal metabolome. BMC Syst Biol 5:70
Laurent P (1995) Contribution à l'étude des protéines régulées par la symbiose chez l'ectomycorhize d'Eucalyptus – Pisolithus. Caractérisation de mannoprotéines pariétales chez le basidiomycete *Pisolithus tinctorius*. PhD thesis, University Henri Poincaré, Nancy
Laurent P, Voiblet C, Tagu D, De Carvalho D, Nehls U, De Bellis R, Balestrini R, Bauw G, Bonfante P, Martin F (1999) A novel class of ectomycorrhiza-regulated cell wall polypeptides in *Pisolithus tinctorius*. Mol Plant Microbe Interact 12:862–871
Le Quéré A, Eriksen KA, Rajashekar B, Schützendübel A, Canbäck B et al (2006) Screening for rapidly evolving genes in the ectomycorrhizal fungus *Paxillus involutus* using cDNA microarrays. Mol Ecol 15:535–550
Leake JR, Johnson D, Donnelly DP, Muckle G, Boddy L, Read D (2004) Networks of power and influence: the role of mycorrhizal mycelium in controlling plant communities and agroecosystem functioning. Can J Bot 82:1016–1045
Linder MB, Szilvay GR, Nakari-Setala T, Penttila ME (2005) Hydrophobins: the protein amphiphiles of filamentous fungi. FEMS Microbiol Rev 29:877–896
Ludwig-Müller J, Bendel U, Thermann P, Ruppel M, Epstein E, Hilgenberg W (1993) Concentrations of indole-3-acetic acid in plants of tolerant and susceptible varieties of Chinese cabbage infected with *Plasmodiophora brassicae* Woron. New Phytol 125:763–769
Ludwig-Müller J, Epstein E, Hilgenberg W (1996) Auxin-conjugate hydrolysis in Chinese cabbage: characterization of an amidohydrolase and its role during infection with clubroot disease. Physiol Plant 97:627–634
Lundeberg G (1970) Utilisation of various nitrogen sources, in particular bound soil nitrogen, by mycorrhizal fungi. Stud For Suec 79:11–17
Luo ZB, Janz D, Jiang X, Gobel C, Wildhagen H, Tan Y, Rennenberg H, Feussner I, Polle A (2009) Upgrading root physiology for stress tolerance by ectomycorrhizas: insights from metabolite and transcriptional profiling into reprogramming for stress anticipation. J Plant Physiol 151:1902–1917
Maier A (2003) Einfluss bakterieller Stoffwechselprodukte auf Wachstum und Proteom des Ektomykorrhizapilzes Amanita muscaria. PhD thesis, University of Tubingen, Tubingen
Martin F, Laurent P, de Carvalho D, Voiblet C, Balestrini R, Bonfante P, Tagu D (1999) Cell wall proteins of the ectomycorrhizal basidiomycete *Pisolithus tinctorius*: identification, function and expression in symbiosis. Fungal Genet Biol 27: 161–174
Martin F (2007) Fair trade in the underworld: the ectomycorrhizal symbiosis. In: Howard RJ, Gow NAR (eds) Biology of the fungal cell, 2nd edn, The mycota VIII. Springer-Verlag, Berlin, Heidelberg, pp 291–308
Martin F, Perotto S, Bonfante P (2007) Mycorrhizal fungi: a fungal community at the interface between soil and roots. In: Pinton R, Varanini Z, Nannipieri P (eds) The rhizosphere. Biochemistry and organic substances at the soil-plant interface. Dekker, New York, pp 263–296

Martin F, Aerts A, Ahrén D, Brun A, Danchin EGJ et al (2008) The genome of *Laccaria bicolor* provides insights into mycorrhizal symbiosis. Nature 452:88–92

Martin F, Kohler A, Murat C, Balestrini R, Coutinho PM et al (2010) Périgord black truffle genome uncovers evolutionary origins and mechanisms of symbiosis. Nature 464:1033–1038

Martin F, Cullen D, Hibbett D, Pisabarro A, Spatafora JW, Baker SE, Grigoriev IV. (2011) Sequencing the Fungal Tree of Life. New Phytologist 190: 818–821

Meixner C, Ludwig-Muller J, Miersch O, Gresshoff P, Staehelin C, Vierheilig H (2005) Lack of mycorrhizal autoregulation and phytohormonal changes in the supernodulating soybean mutant *nts1007*. Planta 222:709–715

Meixner C, Vegvari G, Ludwig-Muller J, Gagnon H, Steinkellner S, Staehelin C, Gresshoff P, Vierheilig H (2007) Two defined alleles of the LRR receptor kinase *GmNARK* in supernodulating soybean govern differing autoregulation of mycorrhization. Physiol Plant 130:261–270

Mengel K, Kirby EA (1987) Plant nutrients. In: Mengel K, Kirby EA (eds) Principles of plant nutrition. International Potash Institute, Geneva, pp 11–24

Mitchell TK, Dean RA (1995) The cAMP dependent protein kinase catalytic subunit is required for appressorium formation and pathogenesis by the rice blast pathogen *Magnaporthe grisea*. Plant Cell 7:1869–1878

Montanini B, Levati E, Bolchi A, Kohler A, Morin E, Tisserant E, Martin F, Ottonello S (2011) Genome-wide search and functional identification of transcription factors in the mycorrhizal fungus *Tuber melanosporum*. New Phytol 189:736–750

Morandi D, Branzanti B, Gianinazzi-Pearson V (1984) Isoflavonoid accumulation in soybean roots infected with vesicular-arbuscular mycorrhizal fungi. Physiol Plant Pathol 24:357–364

Moyersoen B (2006) *Pakaraimaea dipterocarpacea* is ectomycorrhizal, indicating an ancient Gondwanaland origin for the ectomycrrhizal habit in Dipterocarpaceae. New Phytol 172:753–762

Mukhtar MS, Carvunis A-R, Dreze M, Epple P, Steinbrenner J et al (2011) Independently evolved virulence effectors converge onto hubs in a plant immune system network. Science 333:576–601

Müller O, Schreier PH, Uhrig JF (2008) Identification and characterization of secreted and pathogenesis-related proteins in *Ustilago maydis*. Mol Genet Genomics 279:27–39

Narasimhan S, Armstrong MY, Rhee K, Edman JC, Richards FF, Spicer E (1994) Gene for an extracellular matrix receptor protein from *Pneumocystis carinii*. Proc Natl Acad Sci USA 91:7440–7444

Nehls U, Béguiristain T, Ditengou F, Lapeyrie F, Martin F (1998) The expression of a symbiosis-regulated gene in eucalypt roots is regulated by auxins and hypaphorine, the tryptophan betaine of the ectomycorrhizal basidiomycete *Pisolithus tinctorius*. Planta 207:296–302

Nurmiaho-Lassila EL, Timonen S, Haahtela K, Sen R (1997) Bacterial colonization patterns of intact *Pinus sylvestris* mycorrhizospheres in dry pine forest soil: an electron microscopy study. Can J Microbiol 43:1017–1035

Ochsenreiter T, Selezi D, Quaiser A, Bonch-Osmolovskaya L, Schleper C (2003) Diversity and abundance of Crenarchaeota in terrestrial habitats studied by 16S RNA surveys and real time PCR. Environ Microbiol 5:787–797

Parta M, Chang Y, Rulong S, Pinto-DaSilva P, Kwon-Chung KJ (1994) HYP1, a hydrophobin gene from *Aspergillus fumigatus*, complements the rodletless phenotype in *Aspergillus nidulans*. Infect Immun 62:4389–4395

Penalver MC, O'Connor JE, Martinez JP, Gil ML (1996) Binding of human fibronectin to *Aspergillus fumigatus* conidia. Infect Immun 64:1146–1153

Pérez-García A, Olalla L, Rivera E, Del Pino D, Canovas I, De Vicente A, Torés JA (2001) Development of *Sphaerotheca fusca* on susceptible, resistant, and temperature-sensitive resistant cultivars. Mycol Res 105:1216–1222

Pierchbacher MD, Ruoslahti E (1984) Cell attachment activity of fibronectin can be duplicated by small synthetic fragments of the molecule. Nature 309:30–33

Plett JM, Martin F (2011) Blurred boundaries: lifestyle lessons from fungal ectomycorrhizal genomes. Trends Genet 27:14–22

Plett JM, Kemppainen M, Kale SD, Kohler A, Legue V, Brun A, Tyler B, Pardo A, Martin F (2011) A secreted effector protein of *Laccaria bicolor* is required for symbiosis development. Curr Biol 21:1197–1203

Plett JM, Gibon J, Kohler A, Duffy K, Hoegger PJ, Velagapudi R, Han J, Kues U, Grigoriev IV, Martin F (2012) Phylogenetic, genomic organization and expression analysis of hydrophobin genes in the ectomycorrhizal basidiomycete *Laccaria biocolor*. Fungal Genet Biol 49:199–209

Qin M, Wang L-K, Feng X-Z, Yang Y-L, Wang R, Wang C, Yu L, Shao B, Qiao M-Q (2007) Bioactive surface modification of mica and poly(dimethylsiloxane) with hydrophobins for protein immobilization. Langmuir 23:4465–4471

Rajashekar B, Samson P, Johansson T, Tunlid A (2007) Evolution of nucleotide sequences and expression patterns of hydrophobin genes in the ectomycorrhizal fungus *Paxillus involutus*. New Phytol 174:399–411

Rasmussen TB, Skindersoe ME, Bjarnsholt T, Phipps RK, Christensen KB, Jensen PO, Andersen JB, Koch B, Larsen TO, Hentzer M, Eberl L, Hoiby N, Givskov M (2005) Identity and effects of quorum-sensing inhibitors produced by Penicillium species. Microbiology 151:1325–1340

Read DJ, Perez-Moreno J (2003) Mycorrhizas and nutrient cycling in ecosystems – a journey towards relevance? New Phytol 157:475–492

Regenfelder E, Spellig T, Hartmann A, Lauenstein S, Bolker M, Kahmann R (1997) G proteins in *Ustilago maydis*: transmission of multiple signals? EMBO J 16:1934–1942

Rikhvanov EG, Varakina NN, Sozinov DY, Voinikov VK (1999) Association of bacteria and yeasts in hot springs. Appl Environ Microbiol 65:4292–4293

Roby D, Gadelle A, Toppan A (1987) Chitin oligosaccharides as elicitors of chitinase activity in melon plants. Biochem Biophys Res Commun 143:885–892

Ruess RW, Hendrick RL, Burton AJ, Pregitzer KS, Sveinbjornsson B, Allen ME, Maurer GE (2003) Coupling fine root dynamics with ecosystem carbon cycling in black spruce forests of interior Alaska. Ecol Monogr 73:643–662

Ruoslahti E (1996) RGD and other recognition sequences for integrins. Annu Rev Cell Dev Biol 12:697–715

Rupp LA, Mudge KW (1985) Ethephon and auxin induce mycorrhiza-like changes in the morphology of root organ cultures of Mugo pine. Physiol Plant 64:316

Rupp LA, Mudge KW, Negm FB (1989) Involvement of ethylene in ectomycorrhiza formation and dichotomous branching of roots of mugo pine seedlings. Can J Bot 67:477

Scherrer S, Honegger R (2003) Inter- and intraspecific variation of homologous hydrophobin (H1) gene sequences among *Xanthoria* spp. (lichen-forming ascomycetes). New Phytol 158:375–389

Scherrer S, Haisch A, Honegger R (2002) Characterization and expression of XPH1, the hydrophobin gene of the lichen-forming ascomycete *Xanthoria parietina*. New Phytol 154:175–184

Schrey SD, Tarkka MT (2008) Friends and foes: streptomycetes as modulators of plant disease and symbiosis. Antonie Leeuwenhoek 94:11–19

Schrey SD, Schellhammer M, Ecke M, Hampp R, Tarkka MT (2005) Mycorrhiza helper bacterium *Streptomyces* AcH 505 induces differential gene expression in the ectomycorrhizal fungus *Amanita muscaria*. New Phytol 168:205–216

Selosse MA, Richard F, He X, Simard SW (2006) Mycorrhizal networks:des liaisons dangereuses? Trends Ecol Evol 21:621–628

Sharp JA, McNeil M, Albersheim P (1984a) The primary structure of one elicitor active and seven elicitor-inactive hexa (ß-D-glucopyranolsyl)-D-glucitols isolated from the mycelia walls of *Phytophthora megasperma* f.sp. *glycinea*. J Biol Chem 25:11321–11326

Sharp JA, Valent B, Albersheim P (1984b) Purification and partial characterization of a β-glucan fragment that elicits phytoalexin accumulation in soybean. J Biol Chem 25:11312–11320

Singh P, Chauhan N, Ghosh A, Dixon F, Calderone R (2004) *SKN7* of *Candida albicans*: mutant construction and phenotype analysis. Infect Immun 72:2390–2394

Spanu P, Bonfante-Fasolo P (1988) Cell wall-bound peroxidase activity in roots of mycorrhizal *Allium porrum*. New Phytol 109:119–124

Spanu P, Boller T, Ludwig A, Wiemken A, Faccio A, Bonfante-Fasolo P (1989) Chitinase in roots of mycorrhizal *Allium porrum*: regulation and localization. Planta 177:447–455

Splivallo R, Fischer U, Gobel C, Feussner I, Karlovsky P (2009) Truffles regulate plant root morphogenesis via the production of auxin and ethylene. Plant Physiol 150:2018–2029

Sundaram S, Kim SJ, Suzuki H, McQuattie CJ, Hiremath ST, Podila GK (2001) Isolation and characterization of a symbiosis-regulated *ras* from the ectomycorrhizal fungus *Laccaria bicolor*. Mol Plant Microbe Interact 14:618–628

Sunde M, Kwan AHY, Templeton MD, Beever RE, Mackay JP (2008) Structural analysis of hydrophobins. Micron 39:773–784

Tagu D, Nasse B, Martin F (1996) Cloning and characterization of hydrophobin-encoding cDNAs from the ectomycorrhizal basidiomycete *Pisolithus tinctorius*. Gene 168:93–97

Tagu D, Kottke I, Martin F (1998) Hydrophobins in ectomycorrhizal symbiosis: hypothesis. Symbiosis 25:5–18

Tagu D, De Bellis R, Balestrini R, De Vries O, Piccoli G, Stocchi V, Bonfante P, Martin F (2001) Immunolocalization of hydrophobin HYDPt-1 from the ectomycorrhizal basidiomycete *Pisolithus tinctorius* during colonisation of *Eucalyptus globulus* roots. New Phytol 149:127–135

Talbot NJ, Ebbole DJ, Hamer JE (1993) Identification and characterization of MPG1, a gene involved in pathogenicity from the rice blast fungus *Magnaporthe grisea*. Plant Cell 5:1575–1590

Talbot NJ, Kershaw MJ, Wakley GE, de Vries OMH, Wessels JGH, Hamer JE (1996) MPG1 encodes a fungal hydrophobin involved in surface interactions during infection related development by the rice blast fungus *Magnaporthe grisea*. Plant Cell 8:985–999

Tarkka MT, Frey-Klett P (2008) Mycorrhiza helper bacteria. In: Varma A (ed) Mycorrhiza, 3rd edn. Springer, Berlin, pp 113–132

Tarkka MT, Sarniguet A, Frey-Klett P (2009) Inter-kingdom encounters: recent advances in molecular bacterium–fungus interactions. Curr Genet 55:233–243

Taylor AFS, Martin F, Read DJ (2000) Fungal diversity in ectomycorrhizal communities of Norway spruce Picea abies (L.) Karst. and beech (*Fagus sylvatica* L.) along north–south transects in Europe. Ecol Stud 142:343–365

Templeton MD, Rikkerink EHA, Beever RE (1994) Small, cysteine-rich proteins and recognition in fungal–plant interactions. Mol Plant Microbe Interact 7:320–325

Teplitski M, Robinson JB, Bauer WD (2000) Plants secrete substances that mimic bacterial N-acyl homoserine lactone signal activities and affect population density-dependent behaviors in associated bacteria. Mol Plant Microbe Interact 13:637–648

Thau N, Monod M, Cretani B, Rolland C, Tronchin G et al (1994) Rodletless mutants of *Aspergillus fumigatus*. Infect Immun 62:4380–4388

Tibbett M (2000) Roots, foraging and the explication of soil nutrient patches: the role of mycorrhizal symbiosis. Funct Ecol 14:397–399

Tibbett M, Sanders FE, Minto SJ, Dowell M, Cairney JWG (1998) Utilization of organic nitrogen by ectomycorrhizal fungi (Hebeloma spp.) of arctic and temperate origin. Mycol Res 102(12):1525–1532

Tibbett M, Sanders FE, Cairney JWG (1999) Long term storage of ectomycorrhizal basidiomycetes (*Hebeloma* spp.) at low temperature. J Basic Microbiol 39:381–384

Tisserant E, Da Silva C, Kohler A, Morin E, Wincker P, Martin F (2011) Deep RNA sequencing improved the structural annotation of the *Tuber melanosporum* transcriptome. New Phytol 189:883–891

Trembley ML, Ringli C, Honegger R (2002a) Differential expression of hydrophobins DGH1, DGH2 and DGH3 and immunolocalization of DGH1 in strata of the lichenized basidiocarp of *Dictyonema glabratum*. New Phytol 15:185–195

Trembley ML, Ringli C, Honegger R (2002b) Hydrophobins DGH1, DGH2, and DGH3 in the lichen-forming Basidiomycete *Dictyonema glabratum*. Fungal Genet Biol 3:247–259

Tuskan GA, Difazio S, Jansson S, Bohlmann J, Grigoriev I et al (2006) The genome of black cottonwood, *Populus trichocarpa*. Science 313:1596–1604

Urbanek H, Kaczmarek A (1985) Extracellular proteinases of the isolate of *Botrytis cinerea* virulent to apple tissues. Acta Biochim Pol 32:101–109

Urbanek H, Yirdaw G (1984) Hydrolytic ability of acid protease of *Fusarium culmorum* and role in phytopatogenesis. Acta Microbiol Pol 33:131–136

Uroz S, Heinonsalo J (2008) Inactivation of N-acyl homoserine lactone quorum sensing molecules by forest soil fungi. FEMS Microbiol Ecol 65: 271–278

Uroz S, Dangelo D, Carlier A, Faure D, Petit A, Oger P, Sicot C, Dessaux Y (2003) Novel bacteria degrading N-acyl homoserine lactones and their use as quenchers of quorum-sensing regulated functions of plant pathogenic bacteria. Microbiology 149:1981–1989

Vierheilig H (2004) Further root colonization by arbuscular mycorrhizal fungi in already mycorrhizal plants is suppressed after a critical level of root colonization. J Plant Physiol 161:339–341

Vierheilig H, Garcia-Garrido JM, Wyss U, Piché Y (2000a) Systemic suppression of mycorrhizal colonization of barley roots already colonized by AM fungi. Soil Biol Biochem 32:589–595

Vierheilig H, Maier W, Wyss U, Samson J, Strack D, Piché Y (2000b) Cyclohexenone derivative and phosphate-levels in split-root systems and their role in the systemic suppression of mycorrhization in precolonized barley plants. J Plant Physiol 157:593–599

Voiblet C, Duplessis S, Encelot N, Martin F (2001) Identification of symbiosis-regulated genes in *Eucalyptus globulus-Pisolithus tinctorius* ectomycorrhiza by differential hybridization of arrayed cDNAs. Plant J 25:181–192

Walker TS, Bais HP, Grotewold E, Vivanco JM (2003) Root exudation and rhizosphere biology. Plant Physiol 132:44–51

Wang D, Pajerowska-Mukhtar K, Culler AH, Dong X (2007) Salicylic acid inhibits pathogen growth in plants through repression of the auxin signaling pathway. Curr Biol 17:1784–1790

Wang X, Wang H, Huang Y, Zhao Z, Qin X, Wang Y, Miao Z, Chen Q, Qiao M (2010) Noncovalently funtionalized multi-wall carbon nanotubes in aqueous solution using the hydrophobin HFBI and their electroanalytical application. Biosens Bioelectron 26:1104–1108

Wessels JGH (1996) Hydrophobins: proteins that change the nature of the fungal surface. Adv Microb Physiol 38:1–45

Wessels JGH, de Vries OMH, Asgeirsdottir SA, Schuren FHJ (1991) Hydrophobin genes involved in formation of aerial hyphae and fruit bodies in *Schizophyllum*. Plant Cell 3:793–799

Wild A (1988) Plant nutrients in soil. In: Wild A (ed) Russel's soil conditions and plant growth, 11th edn. Longman, Harlow, pp 695–743

Wösten HAB (2001) Hydrophobins: multipurpose proteins. Annu Rev Microbiol 55:625–646

Wösten HAB, Wessels JGH (1997) Hydrophorins, from molecular structure to multiple functions in fungal development. Mycoscience 38:363–374

Yasuda T, Shishido K (1997) Aggregation of yeast cells induced by the Arg-Gly-Asp motif-containing fragment of high-molecular mass cell-adhesion protein MFBA, derived from the basidiomycetous mushroom *Lentinus edodes*. FEMS Microbiol Lett 154:195–200

Zampieri E, Balestrini R, Kohler A, Abbà S, Martin F, Bonfante P (2011) The Perigord black truffle responds to cold temperature with an extensive reprogramming of its transcriptional activity. Fungal Genet Biol 48:585–591

Zhang L, Villalon D, Sin Y, Kazmierczak P, Van Alfen NK (1994) Virus-associated down-regulation of the gene encoding cryparin, an abundant cell surface protein of the chestnut blight fungus *Cryphonectria parasitica*. Gene 139:59-64

Zhao Z-X, Qiao M-Q, Yin F, Shao B, Wu B-Y, Wang Y-Y, Wang X-S, Qin X, Li S, Yu L, Chen Q (2007) Amperometric glucose biosensor based on self-assembly hydrophobin with high efficiency of enzyme utilization. Biosens Bioelectron 22:3021–3027

7 Carbohydrates Exchange Between Symbionts in Ectomycorrhizas

U. Nehls[1], I. Bodendiek[1]

CONTENTS

I. Introduction

Due to their great economic and ecologic importance, forests are of outstanding significance for mankind. To guarantee the optimal function of boreal and temperate forest ecosystems, the mutualistic interaction (ectomycorrhiza) of tree roots with certain soil fungi is essential (Smith and Read 2008). From the plant point of view, this interaction is based on the fact that in natural forests major nutrients (nitrogen, phosphate) are either fixed in the organic layer or contained in microorganisms and lower animals, hence in resources, where trees only have limited access to (Harley and Smith 1983; Smith and Read 1997).

The shortness of easily degradable carbohydrates in forest soils seems to be a major driving force for the fungal partner to become involved in this type of symbiosis. Ectomycorrhizal (ECM) interaction enables both partners to optimize their nutrient supply, thus paving the way for an efficient colonization of nutrient-limited environments. In consequence, these partners are ecologically dependent on each other under most natural conditions. However, under optimal nutritional conditions, which are rare in nature, ECM fungi [in contrast to the obligate biotrophic arbuscular mycorrhizal (AM) fungi] as well as their host plants are independent of each other to a certain extent.

Even though varying with soil depth and host tree species, ECM mycelial biomass seems to correlate with the distribution of tree roots in the respective soil profile (Wallander et al. 2004; Göransson et al. 2006) and can comprise up to 80 % of the total fungal and 30% of the microbial biomass (Wallander et al. 2001; Högberg and Högberg 2002; Wallander 2006). Compared to saprotrophic fungi, ectomycorrhizal fungi are enriched in deeper surface layers, characterized by increased nitrogen to carbon ratio. Here, easily degradable carbohydrate resources are depleted while suitable N sources are still available. Due to their ability to gain plant-derived carbohydrates very efficiently, ectomycorrhizal fungi increase their competitiveness ability in soil (Smith and Read 1997; Leake et al. 2001; Anderson and Cairney 2007) and can fill the niche between litter decomposers, that are dependent on easily degradable soil carbohydrates, and wood/lignin degrading organisms. Direct access to plant carbon sources by mycorrhizal fungi is also responsible for a large, additional carbon input into the soil. Therefore,

[1]Faculty 2, Biology/Chemistry, Botany, University of Bremen, Leobener Str. 2, 28359 Bremen, Germany; e-mail: nehls@uni-bremen.de

Fungal Associations, 2nd Edition
The Mycota IX
B. Hock (Ed.)

ECM fungi are regarded as key elements of forest ecosystem processes, e.g., nutrient cycling and carbon entry (for reviews, see Read et al. 2004; Anderson and Cairney 2007), and improvement of plant nutrition.

Being of such immeasurable value for forest ecosystems, ectomycorrhizal symbiosis stimulated researchers since the initial description by Frank (1885) to investigate their ecological significance in different forest ecosystems and the underlying developmental and functional mechanisms. Due to molecular techniques, finished and ongoing genome sequencing projects, and the capability of genetic manipulation of plant and fungal models, our understanding of the molecular mechanisms underlying ectomycorrhiza formation and function is rapidly evolving, steadily providing new results, and making it difficult to keep track of the latest developments in this research area. For this reason, we start by reviewing the current views on ECM fungal carbohydrate nutrition in symbiosis before trying to deduce some general mechanisms.

II. The Ectomycorrhizal Fungal Colony

Soil growing hyphae, exploring litter or mineral layers for nutrients, can constitute a large part of the ECM fungal colony (Agerer 2001; Anderson and Cairney 2007). When soil-growing hyphae recognize an emerging fine root of a suitable plant partner, they direct their growth towards it (Martin et al. 2001) and colonize the root surface, (often) forming a **sheath or mantle** of hyphae, which encloses the root and isolates it from the surrounding soil (Blasius et al. 1986). Root hairs, which are normally formed by rhizodermal cells, are suppressed by ectomycorrhiza formation. After or parallel to sheath formation, fungal hyphae grow inside the colonized fine root, forming highly branched structures in the apoplast of the rhizodermis or that of the entire root cortex, the so-called **Hartig net** (Kottke and Oberwinkler 1986). Both fungal networks (fungal sheath and Hartig net) have different functions (Harley and Smith 1983; Kottke and Oberwinkler 1986; Smith and Read 2008). The **Hartig net serves as an interface between plant and fungus**, where cells are adapted to the exchange of plant-derived carbohydrates for fungus-derived nutrients. The **fungal sheath serves as an intermediate storage compartment** for: (a) nutrients that are delivered by soil growing hyphae and are intended for the Hartig net and (b) carbohydrates that are taken up by hyphae of the Hartig net and are designated for transport towards the soil-growing mycelium (Jordy et al. 1998). In contrast to fast-growing, saprophytic model fungi like *Aspergillus* or *Neurospora*, large parts of the ECM fungal colonies remain functionally interconnected (Anderson et al. 1998; Agerer 2001) making long-distance nutrient and metabolite exchange within the colony a central element of mycorrhizal fungal physiology. For ECM **fungi two major carbon sources** are available: (a) **soil organic matter** (that is degraded by the extra radical mycelium) and (b) **photosynthates from host trees** (available through ectomycorrhizas). The proportions of both resources on the whole carbon budget of ECM colonies are expected to differ with respect to fungal species, season, soil properties, and microbial communities.

III. Soil Carbohydrates and Fungal C-Nutrition

Although litter and humus layers of forest soils are quite rich in complex carbohydrates (e.g., cellulose and lignin) most basidiomycotic ectomycorrhizal fungi together with a large part of other soil microbes seem to be dependent on **simple, readily utilizable carbohydrates**, such as those contained in living or dead organisms and litter. Decomposition of such organic matter leads to the liberation of suitable carbohydrate sources (see below; Wainwright 1993).

Ericoid (Bajwa et al. 1985; Leake and Read 1997) and certain ectomycorrhizal fungi (Abuzinadah and Read 1986; El-Badaoui and Botton 1989; Zhu 1990; Spägele 1992; Zhu et al. 1994; Bending and Read 1996) are able to utilize proteins as nitrogen and carbon sources. In the genome of *Laccaria bicolor* a number of genes encoding putative extracellular proteases are present (Lucic et al. 2008). Proteases produced by mycorrhizal fungi and characterized so far often reveal acidic pH optima, presumably reflecting the frequently low pH value of forest soils, which is often further decreased by fungal activity. Furthermore, data from *Amanita muscaria* indicate that protease excretion occurs in a pH-dependent manner (Nehls et al. 2001). The protease AmProt1, revealing a narrow optimum around pH 3.0, was mainly released at pH values below pH 5.4. In contrast, a second *Amanita* protease, AmProt2, that revealed a broad optimum at pH 3–6, was only excreted at higher pH values (between 5.4 and 6.3). While AmProt1 is thought to be important for the exploration of protein sources in fragmented litter (with low pH values), proteins from bacterial biofilms (pH 5–6; Fletcher 1996) are the supposed aim for AmProt2 release. *AmProt1* was furthermore expressed in a carbon and nitrogen starvation-dependent manner in addition to the environmental pH (Nehls et al. 2001). While nitrogen starvation alone slightly increased *AmProt1* expression by a factor of about three, the absence of a suitable carbon source increased the transcript level of the gene by a factor of 12, independent on fungal nitrogen nutrition. The authors therefore concluded **that proteins of various sources are a valuable carbon resource** for this ECM fungus.

Amino acids (as the products of protein degradation) are frequently found in forest soils and are thus of great importance for nutrition (Wallenda and Read 1999). Fungal amino acid importer genes have been isolated and functionally characterized to date from *A. muscaria* (Nehls et al. 1999b) and *Hebeloma cylindrosporum* (Wipf et al. 2002). As determined by heterologous expression in yeast, these genes encode high affinity H^+/amino acid symporters with a broad amino acid spectrum. Furthermore, a set of oligopeptide and amino acid importers was identified in the genome of *L. bicolor*, indicating the immense meaning of organic nitrogen sources for fungal nutrition (Lucic et al. 2008).

Other important C-sources of ECM fungi are saccharides. As no indications for a direct import of poly- and oligosaccharides into ECM fungi exist, they are supposed to be hydrolyzed first (Wisser et al. 2000) and taken up as monosaccharides by fungal hyphae. With 15 putative monosaccharide transporter (MST) genes found in the genome of *L. bicolor* (Fajardo Lopez et al. 2008), this basidiomycete contains a similar high number as observed in other fungal genomes. Based on their expression profiles fungal MSTs can be grouped into: (a) constitutively expressed members, (b) genes that are induced in a mycorrhiza-dependent manner, and (c) genes that are mainly expressed under carbon starvation. Due to the usually low hexose concentration in forest soils, the deduced proteins of the latter group are expected to be responsible for hexose uptake by soil-growing hyphae. However, fungal mycelia do contain much higher glucose concentrations than usually found in soil (Wainwright 1993; Wallenda 1996). A reduction of unattended hexose leakage of fungal hyphae by an efficient sugar reimport is thus another supposed function of these hexose importers. With regard to hexose uptake of non-mycorrhizal mycelia, distinct differences in their lifestyles can be seen between *L. bicolor* and other well-investigated basidiomycotic ECM fungal models. While carbon-starved *L. bicolor* mycelia revealed maximal glucose uptake immediately after sugar application, hyphae of *A. muscaria* (Wiese et al. 2000) and *H. cylindrosporum* (Salzer and Hager 1991) needed about one day to reach a maximal glucose import rate. Starting from a similar uptake rate of about 5 μmol glucose h^{-1} g^{-1} hyphal fresh weight under carbon starvation (Wiese et al. 2000; Fajardo Lopez et al. 2008), *A. muscaria* exceeded the uptake rate by *L. bicolor* by a factor of about seven about one day after hexose addition. An explanation for this different behavior is that hexose transporter gene expression is induced in a sugar-dependent manner in *A. muscaria* (with a minimal threshold level of at least 0.5 mM for glucose; Nehls et al. 1998; Nehls et al. 1999a) but not in *L. bicolor* (Fajardo Lopez et al. 2008). Therefore, it can be speculated that *A. muscaria* is more flexible concerning the utilization of local soil saccharide pools than *L. bicolor*.

In contrast to these easily degradable carbon sources, there is ample evidence that pectinolytic activities are rare in basidiomycotic ectomycorrhizal fungi (Lindeberg and Lindeberg 1977) and that cellulose and lignin break-down is feasible but rather slow compared to wood-decaying fungi (Trojanowski et al. 1984). Some species such as *L. bicolor* (Martin et al. 2008) or *A. bisporigera* (Nagendran et al. 2009) are even deprived of genes coding cellulose- and hemicellulose-hydrolyzing enzymes and are also reduced in genes encoding polyphenol degrading proteins. In contrast, the genome sequence of *L. bicolor* also reveals expanded glycosyl hydrolase families and a large set of secreted proteases, chitinases, and glucanases, which are likely involved in the break-down of byproducts from decaying leaf litter and dead organisms (Martin and Nehls 2009).

While these detailed investigations were performed with a limited number of model species, further investigations were also performed at the community level. By using microplate-based activity assays for a series of excreted hydrolytic enzymes, a wide range of degradation capacity of litter polymers was observed for various ECM species (Buee et al. 2005, 2007; Courty et al. 2007). The observed fungal degradation capacity was frequently further enhanced upon carbon and nitrogen starvation (Wisser et al. 2000; Pritsch et al. 2004; Buee et al. 2005; Courty et al. 2005).

In contrast to the well-established saprotrophic capability of ECM fungi, Treseder et al. (2006) reported that **in forest ecosystems less than 2 % of the carbon in ectomycorrhizal biomass originated from litter degradation.** As discussed by the authors, this does not necessarily indicate a low decomposer activity of ECM fungi, as acquired ^{14}C from the litter could be preferentially respired by the fungal mycelium soon after uptake (Högberg et al. 2002; van Hees et al. 2005). Furthermore, as observed by Lindahl et al. (2007), ectomycorrhizal taxa colonize primarily the fragmented litter at lower soil layers and are thus expected to acquire C from soil organic matter rather than from fresh leaf litter. Additionally, studies in temperate forests indicated that ECM fungal decomposer capability is mainly enhanced when photosynthetic rates of their host plants decline (late autumn to spring; Buee et al. 2005, 2007; Courty et al. 2007), pointing towards a seasonal impact on soil carbon utilization driven by carbon starvation. Taken the referred data together, the carbon portions of an ECM fungal colony obtained by degradation of soil organic matter versus by direct carbon support by a host plant remains unclear and needs further elucidation.

IV. The Symbiotic Interface

A. Potential Fungal Carbon Sources

In contrast to soil, **plant root exudates can be rich in simple carbohydrates.** The release of C compounds from roots into the surrounding soil is a ubiquitous phenomenon (Norton et al. 1990; Grayston et al. 1997; Lugtenberg et al. 1999). Studies performed on a broad range of plant species indicated a basal carbon efflux that can reach up to 5 % of the net fixed C (for a review, see Farrar et al. 2003). However, root exudation can be further increased by the plant. Examples where root exudation exceeds the basal efflux rate are the prevention of toxic (e.g., aluminum) compounds from entering the root, the enhanced mobilization of limiting nutrients, responses to biotic and abiotic plant stresses (Sacchi et al. 2000), or a continuous removal of the exudates by the soil microbial community (Cakmak and Marschner 1988).

A strong increase in root carbon exudation is well-documented in ECM symbiosis. One of the first attempts to assay carbon flow from a mycorrhizal plant towards the fungal partner was performed by Melin and Nilsson (1957). They could show that feeding (^{14}C) CO_2 to leaves resulted in the appearance of labeled carbon in the hyphal mantle within one day. Subsequently, these results were confirmed by a number of researchers with different experimental systems estimating that, depending on host plant, ECM fungus, season, plant nutrition, and soil conditions, **between 10 % and 25 % of plant photosynthates are transferred toward the fungal partner** (Lewis and Harley 1965c; Cairney et al. 1989; Söderström 1992; Leake et al. 2001; Högberg and Högberg 2002; Wu et al. 2002; Hobbie 2006). Whether elevated carbon transfer is already occurring prior to physical contact between the partners or only when the entire symbiotic structures are formed is to our knowledge still unknown. There is evidence that **plant root exudates are**

already used by ECM fungal hyphae prior to their contact with the root surface (Herrmann and Buscot 2007). However, **only when ECM fungal hyphae cover the colonized tree fine roots do they have direct and privileged access to these exudates.**

Potential carbon compounds to be delivered by the plant partner in symbiosis are amino acids, carboxylic acids, or soluble sugars. Plant cell wall compounds like pectin, hemicellulose, cellulose, or proteins have also been under discussion (for reviews, see Harley and Smith 1983; Smith and Read 2008).

Indeed, certain basidiomycotic ECM fungi do have some cell wall-degrading activities, but a number of researchers (Trojanowski et al. 1984; Haselwandter et al. 1990; Entry et al. 1991) supposed the rate to be too slow to meet the huge fungal carbohydrate demand and expect these capabilities to be more important for soil-growing hyphae (see above). However, since most studies estimate the cellolytic activities of ECM fungi by using extraradical mycelia these data may have to be interpreted with some care because fungal cell wall-degrading activities might be induced at the symbiotic interface. Nonetheless, the massive gene loss in the plant cell wall-degradation arsenal as observed for *L. bicolor* (Martin et al. 2008) and *A. bisporigera* (Nagendran et al. 2009), where certain classes of hydrolytic enzymes are absent, such as celluloses and pectinases, point towards a prevention of a massive host cell wall degradation in symbiosis and against cell wall compounds as significant fungal carbon sources. Anyway, in contrast to these basidiomycotic fungi, the ascomycotic ECM fungus *Tuber melanosporum* still harbors such wall-degrading enzymes in its genome (Martin et al. 2010).

Amino acids, originating from protein hydrolysis, are rather unlikely as a potential fungal carbon source because net nitrogen nutrition of the plant is necessary for the sustainable formation of ectomycorrhizas. Moreover, fungal proteases and amino acid importers are usually induced upon nitrogen and carbon starvation, conditions that are not supposed for hyphae of the symbiotic interface (Nehls 2004). Nevertheless, a nitrogen cycling model, where amino acids are exported by root cells, imported and degraded by Hartig net hyphae for carbon nutrition and ammonium is given back to the plant, cannot be ruled out. However, neither for *A. muscaria* (Nehls et al. 1999b) nor for *L. bicolor* (Lucic et al. 2008) ectomycorrhizas was a strong increase observed in the expression of potential amino acid importers, making this model rather unlikely.

Organic acids, such as malate, citrate, and oxalate, have been proposed to be involved in many processes taking place in the rhizosphere and the release of organic acids from roots can be driven by environmental stresses (e.g., Al, P and Fe stress, anoxia), nutrient acquisition, mineral weathering and pathogen attraction (for a review, see Jones 1998). Relatively little is known about carboxylic acid excretion by ECM plant roots, but for ECM fungi investigated so far carboxylic acids either serve as a poor carbon source (Litchfield and Arthur 1983) or even inhibit fungal growth (Lindeberg and Lindeberg 1974).

Therefore, from the current point of view the **best candidates for fungal carbohydrate nutrition are soluble sugars.** This assumption is based on a number of facts. Mycelia of many cultivable ECM fungi grow well on simple sugars (Lewis and Harley 1965a; Palmer and Hacskaylo 1970; Salzer and Hager 1991) and uptake experiments with excised ectomycorrhizas indicated elevated sugar uptake of fungal hyphae in symbiosis (Lewis and Harley 1965a, b, c). Moreover, all basidiomycotic ECM fungal models investigated so far do strongly increase their sugar uptake capacity in symbiosis (see below). Based on vesicular mycorrhizas Smith and Smith (1989) developed a model of the plant/fungus interface where either glucose or sucrose is released by plant cortical cells. Sucrose has a long-standing tradition in being favored by researchers, starting from the initial experiments of Lewis and Harley (1965c) who used sucrose for labeling experiments. They based their experiments on the finding that sucrose previously had been identified as the most frequent translocated carbohydrate in angiosperms (Zimmermann 1961) and that early experiments in the 1950s had indicated that sucrose had been the only detectable sugar in sieve tube exudates of beech (Ziegler 1956). Due to the presence of plant-derived cell wall-located invertases in endo- and ectomycorrhizal symbiosis, increased invertase activities in vesicular- and ectomycorrhizas, and transgenic plants revealing modulated invertase activity, **plant sucrose export coupled with cell wall-located invertase activity has been postulated as the most probable mechanism for carbohydrate support of fungal hyphae in symbiosis** (Schaeffer et al. 1997;

Smith and Read 1997; Bücking and Heyser 2003; Tarkka et al. 2005; Nehls 2008; Smith and Read 2008). The following section of this review will discuss this hypothesis in detail by means of new data obtained from studies with poplar and old data based on mycorrhizal labeling experiments.

B. Models for Plant Sugar Export

With few exceptions, the molecular mechanisms of carbohydrate efflux in plant cells are far from being understood. Functionally characterized sugar exporters have mainly been reported from bacteria (e.g., Liu et al. 1999a, b) and animals (Goodyer et al. 1997; Hamill et al. 1999; Uldry and Thorens 2004).

Due to their simultaneous presence in source and sink tissues of plants, it was proposed that phloem unloading might be mediated by the same sucrose-H^+-symporters that are responsible for phloem loading (Truernit and Sauer 1995). This hypothesis was further supported by genetic evidence. *Arabidopsis* mutants, who lack the sucrose-H^+-symporter AtSUC2, were strongly impaired in both phloem loading and unloading of sucrose (Gottwald et al. 2000). A direct proof of the **potential of sucrose-H^+-symporters in sucrose export** was given by Carpaneto et al. (2005), who could demonstrate that ZmSUT1 enables sucrose export after depolarization of the plasma membrane. The authors could show a **direct link between plasma membrane potential and sucrose export** and observed that the smaller the membrane potential was, the higher became the sucrose efflux. Carpaneto et al. (2005) assumed that plasma membrane depolarization is generated by inward rectifying potassium channels. In the phloem ZmSUT1 faces both high sucrose concentrations (>400 mM in the symplast) and a strong sucrose gradient between cytosol and apoplast. Carpaneto et al. (2005) showed for proton coupled sucrose import of ZmSUT1 a $K_m^{(sucrose)}$ of 3.7 mM (membrane potential −180 mV, apoplastic pH 5.6, cytosolic pH 7.5) and for sucrose export a $K_m^{(sucrose)}$ of 161 mM (membrane potential 0 mV, apoplastic pH 5.6, cytosolic pH 7.5). Detectable sucrose efflux by ZmSUT1 was observed under the latter conditions at sucrose concentrations of around 50 mM. With about 10 mM (Schaeffer et al. 1995) the sucrose concentration in non-colonized Norway spruce fine roots is, however, much lower than in the phloem and would be too low for a ZmSUT1-based export mechanism. Nevertheless, lower K_m values for sucrose transporters in ECM plants cannot be ruled out and could enable sucrose efflux. In contrast, in non-mycorrhizal plant fine roots, Jones and Darrah (1996) observed that the basal release of non-charged molecules (e.g., sugars) occurs via a passive mechanism. Facilitators, plasma membrane located proteins that enable the equilibrium of a concentration gradient between the cytosol of plant cells and the apoplast would be one possible explanation. A sucrose gradient between root cell symplast and the apoplast of ECM plants can be expected due to the activity of cell wall-localized acid invertases (see below). Taken together, even when non-colonized and colonized fine roots differ in their physiology, a proton-coupled sucrose efflux mechanism by roots cells as described for phloem unloading by Carpaneto et al. (2005) is less probably for mycorrhizas.

Potential **sucrose efflux facilitators** were indeed found in plants. Zhou et al. (2007) could show that some members of the sucrose transporter gene family of *Pisum sativum* are working as facilitators and not as sucrose-H^+ antiporters as most functionally characterized members of this gene family. The presence of such proteins in plant plasma membrane would thus allow a sucrose efflux from root cells. However, these sucrose facilitators were described only for legume plants yet, where they are involved in seed loading.

What is the situation in ectomycorrhizas? In the genome of the model plant *P. trichocarpa* a total of five putative sucrose transporter genes are present (Nehls et al. unpublished data). When comparing gene expression levels in non-mycorrhized fine roots and mycorrhizas, transcript levels were reduced up to 25 times in symbiosis for four of these genes. The expression level of only one gene was slightly (about twofold) elevated upon ectomycorrhiza formation. Considering the about 20-fold increase in carbon exudation of colonized compared to non-colonized ECM fine roots, the observed sucrose transporter gene expression levels in poplar ectomycorrhizas are therefore far from the increase expected to be necessary for elevated sucrose export in symbiosis. In contrast to plant sucrose transporters, the expression of certain **ECM fungal hexose importers** is strongly elevated upon symbiosis (Nehls et al. 1998; Fajardo Lopez et al. 2008), indicating that the increase in plant sugar export upon symbiosis is, at least at the fungal side, reflected in gene expression. However, in addition to transcriptional regulation **plant sucrose transporters** are also known to be posttranscriptionally regulated at different levels. Redox potential-dependent oligomerization (Krügel et al. 2008) or reverse endocytosis (Liesche et al. 2010) of sucrose transporters has been shown to influence plasma membrane transport activities. Even when glucose but no fructose was detectable in exudates of non-mycorrhizal Norway spruce roots (P. Salzer, personnel communication), excised beech ectomycorrhizas revealed fructose but no glucose enrichment when (artificially) being incubated at low sucrose concentrations (Harley and Jennings 1958). Because ECM fungi but also by the plant partner import preferentially glucose from a 1:1 ratio of glucose and fructose (as created by sucrose hydrolysis), fructose enrichment in the common apoplast of ectomycorrhizas can be expected when sucrose would be exported to serve as fungal carbon source. Taken all data mentioned above together, the possibility of sucrose export by poplar sucrose transporters is not strongly supported but cannot be discarded from the current point of view.

However, Chen et al. (2010) recently discovered a new class of transport proteins in the non-mycorrhizal plant *Arabidopsis thaliana*. The authors could demonstrate that certain of these SWEET called proteins function as plasma membrane facilitators for glucose. Because of the clear preference of ECM fungal hyphae for glucose uptake (Fajardo Lopez et al. 2008), **glucose efflux by plant root cells could be an alternative mechanism for plant-based fungal carbohydrate nutrition in symbiosis.** To support this hypothesis we have identified 26 putative "SWEET"-homologs in the poplar genome from which five members revealed strongly increased transcript levels (up to 250-fold) in mycorrhizas compared to non-colonized poplar fine roots (Nehls et al. unpublished data). Because of the coincidence of high, mycorrhiza-specific induction of certain members of the poplar SWEET gene family and the enormous increase in plant sugar export upon symbiosis these SWEETs are excellent candidates for the long-sought plant carbohydrate exporters in ectomycorrhizal symbiosis and are currently under extensive investigation.

C. Impact of Cell Wall-Bound Acid Invertase on Ectomycorrhizal Symbiosis

Invertases, which hydrolyze sucrose into glucose and fructose, play key roles in primary metabolism and plant development (for a review, see Ruan et al. 2010). Sucrose is the major form of carbohydrates transported from photosynthetically active tissues (e.g., mature leaves) to non-photosynthetic sinks such as roots, where it must be degraded into hexoses for various metabolic and biosynthetic processes. **Sucrolytic enzymes such as invertase have thus been implicated in determination and regulation of the sink strength of plant tissues** (Tymowska-Lalanne and Kreis 1998; Roitsch et al. 2003). In agreement with this, increased activities of cell wall bound invertases are commonly observed upon microbial infection, including fungal (Tetlow and Farrar 1992; Chou et al. 2000; Fotopoulos et al. 2003), bacterial (Bonfig et al. 2006), and viral (Herbers et al. 2000) pathogens but also symbionts (see below). The impact of biotic interaction is best characterized for plant leaves that retain their sink character after maturation or undergo a source to sink transition upon pathogen infection (Horst et al. 2008). Here, hexoses, which are formed by cell wall invertase activities, are thought to aid the pathogen's carbon nutrition (Biemelt and Sonnewald 2006; Seo et al. 2007). In agreement with this, fungal pathogens but also ericoid mycorrhizal fungi (Straker et al. 1992; Parrent et al. 2009) frequently produce their own invertases to ensure/improve their nutritional supply (Walters et al. 1996; Voegele et al. 2001). In contrast to them, **most basidiomycotic ECM fungi investigated so far lack invertase activity** (Lewis and Harley 1965b; Salzer and Hager 1991; Hampp et al. 1995).

Furthermore, investigation of a broad range of agaricomycotic ectomycorrhizal fungi using a set of degenerated primers did not indicate the presence of any invertase genes in these species (Parrent et al. 2009) and invertase genes are also missing in the genome of *L. bicolor* (Martin et al. 2008). Hence, invertase activity, which is usually observed in ectomycorrhizas (see below), is supposed to be of plant origin in ECM symbiosis.

Similar to phytopathogen interaction, increased invertase activities were found in symbiotic arbuscular mycorrhizal (AM) interaction of tomato (Schaarschmidt et al. 2006; García-Rodríguez et al. 2007), common bean, and carrot (Blee and Anderson 2002), or white clover (Wright et al. 1998), but also in ectomycorrhizas of birch (Wright et al. 2000), Norway spruce (Schaeffer et al. 1995), and poplar (Kulmann 2005; Göhringer et al. unpublished data). In agreement with increased enzyme activities, an up to threefold enhanced transcription was observed in ectomycorrhizas for all members of the invertase gene family present in vegetative organs of *Populus trichocarpa* (Göhringer et al. unpublished data). However, with the exception of one of these genes, which revealed its highest transcript level at the developing symbiotic interface, gene expression was highest in the root tip (Göhringer et al. unpublished data), indicating that **invertase activity is mainly affected at the meristem and in the elongation zone upon ectomycorrhiza formation** but not at the active symbiotic interface as observed for Norway spruce (Schaeffer et al. 1995). Moreover, even when increased upon symbiosis, invertase activity does not seem to be a bottleneck of the carbon transfer in AM symbiosis, as mycorrhizal interaction could not be improved by root-specific enhancement of invertase activity in *Nicotiana tabacum* (Schaarschmidt et al. 2007). Nevertheless, the importance of invertase activity for mycorrhizas could be demonstrated for transgenic tobacco plants, revealing a decreased acid invertase activity in roots. These transgenic plants exhibited reduced AM fungal colonization rates together with a lower density of fungal structures (Schaarschmidt et al. 2007). Interestingly, root-specific overexpression of an invertase inhibitor gene in poplar, which is thought to result in a functional reduction of apoplastic invertase activity, did not reveal a decreased number of ectomycorrhizas (Hampp et al. unpublished data).

How can these results be interpreted? In addition to their function in carbohydrate partitioning in sink tissues, **signaling roles of sucrose and hexoses** turned out to be equally important for plant physiology and development (for a review, see Smeekens et al. 2010).

Hexose signals are known to trigger the induction of pathogen related (PR) gene expression and to amplify plant defense reactions (Herbers et al. 2000). A rapid accumulation of soluble sugars caused by cell wall invertase activity after pathogen infection hence leads to the production of well known outputs of plant defense. Observed callose deposition and production of phenolic compounds require large amounts of metabolizable sugars (Herbers et al. 1996; Scharte et al. 2005) and thus promote the successful establishment of host defense reactions (Scharte et al. 2005; Essmann et al. 2008). For arbuscular mycorrhizal symbiosis, Schaarschmidt et al. (2007) therefore suggested that a (only) moderate induction of invertase expression in colonized tomato roots can be interpreted as a plant signal for fine-tuning in the activation of sink metabolism and avoidance of stress-induced defense reactions in symbiosis. In this light the severe increase in invertase gene expression and protein activity observed in different ectomycorrhizal associations can be interpreted in a similar manner as a plant signal to trigger sugar export but not plant defense.

D. Sugar Uptake by Plant Cells at the Symbiotic Interface: Local Control of Fungal Sugar Support?

First hints of a **control of carbon drain towards the fungal partner in ectomycorrhizal symbiosis** came from fertilization experiments, showing that **increased soil nitrogen availability often is associated with decreased production of extra radical hyphae, reduced colonization intensity of fine roots, and decreased fruiting body formation by ECM fungi** (Wallenda and Kottke 1998; Nilsson and Wallander 2003; Nilsson et al. 2005). In agreement with this, Druebert et al. (2009) observed shifts in fungal community composition in response to tree shading or girdling and supposed an alteration of fungal ecosystem function towards less carbon-demanding ECM species.

Because **root exudation of non-charged sugars presumably occurs passively**, driven by the concentration gradient between plant cells and the apoplast (Jones and Darrah 1995), **C-efflux is supposed to be difficult to be controlled by the host plant.** One mechanism to overcome this limitation is sugar re-import by root cells (Jones and Darrah 1996). In the ectomycorrhizal model plant poplar, only one out of five putative sucrose transporter genes was found to be slightly up-regulated (Nehls unpublished data) upon ectomycorrhiza formation, while gene expression of the four other genes was reduced (2–25 times). A substantial sucrose re-uptake from the common apoplast of the symbiotic interface is thus not very likely. However, Grunze et al. (2004) and Nehls et al. (2007) observed a strong increase in the transcript level of one particular *Populus tremula* × *tremuloides* hexose transporter gene upon ectomycorrhiza formation. This finding was interpreted in such a way **that trees may restrict fungal carbohydrate support at the local level**, e.g., in cases when the partner does not deliver sufficient mineral nutrients, but functional analysis of the respective poplar protein was missing in this study. In the genome of *Populus trichocarpa*, 24 putative hexose transporters were found (Göhringer et al. unpublished data), of which transcripts of 13 genes were detectable in vegetative plant organs. The expression of two of them was up-regulated upon ectomycorrhiza formation but a function of the deduced protein as hexose importer could only be confirmed for one of them. The respective monosaccharide importer was capable of importing glucose as well as fructose (Göhringer et al., unpublished data) and revealed a K_m value for glucose import that is similar to that of ECM fungal proteins (Wiese et al. 2000; Polidori et al. 2007; Fajardo Lopez et al. 2008). With a K_m value for fructose import of about 300 μM, this sugar importer would also be able to import fructose originating from sucrose hydrolysis in the common apoplast. With such a K_m value for fructose uptake, fructose concentration in the apoplast would stay high enough to trigger the observed sugar regulated increase in fungal (*A. muscaria*) hexose transporter gene expression (Nehls et al. 1999b) but would not increase to a level that is capable of inhibiting invertase activity (Salzer and Hager 1993) or trigger plant defense. The genome-wide analysis of hexose importers in *P. trichocarpa* therefore supports the hypothesis of Grunze et al. (2004) that plant (poplar) root cells are capable of competing for apoplastic glucose. In agreement with this assumption transgenic poplar plants that are constitutively over-expressing an additional hexose importer gene reveal a reduced mycorrhizal infection capability and an abnormal termination of the symbiosis (Nehls, unpublished data).

E. Fungus-Derived Sink Formation in Colonized Fine Roots

1. Fungal Sugar Uptake at the Symbiotic Interface

In a time course experiment, ectomycorrhizal *Pinus densiflora* seedlings were exposed to C^{14} CO_2 (Wu et al. 2002). Thirteen days after the pulse treatment, 24 % of the label could be detected as sugars within the extraradical mycelium, which proves: (a) **the high sink strength of ectomycorrhizas and** (b) **the fast distribution of plant-derived carbohydrates within the fungal colony**. Different authors calculated that the carbon allocation by entire mycorrhizas (plant plus fungus) is of about 15–19 times higher compared to non-colonized fine roots and that the fungal as well as the plant partner are accountable for about half of the mycorrhiza-allocated carbon (Bevege et al. 1975; Cairney et al. 1989; Wu et al. 2002).

To generate and maintain such a strong C-sink strength, carbohydrate uptake of fungal hyphae in the symbiotic interface has to be significantly increased. As no sucrose uptake by ectomycorrhizal fungi has ever been observed (Harley and Jennings 1958; Salzer and Hager 1991; Chen and Hampp 1993), **hexoses are thought to be the primary carbon source of ECM fungi in symbiosis** (see above). This view is in agreement with current observations in AM symbiosis, where Helber et al. (2011) could demonstrate that the expression of a fungal (*Glomus* sp.) high-affinity monosaccharide transporter (MST2) is crucial for the symbiotic relationship with host plants. Suppression of gene expression leads to fungal growth arrest, an overall reduced degree of mycorrhization, and impairment of arbuscule development. In the ECM fungus *L. bicolor* six out of a total of 15 putative MST genes showed a strong increase in their transcript level upon ectomycorrhiza formation. This observation confirms previous investigations in mycorrhizas of other ECM fungi (Nehls et al. 1998; Nehls 2004; Polidori et al. 2007), together indicating **a strongly enhanced hexose uptake capacity of fungal hyphae as general feature of ectomycorrhizal symbiosis.** However, the mechanisms by which this mycorrhiza-dependent increase in fungal sugar uptake is controlled differ in diverse ECM fungal models. While increased sugar transporter gene expression is controlled in a sugar-dependent manner in *A. muscaria* (Nehls et al. 1998; Nehls 2004), it is developmentally controlled in *L. bicolor* (Fajardo Lopez et al. 2008).

All hexose importers of mycorrhizal fungi investigated so far (Nehls et al. 1998; Nehls 2004; Polidori et al. 2007; Fajardo Lopez et al. 2008) revealed a clear preference for glucose import and a discrimination of fructose uptake, which is also true for ECM fungal mycelia grown in liquid culture (Lewis and Harley 1965b; Chen and Hampp 1993; Stülten 1996; Fajardo Lopez et al. 2008). Fructose was only taken up efficiently from a glucose/fructose mixture when the glucose concentration was below the K_m values of the respective transporters. In addition, no homologs of fructose importers (that are found in certain phytopathogenic fungi; Doehlemann et al. 2005) are present in the genome of *L. bicolor* (Nehls, unpublished data), indicating no potentially mycorrhiza-induced fungal fructose uptake in symbiosis.

2. Fungal Sugar Metabolism in Hartig Net Hyphae

NMR and biochemical investigations revealed that **for the generation and maintenance of a strong carbon sink in symbiosis, monosaccharides are quickly converted into fungal metabolites** either by flux through glycolysis and citric acid cycle or by generation of fungal storage compounds (e.g., Hampp et al. 1995; Kowallik et al. 1998; Martin et al. 1998; Rangel-Castro et al. 2002; Deveau et al. 2008).

The enzyme **phosphofructokinase performs the rate-limiting step in EM fungal glycolysis** (Kowallik et al. 1998) and is activated by fructose 2,6-bisphosphate (F26BP) in *A. muscaria.* Since symbiotic *A. muscaria* hyphae have increased amounts of F26BP (Schaeffer et al. 1996), an increased glycolytic flux in hyphae at the plant/fungus interface can be assumed. Increased flux rates through fungal glycolysis and tricarboxylic acid cycle are also indicated by large-scale expression analysis of established ectomycorrhizas of *Pisolithus microcarpus* (Duplessis et al. 2005) and at the whole genome level with *L. bicolor* (Deveau et al. 2008; Martin et al. 2008). In contrast to data from *Pisolithus*, *A. muscaria*, and *L. bicolor*, ectomycorrhizas formed by *Paxillus involutus* showed a slightly reduced expression of genes involved in sugar uptake and glycolysis (Le Quere et al. 2005; Wright et al. 2005). This could either indicate species-specific differences of ECM fungi upon ectomycorrhiza formation or limited experimental data sets.

Two different pools of storage carbohydrates are supposed in basidiomycotic ECM fungi: (a) short chain carbohydrates like **trehalose and polyols**, and (b) the long chain carbohydrate glycogen.

Besides stress conditions, trehalose accumulation occurs in yeasts and other fast growing saprotrophic fungi only in trehalase-deficient mutants (probably because high trehalase activity normally accompanies trehalose biosynthesis) and trehalose cycling is the important feature in the physiology in these organisms (Wiemken 1990; Francois and Parrou 2001). In contrast, growth of basidiomycotic ECM fungi like *A. muscaria*, *H. cylindrosporum*, or *L. bicolor* on glucose as carbon source resulted in an increased trehalose content over time (Martin et al. 1998; Rangel-Castro et al. 2002; Nehls et al. 2007) until glucose in the growth medium was significantly reduced (Wallenda 1996). Furthermore, trehalose dominated in hyphae of the plant/fungus interface of young Norway spruce ectomycorrhizas (Rieger et al. 1992), indicating this area as the most intense carbon sink and **trehalose as a potential storage carbohydrate in symbiosis.** In accordance with this, the expression of trehalose biosynthetic genes was strongly increased in hyphae of *A. muscaria* upon ectomycorrhiza formation, especially at the symbiotic interface (Fajardo Lopez et al. 2008). **Trehalose and/or polyol (see below) accumulation in ECM fungi was therefore interpreted as a physiological adaptation to a continuous sugar supply by the host plant in symbiosis** (Nehls et al. 2010).

Trehalose biosynthesis can occur via two different pathways in basidiomycotic ECM fungi. Two enzymes, trehalose-6-phosphate synthase and trehalose-6-P phosphatase act together in an energy consuming and therefore irreversible reaction to build trehalose from glucose-6-phosphate and UDP-glucose. Alternatively, trehalose can be formed in a reversible and thus no energy-consuming reaction by a single enzyme, trehalose phosphorylase, from glucose-1-phosphate and glucose (Fajardo Lopez et al. 2008). The expression of both systems was increased upon ectomycorrhiza formation. However, whether the latter enzyme is synthesizing or degrading trehalose depends on the substrate and product concentrations (Eis et al. 1998; Eis and Nidetzky 1999), conditions that are still unknown in ectomycorrhizas. Nevertheless, trehalose accumulation in *A. muscaria* mycelia at the symbiotic interface clearly indicates net biosynthesis under symbiotic conditions (Fajardo Lopez et al. 2008).

In contrast to trehalose, **mannitol is only rarely found** (one exception, *Hygrophorus pustulatus*, is known) in ECM basidiomycetes belonging to the order of agaricales (for a review, see Nehls 2008). In the order of boletales and other orders of ECM basidiomycetes (which are rarely investigated), closely related ECM fungi accumulate mannitol and/or trehalose. In ascomycotic ECM fungi, mannitol (and arabitol) is clearly dominant in substrate mycelia and mycorrhizas while trehalose is rarely found. Hence, it can be concluded that **trehalose and mannitol may have similar functions in different ECM fungi.**

As ECM fungi commonly form an extensive external mycelium, and as soil-growing hyphae are dependent on carbohydrate support by mycorrhizas (Leake et al. 2001; Anderson and Cairney 2007), it is evident that long-distance transport of carbon is of central importance for fungal physiology. Since glycogen is stored in the cytoplasm as large, non-mobile granules, it is (similar to starch in plants) rather unlikely that it serves as the long-distance carbohydrate transport form in fungi. In contrast, **trehalose and polyols are present in large quantities in fungal hyphae, fluctuate** with fungal carbon demand, and are (like sucrose in plants) **highly mobile, predestining them for serving as long-distance transport carbohydrates.** Trehalose and sugar alcohols are therefore supposed to have dual function in ectomycorrhizal fungal colonies: short-term carbon storage and long-distance transport carbohydrates.

In contrast to the trehalose/mannitol content, which fluctuates highly during fungal growth, the **glycogen** content remains relatively stable over a short term. Cultivation of *A. muscaria* mycelia for up to three weeks in the presence or absence of a carbon source had nearly no effect on hyphal glycogen content (Wallenda 1996). However, Jordy et al. (1998) observed changes in glycogen content and localization during ectomycorrhiza formation of *Paxillus involutus*. While glycogen was initially found in Hartig net hyphae (symbiotic interface), it was later observed to be present exclusively in hyphae of the fungal sheath. The initial occurrence of glycogen in the Hartig net hyphae was interpreted (like temporal starch formation in leaves) as a flux control mechanism. When carbohydrate import into hyphae exceeds its export to other parts of the fungal colony, long-term storage pools are filled to ensure continuous fungal carbohydrate sink strength in symbiosis. *Lactarius subdulcis* mycorrhizas collected in a natural forest clearly revealed seasonal changes in their glycogen content (Genet et al. 2000). Glycogen amount was high during winter, declined until summer due to strong fungal propagation, and was restored again in autumn. Taken together these data indicate **glycogen (like starch in plants) as a long-term carbohydrate store, which is only mobilized when the short-term pools (e.g., trehalose and polyols) are empty.**

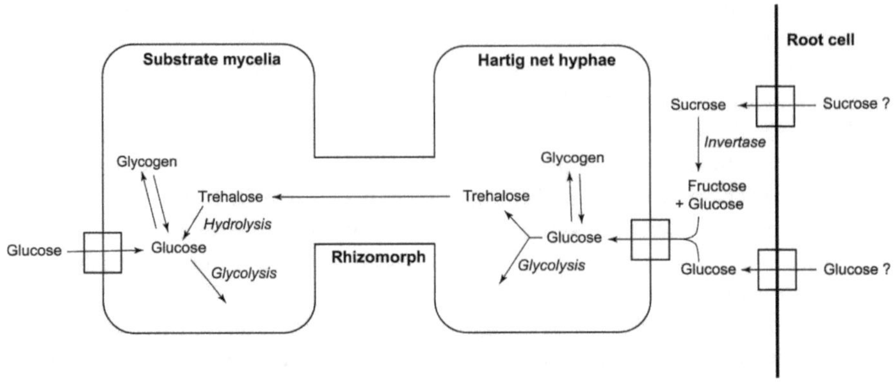

Fig. 7.1. Carbon nutrition of ECM fungal colonies. Shown is a scheme of carbon nutrition and colony-wide carbohydrate exchange in agaric ECM fungal species (soil-growing hyphae and hyphae of the symbiotic interface). Only trehalose metabolism, which is typical for agaric ECM fungi (Nehls 2008), is indicated. The mannitol cycle that is frequently found in other fungal groups has been omitted because of its complex physiology. Furthermore, putative mechanisms of plant carbon export at the functional interface are outlined

V. Conclusions

Ectomycorrhizal fungi have to balance two lifestyles, being on the one hand a good saprotroph in soil, without becoming on the other hand a plant pathogen in symbiotic structures. How this balancing act can be managed is still an open question. Essential for this lifestyle is the fungal carbohydrate support by a host plant. However, the proportion of carbon gained by saprotrophism of soil growing hyphae and hyphae of the symbiotic interface remains unclear. Due to seasonal fluctuations in plant activity, soil temperature, and humidity, etc., the ratio of plant- and soil-gained carbohydrates of a given fungal colony differs. A model summarizing fungal carbon uptake and initial metabolism within an ECM fungal colony (soil-growing hyphae and hyphae of the plant/fungus interface of functional ectomycorrhizas) is outlined in Fig. 7.1.

The huge amount of carbohydrates that is delivered by the plant partner during ectomycorrhizal symbiosis raises the question of how the fungal partner can generate such a strong C-sink strength and maintain it over the ectomycorrhizal lifetime. Efficient glucose uptake in combination with a modulated fungal carbohydrate metabolism (increased flux through glycolysis and conversion into storage carbohydrates) and carbohydrate removal towards other parts of the fungal colony (e.g., soil-growing mycelium) can explain sink formation in symbiosis. Here, trehalose and/or mannitol formation are expected to have a dual function: short-term storage and long-distance transport of carbohydrates. However, a final proof of this hypothesis is still missing. Therefore, the generation of fungal mutants defective in biosynthesis and the degradation of short chain carbohydrates and sugar alcohols will be necessary.

Does the plant partner control the carbohydrate flux towards the fungal partner and, if so, how could it be managed? Essential for addressing this question is the identification of fungal carbon sources delivered by the host plant. In literature, sucrose has been established as putative plant carbon source that is excreted into the common apoplast of the symbiotic interface and hydrolyzed by cell wall-associated, plant-derived invertases. However, the identification of sucrose exporters has not been successful in ECM plant fine roots, yet. In addition, plant invertase gene expression did not increase at the symbiotic interface of poplar ectomycorrhizas. Together with the discovery of a novel class of plant hexose facilitators of which certain members are mycorrhiza-specifically induced in poplar, the function of sucrose as the main carbohydrate for fungal carbon nutrition must therefore be questioned. Instead, glucose might be the plant root exported fungal carbon source in symbiosis. However, further work will be necessary to prove this novel hypothesis.

Mycorrhiza-induced gene expression of putative hexose transporters was observed for different poplar species. The characterized protein of one of these genes revealed glucose import capabilities with a K_m value similar to that of ECM fungal hexose transporters. These findings support the idea of a local control of fungal carbohydrate nutrition by the plant by competing for apoplastic hexoses.

In a nutshell, the genomes of the model plant *P. trichocarpa* and the ECM fungal models *L. bicolor* and *Tuber melanosporum*, which enabled expression analysis at the level of entire gene families and even the whole transcriptome, served as a precious toolbox for molecular investigations that has already resulted in a number of new concepts. With the aid of the JGI genome initiative, targeting 25 mycorrhizal fungal species of different clades (Martin et al. 2011; Plett and Martin 2011), extensive comparative genomics will come into the field and will help to unravel the formation and function of ectomycorrhizal symbiosis. This information will, however, also shine a light on different fungal concepts to enable the tight association with a plant partner.

At this point, it should not be concealed that the expression of genes does not necessarily reflect actual protein activity, a problem that will have to be considered in future experiments. Thus, expression data of entire pathways in combination with analysis of protein function and metabolite content will be necessary to draw sound conclusions. Furthermore, as indicated by symbiotic carbon nutrition, concerted investigation of fungal and plant partners are necessary to unravel not only mycorrhizal organ development but also nutrient and metabolite exchange between both partners.

References

Abuzinadah RA, Read DJ (1986) The role of proteins in the nitrogen nutrition of ectomycorrhizal plants. III. Protein utilisation by *Betula*, *Picea* and *Pinus* in mycorrhizal association with *Hebeloma crustuliniforme*. New Phytol 103:507–514

Agerer R (2001) Exploration types of ectomycorrhizae. Mycorrhiza 11:107–114

Anderson IC, Cairney JWG (2007) Ectomycorrhizal fungi: exploring the mycelial frontier. FEMS Microbiol Rev 31:388–406

Anderson IC, Chambers SM, Cairney JWG (1998) Use of molecular methods to estimate the size and distribution of mycelial individuals of the ectomycorrhizal basidiomycete *Pisolithus tinctorius*. Mycol Res 102:295–300

Bajwa R, Abuarghub S, Read DJ (1985) The biology of mycorrhiza in the ericaceae. X. The utilization of proteins and the production of proteolytic enzymes by the mycorrhizal endophyte and by mycorrhizal plants. New Phytol 101:469–486

Bending GD, Read DJ (1996) Nitrogen mobilization from protein–polyphenol complex by ericoid and ectomycorrhizal fungi. Soil Biol Biochem 28:1603–1612

Bevege DI, Bowen GD, Skinner MF (1975) Comparative carbohydrate physiology of ecto-and endomycorrhizas. In: Sanders FE, Mosse B, Tinker PB (eds) Endomycorrhizas. Academic, London, pp 149–174

Biemelt S, Sonnewald U (2006) Plant-microbe interactions to probe regulation of plant carbon metabolism. J Plant Physiol 163:307–318

Blasius D, Feil W, Kottke I, Oberwinkler F (1986) Hartig net structure and formation of fully ensheated ectomycorrhizas. Nord J Bot 6:837–842

Blee KA, Anderson AJ (2002) Transcripts for genes encoding soluble acid invertase and sucrose synthase accumulate in root tip and cortical cells containing mycorrhizal arbuscules. Plant Mol Biol 50:197–211

Bonfig K, Schreiber U, Gabler A, Roitsch T, Berger S (2006) Infection with virulent and avirulent *P. syringae* strains differentially affects photosynthesis and sink metabolism in *Arabidopsis* leaves. Planta 225:1–12

Bücking H, Heyser W (2003) Uptake and transfer of nutrients in ectomycorrhizal associations: interactions between photosynthesis and phosphate nutrition. Mycorrhiza 13:59–68

Buee M, Vairelles D, Garbaye J (2005) Year-round monitoring of diversity and potential metabolic activity of the ectomycorrhizal community in a beech (*Fagus silvatica*) forest subjected to two thinning regimes. Mycorrhiza 15:235–245

Buee M, Courty PE, Mignot D, Garbaye J (2007) Soil niche effect on species diversity and catabolic activities in an ectomycorrhizal fungal community. Soil Biol Biochem 39:1947–1955

Cairney JWG, Ashford AE, Allaway WG (1989) Distribution of photosynthetically fixed carbon within root systems of *Eucalyptus pilularis* plants ectomycorrhizal with *Pisolithus tinctorius*. New Phytol 112:495–500

Cakmak I, Marschner H (1988) Increase in membrane permeability and exudation in roots of zinc deficient plants. Plant Physiol Biochem 132:356–361

Carpaneto A, Geiger D, Bamberg E, Sauer N, Fromm J, Hedrich R (2005) Phloem-localized, proton-coupled sucrose carrier ZmSUT1 mediates sucrose efflux under the control of the sucrose gradient and the proton motive force. J Biol Chem 280:21437–21443

Chen XY, Hampp R (1993) Sugar uptake by protoplasts of the ectomycorrhizal fungus, *Amanita muscaria* (L. ex Fr.) hooker. New Phytol 125:601–608

Chen L-Q, Hou B-H, Lalonde S, Takanaga H, Hartung ML, Qu X-Q, Guo W-J, Kim J-G, Underwood W, Chaudhuri B, Chermak D, Antony G, White FF, Somerville SC, Mudgett MB, Frommer WB (2010) Sugar transporters for intercellular exchange and nutrition of pathogens. Nature 468:527–532

Chou H, Bundock N, Rolfe S, Scholes J (2000) Infection of *Arabidopsis thaliana* leaves with *Albugo candida* (white blister rust) causes a reprogramming of host metabolism. Mol Plant Pathol 1:99–113

Courty P-E, Pritsch K, Schloter M, Hartmann A, Garbaye J (2005) Activity profiling of ectomycorrhiza communities in two forest soils using multiple enzymatic tests. New Phytol 167:309–319

Courty P-E, Breda N, Garbaye J (2007) Relation between oak tree phenology and the secretion of organic matter degrading enzymes by *Lactarius quietus* ectomycorrhizas before and during bud break. Soil Biol Biochem 39:1655–1663

Deveau A, Kohler A, Frey-Klett P, Martin F (2008) The major pathways of carbohydrate metabolism in the ectomycorrhizal basidiomycete *Laccaria bicolor* S238N. New Phytol 180:379–390

Doehlemann G, Molitor F, Hahn M (2005) Molecular and functional characterization of a fructose specific transporter from the gray mold fungus *Botrytis cinerea*. Fungal Genet Biol 42:601–610

Druebert C, Lang L, Valtanen K, Polle A (2009) Beech carbon productivity as driver of ectomycorrhizal abundance and diversity. Plant Cell Environ 32:992–1003

Duplessis S, Courty PE, Tagu D, Martin F (2005) Transcript patterns associated with ectomycorrhiza development in *Eucalyptus globulus* and *Pisolithus microcarpus*. New Phytol 165:599–611

Eis C, Nidetzky B (1999) Characterization of trehalose phosphorylase from *Schizophyllum commune*. Biochem J 341:385–393

Eis C, Albert M, Dax K, Nidetzky B (1998) The stereochemical course of the reaction mechanism of trehalose phosphorylase from *Schizophyllum commune*. FEBS Lett 440:440–443

El-Badaoui K, Botton B (1989) Production and characterization of exocellular proteases in ectomycorrhizal fungi. Anal Sci For 46:728–730

Entry JA, Donnelly PK, Cromack KJ (1991) Influence of ectomycorrhizal mat soils on lignin and cellulose degradation. Biol Fert Soils 11:75–78

Essmann J, Schmitz-Thom I, Schön H, Sonnewald S, Weis E, Scharte J (2008) RNA interference-mediated repression of cell wall invertase impairs defense in source leaves of tobacco. Plant Physiol 147:1288–1299

Fajardo Lopez M, Dietz S, Grunze N, Bloschies J, Weiss M, Nehls U (2008) The sugar porter gene family of *Laccaria bicolor*: function in ectomycorrhizal symbiosis and soil-growing hyphae. New Phytol 180:365–378

Farrar J, Hawes M, Jones D, Lindow S (2003) How roots control the flux of carbon to the rhizosphere. Ecology 84:827–837

Fletcher M (1996) Bacterial adhesion. In: Fletcher M (ed) Bacterial adhesion: molecular and ecological diversity. Wiley-Liss, New York, pp 59–745

Fotopoulos V, Gilbert MJ, Pittman JK, Marvier AV, Buchanan AJ, Sauer N, Hall JL, Williams LE (2003) The monosaccharide transporter gene, AtSTP4, and the cell-wall invertase, Atfruct1, are induced in *Arabidopsis* during infection with the fungal biotroph *Erysiphe cichoracearum*. Plant Physiol 132:821–829

Francois J, Parrou JL (2001) Reserve carbohydrates metabolism in the yeast *Saccharomyces cerevisiae*. FEMS Microbiol Rev 25:125–145

Frank B (1885) Über die auf Wurzelsymbiose beruhende Ernährung gewisser Bäume durch unterirdische Pilze. Ber Dtsch Bot Ges 3:128–145

García-Rodríguez S, Azcón-Aguilar C, Ferrol N (2007) Transcriptional regulation of host enzymes involved in the cleavage of sucrose during arbuscular mycorrhizal symbiosis. Physiol Plant 129:737–746

Genet P, Prevost A, Pargney JC (2000) Seasonal variations of symbiotic ultrastructure and relationships of two natural ectomycorrhizae of beech (*Fagus sylvatica*/*Lactarius* blennius var. Viridis and *Fagus sylvatica*/*Lactarius subdulcis*). Trees 14:465–474

Goodyer ID, Hayes DJ, Eisenthal R (1997) Efflux of 6-deoxy-D-glucose from *Plasmodium falciparum*-infected erythrocytes via two saturable carriers. Mol Biochem Pathol 84:229–239

Göransson H, Wallander H, Ingerslev M, Rosengren U (2006) Estimating the relative nutrient uptake from different soil depths in *Quercus robur*, *Fagus sylvatica* and *Picea abies*. Plant Soil 286:87–97

Gottwald JR, Krysan PJ, Young JC, Evert RF, Sussman MR (2000) Genetic evidence for the in planta role of phloem-specific plasma membrane sucrose transporters. Proc Natl Acad Sci USA 97:13979–13984

Grayston SJ, Vaughan D, Jones D (1997) Rhizosphere carbon flow in trees, in comparison with annual plants: the importance of root exudation and its impact on microbial activity and nutrient availability. Appl Soil Ecol 5:29–56

Grunze N, Willmann M, Nehls U (2004) The impact of ectomycorrhiza formation on monosaccharide transporter gene expression in poplar roots. New Phytol 164:147–155

Hamill S, Cloherty EK, Carruthers A (1999) The human erythrocyte sugar transporter presents two sugar import sites. Biochemistry 38:16974–16983

Hampp R, Schaeffer C, Wallenda T, Stülten C, Johann P, Einig W (1995) Changes in carbon partitioning or allocation due to ectomycorrhiza formation: biochemical evidence. Can J Bot 73:548–556

Harley JL, Jennings DH (1958) The effect of sugars on the respiratory response of beech mycorrhizas to salts. Proc R Soc Lond 148:403–418

Harley JL, Smith SE (1983) Mycorrhizal symbiosis. Academic, London

Haselwandter K, Bobleter O, Read DJ (1990) Degradation of carbon-14 labelled lignin and dehydropolymer of coniferyl alcohol by ericoid and ectomycorrhizal fungi. Arch Microbiol 153:352–354

Helber N, Wippel K, Sauer N, Schaarschmidt S, Hause B, Requena N (2011) A versatile monosaccharide transporter that operates in the arbuscular mycorrhizal fungus *Glomus* sp. is crucial for the symbiotic relationship with plants. Plant Cell 23: 3812–3823.

Herbers K, Meuwly P, Frommer WB, Metraux JP, Sonnewald U (1996) Systemic acquired resistance mediated by the etopic expression of invertase: possible hexose sensing in the secretory pathway. Plant Cell 8:793–803

Herbers K, Takahata Y, Melzer M, Mock H, Hajirezaei M, Sonnewald U (2000) Regulation of carbohydrate partitioning during the interaction of potato virus Y with tobacco. Mol Plant Pathol 1:51–59

Herrmann S, Buscot F (2007) Cross talks at the morphogenetic, physiological and gene regulation levels between the mycobiont *Piloderma croceum* and oak microcuttings (*Quercus robur*) during formation of ectomycorrhizas. Phytochemistry 68:52–67

Hobbie EA (2006) Carbon allocation to ectomycorrhizal fungi correlates with belowground allocation in culture studies. Ecology 87:563–569

Högberg MN, Högberg P (2002) Extramatrical ectomycorrhizal mycelium contributes one-third of microbial biomass and produces, together with associated roots, half the dissolved organic carbon in a forest soil. New Phytol 154:791–795

Högberg P, Nordgren A, Aren G (2002) Carbon allocation between tree root growth and root respiration in boreal pine forest. Oecologia 132:579–581

Horst R, Engelsdorf T, Sonnewald U, Voll L (2008) Infection of maize leaves with *Ustilago maydis* prevents establishment of C(4) photosynthesis. J Plant Physiol 165:19–28

Jones DL (1998) Organic acids in the rhizosphere: a critical review. Plant Soil 205:25–44

Jones DL, Darrah PR (1995) Influx and efflux of organic acids across the soil–root interface of *Zea mays* L. and its implications in rhizosphere C flow. Plant Soil 173:103–109

Jones DL, Darrah PR (1996) Re-sorption of organic compounds by roots of Zea mays L. and its consequences in the rhizosphere. III Characteristics of sugar influx and efflux. Plant Soil 178:153–160

Jordy MN, Azemar LS, Brun A, Botton B, Pargney JC (1998) Cytolocalization of glycogen, starch, and other insoluble polysaccharides during ontogeny of *Paxillus involutus*/*Betula pendula* ectomycorrhizas. New Phytol 140:331–341

Kottke I, Oberwinkler F (1986) Mycorrhiza of forest trees – structure and function. Trees 1:1–24

Kowallik W, Thiemann M, Huang Y, Mutumba G, Beermann L, Broer D, Grotjohann N (1998) Complete sequence of glycolytic enzymes in the mycorrhizal basidiomycete, *Suillus bovinus*. Z Naturforsch 53:818–827

Krügel U, Veenhoff LM, Langbein J, Wiederhold E, Liesche J, Friedrich T, Grimm B, Martinoia E, Poolman B, Kühn C (2008) Transport and sorting of the *Solanum tuberosum* sucrose transporter SUT1 is affected by posttranslational modification. Plant Cell 20:2497–2513

Kulmann C (2005) Lokalisation und regulation der sauren invertase in mykorrhizierten Kurzwurzeln. University of Bremen, Bremen

Le Quere A, Wright D, Soederstroem B, Tunlid A, Johansson T (2005) Global patterns of gene regulation associated with the development of ectomycorrhiza between birch (*Betula pendula* Roth.) and *Paxillus involutus* (Batsch) Fr. Mol Plant Micobe Interact 18:659–673

Leake J, Read D (1997) Mycorrhizal fungi in terrestrial habitat's. In: Wicklow D, Söderström B (eds) The mycota IV: environmental and microbial relationships. Springer, Berlin, pp 281–301

Leake JR, Donnelly DP, Saunders EM, Boddy L, Read DJ (2001) Rates and quantities of carbon flux to ectomycorrhizal mycelium following ^{14}C pulse labeling of *Pinus sylvestris* seedlings: effects of litter patches and interaction with a wood-decomposer fungus. Tree Physiol 21:71–82

Lewis DH, Harley JL (1965a) Carbohydrate physiology of mycorrhizal roots of beech I. Identity of endogenous sugars and utilization of exogenous sugars. New Phytol 64:224–237

Lewis DH, Harley JL (1965b) Carbohydrate physiology of mycorrhizal roots of beech II. Utilizaton of exogenous sugars by uninfected and mycorrhizal roots. New Phytol 64:238–255

Lewis DH, Harley JL (1965c) Carbohydrate physiology of mycorrhizal roots of beech. III. Movement of sugars between host and fungus. New Phytol 64:265–275

Liesche J, He H-X, Grimm B, Schulz A, Kühn C (2010) Recycling of *solanum* sucrose transporters expressed in yeast, tobacco, and in mature phloem sieve elements. Mol Plant 3:1064–1074

Lindahl BD, Ihrmark K, Boberg J, Trumbore SE, Högberg P, Stenlid J, Finlay RD (2007) Spatial separation of litter decomposition and mycorrhizal nitrogen uptake in a boreal forest. New Phytol 173:611–620

Lindeberg G, Lindeberg M (1974) Effect of short chain fatty acids on the growth of some mycorrhizal and saprophytic hymenomycetes. Arch Microbiol 101:109–114

Lindeberg G, Lindeberg M (1977) Pectinolytic ability of some mycorrhizal and saprophytic hymenomycetes. Arch Microbiol 115:9–12

Litchfield J, Arthur M (1983) Growth of selected ectomycorrhizal fungi in aerated liquid culture. Dev Indust Microb 24:289–294

Liu JY, Miller PF, Gosink M, Olson ER (1999a) The identification of a new family of sugar efflux pumps in *Escherichia coli*. Mol Microbiol 31:1845–1851

Liu JY, Miller PF, Willard J, Olson ER (1999b) Functional and biochemical characterization of *Escherichia coli* sugar efflux transporters. J Biol Chem 274:22977–22984

Lucic E, Fourrey C, Kohler A, Martin F, Chalot M, Brun-Jacob A (2008) A gene repertoire for nitrogen transporters in *Laccaria bicolor*. New Phytol 180:343–364

Lugtenberg BJJ, Kravchenko LV, Simons M (1999) Tomato seed and root exudate sugars: composition, utilization by *Pseudomonas* biocontrol strains and role in rhizosphere colonization. Environ Microbiol 1:439–446

Martin F, Nehls U (2009) Harnessing ectomycorrhizal genomics for ecological insights. Curr Opin Plant Biol 12:508–515

Martin F, Boiffin VV, Pfeffer PE (1998) Carbohydrate and amino acid metabolism in the *Eucalyptus globulus–Pisolithus tinctorius* ectomycorrhiza during glucose utilization. Plant Physiol 118:627–635

Martin F, Duplessis S, Ditengou F, Lagrange H, Voiblet C, Lapeyrie F (2001) Developmental cross talking in the ectomycorrhizal symbiosis: signals and communication genes. New Phytol 151:145–154

Martin F, Aerts A, Ahrén D, Brun A, Duchaussoy F, Kohler A, Lindquist E, Salamov A, Shapiro HJ, Wuyts J, Blaudez D, Buée M, Brokstein P, Canbäck B, Cohen D, Courty PE, Coutinho PM, Danchin EGJ, Delaruelle C, Detter JC, Deveau A, DiFazio S, Duplessis S, Fraissinet-Tachet L, Lucic E, Frey-Klett P, Fourrey C, Feussner I, Gay G, Gibon J, Grimwood J, Hoegger P, Jain P, Kilaru S, Labbé J, Lin Y, Le Tacon F, Marmeisse R, Melayah D, Montanini B, Muratet M, Nehls U, Niculita-Hirzel H, Oudot-Le Secq MP, Pereda VP, Peter M, Quesneville H, Rajashekar B, Reich M, Rouhier N, Schmutz J, Yin T, Chalot M, Henrissat B, Kües U, Lucas S, Van de Peer Y, Podila G, Polle A, Pukkila PJ, Richardson PM, Rouzé P, Sanders I, Stajich JE, Tunlid A, Tuskan G, Grigoriev I (2008) The genome sequence of the basidiomycete fungus *Laccaria bicolor* provides insights into the mycorrhizal symbiosis. Nature 452:88–92

Martin F, Kohler A, Murat C, Balestrini R, Coutinho PM, Jaillon O, Montanini B, Morin E, Noel B, Percudani R, Porcel B, Rubini A, Amicucci A, Amselem J, Anthouard V, Arcioni S, Artiguenave F, Aury J-M, Ballario P, Bolchi A, Brenna A, Brun A, Buee M, Cantarel B, Chevalier G, Couloux A, Da Silva C, Denoeud F, Duplessis S, Ghignone S, Hilselberger B, Iotti M, Marcais B, Mello A, Miranda M, Pacioni G, Quesneville H, Riccioni C, Ruotolo R, Splivallo R, Stocchi V, Tisserant E, Viscomi AR, Zambonelli A, Zampieri E, Henrissat B, Lebrun M-H, Paolocci F, Bonfante P, Ottonello S, Wincker P (2010) Perigord black truffle genome uncovers evolutionary origins and mechanisms of symbiosis. Nature 464:1033–1038

Martin F, Cullen D, Hibbett D, Pisabarro A, Spatafora JW, Baker SE, Grigoriev IV (2011) Sequencing the fungal tree of life. New Phytol 190:818–821

Melin E, Nilsson H (1957) Transport of C^{14}-labelled photosynthate to the fungal associate of pine mycorrhiza. Svensk Bot Tidskr 51:166–186

Nagendran S, Hallen-Adams H, Paper J, Aslam N, Walton J (2009) Reduced genomic potential for secreted plant cell-wall-degrading enzymes in the ectomycorrhizal fungus *Amanita bisporigera*, based on the secretome of *Trichoderma reesei*. Fungal Genet Biol 46:427–435

Nehls U (2004) Carbohydrates and nitrogen: nutrients and signals in ectomycorrhizas. In: Varma A, Abbott L, Werner D, Hampp R (eds) Plant surface microbiology. Springer, Berlin, pp 373–392

Nehls U (2008) Mastering ectomycorrhizal symbiosis: the impact of carbohydrates. J Exp Bot 59:1097–1108

Nehls U, Wiese J, Guttenberger M, Hampp R (1998) Carbon allocation in ectomycorrhizas: identification and expression analysis of an *Amanita muscaria* monosaccharide transporter. Molec Plant Microbe Interact 11:167–176

Nehls U, Ecke M, Hampp R (1999a) Sugar- and nitrogen-dependent regulation of an *Amanita muscaria* phenylalanine ammonium lyase gene. J Bacteriol 181:1931–1933

Nehls U, Kleber R, Wiese J, Hampp R (1999b) Isolation and characterization of a general amino acid permease from the ectomycorrhizal fungus *Amanita muscaria*. New Phytol 144:343–349

Nehls U, Bock A, Einig W, Hampp R (2001) Excretion of two proteases by the ectomycorrhizal fungus *Amanita muscaria*. Plant Cell Environ 24:741–747

Nehls U, Grunze N, Willmann M, Reich M, Küster H (2007) Sugar for my honey: carbohydrate partitioning in ectomycorrhizal symbiosis. Phytochemistry 68:82–91

Nehls U, Göhringer F, Wittulsky S, Dietz S (2010) Fungal carbohydrate support in the ectomycorrhizal symbiosis: a review. Plant Biol 12:292–301

Nilsson LO, Wallander H (2003) Production of external mycelium by ectomycorrhizal fungi in a Norway spruce forest was reduced in response to nitrogen fertilization. New Phytol 158:409–416

Nilsson LO, Giesler R, Baath E, Wallander H (2005) Growth and biomass of mycorrhizal mycelia in coniferous forests along short natural nutrient gradients. New Phytol 165:613–622

Norton JM, Smith JL, Firestone MK (1990) Carbon flow in the rhizosphere of ponderosa pine seedlings. Soil Biol Biochem 22:449–455

Palmer JG, Hacskaylo E (1970) Ectomycorrhizal fungi in pure culture I. Growth on single carbon sources. Physiol Plant 23:1187–1197

Parrent JL, James TY, Vasaitis R, Taylor AFS (2009) Friend or foe? Evolutionary history of glycoside hydrolase family 32 genes encoding for sucrolytic activity in fungi and its implications for plant-fungal symbioses. BMC Evol Biol 9:148

Plett JM, Martin F (2011) Blurred boundaries: lifestyle lessons from ectomycorrhizal fungal genomes. Trends Genet 27:14–22

Polidori E, Ceccaroli P, Saltarelli R, Guescini M, Menotta M, Agostini D, Palma F, Stocchi V (2007) Hexose uptake in the plant symbiotic ascomycete *Tuber borchii* Vittadini: biochemical features and expression pattern of the transporter TBHXT1. Fungal Genet Biol 44:187–198

Pritsch K, Raidl S, Marksteiner E, Blaschke H, Agerer R, Schloter M, Hartmann A (2004) A rapid and highly sensitive method for measuring enzyme activities in single mycorrhizal tips using 4-methylumbelliferone-labelled fluorogenic substrates in a microplate system. J Microbiol Methods 58:233–241

Rangel-Castro JI, Danell E, Pfeffer PE (2002) A ^{13}C-NMR study of exudation and storage of carbohydrates and amino acids in the ectomycorrhizal edible mushroom *Cantharellus cibarius*. Mycologia 94:190–199

Read DJ, Leake JR, Perez-Moreno J (2004) Mycorrhizal fungi as drivers of ecosystem processes in heathland and boreal forest. Can J Bot 82(8):1243–1263

Rieger A, Guttenberger M, Hampp R (1992) Soluble carbohydrates in mycorrhized and non-mycorrhized fine roots of spruce seedlings. Z Naturforsch 47:201–204

Roitsch T, Balibrea ME, Hofmann M, Proels R, Sinha AK (2003) Extracellular invertase: key metabolic enzyme and PR protein. J Exp Bot 54:513–524

Ruan Y-L, Jin Y, Yang Y-J, Li G-J, Boyer JS (2010) Sugar input, metabolism, and signaling mediated by invertase: roles in development, yield potential, and response to drought and heat. Mol Plant 3:942–955

Sacchi GA, Abruzzese A, Lucchini G, Fiorani F, Cocucci S (2000) Efflux and active re-absorption of glucose in roots of cotton plants grown under saline conditions. Plant Soil 220:1–11

Salzer P, Hager A (1991) Sucrose utilization of the ectomycorrhizal fungi *Amanita muscaria* and *Hebeloma crustuliniforme* depends on the cell wall-bound invertase activity of their host *Picea abies*. Bot Acta 104:439–445

Salzer P, Hager A (1993) Characterization of wall-bound invertase isoforms of *Picea abies* cells and regulation by ectomycorrhizal fungi. Physiol Plant 88:52–59

Schaarschmidt S, Roitsch T, Hause B (2006) Arbuscular mycorrhiza induces gene expression of the apoplastic invertase LIN6 in tomato (*Lycopersicon esculentum*) roots. J Exp Bot 57:4015–4023

Schaarschmidt S, Gonzalez MC, Roitsch T, Strack D, Sonnewald U, Hause B (2007) Regulation of arbuscular mycorrhization by carbon. The symbiotic interaction cannot be improved by increased carbon availability accomplished by root-specifically enhanced invertase activity. Plant Physiol 143:1827–1840

Schaeffer C, Wallenda T, Guttenberger M, Hampp R (1995) Acid invertase in mycorrhizal and non-mycorrhizal roots of Norway spruce (*Picea abies* [L.] Karst.) seedlings. New Phytol 129:417–424

Schaeffer C, Johann P, Nehls U, Hampp R (1996) Evidence for an up-regulation of the host and a down-regulation of the fungal phosphofructokinase activity in ectomycorrhizas of Norway spruce and fly agaric. New Phytol 134:697–702

Schaeffer C, Wallenda T, Hampp R, Salzer P, Hager A (1997) Carbon allocation in ectomycorrhizae. In: Rennenberg H, Eschrich W, Ziegler H (eds) Trees – contributions to modern tree physiology. Backhuys, Leiden, pp 393–407

Scharte J, Schön H, Weis E (2005) Photosynthesis and carbohydrate metabolism in tobacco leaves during an incompatible interaction with *Phytophthora nicotianae*. Plant Cell Environ 28:1421–1435

Seo Y, Cho J, Lee SK, Ryu H, Han M, Hahn T, Sonnewald U, Jeon J (2007) Current insights into the primary carbon flux that occurs in plants undergoing a defense response. Plant Stress 1:42–49

Smeekens S, Ma J, Hanson J, Rolland F (2010) Sugar signals and molecular networks controlling plant growth. Curr Opin Plant Biol 13:273–278

Smith SE, Read DJ (1997) Mycorrhizal symbiosis. Academic, London

Smith SE, Read DJ (2008) Mycorrhizal symbiosis. Academic, London

Smith FA, Smith SE (1989) Membrane transport at the biotrophic interface: an overview. Austral J Plant Physiol 16:33–43

Söderström B (1992) The ecological potential of the ectomycorrhizal mycelium. In: Read DJ, Lewis DH, Fitter AH, Alexander IJ (eds) Mycorrhizas in ecosystems. CAB International, Wallingford, pp 77–83

Spägele S (1992) Charakterisierung der intra- und extrazellulären Proteasenaktivitäten des Fliegenpilzes (*Amanita muscaria* [L. ex Fr.] Hooker). Eberhard Karls University, Tübingen

Straker CJ, Schnippenkoetter WH, Lemoine M-C (1992) Analysis of acid invertase and comparison with acid phosphatase in the ericoid mycorrhizal fungus *Hymenoscyphus ericae* (Read) Korf and Kernan. Mycorrhiza 2:63–67

Stülten C (1996) Aufnahme und Metabolisierung von Kohlenhydraten und Aminosäuren durch den Ektomykorrhizapilz *Cenococcum geophilum* Stamm SIV. Eberhard Karls University, Tübingen

Tarkka M, Nehls U, Hampp R (2005) Physiology of Ectomycorrhiza (ECM). In: Lüttge U (ed) Progress in botany. Springer, Berlin, pp 247–276

Tetlow IJ, Farrar JF (1992) Sucrose-metabolizing enzymes from leaves of barley infected with brown rust (*Puccinia hordei* Otth.). New Phytol 120:475–480

Treseder KK, Torn MS, Masiello CA (2006) An ecosystem-scale radiocarbon tracer to test use of litter carbon by ectomycorrhizal fungi. Soil Biol Biochem 38:1077–1082

Trojanowski J, Haider K, Hüttermann A (1984) Decomposition of ^{14}C-labelled lignin, holocellulose and lignocellulose by mycorrhizal fungi. Arch Microbiol 139:202–206

Truernit E, Sauer N (1995) The promotor of the *Arabidopsis thaliana SUC2* sucrose-H^+ symporter gene directs expression of β-glucuronidase to the phloem: evidence for phloem loading and unloading by SUC2. Planta 196:564–570

Tymowska-Lalanne Z, Kreis M (1998) Expression of the *Arabidopsis thaliana* invertase gene family. Planta 207:259–265

Uldry M, Thorens B (2004) The SLC2 family of facilitated hexose and polyol transporters. Eur J Physiol 447:480–489

van Hees P, Jones D, Finlay R, Godbold D, Lundström U (2005) The carbon we do not see-the impact of low molecular weight compounds on carbon dynamics and respiration in forest soils: a review. Soil Biol Biochem 37:1–13

Voegele RT, Struck C, Hahn M, Mendgen K (2001) The role of haustoria in sugar supply during infection of broad bean by the rust fungus *Uromyces fabae*. Proc Natl Acad Sci 98:8133–8138

Wainwright M (1993) Oligotrophic growth of fungi – stress or natural state. In: Jennings DH (ed) Stress tolerance of fungi. Dekker, New York, pp 127–144

Wallander H (2006) External mycorrhizal mycelia – the importance of quantification in natural ecosystems. New Phytol 171:240–242

Wallander H, Nilsson LO, Hagerberg D, Baath E (2001) Estimation of the biomass and seasonal growth of external mycelium of ectomycorrhizal fungi in the field. New Phytol 151:753–760

Wallander H, Göransson H, Rosengren U (2004) Production, standing biomass and natural abundance of ^{15}N and ^{13}C in ectomycorrhizal mycelia collected at different soil depths in two forest types. Oecologia 139:89–97

Wallenda T (1996) Untersuchungen zur Physiologie der Pilzpartner von Ektomykorrhizen der Fichte (*Picea abies* [L.] Karst.). PhD thesis, Eberhard Karls University, Tübingen

Wallenda T, Kottke I (1998) Nitrogen deposition and ectomycorrhizas. New Phytol 139:169–187

Wallenda T, Read DJ (1999) Kinetics of amino acid uptake by ectomycorrhizal roots. Plant Cell Environ 22:179–187

Walters DR, Cowley T, McPherson A, Marshall G, McRoberts N (1996) Sugar transport in the light leaf spot pathogen *Pyrenopeziza brassicae*. FEBS Microbiol Lett 143:285–289

Wiemken A (1990) Trehalose in yeast, stress protectant rather than reserve carbohydrate. Antonie Leeuwenhoek 58:209–217

Wiese J, Kleber R, Hampp R, Nehls U (2000) Functional characterization of the *Amanita muscaria* monosaccharide transporter Am Mst1. Plant Biol 2:1–5

Wipf D, Benjdia M, Tegeder M, Frommer WB (2002) Characterization of a general amino acid permease from *Hebeloma cyclindrosporum*. FEBS Lett 528:119–124

Wisser G, Guttenberger M, Hampp R, Nehls U (2000) Identification and characterization of an extracellular acid trehalase from the ectomycorrhizal fungus *Amanita muscaria*. New Phytol 146:169–175

Wright DP, Read DJ, Scholes JD (1998) Mycorrhizal sink strength influences whole plant carbon balance of *Trifolium repens* L. Plant Cell Environ 21:881–891

Wright DP, Scholes JD, Read DJ, Rolfe SA (2000) Changes in carbon allocation and expression of carbon transporter genes in *Betula pendula* Roth. colonized by the ectomycorrhizal fungus *Paxillus involutus* (Batsch) Fr. Plant Cell Environ 23:39–49

Wright DP, Johansson T, Le Quere A, Soderstrom B, Tunlid A (2005) Spatial patterns of gene expression in the extramatrical mycelium and mycorrhizal root tips formed by the ectomycorrhizal fungus *Paxillus involutus* in association with birch (*Betula pendula*) seedlings in soil microcosms. New Phytol 167:579–596

Wu B, Nara K, Hogetsu T (2002) Spatiotemporal transfer of carbon-^{14}-labelled photosynthate from ectomycorrhizal *Pinus densiflora* seedlings to extraradical mycelia. Mycorrhiza 12:83–88

Zhou Y, Qu H, Dibley KE, Offler CE, Patrick JW (2007) A suite of sucrose transporters expressed in coats of developing legume seeds includes novel pH-independent facilitators. Plant J 49:750–764

Zhu H (1990) Purification and characterization of an extracellular acid proteinase from the ectomycorrhizal fungus *Hebeloma crustuliniforme*. Appl Environ Microbiol 56:837–843

Zhu H, Dancik BP, Higginbotham KO (1994) Regulation of extracellular proteinase production in an ectomycorrhizal fungus *Hebeloma crustuliniforme*. Mycologia 86:227–234

Ziegler H (1956) Untersuchungen Über die Leitung und Sekretion der Assimilate. Planta 47:447–500

Zimmermann M (1961) Movement of organic substances in trees. Sci Total Environ 133:73–79

8 Lipids of Mycorrhizas

Y. Dalpé[1], M. Trépanier[2], A. Lounès-Hadj Sahraoui[3], J. Fontaine[3], M. Sancholle[3]

CONTENTS

I. Introduction

Lipids play a central role in all biological systems; phospholipids (PPLs) and sterols as integral components of membranes and neutral lipids (NLs) such as glycerides as energy storage. With the symbiont way of life of mycorrhizal fungi (MF), a high percentage of the carbon (C) provided by plant partners serves to metabolize lipids and the fungal membranes are privileged sites of bi-directional nutrient exchanges. PPL have specific functions in cellular physiology such as maintenance of cell membrane fluidity, hydrophobic barriers, organelle building, cell signalling, and several elements of metabolic regulation. Lipids represent a non-negligible fraction of fungal biomass, comprising 10–15% of ectomycorrhizal fungi (EMF) basidiocarps and up to 50% of arbuscular mycorrhizal fungi (AMF) spores. First considered as a biochemical tool for the quantitative evaluation of mycorrhizas in a given system, the composition and transformation of plants and fungal lipids during the establishment of symbiosis has been studied, gradually clarifying the role and impact of lipid metabolism in nutrient exchanges, C cycling,

[1]Eastern Cereal and Oilseed Research Centre, Agriculture and Agri-Food Canada, Ottawa, Canada;Yolande.Dalpe@AGR.GC.CA
[2]Faculté des Sciences de l'Agriculture et de l'Alimentation, Département de Phytologie, Université Laval, Québec, Canada;
[3]Unité de Chimie Environnementale et Interactions sur le Vivant, Université du Littoral Côte d'Opale, Calais, France;

Fungal Associations, 2nd Edition
The Mycota IX
B. Hock (Ed.)

plant protection against stresses and toxins, and fungal heterotrophy.

As will be shown by this review, several areas of lipid metabolism in MF have been investigated during the last decade, generating valuable data for a better understanding of their role and functionality in symbioses. The fungi and plants involved in mycorrhizal associations belong to such diverse classes and groups of organisms that, to avoid redundancies, this review is intentionally subdivided into research topics instead of mycorrhizal categories. Firstly, a survey of the lipid composition of MF aims to establish the respective profiles of fungal symbionts highlighting specific particularities associated with fungal classification. The qualitative and quantitative transformations occurring during the establishment of mycorrhizas are described, with an emphasis on the role of lipophilic signals involved in root penetration by hyphae, arbuscle differentiation, phosphorus (P) transportation processes, and C cycling. The impact of P and nitrogen (N) supplementation on storage and membrane lipid synthesis in root and seeds is discussed in light of recent studies. Overviews on fungal lipid changes during the growth process deal mainly with germination, mycelial growth, and fungal maturation. Lipid metabolism in mycorrhizal plants submitted to exogenous additives such as fungicides and pollutants are described in a context of phytoremediation and plant ability to tolerate noxious environments. Finally, an up to date survey provides an overview of the diverse biotechnology tools, such as specific lipids, sterols, and fatty acids (FAs) currently used for the detection and evaluation of MF biomass, root colonization, and for the estimation of the mycorrhizal population in a soil community.

II. Lipid Profiles of Mycorrhizal Fungi

Mycorrhizal fungi symbionts belong to diverse taxonomic groups, including the Ascomycota, Basidiomycota, and Glomeromycota, all sharing root association and plant benefits. The first attempts to study MF lipid composition were performed within the frame of exploratory surveys of the whole fungal diversity world from which acquired knowledge is based on the extracted information actually available.

A. Arbuscular Mycorrhizas

A simple microscopic observation of intact and crushed spores of AMF reveals their richness in lipid material. In fact, **AMF spores are filled with lipids** and thus AMF were designated as "oleaginous fungi" (Gaspar et al. 1994). **Lipids constitute almost 50% of the AMF spore dry weight** (Beilby 1983; Schubert et al. 1992). All AMF propagules such as spores, hyphae, vesicles, arbuscles, including the auxiliary cells of the Gigasporaceae and Scutellosporaceae representatives were found to contain lipids (Selvaraj and Subramanian 1990) in which the NL fraction largely predominates over PL (Jabaji-Hare et al. 1984; Olsson and Wilhelmsson 2000). Total lipids of *Glomus irregulare* (formerly identified by Schenck as *G. intraradices* DAOM 181602; Sokolski et al. 2010) vesicles reached 58% of their dry weight (Jabaji-Hare et al. 1984). The histochemical characterization of AMF propagules by Nemec (1981) suggested that AMF lipids contained phosphoglycerides and glycolipids.

The **major saturated FAs of AMF** are **palmitic (C16:0)** and **stearic acids (C18:0). Monounsaturated acids account for almost 50% of the total FA.** Of the monoenes, **oleic acid (C18:1*n*-9)** usually dominates in gigasporoid spores (Sancholle and Dalpé 1993; Madan et al. 2002; Salvioli et al. 2010), and alpha and gamma linolenic acids (C18:2*n*-6) are found as trace elements. **Hexadecenoic acid (C16:1*n*-5)** makes up to 80% of glomoid spore FA (Nordby et al. 1981; Gaspar et al. 1994; Grandmougin-Ferjani et al. 1997) and is dominant in the membrane and storage lipids found in both PL and NL fractions (Olsson and Wilhelmsson 2000). This unusual FA is specific to AMF and has, to date, not been found in any other fungal groups or species except *Pisolithus tinctorius* (Campos da Rocha et al. 2008) and as traces in *Laccaria bicolor* basidiocarps (Reich et al. 2009). As such it is currently considered a powerful **signature FA** with ecological potential for the evaluation of soil communities and populations (see Sect. VI). The AMF FAs were demonstrated to be stable during mycelium maturation and long-term storage (Olsson and Johansen 2000), heritable through several generations of pot culturing, and not affected by host plant identity (Bentivenga and Morton 1994b). The numerous data available on the total lipid and FA composition of AMF (Sancholle and Dalpé 1993; Gaspar et al. 1994; Grandmougin-Ferjani et al. 1997) easily allow segregation of species at the family level. Lipid distribution and concentration as storage mycelial lipid in intraradical propagules such as vesicles (*Glomus*) and arbuscles versus its accumulation in extraradical mycelium has been successfully demonstrated by van Aarle and Olsson (2003).

The **FA profiles** of gigasporoid spores (*Gigaspora* and *Scutellospora* species) mainly differ from that of glomoid spores (*Glomus* species) by a much lower level of C16:1*n*-5 (40–50% compared to 5–8%) and a high proportion of C16:0 and C18:1*n*-9. *Gigaspora margarita* spores are characterized by their **endophytic associations**, including a Gram-negative type bacterial endophyte (Bianciotto et al. 2003). When the FA profile of a *Gigaspora* line, deprived of its Gram-negative endophyte, was compared to the "wild-type" strain, the total and specific FA contents were found to decrease in bacteria-free spores by a range of 30–70% with, for example, 66.0% and 54.5% less C16:0 and C18:1*n*-9 respectively (Salvioli et al. 2010). Interestingly, the C16:1*n*-5 content also decreased by 53%, but still made up approximately 5% of the total FAs. As Gram-negative bacteria are among the few microorganisms able to synthesize C16:1*n*-5 (Frostegård et al. 1993), such results confirmed that the AMF C16:1n-5 content originated from *Gigaspora* and not from its bacterial endophyte.

The use of **FAME profiles** (fatty acid methyl ester) has been **proposed for the identification of AMF** (Bentivenga and Morton 1994b; Madan et al. 2002). Three FAs may be used to delimit taxonomic boundaries at the family level: C18:0 3OH (0–2% in *Glomeraceae*, absent from *Gigasporaceae*); C20:0 (10–16% in *Glomeraceae*, <1% in *Gigasporaceae*) and *cis* C20:1*n*-9 (2–15% in *Gigasporaceae*, absent from *Glomeraceae*; Bentivenga and Morton 1994a, b). In contrast to bacteria, most fungi (including EMF) contain considerable amounts of C18:2*n*-6 (Müller et al. 1994) which has therefore been used diagnostically (Olsson 1999; Wallander et al. 2001; Bååth et al. 2004). However, as this FA also occurs in large amounts in plants and saprotrophic fungi (Olsson 1999), it should only be used as a "signature FA" for EMF under well defined contexts.

Depending on the studies, **24-ethylcholesterol** (Beilby 1980; Beilby and Kidby 1980; Grandmougin-Ferjani et al. 1997, 1999; Fontaine et al. 2001a, 2004) or **24-methylcholesterol** (Nordby et al. 1981; Fontaine et al. 2001a, 2004; Salvioli et al. 2010) were described as major sterols in AMF spores. Minor sterols detected were cholesterol, Δ5-avenasterol, and 24-ethylcholesta-5,22-dienol. Contradictory results have been published on the presence of ergosterol in AMF propagules. The use of monoxenic cultures allowed confirming the absence of ergosterol in AMF (Fontaine et al. 2001a, 2004). Additionally, α-amyrin, a common vascular plant triterpene, was detected in spores of all AMF species studied (Grandmougin-Ferjani et al. 1999). The close similarity of sterol composition between AMF species revealed the unique metabolic behaviour of these organisms which are currently related taxonomically on the basis of their obligate symbiotic way of life.

B. Ectomycorrhizas

The majority of fungal organisms known for their involvement in ectomycorrhizal (EM) symbioses belong to the classes Ascomycota and Basidiomycota, with a few representatives within Zygomycota. The available data on species lipid profiles are principally related to fruiting bodies, with surprisingly little information on the fungal lipid composition of the mycelium that makes the specialized filamentous mantle that surrounds tree roots.

The species most frequently studied for their lipid composition belong to the macroscopic EMF of the genera *Amanita*, *Boletus*, *Cantharellus*, *Lactarius*, *Russula*, and *Tricholoma* (Basidiomycota) and the truffle species, *Tuber* and *Terfezia* (Ascomycota; Table 8.1). **Total lipid contents of dried fructifications**, even though variable among species, ranged

Table 8.1. Fatty acid profiles (%) of selected ectomycorrhizal fungi fruiting bodies

	C12:0	C14:0	C15:0	C16:0	C16:1	C16:2	C17:0	C18:0	C18:1	C18:2	C18:3	C18:X	C20–C24	Other	References
Ascomycetes															
Cenococcum graniforme				32.0					37.0	31.0					Martin et al. (1984b)
Tuberaceae															
Terfezia claveryii		2.1	1.3	17.0	1.4		1.4	4.5	6.9	45.4	5.8		3.9	10.0	Murcia et al. (2003)
Tuber indicum		0.1	5.2	0.1	tr			7.7	17.7	67.1		0.2	1.6		Tang et al. (2011)
T. melanosporum		6.3		18.9				4.7	20.6	47.9	1.6			1.2	Harki et al. (2006)
Basidiomycetes															
Amanitaceae															
Amanita muscaria		0.1	0.6	4.4	0.5		0.2	7.1	12.2	73.0				1.8	Karlinski et al. (2007)
Boletaceae															
Boletus edulis				8.9				3.0	30.8	52.9				4.4	Hanus et al. (2008)
Suillus grevillei		0.1	1.7	12.8			4.9	0.6		67.1	1.4			11.3	Karlinski et al. (2007)
Cantharellaceae															
Cantharellus cibarius				6.3	0.6			1.2	18.0	41.2		32.3			De Kok et al. (1982)
Cortinariaceae															
Cortinarius nemorensis				15.0				3.0	55.4	26.6					Brondz et al. (2004)
Hericiaceae															
Hydnum rufescens		1.2		24.7	7.9	3.7		2.0	54.3	6.2					Prostenik et al. (1978)
Russulaceae															
Lactarius deliciosus				12.1				25.3	41.3	17.1	0.3				Barros et al. (2007)
Russula rosea				13.6			3.4	8.9	9.4	61.3				3.2	Karlinski et al. (2007)
Sclerodermataceae															
Pisolithus tinctorius (mycelium)		0.4	0.8	19.5	0.8	0.5	0.9	11.7	65.0				tr	tr	Laczko et al. (2003)
P. tinctorius (peridioles)			0.5	17.1				1.0	40.7	34.1					Campos da Rocha et al. (2008)
P. tinctorius (free spores)			0.6	20.7	12.1			1.5	50.5	13.5					Campos da Rocha et al. (2008)
Tricholomataceae															
Tricholoma portentosum				5.6				2.3	58.4	35.4	0.2				Barros et al. (2007)

between 1.8% and 21% (Dembitsky and Pechenkina 1991; Pedneault et al. 2006; Hanus et al. 2008). The **major polar lipid (PL)** component of EMF is **phosphatidylcholine (PC)** (Weete 1989) and, interestingly, several *Boletus* species metabolize high amounts of a **betaine lipid, diacylglyceryltrimethylhomoserine**, which together with PC makes up 73–92% of the total PLs (Hanus et al. 2008). Glycolipids, cell membrane components involved in cell signalling, were detected in a diversity of EMF and found to belong to inositolphosphoceramids and glucosylphosphoceramids (Jennemann et al. 2001; Olsson and Jantzen 2007).

For the majority of EMF, as for saprotrophic mushrooms, the **major FAs** belong to the **C18 di- and mono-unsaturated acids (C18:2*n*-6 and C18:1*n*-9)** and **palmitic acid (C16:0)**. When combined, these FAs can make up to 90% of the total fruit body FAs (Table 8.1).

Significant quantitative but not qualitative differences were observed between the FA profiles of EMF species (Prostenik et al. 1983; Dembitsky et al. 2010). PLs were more unsaturated than NL, linoleic acid being the dominant polar lipid fatty acid (PLFA), with C16:0 more abundant in NL fractions. In *Terfezia boudieri*, profiles were reversed (Al-Saadi et al. 2005). Unusual FAs have been detected in EMF species such as a trihydroxy FA isolated from *Tuber indicum* (Gao et al. 2001). Neo acids heptadecenoic (C17:1*n*-5) and elaidic acids (C18:1*n*-9t) were found in basidiomycetous fungi (Dembitsky 2006; Pedneault et al. 2006, 2008). Interestingly, small amounts of hexadecenoic acid (C16:1*n*-5), considered a biomarker for AMF, were detected in *Laccaria bicolor* (Reich et al. 2009) as well as in the developing basidiospores of *Pisolithus* sp. (Campos da Rocha et al. 2008). Surprisingly, one major FA of *Amanita rubescens* is a C13:0 2OH making up 37.5% of the total FAs (Karlinski et al. 2007). Important changes in FA concentrations were observed during ascocarp maturation of *T. melanosporum* (Harki et al. 2006) and during basidiospore differentiation and maturation of *Pisolithus* sp. (Campos da Rocha et al. 2008).

Free sterols may represent **up to 40% of the total lipid content**, as in *Boletus edulis* basidiocarps (Hanus et al. 2008). **Ergosterol** remains the predominant sterol of EMF, often making up more than 50% of the total sterol fraction (Senatore et al. 1988; De Simone et al. 1979; Cerri et al. 1981; Solberg 1989). A major exception comes from truffles where **brassicasterol** was found to make up 98% of the sterol fraction of a commercial *Terfezia* (Weete et al. 1985). Brassicasterol has also been found to be quite abundant (26–44% of total sterols) in the black truffle *Tuber melanosporum* and in *T. brumale* (Weete et al. 1985), comprising, when combined with ergosterol, 90% of the total sterols of *T. melanosporum* (Harki et al. 1996). Interestingly, ascocarps of *Morchella esculenta* also contain large amounts of brassicasterol (Yokokawa 1994). The only basidiomycetous EMF found to contain brassicasterol (5–9% of total sterols) belonged to *Tricholoma* species (Senatore et al. 1988). Solberg (1989) and Weete et al. (2010) published good reviews on the lipid constituents of fungi, including several EM species.

The lipid profiles of fungal mycelia particularly caught our attention, as hyphae constitute the basic link between plant and fungal symbionts. Unfortunately, little documentation is available in the literature, either for mycelium generated by fungal pure cultures or hyphal phases extracted from natural, or for artificial mycorrhizal plant–fungi associations. The lipid droplets observed along the hyphae of pure cultures of *Cenococcum graniforme* and *Hebeloma crustuliniforme* mycelial cultures contained high amounts of triacylglycerol (Martin et al. 1984a). Total lipid mycelium content is estimated to make up 5–10% of mycelium dry weight and is considered to be involved in C storage (Martin et al. 1984a, b).

In a comparative analysis of the FA content of *Tuber* spp., Tang et al. (2011) found that the FAME profiles of *T. indicum* and *T. aestivum* fruit bodies and mycelium were nearly identical, even though influenced by environmental conditions. Similar results were obtained by Karlinski et al. (2007) working on 13 basidiomycetous EMF species.

C. Other Mycorrhizas

Little or no information is available in the scientific literature on lipid profiles of ectendo, ericoid, orchid, and sebacinoid MF. As for the whole filamentous fungal community, lipid droplets are currently observed in the cytoplasm of hyphal endophytes (Werner et al. 2000). Sudan IV and Sudan black staining allow current detection of lipids in the cytoplasm of non-pelotonic hyphae of *Sphatoglottis plicata* (Senthilkumar and Britoo 2001). In the only reference on orchid mycorrhizal fungal sterol composition (Barroso et al. 1987), no sterols were recovered from the diethyl ether extraction of a mycelium endophyte of *Ophrys lutea* grown in pure culture even though the sterol content of the colonized orchid tubers underwent an important transformation. The lipid composition of mycorrhizal *Piriformospora indica* species (Sebacinales) has not yet been investigated.

III. Plant Lipid Transformations Induced by Mycorrhizas

The close interaction between plant and fungi mycorrhizal symbionts influences their respective metabolism and biology, because mechanical contacts, signalling, and material exchanges occur in a continuous process. It is likely that major changes in plant lipid metabolism take place at symbiosis, influenced by mineral nutrition and P and N inputs, with repercussions on C cycling.

A. Impact of Mycorrhizas on Plant Lipids

1. Quantitative Changes, Colonized Versus Non-Colonized Roots

a) Lipids and Fatty Acids

The accumulation of lipids in AMF-colonized roots has been found to be directly proportional to the number of intraradical AMF vesicles (Pacovsky and Fuller 1987; Johansen et al. 1996). As a consequence, **colonized roots contain generally more lipids than non-colonized ones** (Graham et al. 1996; Gaspar et al. 1997).

Depending upon P availability, such content can be either increased (Bethlenfalvay et al. 1994), decreased (Bethlenfalvay et al. 1997), or in some cases remain unchanged (Campagnac et al. 2010) and be dependent on the fungal symbiont and related to its aggressiveness (Nordby et al. 1981; Graham et al. 1996). The increase in lipid content was reported to be much smaller with Gigasporaceae- than Glomeraceae-colonized plants. Such a result may be explained by the fact that gigasporoid species do not differentiate intraradical vesicles, which are particularly lipid-rich propagules. The accuracy of this hypothesis was indirectly supported by van Aarle and Olsson (2003) in a comparative study of the FA composition of PLs and NLs of a glomoid and a gigasporoid spore species. Abundant oil droplets were detected in root cells of *Betula verrucosa* and *Pinus radiata* mycorrhiza associated respectively with *Lactarius pubescens* and *Rhizopogon luteolus* (Bevege et al. 1975). However no studies have evaluated quantitative root lipid changes in EM-colonized roots.

b) Sterols

An **increase in sterol content was measured during AM establishment** in roots of maize and clover colonized by a diversity of AMF (Cooper and Lösel 1978). Similar results were observed by Fontaine et al. (2001a, 2004) working with AMF monoxenic root culture. However, in many other similar trials, no significant differences in sterol content were observed between mycorrhizal and non-mycorrhizal roots (Cooper and Lösel 1978; Nagy et al. 1980; Campagnac et al. 2008).

Few data are available on the content and sterol composition of EM roots. In a number of studies, only the ergosterol measurement is provided as its value correlates with root colonization levels (see Sect. VI). However, Laczko et al. (2003) reported that, during the colonization of *Pinus sylvestris* roots by *Pisolithus tinctorius*, the sterol content remained stable, in contrast to the neutral lipid fatty acid (NLFA) fraction. Reich et al. (2009), when analysing membranes of the AMF *Laccaria bicolor*, showed that the glycolipid fraction was composed mainly of acylated and non-acylated sterol glycosides, with NLs containing free sterols and sterol esters.

2. Qualitative Changes, Colonized Versus Non-Colonized Roots

a) Lipids and Fatty Acids

The FA composition of AM-colonized roots differs from non-mycorrhizal ones in the presence of C16:1*n*-5, which is a specific and major FA for most *Glomus* isolates (see Sect. I). Indeed, **the AMF-specific FA C16:1*n*-5 was totally absent from non-mycorrhizal roots** (Olsson et al. 1997; Stumpe et al. 2005; Campagnac et al. 2010), supporting its value as a biomarker for glomalean fungi (see Sect. VI). *Glomus* species were shown to be particularly rich in C16:1*n*-5 (Gaspar et al. 1994; Grandmougin-Ferjani et al. 2005), making up 34–78% of the neutral FA fraction as storage lipids, with a minor occurrence in membranes representing 13–23% of the PFLA (Olsson and Johansen 2000; Van Aarle and Olsson 2003). The ratio between PPL levels (related to arbuscular colonization) and the NLs (related to vesicle abundance) has been successfully applied to estimate the physiological state of the fungus in mutualistic symbiosis (Olsson et al. 1997; Van Aarle and Olsson 2003).

AM root colonization levels can be estimated either cytologically (Brundrett et al. 1994), by quantification of PCR fungus-specific DNA (Filion et al. 2003), by fungus-specific quantitative PCR RNA (Isayenkov et al. 2004), or by analysis of fungal specific esterified FA profiles. Not surprisingly, this latter method was as efficient as PCR, with the added advantage of being much faster and less laborious. With ***Glomus*** **mycorrhizas**, the **amount of C16:1*n*-5 is well correlated with root colonization levels in a variety of plant hosts** (Graham et al. 1995; Olsson et al. 1997; Fontaine et al. 2004; Stumpe et al. 2005; Campagnac et al. 2010). In *Medicago truncatula* colonized roots, in addition to C16:1*n*-5, esterified palmitic acid (C16:0) and oleic acid (C18:1*n*-9) contents also increased compared to control roots (Stumpe et al. 2005). Fester et al. (2001) suggested that, due to the proliferation of plastids in arbuscle-containing cells, the plastidic pathway to FA biosynthesis may be involved. However, when the colonizing organism belongs to the genus *Gigaspora*, there is an increase in C18:1*n*-9 which comprises 35–55% of the total profile whereas C16:1*n*-5 represents less than 2%. Indeed, the FA profile of *Gi. margarita* spores showed a clear predominance of C18:1*n*-9 (Bentivenga and Morton 1994b; Salvioli et al. 2010).

In herbaceous plants, PC dominates (Cowan 2006) when PC and phosphatidylethanolamine (PE) are the major PPLs in AMF. Cooper and Lösel (1978) reported an increase of all molecular types of PPLs in mycorrhizal roots. In particular, increases in phosphatidic acid and phosphatidylinositol (PI) in mycorrhizal roots are of interest since these molecules were not detected in the isolated AM mycelium. In plant–pathogen interactions, PI is considered to play an important role in plant cell signalling; this may explain its increase in colonized roots. Drissner et al. (2007) highlighted the importance of lyso-phosphatidylcholine as a plant signal, particularly in AM symbiosis. Root extracts of mycorrhizal plants harbouring the lipophilic signal induced phosphate transporter genes.

To study EM formation and function, model systems using axenically raised tree seedlings have been used for studies of signal exchange (Martin et al. 2001), gene expression (Voiblet et al. 2001), N and C metabolism (Plassard et al. 2000; Nehls et al. 2001), and lipid metabolism (Laczko et al. 2003; Reich et al. 2009). EM roots showed a more diverse FA pattern than free-living mycelium, potentially due to the presence of additional FAs of plant origin (Laczko et al. 2003; Reich et al. 2009).

In *P. tinctorius* and *L. bicolor* fungi, the most abundant FAs were palmitic acid (C16:0), oleic acid (C18:1*n*-9), and linoleic acid (C18:2*n*-6). Reich et al. (2009), when comparing the relative abundance of FAs in different root tissues, found that EM were significantly enriched in C16:1*n*-5 by 5.4-fold and in C18:2*n*-6 by 2.2-fold. Linoleic acid, the most abundant FA, shows a strong decline with age. In contrast, C16:1*n*-5 occurred as traces in roots (0.05%) but displayed the highest absolute amounts and relative enrichment factors in EM and showed no age-dependent changes (Reich et al. 2009). In *P. tinctorius*/*Pinus sylvestris* EM, PLFAs (54%) dominated over NLFAs (38%) and sterols (8%), but the level of NLFAs was more than three times higher than in axenically grown roots, indicating that mycorrhiza formation should have a major stimulatory effect on NL accumulation (Laczko et al. 2003). In a single day period, the saturated FA concentration in the

NLFA fraction first increased and then slowly decreased. Saturated FAs were detected in the extraradical fungal mycelium, indicating that the participation of NLs in C transfer from plant to fungus occurs prior to carbohydrate involvement, suggesting the presence of an unknown signal and a probable specific role of NLFAs in the root itself (Laczko et al. 2003). In plant–pathogen interactions, host plants respond to fungal infection by the synthesis of FAs, in particular, the accumulation of epicuticular waxes (C20, C22) on shoot and root surfaces (Zhukov et al. 2001). A similar defence reaction has been observed at the first contact of symbionts (Bonanomi et al. 2001). Laczko et al. (2003) suggested that a similar process may exist in EM as well.

Laczko et al. (2003) also observed, in the initial stages of mycorrhizas establishment, a significant increase in the saturated PLFA C18:0 at the expense of the unsaturated PLFAs C18:2*n*-6 and C18:1*n*-6 in the extraradical mycelium, potentially linked to a structural reorganization of the membrane concomitant with the adaptation to symbiosis conditions, and possibly related to the motile tubular vacuolar systems (Cole et al. 1998; Weber et al. 2001).

b) Sterols

Changes in sterols in mycorrhizal plants can be attributed to a metabolic response to intraradical fungal root colonization. Gaude et al. (2004) showed an accumulation of the galactolipid digalactosylglycerol in the peribacteroid membrane of the N-fixing nodules of soybean and lotus. It is likely that such changes also take place during the establishment of mycorrhizal symbiosis.

Non-colonized herbaceous roots contain sitosterol, campesterol, and stigmasterol as the major sterols; **in AM-colonized roots, sitosterol and stigmasterol levels decreased while campesterol (24-methylcholesterol) increased concomitantly** (Nagy et al. 1980; Grandmougin-Ferjani et al. 1995). Similar results were obtained on in vitro carrot—and chicory—*G. intraradices* root cultures (Fontaine et al. 2001a, 2004; Campagnac et al. 2008). The sterols 24-methyl- and 24-ethylcholesterol were found to predominate in spores either produced in pot cultures (Beilby 1980; Beilby and Kidby 1980; Grandmougin-Ferjani et al. 1999) or differentiated in monoxenic cultures (Declerck et al. 2000; Fontaine et al. 2001a, 2004; Campagnac et al. 2009). No ergosterol was detected in the spores and mycelium of AMF (Beilby 1980; Nordby et al. 1981; Grandmougin-Ferjani et al. 1999; Salvioli et al. 2010). The increase in campesterol in mycorrhizal roots is attributed to the presence of AMF in the roots, although a plant metabolic response to root colonization cannot completely be excluded. 24-Methylenecholesterol, a sterol not easily separated from 24-methylcholesterol by gas chromatography, was also detected in colonized roots (Schmitz et al. 1991; Grandmougin-Ferjani et al. 1995). In the light of those results, **the 24-methyl/methylene cholesterol mixture has been proposed as a potential indicator of AM colonization** (Schmitz et al. 1991). Fontaine et al. (2004) found an additional sterol in colonized carrot and chicory transformed roots, 24-methyldesmosterol, and a valuable correlation was established between the content of the sum of 24-methylcholesterol, 24-methylenecholesterol, and 24-methyldesmosterol and AM colonization. These authors noticed that the 24-methyl/methylene sterol increase was detected on well established symbioses, and the sterols were found in both symbiont tissues.

The establishment of functional symbiosis implies complex molecular signalling between mycorrhizal partners, and recent studies strongly implicate lipids in the process (Bucher 2010). **Sterols** were found to **induce plant genes involved in root penetration by hyphae and arbuscle differentiation**, such as the MtMSBS1 gene, coding for a membrane sterol binding protein (Kuhn et al. 2010). This protein, once fixed to sterols, controls their syntheses by regulating P450 cytochromes located in the endoplasmic reticulum. Sterols play a major role in the plasma membrane invagination process and consequently the propagation of the fungal symbiont in the root cortex.

Despite the frequent use of ergosterol analyses in the study of EM ecosystems (see Sect. VI), knowledge of the composition and functional significance of sterols during the establishment of the EM symbioses remains very limited. Laczko et al. (2003) showed that colonization of *Pinus sylvestris* roots by *Pisolithus tinctorius* induced the synthesis of ergosterol and 24-ethyllanosta-8, 24(24′)-

diene-3-β,22-ζ-diol, both recognized as major sterols in EMF. The authors concluded that, although ergosterol is often used as a marker for fungi in EM research, other fungal-specific sterols can be quantitatively more important.

In a comparison of colonized and non-colonized *Ophrys lutea* tubers, Barroso et al. (1987) demonstrated that free sterols are synthesized during the infection process and noted an accumulation of sitosterol and campesterol in colonized tubers which correlated to an interesting decrease of ergosterol and 24-methylcholesterol.

When studying the effects of N fertilization on mycorrhizal infection in *Calluna vulgaris*, Caporn et al. (1995) found that concentrations of ergosterol were significantly greater in fine hair ericaceous roots than in thicker roots and were also higher in the surface horizons than deep in the soil core. Indeed, intracellular fungal coils occur almost exclusively in hairy root cortical cells.

3. Shoots and Seeds in Colonized and Non-Colonized Plants

There are very few data on the lipid metabolism and changes in aerial parts of mycorrhizal plants compared to non-mycorrhizal ones. The first attempt examined changes in FA composition in the leaves of *Citrus* colonized by *G. mosseae* which were found to contain more (C16+C17) and less C18 FAs than leaves of non-colonized plants (Nordby et al. 1981). Leaves of soybeans whose roots were inoculated with *G. fasciculatum* contained more lipids per unit of dry weight than leaves of non-inoculated plants, the increase being higher when plants were dually inoculated with *G. fasciculatum*+*Bradyrhizobium* (Pacovsky and Fuller 1987).

On a qualitative basis, leaves of mycorrhizal soybean contained higher levels of C18:2 and C18:1 than leaves of non-mycorrhizal plants (Pacovsky and Fuller 1987) when C16:0 accumulated in the pods. These results were corroborated by those of Grandmougin-Ferjani et al. (1995) who found that increases of C18:2 and C18:3 correlated to a decrease of C16:0 in leaves of leek plants colonized by *G. mosseae*. In leaves of maize colonized by *G. etunicatum*, there was a general decrease in the FA content, while leaves of dually inoculated plants (*Glomus*+*Azospirillum*) contained more C18:3 and C16:0 than the controls (Pacovsky 1989). Changes can also affect the sterol composition of the leaves of mycorrhizal plants, such as with flax leaves where sitosterol and stigmasterol accumulated in plants with *G. intraradices*-colonized roots but remained unchanged in the leaves of non-colonized plants (Dugassa et al. 1996).

Because in mycorrhizal plants the presence of the mycosymbiont is restricted to the root tissues, the observed changes in the FA and/or sterol contents of leaves cannot be attributed to the presence of substances coming from the mycelium. There is thus evidence that **root colonization may deeply alter the metabolism of the host plant.**

While studying seed nutrient dynamics as influenced by mycorrhizal infection of the maternal generation, Lu and Koide (1991) suggested that seeds produced by mycotrophic plants may be modified specifically in relation to the relative abundance of some of their main storage products: proteins and oils. The lipid content of pea seeds colonized by *G. mosseae* was found to be higher than in seeds from non-mycorrhizal plants (Bethlenfalvay et al. 1994). A similar observation was made with mycorrhizal durum wheat, whose kernels contained more lipid than kernels of non-mycorrhizal wheat (Al-Karaki et al. 1998). In contrast, the opposite results were obtained with soybean seeds of plant roots colonized by *G. mosseae*, *G. etunicatum*, or *Gi. rosea*. Although the seed protein contents remained unchanged, a higher seed protein/lipid ratio was found in AM plants (Bethlenfalvay et al. 1997). All these studies and observations showed that the **lipid metabolism of the host plants can be deeply altered by the presence of the mycosymbiont**, probably through the modified supply of mineral nutrients like N or P.

B. Impact of Mineral Nutrition on Plant Lipid Metabolism

Fungi are well known as mineral scavengers. When involved in mycorrhizal associations operating as extensions of the root system, they can be very efficient in ion uptake, particularly enhancing the uptake of P by the host plant (Rolin et al. 1984). The amount of P uptake per unit of C allocated below ground is thus expected to be higher in mycorrhizal than in non-mycorrhizal roots (Jakobsen 1991). In legumes, AM symbiosis has been shown to increase photosynthetic rates by 14%, which represents more C than the C costs for symbiosis (Kaschuk et al. 2009). As such, the legume received the considerable advantage of the nutrients supplied by the fungus without compromising the amount of C available for its growth. Non-mycorrhizal *Pinus sylvestris* does not respond to P limitation by a proportional increase of root production. This might indicate an obligate dependency on mycorrhiza for effective P uptake, since the production of extramatrical biomass of the mycosymbiont *Paxillus involutus* peaked when P was low and other nutrients were high (Ekblad et al. 1995). As **P is required for the synthesis of PLs**, the **P uptake by MF logically alters their own lipid metabolism as well as that of their host plants**. Other mineral nutrients including N are involved as well, acting in close relationship with P.

1. Repercussions of Phosphorus Uptake on Lipid Metabolism

a) Roots

The amount of P available in the soil may act directly by either limiting or exhausting fungal growth. It may also act indirectly by altering fungal lipid metabolism.

In AM *Citrus* roots, application of P to the soil reduced the total fungal FA content (Peng et al. 1993). Even if the AM colonization of *Citrus* roots was not significantly reduced by a high P supply, the relative content of C16:1*n*-5 acid was found to be significantly lower in plants grown under high P concentration (Graham et al. 1997). Olsson et al. (1997) demonstrated that P applications resulted in an increase in P concentration in both the shoots and the roots of non-mycorrhizal plants of *Cucumis sativa* while *G.caledonium*-colonized plants showed a decrease in P concentration in the shoots but not in the roots. The root colonization level decreased as the applications of P increased and the C16:1*n*-5 concentration decreased in roots and soil as P application increased. P supplementation negatively influenced lipid metabolism, particularly the synthesis of storage lipids, as shown by the dramatic decrease of the NLFA/PLFA lipid ratio following the application of 15 mg $P kg^{-1}$. The P available for roots controls the C flow to the fungus. However, if P is low, the C flow from root to fungus will be high, up to the moment that P is no longer a limiting factor for the root, after which the C flow to the fungus diminishes (Olsson et al. 2006).

High P content induces an increase in membrane PPL content, reducing permeability and resulting in a **reduced leakage of sugars and amino acids** which are essential for the establishment of the mycorrhiza (Ratnayake et al. 1978).

There probably is a need for the regulation of P and N contents, since roots dually inoculated (*Azospirillum brasilense* + *G. etunicatum*) contained more specific FAs than those inoculated with only one or none of the symbionts (Pacovsky 1989). Schliemann et al. (2008) published an extensive study on the metabolite profiling of mycorrhizal roots of *Medicago truncatula*. With respect to lipids, they observed an increase of some FAs (C16:0, C18:1*n*-6) during the most active stages of root colonization, indicating a mycorrhiza-specific activation of plastidial metabolism. Accumulation of fungus-specific FAs (C16:1*n*-5, C18:1*n*-7) designated as markers for colonized roots or soil fungi was also observed (Olsson 1999; van Aarle and Olsson 2003; Olsson et al. 2005b).

Working with the EMF *Hebeloma cylindrosporium*, Rolin et al. (1984) showed that P uptake by the fungus was stored as orthophosphate, an easily mobilized form for DNA and PL synthesis. Unfortunately, these authors did not perform analyses of lipid content in the roots or in the seeds.

b) Seeds

The physiological effects of N and P supplementation were studied on *Glycine max* colonized either by *Glomus fasciculatum*, or *Bradyrhizobium japonicum*, or by both symbionts. The response to root colonization is particularly spectacular in the seed lipid fraction which is twice as high in dually inoculated plants compared to non-inoculated plants supplied with N or P fertilizers (Pacovsky and Fuller 1987). This result demonstrates that the **regulation of P uptake by mycorrhizas strongly influences the synthesis of storage lipids in seeds.** The lipid content of seeds produced by mycorrhizal *Pisum sativum* is also highly correlated to seed P content. The level of AM root colonization has been shown to affect both the protein and lipid composition of pea seeds, acting in the same way as changes in soil P availability (Bethlenfalvay et al. 1994). A **negative correlation between seed P and lipid concentration** was clearly observed with *Glycine max* (Bethlenfalvay et al. 1997). Lipid concentration is more affected by P supplementation or AM root colonization than protein and tends to decrease with higher levels of root colonization. This results in higher protein/lipid ratios in the seeds of highly colonized plants. In contrast, *Glomus*-inoculated soybeans contained more lipids per unit of dry weight in the leaves and roots than P-fertilized non-inoculated soybeans. However, the highest lipid contents are obtained with dually (*Bradyrhizobium* + *Glomus* sp.) inoculated plants (Pacovsky and Fuller 1987).

Without supplemental P, durum wheat inoculated with *G. mosseae* had a significantly higher lipid concentration in the seed, but the protein concentration was lower when compared to non-inoculated plants, resulting in a lower protein/lipid concentration ratio in AM plants. The addition of P to non-AM plants does not result in a modification of this ratio, indicating that the presence of the AM fungus affects the production of seed protein and lipid with a mechanism distinct from those mediated by P nutrition (Al-Karaki and Clark 1999). In oil flax, inoculation with *G. mosseae* or *G. intraradices* increased the shoot P concentration, the number of stem branches, and the number of seed bolls per flower, but caused a slight diminution in the total unsaturated FAs FAs (Rydlov et al. 2011). Thus, AMF may modify the composition of FAs in the seeds of oil flax. Working with legume species, Kaschuk et al. (2010) obtained higher yields with AMF and an improvement of 14% in the seed protein mass fraction, but observed no discernable effects of AMF on seed lipid mass fractions. This suggests that yield increases due to fungi resulted from an increase in the seed proteins.

2. Effects of Nitrogen on Lipid Content

The availability of N has been shown to alter the lipid metabolism of AM-colonized plants. Changes can be induced, not only by the level of available N, but also by the type of N supplied. The amount of N available to a plant may modify the mycorrhizal relationship between symbionts. Working with monoxenic AMF root-organ culture, Olsson et al. (2005a) observed that **when N is high, the C-flow to the fungus is reduced**, as demonstrated by a significant reduction in fungal C16:1*n*-5 FA in the root and in the extraradical mycelium. This **decrease in fungal lipid storage in roots when N is applied** has also been observed for *Acer saccharum* growing in northern hardwood forests (van Diepen et al. 2007). One hypothesis is that host C allocation to AMF may be reduced with N addition. Another hypothesis is that N addition could lead to modification of the AMF community structure by the selection of less beneficial fungi in some ecosystems (van Diepen et al. 2011).

In a similar way, when nitrates are supplied up to an optimum, root colonization by AM fungus is reduced. Nitrite NO reductase in the plant plasma membrane vesicles of root cells increased in parallel with nitrate supply in the AM root, but not in the roots of control plants. The enzyme, dependent on nitrate availability, induces a defence response limiting the colonization of roots by AMF. However, under very high levels of applied N, nitrite NO reductase was strongly reduced and root colonization increased, indicating that the defence system of the plant was impaired (Moche et al. 2010).

With EM, ergosterol seems to be particularly sensitive to changes in N sources. As with P, an increased supply of mineral N results in a decrease of ergosterol content in roots of *Picea abies* colonized by *Thelephora terrestris* and *Cenococcum geophilum*. This is true when inorganic N is supplied (ammonium nitrate), but not with keratin, an organic form (Wallenda et al. 1996). With *Pisolithus tinctorius*, total FA levels were found to decrease under extreme N inputs, but their concentration increased with N increases (Melhuish and Janerette 1979). These results indicate that the type of N supplied can influence both the levels and the profile of fungal lipids. In *Calluna vulgaris* mycorrhizal plants, the maximum ergosterol content was found at 80 kg ha^{-1} $year^{-1}$ NH_4NO_3 application, but was found variable through the growing season (Caporn et al. 1995).

3. Influence of Other Minerals on Lipid Content

Other minerals, K, Ca, and Mg, may also influence the lipid composition of mycorrhizal plants (Lynd and Ansman 1989). Strullu et al. (1983) showed that the vesicles of AMF contain K, Ca, Mg, and Na as principal elements, in addition to lipids. Aluminium was found to affect the membrane fluidity of the mycorrhizal fungus *Amanita muscaria* through the ordering and dynamics of lipid molecules in the plasmalemma (Žel et al. 1993). Even if they are not directly incorporated into the lipid molecules, mineral nutrients may influence the lipid content of the MF by interfering with each other.

C. Carbon Cycling

In most mycorrhizas, the host plant represents the major C source for the fungus which is heterotrophic for such nutrient sources. The inability to culture AMF saprophytically (Jakobsen 1991) has brought research to focus on physiological studies of symbiotic fungi and their interactions with plants. Fungal survival and growth depend on the plant for specific nutrients. Several observations and experiments point to the conclusion that precursors for the metabolism of the mycosymbiont undoubtedly come from the host plant, despite the fact that enzymes for the dark fixation of CO_2, such as pyruvate carboxylase and phosphoenol-pyruvate carboxykinase, have been detected in filamentous fungi (reported by Jakobsen 1991). Using labelled ^{13}C, Martin and Canet (1986) demonstrated the presence of these enzymes in EMF, while Bago et al. (1999) obtained the same results with AMF. This direct CO_2 assimilation is certainly not the main metabolic pathway for MF development.

Carbon cycling through the host plant seems to be a promising research avenue. Symbiotic fungi play an important role in nutrient cycling through the uptake, transformation, and release of nutrients. Of course, the host photosynthetic plants provide the mycosymbionts with enough variety and quantity of photosynthate to cover their C requirements (Ho and Trappe 1973; Bevege et al. 1975; Cox et al. 1975; Lösel and Cooper 1979). As the mycorrhizal association improves the plant's access to soil water and minerals, C fixation activities are consequently stimulated (Paul and Kucey 1981; Trent et al. 1989). Several authors have demonstrated that **mycorrhizas may increase below-ground C allocation** by 4–20% (Pang and Paul 1980; Snellgrove et al. 1982; Douds et al. 1988; Wang et al. 1989; Jakobsen and Rosendahl 1990; Graham 2000). When studying the C economy of sour orange in relation to mycorrhizal colonization, Eissenstat et al. (1993) concluded that mycorrhizas increased the root biomass fraction, root length/leaf area ratio, and the percentage of ^{14}C recovered from below-ground components. All of these results showed clearly that the **presence of mycorrhizas stimulates the transfer of photosynthate to the roots**. Moreover, **up-regulation of photosynthesis by AMF** has been observed, leading to a direct net C gain for the host (Miller 2002). Using soybean–*Rhizobium*–*Glomus* associations, Harris et al. (1985) showed that the rate of specific ^{14}C uptake was greater in symbiotic plants, but the starch content of their leaves was reduced by half compared to that of control plants. One can suppose that most of the missing starch is used as a C supply by the microbial symbionts. There is no doubt that the development of a functional mycorrhiza simultaneously depends on the availability of photosynthates and enhances their availability to the host plant (Hampp et al. 1995). Provided there is full compensation for the C drain by the mycobiont in the form of an increased photosynthetic rate, plant growth should not be affected by the C drain (Jakobsen 1991).

For a temperate woody plant such as grapevine, the C stored in the stem is necessary for initial spring growth. Mortimer et al. (2005)

observed a C decrease in the stem of mycorrhizal plants and reduced shoot growth during the first 67 days, corresponding with the phase of rapid root colonization. After this period, root colonization stabilized, normal shoot growth was observed and C content in stem was replenished. Graham et al. (1997) showed that colonized *Citrus* plants contained more sucrose and less starch than non-mycorrhizal plants. Sucrose concentration is consistently lower in the root tissues of P-fed mycorrhizal than non-mycorrhizal plants, indicating a substantial utilization of sucrose by the fungus (Peng et al. 1993). These authors found significantly higher FA concentrations in mycorrhizal compared to non-mycorrhizal fibrous roots. There was a 37% difference in daily total root/soil respiration, two-thirds of which was associated with the construction of lipid-rich roots, the maintenance of the root fungal tissues, and the growth and maintenance of extramatrical hyphae. An increased capacity for new sucrose synthesis was also observed in mycorrhizal seedlings of spruce (Hampp et al. 1995).

The limitation of phloem loading of photoassimilates in tobacco transgenic plants led to a significant decrease in root sugar content and root colonization (Schaarschmidt et al. 2007). The same authors also found that inhibition of root invertase activity decreased root colonization, but over-expression of this enzyme in the root did not improve root colonization, suggesting that the optimal level of hexoses is already present in wild-type tobacco. In the presence of a sufficient amount of P, shading of the plant resulted in a reduced fungal lipid accumulation in the intraradical mycelium, but if P availability is low, shading of plant did not reduce C transfer to the fungus (Olsson et al. 2010). The authors concluded that, for shaded plants, more C is translocated to the extraradical mycelium with the strategy of finding a new host with a higher C allocation capacity. Moreover, **under sufficient nutrient levels, a host plant with low C status transfers less C to the fungus, which diminishes fungal nutrient transfer to the plant and increases accumulation in fungal spores and hyphae** (Hammer et al. 2011).

Even if a proportion of this C is respired by the fungus, Olsson and Johnson (2005) found that most C remained in the AMF for the duration of their experiment (32 days). In a context of **global warming**, AMF could dramatically improve C sequestration in soils, especially in nutrient-limited environments (Zhu and Miller 2003). In response to elevated atmospheric CO_2, increased photosynthetic activity is expected, especially for C-3 plants. This newly fixed C is transferred in higher quantity to AMF which then transfer it to their associated soil microbial communities. Under elevated atmospheric CO_2, Drigo et al. (2010) observed marked changes in bacterial diversity and activity, which favoured the specific population best adapted to this enriched nutrient mycorhizosphere.

The review of Smith and Gianinazzi-Pearson (1988) presented a good overview of C distribution in the symbionts and concluded that the **fungi can convert host metabolites into specific fungal compounds such as lipids** which are particularly abundant in AM fungal vesicles. The **AMF vesicles possess high lipid contents** and may act as reservoirs for many elements (Strullu et al. 1983); this is consistent with Bevege et al. (1975), who considered that lipid fractions in the hyphae may have an important storage function. The analysis of root FAMEs not only provided a measure of colonization development, but also served as an index of C allocated to intraradical fungal growth and lipid storage. Moreover, ^{13}C-labelling of the specific NLFA 16:1*n*-5 is now a useful tool to study the distribution of plant C allocations in response to applied treatments (Gavito and Olsson 2003; Olsson et al. 2005b).

With respect to C transfer via AMF between two host plants, some recent studies concluded that C given by a host plant can move through the extraradical mycelium to reach the root of a second plant, but the transferred C remained in the intraradical mycelium of the fungus and was never transferred to the plant (Pfeffer et al. 2004; Voets et al. 2008). Moreover, in a system where many root systems are connected with a mutual AM network, more lipid accumulation takes place in the intraradical mycelium of the C-limited root than in the C-sufficient root, suggesting that AM colonization may poorly reflect host quality for C allocation (Lekberg et al. 2010).

The first studies on the **transfer of C from carbohydrates into lipids** were performed during the 1980s using ^{14}C-labelled precursors. These studies demonstrated the incorporation of a substantial proportion of host photosynthate into fungal lipids, providing short- or long-term C sinks and membrane components for an often extensive inter- and intra-cellular fungal thallus (Lösel 1980). ^{14}C-Labelled photosynthates were found to accumulate in the lipid-filled spores (Ho and Trappe 1973) as well as in hyphae, vesicles, and arbuscles (Cox et al. 1975). The authors suggested that the **synthesis of lipid by MF could provide an alternative sink for photosynthate.** Studying the distribution of ^{14}C-labelled photosynthate in colonized and non-colonized roots, Bevege et al. (1975) concluded that **a high proportion of the ^{14}C-lipid fraction was found in the extraradical fungal hyphae.**

In the case of EMF, the proportion of total host photosynthate incorporated into the root lipid phase was not found to be significantly different, based on the mycorrhizal status of the plant, although mycorrhizal plants assimilated more ^{14}C through photosynthesis (Lösel 1980). Studying carbohydrate storage by ^{13}C NMR, Martin et al. (1984a) showed that triacylglycerols represented only 20% of the accumulated C. More recently, Laczko et al. (2003) observed an increased in saturated NLFAs in roots just after inoculation with *P. tinctorius* in *Pinus sylvestris* and, subsequently, saturated NLFAs appeared in the extraradical mycelium. They concluded that, in response to the fungus, plants produce increased amounts of NLs that are transferred to the fungus. This type of lipid transfer was proposed by Trépanier et al. (2005) for AMF but, to date, has not been proved. C allocation to *Pisolithus tinctorius* by *Pinus pinaster* may be related to N supply. Corrêa et al. (2011) found that, in early mycorrhizal stages, C allocation remained independent of N supply but, after 10 days, C allocation became higher if N supply was low. It was concluded that N availability in soil could control the below-ground C investment of EM plants.

Recent studies on the lipid metabolism of AMF have been performed using compartmentalized monoxenic cultures of *G. intraradices*

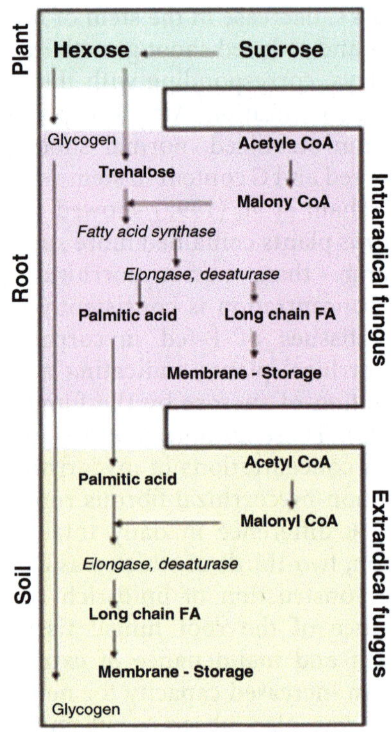

Fig. 8.1. Metabolic pathways of fatty acid synthesis in a functional arbuscular mycorrhizal symbiosis

grown on Ti-DNA transformed carrot roots. **Arbuscles are proposed to be the main site of C transfer between the root and the hyphae of AMF** (Blee and Anderson 1998) although some minor transfers have also been observed from the fungal structures in the epidermis and hypodermis (Manjarrez et al. 2008). Excreted into the apoplasm by the root cells, hexoses are absorbed by the fungus via a sugar-transporter (Schüβler et al. 2006). Inside the intraradical hyphae, this hexose is converted into glycogen and palmitic acid, which is incorporated into triacylglycerol (TAG), the main C reserve of AMF. No synthesis of C16:0, the main component of TAG, is possible inside the extraradical hyphae (Trépanier et al. 2005; Fig. 8.1). The TAG and glycogen can both be translocated to extraradical hyphae (Pfeffer et al. 1999; Bago et al. 2003). Inside the intraradical or the extraradical hyphae, the C16:0 FA acids can be elongated to C18 or C20 FAs using acetate coming from the citric acid cycle or from the media (Trépanier et al. 2005). In the extraradical hyphae, glycogen can be hydrolysed to hexoses and trehalose; when TAG is oxidized to acetyl-CoA, this acetyl-CoA can be further utilized in the glyoxylate cycle to produce hexoses (Lammers et al. 2001) which can be used for trehalose, chitin or nucleic acid synthesis. In the extraradical hyphae, AMF are also able to incorporate acetyl-CoA into sterols and their precursors (Fontaine 2001; Fontaine et al. 2001a, b). All these works clearly demonstrated transfer of photosynthates from the host plant to fungal lipids. Two main mysteries remain: why **only intraradical hyphae can absorb hexoses**, and why **FA synthesis only takes place inside the intraradical hyphae.** Molecular biology methods represent a promising avenue to elucidate these fundamental points.

IV. Fungal Lipid Changes Through the Growth Process

Every key step in the mycorrhizal fungal life cycle, from spore germination to mycelium differentiation, root colonization, symbiosis establishment, and sporulation, involves lipids without which propagation, symbiotic functioning, and species survival could not be achieved.

A. Spore and Mycelium

1. Arbuscular Mycorrhizas

As detailed in Sect. II, AMF spores and fungal propagules possess unique lipid profiles. **AMF spores, totally filled with lipids, contain enough reserves to support germination and subsequent hyphal growth and development** (Bago et al. 2000). Lipid behaviour has been interestingly studied on pre-germinated, germinating, and post-germinated spores submitted to ^{14}C acetate (Beilby 1980, 1983; Beilby and Kidby 1980). During the hydration period, added ^{14}C acetate was first located in the triacylglycerides and the free FA fractions and then, at spore germination, was recovered in the PPL and the sterol fractions. *De novo* lipid synthesis was demonstrated to occur simultaneously with lipid breakdown (Beilby 1983). FA, sterol, diacylglyceride, and PPL concentrations increased during the 7 days following spore germination. PC was found to be the major PL in ungerminated spores, while PE accumulated in germinated spores. Gaspar et al. (1994) clearly demonstrated **the consumption of triacylglycerides during the first stages of germination.** The decrease of total and NLs in germinated spores was evaluated at 75% and 50%, respectively, with a concomitant gradual production of long-chain FAs and PPLs. Hydrolytic enzymes degraded triacylglycerides and released C, which was used for the synthesis of PPLs. Lipolytic enzymes degrade triacylglycerols, and PPL and sterol syntheses insure membrane differentiation and consequently the elongation of germinating hyphae (Beilby and Kidby 1980; Gaspar et al. 1994; Bago et al. 1999; Fontaine 2001). The bi-compartimented Petri dish system for in vitro root culture propagation of AMF (Fortin et al. 2002) allows the separation of root and fungal phases, providing pure fungal material particularly suitable for studies on lipid changes during all stages of fungal development. Using such a system, Trépanier et al. (2005) demonstrated that **germinating spores and extraradical mycelium do not have the potential to proceed to the de novo synthesis of FAs, but can perform FA chain elongation and desaturation on preexisting C16 FAs** (see Sect. III.C). According to these authors, the genetic mechanisms involved in FA synthesis were expressed solely in the intraradical mycorrhiza and such a regulation was hypothesized to be related to the AMF biotrophy. **De novo synthesis of sterols** was demonstrated by Fontaine et al. (2001a, b) in both symbiotic and non-symbiotic phases. ^{14}C sodium acetate, a precursor of sterols, was found to be incorporated into sterols by both quiescent and germinating spores of *G. irregulare* (DAOM 197198) as well as by the fungus in the symbiotic phase. In all three systems studied, ^{14}C radioactivity was detected mainly in 24-methylcholesterol and 24-ethylcholesterol, the two most abundant AMF sterols.

According to Olsson and Johansen (2000) the FA profiles of AMF mycelium and spores, measured as C16:1*n*-5 concentrations, remained unchanged with age. However, measurements were made for only a 2-month period, essentially representing only the phase of mycorrhizal establishment. In our opinion, their results did not encompass a sufficient time period to adequately cover changes during maturation and aging processes. Nemec (1981) previously detected dehydrogenase activity in young arbuscles as well as in intraradical hyphae, and he detected peroxidase and catalase activities in senescing arbuscles which were tentatively attributed to FA oxidation. As a whole, total lipid contents of AM roots were found to be proportional to root colonization levels (Graham et al. 1996; Johansen et al. 1996; Pacovsky and Fuller 1987) and the number of spores in soil correlated well with AMF NL content (Olsson et al. 1997).

2. Ectomycorrhizas

To our knowledge, the most detailed lipid study on the sporogenesis of EMF was performed by Campos da Rocha et al. (2008) working with *Pisolithus tinctorius*; mature peridioles contained more than double the total FA compared to hymenium, young peridioles, and mature spore structures. These data indicated lipid accumulation in the peridiole prior to transfer into the spore. Moreover, short-chain saturated FAs found specifically in the early basidiogenesis stage were absent from mature spore profiles. C18:1*n*-9 FA concentration increased with spore maturation, becoming the major spore FA, in contrast to a majority of EMF. Some unusual FAs were detected only in the peridioles (C17:1*n*-9c) and in the spores (C16:1*n*-5), the latter FA being rarely found in fungi other than AMF.

Very little is actually known about lipid transfer from roots to hyphae except the confirmation of a decrease in the content of root FAs and their simultaneous accumulation in the extraradical hyphae (Laczko et al. 2003). Other lipid transformations that occur during EM development deal with ergosterol, which is by far the major sterol of EMF. Ekblad et al. (1998) observed the ergosterol content of *Pinus sylvestris–Paxillus involutus* mycorrhiza to decrease substantially with ageing. As ergosterol is correlated to living biomass, its direct measurements may serve to estimate the inoculum fungal viability and senescence, for example, of commercial bio-products.

B. Storage Lipids

Storage occurs mainly in the form of sugars, such as trehalose and glycogen, and lipids. Since early research on AMF, the lipidic globules observed inside AMF spores and hyphae have been considered as storage compounds playing the same role as carbohydrates play for other fungi. **AMF accumulate significant amounts of lipids in hyphae as well as in fructifications and spores.** These **lipids serve as C and energy sources during starvation and spore germination** (Laczko et al. 2003; Trépanier et al. 2005). AMF storage lipids are essentially located inside the oldest parts of the infection units, near the entry points, in both intercellular and external hyphae (Cooper and Lösel 1978). As the mycelium ages, a larger proportion of the biomass is represented by persistent lipid storage structures. No visible lipids were detected in young arbuscles, but lipids were observed in the large hyphae and older branches of the arbuscles as well as in the mycelium and the intraradical *Glomus* spores (Selvaraj and Subramanian 1990) and in the non-pelotonic hyphae of ericoid mycorrhizal fungi (Senthilkumar and Britoo 2001). Fontaine (2001) underlined the lipid richness of *G. intraradices* spores using Fluorol yellow 088, a lipid-specific fluorochrome (Brundrett et al. 1991). Depending on the species, **lipids represented between 45% and 72% of the fungal biomass**, which corresponds to 45–95% of the C pool (Beilby and Kidby 1980; Beilby 1983; Jabaji-Hare et al. 1984; Schubert et al. 1992). The C16:1*n*-5 hexadecenoic FA, extracted exclusively from mycorrhizal propagules, has been attributed to the root-associated AMF symbiont (Nordby et al. 1981; Gaspar and Pollero 1994; Gaspar et al. 1994; Grandmougin-Ferjani et al. 1997). This FA accumulates in large amounts in both the neutral and PPL fractions and is currently used as the signature FA in AMF studies (see Sects. II and VI).

EMF lipid reserves are usually located in the acylglyceride and FA fraction of NLs and found to vary between fungal growth phases and fungal structures (Martin et al. 1984b). *Paxillus involutus* and *Pisolithus tinctorius*, when grown either in pure cultures or in association with *Pinus strobus* were found to accumulate large amounts of lipid in their sclerotia cortical and medullar zones (Grenville et al. 1985a, b; Moore et al. 1991) as well as in mycelium and spores (Laczko et al. 2003). Triglyceride levels increased with hyphal maturity and remained constant until senescence. Histochemical tests revealed the NL nature of the - sclerotium-like bodies developed in *Hebeloma sacchariolens–Betula* sp. mycorrhiza (Fox 1986). Lipids also accumulated in the hyphae of the outer mantle (Dell et al. 1990), in rhizomorphs (Franz and Acker 1995), and in Hartig net cells (Bonfante-Fasolo and Scannerini 1977; Herr and Peterson 1996).

V. Impact of Exogenous Additives on Mycorrhizas

Based on the ecological importance of mycorrhizal symbioses in the environment and on their continuous exposure in the soil, the responses of these organisms to long-term soil and air contamination could give important information for assessing associated environmental risks. The efficiency of phytoremediation depends, in addition to substrate type, on several plant characteristics, such as the ability to eliminate the pollutant and to tolerate pollutant toxicity. In many instances, mycorrhizal symbioses help plants to sustain such challenges where lipid metabolism is currently involved.

A. Impact of Fungicides on Lipid Metabolism

Most fungicides act directly on essential fungal functions such as respiration, cell division, or lipid synthesis (Leroux 2003). **Fungicides of the sterol biosynthesis inhibitor (SBI) family** have been shown to exhibit **undesirable side effects on non-target organisms such as mycorrhizas** (Sukarno et al. 1996; Zocco et al. 2008). Numerous studies performed on the effects of SBI fungicides on mycorrhizal plants have generated contradictory results dealing with plant growth, AMF development, and symbiosis functioning (Dodd and Jeffries 1989; von Alten et al. 1993; Kling and Jakobsen 1997; Kjoller and Rosendahl 2000; Schweiger et al. 2001). The diversity of experimental approaches taken (plant species, growth conditions, fungicide formulation, application methods, micro-organisms of soil) did not allow comparisons to be made between the published results nor to draw clear conclusions on the impact of SBI fungicides on AMF.

Monoxenic cultures therefore seemed more appropriate tools to monitor the impact of fungicides and their toxicity mechanisms on AMF (Wan et al. 1998) as they enable non-destructive observations (Fortin et al. 2002), standardize and insure reproducibility of experiments, and allow comparisons between studies (Hillis et al. 2008; Zocco et al. 2008; Campagnac et al. 2010). Furthermore, with such cultures, large quantities of biological material free of contaminants can be produced, thus giving access to a diversity of biochemical and molecular investigations that were previously inconceivable. The fungitoxicity of fenpropimorph against *G. intraradices* was explained by a decrease of sterol content, in addition to an increase amount of a precursor, the squalene (Campagnac et al. 2009).

These data suggested that the sterol pathway is severely slowed down or that squalene epoxidase, an unusual target enzyme in fungi, was inhibited by fenpropimorph. Furthermore, after characterization of the first gene involved in the sterol biosynthetic pathway encoding a C-4 sterol methyl oxidase *Gint*SMO, Oger et al. (2009) showed changes in *Gint*SMO transcript levels on in vitro grown *G. intraradices* extraradical structures exposed to the SBI fungicide fenpropimorph. Gonzlez-Guerrero et al. (2010) then suggested that the antifungal effects of fenpropimorph might not only be due to interference with sterol metabolism, but also to perturbation in the production and scavenging of reactive oxygen species (ROS). Indeed, it was demonstrated that GintSOD1 (a gene encoding CuZn superoxide dismutase (SOD)) expression is induced by the well known ROS-inducing agents paraquat and copper, and also by fenpropimorph. These results suggested that GintSOD1 is involved in the detoxification of ROS generated from metabolic processes and by external agents. Interestingly, *G. intraradices* colonization was shown to partially protect roots from oxidative stress induced by fenpropimorph (Campagnac et al. 2010), suggesting that a fungal antioxidant, including SOD activity, could potentially be involved in the protective effect. This observation was reinforced by the concomitant decrease of C18:3 levels and high malondialdehyde (MDA) content produced in non-AM roots exposed to this fungicide in comparison to the control.

These observations may therefore be related to the direct **reaction of oxygen free radicals with unsaturated lipids leading to MDA formation**, one of the most common reactive aldehydes resulting from the oxidative degradation of polyunsaturated FA. Moreover, the increase in the ratio of saturated/unsaturated FAs in non-mycorrhizal roots exposed to fenpropimorph suggested a modification in membrane fluidity (Campagnac et al. 2010). Likewise, Calonne et al. (2010) demonstrated that the drastic decrease in *G. irregulare* development observed in the presence of propiconazole was linked to both the toxicity of MDA accumulation and the disruption of lipid metabolism. While drastic increases in PC and its PLFA (16:0, 18:0, 18:3) quantities were observed in *G. irregulare* grown under propiconazole treatment, significant reductions in the sterol end-products 24-ethylcholesterol and 24-methylcholesterol, and an accumulation of the precursor 24-methylenedihydrolanosterol, were detected. The latter findings suggest that the SBI fungicide affected the sterol biosynthesis pathway of the non-target organism *G. irregulare* by inhibiting a key enzyme of sterol biosynthesis, the sterol 14-α-demethylase, as observed in phytopathogenic fungi (Calonne et al. 2012).

Campagnac et al. (2008) pointed out differential effects of the two SBI fungicides (fenpropimorph, fenhexamid) on sterol metabolism of the AM symbiosis grown in vitro. Whereas **fenpropimorph** application at increasing concentrations drastically reduced mycorrhizal root growth and root colonization, fenhexamid had no effect. A good correlation was established between decreases in the glomalean lipid marker FA C16:1*n*-5 and the negative effect of fenpropimorph on chicory root colonization (Campagnac et al. 2010). In mycorrhizal roots treated with fenpropimorph, the usual Δ^5-sterols were replaced by unusual compounds such as 24-methylpollinastanol, ergosta-8,14-dienol, stigmasta-8,14-dienol, Δ^8 sitosterol, and ergosta-7,22-dienol. In contrast, fenhexamid treatment did not modify the sterol profiles of AM carrot roots. Comparison of the effects caused by the two tested fungicides indicated **the involvement of phytosterols in the development of symbiosis.** Indeed, observed modifications of the root sterol composition could explain the high fenpropimorph toxicity to AM symbiosis (Campagnac et al. 2008). Many reports have studied the effects of a diversity of fungicides and herbicides commonly used in nurseries and afforestation practices on the growth of the EMF *Lactarius deliciosus* and *Pisolithus tinctorius* (Díaz et al. 2003; Carrillo et al. 2011), but no information is available regarding their impact on lipid metabolism.

B. Impact of Pollutants on Lipid Metabolism

The effects of pollution on the constitutive and adaptive mechanisms of mycorrhizas in contaminated soils, and the potential contribution of the mycorrhizal symbiosis to the phytoremediation of polluted soils, have been reviewed regularly and in depth (Meharg and Cairney 2000; Meharg 2003; Göhre and Paszkowski 2006; Hildebrandt et al. 2007; Ferrol et al. 2009; Gonzlez-Guerrero et al. 2010). In spite of this, very little research has been performed to date on the influence of pollutants on the lipid metabolism of mycorrhizas.

Rabie (2005a) studied the role of AMF in plant-based remediation strategies of soils highly contaminated with heavy metals. Working with *G. mosseae*—red kidney bean and wheat symbioses in soils artificially contaminated with high concentrations of Zn, Cu, Pb, and Cd, he demonstrated that mycorrhizal plants could accumulate relatively high metal concentrations in their roots and shoots compared to non-mycorrhizal plants, suggesting the involvement of internal detoxification as metal tolerance mechanisms. This author hypothesized that **AMF can protect plants against heavy metal accumulation by increasing lipid production**, which compensates for lipid destruction by peroxidation. Highly significant positive correlations were found between mycorrhizal root colonization, plant water content, root lipids, oxido-reductase enzyme activities, total microbial counts in the soil rhizosphere, and dissipation of polycyclic aromatic hydrocarbons (PAH), and these were suggested to be indicative of plant resistance and acclimation to contaminants in the environment (Rabie 2005b). Mycorrhizal mung bean plants were the most tolerant to and effective at dissipating PAH, with higher water and root lipid contents, suggesting a higher dilution of toxic material in plant tissues and fewer detrimental effects. When the root water phase was demonstrated to be the reservoir for water-soluble contaminants, lipids in roots, even in small amounts, were considered to be a reservoir for liposoluble contaminants (Chiou et al. 2001). Flagging root lipid content as a good predictor for root accumulation of PAH is consistent with the observations of Paterson and Mackay (1994), Simonich and Hites (1995), and Petersen et al. (2002).

Using Sudan red 7B (a lipid-specific dye) and the PAH fluorescent property, it was demonstrated that **PAH, like anthracene, accumulated in the lipid bodies of AMF** and in cortical cells of transformed chicory roots (Verdin et al. 2006). These findings indicated that AM symbiosis could be involved in the biodegradation and intracellular storage of the pollutant via the interaction of lipid bodies with the endoplasmic reticulum where membrane-bound cytochrome P450 enzymes, known to be involved in PAH metabolism, were localized (Müllner and Daum 2004). There is increasing evidence that lipid bodies may play important roles in various aspects of lipid trafficking (Murphy 2001). **Bidirectional movements of lipid bodies along coenocytic AMF hyphae and between intra- and extraradical hyphae** (Bago et al. 2002) strongly suggest their involvement in the **intrahyphal transportation of anthracene.** Verdin et al. (2006) and Debiane et al. (2008, 2009) noted that the adverse impacts of PAH on root growth were less marked on colonized roots than on non-colonized ones, thereby suggesting a protective effect of AM colonization. The lipid peroxidation biomarker, MDA, was increased by PAH addition in non-colonized roots compared to colonized roots. This result suggests that **the protective role of AMF is associated with a reduction in PAH-induced oxidative stress damage**, as an increase in MDA indicates the production of ROS and oxidation of unsaturated membrane PPLs (Debiane et al. 2008, 2009).

Increases in membrane lipid peroxidation have been shown to increase membrane permeability, exosmosis of electrolytes, and finally injuries to the cell membrane system (Zhu et al. 2010). Several authors have reported lipid disruption in plants in response to pollutants (Harwood 1996; Berglund et al. 2002; Ben Youssef et al. 2005). Data on the impact of pollutants on membrane lipid composition of mycorrhizas remain scarce. Debiane et al. (2012) examined the impact of benzo[a]pyrene, a high molecular weight PAH, on the lipid metabolism of the symbiosis (chicory roots—*G. irregulare*) grown in vitro. Whereas no changes were observed in the lipids of mycorrhizal roots, decreases in PPLs (PE, PC) and stigmasterol were observed in non-mycorrhizal roots. These findings suggested that the disturbances in lipid metabolism and toxic effects in AM roots under PAH pollution could be assigned to depletions of the main membrane constituents as well as to MDA accumulation (Debiane et al. 2011). Similarly, Kirk et al. (2005) suggested that interference with fungal membranes may explain the reduced extraradical hyphal growth of AMF (*Glomus* species) in petroleum hydrocarbon.

There is very little information available about the effects of pollutants on ectomycorrhizas, with regard to the lipid metabolism of the symbiotic fungus and/or the host plant. Collin-Hansen et al. (2005) showed an increased frequency of apurinic/apyrimidinic sites in DNA and a concentration of lipid hydroperoxides in the fruit bodies of *Boletus edulis* growing in soils polluted with a mixture of heavy metals. The authors pointed out significant positive correlations between concentrations of Cd, Zn, or Cu and the degree of oxidative damage to DNA and lipids.

C. Impact of Exogenous Lipids on Mycorrhizas

A diversity of fungi has demonstrated their capacity to use lipids as a C source via **extracellular lipase synthesis**, in replacement of the currently available carbohydrates (Caldwell et al. 1991; Sancholle and Lösel 1995). In some cases lipids, when incorporated in culture media, can stimulate growth and induce the reproduction of fungi (Lösel 1988).

1. Arbuscular Mycorrhizas

Root exudates from mycotrophic herbaceous plants were demonstrated to **contain lipids in the form of FAs, PPLs, and sterols** (Hale et al. 1981; Thompson and Hale 1983; Bozhkov et al. 1998; Nagahashi and Douds 2011). Semi-purified extracts of transformed carrot roots used for the in vitro propagation of AMF were suspected to contain two hydroxy FAs of C12 and C14 chain length. When testing equivalent chemical commercial products on *Gigaspora* spores, significant stimulation was observed at the levels of germ tube elongation, branching, and branching elongation. Similar tests made with the non-hydroxylized palmitic and stearic acids (C14:0, C16:0 FAs) had no effect on hyphal growth or ramification. The lipid identity of carrot root exudates remains to be confirmed.

Hyphal growth of AMF appeared to be slightly more stimulated by pea root exudates than by water (Samra et al. 1996). Root exudates extracted from plants grown in low P conditions were clearly shown to promote hyphal growth and to induce significantly more hyphal branching than exudates coming from plants grown in high P plant conditions (Nagahashi et al. 1996; Tawaraya et al. 1996). In a study on the relationship between root exudation and plant P supply, Ratnayake et al. (1978) demonstrated that a decrease in root P content corresponded to an increase in root membrane permeability and consequently a decrease in the PPL content of root tissues. These observations unfortunately do not provide direct information about the possible involvement of lipid components in the exudates of AMF roots; however, they clearly demonstrate the impact of P nutrition on the composition of root exudates.

2. Ectomycorrhizas

Hutchinson (1990) demonstrated the capacity of axenically grown *Amanita* strains to use polyethylene sorbitol esterified FAs (Tweens) as a sole source of C. **Short-chain FAs** such as butyric acid were found to **induce basidiospore germination in several EMF species, favouring fungal survival and propagation** (Ohta 1988). The EMF *Boletus*, when cultivated in pure cultures in the presence of safflower oil, a rich source of oleic acid, showed a mycelial growth increase (Schisler and Volkoff 1977); otherwise the growth of *B. variegatus* in the presence of short-chain FAs was inhibited by a leakage of its cell material and blockage of respiration (Pedersen 1970). The inhibitory effect of shorter chain FAs has been attributed to their interaction with membrane components, as discussed by Lösel (1988). Among the ectendomycorrhizal fungi, *Phialophora finlandia* and *Phialocephala fortinii* were found to be able to use Tween 40 (a FA ester) as a C source (Caldwell et al. 2000). Such benefits were hypothesized to allow the fungus better access to plant nutrients.

Tree root exudates are known to stimulate the spore germination of EMF (Birraux and Fries 1981; Xiaomei et al. 1995). Fries et al. (1985) demonstrated the lipidic nature of some of these germination inducers, but without determining with precision the identity of the active substances. Exudates from 6-month-old *Pinus sylvestris* roots were found to contain lipids at the level of 2.1% of the fresh weight of the seedling roots. The FA content of pine root exudates was revealed to be predominantly linoleic (42%), oleic (23%), stearic (18%), and palmitic (14%) acids. Once added to the culture media of axenically grown EMF, **the lipid fraction of root exudates increased the mycelial growth** of *Laccaria bicolor* and *L. amethystina* strains by 150–250% but had a much smaller growth-promoting effect (17–21%) on *Leccinum aurantiacum*. Such differential fungal responses suggested the concept of possible selection of fungal symbiont partners by the host through root exudates. Later, Hua et al. (1995), in a study on the improvement of mycorrhizal symbiosis in pine nursery, demonstrated that pine root exudates can stimulate the germination and growth of MF. Unfortunately, the root exudate factors that may be responsible for this stimulation were not identified.

VI. Lipids as Indicators of Mycorrhizal Fungi

The numerous and original evaluation tools currently available to characterize and quantify mycorrhizas and mycorrhizal soil communities have been developed through the gradual acquisition of knowledge on fungal lipid profiles and on mycorrhizal functioning. Such tools may be either fungal group- or genus-specific, using lipid categories or specific sterols or FAs. Reliable and user friendly lipid-based methodologies have now reached workers of all disciplines, from fundamental research in fungal physiology to field ecology.

A. Specific Sterols

1. Arbuscular Mycorrhizas

Microscopic evaluation of AMF colonization levels on stained roots is a tedious and time-consuming process (Sylvia et al. 1993). Several biochemical assays have been developed and tested through time. In the search for potential biomarkers for AMF, the major fungal sterols were found to be 24-ethylcholesterol and 24-methylcholesterol (Beilby 1980; Beilby and Kidby 1980; Grandmougin-Ferjani et al. 1995; Grandmougin-Ferjani et al. 1996; Grandmougin-Ferjani et al. 1999). AM-colonized roots were also reported to contain more campesterol (24-methylcholesterol) and 24-methylene-cholesterol than non-mycorrhizal roots (Nagy et al. 1980; Nordby et al. 1981; Schmitz et al. 1991; Grandmougin-Ferjani et al. 1995). Actually, 24 methyl/methylene sterols are considered the most appropriate to quantify root colonization, at least with AM-colonized transformed roots (Fontaine et al. 2004).

2. Ectomycorrhizas

As presented in Sect. II, **most EMF contain ergosterol as their major sterol** with only a few known exceptions such as truffles and *Pisolithus tinctorius* (Laczko et al. 2003; see Sect. I). Its concentration was found to correlate with fungal hyphal length (West et al. 1987; Koo et al. 2009) and to vary with environmental conditions (Bermingham et al. 1995; Ekblad et al. 1998) as well as with fungal species and the physiological state of the fungi (Bermingham et al. 1995; Domenech et al. 2004). Being a regular component of fungal membranes, **ergosterol** is considered to be a **good measure of the metabolically active fungal biomass** (Nylund and Wallander 1989; Ekblad et al. 1998; Parsi and Gorecki 2006) based on the assumption that it is unstable and rapidly degraded after cell death (Ekblad et al. 1998; Montgomery et al. 2000). The ratio ergosterol/sistosterol, where ergosterol content refers to the fungal biomass and sitosterol to the host plant biomass, was proposed as a potential measurement for the evaluation of fungal biomass inside plant tissues but was later considered not appropriate as such a ratio may differ according to plant/fungi partners and the metabolic status of the symbionts.

Chitin, even though less fungus-specific, was also used until recently to determine the level of root fungal colonization (Probanza et al. 2001). However, the chitin assay was also revealed to be quantitatively less sensitive than ergosterol measurements (Johnson and McGill 1990) and more sensitive to chemical interference (Matcham et al. 1985). The chitin content reflects all fungal biomass, living and dead, and thus can be considered an indicator of the integration of the fungus over the life span of the root system (Ekblad et al. 1998; Probanza et al. 2001; Domenech et al. 2004). As suggested by Ekblad et al. (1998) both markers can be used in combination (ergosterol/chitin ratio) to estimate the living fraction of fungal biomass.

3. Ericoid Mycorrhizas

The ergosterol content in ericoid mycorrhizal fungi was estimated by Padgett and Posey (1993) to be 2.3 mg g^{-1}, a level similar to the amount generally found in Ascomycetes and Basidiomycetes. Accordingly, **ergosterol** is currently used as a **biomarker for ericoid mycorrhizal colonization** (Caporn et al. 1995; Genney et al. 2000; Kasurinen et al. 2001; Olsrud et al. 2004, 2007) and found at rates similar to those observed with EMF. However, Olsrud et al. (2007) showed that ergosterol content in the ericaceous fine roots did not correlate with fungal colonization levels and recommended not to use the sterol as a biomarker, at least with roots grown in soil.

B. Specific Fatty Acids

1. Arbuscular Mycorrhizas

The hexadecenoic acid, C16:1*n*-5 is a monounsaturated FA found in a variety of soil micro-organisms such as Gram negative bacteria, filamentous bacteria, spirochaetes (Livermore et al. 1969; Nichols et al. 1986; Frostegård et al. 1993), and also in primitive vascular plants belonging to the Proteaceae, Ginkoaceae, and Euphorbiaceae (Gellerman and Schlenk 1963; Cronquist 1968; Thorne 1992). Peculiarly, this unsaturated mono-ene occurs as the main FA of several Glomeromycetes spores, including *Acaulospora*, *Gigaspora*, *Glomus*, and *Scutellospora* species (Sancholle and Dalpé 1993; Gaspar and Pollero 1994; Madan et al. 2002) as well as in the AMF-colonized roots (Pacovsky and Fuller 1987; Olsson et al. 1995). **The unique distribution of C16:1*n*-5 makes this molecule the signature FA of AMF and the best biomarker tool for the evaluation of root fungal biomass.** During the last decade, C16:1*n*-5 as a tool gained popularity and demonstrated high performance in a variety of trials (Van Aarle and Olsson 2003; Fontaine et al. 2004; Stumpe et al. 2005; Campagnac et al. 2008, 2010). It has been applied to estimate AMF biomass in soil samples (Van Aarle and Olsson 2003; Stumpe et al. 2005) and spore-specific biomass (Madan et al. 2002), to characterize microbial communities (Olsson 1999; Bååth 2003; Van Aarle and Olsson 2003) and soil food webs (Ruess et al. 2002). Tools to quantify root fungal biomass were described early by Graham et al. (1995) using *Citrus* roots where the amount of 16:1*n*-5 was found to correlate with the level of root colonization but not with fungal dispersion inside roots. Gaspar et al. (2001) suggested that 16:1*n*-5 FA concentration could be used to estimate the amount of C allocated to fungal growth and to lipid storage within roots. Van Aarle and Olsson (2003), when studying C16:1*n*-5 distribution within fungal lipid fractions, proposed to use the NLFA/PLFA ratio for direct evaluation of the storage status of AMF. This measurement, applied to AMF cultured in vitro, is becoming a reliable tool for the estimation of mycorrhizal efficiency of fungal strains, as well as for the determination of fungal biomass in excised colonized roots.

2. Ectomycorrhizas

The most common FA profiles found in EMF include unsaturated C12–20 chain length and C18 di- and monoene unsaturated FAs. Linolenic acid (C18:2*n*-6) is the most abundant acid, followed by oleic acid (C18:1*n*-9) and palmitic acid (C16:0), distributed in the PL fraction involved in membrane differentiation and in the NL fraction mainly as storage material (Pedneault et al. 2006, 2008; Kalac 2009). Some specific profiles were observed: *Cantharellus cibarius* and *Hydnum rufescens* fructifications revealed a particularly high content of polyunsaturated C18 FAs not found in other EMF (Table 8.1). Tuberaceae species, from the genera *Terfezia*, *Tirmania*, and *Tuber* contained long-chain FAs from C20 to C24. Even though found at low concentrations, these FAs seemed to be truffle-specific and were not found, even in trace amounts, in any of the other EMF species studied to date. Among other specific FAs of EMF, basidiocarps of *Lactarius chrysorrheus* were found to synthesize 6 oxo-decanoic acid (Hiroi 1978) and elaidic acid (C18:1*n*-9t) has been detected in *Amanita*, *Boletus*, and *Suillus* species (Pedneault et al. 2008). Thus far, none of these specific FAs has yet been assayed for the detection and estimation of fungal biomass.

C. Soil Community

As described above, FAME analyses of the PLFA and NLFA fractions are currently used as lipid biomarkers in studies on microbial ecology and soil mycorrhizal fungal populations (Bååth and Anderson 2003; Nilsson et al. 2005). PLFAs being bound to membranes and their profiles recognized as being specific to groups of organisms, their evaluation may provide an opportunity to distinguish between microbial communities (Thoms et al. 2010). FAMEs of the NL fraction, associated with storage lipids, but also found in spores and mycelium, is to some extent considered a reliable tool for the evaluation of the total fungal biomass of targeted organisms (Bååth 2003; Balser et al. 2005; Nilsson et al. 2005; Ngosong et al. 2010).

FA values help to quantify some of the components of microbial soil communities and to gain insight into the diversity of community structure. C18:2*n*-6 is used to evaluate the rhizosphere fungal biomass of ericoid mycorrhizal fungi (Fritze and Bååth 1993) and EMF (Frostegård and Bååth 1996; Olsson 1999; Nilsson et al. 2005). However, linoleic acid is also a major FA in a large number of saprotrophic macro and microfungi found in soil, thus weakening data accuracy when only the MF population is evaluated. An unusual PL component of higher fungi, diacylglycerohomoserine, has been found in *Boletus edulis* (Vaskovsky et al. 1991) and was found to be, after PC, the second major PL of several *Boletus* species (Hanus et al. 2008). Targeting such PLs could become a valuable tool for chemotaxonomic recognition of family-specific EM.

C16:1*n*-5, the **signature FA** of AMF can be **used to estimate soil AMF total biomass** (Balser et al. 2005; Nilsson et al. 2005; Ngosong et al. 2010). Moreover, Olsson et al. (1995) reported an interesting correlation between the AM hyphal length in soil and the combined levels of the C16:1*n*-5 and the polyunsaturated C20:5 FA. The ratio between C16:1*n*-5 FA and total FA in a given substrate allows quantitative and comparative analyses of soil microbial communities (Frostegård and Bååth 1996; Madan et al. 2002; Bååth and Anderson 2003; Balser et al. 2005). Because C16:1*n*-5 FA is also commonly isolated from soil Gram negative bacteria (Frostegård et al. 1993; Olsson et al. 1995), its use as a lipid indicator may face some limitations when used under natural conditions.

As **ergosterol** is not present in either procaryotic cells or in vascular plants (Parsi and Gorecki 1996; Hannich et al. 2011), but is recognized as a **fungus-specific component of the great majority of EMF**, this sterol can be **used as a reliable root fungal biomass indicator** (Nylund and Wallander 1989; Ekblad et al. 1998). When compared to other methods used for the evaluation of EMF inside colonized roots, the ergosterol assay allows quantitative measurements not available through the microscopic observation of stained roots (Sylvia et al. 1993).

VII. Future Research Avenues

Mycorrhizal symbiosis is a complex process which is programmed to perfection by the two partners. The transfer of C is the basis of this relationship. With AMF, C is largely converted into lipids and used for both structural material and energy storage. The last decade has seen major advances in the world of mycorrhizal symbiosis, especially with respect to the various signals exchanged between the plant and the fungus before and during colonization, some of these signals involving lipids. Due to their specificity and ease of analysis, some FAs and sterols are now commonly used to assess the population of mycorrhizas in soil, or the degree of root colonization. Before these discoveries, the exploration and quantification of the extramatrical network was often random, time-consuming, or inaccurate.

In agriculture, where an input as important as P has undergone a dramatic increase in cost, the large-scale use of mycorrhizal inoculants is now economically viable for farmers, resulting in thousands of hectares being inoculated annually. In a context of global warming and multiple attempts to capture atmospheric CO_2, mycorrhizas could play a prominent role, as yet unknown, in increasing the rate of photosynthesis by about 15%. This C ends up stored in the ground, initially as lipid, and contributes to the maintenance of an efficient and functional soil ecosystem in both agricultural and forestry environments. The demonstrated remediation potential of mycorrhizas allows their use to be foreseen as a green and sustainable alternative to soil depollution using lipid markers as tools to measure pollutant ecotoxicity.

Tomorrow's world will be turned to the production of alternative fuels such as biodiesel. Made from various vegetable oils, the increased production of these oils favoured by MF justifies further agricultural use. The present craze for nutraceutical foods provides more reasons to consume EMF because they contain non-negligible amounts of lipid compounds of great interest for health, such as sterols, which may play a role in preventing cardiovascular disease.

With respect to fundamental research, mysteries remain to be elucidated. Why are AMF unable to synthesize FAs outside the root? Is it because lipids are more easily exported over long distances than sugar? Would this incapacity be related to the obligate mycotrophy of several MF? The genes responsible for FA synthesis in fungal AMF have not yet been identified. The discovery of factors regulating the transcription of these genes only in the root may represent a considerable advance in the elucidation of the obligate biotrophy of these fungi. In our opinion, this point is fundamental and represents the research avenue that must be prioritized in the coming years. Novel technologies in molecular biology should considerably simplify and accelerate such research.

Acknowledgement The authors wish to thank Dr. Shea Miller from Eastern Cereal and Oilseed Research Centre for her useful suggestions and revision of the manuscript.

References

Al-Karaki GN, Clark RB (1999) Mycorrhizal influence on protein and lipid of durum wheat grown at different soil phosphorus level. Mycorrhiza 9:97–101

Al-Karaki GN, Al-Raddad A, Clark RB (1998) Water stress and mycorrhizal isolate effect on growth and nutrient acquisition of wheat. J Plant Nutr 21:891–902

Al-Saadi AH, Benkayal FA, Al-Saadi MH (2005) Libyan truffles *Terfezia boudieri* Chatin: characterization of lipids and total protein electrophoretic pattern. Dirasat Online Pure Sci 32:269–277

Bååth E (2003) The use of neutral lipid fatty acids to indicate physiological conditions of soil fungi. Microb Ecol 45:373–383

Bååth E, Anderson TH (2003) Comparison of soil fungal/bacterial ratios in a pH gradient using physiological and PLFA-based techniques. Soil Biol Biochem 35:955–963

Bååth E, Nilsson LO, Göransson H, Wallander H (2004) Can the extent of degradation of soil fungal mycelium during soil incubation be used to estimate ectomycorrhizal biomass in soil? Soil Biol Biochem 36:2105–2109

Bago B, Pfeffer PE, Douds DD, Brouilette J, Bécard G, Shachar-Hill Y (1999) Carbon metabolism in spores of the arbuscular mycorrhizal fungus *Glomus intraradices* as revealed by nuclear magnetic resonance spectroscopy. Plant Physiol 121:263–271

Bago B, Shachar-Hill Y, Pfeffer PE (2000) Carbon metabolism and transport in arbuscular mycorrhizas. Plant Physiol 124:949–957

Bago B, Zipfel W, Williams RM, Jun J, Arreola R, Lammers PJ, Pfeffer PE, Shachar-Hill Y (2002) Translocation and utilization of fungal storage lipid in the arbuscular mycorrhizal symbiosis. Plant Physiol 128:108–124

Bago B, Pfeffer PE, Abubaker J, Jun J, Allen JW, Brouillette J, Douds DD, Lammers PJ, Shachar-Hill Y (2003) Carbon export from arbuscular mycorrhizal roots involves the translocation of carbohydrate as well as lipid. Plant Physiol 131:1496–1507

Balser TC, Treseder KK, Ekenler M (2005) Using lipid analysis and hyphal length to quantify AM and saprotrophic fungal abundance along a soil chronosequence. Soil Biol Biochem 37:601–604

Barros L, Baptista P, Correia DM, Casal S, Oliveira B, Ferreira ICFR (2007) Fatty acid and sugar compositions, and nutritional value of five wild edible mushrooms from northeast Portugal. Food Chem 105:140–145

Barroso J, Neves HC, Pais MSS (1987) Production of free sterols by infected tubers of *Ophrys lutea* Con.: identification by gas chromatography–mass spectrometry. New Phytol 106:147–152

Beilby JP (1980) Fatty acid and sterol composition of ungerminated spores of the vesicular-arbuscular mycorrhizal fungus, *Acaulospora laevis*. Lipids 15:949–952

Beilby JP (1983) Effects of inhibitors on early protein, RNA, and lipid synthesis in germinating vesicular arbuscular mycorrhizal fungal spores of *Glomus caledonium*. Can J Microbiol 29:96–601

Beilby JP, Kidby DK (1980) Biochemistry of ungerminated and germinated spores of the vesicular arbuscular mycorrhizal fungus, *Glomus caledonium*: changes in neutral and polar lipids. J Lipid Res 21:739–750

Bentivenga SP, Morton JB (1994a) Stability and heritability of fatty acid methyl ester profiles of glomalean endomycorrhizal fungi. Mycol Res 98:1419–1426

Bentivenga SP, Morton JB (1994b) Congruence of fatty acid methyl ester profiles and morphological characters of arbuscular mycorrhizal fungi in Gigasporaceae. Proc Natl Acad Sci USA 93:659–662

Ben Youssef N, Nouairi I, Ben Temime S, Taamalli W, Zarrouck M, Ghorbal MH, Ben Miled Daoud D (2005) Effets du cadmium sur le métabolisme des lipides de plantules de colza (*Brassica napus* L.). C R Biol 328:745–757

Berglund AH, Quatacci MF, Calucci L, Navarri-Izzo F, Pinzino C, Liljenberg C (2002) Alterations of wheat root plasma membrane lipid composition induced by copper stress result in changed physicochemical properties of plasma membrane lipid vesicles. Biochim Biophys Acta Biomembr 1564:466–472

Bermingham S, Maltby L, Cooke RC (1995) A critical assessment of the validity of ergosterol as an indicator of fungal biomass. Mycol Res 99:479–484

Bethlenfalvay GJ, Mihara KL, Schreiner RP (1994) Mycorrhizae alter protein and lipid contents and yield of pea seeds. Crop Sci 34:998–1003

Bethlenfalvay GJ, Schreiner RP, Mihara KL (1997) Mycorrhizal fungi effect on nutrient composition and yield of soybean seeds. J Plant Nutr 20:581–591

Bevege DI, Bowen GD, Skinnner MF (1975) Comparative carbohydrate physiology of ecto and endomycorrhizas. In: Sanders FE, Mosse B, Tinker PB (eds) Endomycorrhizas. Academic, London, pp 149–174

Bianciotto V, Lumini E, Bonfante P, Vandamme P (2003) '*Candidatus* Glomeribacter gigasporarum' gen. nov., sp nov., an endosymbiont of arbuscular mycorrhizal fungi. Int J Syst Evol Microbiol 53:121–124

Birraux D, Fries N (1981) Germination of *Thelephora terrestris* basidiospores. Can J Bot 59:2062–2064

Blee KA, Anderson AJ (1998) Regulation of arbuscule formation by carbon in the plant. Plant J 16:523–530

Bonanomi A, Wiemken A, Boller T, Salzer P (2001) Local induction of a mycorrhiza-specific class III chitinase gene in cortical root cells of *Medicago truncatula* containing developing or mature arbuscules. Plant Biol 3:194–200

Bonfante-Fasolo P, Scannerini S (1977) Cytological observations on the mycorrhiza *Endogone flammicorona–Pinus strobus*. Allionia 22:23–54

Bozhkov AI, Menzyanova NG, Leontovich VP, Belous AM (1998) Dynamics of the lipid composition of root exudates during growth of *Triticum aestivum*. L Dop Nat Akad Nauk Ukrayiny 1998:166–170

Brondz I, Hoiland K, Ekeberg D (2004) Multivariate analysis of fatty acids in spores of higher basidiomycetes: a new method for chemotaxonomical classification of fungi. J Chromatogr B 800:303–307

Brundrett MC, Kendrick B, Peterson CA (1991) Efficient lipid staining in plant material with Sudan red 7B or Fluoral yellow 088 in polyethylene glycol glycerol. Biotechnol Histochem 66:111–116

Brundrett MC, Melville L, Peterson RL (1994) Estimation of root length and colonization by mycorrhiza fungi. Practical methods in mycorrhizal research. Mycologue, Montreal

Bucher M (2010) A novel lipid signal in the arbuscular mycorrhizal symbiosis within eyesight? New Phytol 185:593–595

Caldwell BA, Castellano MA, Griffiths RP (1991) Fatty acid esterase production by ectomycorrhizal fungi. Mycologia 83:233–236

Caldwell BA, Jumpponen A, Trappe JM (2000) Utilization of major detrital substrates by dark-septate, root endophytes. Mycologia 92:230–232

Calonne M, Fontaine J, Debiane D, Laruelle F, Grandmougin-Ferjani A, Lounès-Hadj Sahraoui A (2010) Propiconazole toxicity on the non-target organism, the arbuscular mycorrhizal fungus, *Glomus irregulare*. In: Carisse O (ed) Fungicides. InTech, Rijeka, pp 325–346

Calonne C, Lounès-Hadj Sahraoui A, Campagnac E, Debiane D, Laruelle F, Grandmougin-Ferjani A, Fontaine J (2012) Propiconazole inhibits the sterol 14α-demethylase in *Glomus irregulare* like in phytopathogenic fungi. Chemosphere 87:376–383

Campagnac E, Fontaine J, Lounès-Hadj Sahraoui A, Laruelle F, Durand R, Grandmougin-Ferjani A (2008) Differential effects of fenpropimorph and fenhexamid, two sterol biosynthesis inhibitor fungicides, on arbuscular mycorrhizal development and sterol metabolism in carrot roots. Phytochemistry 69:2912–2919

Campagnac E, Fontaine J, Lounès-Hadj Sahraoui A, Laruelle F, Durand R, Grandmougin-Ferjani A (2009) Fenpropimorph slows down the sterol pathway and the development of the arbuscular mycorrhizal fungus *Glomus intraradices*. Mycorrhiza 20:167–178

Campagnac E, Lounès-Hadj Sahraoui A, Debiane D, Fontaine J, Laruelle F, Garçon G, Verdin A, Durand R, Shirali P, Grandmougin-Ferjani A (2010) Arbuscular mycorrhiza partially protect chicory roots against oxidative stress induced by two fungicides, fenpropimorph and fenhexamid. Mycorrhiza 20:167–178

Campos da Rocha AN, Costa MD, Totola MR, Borges AC (2008) Total lipid and fatty acid accumulation during basidiospore formation in the ectomycorrhizal fungus *Pisolithus* sp. Rev Bras Ciênc Solo 32:1531–1540

Caporn SJM, Song W, Read DJ, Lee JA (1995) The effect of repeated nitrogen fertilization on mycorrhizal infection in heather *Calluna vulgaris* (L.)Hull. New Phytol 129:605–609

Carrillo C, Díaz G, Honrubia M (2011) Testing the effect of routine fungicide application on ectomycorrhiza formation on *Pinus halepensis* seedlings in a nursery. Forest Pathol 41:70–74

Cerri R, De Simone F, Senatore F (1981) Sterols of three *Lactarius* species. Biochem Syst Ecol 9:247–248

Chiou CT, Sheng G, Manes M (2001) A partition-limited model for the plant uptake of organic contaminants from soil and water. Environ Sci Technol 35:1437–1444

Cole L, Orlovich DA, Ashford AE (1998) Structure, function and motility of vacuoles in filamentous fungi. Fungal Genet Biol 24:86–100

Collin-Hansen N, Andersen RA, Stennes E (2005) Damage to DNA and lipids in *Boletus edulis* exposed to heavy metals. New Phytol 12:1386–1396

Cooper KM, Lösel DM (1978) Lipid physiology of vesicular-arbuscular mycorrhiza. I. Composition of lipids in roots of onion, clover and ryegrass infected with *Glomus mosseae*. New Phytol 80:143–151

Corrêa A, Hampp R, Magel E, Martins-Loução MA (2011) Carbon allocation in ectomycorrhizal plants at limited and optimal N supply: an attempt at unraveling conflicting theories. Mycorrhiza 21:35–51

Cowan AK (2006) Phospholipids as plant growth regulators. Plant Growth Regul 48:97–109

Cox G, Sanders FE, Tinker PB, Wild JA (1975) Ultrastructural evidence relating to host-endophyte transfer in a vesicular-arbusular mycorrhiza. In: Sanders FE, Mosse B, Tinker PB (eds) Endomycorrhizas. Academic, London, pp 297–311

Cronquist A (1968) The evolution and classification of flowering plants, Riverside studies in biology. Riverside, London

Debiane D, Garçon G, Verdin A, Fontaine J, Durand R, Grandmougin-Ferjani A, Shirali P, Lounès-Hadj Sahraoui A (2008) In vitro evaluation of the oxidative stress and genotoxic potentials of anthracene on mycorrhizal chicory roots. Environ Exp Bot 64:120–127

Debiane D, Garçon G, Verdin A, Fontaine J, Durand R, Shirali P, Grandmougin-Ferjani A, Lounès-Hadj Sahraoui A (2009) Mycorrhization alleviates benzo [a]pyrene-induced oxidative stress in an in vitro chicory root model. Phytochem 70:1421–1427

Debiane D, Calonne M, Fontaine J, Laruelle F, Grandmougin-Ferjani A, Lounès-Hadj Sahraoui A (2011) Lipid content disturbance in the arbuscular mycorrhizal, *Glomus irregulare* grown in monoxenic conditions under PAHs pollution. Fungal Biol 115:782–792

Debiane D, Calonne M, Fontaine J, Laruelle F, Grangmougin-Ferjani A, Lounès-Hadj Sahraoui A (2012) Benzo[a]pyrene induced lipid changes in the monoxenic arbuscular mycorrhizal chicory roots. J Hazard Mater 209(210):18–26

Declerck S, Cranenbrouck S, Dalpé Y, Séguin S, Grandmougin-Ferjani A, Fontaine J, Sancholle M (2000) *Glomus proliferum sp. nov.*: a description based on morphological, biochemical, molecular and monoxenic cultivation data. Mycologia 92:1178–1187

De Kok LJ, Kuiper PJC, Bruins AP (1982) A polyunsaturated octadecanoic acid derivative, a major fatty acid in sporophores of *Cantharellus cibarius* Fr. In: Wintermans JFGM, Kuiper PJC (eds) Biochemistry and metabolism of plant lipids. Elsevier Biomedical, Amsterdam

Dell B, Malajczuk N, Thompson GT (1990) Ectomycorrhiza formation in *Eucalyptus*: V. A tuberculate ectomycorrhiza of *Eucalyptus pilularis*. New Phytol 114:633–640

Dembitsky VM (2006) Natural neo acids and neo alkanes: their analogs and derivatives. Lipids 41:309–340

Dembitsky VM, Pechenkina EE (1991) Phospholipid and fatty acid compositions of higher fungi. Chem Nat Comp 27:155–156

Dembitsky VM, Terent AO, Levitsky DO (2010) Amino and fatty acids of wild edible mushrooms of the genus *Boletus*. Rec Nat Prod 4:218–223

De Simone F, Senatore F, Sica D, Zollo F (1979) Sterols from Badisiomycetes. Phytochemistry 18:1572–1573

Díaz G, Carrillo C, Honrubia M (2003) Differential responses of ectomycorrhizal fungi to pesticides in pure culture. Cryptogamie Mycol 24:199–211

Dodd JC, Jeffries P (1989) Effects of fungicides on three vesicular-arbuscular mycorrhizal fungi associated with winter wheat (*Triticum aestivum* L.). Biol Fertil Soils 7:120–128

Domenech J, Ramos-Solano B, Probanza A, Lucas-García JA, Colón JJ, Gutiérrez-Mañero FJ (2004) *Bacillus spp.* and *Pisolithus tinctorius* effects on *Quercus ilex* ssp. ballot: a study on tree growth, rhizosphere community structure and mycorrhizal infection. For Ecol Manage 194:293–303

Douds DD Jr, Johnson CR, Koch KE (1988) Carbon cost of the fungal symbiont relative to net leaf P accumulation in a split-root VA mycorrhizal symbiosis. Plant Physiol 86:491–496

Drigo B, Pijl AS, Duyts H, Kielak AM, Gamper HA, Houtekamer MJ, Boschker HTS, Bodelier PLE, Whiteley AS, van Veen JA, Kowalchuk GA (2010) Shifting carbon flow from roots into associated microbial communities in response to elevated atmospheric CO_2. Proc Natl Acad Sci USA 107:10938–10942

Drissner D, Kunze G, Callewaert N, Gehrig P, Tamasloukht M, Boller T, Felix G, Amrhein N, Bucher M (2007) Lyso-phosphatidylcholine is a signal in the arbuscular mycorrhizal symbiosis. Science 318:265–268

Dugassa GD, Von Alten H, Schönbeck F (1996) Effects of arbuscular mycorrhiza (AM) on health of *Linum usitatissimum* L. infected by fungal pathogens. Plant Soil 185:173–182

Eissenstat DM, Graham JH, Syvertsen JP, Drouillard DL (1993) Carbon economy of sour orange in relation to mycorrhizal colonization and phosphorus status. Ann Bot 71:1–10

Ekblad ALF, Wallander H, Carlsson R, Huss-Danell K (1995) Fungal biomass in roots and extramatrical mycelium in relation to macronutrients and plant biomass of ectomycorrhizal *Pinus sylvestris* and *Alnus incana*. New Phytol 131:443–451

Ekblad A, Wallander H, Näsholm T (1998) Chitin and ergosterol combined to measure total and living fungal biomass in ectomycorrhizas. New Phytol 138:143–149

Ferrol N, Gonzlez-Guerrero M, Valderas A, Benabdellah K, Azcón-Aguilar C (2009) Survival strategies of arbuscular mycorrhizal fungi in Cu-polluted environments. Phytochem Rev 8:551–559

Fester T, Strack D, Hause B (2001) Reorganization of tobacco root plastids during arbuscule development. Planta 213:864–868

Filion M, St-Arnaud M, Jabaji-Hare SH (2003) Direct quantification of fungal DNA from soil substrate using real-time PCR. J Microbiol Methods 53:67–76

Fontaine J (2001) Etude du métabolisme lipidique d'un champignon endomycorhizien à vésicules et arbuscules: *Glomus intraradices*. PhD thesis, Université du Littoral Côte d'Opale, Calais, France

Fontaine J, Grandmougin-Ferjani A, Hartmann MA, Sancholle M (2001a) Sterol biosynthesis by *Glomus intraradices*. Lipids 36:1357–1363

Fontaine J, Grandmougin-Ferjani A, Sancholle M (2001b) Métabolisme lipidique du champignon endomycorhizien: *Glomus intraradices*. C R Acad Sci Ser III Sc Vie 324:847–853

Fontaine J, Grandmougin-Ferjani A, Glorian V, Durand R (2004) 24-Methyl/methylene sterols increase in monoxenic roots after colonization by arbuscular mycorrhizal fungi. New Phytol 163:159–167

Fortin JA, Bécard G, Declerck S, Dalpé Y, St Arnaud M, Coughlan AP, Piché Y (2002) Arbuscular mycorrhiza on root-organ cultures. Can J Bot 80:1–20

Fox FM (1986) Ultrastructure and infectivity of sclerotium-like bodies of the ectomycorrhizal fungus *Hebeloma sacchariolens*, on Birch (*Betula spp.*). Trans Br Mycol Soc 87:359–369

Franz F, Acker G (1995) Rhizomorphs of *Picea abies* ectomycorrhizae: ultrastructural aspects and elemental analysis (EELS and ESI) on hyphal inclusions. Nova Hedwigia 60:253–267

Fries N, Bardt M, Serck-Hanssen K (1985) Growth of ectomycorrhizal fungi stimulated by lipids from a pine root exudate. Plant Soil 86:287–290

Fritze H, Bååth E (1993) Microfungal species composition and fungal biomass in a coniferous forest soil polluted by alkaline deposition. Microb Ecol 25:83–92

Frostegård A, Bååth E (1996) The use of phospholipid fatty acid analysis to estimate bacterial and fungal biomass in soil. Biol Fertil Soils 22:59–65

Frostegård A, Bååth E, Tunlid A (1993) Shifts in the structure of soil microbial communities in limed forests as revealed by phospholipids fatty analysis. Soil Biol Biochem 25:723–730

Gao JM, Wang CY, Zang AL, Liu JK (2001) A new trihydroxy fatty acid from the Ascomycetes, Chinese truffle *Tuber indicum*. Lipids 36:1365–1370

Gaspar ML, Pollero RJ (1994) *Glomus antarcticum*: the lipids and fatty acid composition. Mycotaxon 51:129–136

Gaspar ML, Pollero RJ, Cabello MN (1994) Triacylglycerol consumption during spore germination of vesicular-arbuscular mycorrhizal fungi. J Am Oil Chem Soc 71:449–452

Gaspar ML, Pollero R, Cabello M (1997) Variations in the lipid composition of alfalfa roots during colonization with the arbuscular mycorrhizal fungus *Glomus versiforme*. Mycologia 89:37–82

Gaspar ML, Cabello MN, Pollero R, Aon MA (2001) Fluorescein diacetate hydrolysis as a measure of fungal biomass in soil. Curr Microbiol 42:339–344

Gaude N, Tippmann H, Flemetakis E, Katinakis P, Udvardi M, Dörmann P (2004) The galactolipid digalactosyldiacylglycerol accumulates in the peribacteroid membrane of nitrogen-fixing nodules of soybean and lotus. J Biol Chem 279:34624–34630

Gavito ME, Olsson PA (2003) Allocation of plant carbon to foraging and storage in arbuscular mycorrhizal fungi. FEMS Microbiol Ecol 45:181–187

Gellerman JL, Schlenk H (1963) A group of fatty acids with 'Tetramethylene interruption' in the double bond system. Experientia 19:522–523

Genney DR, Alexander IJ, Hartley SE (2000) Exclusion of grass roots from soil organic layers by *Calluna*: the role of ericoid mycorrhizas. J Exp Bot 51:1117–1125

Göhre V, Paszkowski U (2006) Contribution of the arbuscular mycorrhizal symbiosis to heavy metal phytoremediation. Planta 223:1115–1122

Gonzlez-Guerrero M, Oger E, Benabdellah K, Azcón-Aguilar C, Lanfranco L, Ferrol N (2010) Characterization of a CuZn superoxide dismutase gene in the arbuscular mycorrhizal fungus *Glomus intraradices*. Curr Genet 56:265–274

Graham JH (2000) Assessing the cost of arbuscular mycorrhizal symbiosis in agroecosystems. In: Podila GK, Douds DD (eds) Current advances in mycorrhizal research. The American Phytopathological Society, St. Paul, pp 127–140

Graham JH, Hodge NC, Morton JB (1995) Fatty acid methyl ester profiles for characterization of glomalean fungi and their endomycorrhizae. Appl Environ Microbiol 61:58–64

Graham JH, Drouillard DL, Hodge NC (1996) Carbon economy of sour orange in response to different *Glomus* spp. Tree Physiol 16:1023–1029

Graham JH, Duncan LW, Eissenstat DM (1997) Carbohydrate allocation patterns in citrus genotypes as affected by phosphorus nutrition, mycorrhizal colonization and mycorrhizal dependency. New Phytol 135:335–343

Grandmougin-Ferjani A, Dalpé Y, Veignie E, Hartmann MA, Rafin C, Sancholle M (1995) Infection by arbuscular mycorrhizal fungus *Glomus mosseae* of leek plants (*Allium porrum* L.) effects of lipids. In: Kader JC, Mazliak P (eds) Plant lipid metabolism. Kluwer, Dordrecht, pp 444–446

Grandmougin-Ferjani A, Dalpé Y, Hartmann MA, Laruelle F, Couturier D, Sancholle M (1996) Taxonomic aspect of the sterol and Δ11-hexadecenoic acid (C16:1Δ11) distribution in arbuscular mycorrhizal spores. In: Williams JP, Kahn MU, Lem NW

(eds) Physiology, biochemistry and molecular biology of plant lipids. Kluwer, Dordrecht, pp 195–197
Grandmougin-Ferjani A, Dalpé Y, Hartmann MA, Laruelle F, Couturier D, Sancholle M (1997) Taxonomic aspect of the sterol and Δ11-hexadecenoic acid (C16:1Δ11) distribution in arbuscular mycorrhizal spores. In: Williams JP, Khan MU, Lem NW (eds) Physiology, biochemistry and molecular biology of plant lipids. Kluwer, Dordrecht, pp 195–197
Grandmougin-Ferjani A, Dalpé Y, Hartmann MA, Laruelle F, Sancholle M (1999) Sterol distribution in arbuscular mycorrhizal fungi. Phytochemistry 50:1027–1031
Grandmougin-Ferjani A, Fontaine J, Durand R (2005) Carbon metabolism, lipid composition and metabolism in arbuscular mycorrhizal fungi. In: Declerck S, Strullu DG, Fortin JA (eds) In vitro culture of mycorrhizas. Springer, Berlin, pp 159–180
Grenville DJ, Peterson RL, Piché Y (1985a) The development, structure, and histochemistry of sclerotia of ectomycorrhizal fungi: I. *Pisolithus tinctorius*. Can J Bot 63:1402–1411
Grenville DJ, Peterson RL, Piché Y (1985b) The development, structure, and histochemistry of sclerotia of ectomycorrhizal fungi: II. *Paxillus involutus*. Can J Bot 63:1412–1417
Hale MG, Moore LD, Orcutt DM (1981) Effects of gibberellin A3 and 2,4-D on plant and root exudate lipids and susceptibility to *Pythium muriotylum*. Soil Biol Biochem 13:395–399
Hammer EC, Pallon J, Wallander H, Olsson PA (2011) Tit for tat? A mycorrhizal fungus accumulates phosphorus under low plant carbon availability. FEMS Microbiol Ecol 76:236–244
Hampp R, Schaeffer C, Wallenda T, Stulten C, Johann P, Einig W (1995) Changes in carbon partitioning of allocation due to ectomycorrhiza formation biochemical evidence. Can J Bot 73:548–556
Hannich JT, Umebayashi K, Riezman H (2011) Distribution and functions of sterols and sphingolipids. Cold Spring Harb Perspect Biol 2011:a004762
Hanus LO, Shkrob I, Dembitsky VM (2008) Lipids and fatty acids of wild edible mushrooms of the genus *Boletus*. J Food Lipids 15:370–383
Harki E, Klaebe A, Talou T, Dargent R (1996) Identification and quantification of *Tuber melanosporum* Vitt. sterols. Steroids 61:609–612
Harki E, Bouya D, Dargent R (2006) Maturation-associated alterations of the biochemical characteristics of the black truffle *Tuber melanosporum* Vitt. Food Chem 99:394–400
Harris D, Pacovsky RS, Paul EA (1985) Carbon economy of soybean-*Rhizobium*-*Glomus* associations. New Phytol 101:427–440
Harwood JL (1996) Recent advances in the biosynthesis of plant fatty acids. Biochim Biophys Acta Lipids Lipid Metab 1301:7–56
Herr DG, Peterson RL (1996) Morphology, anatomy and histochemistry of *Fagus grandifolia* Ehrh. (North American Beech) ectomycorrhizas. Bot Acta 109:64–71
Hildebrandt U, Regvar M, Bothe H (2007) Arbuscular mycorrhiza and heavy metal tolerance. Phytochemistry 68:139–146
Hillis DG, Antunes P, Sibley PK, Klironomos JN, Solomon KR (2008) Structural responses of *Daucus carota* root-organ cultures and the arbuscular mycorrhizal fungus, *Glomus intraradices*, to 12 pharmaceuticals. Chemosphere 73:344–352
Hiroi M (1978) Identification of 6-oxooctadecanoic acid in mushroom, *Lactarius chrysorheus* Fr. lipid. J Agric Chem Soc Jpn 52:351–353
Ho I, Trappe JM (1973) Translocation of 14C from *Festuca* plants to their endomycorrhizal fungi. Nature 244:30
Hua X, Liu G, Zhang X, Yu L, Zeng P, Huang D (1995) Study on mycorrhization of pine in nursery and field by cutting primary root apex of young seedlings off. For Res 8:535–543
Hutchinson LJ (1990) Studies on the systematics of ectomycorrhizal fungi in axenic culture: II. The enzymatic degradation of selected carbon and nitrogen compounds. Can J Bot 68:1522–1530
Isayenkov I, Fester T, Hause B (2004) Rapid determination of fungal colonization and arbuscule formation in roots of *Medicago truncatula* using real-time (RT) PCR. J Plant Physiol 161:1379–1383
Jabaji-Hare SH, Deschene A, Kendrick B (1984) Lipid content and composition of vesicles of a vesicular arbuscular mycorrhizal fungus. Mycologia 76:1024–1030
Jakobsen I (1991) Carbon metabolism in mycorrhiza. Methods Microbiol 23:149–180
Jakobsen I, Rosendahl L (1990) Carbon flow into soil and external hyphae from roots of mycorrhizal cucumber plants. New Phytol 115:77–83
Jennemann R, Geyer R, Sandhoff R, Gschwind RM, Levery SB, Groene HJ, Wiegandt H (2001) Glycoinositolphosphosphingolipids (basidiolipids) of higher mushrooms. Eur J Biochem 268:1190–1205
Johansen A, Finlay RD, Olsson PA (1996) Nitrogen metabolism of external hyphae of the arbuscular mycorrhizal fungus *Glomus intraradices*. New Phytol 133:705–712
Johnson BN, McGill WB (1990) Comparison of ergosterol and chitin as quantitative estimate of mycorrhizal infection and *Pinus contorta* seedling response to inoculation. Can J For Res 20:1125–1131
Kalac P (2009) Chemical composition and nutritional value of European species of wild growing mushrooms: a review. Food Chem 113:9–16
Karlinski L, Ravnskov S, Kieliszewska-Rokicka B, Larsen J (2007) Fatty acid composition of various ectomycorrhizal fungi and ectomycorrhizas of Norway spruce. Soil Biol Biochem 39:854–866

Kaschuk G, Kuyper TW, Leffelaar PA, Hungria M, Giller KE (2009) Are the rates of photosynthesis stimulated by the carbon sink strength of rhizobial and arbuscular mycorrhizal symbioses? Soil Biol Biochem 41:1233–1244

Kaschuk G, Leffelaar PA, Giller KE, Alberton O, Hungria M, Kuyper TW (2010) Responses of legumes to rhizobia and arbuscular mycorrhizal fungi: a meta-analysis of potential photosynthate limitation of symbioses. Soil Biol Biochem 42:125–127

Kasurinen A, Holopainen T, Anttonen S (2001) Mycorrhizal colonisation of highbush blueberry and its native relatives in central Finland. Agric Food Sci Finland 10:113–119

Kirk JL, Moutoglis P, Klironomos J, Lee H, Trevors JT (2005) Toxicity of diesel fuel to germination, growth and colonization of *Glomus intraradices* in soil and in vitro transformed carrot root cultures. Plant Soil 270:23–30

Kjoller R, Rosendahl S (2000) Effects of fungicides on arbuscular mycorrhizal fungi: differential responses in alkaline phosphatase activity of external and internal hyphae. Biol Fertil Soils 31:361–365

Kling M, Jakobsen I (1997) Direct application of carbendazim and propiconazole at field rates to the external mycelium of three arbuscular mycorrhizal fungi species: effect on 32P transport and succinate dehydrogenase activity. Mycorrhiza 7:33–37

Koo CD, Lee DH, Park YW, Lee YN, Ka KH, Park H, Bak WC (2009) Ergosterol and water changes in tricholoma matsutake soil colony during the mushroom fruiting season. Mycobiology 37:10–16

Kuhn H, Ku H, Requena N (2010) Membrane steroid-binding protein 1 induced by a diffusible fungal signal is critical for mycorrhization in *Medicago truncatula*. New Phytol 185:716–733

Laczko E, Boller T, Wiemken V (2003) Lipids in roots of *Pinus sylvestris* seedlings and in mycelia of *Pisolithus tinctorius* during ectomycorrhiza formation: changes in fatty acid and sterol composition. Plant Cell Environ 27:27–40

Lammers PJ, Jun J, Abubaker J, Arreola R, Gopalan A, Bago B, Hernandez-Sebastia C, Allen JW, Douds DD, Pfeffer PE, Shachar-Hill Y (2001) The glyoxylate cycle in an arbuscular mycorrhizal fungus. Carbon flux and gene expression. Plant Physiol 127:1287–1298

Lekberg Y, Hammer ED, Olsson PA (2010) Plants as resource islands and storage units—adopting the mycocentric view of arbuscular mycorrhizal networks. FEMS Microbiol Ecol 74:336–345

Leroux P (2003) Mode of action of agrochemicals towards plant pathogen. C R Acad Sci 326:9–21

Livermore BP, Johnson RC, Jenkin HM (1969) Isolation of an unusual positional isomer of Hexadecenoic acid from a parasitic Leptospire. Lipids 4:166

Lösel DM (1980) The effect of biotrophic fungal infection on the lipid metabolism of green plants. In: Mazliak P, Benveniste P, Costes C, Douce R (eds) Biogenesis and function of plant lipids. Elsevier/North Holland Biomedical, New York, pp 263–268

Lösel DM (1988) Fungal lipids. In: Ratledge C, Wilkinson SG (eds) Microbial lipids, vol 1. Academic, London, pp 699–806

Lösel DM, Cooper KM (1979) Incorporation of 14C-labelled substrates by uninfected and VA mycorrhizal roots of onion. New Phytol 83:415–426

Lu X, Koide RT (1991) *Avena fortma* L. seedling nutrient dynamics as influenced by mycorrhizal infection of natural generation. Plant Cell Environ 14:931–939

Lynd JQ, Ansman TR (1989) Effects of phosphorus, calcium with four potassium levels on nodule histology, nitrogenase activity and improve Spanco peanut yields. J Plant Nutr 12:65–84

Madan R, Pankhurst C, Hawke B, Smith S (2002) Use of fatty acids for identification of AM fungi and estimation of the biomass of AM spores in soil. Soil Biol Biochem 34:125–128

Manjarrez M, Smith FA, Marschner P, Smith SE (2008) Is cortical root colonization required for carbon transfer to arbuscular mycorrhizal fungi? Evidence from colonization phenotypes and spore production in the reduced mycorrhizal colonization (rmc) mutant of tomato. Botany 86:1009–1019

Martin F, Canet D (1986) Biosynthesis of amino acids during carbon-13 glucose utilization by the ectomycorrhizal ascomycete *Cenococcum geophilum* monitored by carbon-13 NMR. Physiol Veg 24:209–218

Martin F, Canet D, Marchal JP (1984a) In vivo natural abundance of carbon 13 NMR studies of the carbohydrate storage in ectomycorrhizal fungi. Physiol Veg 22:733–744

Martin F, Canet D, Marchal JP, Brondeau J (1984b) In vivo natural abundance 13C nuclear magnetic resonance studies of living ectomycorrhizal fungi. Plant Physiol 75:151–153

Martin M, Duplessis S, Ditengou F, Lagrange H, Voiblet C, Lapeyrie F (2001) Developmental cross talking in the ectomycorrhizal symbiosis: signals and communication genes. New Phytol 151:145–154

Matcham SE, Jordan BR, Wood DA (1985) Estimation of fungal biomass in a solid substrate by three independent methods. Appl Microbiol Biotechnol 21:108–112

Meharg AA (2003) The mechanistic basis of interactions between mycorrhizal associations and toxic metal cations. Mycol Res 107:1253–1265

Meharg AA, Cairney JWG (2000) Ectomycorrhizas—extending the capabilities of rhizosphere remediation? Soil Biol Biochem 32:1475–1484

Melhuish JH, Janerette CA (1979) The effects of the carbon-nitrogen ratios on carbohydrate, protein, lipid and fatty acid production in *Pisolithus tinctorius*. Proc N Am Conf Mycorrhizae 4:67

Miller RM (2002) Mycorrhizal mediated feedbacks influence net carbon gain and nutrient uptake in *Andropogon gerardii* Vitman. New Phytol 155:149–162

Moche M, Stremlau S, Hecht L, Göbel C, Feussner I, Stöhr C (2010) Effect of nitrate supply and mycorrhizal inoculation on characteristics of tobacco root plasma membrane vesicles. Planta 231:425–436

Montgomery HJ, Monreal CM, Young JC, Seifert KA (2000) Determination of soil fungal biomass from soil ergosterol analyses. Soil Biol Biochem 32:1207–1217

Moore AEP, Ashford AE, Peterson RL (1991) Reserve substances in *Paxillus involutus* sclerotia: determination by histochemistry and X-ray microanalysis. Protoplasma 163:67–81

Mortimer PE, Archer E, Valentine AJ (2005) Mycorrhizal C costs and nutritional benefits in developing grapevines. Mycorrhiza 15:159–165

Müller MM, Kantola R, Kitunen V (1994) Combining sterol and fatty acid profiles for the characterization of fungi. Mycol Res 98:593–603

Müllner H, Daum G (2004) Dynamics of neutral lipid storage in yeast. Acta Biochim Pol 51:323–349

Murcia MA, Martínez-Tomé M, Vera A, Morte A, Gutierrez A, Honrubia M, Jiménez AM (2003) Effects of industrial processing on desert truffles *Terfezia claveryi* Chatin and *Picoa juniperi* Vittandini: proximate composition and fatty acids. J Sci Food Agric 83:535–541

Murphy DJ (2001) The biogenesis and functions of lipid bodies in animals, plants and microorganisms. Prog Lipid Res 40:325–438

Nagahashi G, Douds DD Jr (2011) The effects of hydroxy fatty acids on the hyphal branching of germinated spores of AM fungi. Fungal Biol 115:351–358

Nagahashi G, Douds DD Jr, Abney GD (1996) Phosphorus amendment inhibits hyphal branching of the VAM fungus *Gigaspora margarita* directly and indirectly through its effect on root exudation. Mycorrhiza 6:403–408

Nagy S, Nordby HE, Nemec S (1980) Composition of lipids in roots of six *Citrus* cultivars infected with the vesicular-arbuscular mycorrhizal fungus, *Glomus mosseae*. New Phytol 85:377–384

Nehls U, Mikolajewski S, Magel E, Hampp R (2001) Carbohydrate metabolism in ectomycorrhizas: gene expression, monosaccharide transport and metabolic control. New Phytol 150:533–541

Nemec S (1981) Histochemical characteristics of *Glomus etunicatum* infection of *Citrus limon* fibrous roots. Can J Bot 59:609–617

Ngosong C, Jarosch M, Raupp J, Neumann E, Ruess L (2010) The impact of farming practice on soil microorganisms and arbuscular mycorrhizal fungi: crop type versus long-term mineral and organic fertilization. Appl Soil Ecol 46:134–142

Nichols P, Stulp BK, Jones JG, White DC (1986) Comparison of fatty acid content and DNA homology of the filamentous gliding bacteria *Vitreoscilla, Felxibacter, Filibacter*. Arch Microbiol 146:1–6

Nilsson LO, Giesler R, Bååth E, Wallander H (2005) Growth and biomass of mycorrhizal mycelia in coniferous forests along short natural nutrient gradients. New Phytol 165:613–622

Nordby HE, Nemec S, Nagy S (1981) Fatty acids and sterols associated with *Citrus* root mycorrhizae. J Agric Food Chem 29:396–401

Nylund JE, Wallander H (1989) Effects of ectomycorrhiza on host growth and carbon balance in a semihydroponic cultivation system. New Phytol 112:389–398

Oger E, Ghignone S, Campagnac E, Fontaine J, Grandmougin-Ferjani A, Lanfranco L (2009) Functional characterization of a C-4 sterol methyl oxidase from the endomycorrhizal fungus *Glomus intraradices*. Fungal Genet Biol 46:486–495

Ohta A (1988) Effects of butyric acid and related compounds on basidiospore germination of some mycorrhizal fungi. Trans Mycol Soc Jpn 29:375–382

Olsrud M, Melillo JM, Christensen TR, Michelsen A, Wallander H, Olsson P (2004) Response of ericoid mycorrhizal colonization and functioning to global change factors. New Phytol 162:459–469

Olsrud M, Michelsena A, Wallanderb H (2007) Ergosterol content in ericaceous hair roots correlates with dark septate endophytes but not with ericoid mycorrhizal colonization. Soil Biol Biochem 39:1218–1221

Olsson I, Jantzen E (2007) Sphingolipids in bacteria and fungi. Anaerobe 7:103–112

Olsson PA (1999) Signature fatty acids provide tools for determination of the distribution and interactions of mycorrhizal fungi in soil. FEMS Microbiol Ecol 29:303–310

Olsson PA, Johansen A (2000) Lipid and fatty acid composition of hyphae and spores of arbuscular mycorrhizal fungi at different growth stages. Mycol Res 104:429–434

Olsson PA, Johnson NC (2005) Tracking carbon from the atmosphere to the rhizosphere. Ecol Lett 8:1264–1270

Olsson PA, Wilhelmsson P (2000) The growth of external AM fungal mycelium in sand dunes and in experimental systems. Plant Soil 226:161–169

Olsson PA, Bååth E, Jacobsen I, Söderstrom B (1995) The use of phospholipid and neutral lipid fatty acids to estimate biomass of arbuscular mycorrhizal fungi in soil. Mycol Res 99:623–629

Olsson PA, Bääth E, Jakobsen I (1997) Phosphorus effects on the mycelium and storage structures of an arbuscular mycorrhizal fungus as studied in the soil and roots by analysis of fatty acid signatures. Appl Environ Microbiol 63:3531–3538

Olsson PA, Burleigh SH, van Aarle IM (2005a) The influence of external nitrogen on carbon allocation to *Glomus intraradices* in monoxenic arbuscular mycorrhiza. New Phytol 168:677–686

Olsson PA, van Aarle IM, Gavito ME, Bengtson P, Bengtsson G (2005b) ^{13}C incorporation into signature fatty acids as an assay for carbon allocation in arbuscular mycorrhiza. Appl Environ Microbiol 71:2592–2599

Olsson PA, Hansson MC, Burleigh SH (2006) Effect of P availability on temporal dynamics of carbon allocation and *Glomus intraradices* high-affinity P transporter gene induction in arbuscular mycorrhiza. Appl Environ Microbiol 72:4115–4120

Olsson PA, Rahm J, Aliasgharzad N (2010) Carbon dynamics in mycorrhizal symbioses is linked to carbon costs and phosphorus benefits. FEMS Microbiol Ecol 72:123–131

Pacovsky RS (1989) Metabolic difference in *Zea-Glomus-Azospirillum* symbioses. Soil Biol Biochem 21:953–960

Pacovsky RS, Fuller G (1987) Lipids of soybean inoculated with microsymbionts. In: Stumpf PK et al (eds) The metabolism structure and function of plant lipids. Plenum, London, pp 349–351

Padgett DE, Posey MH (1993) An evaluation of the efficiencies of several ergosterol extraction techniques. Mycol Res 97:1476–1480

Pang PC, Paul EA (1980) Effects of vesicular-arbuscular mycorrhiza on C and N distribution in nodulated fababeans. Can J Soil Sci 60:241–250

Parsi Z, Gorecki T (2006) Determination of ergosterol as an indicator of fungal biomass in various samples using non-discriminating flash pyrolysis. J Chromatogr A 1130:145–150

Paterson S, Mackay D (1994) A model of organic chemical uptake by plants from soil and the atmosphere. Environ Sci Technol 28:2259–2266

Paul EA, Kucey RMN (1981) Carbon flow in microbial associations. Science 213:473–474

Pedersen TA (1970) Effects of fatty acids and methyl octanoate on resting mycelium of *Boletus variegatus*. Physiol Plant 23:654–666

Pedneault K, Angers P, Gosselin A, Twedell RJ (2006) Fatty acid composition of lipids from mushrooms belonging to the family Boletaceae. Mycol Res 110:1179–1183

Pedneault K, Angers P, Gosselin A, Twedell RJ (2008) Fatty acid profiles of polar and neutral lipids of ten species of higher basidiomycetes indigenous to eastern Canada. Mycol Res 112:1428–1434

Peng S, Eissenstat DM, Graham JH, Williams K, Hodge NC (1993) Growth depression in mycorrhizal *Citrus* at high phosphorus supply. Analysis of carbon costs. Plant Physiol 101:1063–1071

Petersen LS, Larsen EH, Larsen PB, Bruun P (2002) Uptake of trace elements and PAHs by fruit and vegetables from contaminated soils. Environ Sci Technol 36:3057–3063

Pfeffer PE, Douds DD, Bécard G, Shachar-Hill Y (1999) Carbon uptake and the metabolism and transport of lipids in an arbuscular mycorrhiza. Plant Physiol 120:587–598

Pfeffer PE, Douds DD, Bücking H, Schwartz DP, Shachar-Hill Y (2004) The fungus does not transfer carbon to or between roots in an arbuscular mycorrhizal symbiosis. New Phytol 163:617–627

Plassard C, Bonafos B, Touraine B (2000) Differential effects of mineral and organic N sources, and of ectomycorrhizal infection by *Hebeloma cylindrosporum*, on growth and N utilization in *Pinus pinaster*. Plant Cell Environ 23:1195–1205

Probanza A, Mateos JL, Lucas Garcia JA, Ramos B, de Felipe MR, Gutierrez Manero FJ (2001) Effects of inoculation with PGPR *Bacillus* and *Pisolithus tinctorius* on *Pinus pinea* L. growth, bacterial rhizosphere colonization, and mycorrhizal infection. Microb Ecol 41:140–148

Prostenik M, Burcar I, Castek A, Cosovic C, Golem J, Jandric Z, Kljaic K, Ondrusek V (1978) Lipids of higher fungi. III The fatty acids and 2-hydroxy-fatty acids in some species of Basidiomycetes. Chem Phys Lipids 22:97–103

Prostenik M, Castek A, Cosovic C, Gospocic C, Jandric Z, Kljaic K, Ondrusek V (1983) The patterns of the constituent fatty acids of glycerophospholipids in mushrooms: lipids of higher fungi. Exp Mycol 7:74–81

Rabie GH (2005a) Contribution to arbuscular mycorrhizal fungus to red kidney and wheat plants tolerance grown in heavy metal-polluted soil. Afr J Biotechnol 4:332–345

Rabie GH (2005b) Role of arbuscular mycorrhizal fungi in phytoremediation of soil rhizosphere spiked with poly aromatic hydrocarbons. Microbiology 33:41–50

Ratnayake M, Leonard RT, Menge JA (1978) Root exudation in relation to supply of phosphorus and its possible relevance to mycorrhizal formation. New Phytol 81:543–552

Reich M, Göbel C, Kohler A, Buée M, Martin F, Feussner I, Polle A (2009) Fatty acid metabolism in the ectomycorrhizal fungus *Laccaria bicolor*. New Phytol 182:950–964

Rolin D, Le Tacon F, Larher F (1984) Characterization of the different forms of phosphorus in the mycelium of the ectomycorrhizal fungus *Hebeloma cylindrosporum* cultivated in pure culture. New Phytol 98:335–344

Ruess L, Häggblom MM, Garzía Zapata EJ, Dighton J (2002) Fatty acids of fungi and nematodes—possible biomarkers in the soil food chain? Soil Biol Biochem 34:745–756

Rydlov J, Püschel D, Sudov R, Gryndler M, Mikanov O, Vostka M (2011) Interaction of arbuscular mycorrhizal fungi and rhizobia: effects on flax yield in spoil-bank clay. J Plant Nutr Soil Sci 174:128–134

Salvioli A, Chiapello M, Fontaine J, Lounes Hadj-Sahraoui A, Grandmougin-Ferjani A, Lanfranco L,

Bonfante P (2010) Endobacteria affect the metabolic profile of their host *Gigaspora margarita*, an arbuscular mycorrhizal fungus. Environ Microbiol Rep 12:2083–2095

Samra A, Dumas-Gaudot E, Gianinazzi-Pearson V, Gianinazzi S (1996) Soluble proteins and polypeptide profiles of spores of arbuscular mycorrhizal fungi. Interspecific variability and effects of host (myc+) and non-host (myc–) *Pisum sativum* root exudates. Agronomie 16:709–719

Sancholle M, Dalpé Y (1993) Taxonomic relevance of fatty acids of arbuscular mycorrhizal fungi and related species. Mycotaxon 49:187–193

Sancholle M, Lösel DM (1995) Lipids in fungal biotechnology. In: Kück B (ed) The Mycota II. Genetics and biotechnology. Springer, Berlin, pp 339–367

Schaarschmidt S, Gonzlez M-C, Roitsch T, Strack D, Sonnewald U, Hause B (2007) Regulation of arbuscular mycorrhization by carbon. The symbiotic interaction cannot be improved by increased carbon availability accomplished by root-specifically enhanced invertase activity. Plant Physiol 143:1827–1840

Schisler LC, Volkoff O (1977) The effects of safflower oil on mycelial growth of Boletaceae in submerged liquid cultures. Mycologia 69:118–125

Schliemann W, Ammer C, Strack D (2008) Metabolite profiling of mycorrhizal roots of *Medicago truncatula*. Phytochemistry 69:112–146

Schmitz O, Danneberg G, Hundeshagen B, Klingner A, Bothe H (1991) Quantification of vesicular arbuscular mycorrhiza by biochemical parameters. J Plant Physiol 139:106–114

Schubert A, Wys P, Wiemken A (1992) Occurrence of trehalose in vesicular-arbuscular mycorrhizal fungi and in mycorrhizal roots. J Plant Physiol 140:41–45

Schüßler A, Martin H, Cohen D, Fitz M, Wipf D (2006) Characterization of a carbohydrate transporter from symbiotic glomeromycotan fungi. Nature 444:933–936

Schweiger PF, Spliid NH, Jakobsen I (2001) Fungicide application and phosphorus uptake by hyphae of arbuscular mycorrhizal fungi into field-grown peas. Soil Biol Biochem 33:1231–1237

Selvaraj T, Subramanian G (1990) Phenols and lipids in mycorrhizal and non mycorrhizal roots of *Sesamum indicum*. Curr Sci 59:471–473

Senatore F, Dini A, Marino A, Schettino O (1988) Chemical constituents of some Basidiomycetes. J Sci Food Agric 45:337–345

Senthilkumar S, Britoo SJ (2001) Morphological and cytochemical study of the pelotonic and non-pelotonic hyphae of the orchid *Spathoglottis plicata*. Mycorrhiza News 13:16–19

Simonich SL, Hites RA (1995) Organic pollutant accumulation in vegetation. Environ Sci Technol 29:2095–2103

Smith SE, Gianinazzi-Pearson V (1988) Physiological interactions between symbionts in vesicular arbuscular mycorrhizal plants. Annu Rev Plant Physiol Plant Mol Biol 39:221–244

Snellgrove RC, Splittstoesser WE, Stribley DP, Tinker PB (1982) The distribution of carbon and the demand of the fungal symbiont in leek plants with vesicular-arbuscular mycorrhizae. New Phytol 92:75–87

Sokolski S, Dalpé Y, Séguin S, Khasa D, Lévesque CA, Piché Y (2010) Conspecificity of DAOM 197198, the model arbuscular mycorrhizal fungus, with *Glomus irregulare*: molecular evidence with three protein-encoding genes. Botany 88:829–838

Solberg Y (1989) A literature review of the lipid constituents of higher fungi new investigations of Agaricales species. Int J Mycol Lichenol 4:137–154

Strullu DG, Chamel A, Eloy JF, Gourret JP (1983) Ultrastructure and analysis, by laser probe mass spectrography, of the mineral composition of the vesicles of *Trifolium pratense* endomycorrhizas. New Phytol 94:81–88

Stumpe M, Carsjens JG, Stenzel I, Göpel C, Lang I, Pawlowski K, Hause B, Feussner I (2005) Lipid metabolism in arbuscular mycorrhizal roots of *Medicago truncatula*. Phytochemistry 66:781–791

Sukarno N, Smith SE, Scott ES (1996) The effect of fungicides on vesicular-arbuscular mycorrhizal symbiosis. New Phytol 125:139–147

Sylvia DM, Wilson DO, Graham JH, Maddo JJ, Millner P, Morton JB, Skipper HD, Wright SF, Jarstfer AG (1993) Evaluation of vesicular-arbuscular mycorrhizal fungi in diverse plants and soils. Soil Biol Biochem 25:705–713

Tang Y, Li YY, Li HM, Wan DJ, Tang YJ (2011) Comparison of lipid content and fatty acid composition between tuber fermentation mycelia and natural fruiting bodies. J Agric Food Chem 59:4736–4742

Tawaraya K, Watanabe S, Yoshida E, Wagatsuma T (1996) Effect of onion (*Allium cepa*) root exudates on the hyphal growth of *Gigaspora margarita*. Mycorrhiza 6:57–59

Thompson LK, Hale MG (1983) Effect of kinetin in the rooting medium on root exudation of free fatty acids and sterols from roots of *Arachis hypogaea* L. 'Argentine' under axenic conditions. Soil Biol Biochem 15:125–126

Thoms C, Gattinger A, Jacob M, Thomas FM, Gleixner G (2010) Direct and indirect effects of tree diversity drive soil microbial diversity in temperate deciduous forest. Soil Biol Biochem 42:1558–1565

Thorne RT (1992) Classification and geography of the flowering plants. Bot Rev 58:225–348

Trent JD, Svejcar TJ, Christiansen S (1989) Effects of fumigation on growth, photosynthesis water relations and mycorrhizal development of winter wheat in the field. Can J Plant Sci 69:535–540

Trépanier M, Bécard G, Moutoglis P, Willemot C, Gagné S, Avis TJ, Rioux JA (2005) Dependance of arbuscular-mycorrhizal fungi on their plant host for palmitic acid synthesis. Appl Environ Microbiol 71:5341–5347 van Aarle IM, Olsson PA (2003) Fungal lipid accumulation and development of mycelial structures by two arbuscular mycorrhizal fungi. Appl Environ Microbiol 69:6762–6767

van Diepen LTA, Lilleskov EA, Pregitzer KS, Miller RM (2007) Decline of arbuscular mycorrhizal fungi in northern hardwood forests exposed to chronic nitrogen additions. New Phytol 176:175–183

van Diepen LTA, Lilleskov EA, Pregitzer KS (2011) Simulated nitrogen deposition affects community structure of arbuscular mycorrhizal fungi in northern hardwood forests. Mol Ecol 20:799–811

Vaskovsky VE, Khotimchenko SV, Benson AA (1991) Identification of diacylglycero-4′-o-(N,N, N-trimethyl) homoserine in mushrooms. Lipids 326:254–256

Verdin A, Lounès-Hadj Sahraoui A, Fontaine J, Grandmougin-Ferjani A, Durand R (2006) Effects of anthracene on development of an arbuscular mycorrhizal fungus and contribution of the symbiotic association to pollutant dissipation. Mycorrhiza 16:397–405

Voets L, Gougau I, Olsson PA, Merckx R, Declerk S (2008) Absence of carbon transfer between *Medicago truncatula* plants linked by a mycorrhizal network, demonstrated in an experimental microcosm. FEMS Microbiol Ecol 65:350–360

Voiblet C, Duplessis S, Encelot N, Martin F (2001) Identification of symbiosis-regulated genes in *Eucalyptus globulus-Pisolithus tinctorius* ectomycorrhiza by differential hybridization of arrayed cDNAs. Plant J 25:181–191

von Alten H, Lindermann A, Schonnbeck F (1993) Stimulation of vesicular-arbuscular mycorrhiza by fungicides or rhizosphere bacteria. Mycorrhiza 2:167–173

Wallander H, Nilsson LO, Hagerberg D, Bååth E (2001) Estimation of the biomass and seasonal growth of external mycelium of ectomycorrhizal fungi in the field. New Phytol 151:753–760

Wallenda T, Schaeffer C, Einig W, Wingler A, Hampp U, Seith B, George E, Marschner H (1996) Effects of varied soil nitrogen suupply on Norway spruce (*Picea abies* [L.] Karst.). Plant Soil 186:361–369

Wan MT, Rahe JE, Watts RG (1998) A new technique for determining the sublethal toxicity of pesticides to the vesicular-arbuscular mycorrhizal fungus *Glomus intraradices*. Environ Toxicol Chem 17:1421–1428

Wang GM, Coleman DC, Freckman DW, Dyer MI, McNaughton SJ, Acra MA, Goeschl JD (1989) Carbon partitioning patterns of mycorrhizal versus non-mycorrhizal plants: real-time dynamic measurements using CO_2. New Phytol 112:489–493

Weber RWS, Wakley GE, Thines E, Talbot NJ (2001) The vacuole as central element of the lytic system and sink for lipid droplets in maturing appressoria of *Magnaporthe grisea*. Protoplasma 216:101–112

Weete JD (1989) Structure and function of sterols in fungi. Adv Lipid Res 23:115–167

Weete JD, Kulifaj M, Montant C, Nes WR, Sancholle M (1985) Distribution of sterols in fungi II Brassicasterol in *Tuber* and *Terfezia* species. Can J Microbiol 31:1127–1130

Weete JD, Abril M, Blackwell M (2010) Phylogenetic distribution of fungal sterols. PLoS One 5:e10899

Werner A, Idzikowska K, Napierala-Filipiak A (2000) Ectendotrophic mycorrhiza of Scots pine (*Pinus sylvestris* L.) [in Polish]. Sylwan 144:53–67

West AW, Grant WD, Sparling GP (1987) Use of ergosterol, diaminopimelic acid and glucosamine contents of soil to monitor changes in microbial populations. Soil Biol Biochem 19:607–612

Xiaomei H, Guolong L, Xiaolin Z, Lianfu Y, Pingsheng Z, Dongqing H (1995) Study on mycorrhization of pine in nursery and field by cutting primary root apex of young seedlings off. For Res 8:535–543

Yokokawa H (1994) Sterol compositions of the fruit-bodies of higher fungi. Int Mycol Congr Abstr 5:250

Žel J, Svetek J, Črne H, Schara M (1993) Effects of aluminium on membrane fluidity of the mycorrhizal fungus *Amanita muscaria*. Physiol Plant 89:172–176

Zhu X, Song F, Xu H (2010) Influence of arbuscular mycorrhiza on lipid peroxidation and antioxidant enzyme activity of maize plants under temperature stress. Mycorrhiza 20:325–332

Zhu YG, Miller RM (2003) Carbon cycling by arbuscular mycorrhizal fungi in soil–plant systems. Trends Plant Sci 8:407–409

Zhukov AV, Lebedeva NI, Vereshchagin AG (2001) Effect of Helminthosporic root rot on the lipid content in wheat seedlings. Russ J Plant Physiol 48:111–115

Zocco D, Fontaine J, Lozanova E, Renard L, Bivort C, Durand R, Grandmougin-Ferjani A, Declerck S (2008) Influence of two sterol biosynthesis inhibitor fungicides (fenpropimorph and fenhexamid) on the development of an arbuscular mycorrhizal fungus. Mycol Res 112:592–601

9 The Role of the Stonesphere for the Interactions Between Mycorrhizal Fungi and Mycorrhizosphere Bacteria During Mineral Weathering

N. Koele[1]

CONTENTS

I. The Stonesphere

The stonesphere encompasses rock fragments in the rooted zone of the soil that interact with the soil environment physically, chemically or biologically. It includes the surface and cracks of rocks as a physical niche for microbial communities (Certini et al. 2004), providing a rich environment with lower acidity, more available nutrients and organic compounds as an energy source for non-symbiotic microorganisms (Koele et al. 2010). The stonesphere has been largely overlooked as a nutrient source and habitat for micro-organisms, but more and more evidence is emerging, showing this sphere in soils **provides valuable nutrient elements** (e.g. Corti et al. 2002; Heisner et al. 2004) and **a niche or refuge for microorganisms** (Certini et al. 2004; Koele et al. 2010). In addition, "common sense" and practitioners' experiences (Poesen and Lavee 1994) imply that, when plant roots and their symbionts are in such close contact as depicted in Fig. 9.1, interactions between the mineral and biological components are likely to take place. To understand biological processes in the stonesphere, however, its chemical and physical properties need to be defined first.

A. Functions of Stones in the Soil

Traditionally soil analyses are performed on the soil fraction <2 mm. This distinction is based on the assumption that the coarse-soil fraction, typically rock fragments >2 mm, is inert for the

[1]Landcare Research, PO Box 40, Lincoln 7640, New Zealand; e-mail: koelen@landcareresearch.co.nz

Fungal Associations, 2nd Edition
The Mycota IX
B. Hock (Ed.)

Fig. 9.1. The stonesphere in a broad sense: a fallen tree shows rock fragments of different size classes (<2 mm to >50 cm) that are largely incorporated into the (mycor)rhizosphere

Struchtemeyer 1966), rock fragments in soil have been recognized to affect physical soil properties (reviewed by Poesen and Lavee 1994). **Rock fragments at the soil surface affect erosion** at multiple scales. Individual fragments intercept rain drops, reducing infiltration but also preventing splash erosion. At larger scales rock fragments can have variable effects but tend to decrease soil erosion. This is mostly due to the protective cover of stones and the smaller flow velocities caused by rock fragments. Without a soil crust the rock fragments on top of the soil and embedded in the soil enhance infiltration and reduce erosion. With a soil crust, however, embedded rock fragments will cause overland flow concentration and erosion (Poesen and Ingelmo-Sanchez 1992). The size of the rock fragments plays an important role: generally the smaller size fractions distribute the flow better and enhance infiltration, whereas larger fractions cause concentrated overland flow (Lavee and Poesen 1991).

Table 9.1. Some effects of stones on soil properties with references to relevant studies

Effects of stones on soil properties	Physical/chemical/biological	Key reference(s)
Increased/decreased erosion	Physical	Poesen and Lavee (1994)
Increased/decreased water infiltration	Physical	Lavee and Poesen (1991)
Increased soil structure	Physical	Agnelli et al. (2000)
Decreased water holding capacity	Physical	Brakensiek and Rawls (1994)
Temperature	Physical	Childs and Flint (1990)
Water evaporation	Physical	Van Wesemael et al. (1996)
Increased fertility	Chemical	Heisner et al. (2004), Kohler et al. (2000), Ugolini et al. (2001)
Increased buffer against acidification	Chemical	Deutschmann (1998)
Organic matter composition	Chemical/biological	Agnelli et al. (2000), Bornemann et al. (2011)
Root branching	Biological	Jackson et al. (2011)
Microbial community structure	Biological	Certini et al. (2004)

most critical soil processes and therefore dilutes the active soil fraction. However, the important arguments that rock fragments generally do not contribute to the soil's water holding capacity and nutrient pool, and thus cannot sustain a microbial community, have been countered. Table 9.1 summarizes the effects of rock fragments in soil on physical, chemical and biological processes.

1. Influence of Stones on Soil Physical Properties

Especially in (semi)arid ecosystems where stony soils make up approximately 80% (Epstein and

Rock fragments within the soil profile generally decrease the water holding capacity of the soil, although this depends on the type of the rock, the weathering state and its size; some types can actually increase water holding capacity. In the soil profile rock fragments also reduce the amount of pores for waterflow and increase the tortuosity (Brakensiek and Rawls 1994), important for air-filled pores.

Most studies on the physical properties of rock fragments in the soil focus on agricultural land and/or non-vegetated areas. Therefore it is not certain that the above-outlined processes are also of significance in forest soils, where plant cover and roots also play an important role in reducing erosion, and the role of rock fragments is likely coupled to this.

2. Influence of Stones on Soil Chemical Properties

Traditionally the coarse-soil fraction is regarded as chemically inert. This assumption stems from the fact that rock (fragments) weathers so slowly that this is not a considerable nutrient source on a medium- to short-term basis. **Weathering releases nutrient cations from the mineral structures of the rock.** In long-term calculations, weathering is considered an important nutrient source, especially of the base cations (Mg, Ca, K), but on short- to medium-term time-scales, nutrient cycling from litter and atmospheric inputs are the major nutrient inputs into the ecosystem (Kimmins 1987; Prescott 2002). Calculated from the mineral composition and the kinetics of protolytic mineral weathering, the input of nutrients via chemical weathering is indeed slow and should be only considered on time-scales of years or decades (e.g. Sverdrup 1990).

Several studies however reported that rock fragments contain similar amounts of plant-available nutrients compared to the fine earth (Rivard and de Kimpe 1980; Corti et al. 1997; Schüler and Butz-Braun 2001; Ugolini et al. 2001; Heisner et al. 2004).

Kohler et al. (2000) and Martín-García et al. (1999) argued that **rock fragments can contain nutrients in weathering cracks.** These cracks have filled up with in situ weathering products such as vermiculites and chlorites, which have been protected from leaching from the bulk soil solution by the stone matrix. Certain rock fragments are used as fertilizers, which is another sign that the nutritional value of rock fragments is known by knowledge of experience. In one of the first stone-related studies, de Turk (1919) investigated the use of silicate minerals as a K fertilizer for agricultural crops. This concept is still being applied, although with crushed rocks rather than actual coarse fractions, and with mixed results. In agriculture, crushed rock fragments are often considered too slow in releasing K (and other cations).

II. Mineral Weathering in the Mycorrhizosphere

According to Calvaruso et al. (2007), mycorrhizal fungi modify the physical, chemical and microbiological characteristics of the surrounding soil and create a special environment called the mycorrhizosphere, in which microbial communities differ from those of the rhizosphere soil.

It has been long known that plants and associated micro-organisms can increase their nutrient uptake via localized weathering in the rhizosphere (Boyle and Voigt 1973). **Plant roots, (mycorrhizal) fungi and bacteria can excrete low molecular weight organic acids** (LMWOAs) and other substances that **dissolve the mineral structure and consequently release nutrient elements** (e.g. Bormann et al. 1998; Kelly et al. 1998). The interactions between these actors in the mycorrhizosphere probably depend on spatial and temporal nutrient availability.

A. Mycorrhizal Weathering

Landeweert et al. (2001) described the mechanisms of ectomycorrhizal weathering. Excreting organic acids, ectomycorrhizal hyphae are able to dissolve mineral structures and obtain the nutrients from within that mineral structure. This mechanism has been shown in numerous laboratory experiments (e.g. Leyval and Berthelin 1989; Paris et al. 1996; Wallander et al. 1997; Glowa et al. 2003). The general conclusion of these studies was that **especially in nutrient-deficient environments plant roots and associated ectomycorrhizal fungi take up these nutrients from the mineral structure of the rock minerals.** However these studies with very simplified conditions could not be scaled up to the ecosystem level.

Research on mycorrhizal weathering almost exclusively focuses on the ectomycorrhizal symbiosis. Ectomycorrhizal fungi have been repeatedly shown to be able to produce strong weathering agents such as organic acids, and they produce a vast network of mycelia that explores the soil. In contrast, arbuscular mycorrhizal fungi, although important in plant nutrient uptake, are not generally considered to be important in mineral weathering. However, a study by Arocena et al. (2012) showed that arbuscular mycorrhizal fungi are capable of biotite alteration in order to obtain the nutrient element K.

B. Weathering by Bacteria in the Mycorrhizosphere

Bacterial subsurface weathering is less studied than mycorrhizal weathering in soil. For surface rock weathering Gleeson et al. (2006) found that **bacterial communities preferentially inhabit minerals with specific inorganic nutrient contents such as muscovite, plagioclase and K-feldspar (rich in Mg, Ca, K).** For bacteria isolated from the subsurface Barker et al. (1998) demonstrated that these bacteria are capable of weathering the nutrient-rich mineral grains biotite and plagioclase feldspars. Uroz et al. (2009) provided a list of bacteria identified as capable of weathering. Specifically strains from the mycorrhizosphere seem to be more proficient in releasing Fe, Si and nutrient cations from minerals than strains isolated from the bulk soil (Uroz et al. 2007), so it is not surprising that especially *Burholderia* and *Pseudomonas* isolated from the mycorrhizosphere have been cited multiple times to have good weathering abilities (Uroz et al. 2009; and see references therein). The mechanisms of bacterial weathering are in principle the same as for ectomycorrhizal weathering: the release of protons and siderophores to reduce the pH and dissolve minerals.

C. Interactions Between Mycorrhizal Fungi and Bacteria

Ever since several mycorrhizal fungi and rhizosphere bacteria were shown to increase plant nutrient uptake and growth, possible interactions have been hypothesized. Co-inoculation of plant roots with both mycorrhizal fungi and bacteria had various effects, and so far no conclusive trend has been found. The inconsistent effect of co-inoculation shows the interaction between these organisms is not a simple additive effect of the two organisms. **The interactions are complex and depend on various factors in the soil environment such as nutrient status and moisture content.** Koele et al. (2009) hypothesize that **the tips of ectomycorrhizal hyphae are a niche for weathering bacteria that are enabled by the fungi through exudation of specific carbohydrates** (Frey et al. 1997).

III. The Stonesphere as Attractor for the Mycorrhizosphere

A. Rock Fragments Analogous to Soil Aggregates: Fungal Hyphae and Soil Structure Feedbacks

Soil has been often regarded as a black box, especially with regard to microbial processes. Regarding soil as one entity and neglecting its structure and spatial distribution of nutrients, water, oxygen, roots, macro- and microfauna can lead to serious over- or underestimates of nutrient analyses and more importantly soil-forming processes. For instance Hildebrand (1990) **showed that soil structure is important for the spatial distribution of nutrients.** Macropores and medium-sized pores in contact with the bulk soil solution are where most active nutrient exchange with plant roots and micro-organisms take place. Similarly the standard procedure for nutrient analyses to homogenize the soil does not reflect the spatial area in which micro-organisms live. Soil structure is influenced by and influences fungal hyphae: Tisdall et al. (1997) showed that **saptrotrophic, arbuscular, ericoid and ectomycorrhizal fungi are capable of forming stable soil aggregates.** Arbuscular mycorrhizal fungi have been shown to react to their physical environment by excreting increased amounts of glomalin, a substance that enhances soil aggregation, in a fine-grained medium (Rillig and Steinberg 2002). **Stable soil aggregates increase the formation of macropores in the soil.** Macropores influence soil aeration and water infiltration, and Schack-Kirchner et al. (2000) showed that fungal **hyphae (ectomycorrhizal and saprotrophic) preferentially grow in or near macropores because of their high oxygen demand.** As shown by Read and Armstrong (1972) ectomycorrhiza cannot form when not in contact with the atmosphere or without association to an oxygen-providing root. Indeed Agnelli et al. (2000) assumed that highly weathered rock fragments are ecological analogies to soil aggregates, but are more stable. Koele et al. (2010) suggested that rock fragments act as a refuge for micro-organisms in local or temporal extreme environmental conditions, such as droughts or acidification.

B. Minerals Attracting Fungal Hyphae Attracting Bacteria?

1. Mycorrhizal Hyphae Biosensing Nutrient Hotspots

Considering that the **stonesphere** contains large amounts of weatherable mineral nutrients it is not surprising that this zone has been called a **nutrient "hotspot" for micro-organisms** (Kohler and Hildebrand 2004). Findings by Certini et al. (2004) that a different microbial community inhabits the stonesphere compared to the surrounding bulk soil support the idea that specialized micro-organisms actively harvest those nutrients. Indeed a positive relationship between the growth of ectomycorrhizal hyphae and nutrient-rich minerals has been established, both under laboratory conditions (Rosling et al. 2004; Leake et al. 2008) and in the field (Hagerberg et al. 2003). According to Leake et al. (2008) **some ectomycorrhizal fungi act as biosensors** and send out fungal hyphae where nutrient-rich substrate is encountered. Finlay and Read (1986a, b) and Rosling et al. (2004) furthermore demonstrated that more photosynthates are allocated to the ectomycorrhizal mycelium that encounters nutrient-rich patches, and in addition the hyphae at the nutrient-rich substrate proliferate locally to form patches of mycelium on the nutrient-rich substrate.

A possible explanation beyond nutrient elements within the mineral structures may be a "mantle" of fine particles surrounding rock fragments in soil: Kern et al. (2006) showed that the nutrient cations within this mantle material are significantly higher than in bulk soil fine earth. Similarly Fernandez-Sanjurjo et al. (2011) found higher organic C, total N and available P contents, as well as higher exchangeable cations in fine particles surrounding a volcanic skeleton than in the other soil fractions. Although the mass percentage in soil of this fine earth mantle fraction is small, so that the overall contribution of this fraction to plant nutrition seems insignificant at first sight, it may however act as an attractor for microbial communities.

2. Selection of Bacteria by Ectomycorrhizal Fungi

Both ectomycorrhizal fungi and certain bacteria occupy specific nutrient hotspots in the soil, such as the stonesphere. Mycorrhizal biosensing of such hotspots provides a good explanation of the mycorrhizal presence, because mycorrhizal fungi obtain their energy needs from the host plant and are thus not dependent on energy sources in the soil. Ectomycorrhizal fungi occupy most of the root system, so that they become in fact the sole surface in the mycorrhizosphere for the uptake and release of elements. Therefore **the microbial activity in the mycorrhizosphere is likely to be predominantly controlled by ectomycorrhizal fungi.** Indeed Frey-Klett et al. (2004) and Calvaruso et al. (2007) demonstrated that ectomycorrhizal fungi select bacteria in the mycorrhizosphere in nutrient-poor forest soils, especially those bacteria capable of mineral weathering. According to Frey et al. (1997) **the specific carbohydrates exuded by ectomycorrhizal hyphae can both attract specific bacteria and inhibit others (e.g. pathogens).**

However there might be a difference in the ectomycorrhizal selection of bacteria based on the availability of organic and inorganic nutrients (Brooks et al. 2011). Furthermore Koele et al. (2009) suggested that some ectomycorrhizal fungi may select bacteria for weathering while the fungus itself mainly increases the uptake ability of the host plant, benefitting from increased release of nutrients by bacterial weathering.

3. Biofilms

Balogh-Brunstad et al. (2008) studied the interactions of ectomycorrhizal fungi and bacteria on mineral weathering and acknowledged the importance of biofilms. They **define biofilms as "micro-organisms concentrated on the root—hypha—mineral interface and surrounded by extracellular polymers they produce, utilizing plant and fungal exudates in soils"**. Biofilms provide a **local environment for the exchange of elements protected from bulk soil leaching** (Banfield et al. 1999; Gadd 2010). Furthermore biofilms probably help prevent desiccation and thus provide a longer residence time of water and associated diffusion potential, thus enhancing the microbial weathering (Barker et al. 1998). Because the biofilm can be much larger than the area micro-organisms occupy, the potential area where weathering can occur is also increased (Barker et al. 1998).

C. Different Microbial Communities and Hyphal Length in the Stonesphere

Certini et al. (2004) found a **different microbial community in the stonesphere than in the surrounding bulk soil** and attributed this to a favourable physical environment in the stonesphere. Rock fragments have also been shown to contain humic substances chemically different from the fine earth, indicating an active and different microbial community in rock fragments (Agnelli et al. 2000). Additionally Koele et al. (2011) suggested that a combination of physical and chemical characteristics of the stonesphere might influence the distribution of fungal hypha. Weathering cracks filled with in situ weathered material and infillings from the surrounding bulk soil will become less easily depleted upon soil acidification due to the protective nature of the rock fragment (Kohler et al. 2000). These weathering fissures are usually medium- to micro-sized pores that are not in diffusive contact with the bulk soil solution, preventing rapid depletion of nutrient elements. Simultaneously especially coarser rock fragments and soil aggregates (>2 mm diameter) will ensure the formation of larger pores in the matrix, providing oxygen for microbial respiration. Koele et al. (2011) measured the hyphal length of rock fragments and fine earth in an acidic soil profile and noticed that fungal hyphae are distributed differently in fine earth and rock fragments, possibly linked to nutrient sources.

D. Soil Depth Influence on Community Structure in the Mycorrhizosphere

Mycorrhizal roots likely occur throughout the rooted zone of the soil. However, with depth the soil characteristics change, and thus the amount, form and availability of nutrients, water and oxygen changes. It is therefore not surprising that the microbial communities also change. Even within the organic forest floor layers Dickie et al. (2002) found a **vertical differentiation of ectomycorrhizal fungi**. They point out a possible specialization of mycorrhizal fungi in different niches in the soil, similar to the spatial niche differentiation of bacteria found by Uroz et al. (2007). Throughout a complete soil profile including mineral soil, Rosling et al. (2003) reported **ectomycorrhizal root tips to a depth of 54 cm, with half the taxa occurring in mineral soil**. This highlights the specialization of specific taxa for organic (forest floor) and inorganic (mineral soil) nutrient acquisition. For bacterial communities Calvaruso et al. (2007) showed varying functional diversity with soil depth, with high-weathering

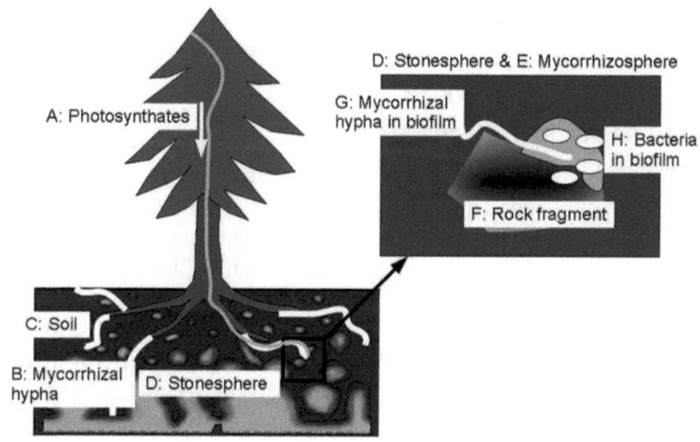

Fig. 9.2. Conceptual visualization of processes in the stonesphere and the mycorrhizosphere during mineral weathering. (**A**) Photosynthates are assimilated in foliage and transported via the roots to mycorrhizal hyphae in the soil. (**B**) Mycorrhizal fungi are associated with plant roots to increase nutrient uptake surface area and the exploration through the soil. (**C**) In the bulk soil mycorrhizal hyphae explore for resources needed by the host plant such as water and nutrients. Mycorrhizal hyphae use biosensing to find specific nutrients. (**D**) The stonesphere is potentially rich in nutrient bearing minerals, and a buffer during acidification or drought. (**E**) The mycorrhizosphere is directly influenced by the mycorrhizal hypha and can be altered (for instance by formation of stable aggregates) to facilitate processes such as nutrient uptake. (**F–H**) A single rock fragment (**F**) can become part of the mycorrhizosphere (**E**) especially when mycorrhizal hyphae and bacteria form a biofilm (**G, H**) in which controlled conditions allow localized nutrient uptake from the rock surface or weathering cracks

bacterial strains found predominantly in the mycorrhizosphere of deeper mineral soil layers.

IV. Ecological Significance of Mineral Weathering in the Mycorrhizosphere

Figure 9.2 gives a conceptual overview of processes through the biosphere into the mycorrhizosphere that are important for micro-organisms interacting in the stonesphere. Considering the nutritional value of the coarse soil fraction, the ability of micro-organisms to access this nutrient source through biological weathering, and the close interactions between micro-organisms and the substrate, **the stonesphere can be regarded as an ecological "buffer"**. Indeed Deutschmann (1998) calculated that, when rock fragment exchangeable cations are incorporated into the available nutrient stock, 200 instead of 8 years (fine earth only) are needed to deplete the exchangeable cation bases if the site-specific acidification rates continue. The **stonesphere attracts mycorrhizal hyphae and mycorrhizosphere bacteria** because of its favourable nutrient status and buffer capacity compared to the surrounding bulk soil, especially in acidified soils. Plant roots or mycorrhizal hyphae select bacteria to perform functions such as mineral weathering and mycorrhizal infection (Frey-Klett et al. 2007) as needed. The stonesphere sustains different microbial communities than the bulk soil and depending on environmental conditions (e.g. drought, acidification, pathogen attack) micro-organisms with specific capabilities are favoured to cope with those conditions.

Although all the above-mentioned studies present evidence for a role of bacteria and fungi, in addition to plant roots, for mineral weathering in the stonesphere, its ecological significance has not yet been quantified. Upscaling to ecosystem scale remains a problem with laboratory experiments, although field observations (e.g. through mineral amended mesh bags; Wallander et al. 2006) support the idea of mycorrhizosphere weathering of minerals. Furthermore researchers continue to ignore the deeper mineral soil horizons and the fraction coarser than 2 mm in both soil nutrient and microbial analyses. It is true that analysing all size fractions separately is time-consuming, but one should consider carefully the goal of the research and selectively sample the area of interest, e.g. the rhizosphere (including rock fragments), the mycorrhizosphere and, as control, the bulk soil.

It has not yet been clarified sufficiently whether fungal and bacterial weathering in the stonesphere are a commonplace behaviour, or whether it is restricted to spatial and temporal resource (nutrient) deficiencies. An interesting observation by Arocena et al. (2012) is that mycorrhizal fungi (*Glomus*) selectively extract K from biotite for immediate use and at the same time leave unaltered biotite for future extractions. If this can be extrapolated to more mycorrhizal species and minerals, interactions in the mycorrhizosphere during mineral weathering can be considered a dynamic responsive system to adapt to changing conditions.

Acknowledgement The author was partially supported by funding from the Royal Society of New Zealand Marsden Fund.

References

Agnelli A, Celi L, Degl'Innocenti A, Corti G, Ugolini FC (2000) Chemical and spectroscopic characterization of the humic substances from sandstone-derived rock fragments. Soil Sci 165:314–327

Arocena JM, Velde B, Robertson SJ (2012) Weathering of biotite in the presence of arbuscular mycorrhizae in selected agricultural crops. Appl Clay Sci. doi:10.1016/j.clay.2011.06.013

Balogh-Brunstad Z, Keller CK, Gill RA, Bormann BT, Li CY (2008) The effect of bacteria and fungi on chemical weathering and chemical denudation fluxes in pine growth experiments. Biogeochemistry 88:153–167

Banfield JF, Barker WW, Welch SA, Taunton A (1999) Biological impact on mineral dissolution: application of the lichen model to understanding mineral weathering in the rhizosphere. Proc Natl Acad Sci USA 96:3404–3411

Barker W, Welch S, Chu S, Banfield J (1998) Experimental observations of the effects of bacteria on aluminosilicate weathering. Am Mineral 83:1551

Bormann BT, Wang D, Snyder MC, Bormann FH, Benoit G, April R (1998) Rapid, plant-induced weathering in an aggrading experimental ecosystem. Biogeochemistry 43:129–155

Bornemann L, Herbst M, Welp G, Vereecken H, Amelung W (2011) Rock fragments control size and saturation of organic carbon pools in agricultural topsoil. Soil Sci Soc Am J 75:1898–1907

Boyle J, Voigt G (1973) Biological weathering of silicate minerals. Plant Soil 38:191–201

Brakensiek D, Rawls W (1994) Soil containing rock fragments: effects on infiltration. Catena 23:99–110

Brooks DD, Chan R, Starks ER, Grayston SJ, Jones MD (2011) Ectomycorrhizal hyphae structure components of the soil bacterial community for decreased phosphatase production. FEMS Microbiol Ecol 76:245–255

Calvaruso C, Turpault M, Leclerc E, Frey-Klett P (2007) Impact of ectomycorrhizosphere on the functional diversity of soil bacterial and fungal communities from a forest stand in relation to nutrient mobilization processes. Microbial Ecol 54:567–577

Certini G, Campbell C, Edwards A (2004) Rock fragments in soil support a different microbial community from the fine earth. Soil Biol Biochem 36:1119–1128

Childs SW, Flint AL (1990) Physical properties of forest soils containing rock fragments. Sustained productivity of forest soils. In: Gessel SP, Lacate DS, Weetman GF, Powers RF (eds) Sustained productivity of forest soils. University of British Columbia, Vancouver, pp 95–121

Corti G, Agnelli A, Ugolini F (1997) Release of Al by hydroxy interlayered vermiculite and hydroxy interlayered smectite during determination of cation exchange capacity in fine earth and rock fragments fractions. Eur J Soil Sci 48:249–262

Corti G, Ugolini F, Agnelli A, Certini G, Cuniglio R, Berna F, Fernandez-Sanjurjo MJ (2002) The soil skeleton, a forgotten pool of carbon and nitrogen in soil. Eur J Soil Sci 53:283–298

de Turk E (1919) Potassium-bearing minerals as a source of potassium for plant growth. Soil Sci 8:269–301

Deutschmann G (1998) New aspects of buffering processes in stony soils. Chemosphere 36:1149–1154

Dickie I, Xu B, Koide R (2002) Vertical niche differentiation of ectomycorrhizal hyphae in soil as shown by T-RFLP analysis. New Phytol 156:527–535

Epstein EG, Struchtemeyer W (1966) Effects of stones on runoff, erosion, and soil moisture 1. Soil Sci Soc Am J 30:638

Fernandez-Sanjurjo MJ, Corti G, Agnelli A (2011) Genesis and role of the skeleton water-extractable fines in volcanic soils. Soil Sci Soc Am J 75:1019–1031

Finlay RD, Read DJ (1986a) The structure and function of the vegetative mycelium of ectomycorrhizal plants. I. Translocation of 14C-labelled carbon between plants interconnected by a common mycelium. New Phytol 103:143–156

Finlay RD, Read DJ (1986b) The structure and function of the vegetative mycelium of ectomycorrhizal plants. II. The uptake and distribution of phosphorus by mycelial strands interconnecting host plants. New Phytol 103:157–165

Frey P, Frey-Klett P, Garbaye J, Berge O, Heulin T (1997) Metabolic and genotypic fingerprinting of fluorescent pseudomonads associated with the Douglas Fir-*Laccaria bicolor* mycorrhizosphere. Appl Environ Microbiol 63:1852–1860

Frey-Klett P, Chavatte M, Clausse M-L, Courrier S, le Roux C, Raaijmakers J, Martinotti MG, Pierrat J-C, Garbaye J (2004) Ectomycorrhizal symbiosis affects functional diversity of rhizosphere fluorescent pseudomonads. New Phytol 165:317–328

Frey-Klett P, Garbaye J, Tarkka M (2007) The mycorrhiza helper bacteria revisited. New Phytol 176:22–36

Gadd GM (2010) Metals, minerals and microbes: geomicrobiology and bioremediation. Microbiology 156:609–643

Gleeson D, Kennedy N, Clipson N, Melville K, Gadd G, McDermott F (2006) Characterization of bacterial community structure on a weathered pegmatitic granite. Microbial Ecol 51:526–534

Glowa K, Arocena J, Massicotte H (2003) Extraction of potassium and/or magnesium from selected soil minerals by Piloderma. Geomicrobiol J 20: 99–111

Hagerberg D, Thelin G, Wallander H (2003) The production of ectomycorrhizal mycelium in forests: Relation between forest nutrient status and local mineral sources. Plant Soil 252:279–290

Heisner U, Raber B, Hildebrand EE (2004) The importance of the soil skeleton for plant-available nutrients in sites of the Southern Black Forest, Germany. Eur J For Res 123:249–257

Hildebrand E (1990) The spatial heterogeneity of chemical properties in acid forest soils and its importance for tree nutrition. Water Air Soil Pollut 54:183–191

Jackson LP, Hall IV, Aalders LE (2011) Lowbush blueberry seedling growth as affected by soil type. Can J Soil Sci 52:113–115

Kelly EF, Chadwick OA, Hilinski TE (1998) The effect of plants on mineral weathering. Biogeochemistry 42:21–53

Kern M, Raber B, Hildebrand EE (2006) Verfahren zur Ermittlung des Nährelementpotenzials des Gesamtbodens unter besonderer Berücksichtigung des Bodenskeletts und deren Integration in die Bodenzustandserhebung im Wald (BZE II). Abschlussbericht Institut für Bodenkunde und Waldernährungslehre Universität Freiburg, Freiburg, p 97

Kimmins J (1987) Forest ecology. Macmillan, New York

Koele N, Turpault MP, Hildebrand EE, Uroz S, Frey-Klett P (2009) Interactions between mycorrhizal fungi and mycorrhizosphere bacteria during mineral weathering: budget analysis and bacterial quantification. Soil Biol Biochem 41:1935–1942

Koele N, Hildebrand EE, Schack-Kirchner H (2010) Effects of weathering state of coarse-soil fragments on tree-seedling nutrient uptake. J Plant Nut Soil Sci 173:245–251

Koele N, Storch F, Hildebrand EE (2011) The coarse-soil fraction is the main living space of fungal hyphae in the BhBs horizon of a podzol. J Plant Nutr Soil Sci 174:750–753. doi:10.1002/jpln.201000296

Kohler M, Hildebrand E (2004) New aspects of element cycling and forest nutrition. Towards the sustainable use of Europe's forests—forest ecosystem and landscape research. Sci Chall Opport 49:171–180

Kohler M, Wilpert KV, Hildebrand EE (2000) The soil skeleton as a source for the short-term supply of "base cations" in forest soils of the Black Forest (Germany). Water Air Soil Poll 122:37–48

Landeweert R, Hoffland E, Finlay RD, Kuyper TW, van Breemen N (2001) Linking plants to rocks: ectomycorrhizal fungi mobilize nutrients from minerals. Trends Ecol Evol 16:248–254

Lavee H, Poesen JWA (1991) Overland flow generation and continuity on stone-covered soil surfaces. Hydrol Process 5:345–360

Leake J, Duran A, Hardy K, Johnson I, Beerling DJ, Banwart SA, Smits MM (2008) Biological weathering in soil: the role of symbiotic root-associated fungi biosensing minerals and directing photosynthate-energy into grain-scale mineral weathering. Mineral Mag 72:85–89

Leyval C, Berthelin J (1989) Interactions between laccaria laccata, agrobacterium radiobacter and beech roots: influence on P, K, Mg, and Fe mobilization from minerals and plant growth. Plant Soil 117:103–110

Martín-García J, Delgado G, Parraga J, Gamiz E, Delgado R (1999) Chemical, mineralogical and micro morphological study of coarse fragments in Mediterranean red soils. Geoderma 90:23–47

Paris F, Botton B, Lapeyrie F (1996) In vitro weathering of phlogopite by ectomycorrhizal fungi. Plant Soil 179:141–150

Poesen J, Ingelmo-Sanchez F (1992) Runoff and sediment yield from topsoils with different porosity as affected by rock fragment cover and position. Catena 19:451–474

Poesen J, Lavee H (1994) Rock fragments in top soils: significance and processes. Catena 23:1–28

Prescott CE (2002) The influence of the forest canopy on nutrient cycling. Tree Physiol 22:1193–1200

Read D, Armstrong W (1972) A relationship between oxygen transport and the formation of the ectotrophic mycorrhizal sheath in conifer seedlings. New Phytol 71:49–53

Rillig MC, Steinberg PD (2002) Glomalin production by an arbuscular mycorrhizal fungus: a mechanism of habitat modification? Soil Biol Biochem 34:1371–1374

Rivard R, De Kimpe C (1980) Propriétés de quelques sols riches en graviers dans la région de Québec. Can J Soil Sci 60:263–273

Rosling A, Landeweert R, Lindahl B, Larsson KH, Kuyper TW, Taylor AFS, Finlay RD (2003) Vertical distribution of ectomycorrhizal fungal taxa in a podzol soil profile. New Phytol 159:775–783

Rosling A, Lindahl BD, Finlay RD (2004) Carbon allocation to ectomycorrhizal roots and mycelium colonising different mineral substrates. New Phytol 162:795–802

Schack-Kirchner H, von Wilpert K, Hildebrand EE (2000) The spatial distribution of soil hyphae in structured spruce-forest soils. Plant Soil 224: 195–205

Schüler G, Butz-Braun R (2001) Nährstoffversorgung aus dem Bodenskelett eines Buntsandstein-Standortes im Pfälzerwald in Abhängigkeit von Tiefenstufen und Gesteinsgefüge. Mitteilungen der Deutschen Bodenkundlichen Gesellschaft, pp 693–694

Sverdrup H (1990) The kinetics of base cation release due to primary silicate weathering. Lund University Press, Lund, p 246

Tisdall J, Smith S, Rengasamy P (1997) Aggregation of soil by fungal hyphae. Austral J Soil Res 35:55–60

Ugolini FC, Corti G, Dufey JE, Agnelli A, Certini G (2001) Exchangeable Ca, Mg, and K of rock fragments and fine earth from sandstone and siltstone derived soils and their availability to grass. J Plant Nutr Soil Sci 164:309–315

Uroz S, Calvaruso C, Turpault M, Pierrat JC, Mustin C, Frey-Klett P (2007) Effect of the mycorrhizosphere on the genotypic and metabolic diversity of the bacterial communities involved in mineral weathering in a forest soil. Appl Environ Microbiol 73:3019–3027

Uroz S, Calvaruso C, Turpault M, Frey-Klett P (2009) Mineral weatheringby bacteria: ecology, actors and mechanisms. Trends Microbiol 17:378–387

van Wesemael B, Poesen J, Kosmas CS, Danalatos NG, Nachtergale J (1996) Evaporation from cultivated soils containing rock fragments. J Hydrol 182:65–82

Wallander H, Wickman T, Jacks G (1997) Apatite as a P source in mycorrhizal and non-mycorrhizal *Pinus sylvestris* seedlings. Plant Soil 196:123–131

Wallander H, Hagerberg D, Åberg G (2006) Uptake of 87Sr from microcline and biotite by ectomycorrhizal fungi in a Norway spruce forest. Soil Biol Biochem 38:2487–2490

10 Context-Dependent Interaction Hierarchies and the Organization of Ectomycorrhizal Fungal Communities

R.T. Koide[1]

CONTENTS

I. Introduction

This chapter concerns interactions among species within communities of ectomycorrhizal fungi, but if the principles discussed herein are relevant to those communities, they should be relevant to all biological communities. At the outset of this contribution I feel I must define what is meant by "interactions", explain my view of ectomycorrhizal fungal communities, and describe why I do not include a discussion of intraspecific interactions. It is not often that this degree of specification becomes necessary in a chapter of this sort, but it is important here mainly because "interactions" and "communities" are terms that are frequently used in a variety of ways and, because individuals of mycelial organisms are difficult to define, the concept of intraspecific interaction is fraught with difficulty.

In one sense, to say that species of ectomycorrhizal fungi in a single community interact is trivial. By some definitions of community, interaction of some sort is a prerequisite to membership. Nevertheless, settling on what actually comprises "interaction" is not trivial. The best studied and the most frequently written about interaction is **competition**, and an excellent, recent review of this subject is available (Kennedy 2010). Other important contributions that have discussed competition at some length include Bruns (1995), Jumpponen and Egerton-Warburton (2005), and Koide et al. (2011). But competition is only one, frequently negative, kind of interspecific interaction. Other types of negative interaction include **allelopathy** and **parasitism**. Interspecific interactions may also be positive, and some cases of this are mentioned below. Because little is known about species interactions other than those assumed to be competitive, this chapter is, of necessity, primarily focused on competition and interactions that are phenomenologically indistinguishable from it.

The term "community" is frequently interpreted differently by different people. In the most inclusive case, a community comprises all organisms that live together in the same place at the same time. Each member of such a community interacts with the others in some fashion by virtue of its contemporaneity and physical proximity. Plants, animals, fungi, and bacteria coexisting in a forest are, in this sense, members of the same Community (with a capital C). This most inclusive concept of community is particularly useful in studies of carbon and nutrient cycling through food webs.

[1]The Pennsylvania State University, University Park, PA 16802, USA; e-mail: rxk13@psu.edu

Fungal Associations, 2nd Edition
The Mycota IX
B. Hock (Ed.)

Subsets of the Community can also be considered communities in their own right. For example, the phylum Basidiomycota comprises a subset of organisms in forest communities. Clearly, basidiomycotan fungi interact strongly with members of many other phyla including ascomycotan fungi possessing complementary suites of enzymes (Wu et al. 2005), soil arthropods from which they may obtain nitrogen (Klironomos and Hart 2001), bacterial communities that may be nourished by their exudates (Olsson and Wallander 1998), and plants that serve as their sources of reduced carbon. Despite frequent and ecologically important interactions with species in such disparate phyla, the concept of a basidiomycotan community is still meaningful; the organization of the basidiomycotan community reveals patterns of adaptive radiation and niche differentiation. For example, closely related basiodomycotan fungi are alternately mycorrhizal (biotrophic) or saprotrophic (Hibbett et al. 2000), and specialization occurs with respect to forest successional stage and season (Dighton et al. 1986).

Another conceptualization is one in which the community comprises trophically similar species. Hubbell's "unified neutral theory" (Hubbell 2001), for example, restricts itself to within-trophic-level diversity. According to this view, saprotrophic and biotrophic members of the Basidiomycota are treated as members of separate communities, which is the approach taken by most ectomycorrhizal fungal ecologists and the approach taken herein. This within-trophic-level approach to communities is rather traditional; many of community ecology's roots lie in the myriad studies of plant communities comprising trophically similar (photosynthetic) higher plants. In communities of a single trophic level, resource acquisition among members is expected to be very similar overall and identical in terms of carbon capture. Thus, competition and factors that modify competition are likely to be very important in structuring these communities (Fleming 1985). If we consider a community to consist of interacting or potentially interacting species then, practically speaking, ectomycorrhizal fungal communities can be defined by a single stand of trees, a single tree, or a portion of a single root system.

The reason why intraspecific interactions are not considered here stems from the fact that the concept of interaction assumes that individual organisms are the players in the interaction. In the case of interspecific interactions it is obvious that at least two distinct entities are interacting. In the case of intraspecific interactions one must also be able to clearly identify separate individuals, but this is difficult with fungi because they are mycelial. It is not clear with mycelial organisms that the concept of individuals is frequently relevant. A single genet of some fungal species may break apart into physically distinct ramets (Detman and van de Kamp 2001), and smaller ramets may anastamose to form larger ramets (Buchalo et al. 1989). Because the individual is ill-defined for most fungi, the concept of intraspecific interactions is difficult conceptually, let alone experimentally, and I do not consider it further here.

II. Interactions Between Ectomycorrhizal Fungi and Other Kinds of Organisms

Ectomycorrhizal fungi obviously interact with more than each other. Interactions are unavoidable with the many other organisms that coexist with them either in or on roots, or in the soil. While the focus of this chapter is interactions among species of ectomycorrhizal fungi, I would be remiss if I did not at least call attention to the fact that, within a Community, a "community" of ectomycorrhizal fungi does not live apart from "communities" of other kinds of organisms. Obviously ectomycorrhizal fungi must interact with their **host plants** to obtain photosynthate, and varying degrees of host specificity (Molina et al. 1992; Ishida et al. 2007) suggest that such interactions can be very highly controlled. Moreover, because some host plants form both arbuscular mycorrhizas and ectomycorrhizas, ectomycorrhizal fungi have the potential to compete with arbuscular mycorrhizal fungi in the colonization of roots

(Lapeyrie and Chilvers 1985; Chen et al. 2000). Ectomycorrhizal fungi also interact with **prokaryotes** (Olsson and Wallander 1998; Schrey et al. 2007), the best studied of which are the so-called helper bacteria (Frey-Klett et al. 2007), with **saprotrophic fungi** (Leake et al. 2001; Wu et al. 2005; Zadworny et al. 2007), with **pathogenic fungi** (Chakravarty et al. 1991; Duchesne et al. 1988; Malajczuk 1988), and with **soil arthropods** (Klironomos and Hart 2001). Because species of ectomycorrhizal fungi are functionally distinct, interactions between them and other soil-dwelling organisms are also likely to influence interactions among ectomycorrhizal fungi.

III. Interactions Among Species of Ectomycorrhizal Fungi

A. Non-Competitive Interactions

The vast majority of interactions documented among ectomycorrhizal fungal species are competitive or phenomenologically indistinguishable from competitive interactions. However, in a small number of cases interactions that are not competitive have been documented. For example, positive correlations in space and time occur among species of ectomycorrhizal fungi. There are a number of possible reasons for these. Fleming (1985), Mamoun and Olivier (1993a, b), and Shaw et al. (2003), for example, observed that the presence of one ectomycorrhizal fungal species on a root may facilitate colonization by another. Koide et al. (2005) also found a number of positive correlations of occurrence of hyphae of certain pairs of species of ectomycorrhizal fungi in space and time within the forest floor of a *Pinus resinosa* plantation. The causes of spatial correlations of hyphae of different species are not known, but they could derive from functional complementarity resulting in a more efficient capture of soil-borne resources, which has been noted among saprotrophic fungi (Hättenschwiler et al. 2005). It must be noted, however, that positive correlations in the occupancy of space are not necessarily indicative of positive interactions among species. The co-occurrence of members of the Gomphidiaceae and suilloid fungi (Agerer 1990, 1991) may be due to a parasite–host relationship (Olsson et al. 2000).

B. Competition

Competition and a number of other types of negative interaction, including allelopathy, while quite different mechanistically, often have similar consequences and are thus difficult to tell apart phenomenologically. Kennedy (2010), therefore, took what he called an "outcome-based" definition of competition, which is to say that he lumped competition with other interactions that looked like it. I agree that, practically speaking, such is the approach we must take for now simply because, as is often the case with plant "competition", it is very difficult to distinguish among interactions that produce similar results. Nevertheless, while competition and allelopathy, for example, may have the same appearance, the mechanisms that give rise to the phenomena can be very different. For completeness, I describe here the conceptual distinction between competition and allelopathy.

In competition, an organism takes a resource from a common pool and, as a consequence, reduces the amount of that resource available to others. We might refer to this as direct competition when the organisms are present at the same time, or pre-emptive competition if they are not. In either case, if the resource is limiting, the consequence of competition may be some degree of reduced vigor (growth or reproduction or some other measure of success) of one or more of the organisms involved. The competition need not be symmetric, in which case the degree of reduced vigor need not be the same for all competitors and, in the most extreme case of asymmetry, one of the "competitors" may not suffer from any reduction of vigor at all. However, competition may be inconsequential if the resource in question is not limiting, at least for the time being. As time progresses, however, a formerly non-limiting resource can become limiting, at which time the previously inconsequential competition becomes consequential.

Allelopathy, by contrast, is the case in which one organism produces chemicals that result in reduced vigor of other organisms. In the strictest definition of allelopathy, reduced vigor is the result ***whether or not resources are limiting***. In truth, allelochemicals produced by one organism could theoretically reduce the capacity of another organism to acquire resources. We might rightfully refer to this as "allelopathy-induced competition", but that is splitting hairs. In the end, it might be best to refer to any interaction involving the production of chemicals that reduce the vigor of nearby organisms as allelopathy and any interaction involving removal of resources from a common pool as competition, conceding that both may take place at the same time.

Competition and allelopathy can be important forces in nature. For example, competition can influence the structure of biological communities (Bengtsson et al. 1994), leading to reduced evenness (Lamb et al. 2009) or, ultimately, to "competitive" exclusion (Grace and Wetzel 1981) in the absence of forces that prevent exclusion. Over the course of succession, for example, rapidly growing, early successional ectomycorrhizal fungal species may eventually be outcompeted by later successional species (Visser 1995). However, some studies of succession of ectomycorrhizal fungi demonstrate that communities change over time mainly as a consequence of species accretion rather than competitive exclusion (Visser 1995; Nara et al. 2003a, b; Shaw et al. 2003). Therefore the coexistence of species requires further discussion.

Coexistence may occur simply because species interactions are not stable in space or time, so that the ultimate result of negative species interactions is never realized. Such instability could be the result of various kinds of "disturbance", including the death of individuals (Zhou and Hogetsu 2002), fire (Stendell et al. 1999; Grogan et al. 2000), physical soil disruption (Lilleskov and Bruns 2003), or a change in resource availability such as with the production of new roots (Koide et al. 2011). Even in the absence of disturbance, habitat partitioning can ultimately result in stable coexistence. For example, in the red pine (*Pinus resinosa*) plantation where we have worked, both spatial and temporal partitioning occur (Dickie et al. 2002; Koide et al. 2005, 2007). The hyphae of *Cenococcum geophilum* and *Clavulina cinerea* were primarily found in a different layer of the forest floor from that occupied by *Lactarius* sp. and *Tylopilus felleus*, which was different again from the layer occupied by *Russula* sp. Moreover, while *Ce. geophilum* and *Cl. cinerea* were found mainly in the same forest floor layer, they generally occupied different small volumes of that layer (Koide et al. 2005). We also found that *Lactarius* sp. and *T. felleus* exhibited similar temporal patterns of occurrence, but those were different from another temporal pattern exhibited by *Ce. geophilum*, *Cl. cinerea*, and *Russula* sp. (Koide et al. 2007).

Despite the existence of many mechanisms that promote coexistence, **competition certainly must occur**. Indeed, the expression of habitat partitioning of the kind discussed above is presumably the result of community sorting resulting from competition. This will be discussed below in greater detail. It is sufficient here to state that there is plenty of evidence for competition or, at least, interactions that produce the same phenomena. For example, the existence of multiple species of ectomycorrhizal fungi on a single root tip (Villeneuve et al. 1991; Mamoun and Olivier 1993a, b; Wu et al. 1999; Koide et al. 2005) or in a small volume of soil (Koide et al. 2005) suggests that direct competition for a root or resources within a small volume of soil is possible. Negative correlations in space strongly suggest that competition (or allelopathy) has operated, and the actual displacement of one species of fungus from a root tip or volume of soil by another serves as direct evidence for competition or allelopathy, and these have been noted multiple times (Benecke and Göbl 1974; Bledsoe et al. 1982; Villeneuve et al. 1991; Wu et al. 1999; Koide et al. 2005; Iotti et al. 2010). Moreover, competition for roots is expressed when a previously colonizing species prevents colonization by a latecomer (Villeneuve et al. 1991; Kennedy and Bruns 2005; Kennedy et al.

2009), or when a latecomer overruns the previous colonizer (Wu et al. 1999).

Unless saprotrophy is significant, the ability to colonize a root, extract photosynthate from it and prevent others from taking over must be important in establishing and maintaining competitive hierarchies among ectomycorrhizal fungal species. With respect to N, P or water, the ability that hyphae have to colonize soil and maintain possession of it must be similarly important in the determination of competitive hierarchies for those resources. To illustrate some of these concepts, consider the competition between two species of ectomycorrhizal fungi, *Pisolithus tinctorius* and an unnamed isolate Tanashi 01, for uncolonized roots of *Pinus densiflora* (Wu et al. 1999). Both species need to colonize roots in order to acquire photosynthate. However, while both species appear to be early successional (K. Nara, personal communication), they exhibit very different characteristics, which undoubtedly have large consequences in competitive interactions. In one experiment, *Pis. tinctorius* exhibited more rapid hyphal extension than Tanashi 01 but, when the hyphae came into contact, Tanashi 01 was able to suppress and eliminate *Pis. tinctorius* hyphae from the soil. Moreover, when the hyphae of Tanashi 01 reached the roots previously colonized by *Pis. tinctorius*, Tanashi 01 was able to successfully colonize those roots. Thus, *Pis. tinctorius* was not able to maintain its colonization of roots or soil in the face of competition with Tanashi 01 under the specific conditions of that experiment. The loss of control of soil volume or a root tip by one fungus to another can occur because of the death of the former, because the second can extract resources to a concentration that is too low to be acquired by the first, or because the second produces allelochemicals, which are known from some ectomycorrhizal fungi such as *Cenococcum* sp. (Krywolap 1964).

Wu et al. (1999) were careful to note that, had the environmental conditions been different, there might have been a different outcome. For example, had the soil been repeatedly physically disturbed, it is possible that the more rapid growth by *Pis. tinctorius* would have made it competitively superior to Tanashi 01. This suggests that the hierarchies established by species interactions may be "context dependent".

IV. Context-Dependent Interactions— The Interplay of Species Traits and Environmental Heterogeneity

One can envision multiple phenomena that can be explained by context-dependent interactions. **Habitat partitioning** is the case in which species exploit distinct portions of a heterogeneous habitat. As mentioned above, habitat partitioning by members of an ectomycorrhizal fungal community (e.g., see Dickie et al. 2002) can be the result of the sorting of species as a consequence of multiple, context-dependent interaction hierarchies. Let us imagine a scenario in which the upper layers of the forest floor are relatively dry (experience low water potential) and where deeper layers are well-supplied with water. Thus, organisms confined to the upper layers, but not the lower layers, will be frequently limited by water shortage. Now let us assume that ectomycorrhizal fungal species A is superior to species B in acquiring water at low water potentials (see Coleman et al. 1989). As a consequence of this physiological difference, species A becomes the dominant species in the upper layers of the forest floor as it outcompetes species B, which remains rare. However, if species B colonizes far more roots than species A when moisture is not limiting, then in the deeper layers of the forest floor where moisture is more reliably available, species B will become the dominant species as it outcompetes the rarer species A for roots. Thus, habitat partitioning can result from environmental heterogeneity of the habitat (in space or time) and competitive hierarchies (or, in the more general case, interaction hierarchies) that depend on environmental conditions. Thus, context-dependent interaction hierarchies are caused by idiosyncratic preferences for or tolerances to various environmental conditions. While it might appear that habitat partitioning reduces competition, habitat portioning might instead be the result of competition.

Host preference (DeBellis et al. 2006; Durall et al. 2006; Ishida et al. 2007; Iotti et al. 2010) can also be explained by context-dependent interactions among species. For example, let us assume that fungal species A and B each colonize a single root of host 1. Furthermore, let us assume that from those single colonized roots, species A is able to derive twice as much photosynthate as species B. The possibility for this is suggested by the highly variable degree of ^{13}C labeling of ectomycorrhizal root tips of *Pinus sylvestris* (Högberg et al. 2008). If photosynthate limits the vigor of the fungi equally, then on host 1 species A might grow more vigorously than species B and, all else being equal, species A might be expected to outcompete species B in subsequent root colonization attempts. On host 2, however, if species B is able to derive more photosynthate per colonized root than species A, a competitive superiority of species B over species A might be expected, leading to a domination of that host by species B. The perception of host specificity, documented as vastly different frequencies of occurrence, may thus result from distinct hierarchies of competition for photosynthate that depend on host species.

The degree of **disturbance** is also known to influence communities (Taylor and Bruns 1999) and this may also occur via context-dependent interaction hierarchies. In the absence of disturbance (an environmental condition), variation among species in ramet longevity (see Zhou and Hogetsu 2002) may result in the eventual domination of the community by the longest-lived species over shorter-lived species because of a differential maintenance of space; as ramets of the shorter-lived species die, the space they formerly occupied can be slowly occupied by ramets of the longer-lived species. However, in the presence of disturbance, shorter-lived species may be competitively superior if they are also more rapidly growing.

These examples illustrate how differences among ectomycorrhizal fungal species in key functional traits produce interaction hierarchies that depend on environmental conditions. When the conditions change, the hierarchy is modified and species interactions produce different outcomes. The literature is replete with examples of environmental factors that differentially affect the relative frequencies of species of ectomycorrhizal fungi, presumably because of inherent differences in physiology.

The following is an incomplete list of such environmental factors: forest successional status (Twieg and Durall 2007), litter quality (Goodman and Trofymow 1998; Conn and Dighton 2000) and host nutritional status, growth rate, size or health (Nara et al. 2003a; Cullings et al. 2005; Korkama et al. 2006; Blom et al. 2009; Leski et al. 2010), slope, aspect and openness of the habitat (Kranabetter and Friesen 2002; Scattolin et al. 2007; Di Marino et al. 2009), microsite (Tedersoo et al. 2008, 2009) and factors related to microsite variation including temperature (Rygiewicz et al. 2000), moisture (Gehring et al. 1998; O'Dell et al. 1999; Swaty et al. 2004; Cavender-Bares et al. 2009; Di Marino et al. 2009), N availability (Sagara 1995; Fransson et al. 2000; Lilleskov et al. 2001, 2002a, 2008; Toljander et al. 2006; Taniguchi et al. 2007; Wright et al. 2009), P availability (van der Heijden et al. 1999; Brearley et al. 2007; Wright et al. 2009), and pH (van der Heijden et al. 1999; Scattolin et al. 2007), edaphic factors including soil type (Gehring et al. 1998; Brearley 2006) and soil horizon (Malajczuk and Hingston 1981; Taylor and Bruns 1999; Dickie et al. 2002; Landeweert et al. 2003; Rosling et al. 2003; Baier et al. 2006; Courty et al. 2008; Genney et al. 2006; Buée et al. 2007; Scattolin et al. 2008), season (Buée et al. 2005; Koide et al. 2007; Courty et al. 2008; Wang and Guo 2010), fire (Smith et al. 2004, 2005; de Romn and de Miguel 2005), which may reflect vertical stratification of ectomycorrhizal fungal communities in the forest floor (Stendell et al. 1999), silvicultural practices (thinning, clear cutting) presumably due to changes in various environmental factors (Shaw et al. 2003; Durall et al. 2006; Heinonsalo et al. 2007; Mosca et al. 2007), response to disturbance (Taylor and Bruns 1999), and response to change in host carbon status (Kranabetter and Wylie 1998; Pena et al. 2010; Pestaña and Santolamazza-Carbone 2011). Species-specific interactions with other soil-dwelling organisms such as "helper bacteria" (Frey-Klett et al. 2007) would also result in context-dependent interactions among ectomycorrhizal fungi.

Relatively little of what might be termed the "comparative physiological ecology" of ectomycorrhizal fungi has been investigated. Therefore, descriptions of variation among species in ecologically relevant traits are relatively rare. Nevertheless, the few published studies in this area have demonstrated that significant variation among species occurs for traits that I hypothesize do influence interaction hierarchies. For example, we know species vary in growth at low water potentials (Coleman et al. 1989), ramet longevity (Zhou and Hogetsu 2002), growth rate of hyphae or rate of extension into unexploited soil (Wu et al. 1999), hyphal exploration strategy (Agerer 2001), propensity of hyphae to colonize root tips (Wu et al. 1999), the ability of hyphae to stimulate new root production (Niemi et al. 2002), preference for inorganic versus organic N source (Keizer and Arnolds 1994; Dickie et al. 1998; Lilleskov et al. 2002b), pattern of allocation between fruiting bodies and vegetative tissues (Gardes and Bruns 1996), salinity tolerance (Matsuda 2007), and in ability of spores to disperse to new habitats (Ashkannejhad and Horton 2006), persist in the environment (Taylor and Bruns 1999), germinate (Ishida et al. 2008), or colonize roots (Ishida et al. 2008).

Let us consider a hypothetical case in which significant variation in protease activity occurs between two species of ectomycorrhizal fungi. Protease enzymes allow fungi to hydrolyze proteins into amino acids that can be taken up and used as sources of nitrogen (N). Let us assume that the two species are able to disperse across an environmental gradient of N mineralization rate. Keep in mind that this gradient will map to locations in real space but not necessarily to locations that are spatially contiguous. High protease activity allows species H to grow well even when N mineralization rates are low as long as protein comprises some fraction of the soil total N pool (Fig. 10.1a). When N mineralization rates are high the production of protease is likely to be unnecessary as the uptake of mineral N can occur. In contrast, low protease activity (exhibited by species L) is assumed to be disadvantageous when N mineralization rates are low because the species will be limited by N (Fig. 10.1b). The success in colonizing a root system or a volume of soil is certain to be at least partly a function of the preferences or tolerances to environmental factors (its traits) by component species of ectomycorrhizal fungi even in the absence of species interactions such as competition. However, success is also likely to be an expression of the interaction hierarchy, which is determined by the traits interacting with the environment and which may be context-dependent. Therefore competition favors species H at lower N mineralization rates, but at some level of N mineralization competition will become balanced, and eventually competition could favor species L due to the use of energy by species H to produce protease, resulting in the pattern of root colonization illustrated in Fig. 10.1c. Figure 10.1c illustrates a case of context-dependent competitive hierarchy in which species H is competitively superior at low N mineralization rates but competitively inferior at the highest N mineralization rates. In the absence of competing species, 100% of colonized roots will be colonized by either species H (Fig. 10.1d) or species L (Fig. 10.1e).

General principles of community organization are derived by relating physiological differences among species to dominance hierarchies in the following way. Let us assume the existence of a gradient of N mineralization rate along which five species (*A, B, C, D, E*) occur (Fig. 10.2a). Let us further assume that at two portions of the gradient we find the same five species arranged into two distinct communities (*X* and *Y*); they possess distinct relative frequencies. We hypothesize that for community *X* protease activity will be positively related to relative frequency, but that for community *Y* protease activity will be negatively related to relative frequency. In this example, significant variability in relative frequency is, in fact, explained by a variation in protease activity for both communities (Fig. 10.2b, d). Let us assume here that the N mineralization rate is at least partly related to prevailing temperature, so we further hypothesize that the ability to grow at cold temperature should be positively related to relative frequency for community *X* and, perhaps, negatively related to relative frequency in community *Y*. We find, here, that significant variability in relative frequency is also explained by the ability to grow at cold temperatures (Fig. 10.2c, e). Thus, we conclude that both protease activity and the ability to grow at cold temperatures help to structure these communities.

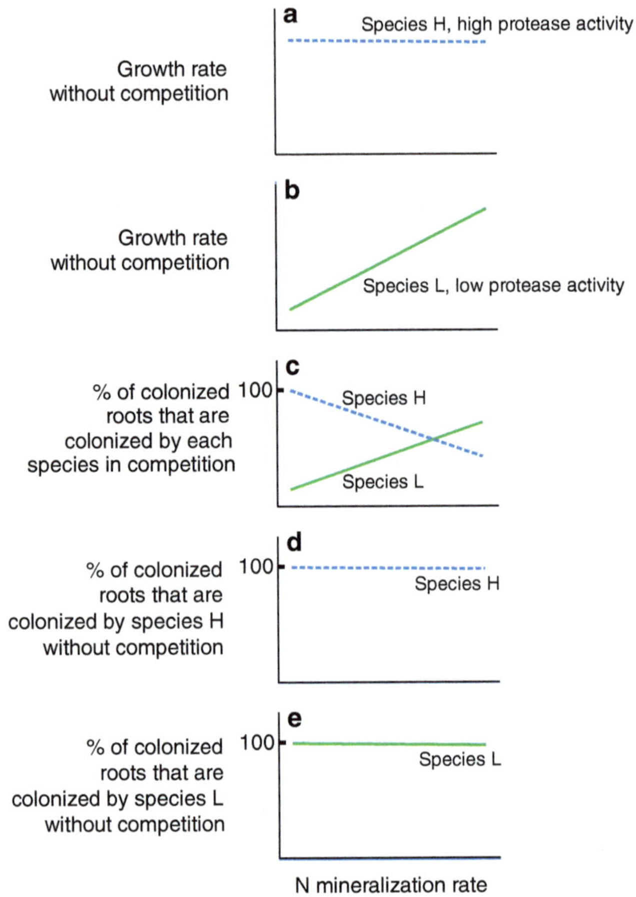

Fig. 10.1. The relationship of a key trait (protease activity) to species interactions along an environmental gradient (N mineralization rate) for a species with inherently high protease activity (species *H*) and another with inherently low protease activity (species *L*). The assumption is that the two species are able to disperse across an environmental gradient of N mineralization rate. This gradient will map to locations in real space but not necessarily to locations that are spatially contiguous. High protease activity allows species *H* to grow well even when N mineralization rates are low as long as protein comprises some fraction of the soil total N pool (**a**). When N mineralization rates are high the production of protease is likely to be unnecessary as the uptake of mineral N can occur. In contrast, low protease activity (exhibited by species *L*) is assumed to be disadvantageous when N mineralization rates are low because the species will be limited by N (**b**). Therefore competition favors species *H* at lower N mineralization rates, but at some level of N mineralization competition is balanced, and at higher rates of N mineralization competition favors species *L* due to the use of energy by species *H* to produce protease, resulting in the pattern of root colonization illustrated in (**c**). In the absence of competing species, 100 % of colonized roots are colonized by either species *H* (**d**) or species *L* (**e**)

This approach allows one to determine the importance of traits of species to their performance in any community. Of course no single trait such as protease activity or growth at cold temperatures is expected to account for all variability in fungal performance. However, it is likely that certain key traits will have measurable impacts on performance, as has been demonstrated for protease activity along mineral N availability gradients (Taylor et al. 2000; Lilleskov et al. 2002a, b). This approach should be relevant to a number of phenomena. For example, successful invasion of a community by a novel species should be determined by the relevant traits possessed by the species; the likelihood of community *X* (Fig. 10.2) being invaded

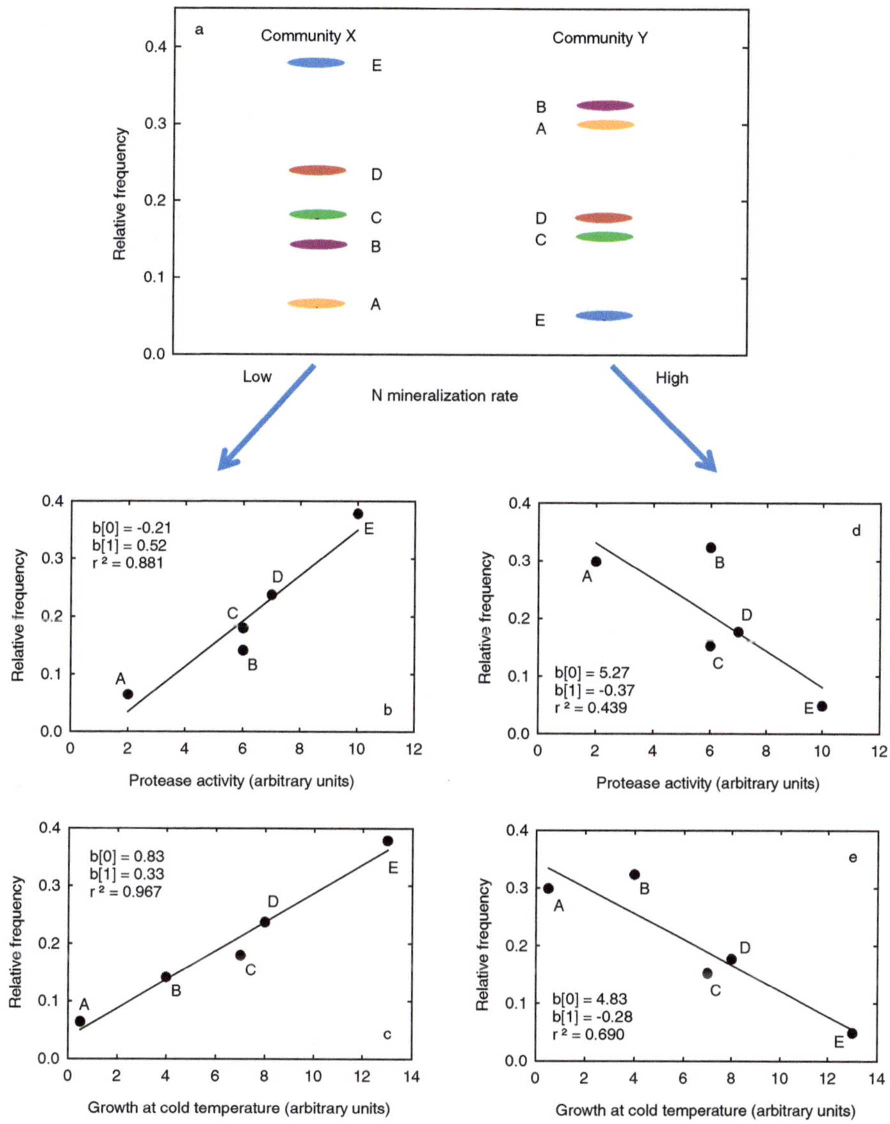

Fig. 10.2. (**a**) In order to illustrate how general principles of community organization can be gleaned from traits that determine environmentally context-dependent interaction hierarchies, assume, as for Fig. 10.1, a gradient of N mineralization rate along which five species occur (*A–E*). Further, assume that at two portions of the gradient we find the same five species, but in two distinct communities (*X*, *Y*); they possess distinct relative frequencies. For community *X* we hypothesize that protease activity will be positively related to relative frequency, but that for community *Y* protease activity will be negatively related to relative frequency. In this hypothetical case, significant variabiltiy in relative frequency is, in fact, explained by variation in protease activity for both communities (**b**, **d**). Because we suspect that the mineralization rate is at least partly related to prevailing temperature, we further hypothesize that the ability to grow at cold temperature should be positively related to relative frequency for community *X* and, perhaps, negatively related to relative frequency in community *Y*. We find, here, that significant variability in relative frequency is also explained by the ability to grow at cold temperatures (**c**, **e**). Thus, we conclude that both protease activity and the ability to grow at cold temperatures help to structure these communities

is greater for a species with high protease activity than for a species with low protease activity.

V. Conclusions

I believe an **understanding of ecologically relevant traits that determine environmentally context-dependent interaction hierarchies is the key to elucidating general principles that structure biological communities.** I will admit, however, that stating that context-dependent interaction hierarchies exist is really just a fancy way of indicating that the nature of competition, allelopathy, and other interactions among species of ectomycorrhizal fungi depends on environmental conditions. This is nothing more than the classical "interaction" in an analysis of variance. There is certainly nothing new to the appreciation of statistical interactions. Many years ago my postdoctoral supervisor, Harold (Hal) Mooney, once asked "how do you spell interactions?" An excellent professor, Hal answered the question himself by spelling the word "e-c-o-l-o-g-y". In his estimation ecology is the study of interactions among factors and I would have to agree. In this case, understanding the statistical interaction between environment and competitive hierarchies may lead to profound insights into community organization.

Context-dependent interspecific interactions occur because different species react differently to the very same set of conditions. Thus, as environmental conditions change, competitive superiority shifts from one species to the next and communities of distinct structure are established. These concepts relating species traits to environmental variability (including climate change) exist at the interface between physiological ecology and community ecology, one of the most exciting and relevant areas of research today. We need to build on the growing body of information relating the ecologically relevant traits of species to the environmental heterogeneity of the habitat in order to better understand the general principles that shape communities of ectomycorrhizal fungi. In particular if we focus on traits that influence species interactions, and learn how these interactions are influenced by environmental heterogeneity, we will have come a long way toward discerning the general principles by which communities are organized and by which succession proceeds.

Some of the dissatisfaction with traditional community ecology (Pianka 1992; Lawton 1999) stems from the lack of general rules for community assembly and our inability to predict which species will occur where and when. Lawton (1999) complained of "too much contingency" and that "important processes and resulting community dynamics differ, often markedly, from system to system...". That is the point, is it not? Knowing how interaction hierarchies are modified by shifts in environmental conditions reveals much of what is important. The approach advocated herein is incapable of predicting which species will be present in any community but that is not the goal. Its utility lies in our ability to extract from it general principles of community organization. I suppose if enough traits were known for all species that could possibly migrate to a particular place, it might then be possible to predict species distributions. But it does not seem even remotely possible to catalog enough of the relevant traits of enough species to ever make community ecology predictive in that way. You may recall the theoretical brilliance but utter impracticality of the *n*-dimensions of Hutchinson's "hypervolume" (Hutchinson 1957). Instead, much of the structuring of communities can probably be explained by relating key traits to interaction hierarchies along environmental gradients. While this approach may not be recognizable as traditional community ecology, it would appear to be far more likely to produce general principles of community structuring than more traditional approaches.

Acknowledgements I thank Chris Fernandez, Kristin Haider, Matt Peoples, and Krittika Petprakob for helpful suggestions during the development of this chapter. I also acknowledge financial support from the Northeast Sun Grant Initiative and the USDA NIFA AFRI.

References

Agerer R (1990) Studies on ectomycorrhizae XXIV – ectomycorrhizae of *Chroogomphus helveticus* and *C. rutilus* (Gomphidiaceae, Basidiomycetes) and their relationship to those of *Suillus* and *Rhizopogon*. Nova Hedwigia 50:1–63

Agerer R (1991) Studies on ectomycorrhizae XXXIV – mycorrhizae of *Gomphidius glutinosus* and of *G. roseus* with some remarks on Gomphidiaceae (Basidiomycetes). Nova Hedwigia 53:127–170

Agerer R (2001) Exploration types of ectomycorrhizae. Mycorrhiza 11:107–114

Ashkannejhad S, Horton TR (2006) Ectomycorrhizal ecology under primary succession on coastal sand dunes: interactions involving *Pinus contorta*, suilloid fungi and deer. New Phytol 169:345–354

Baier R, Ingenhaag J, Blaschke H, Göttlein A, Agerer R (2006) Vertical distribution of an ectomycorrhizal community in upper soil horizons of a young Norway spruce (*Picea abies* [L.] Karst.) stand of the Bavarian Limestone Alps. Mycorrhiza 16:197–206

Benecke U, Göbl F (1974) The influence of different mycorrhizae on growth, nutrition and gas-exchange of *Pinus mugo* seedlings. Plant Soil 40:21–32

Bengtsson J, Fagerstrom T, Rydin H (1994) Competition and coexistence in plant communities. Trends Ecol Evol 9:246–250

Bledsoe CS, Tennyson K, Lopushinsky W (1982) Survival and growth of outplanted Douglas-fir seedlings inoculated with mycorrhizal fungi. Can J For Res 12:720–723

Blom JM, Vannini A, Vettraino AM, Hale MD, Godbold DL (2009) Ectomycorrhizal community structure in a healthy and a *Phytophthora*-infected chestnut (*Castanea sativa* Mill.) stand in central Italy. Mycorrhiza 20:25–38

Brearley FQ (2006) Differences in the growth and ectomycorrhizal community of *Dryobalanops lanceolata* (Dipterocarpaceae) seedlings grown in ultramafic and non-ultramafic soils. Soil Biol Biochem 38:3407–3410

Brearley FQ, Scholes JD, Press MC, Palfner G (2007) How does light and phosphorus fertilisation affect the growth and ectomycorrhizal community of two contrasting dipterocarp species? Plant Ecol 192:237–249

Bruns TD (1995) Thoughts on the processes that maintain local species diversity of ectomycorrhizal fungi. Plant Soil 170:63–73

Buchalo AS, Šašek V, Zakordonec OA (1989) Scanning electron microscopic study of vegetative mycelium of higher basidiomycetes. Folia Microbiol 34:146–150

Buée M, Vairelles D, Garbaye J (2005) Year-round monitoring of diversity and potential metabolic activity of the ectomycorrhizal community in a beech (*Fagus silvatica*) forest subjected to two thinning regimes. Mycorrhiza 15:235–245

Buée M, Courty PE, Mignot D, Garbaye J (2007) Soil niche effect on species diversity and catabolic activities in an ectomycorrhizal fungal community. Soil Biol Biochem 39:1947–1955

Cavender-Bares J, Izzo A, Robinson R, Lovelock CE (2009) Changes in ectomycorrhizal community structure on two containerized oak hosts across an experimental hydrologic gradient. Mycorrhiza 19:133–142

Chakravarty C, Peterson RL, Ellis BE (1991) Interaction between the ectomycorrhizal fungus *Paxillus involutus*, damping-off fungi and *Pinus resinosa* seedlings. J Phytopathol 132:207–218

Chen YL, Brundrett MC, Dell B (2000) Effects of ectomycorrhizas and vesicular-arbuscular mycorrhizas, alone or in competition, on root colonization and growth of *Eucalyptus globulus* and *E. urophylla*. New Phytol 146:545–555

Coleman MD, Bledsoe CS, Lopushinsky W (1989) Pure culture response of ectomycorrhizal fungi to imposed water stress. Can J Bot 67:29–39

Conn C, Dighton J (2000) Litter quality influences on decomposition, ectomycorrhizal community structure and mycorrhizal root surface acid phosphatase activity. Soil Biol Biochem 32:489–496

Courty PE, Franc A, Pierrat JC, Garbaye J (2008) Temporal changes in the ectomycorrhizal community in two soil horizons of a temperate oak forest. Appl Environ Microbiol 74:5792–5801

Cullings K, Raleigh C, Vogler DR (2005) Effects of severe dwarf mistletoe infection on the ectomycorrhizal community of a *Pinus contorta* stand in Yellowstone Park. Can J Bot 83:1174–1180

de Romn M, de Miguel AM (2005) Post-fire, seasonal and annual dynamics of the ectomycorrhizal community in a *Quercus ilex* L. forest over a 3-year period. Mycorrhiza 15:471–482

DeBellis T, Kernaghan G, Bradley R, Widden P (2006) Relationships between stand composition and ectomycorrhizal community structure in boreal mixed-wood forests. Microbial Ecol 52:114–126

Detman JR, van de Kamp BJ (2001) The population structure of *Armillaria ostoyae* and *Armillaria sinapina* in the central interior of British Columbia. Can J Bot 79:600–611

Di Marino E, Montecchio L, Scattolin L, Abs C, Agerer R (2009) The ectomycorrhizal community structure in European beech forests differing in coppice shoot age and stand features. J For 107:250–259

Dickie I, Koide R, Stevens C (1998) Tissue density and growth response of ectomycorrhizal fungi to nitrogen source and concentration. Mycorrhiza 8:145–148

Dickie IA, Xu B, Koide RT (2002) Vertical niche differentiation of ectomycorrhizal hyphae in soil as shown by T-RFLP analysis. New Phytol 156:527–535

Dighton J, Poskitt JM, Howard DM (1986) Changes in occurrence of basidiomycete fruit bodies during forest stand development: with specific reference to mycorrhizal species. Trans Br Mycol Soc 87:163–171

Duchesne LC, Peterson R, Ellis BE (1988) Interaction between the ectomycorrhizal fungus *Paxillus involutus* and *Pinus resinosa* induces resistance to *Fusarium oxysporum*. Can J Bot 66:558–562
Durall DM, Gamiet S, Simard SW, Kudrna L, Sakakibara SM (2006) Effects of clearcut logging and tree species composition on the diversity and community composition of epigeous fruit bodies formed by ectomycorrhizal fungi. Can J Bot 84:966–980
Fleming LV (1985) Experimental study of sequences of ectomycorrhizal fungi on birch (*Betula* sp.) seedling root systems. Soil Biol Biochem 17:591–600
Fransson PM, Taylor AFS, Finlay RD (2000) Effects of continuous optimal fertilization on belowground ectomycorrhizal community structure in a Norway spruce forest. Tree Physiol 20:599–606
Frey-Klett P, Garbaye J, Tarkka M (2007) The mycorrhiza helper bacteria revisited. New Phytol 176:22–36
Gardes M, Bruns TD (1996) Community structure of ectomycorrhizal fungi in a *Pinus muricata* forest: above- and below-ground views. Can J Bot 74:1572–1583
Gehring CA, Theimer TC, Whitham TG, Keim P (1998) Ectomycorrhizal fungal community structure of pinyon pines growing in two environmental extremes. Ecology 79:1562–1572
Genney DR, Anderson IC, Alexander IJ (2006) Fine-scale distribution of pine ectomycorrhizas and their extramatrical mycelium. New Phytol 170:381–390
Goodman DM, Trofymow JA (1998) Distribution of ectomycorrhizas in microhabitats in mature and old-growth stands of Douglas-fir on southeastern Vancouver Island. Soil Biol Biochem 30:2127–2138
Grace JB, Wetzel RG (1981) Habitat partitioning and competitive displacement in cattails (*Typha*): experimental field studies. Am Nat 118:463–474
Grogan P, Baar J, Bruns TD (2000) Below-ground ectomycorrhizal community structure in a recently burned bishop pine forest. J Ecol 88:1051–1062
Hättenschwiler S, Tiunov AV, Scheu S (2005) Biodiversity and litter decomposition in terrestrial ecosystems. Annu Rev Ecol Evol Syst 36:191–218
Heinonsalo J, Koskiahde I, Sen R (2007) Scots pine bait seedling performance and root colonizing ectomycorrhizal fungal community dynamics before and during the 4 years after forest clear-cut logging. Can J For Res 37:415–429
Hibbett DS, Gilbert LB, Donoghue MJ (2000) Evolutionary instability of ectomycorrhizal symbioses in basidiomycetes. Nature 407:506–508
Högberg P, Högberg M, Göttlicher S, Betson N, Keel S, Metcalfe D, Campbell C, Schindlbacher A, Hurry V, Lundmark T et al (2008) High temporal resolution tracing of photosynthate carbon from the tree canopy to forest soil microorganisms. New Phytol 177:220–228
Hubbell SP (2001) The unified neutral theory of biodiversity and biogeography. Princeton University Press, Princeton
Hutchinson GE (1957) Concluding remarks. Cold Spring Harbor Symp Quant Biol 22:415–427
Iotti M, Lancellotti E, Hall I, Zambonelli A (2010) The ectomycorrhizal community in natural *Tuber borchii* grounds. FEMS Microbiol Ecol 72:250–260
Ishida TA, Nara K, Hogetsu T (2007) Host effects on ectomycorrhizal fungal communities: insight from eight host species in mixed conifer-broadleaf forests. New Phytol 174:430–440
Ishida TA, Nara K, Tanaka M, Kinoshita A, Hogetsu T (2008) Germination and infectivity of ectomycorrhizal fungal spores in relation to their ecological traits during primary succession. New Phytol 180:491–500
Jumpponen A, Egerton-Warburton LM (2005) Mycorrhizal fungi in successional environments: a community assembly model incorporating host plant, environmental, and biotic filters. In: Dighton J, White JF, Oudemans P (eds) The fungal community: its organization and role in the ecosystem, 3rd edn. Taylor & Francis, Boca Raton, pp 139–168
Keizer P, Arnolds E (1994) Succession of ectomycorrhizal fungi in roadside verges planted with common oak (*Quercus robur* L.) in Drenthe, The Netherlands. Mycorrhiza 4:147–159
Kennedy P (2010) Ectomycorrhizal fungi and interspecific competition: species interactions, community structure, coexistence mechanisms, and future research directions. New Phytol 187:895–910
Kennedy PG, Bruns TD (2005) Priority effects determine the outcome of ectomycorrhizal competition between two Rhizopogon species colonizing *Pinus muricata* seedlings. New Phytol 166:631–638
Kennedy PG, Peay KG, Bruns TD (2009) Root tip competition among ectomycorrhizal fungi: are priority effects a rule or an exception? Ecology 90:2098–2107
Klironomos J, Hart M (2001) Animal nitrogen swap for plant carbon. Nature 410:651–652
Koide RT, Xu B, Sharda J, Lekberg Y, Ostiguy N (2005) Evidence of species interactions within an ectomycorrhizal fungal community. New Phytol 165:305–316
Koide RT, Shumway DL, Xu B, Sharda JN (2007) On temporal partitioning of a community of ectomycorrhizal fungi. New Phytol 174:420–429
Koide RT, Fernandez C, Petprakob K (2011) General principles in the community ecology of ectomycorrhizal fungi. Ann For Sci 68:45–55
Korkama T, Pakkanen A, Pennanen T (2006) Ectomycorrhizal community structure varies among Norway spruce (*Picea abies*) clones. New Phytol 171:815–824
Kranabetter J, Friesen J (2002) Ectomycorrhizal community structure on western hemlock (*Tsuga heterophylla*) seedlings transplanted from forests into openings. Can J Bot 80:861–868
Kranabetter J, Wylie T (1998) Ectomycorrhizal community structure across forest openings on naturally

regenerated western hemlock seedlings. Can J Bot 76:189–196

Krywolap GN (1964) *Cenococcum graniforme* antibiotics: elaboration in pure culture and during mycorrhizal association. PhD thesis, Pennsylvania State University, University Park

Lamb EG, Kembel SW, Cahill JF (2009) Shoot, but not root, competition reduces community diversity in experimental mesocosms. J Ecol 97:155–163

Landeweert R, Leeflang P, Kuyper TW, Hoffland E, Rosling A, Wernars K, Smit E (2003) Molecular identification of ectomycorrhizal mycelium in soil horizons. Appl Environ Microbiol 69:327–333

Lapeyrie F, Chilvers G (1985) An endomycorrhiza-ectomycorrhiza succession associated with enhanced growth of *Eucalyptus dumosa* seedlings planted in a calcareous soil. New Phytol 100:93–104

Lawton JH (1999) Are there general laws in ecology? Oikos 84:177–192

Leake JR, Donnelly DP, Saunders EM, Boddy L, Read DJ (2001) Rates and quantities of carbon flux to ectomycorrhizal mycelium following 14C pulse labeling of *Pinus sylvestris* seedlings: effects of litter patches and interaction with a wood-decomposer fungus. Tree Physiol 21:71–82

Leski T, Aucina A, Skridaila A, Pietras M, Riepsas E, Rudawska M (2010) Ectomycorrhizal community structure of different genotypes of Scots pine under forest nursery conditions. Mycorrhiza 20:473–481

Lilleskov EA, Bruns TD (2003) Root colonization dynamics of two ectomycorrhizal fungi of contrasting life history strategies are mediated by addition of organic nutrient patches. New Phytol 159:141–151

Lilleskov EA, Fahey T, Lovett G (2001) Ectomycorrhizal fungal aboveground community change over an atmospheric nitrogen deposition gradient. Ecol Appl 11:397–410

Lilleskov EA, Fahey T, Horton T (2002a) Belowground ectomycorrhizal fungal community change over a nitrogen deposition gradient in Alaska. Ecology 83:104–115

Lilleskov EA, Hobbie E, Fahey TJ (2002b) Ectomycorrhizal fungal taxa differing in response to nitrogen deposition also differ in pure culture organic nitrogen use and natural abundance of nitrogen isotopes. New Phytol 154:219–231

Lilleskov EA, Wargo PM, Vogt KA, Vogt DJ (2008) Mycorrhizal fungal community relationship to root nitrogen concentration over a regional atmospheric nitrogen deposition gradient in the northeastern USA. Can J For Res 38:1260–1266

Malajczuk N (1988) Interaction between *Phytophthora cinnamomi* zoospores and micro-organisms on non-mycorrhizal and ectomycorrhizal roots of *Eucalyptus marginata*. Trans Br Mycol Soc 90:375–382

Malajczuk N, Hingston FJ (1981) Ectomycorrhizae associated with Jarrah. Austral J Bot 29:453–462

Mamoun M, Olivier J (1993a) Competition between *Tuber melanosporum* and other ectomycorrhizal fungi under two irrigation regimes I. Competition with *Tuber brumale*. Plant Soil 149:211–218

Mamoun M, Olivier JM (1993b) Competition between *Tuber melanosporum* and other ectomycorrhizal fungi under two irrigation regimes. II. Comparison of soils artificially infested with *T. melanosporum* and *T. brumale*. Plant Soil 149:211–218

Matsuda Y (2007) Studies on ectomycorrhizal community structure in an *Abies firma* forest and techniques for restoring natural ecosystems with mycorrhizal associations. Nippon Kingakukai Kaiho 48:32–41

Molina R, Massicotte HB, Trappe JM (1992) Ecological role of specificity phenomena in ectomycorrhizal plant communities: potentials for interplant linkages and guild development. In: Read DJ, Lewis DH, Fitter AH, Alexander IJ (eds) Mycorrhizas in ecosystems. CAB International, Wallingford, pp 106–112

Mosca E, Montecchio L, Sella L, Garbaye J (2007) Short-term effect of removing tree competition on the ectomycorrhizal status of a declining pedunculate oak forest (*Quercus robur* L.). For Ecol Manage 244:129–140

Nara K, Nakaya H, Hogetsu T (2003a) Ectomycorrhizal sporocarp succession and production during early primary succession on Mount Fuji. New Phytol 158:193–206

Nara K, Nakaya H, Wu B, Zhou Z, Hogetsu T (2003b) Underground primary succession of ectomycorrhizal fungi in a volcanic desert on Mount Fuji. New Phytol 159:743–756

Niemi K, Häggman H, Sarjala T (2002) Effects of exogenous diamines on the interaction between ectomycorrhizal fungi and adventitious root formation in Scots pine in vitro. Tree Physiol 22:373–381

O'Dell TE, Ammirati JF, Schreiner EG (1999) Species richness and abundance of ectomycorrhizal basidiomycete sporocarps on a moisture gradient in the *Tsuga heterophylla* zone. Can J Bot 77:1699–1711

Olsson PA, Wallander H (1998) Interactions between ectomycorrhizal fungi and the bacterial community in soils amended with various primary minerals. FEMS Microbiol Ecol 27:195–205

Olsson PA, Münzenberger B, Mahmood S, Erland S (2000) Molecular and anatomical evidence for a three-way association between *Pinus sylvestris* and the ectomycorrhizal fungi *Suillus bovinus* and *Gomphidius roseus*. Mycol Res 104:1372–1378

Pena R, Offermann C, Simon J, Naumann PS, Gessler A, Holst J, Dannenmann M, Mayer H, Kögel-Knabner I, Rennenberg H, Polle A (2010) Girdling affects ectomycorrhizal fungal (EMF) diversity and reveals functional differences in EMF community

composition in a beech forest. Appl Environ Microbiol 76:1831–1841
Pestaña M, Santolamazza-Carbone S (2011) Defoliation negatively affects plant growth and the ectomycorrhizal community of *Pinus pinaster* in Spain. Oecologia 165:723–733
Pianka ER (1992) The state of the art in community ecology. In: Adler K (ed) Herpetology: current research on the biology of amphibians and reptiles. Proceedings of the first world congress of herpetology, Society for the Study of Amphibians and Reptiles, Oxford, OH, pp 141–162
Rosling A, Landeweert R, Lindahl BD, Larsson K-H, Kuyper TW, Taylor AFS, Finlay RD (2003) Vertical distribution of ectomycorrhizal fungal taxa in a podzol soil profile. New Phytol 159:775–783
Rygiewicz PT, Martin KJ, Tuininga AR (2000) Morphotype community structure of ectomycorrhizas on Douglas fir (*Pseudotsuga menziesii* Mirb. Franco) seedlings grown under elevated atmospheric CO2 and temperature. Oecologia 124:299–308
Sagara N (1995) Association of ectomycorrhizal fungi with decomposed animal wastes in forest habitats a cleaning symbiosis? Can J Bot 73:S1423–S1433
Scattolin L, Montecchio L, Agerer R (2007) The ectomycorrhizal community structure in high mountain Norway spruce stands. Trees 22:13–22
Scattolin L, Montecchio L, Mosca E, Agerer R (2008) Vertical distribution of the ectomycorrhizal community in the top soil of Norway spruce stands. Eur J For Res 127:347–357
Schrey SD, Salo V, Raudaskoski M, Hampp R, Nehls U, Tarkka MT (2007) Interaction with mycorrhiza helper bacterium Streptomyces sp. AcH 505 modifies organisation of actin cytoskeleton in the ectomycorrhizal fungus *Amanita muscaria* (fly agaric). Curr Genet 52:77–85
Shaw PJA, Kibby C, Mayes J (2003) Effects of thinning treatment on an ectomycorrhizal succession under Scots pine. Mycol Res 107:317–328
Smith J, McKay D, Niwa C, Thies W, Brenner G, Spatafora J (2004) Short-term effects of seasonal prescribed burning on the ectomycorrhizal fungal community and fine root biomass in ponderosa pine stands in the Blue Mountains of Oregon. Can J For Res 34:2477–2491
Smith JE, McKay D, Brenner G, Mciver J, Spatafora JW (2005) Early impacts of forest restoration treatments on the ectomycorrhizal fungal community and fine root biomass in a mixed conifer forest. J Appl Ecol 42:526–535
Stendell E, Horton T, Bruns T (1999) Early effects of prescribed fire on the structure of the ectomycorrhizal fungus community in a Sierra Nevada ponderosa pine forest. Mycol Res 103:1353–1359
Swaty RL, Deckert RJ, Whitham TG, Gehring CA (2004) Ectomycorrhizal abundance and community composition shifts with drought: predictions from tree rings. Ecology 85:1072–1084
Taniguchi T, Kanzaki N, Tamai S, Yamanaka N, Futai K (2007) Does ectomycorrhizal fungal community structure vary along a Japanese black pine (*Pinus thunbergii*) to black locust (*Robinia pseudoacacia*) gradient? New Phytol 173:322–334
Taylor DL, Bruns TD (1999) Community structure of ectomycorrhizal fungi in a *Pinus muricata* forest: minimal overlap between the mature forest and resistant propagule communities. Mol Ecol 8:1837–1850
Taylor AFS, Martin F, Read DJ (2000) Fungal diversity in ectomycorrhizal communities of Norway spruce [*Picea abies* (L.) Karst.] and beach (*Fagus sylvatica* L.) along north–south transects in Europe. In: Schulze E-D (ed) Carbon and nitrogen cycling in European forest ecosystems. Springer, Berlin Heidelberg New York, pp 343–365
Tedersoo L, Suvi T, Jairus T, Kõljalg U (2008) Forest microsite effects on community composition of ectomycorrhizal fungi on seedlings of *Picea abies* and *Betula pendula*. Environ Microbiol 10:1189–1201
Tedersoo L, Gates G, Dunk CW, Lebel T, May TW, Kõljalg U, Jairus T (2009) Establishment of ectomycorrhizal fungal community on isolated *Nothofagus cunninghamii* seedlings regenerating on dead wood in Australian wet temperate forests: does fruit-body type matter? Mycorrhiza 19:403–416
Toljander JF, Eberhardt U, Toljander YK, Paul LR, Taylor AFS (2006) Species composition of an ectomycorrhizal fungal community along a local nutrient gradient in a boreal forest. New Phytol 170:873–884
Twieg B, Durall D (2007) Ectomycorrhizal fungal succession in mixed temperate forests. New Phytol 176:437–447
van der Heijden EW, Vries FWD, Kuyper TW (1999) Mycorrhizal associations of *Salix repens* L. communities in succession of dune ecosystems. I. Above-ground and below-ground views of ectomycorrhizal fungi in relation to soil chemistry. Can J Bot 77:1821–1832
Villeneuve N, Tacon F, Bouchard D (1991) Survival of inoculated *Laccaria bicolor* in competition with native ectomycorrhizal fungi and effects on the growth of outplanted Douglas-fir seedlings. Plant Soil 135:95–107
Visser S (1995) Ectomycorrhizal fungal succession in jack pine stands following wildfire. New Phytol 129:389–401
Wang Q, Guo L-D (2010) Ectomycorrhizal community composition of *Pinus tabulaeformis* assessed by ITS-RFLP and ITS sequences. Botany 88:590–595
Wright SH, Berch SM, Berbee ML (2009) The effect of fertilization on the below-ground diversity and community composition of ectomycorrhizal fungi associated with western hemlock (*Tsuga heterophylla*). Mycorrhiza 19:267–276
Wu B, Nara K, Hogetsu T (1999) Competition between ectomycorrhizal fungi colonizing *Pinus densiflora*. Mycorrhiza 9:151–159

Wu T, Kabir Z, Koide R (2005) A possible role for saprotrophic microfungi in the N nutrition of ectomycorrhizal *Pinus resinosa*. Soil Biol Biochem 37:965–975

Zadworny M, Smolinski D, Idzikowska K, Werner A (2007) Ultrastructural and cytochemical aspects of the interaction between the ectomycorrhizal fungus *Laccaria laccata* and the saprotrophic fungi, *Trichoderma harzianum* and *T. virens*. Biocontrol Sci Technol 17:921–932

Zhou Z, Hogetsu T (2002) Subterranean community structure of ectomycorrhizal fungi under *Suillus grevillei* sporocarps in a *Larix kaempferi* forest. New Phytol 154:529–539

11 Ectendomycorrhizas: Occurrence, Structural Characteristics, and Possible Roles

R.L. Peterson[1]

CONTENTS

I. Introduction

Plants interact with a diverse assemblage of soil fungi, some of which colonize roots to form mycorrhizas. Mycorrhizas have traditionally been categorized partially on the identity of the fungal partner(s) but mainly on the modifications of these fungi and their associated roots that lead to the distinct mature structure (Brundrett 2004; Peterson and Massicotte 2004; Smith and Read 2008). **Ectendomycorrhiza,** one of approximately seven currently recognized mycorrhiza categories (Peterson et al. 2004; Smith and Read 2008), has been defined either broadly (Smith and Read 2008) or narrowly (Yu et al. 2001; Peterson et al. 2004). The **broad classification** includes arbutoid mycorrhizas that are characteristic of several genera in the large family, Ericaceae (Smith and Read 2008). The **narrow view** confines the term ectendomycorrhizas to specialized associations that form primarily with two conifer genera, *Pinus* and *Larix*, and a limited number of ascomycete fungal species (Yu et al. 2001). Yu et al. (2001) summarized the *Pinus* and *Larix* species as well as several angiosperm species that have been reported to have ectendomycorrhizas. They pointed out, however, that the evidence for some of these reports is not particularly strong because observations were based on field-collected material of unknown age and the fungal symbionts were not identified.

There has been debate as to whether ectendomycorrhiza should be considered a separate category or a developmental phase or evolutionary stage of ectomycorrhizas (Egger and Fortin 1988). More recently, Brundrett (2004) argued that since ectendomycorrhizas do not occur in a distinct and separate plant lineage, they should be considered as a 'fungal morphotype' included within the category, ectomycorrhiza. The examination of root systems and their associated symbiotic fungi of plant species in parts of the world that until now have been poorly studied may lead to a reassessment, not only of the category, ectendomycorrhiza, but also of other categories of mycorrhizas.

Beck et al. (2005) have used the term to describe the morphological features of mycorrhizas formed by two members of the Glomeromycota with the tropical tree species (*Alzatea verticillata*). The evidence cited for designating the mycorrhiza as an ectendomycorrhiza is the occurrence of highly branched intercellular hyphae resembling Hartig net hyphae in ectomycorrhizas as well as the presence of intracellular structures, in this case typical arbuscules.

[1]Department of Molecular and Cellular Biology, University of Guelph, Guelph, ON N1G 2W1, Canada; e-mail: lpeterso@uoguelph.ca

Fungal Associations, 2nd Edition
The Mycota IX
B. Hock (Ed.)

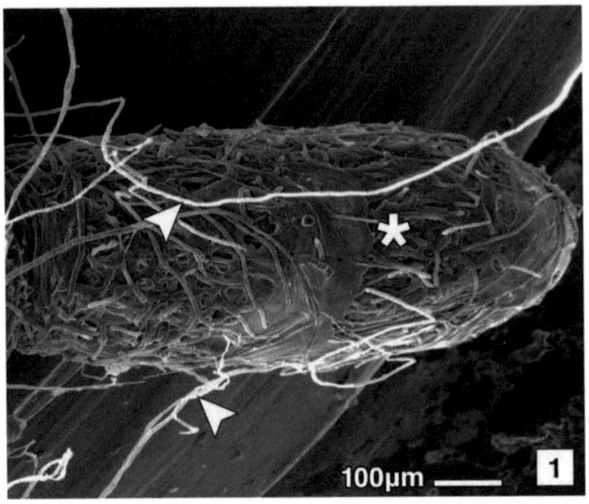

Fig. 11.1. Scanning electron microscopy image of a *Pinus banksiana* monopodial short root colonized by *Wilcoxina mikolae* var. *mikolae*. The mantle (*) consists of loosely arranged hyphae. A few extraradical hyphae (*arrowheads*) are present

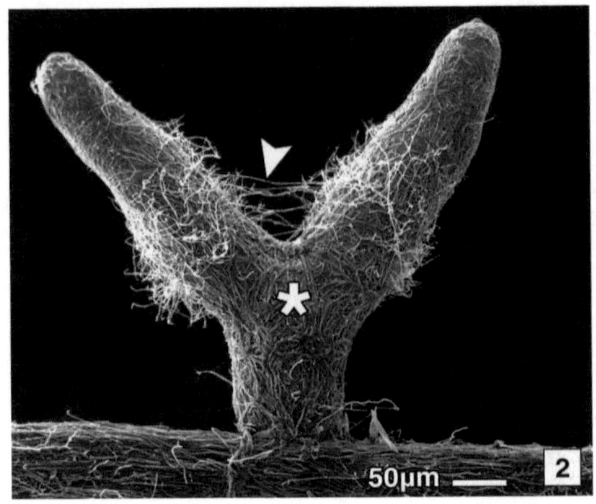

Fig. 11.2. Scanning electron microscopy image of a *Pinus banksiana* dichotomous short root colonized by *Wilcoxina mikolae* var. *mikolae*. The mantle (*) consists of loosely arranged hyphae. A few extraradical hyphae (*arrowhead*) are present

In this chapter, ectendomycorrhizas will be considered from a narrow perspective keeping in mind that, with new evidence, this category may not be maintained (Brundrett 2004) or may be expanded to include a broader range of plant species and fungal symbionts. Arbutoid mycorrhizas and similar mycorrhizas of members of the Ericaceae (see Setaro et al. 2006) will not be discussed in this chapter since they are covered elsewhere in this volume.

II. Structural Characteristics of Ectendomycorrhizas

The short roots of pine species colonized by ectendomycorrhizal fungal species are usually monopodial (Fig. 11.1) but dichotomies may occur (Fig. 11.2). **Ectendomycorrhizas share two important characteristics with ectomycorrhizas:** a **fungal mantle** and **Hartig net**. The mantle, however, may be poorly developed (Figs. 11.1, 11.2, 11.3) or in some cases absent (Smith and Read 2008). The mantle of an E-strain fungal species associated with *Pinus strobus* has been described, using laser scanning confocal microscopy, as a loosely organized net prosenchyma (Schelkle et al. 1996). Mantle hyphae are frequently embedded in surface mucigel (Scales and Peterson 1991b; Fig. 11.3). Hartig net hyphae develop between the epidermal and cortical cells in *Pinus* spp. (Mikola 1965; Wilcox 1969; Wilcox and Ganmore-Neumann 1974; Piché et al. 1986; Ivory and Pearce 1991; Scales and Peterson 1991a; Ursic and Peterson 1997; Peterson et al. 2004; Figs. 11.3 and 11.4) and in *Larix occidentalis* (Laiho 1965). An additional feature, **intracellular hyphae within root epidermal and cortical cells**, distinguishes ectendomycorrhizas from typical ectomycorrhizas (Yu et al. 2001; Peterson and Massicotte 2004; Smith and Read 2008; Figs. 11.3 and 11.4). These hyphae are formed by the ingress of Hartig net hyphae through epidermal and cortical cell walls and, once within these cells, the hyphal diameter increases substantially (Piché et al. 1986; Scales and Peterson 1991a; Figs. 11.4 and 11.5). **Ultrastructural details** of the intracellular hyphae of *Wilcoxina mikolae* var. *mikolae* within *Pinus banksiana* root cells close to the apical meristem showed that they are rich in cytoplasmic organelles and are separated from host cell cytoplasm by the development of a host-derived plasma membrane and interfacial matrix material (Scales and Peterson 1991a; Fig. 11.6). The nature of the matrix material

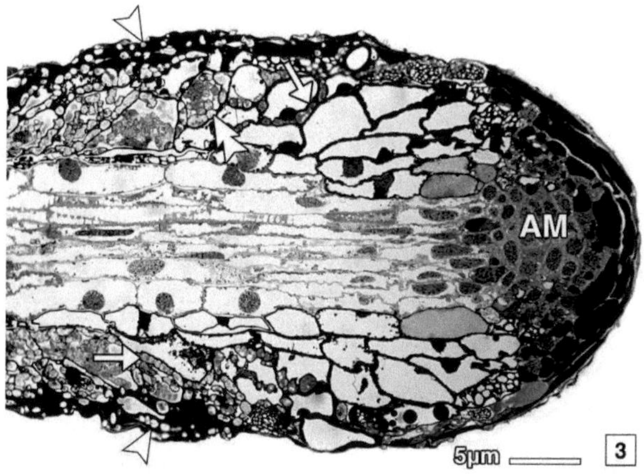

Fig. 11.3. Light microscopy of a longitudinal section of a monopodial short root of *Pinus resinosa* colonized by *Wilcoxina mikolae* var. *mikolae*. Mantle hyphae (*arrowheads*) are embedded in mucilage. Hartig net hyphae (*arrows*) and intracellular hyphae (*double arrowhead*) are present. Colonization has occurred close to the root apical meristem (*AM*)

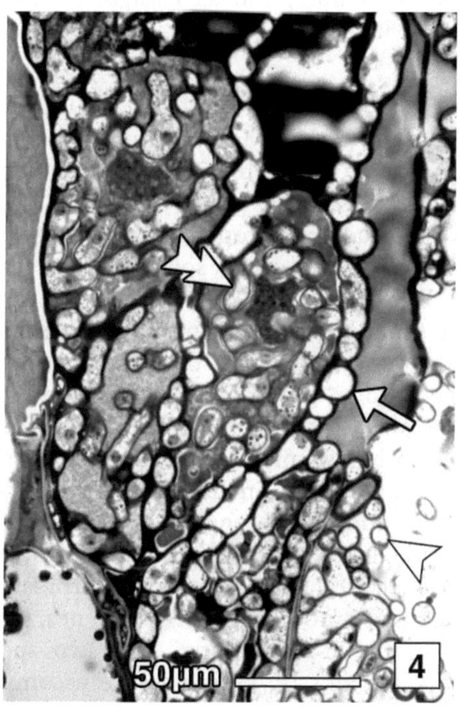

Fig. 11.4. Light microscopy of a longitudinal section of a monopodial short root of *Pinus resinosa* colonized by *Wilcoxina mikolae* var. *mikolae*. Few mantle hyphae (*arrowhead*), Hartig net hyphae (*arrow*) and intracellular hyphae (*double arrowhead*) are evident

has not been determined but its presence, along with the host-derived plasma membrane, suggests that intracellular hyphae may play a role in nutrient exchange between the symbionts; this has not been shown experimentally. Microtubules are closely associated with the intracellular hyphae and these may be involved in the formation of the interface between the symbionts (Kuga-Uetake et al. 2004).

The same isolate of *W. mikolae* var. *mikolae* that forms ectendomycorrrhizas with *P. banksiana* (Scales and Peterson 1991a) forms typical ectomycorrhizas with *Picea mariana* and *Betula alleghaniensis* (Scales and Peterson 1991b), indicating the importance of the host genome in the type of mycorrhiza formed. Similar results were reported previously with other E-strain fungi (Laiho 1965) and these observations led Molina et al. (1992) to conclude that ectendomycorrhizal fungi show broad host range responses with species in which ectomycorrhizas typically form but intermediate specificity with *Pinus* and *Larix* species in which ectendomycorrhizas form. This observation was confirmed experimentally by Massicotte et al. (1999) by growing seedlings of three conifer species (*Abies grandis*, *Pseudotsuga menziesii*, *Pinus ponderosa*) and two angiosperm species (*Lithocarpus densiflora*, *Arbutus menziesii*) in soil collected from three forest sites in southwestern Oregon, United States. The authors suggested that one of the morphotypes formed on all species was consistent with descriptions of that formed by *W. mikolae*.

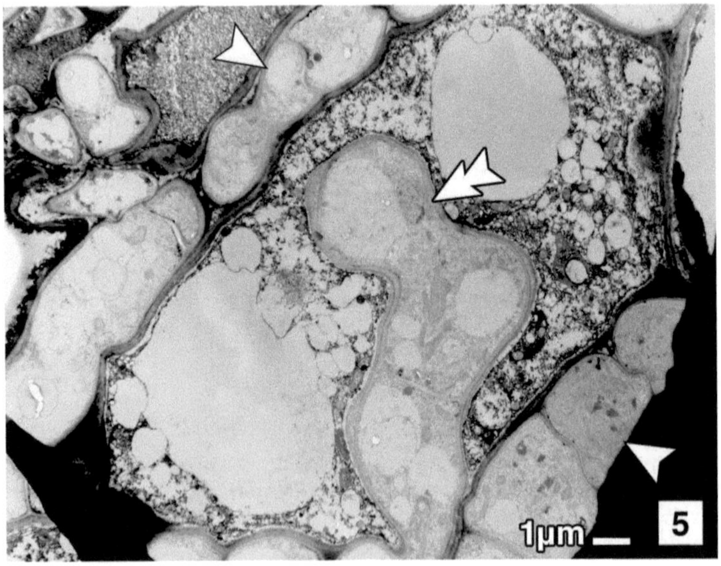

Fig. 11.5. Transmission electron micrograph of cortical cells in of a *Pinus banksiana* short root colonized by *Wilcoxina mikolae* var. *mikolae* showing Hartig net hyphae (*arrowheads*) and an enlarged intracellular hypha (*double arrowhead*)

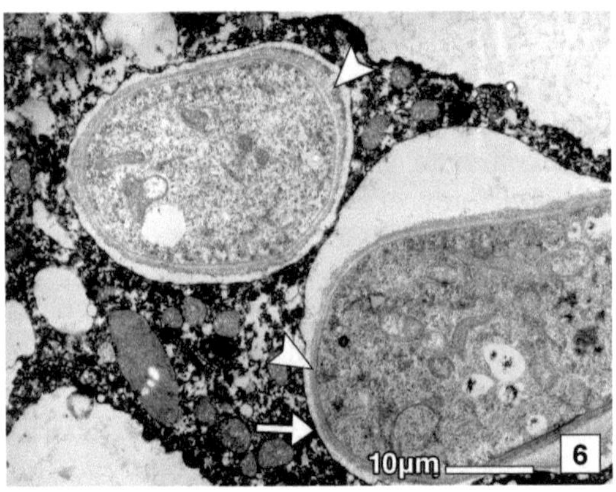

Fig. 11.6. Transmission electron micrograph of a cortical cell in a *Pinus banksiana* short root colonized by *Wilcoxina mikolae* var. *mikolae* showing enlarged intracellular hyphae each surrounded by interfacial material (*arrowheads*) and a host-derived plasma membrane (*arrow*). Some plasmolysis has occurred during fixation

III. Fungal Species Involved

A. Systematics

A detailed discussion of the history of the classification of fungi reported to be involved in the formation of ectendomycorrhizas can be found in Yu et al. (2001). Many of the early studies identified the fungi based on morphological characteristics either of sterile hyphae of isolates cultured from roots of various gymnosperm species or from soil-borne mycelium (e.g. Wilcox et al. 1974; Danielson 1982). These fungi comprised a number of problematic taxonomic isolates that were originally placed into a broad group, 'E-strain' fungi (Laiho and Mikola 1964). They were determined to be ascomycetes based on diagnostic features of hyphae, including the presence of Woronin bodies and regular septation (Danielson 1982). Later studies (Egger and Fortin 1990; Egger et al. 1991) comparing poymorphisms in nuclear and mitochondrial rRNA genes, placed most of the E-strain fungi into two species *W. mikolae* and *W. rehmii* in the ascomycete order Pezizales. Sequence analysis of *W. mikolae* identified two varieties *W. mikolae* var. *mikolae* and var. *tetraspora* (Egger 1996).

Other fungal species that are known to form ectendomycorrhizas under some conditions include *Phialophora finlandia* and *Chloridium paucisporum* (Wang and Wilcox 1985; Wilcox and Wang 1987a, b). In a study of the colonization of *Pinus contorta* (lodgepole pine) roots by a number of post-fire ascomycetes in the Pezizales, Egger and Paden (1986) found that one species, *Sphaerosporella brunnea* formed typical ectendomycorrhizas. This was confirmed for this host species as well as for *P. banksiana* (Jack pine; Iwanyzki 1992). The mycorrhizal status of *S. brunnea* may, however, vary depending on the host species and conditions during mycorrhiza formation (Danielson 1984).

B. Physiology

Physiological aspects of ectendomycorrhizal fungi have not been studied to the same extent as ectomycorrhizal fungi. Mikola (1965) determined the **carbon, nitrogen, and pH requirements** for a number of E-strain isolates from pine and observed that none could use cellulose as a carbon source. However, Caldwell et al. (2000) found that an isolate of *Phialophora finlandia*, when grown in vitro, utilized cellulose, laminarin, starch, and xylan as a carbon source. This isolate also was capable of hydrolyzing protein, ribonucleic acids, and a fatty acid ester. Redlak et al. (2001) found that *Wilcoxina* spp. produced cellulolytic, pectolytic, proteolytic and chitinolytic activity in culture medium but at very low levels. Phenolic compounds had various effects on enzyme production in three ectendomycorrhizal fungal species; this depended on the particular phenolic compound and the enzyme being assayed (Dahm and Redlak 2000). For example, there was no effect on the production of β-glucosidases or pectolytic enzymes by any of the phenolic compounds but endocellulases were inhibited. It is not clear how these results relate to the colonization of roots by ectendomycorrhizal fungi.

Sphaerosporella brunnea is able to hydrolyze complex compounds such as gelatin, cellulose, oil, and pectins, depending on the pH of the medium (Egger 1986). It is not known whether carbon compounds resulting from the breakdown of complex organic compounds could be transported to host roots under certain conditions.

Martin et al. (1988), using ^{13}C-labeled glucose, found evidence for a direct pathway from glucose to mannitol, the main carbohydrate reserve substance in *S. brunnea*, as well as evidence for a lesser accumulation of glycogen and trehalose. In addition, as much as 40% of the ^{13}C-labeled glucose ended up in the free amino acid pools in mycelium.

Ectendomycorrhizal fungi, like other mycorrhizal symbionts, are likely to **benefit plant species by the increased uptake of various nutrients from the substrate.** Mycorrhizal fungi have access to inorganic and organic nitrogen sources in various ecosystems (Smith and Read 2008), an important feature since nitrogen is often limiting to plant growth in many of these ecosystems. Rudawska et al. (1994) confirmed that the ectendomycorrhizal fungal isolate MrgX obtained originally from roots of *Pinus sylvestris* (Pachlewski 1983), grew on medium with ammonium as the source of nitrogen and that this fungus possessed ammonium assimilation enzymes with the glutamine synthetase (GS-GOGAT) pathway being the most important. Prabhu et al. (1995) provided the first evidence for the presence of a NADPH-specific nitrate reductase which catalyzes the first step in nitrate assimilation, in an ectendomycorrhizal ascomycete fungal species, *W. mikolae* var. *mikolae*. By using urea, a neutral nitrogen source, in the culture medium Prabhu et al. (1996a) showed that this enzyme was induced by nitrate and repressed by ammonium.

Two isolates of *W. mikolae* and one of *W. rehmii* were shown to produce the siderophore, ferricrocin (Prabhu et al. 1996b). Siderophores act as chelating agents solubilizing ferric iron and therefore increasing iron absorption by mycorrhizal plants (Haselwandter 1995).

The ectendomycorrhizal fungal isolate MrgX is capable of synthesizing a number of indole compounds (Rudawska et al. 1992). Of these, indole-3-acetic acid (auxin) and indole-3-carboxyl acid are produced in the greatest amounts.

IV. Factors Affecting Mycorrhiza Formation

Temperature has been shown to **affect growth of E-strain fungal isolates** in culture with isolates from northern United States growing better at 20 °C and isolates from southern United States growing better at 24 °C (Wilcox et al. 1983). Northern isolates showed some growth at 4 °C but southern isolates failed to grow at this temperature. Northern isolates formed ectendomycorrhizas with *Pinus resinosa* whereas a southern isolate formed ectomycorrhizas with the same pine species (Wilcox et al. 1983).

Exposure of *P. halepensis* to the **atmospheric pollutants**, ozone and sulfur dioxide in combination, resulted in a decrease in the percentage of mycorrhizal colonization and a change in the morphological appearance of mycorrhizas, with fewer coralloid morphotypes and more simple morphotypes formed (Díaz et al. 1996). Although the fungal symbionts were not identified, structural features of the latter morphotype included a thin mantle, Hartig net hyphae, and intracellular hyphae, typical of ectendomycorrhizas.

Pine nurseries are frequently treated with **herbicides** as a weed control measure and, in a *P. resinosa* nursery in Victoria, Australia, two herbicides, propazine and chlorthal dimethyl are widely used (Marks and Becker 1990). These authors showed that both herbicides suppressed mycorrhiza formation in greenhouse experiments and that in both control and herbicide treatments only two unidentified morphotypes formed, an ectomycorrhiza and an ectendomycorrhiza. The ratio of the ectomycorrhiza morphotype to the ectendomycorrhiza morphotype was reduced in both herbicide treatments. The structure of the ectendomycorrhiza morphotype was modified in the chlorthal dimethyl treatment with only the development of a Hartig net without an evident mantle or intracellular hyphae.

Pathogenic fungi are often problematic in conifer nurseries leading to the use of various **fungicides** to minimize seedling loss. The effect of two fungicides, benomyl and oxine benzoate, on mycelial growth of four ectendomycorrhizal fungal species was studied by Chakravarty et al. (1990). Treatments with both fungicides at 50 ppm and above significantly reduced mycelial growth, indicating that the much higher recommended field rates of both fungicides are likely detrimental to these fungi in nurseries.

There is increasing evidence that the mycorrhizosphere and the hyphosphere host a variety of bacteria (Bending et al. 2006; Smith and Read 2008), some of which have been designated as **'mycorrhiza helper bacteria'** since they have a positive effect on the formation of ectomycorrhizas (Garbaye 1994). Bending et al. (2006) provide a thorough discussion of the diverse interactions that occur between bacteria and the two most prevalent mycorrhizas: ectomycorrhizas and arbuscular mycorrhizas. Research on interactions between bacteria and ectendomycorrhizal fungi is limited. Chanway and Holl (1991) determined the effect of a plant growth promoting *Bacillus* isolate, either alone or in combination with *W. mikolae*, on the growth and nutrient status of *P. contorta* seedlings. Treatment with *Bacillus* alone had no effect on shoot and root biomass or total leaf nitrogen content, whereas treatment with *W. mikolae* alone reduced shoot biomass and total leaf nitrogen content. Inoculation with both resulted in higher root and shoot biomass but lower foliar nitrogen content compared with controls.

Chitinase genes have been **inserted into a number of plant species** in attempts to increase their resistance to pathogenic fungi. The endochitinase gene ech-42 has been transferred into *Picea glauca* (white spruce) and subsequently the transformed seedlings were shown to be more resistant to the root fungal pathogen *Cylindrocladium floridanum* than controls (Noël et al. 2005). Transformed white spruce with the same inserted gene was recently tested for mycorrhiza formation by ectendomycorrhizal *Wilcoxina* spp. (Stefani et al. 2010). The authors showed that mycorrhization was not affected, with roots of transformed seedlings developing a Hartig net and intracellular hyphae. It is of interest however, that *Wilcoxina* spp. usually form ectomycorrhizas with *Picea* spp. (Mikola 1988; Scales and Peterson 1991b).

V. Occurrence and Ecological Considerations

Early studies by Laiho (1965) and Mikola (1965) and later by Mikola (1988) and Lehto (1989) established that ectendomycorrhizas formed on pine seedlings by E-strain fungi were common in Finnish and other European nurseries. A number of other reports confirm the **prevalence of ectendomycorrhizas in pine nurseries** in Canada (Danielson and Visser 1989a; Ursic and Peterson 1997; Ursic et al. 1997), the United States (Laiho 1965; Wilcox 1971; Wilcox et al. 1983), several African countries, New Zealand, and Australia (Mikola 1980). Ectendomycorrizas have also been reported in pine plantations in the United States (Menge and Grand 1981).

There is some evidence that **pine seedlings colonized with ectendomycorrhizal fungi** are more **resistant to harsh environments** than those colonized by ectomycorrhizal fungi. For example, *Pinus resinosa* seedlings inoculated with either the E-strain fungus BDG-58 or *Phialophora finlandia*, both shown to produce typical ectendomycorrhizas with this pine species, had better survival rates than seedlings inoculated with two ectomycorrhizal fungal species when grown on iron tailings (LoBuglio and Wilcox 1988). In a study of survival of *Pinus banksiana* seedlings inoculated with 11 mycorrhizal fungal species and outplanted to oil-sands tailings, only E-strain ectendomycorrhizas were present in substantial numbers after 3 years (Danielson and Visser 1989b).

In a study of successional changes in mycorrhizas of a chronological sequence of *P. banksiana* stands following a wild fire, **E-strain fungi** were **prevalent as early-stage fungi** (Visser 1995) in 6-year-old plantations; these were replaced in older stands, primarily by ectomycorrhizal basidiomycete species. In bioassays with *P. halepensis* seedlings grown in soil collected from two sites in which fire had killed all conifer and shrub species, E-strain morphotypes were up to 20 times more frequent than all ectomycorrhizal morphotypes combined after 1 and 2 years growth (Torres and Honrubia 1997). Also, *W. rehmii*, a known ectendomycorrhizal fungus on pine species, was the most common ascomycete identified by analysis of 18S rDNA and ITS 1 data collected from colonized root tips of *P. ponderosa* after a prescribed burn in eastern Oregon, United States (Fujimura et al. 2005).

The **widespread occurrence of ectendomycorrhizas** geographically, their occurrence in pine seedlings under nursery conditions, and the resilience of ectendomycorrhizal fungal species following various disturbance events, suggests that they **may play important roles in conifer seedling establishment**. However, more controlled experiments are needed before decisions are made to use these fungal species as inoculum for outplanted seedlings.

References

Beck A, Kottke I, Oberwinkler F (2005) Two members of the Glomeromycota form distinct ectendomycorrhizas with *Alzatea verticillata*, a prominent tree in the mountain rain forest of southern Ecuador. Mycol Prog 4:11–22

Bending GD, Aspray TJ, Whipps JM (2006) Significance of microbial interactions in the mycorrhizosphere. Adv Appl Microbiol 60:97–132

Brundrett M (2004) Diversity and classification of mycorrhizal associations. Biol Rev 79:473–495

Caldwell BA, Jumpponen A, Trappe JM (2000) Utlilization of major detrital substrates by dark-septate, root endophytes. Mycologia 92:230–232

Chakravarty P, Jacobs PF, Peterson RL (1990) Effect of the fungicides benomyl and oxine benzoate on the mycelial growth of four isolates of E-strain fungi in vitro. Eur J For Pathol 20:381–385

Chanway CP, Holl FB (1991) Biomass increase and associative nitrogen fixation of mycorrhizal *Pinus contorta* seedlings inoculated with a plant growth promoting *Bacillus* strain. Can J Bot 69:507–511

Dahm H, Redlak K (2000) Effect of phenolic compounds on the enzymatic activity of ectendomycorrhizal fungi. Sylwan 144:77–85

Danielson RM (1982) Taxonomic affinities and criteria for identification of the common ectendomycorrhizal symbiont of pines. Can J Bot 60:7–18

Danielson RM (1984) Ectomycorrhiza formation by the operculate discomycete *Sphaerosporella brunnea* (Pezizales). Mycologia 76:454–461

Danielson RM, Visser S (1989a) The mycorrhizal status and nodulation status of container-grown trees and shrubs reared in commercial nurseries. Can J For Res 20:609–614

Danielson RM, Visser S (1989b) Host response to inoculation and behaviour of introduced and indigenous ectomycorrhizal fungi of jack pine grown on oil-sands tailings. Can J For Res 19:1412–1421

Díaz G, Barrantes O, Honrubia M, Gracia C (1996) Effect of ozone and sulphur dioxide on mycorrhizae of *Pinus halepensis* Miller. Ann Sci For 53:849–856

Egger KN (1986) Substrate hydrolysis patterns of post-fire ascomycetes (Pezizales). Mycologia 78:771–780

Egger KN (1996) Molecular systematics of E-strain mycorrhizal fungi: *Wilcoxina* and its relationship to *Tricharina* (Pezizales). Can J Bot 74:773–779

Egger KN, Fortin JA (1988) Ectendomycorrhizae: diversity and classification. In: Lalonde M, Piché Y (eds) Canadian workshop on mycorrhizae in forestry. CRBF, Université Laval, Ste-Foy, pp 113–114

Egger KN, Fortin JA (1990) Identification of taxa of E-strain mycorrhizal fungi by restriction fragment analysis. Can J Bot 68:1482–1488

Egger KN, Paden JW (1986) Biotrophic associations between lodgepole pine seedlings and postfire ascomycetes (Pezizales) in monoxenic culture. Can J Bot 64:2719–2725

Egger KN, Danielson RM, Fortin JA (1991) Taxonomy and population structure of E-strain mycorrhizal fungi inferred from ribosomal and mitochondrial DNA polymorphisms. Mycol Res 95:866–872

Fujimura KE, Smith JE, Horton TR, Weber NS, Spatafora JW (2005) Pezizalean mycorrhizas and sporocarps in ponderosa pine (*Pinus ponderosa*) after prescribed fires in eastern Oregon, USA. Mycorrhiza 15:79–86

Garbaye J (1994) Helper bacteria: a new dimension to the mycorrhizal symbiosis. New Phytol 128:197–210

Haselwandter K (1995) Mycorrhizal fungi: siderophore production. Crit Rev Biotechnol 15:287–291

Ivory MH, Pearce RB (1991) *Wilcoxina mikolae* newly identified as a mycorrhizal fungus on pines in Africa. Mycol Res 95:250–253

Iwanyzki S (1992) Structure and development of mycorrhizae formed between two pine species (*Pinus banksiana, Pinus contorta*) and the postfire ascomycete *Sphaerosporella brunnea*. MSc thesis, University of Guelph, Guelph

Kuga-Uetake Y, Purich M, Massicotte HB, Peterson RL (2004) Host microtubules in the Hartig net region of ectomycorrhizas, ectendomycorrhizas, and monotropoid mycorrhizas. Can J Bot 82:938–946

Laiho O (1965) Further studies on the ectendotrophic mycorrhiza. Acta For Fenn 79:1–35

Laiho O, Mikola P (1964) Studies on the effect of some eradicants on mycorrhizal development in forest nurseries. Acta For Fenn 77:1–33

Lehto T (1989) Mycorrhizal status of Scots pine nursery stock in Finland. Folia For 726:3–15

LoBuglio KF, Wilcox HE (1988) Growth and survival of ectomycorrhizal and ectendomycorrhizal seedlings of *Pinus resinosa* on iron tailings. Can J Bot 66:55–60

Marks GC, Becker SL (1990) Influence of propazine and chlorthal dimethyl on mycorrhizal development in *Pinus radiata* seedlings. Austral J Bot 38:341–349

Martin F, Ramstedt M, Söderhäll K, Canet D (1988) Carbohydrate and amino acid metabolism in the ectomycorrhizal ascomycete *Sphaerosporella brunnea* during glucose utilization. Plant Physiol 86:935–940

Massicotte HB, Molina R, Tackaberry LE, Smith JE, Amaranthus MP (1999) Diversity and host specificity of ectomycorrhizal fungi retrieved from three adjacent forest sites by five host species. Can J Bot 77:1053–1076

Menge JA, Grand LF (1981) Mycorrhizae of 11-year old *Pinus taeda* L. plantations in North Carolina. J Elisha Mitchell Sci Soc 97:55–66

Mikola P (1965) Studies on the ectendotrophic mycorrhiza of pine. Acta For Fenn 79:1–56

Mikola P (1980) Mycorrhizae across the frontiers. In: Mikola P (ed) Tropical mycorrhiza research. Clarendon, Oxford, pp 1–10

Mikola P (1988) Ectendomycorrhiza of conifers. Silva Fenn 22:19–27

Molina R, Massicotte H, Trappe JM (1992) Specificity phenomena in mycorrhizal symbioses: community-ecological consequences and practical implications. In: Allen MF (ed) Mycorrhizal functioning. An integrative plant–fungal process. Chapman and Hall, New York, pp 357–423

Noël A, Levasseur C, Le VQ, Séguin A (2005) Enhanced resistance to fungal pathogens in forest trees by genetic transformation of black spruce and hybrid poplar with a *Trichoderma harzianum* endochitinase gene. Physiol Mol Plant Pathol 67:92–99

Pachlewski R (1983) Symbiotic fungi and mycorrhiza of Scotch pine (*Pinus sylvestris* L.). Pr Inst Badaw Lesn 615:3–132

Peterson RL, Massicotte HB (2004) Exploring structural definitions of mycorrhizas, with emphasis on nutrient-exchange interfaces. Can J Bot 82:1074–1088

Peterson RL, Massicotte HB, Melville LM (2004) Mycorrhizas: anatomy and cell biology. NRC Research Press, Ottawa

Piché Y, Ackerley CA, Peterson RL (1986) Structural characteristics of ectendomycorrhizas synthesized between roots of *Pinus resinosa* and the E-strain fungus, *Wilcoxina mikolae* var. *mikolae*. New Phytol 104:447–452

Prabhu V, Wilcox HE, Boyer GL (1995) Properties of nitrate reductase from the mycorrhizal ascomycete *Wilcoxina mikolae* var. *mikolae*. Mycol Res 99:1356–1360

Prabhu V, Wilcox HE, Boyer GL (1996a) Regulation of nitrate reductase in the mycorrhizal ascomycete *Wilcoxina mikolae* var. *mikolae*. Mycol Res 100:333–336

Prabhu V, Biolchini PF, Boyer GL (1996b) Detection and identification of ferricrocin produced by ectendomycorrhizal fungi in the genus *Wilcoxina*. Biometals 9:229–234

Redlak K, Dahm H, Ciesielska A, Strzelczyk E (2001) Enzymatic activity of ectendomycorrhizal fungi. Biol Fertil Soils 33:83–90

Rudawska M, Bernillon J, Gay G (1992) Indole compounds released by the ectendomycorrhizal fungal strain MrgX isolated from a pine nursery. Mycorrhiza 2:17–23

Rudawska M, Kieliszewska-Rokicka B, Debaud J-C, Lewandowski A, Gay G (1994) Enzymes of ammonium metabolism in ectendomycorrhizal and ectomycorrhizal symbionts of pine. Physiol Plant 92:279–285

Scales PF, Peterson RL (1991a) Structure and development of *Pinus banksiana* – *Wilcoxina* ectendomycorrhizae. Can J Bot 69:2135–2148

Scales PF, Peterson RL (1991b) Structure of ectomycorrhizae formed by *Wilcoxina mikolae* var. *mikolae* with *Picea mariana* and *Betula alleghaniensis*. Can J Bot 69:2149–2157

Schelkle M, Ursic M, Farquhar M, Peterson RL (1996) The use of laser scanning confocal microscopy to characterize mycorrhizas of *Pinus strobus* L. and to localize associated bacteria. Mycorrhiza 6:431–440

Setaro S, Weiss M, Oberwinkler F, Kottke I (2006) Sebacinales form ectendomycorrhizas with *Cavendishia nobilis*, a member of the Andean clade of Ericaceae, in the mountain forest of southern Ecuador. New Phytol 169:355–365

Smith SE, Read DJ (2008) Mycorrhizal symbiosis, 3rd edn. Academic, New York

Stefani FOP, Tanguay P, Pelletier G, Piché Y, Hamelin RC (2010) Impact of endochitinase-transformed white spruce on soil fungal biomass and ectendomycorrhizal symbiosis. Appl Environ Microbiol 76:2607–2614

Torres P, Honrubia M (1997) Changes and effects of a natural fire on ectomycorrhizal inoculum potential of soil in a *Pinus halepensis* forest. For Ecol Manage 96:189–196

Ursic M, Peterson RL (1997) Morphological and anatomical characterization of ectomycorrhizas and ectendomycorrhizas on *Pinus strobus* seedlings in a southern Ontario nursery. Can J Bot 75:2057–2072

Ursic M, Peterson RL, Husband B (1997) Relative abundance of mycorrhizal fungi and frequency of root rot on *Pinus strobus* seedlings in a southern Ontario nursery. Can J For Res 27:54–62

Visser S (1995) Ectomycorrhizal fungal succession in jack pine stands following wildfire. New Phytol 129:389–401

Wang CJK, Wilcox HE (1985) New species of ectendomycorrhizal and pseudo mycorrhizal fungi: *Phialophora finlandia*, *Chloridium paucisporum*, and *Phialocephela fortinii*. Mycologia 77:951–958

Wilcox HE (1969) Morphology of ectendomycorrhizae in *Pinus resinosa*. In: USDA (ed) Proceedings of the first North American Conference on mycorrhizae. USDA miscellaneous publication 1189. USDA Forest Service, Washington, DC, pp 54–68

Wilcox HE (1971) Morphology of ectendomycorrhizae in *Pinus resinosa*. In: Hacskylo E (ed) Mycorrhizae. USDA, Washington, DC, pp 54–68

Wilcox HE, Ganmore-Neumann R (1974) Ectendomycorrhizae in *Pinus resinosa* seedlings. I. Characteristics of mycorrhizae produced by a black imperfect fungus. Can J Bot 52:2145–2155

Wilcox HE, Wang CJK (1987a) Ectomycorrhizal and ectendomycorrhizal associations of *Phialophora finlandia* with *Pinus resinosa*, *Picea rubens*, and *Betula alleghaniensis*. Can J For Res 17:976–990

Wilcox HE, Wang CJK (1987b) Mycorrhizal and pathological associations of dematiaceous fungi in roots of 7-month-old tree seedlings. Can J For Res 17:884–889

Wilcox HE, Ganmore-Neumann R, Wang CJK (1974) Characteristics of two fungi producing ectendomycorrhiza in *Pinus resinosa*. Can J Bot 52:2279–2282

Wilcox HE, Yang CS, LoBuglio KF (1983) Responses of pine roots to E-strain ectendomycorrhizal fungi. Plant Soil 71:293–297

Yu EJ-C, Egger KN, Peterson RL (2001) Ectendomycorrhizal associations – characteristics and functions. Mycorrhiza 11:167–177

12 Orchid Mycorrhizas: Molecular Ecology, Physiology, Evolution and Conservation Aspects

J.D.W. Dearnaley[1], F. Martos[2], M.-A. Selosse[3]

CONTENTS

[1]Australian Centre for Sustainable Catchments and Faculty of Sciences, University of Southern Queensland, Toowoomba 4350, Australia; e-mail: john.dearnaley@usq.edu.au
[2]School of Life Sciences, University of KwaZulu-Natal, Private Bag X01 Scottsville, Pietermaritzburg 3209, South Africa
[3]Centre d'Ecologie Fonctionnelle et Evolutive (CNRS, UMR 5157), 1919 Route de Mende, 34293 cedex 5 Montpellier, France

I. Introduction

Since the pioneering works by Noël Bernard (1874–1911; Boullard 1985), whose centenary of death was celebrated in 2011, scientific interest in orchid mycorrhizas has continued to grow. This might seem somewhat surprising as the association concerns a single angiosperm family and is less widespread among land plants than the ectomycorrhizal (ECM) or arbuscular mycorrhizal (AM) symbioses (Smith and Read 2008). Impetus for research into orchid mycorrhizas has been multifaceted. As the earth's largest flowering plant family with 27,135 accepted species (The Plant List 2010), accumulating basic biological knowledge of what represents approximately 10 % of the botanical kingdom diversity is justified. For the mycologist, these plants that shifted from an ancestral AM symbiosis with Glomeromycetes to an original symbiosis with new fungal partners (Yukawa et al. 2009) offer a window on mycorrhizal abilities in numerous, and sometimes unexpected, fungal lineages. Moreover, many species, such as aromatic *Vanilla* spp. or many ornamental species, are of economic value. Illuminating studies of orchid mycorrhizas have shown that some non-chlorophyllous plants live as parasites on ECM interactions or saprotrophic fungi and that some green orchids are partially heterotrophic as adults. Recent research has given insights into the nature of the mycorrhizal association of autotrophic orchids, suggesting that the association may be mutualistic (Cameron et al. 2006, 2008). The mycorrhizas of orchids also offer general perspectives on the evolution of specificity and

Fungal Associations, 2nd Edition
The Mycota IX
B. Hock (Ed.)

mycorrhizal networks among plants. Finally, as orchids require the presence of suitable fungal partners for seed germination and seedling establishment, a more complete understanding of the mycorrhizal biology of the many threatened orchid species is required for conservation action plans.

Orchids have historically been divided into three main types on the basis of lifestyle, i.e. terrestrial (soil dwelling), epiphytic (plant surface dwelling) and lithophytic (rock surface dwelling) species. Recent literature (e.g. Gebauer and Meyer 2003; Selosse et al. 2004; for reviews, see Merckx et al. 2009; Selosse and Roy 2009) has suggested a division of orchids into three physiological types based on carbon nutrition. Fully autotrophic species (the majority of taxa) are those that are chlorophyllous and, as adults, obtain their carbon compounds via photosynthetic pathways. Fully mycoheterotrophic (MH) species (approximately 200 species worldwide; Leake 1994, 2004) are dependent on fungal carbon throughout their life cycle. A third type, the partial MH species or mixotrophs (Julou et al. 2005; Selosse and Roy 2009) are intermediate, carrying out some photosynthetic carbon fixation as well as receiving fungal carbon.

Regardless of their carbon nutrition at adult stage, all orchids produce minute, endosperm-lacking seeds and are dependent on fungal colonization for germination and growth into an underground heterotrophic, achlorophyllous stage called a protocorm (Rasmussen 1995; Smith and Read 2008). Environmental fungi colonize though embryo suspensor tissues or epidermal hairs and enter the cortical cells. Colonizing fungal hyphae do not breach the cortical cell membrane but ramify in the space between cell wall and membrane, forming elaborate coiled structures known as pelotons (Fig. 12.1) that collapse at later stages as a result of plant digestion. Intact fungal coils and not collapsed pelotons are likely the site of nutrient exchange between plant and fungus, as indicated by the fact that in vitro grown protocorms show a growth response **before** peloton collapse (Hadley and Williamson 1971) and that the nutrient fluxes after labelling pulses occur too rapidly to be accounted for by hyphal digestion (Cameron et al. 2008).

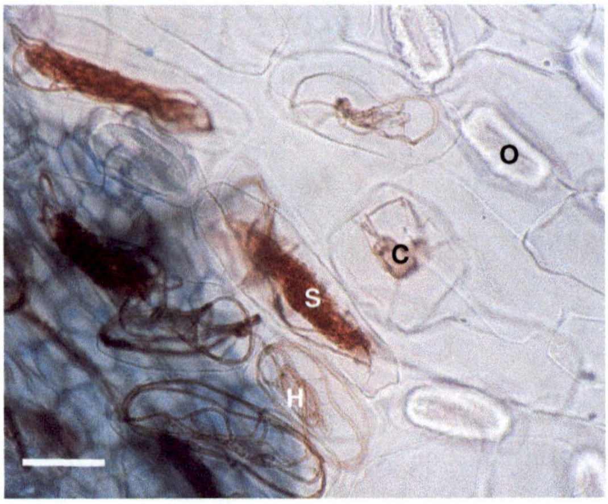

Fig. 12.1. Light microscopy image showing healthy (*H*), slightly degraded (*S*) and collapsed (*C*) fungal pelotons of a *Thelephora* sp. in *Cephalanthera longifolia* roots (Ülle Püttsepp, unpublished micrograph; plant investigated in Abadie et al. (2006)). O Oxalate crystal. *Bar* 50 μm

This chapter will focus on current understanding of the mycorrhizal associations of these three physiological orchid types. For a more comprehensive review of the colonization process and anatomy of orchid mycorrhizas readers are directed to Smith and Read (2008). Here it is intended to update the 10-year-old excellent reviews by Rasmussen (2002) and Taylor et al. (2002), to provide an overview of the contemporary approaches to studying these interactions, to elaborate on what has been gleaned from these studies with regards to the ecology, physiology, evolution and conservation aspects of orchid mycorrhizas and to highlight areas of the association that need further exploration.

II. New Techniques for Studying Orchid Mycorrhizas

A. Molecular Barcoding Approaches

Historically, much knowledge about orchid mycorrhizas has been acquired from **in vitro isolation of fungi**. This has allowed basic fungal identification and simple in vitro seed germination experiments conducted with some root-isolated fungi (e.g. Warcup 1971;

Clements 1988). Indeed, the orchid mycorrhizal association represents possibly one of the easiest symbiotic systems to manipulate under laboratory conditions as both partners can be cultured axenically, at least in the case of the early stages of the fully autotrophic orchids. A hurdle in these types of investigations has been an inability to accurately identify the isolated fungal partners and this has been especially critical to orchid conservation procedures involving restorative work; moreover, isolation often provided mostly contaminants or endophytes (i.e. fungi that for all or part of their life cycle inhabit living plant tissues but do not form pelotons nor cause any obvious disease symptoms; Wilson 1995).

Molecular taxonomy approaches have enhanced fungal taxonomy, especially by isolating fungal DNA and sequencing the nuclear ribosomal DNA (Seiffert 2009). The fungal partners of orchid mycorrhizas can be more accurately and routinely identified from cultured fungi or directly from orchid protocorms, roots, tubers and rhizomes (e.g. Bougoure et al. 2005; Martos et al. 2009; Swarts et al. 2010). For mycobionts recalcitrant to axenic growth, PCR amplification of colonized orchid tissues using fungus-specific primers is commonly used (Dearnaley and Le Brocque 2006; Dearnaley and Bougoure 2010).

Sequencing of the internal transcribed spacer (ITS) of the nuclear ribosomal DNA after PCR amplification using a variety of primer combinations (White et al. 1990; Gardes and Bruns 1993) has been the method of choice for identifying orchid mycobionts over the past decade. One problem is the amplification recalcitrance of Tulasnellaceae, a frequent orchid mycorrhizal taxon (see Sect. III.A), to the 'universal' fungal ITS primers, because they have highly derived nuclear ribosomal DNA sequences. This entailed the need for additional PCR amplifications using Tulasnellaceae-specific PCR primers (Bidartondo et al. 2004; Selosse et al. 2004). Suarez et al. (2006) introduced the Tulasnellaceae-specific primer 5.8S-Tul to amplify the 5' part of 28S rDNA, which was used by Martos et al. (2012). This primer works well on a wide range of clades of Tulasnellaceae and is expected to be frequently used in future studies of orchid mycorrhizal fungi because of the high heterogeneity of the ITS alignment. Recently, some primer pairs specifically devoted to orchid mycorrhizal fungi were described (Taylor and McCormick 2007), but the constantly growing number of fungal taxa (see Sect. III) questions their relevance in the new orchid lineages to be explored. Sequencing of cloned ITS PCR products is often carried out with orchids displaying low fungal specificity (e.g. Selosse et al. 2002; Dearnaley 2006; Liebel et al. 2010; Martos et al. 2012). Sequencing of the large subunit (LSU) of the nuclear ribosomal DNA of the Sebacinales, common orchid mycobionts worldwide, is necessary for higher resolution separation of groups A and B, two major clades in this group (Weiβ et al. 2004, 2011; Selosse et al. 2009). Huynh et al. (2009) also recently showed that ITS sequencing may not sufficiently distinguish isolates of the '*Sebacina vermifera*' complex (Sebacinales group B), common mycobionts of spider orchids in Australia. ITS sequencing and cloning may also reveal many endophytes (Bidartondo et al. 2004; Julou et al. 2005; Abadie et al. 2006; Roy et al. 2009a). In Martos et al. (2012), a wide range of ascomycete and basidiomycete endophytes was identified, perhaps more than in any study of orchid associated fungi to date.

Fungal endophytes are very frequently selected during in vitro isolation or PCR amplification from orchid tissues with fungus-specific primers: dissecting single fungal pelotons from roots tissues before in vitro isolations (Zhu et al. 2008) or PCR amplifications (Rasmussen 1995; Kristiansen et al. 2001) is thus strongly recommended in future work to avoid endophytes. The important diversity of endophytic fungi, mainly from the Helotiales (e.g. *Phialophora*, *Leptodontidium* or *Bisporella* spp.) or Xylariales, will not be discussed in detail here (for a review, see Bayman and Otero 2006), while their effect on orchid growth (potentially deleterious in some species; Bayman et al. 2002) and physiology deserves further study. Chaetotyriales are very common orchid endophytes, at least in tropical areas. Capnodiales are also common in epiphytic taxa, but they might be involved in lichenic symbioses. Many epiphytic orchids root in bryophytes or lichens.

Recently, Jacquemyn et al. (2010) and Lievens et al. (2010) introduced **DNA array technologies** for the identification of orchid fungal partners: oligonucleotides were prepared from a preliminary exploration of fungal diversity in a limited number of individuals (Lievens et al. 2010), and the array was successfully used to investigate the fungal partners of three closely related *Orchis* species and their hybrids (Jacquemyn et al. 2011). This method allows fast and efficient handling of numerous samples, especially compared to the cloning of PCR products. However, some fungal partners may remain overlooked when using this procedure because preliminary exploration overlooks rare fungal taxa that may not be targeted during further investigation (such as taxon 8 and 9 from the Thelephoraceae and Cortinariaceae, respectively, in Lievens et al. 2010; Jacquemyn et al. 2011).

B. Stable and Radioactive Isotopes

A common, but indirect approach to determine the mode of nutrition of individual orchid taxa is **mass spectrometric analysis** of natural C and N isotope abundances (Bidartondo et al. 2004; Julou et al. 2005; Abadie et al. 2006; Zimmer et al. 2007; Ogura-Tsujita et al. 2009; Martos et al. 2009). Fully MH species have been identified to have ^{13}C signatures similar to those of their mycorrhizal partners (Gebauer and Meyer 2003; Trudell et al. 2003) and similar or higher ^{15}N abundance than their mycorrhizal fungi, suggesting a limited trend to ^{15}N accumulation along the food chain (Trudell et al. 2003). As expected, mixotrophs have stable isotope signatures intermediate between fully MH and autotrophic species (Julou et al. 2005; Abadie et al. 2006). Some fully autotrophic species such as *Goodyera* spp. have even lower amounts of these natural isotopes as expected for plants less reliant on nutrient acquisition from fungi (Gebauer and Meyer 2003; Bidartondo et al. 2004). The strength of this method is that abundances oversee the long-term metabolism of the plants, with little interference from the observer.

On the fungal side, Latalova and Balaz (2010) showed that a *Tulasnella* species associated in vitro with the orchid *Serapias strictiflora* was able to mix carbon from the orchid (a C_3 plant) and dead maize roots (a C_4 plant enriched in ^{13}C). The fungus was able to grow with the orchid alone, with ^{13}C abundance close to its host, while addition of dead maize roots resulted in an isotopic shift, so that the latter source furnished ca. 30 % of the fungal biomass.

Experiments tracing **the movement of isotopically labelled compounds** to orchid mycorrhizas have been especially revealing. Although they only provide snapshot views of the metabolism at the time of pulse, they allow tracking of exchanges between symbionts. McKendrick et al. (2000) provided the first clear demonstration of movement of ^{14}C-labelled photosynthates from tree species to the fully MH orchid *Corallorhiza trifida* via ECM fungi. Bougoure et al. (2010) recently demonstrated the flow of ^{13}C-labelled carbon from *Melaleuca scalena* to the fully MH orchid *Rhizanthella gardneri* via an ECM fungal conduit. *R. gardneri* also obtained nitrogen from its fungal partner as indicated by adding $^{13}C+^{15}N$-labelled glycine to hyphae and surrounding soil. Labelling experiments also demonstrated that the fully autotrophic orchid *Goodyera repens* acquires carbon, nitrogen and phosphorous from its fungal partner (Cameron et al. 2006, 2007, 2008). Notably, *G. repens* also transfers significant amounts of photosynthate (likely greater than 3 % of its photosynthetic carbon) back to its *Ceratobasidium* mycobiont – the first direct demonstration of a net carbon flow from orchid to fungi (Cameron et al. 2006, 2008).

C. Other Approaches

In contrast to other mycorrhizal symbioses, such as ECM and AM associations, **gene expression studies** in orchid mycorrhizas have largely been neglected. Watkinson and Welbaum (2003) analysed gene expression in the mycorrhizal association of *Cypripedium parviflorum* var. *pubescens* via differential mRNA display. A trehalose phosphate phosphatase was downregulated in the association, indicating changes to orchid carbohydrate transport. Upregulation of a nucleotide binding protein possibly indicated increased cytokinesis during orchid colonization. As indicated by Dearnaley (2007), modern gene expression techniques such as microarrays, RT-PCR and in situ hybridization may provide additional understanding of the molecular functioning of orchid mycorrhizas. In particular, whole-genome sequencing and transcript profiling of orchid mycorrhizal

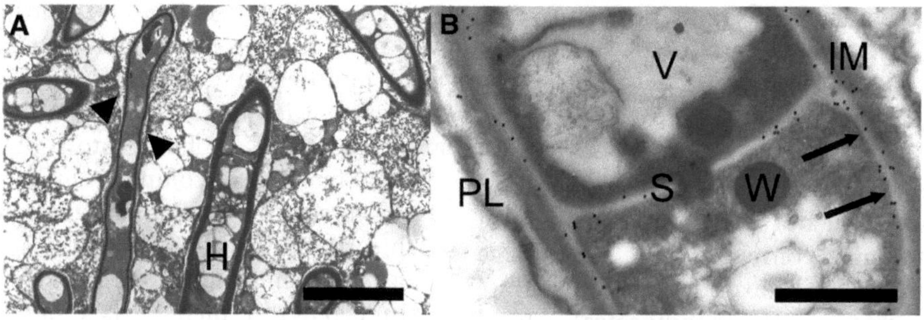

Fig. 12.2. TEM images of *Epipactis microphylla* cells colonized by truffles. (**A**) Truffle septate hyphae (*H*) inside an orchid host cell, where the host plasma membrane (*arrowheads*) tightly surrounds the fungus (*bar* 4 μm). (**B**) Gold granules are regularly distributed (*arrows*) on the longitudinal and septum wall of the fungus after immunogold reaction with the anti-Tbsp1 antibody specific for a truffle phospholipase. *IM* Interfacial material, *PL* plasma membrane of the host cell, *S* septum, *V* vacuole, *W* Woronin bodies. *Bar* 0.6 μm (Modified from Selosse et al. (2004), reproduced with permission of the publisher)

fungi, both free-living and *in planta*, may reveal fungal genes that are upregulated in the symbiosis (Martin et al. 2008).

The use of **electron microscopy** to investigate fungal symbionts in orchid mycorrhizas had somewhat of a rebirth in the past decade (e.g. Pereira et al. 2003; Selosse et al. 2004; Suarez et al. 2008; Martos et al. 2009; Kottke et al. 2010; Schatz et al. 2010; Martos et al. 2012). First, features of the fungal cell wall as well as septal structure, e.g. dolipore and parenthesomes, allow a distinction of the three major mycorrhizal taxa encompassed under the name 'rhizoctonia' (see Sect. III.A; Moore 1987). Moreover, it has been used to confirm how some unexpected taxa do form pelotons and thus are mycorrhizal. Kottke et al. (2010) has given support to molecular data suggesting that Atractiellomycetes, members of the rust lineage (Pucciniomycotina), are mycorrhizal in some neotropical orchids. Selosse et al. (2004) corroborated molecular identification of ascomyceteous *Tuber* spp. as the main mycorrhizal partners in *Epipactis microphylla* by using transmission electron microscopy to check for the presence of Woronin bodies in pelotons and immunogold reactions using antibodies specifically raised against a truffle phospholipase A2 (Fig. 12.2) - interestingly, in this study, basidiomycetes that were found by molecular means were never seen by microscopy. Immunolabelling transmission electron microscopy has been used to demonstrate pectin deposition in the interfacial matrix around *Ceratobasidium* hyphae, but not *Russula* hyphae, in adjacent mycorrhizal root cells of *Limodorum abortivum*, highlighting an orchid's exquisite capability to react distinctly to different fungal symbionts (Paduano et al. 2011). Finally, Huynh et al. (2004) used scanning electron microscopy imaging of stems and protocorms to determine the most effective fungal isolates for conservation of the threatened *Caladenia formosa*.

Other valuable new approaches include: (1) an orchid root peloton isolation and culturing method that maximizes the number of mycorrhizal fungi obtained but minimizes contamination from non-mycorrhizal fungi and bacteria (Zhu et al. 2008) and (2) a modification of the seed packet burial technique originally conceived by Rasmussen and Whigham (1993) which involves removal of site soil and monitoring of symbiotic seed germination under laboratory conditions (Brundrett et al. 2003). (3) Another method of orchid mycorrhizal fungal identification was proposed by Kristiansen et al. (2001), that is, PCR amplification from single pelotons. Now that large-scale environmental detection of fungi is possible through such approaches as t-RFLP (Dickie and FitzJohn 2007), DGGE (Bougoure and Cairney 2005), pyrosequencing (Dumbrell et al. 2011) and DNA microarrays (Lievens et al. 2010), it will be intriguing to see how populations of orchid mycobiota change with time, orchid life stage and environmental conditions.

III. The Diversity of Orchid: Fungus Associations

A. The Diversity of the 'Rhizoctonias'

For many years orchids were considered to interact largely, if not only, with members of the '**rhizoctonia**' complex. This assemblage contains three now taxonomically disparate Agaricomycetes (=Hymenomycetes) taxa: **Sebacinales, Ceratobasidiaceae** and **Tulasnellaceae** (Table 12.1). None of them actually fit the exact definition of the asexual genus *Rhizoctonia* by De Candolle (1815), i.e. the absence of sporulation and formation of sclerotia, so that the name 'rhizoctonia' will be used here not in a taxonomic way, but only to conveniently encompass the three above-mentioned taxa, which are common orchid partners. 'Rhizoctonias' have also been divided into two asexual genera, namely *Ceratorhiza* and *Epulorhiza* (Table 12.1), but this approach is uncomfortable to non-mycologists and, given the current trend to abandon asexual classification, we recommend no longer using these names.

Recent research has highlighted the diverse ecology of these three 'rhizoctonia' taxa. While some species are known to be parasitic, such as in the Ceratobasidiaceae, or are suspected to be saprotrophic, e.g. due to their cultivability in vitro on organic substrates, this classical view (Smith and Read 2008) is now challenged at least for some species. Sebacinales encompasses two major groups (Weiβ et al. 2011) that both occur as endophytes in the roots of many plant species (Selosse et al. 2009): group B additionally forms mycorrhizae with green orchids and Ericaceae, while group A forms ECM on trees and is also associated with some MH orchids (see Sect. III.B; group A is usually not encompassed in 'rhizoctonias'). In an interesting example of convergent evolution, group B is involved in symbiotic germination of *Pyrola* spp. (Ericaceae), another taxon with dust-seeds and MH germination (Hashimoto et al. 2012). ECM clades may exist within the Tulasnellaceae (Bidartondo et al. 2003) and Ceratobasidiaceae (Yagame et al. 2008, 2012; Collier and Bidartondo 2009), and noteworthy MH orchids were instrumental in establishing ECM abilities in these taxa (see Sect. III.B). However, it is unlikely that ECM 'rhizoctonias' are mycorrhizal in fully autotrophic orchids. We are far from a complete understanding of the diversity of nutritional strategies (out of orchid roots) for the Tulasnellaceae and Ceratobasidiaceae: at least, we suspect that their main ecological niche exists out of orchids roots.

Molecular taxonomic identification of orchid mycobionts has now revealed that the **diversity of orchid associates** is much more complex and that other basidiomycetes and even ascomycetes can be involved in orchid mycorrhizas (Table 12.1). The recent overall picture (discussed by Motomura et al. 2010) is that autotrophic orchids largely associate with 'rhizoctonias' worldwide. However, in tropical regions, Atractiellomycetes (Pucciniomycotina) may be common mycorrhizal partners of some epiphytic and terrestrial autotrophic orchids, as shown in the neotropics (Kottke et al. 2010) and in the paleotropics (Martos et al. 2012). The study of South African Diseae (*Pterygodium* and *Corycium* spp.) revealed ECM Ascomycetes such as *Tricharina* and *Peziza* (Waterman et al. 2011), although no direct visualization was obtained. One may expect this list of mycobionts to enlarge in the future. Nevertheless, the study of the earliest-diverging orchid lineages and distribution of fungal associates across orchid phylogeny support that the ancestral state is an association to the three 'rhizoctonia' lineages (Yukawa et al. 2009). Interestingly, Tulasnellaceae turn out to be the most frequently found 'rhizoctonias', in both temperate and tropical regions (Rasmussen 1995; Yuan et al. 2010): in a survey of 77 orchid species from La Réunion island (Indian Ocean), Martos et al. (2012) found them in 88 % of the investigated species (versus 42 % for Sebacinales and 18 % for the Ceratobasidiaceae). By contrast, mixotrophic or fully MH orchids revealed associations with more diverse fungal lineages.

Table 12.1. Summary of the fungal genera forming orchid mycorrhizas. Examples of studies which have identified mycorrhizal genera are given. Taxa in bold are the three groups usually named 'rhizoctonias' in the orchid literature (see text; including the asexual genera *Ceratorhiza*[b] and *Epulorhiza*[a])

- *Phylum Basidiomycota*
  - Sub phylum Pucciniomycotina
    - Class Atractiellomycetes (e.g. Kottke et al. 2010)
  - Sub phylum Agaricomycotina
    - Class Agaricomycetes
      - Order Agaricales
        - *Armillaria* (e.g. Kikuchi et al. 2008)
        - *Campanella* (e.g. Dearnaley and Bougoure 2010)
        - *Coprinus* (e.g. Yagame et al. 2007)
        - *Gymnopus* (e.g. Dearnaley 2006)
        - *Hymenogaster* (e.g. Julou et al. 2005)
        - *Inocybe* (e.g. Roy et al. 2009b)
        - *Marasmius* (e.g. Burgeff 1959)
        - *Mycena* (e.g. Ogura-Tsujita et al. 2009)
        - *Psathyrella* (e.g. Yamato et al. 2005)
      - *Order Cantharellales*
        - ***Tulasnella***[a] (e.g. Jacquemyn et al. 2010)
        - *Clavulina* (e.g. Selosse, unpublished data)
        - *Ceratobasidium*[b] (e.g. Otero et al. 2002)
        - *Thanatephorus*[b] (e.g. Warcup 1991)
      - *Order Russulales*
        - *Gymnomyces* (e.g. Dearnaley and Le Brocque 2006)
        - *Russula* (e.g. Taylor et al. 2004)
      - *Order Hymenochaetales*
        - *Erythromyces* (e.g. Umata 1995)
        - *Resinicium* (e.g. Martos et al. 2009)
      - *Order Sebacinales*
        - *Sebacina* group A (e.g. McKendrick et al. 2002)
        - ***Sebacina* group B**[a] (e.g. Bougoure et al. 2005)
      - Order Thelephorales
        - *Thelephora/Tomentella* (e.g. Bidartondo et al. 2004)
- *Phylum Ascomycota*
  - *Sub phylum Pezizomycotina*
    - Class Pezizomycetes
      - Order Pezizales
        - *Tuber* (e.g. Selosse et al. 2004)
        - *Tricharina* (e.g. Waterman et al. 2011)
        - *Peziza* (e.g. Waterman et al. 2011)

[a]The unrelated sexual genera *Tulasnella* and *Sebacina* encompass species from the asexual genus *Epulorhiza*.
[b]The sexual genera *Ceratobasidium* and *Thanatephorus* encompass species from the asexual genus *Ceratorhiza*.

B. Fully Mycoheterotrophic Orchids and Ectomycorrhizal Fungi

MH orchids are achlorophyllous and receive all their carbon from their mycorrhizal fungi. In this way, they are paedomorphic, i.e. preserving a juvenile trait (heterotrophy, that is a feature of protocorms **only** in other orchid species) during the adult stage. Since the key studies of *Corallorhiza* and *Cephalanthera* species by Taylor and Bruns (1997) and McKendrick et al. (2000), a large number of works indicate that many other fully MH orchids receive carbon from the ECM associations of autotrophic plants in temperate regions (e.g. Selosse et al. 2002; Taylor et al. 2004; Dearnaley and Le Brocque 2006; Roy et al. 2009a) and in some tropical forests (Roy et al. 2009b). Two studies (Taylor and Bruns 1997; Selosse et al. 2002) provided evidence that the same fungal individual was present on MH orchids and surrounding ectomycorrhizal tree roots in situ: although this relied on the polymorphism of a single genetic marker (nuclear ribosomal DNA), it supports that **hyphal connection** can transfer sufficient carbon from surrounding trees to the MH plants to support their growth, as was more recently supported by ex situ resynthesis experiments (Bougoure et al. 2010). While the association is always specific in temperate regions, a recent study showed that, at least in some tropical areas, some *Aphyllorchis* MH species harboured several different ECM fungi in their roots (Roy et al. 2009b).

In most cases, the Russulaceae, Sebacinales and Thelephoraceae are the most frequently involved taxa (Kennedy et al. 2011); Clavulinaceae may also occur in some *Gastrodia* species (M.-A. Selosse, unpublished data) – interestingly, they also belong to the most frequent taxa in ECM communities (Tedersoo and Nara 2010). For the less specific tropical orchids, lack of specificity may ensure the finding of suitable partners at most sites, and one can speculate that associating to less frequent partners may be evolutionarily risky. Nevertheless, some orchids do associate with rarer taxa, such as *Inocybe* spp. (Roy et al. 2009a; Liebel and Gebauer 2011) or the ECM Ceratobasidiaceae (Yagame et al. 2008, 2012; Bougoure et al. 2009, 2010): the latter are so rare in ECM communities (Collier and Bidartondo 2009) that MH orchids were instrumental in confirming their ECM status (Bougoure et al. 2010; see also Yagame et al. 2012). The common features of ECM fungi supporting MH orchids are unclear, as they are dissimilar in phylogenetic position, ecological preferences and mycelial morphology (shape of mycorrhiza and soil exploration type; R. Agerer, personal communication).

C. Fully Mycoheterotrophic Orchids and Saprotrophic Agaricomycetes

Shifts of fungal partners from non-ECM 'rhizoctonia' to various ECM fungi during MH orchid evolution are considered to give orchids a more continuous carbon supply than that provided by the putatively saprotrophic 'rhizoctonias' (Taylor and Bruns 1997). However, some tropical MH orchids live in forests that are devoid of ECM fungal communities (Smith and Read 2008). Other investigations using molecular fungal identification and stable isotope analyses have now shown **associations to non-'rhizoctonia' saprotrophic partners.**

In the fully MH orchid genus *Gastrodia* the main mycobionts are related to *Marasmius* (Martos et al. 2009; Dearnaley and Bougoure 2010), *Mycena* (Martos et al. 2009; Ogura-Tsujita et al. 2009), *Resinicium* (Martos et al. 2009), or *Armillaria* (Kikuchi et al. 2008), depending on the species. Wood-decaying *Erythromyces* occur in *Galeola* species (Umata 1995) and litter-decaying *Mycena* in *Wullschlaegelia aphylla* (Martos et al. 2009). In both *Epipogium roseum* and *Eulophia zollingeri*, the mycobionts involved are saprotrophic Coprinaceae (Yamato et al. 2005; Yagame et al. 2007; Ogura-Tsujita and Yukawa 2008a). There are also a number of pre-molecular, morphological studies identifying diverse saprobic fungal taxa in fully MH orchids (including *Lycoperdon*; for a review, see Ogura-Tsujita and Yukawa (2008a) that need to be revisited by modern molecular tools.

Although direct data (e.g. isotope tracer studies) that tropical orchids receive carbon from decomposing plant matter via a hyphal conduit is still lacking (Selosse et al. 2010), fungal rhizomorphs linking dead organic matter to the orchid mycorrhizal roots can sometimes be visualized (Kusano 1911; Martos et al. 2009; Fig. 12.3). Moreover, the stable isotope abundance signatures of these orchids are distinctive: they often have slightly higher ^{13}C abundance but substantially lower ^{15}N than ECM-associating plants (Ogura-Tsujita et al. 2009), reflecting the higher ^{13}C and lower ^{15}N abundance of saprotrophic fungi as compared with ECM fungi (Hobbie et al. 2001).

Typically fungal colonization is sparse in fully MH orchids that rely on saprotrophic fungi than on ECM fungi (Dearnaley 2006; Dearnaley and Bougoure 2010), or even not continuous over the year for *Wullschlaegelia aphylla* (Martos et al. 2009), suggesting that the mechanism of obtaining carbon is possibly more efficient than with ECM fungi, but this requires further study. Additionally, hyphae colonize some dead cortical root cells in *Wullschlaegelia aphylla* (Martos et al. 2009), while a complicated pattern of colonization exists in *Gastrodia* roots, with passage cells where hyphae enter the root, in host cells that are permanently colonized and in digestion cells where a carbon flux may occur

Fig. 12.3. Fungal rhizomorphs (*arrows*) of *Mycena* linking dead leaves to mycorrhizal roots (*arrowhead*) of the fully MH orchid *Wullschlaegelia aphylla*, in which the fungus is mycorrhizal (*bar* 1 cm). *Inset* Transverse section of a rhizomorph, with a central hole, and hyphae with thicker, melanized walls at the external border (*bar* 100 μm). F. Martos, unpublished micrograph

(Kusano 1911; Wang et al. 1997). Although their raison d'être remains unclear, these patterns may be evolutionarily derived, emphasizing the secondary evolution of this kind of mycoheterotrophy. Martos et al. (2009) and Selosse et al. (2010) speculated that the shift of fungal partners to various saprotrophic fungi during MH orchid evolution might have occurred in tropical and wet temperate regions, because these environmental conditions stimulate decomposing activity by fungi and might allow higher carbon gain for the plant. Indeed the need to support the large carbon requirement of plants may explain why non-ECM 'rhizoctonias' are rarely found in MH orchids, although they support MH germination in many orchid species: they may simply be too C-limited to fulfil the plant's needs beyond the protocorm stage.

D. Mixotrophs: Green Orchids that Obtain Carbon from Fungi

Stable isotope investigation of an increasingly large number of green orchids, e.g. in the genera *Cephalanthera*, *Epipactis* or *Cymbidium*, has revealed natural abundances of ^{13}C and ^{15}N higher than surrounding autotrophic plants but less than that of fully MH orchids (Gebauer and Meyer 2003; Julou et al. 2005; Abadie et al. 2006). Such intermediate values suggest that these orchids obtain part of their carbon via photosynthesis and part through their mycorrhizal fungi – that is, these plants are mixotrophic (Julou et al. 2005). **Mixotrophy** is thought to be an intermediate step in the evolution of full mycoheterotrophy (Bidartondo et al. 2004; Selosse et al. 2004; Abadie et al. 2006; Motomura et al. 2010). Identifying photosynthesis inefficiency in many chlorophyll-containing orchids may also uncover cryptic mixotrophic orchids (e.g. Girlanda et al. 2006).

These orchids are rarely specific in their mycorrhizal associations and associate with several ECM fungi, with few exceptions, such as *Platanthera minor* that is specific to an ECM *Ceratobasidium* (Yagame et al. 2012). *Epipactis* spp. associate with truffles and related ECM Pezizales, as one of the rare Ascomycete-associated orchid clades known so far (Fig. 12.2; Selosse et al. 2004; Bidartondo and Read 2008; Ogura-Tsujita and Yukawa 2008b; Shefferson et al. 2008). *Tuber* spp. also occur as rare mycobionts in the closely related mixotrophic *Limodorum abortivum* (Girlanda et al. 2006). Mycorrhizal associations with terrestrial orchid species were documented for 13 *Tuber* species that belong to five of the nine main *Tuber* clades (the Excavatum, Aestivum, Rufum, Maculatum and Puberulum clades; Bonito et al. 2010). In a thought-provoking paper based on field data collection in Hungary, Ouanphanivanh et al. (2008) showed that *Epi-*

pactis spp. co-occurred more often than at random with truffle stands (a similar situation was shown for *Cephalanthera* and *Hymenogaster*) and could indicate truffle habitats. A noteworthy feature is the presence of some 'rhizoctonias' in mixotrophic orchids, together with the dominant ECM fungi (Bidartondo et al. 2004; Julou et al. 2005; Abadie et al. 2006; Motomura et al. 2010; Paduano et al. 2011). This feature also supports that these orchids are an intermediate step in the evolution of full mycoheterotrophy, deriving from autotrophic orchid ancestors associated with 'rhizoctonias'.

Some ECM fungi are found from time to time in autotrophic orchids where 'rhizoctonia' are dominant: Russulaceae occur in some *Cypripedium* spp. (Shefferson et al. 2007) and *Pterostylis nutans* (Irwin et al. 2007), *Thelephora* and *Cortinarius* in *Orchis* spp. (Lievens et al. 2010) and putatively ECM Pezizomycetes and Helotiales in *Gymnadenia conopsea* (Stark et al. 2009). ECM Ascomycetes (Waterman et al. 2011) even dominate in some South African Diseae (*Pterygodium* and *Corycium* spp.). Although the exact interaction with the orchid is unknown, the detection of these fungi is highly unexpected as they are not usually contaminants of soil samples. We speculate that their presence in root tissues may give opportunity for their evolution into mycorrhizal partners, accompanying the emergence of mixotrophy. The same may apply for evolution of mycoheterotrophy based on saprotrophic fungi: *Mycena*-related fungi, that are sometimes associated with MH orchids as mentioned above (Sect. III.C), can sometime be found in roots of some green, 'rhizoctonia'-associated orchids (Fan et al. 1996; Guo et al. 1997).

E. Epiphytic Orchids and '*Rhizoctonias*'

Although epiphytic species represent the largest number of orchids worldwide (Jones 2006) they are surprisingly less well studied with regards to their mycorrhizal associations. Mycorrhizal fungi sparsely colonize **roots of epiphytic orchids** in comparison with terrestrial orchids (Boddington and Dearnaley 2008; Smith and Read 2008; Graham and Dearnaley 2012; Martos et al. 2012) but molecular identification of the mycobionts present reveals them as the typical 'rhizoctonias' of green orchids. This has included members of the Ceratobasidiaceae (Otero et al. 2002, 2004, 2005, 2007; Pereira et al. 2005; Gowland et al. 2007; Graham and Dearnaley 2012; Martos et al. 2012), the Tulasnellaceae (Pereira et al. 2003; Suarez et al. 2006; Kottke et al. 2008; Martos et al. 2012) and the Sebacinales (Suarez et al. 2008; Martos et al. 2012).

In the largest survey of orchid mycorrhizas conducted from 77 orchid species from La Réunion island, Martos et al. (2012) found that communities of 'rhizoctonias' significantly differed between epiphytic and terrestrial orchid communities in terms of OTUs, whereas the three 'rhizoctonia' taxa did not differ in frequency. This may reflect that different fungal species are available in soil and on tree bark, but we lack information about the diversity and ecology of 'rhizoctonias' in soil **versus** bark environments.

The dependency of epiphytic orchids on mycorrhizal fungi throughout the life cycle is not surprising as the mycobionts, with their increased surface area, may improve access to water and minerals for plants which can be especially limiting in the epiphytic state (Zotz and Schmidt 2006; Osorio-Gil et al. 2008). The habitat of many epiphytic species that live in the shade of dense forest canopy is typified by low irradiance and it is possible that species will be soon identified as mixotrophic with a dependence on external supplied carbon as well as photosynthesis.

IV. Nutrient Exchanges Between Orchid and Mycobiont

The minute seeds of orchids lack food reserves and colonization by a suitable fungus is necessary for further development under natural conditions (Smith and Read 2008). Both organic and inorganic nutrients have been shown to be transferred from mycorrhizal fungus to protocorms. Experiments using **split-plate systems** and labelled glucose accessible only to the fungal partner have demonstrated carbon flow to orchid protocorms (Purves and Hadley 1975; Alexander and Hadley 1985). A similar split-plate system indicated that phosphate (labelled by ^{32}P) is passed from fungus to protocorms of *Dactylorhiza purpurella* (Smith 1966).

Adult mycorrhizal orchids continue to receive both organic and inorganic nutrients from their fungal partners. When the *Ceratobasidium* partner of *Goodyera repens* is supplied with ^{14}C-labelled glycine in the substrate, labelled carbon is passed to the orchid seedlings (Cameron et al. 2006). The addition of $^{13}C/^{15}N$-labelled glycine to the fungal compartment also demonstrated a transfer of nitrogen to orchid seedlings. Cameron et al. (2007) have shown that adult *Goodyera repens* also receive phosphate from their fungal partner under experimental conditions. Mycorrhizal fungi may be key to ensuring optimal water uptake from the environment: mycorrhizal *Plantanthera integrilabia* and *Epidendrum conopseum* both had higher water content than uncolonized controls (Yoder et al. 2000).

Orchid mycorrhizas have often been considered to be atypical mycorrhizal associations, with the fungus deriving little benefit from the orchid host (Dearnaley 2007; Smith and Read 2008). Key to this assumption was research conducted by Hadley and Purves (1974) and Alexander and Hadley (1985) on mycorrhizal *Goodyera repens*. In their experiments, the orchid was exposed to $^{14}CO_2$ and no subsequent passage of labelled carbon to the fungal partner was detected. These experiments were more recently repeated by Cameron et al. (2006, 2008) using more naturally equivalent conditions (e.g. moderate temperature, lighting, humidity) with contrasting results. In the latter experiments approximately 0.4–3.0 % of the carbon label originally provided to the orchid was passed to the fungal partner (Cameron et al. 2006, 2008). Adult orchid mycorrhizas thus potentially represent a truly mutualistic interaction similar to ECM and AM associations.

Rasmussen and Rasmussen (2007) and Hynson et al. (2009) rightly note that adult orchid mycorrhizal systems other than *Goodyera repens* should be investigated similarly, as well as under field conditions. The importance of the carbon acquired from the orchids in the whole nutritional budget of the fungus has also been questioned (Rasmussen and Rasmussen 2009). However, using a clever experimental design where the carbon source have different ^{13}C abundances, Latalova and Balaz (2010) showed that a *Tulasnella* species received 70 % of its carbon from its host, *Serapias strictiflora* and 30 % from dead maize roots added to the system, but these experiments were again carried out in vitro.

Most importantly, from an evolutionary point of view, the reciprocation in a mutualism cannot only be quantified in terms of nutrient flow, but should result in a **fitness improvement**. Fitness is notoriously difficult to measure in fungi (Pringle and Taylor 2002), and there is currently no evidence that fungi reproduce better with the orchid than without.

Intriguing indirect evidence for mutualism arose recently from the analysis of the **architecture of interaction networks** between autotrophic orchids and their ‘rhizoctonias’. These may vary according to the nature of the interaction, especially when comparing mutualistic and trophic interactions (Thébault and Fontaine 2010). Recent analyses (Jacquemyn et al. 2010 and unpublished data) concluded that orchid mycorrhizal interaction networks displayed a significantly nested structure, i.e. specialized (fungus-specific) orchid species tended to associate with ‘rhizoctonia’ species that themselves associated with more generalized (not fungus-specific) orchid species, and vice versa. Conversely, Martos et al. (2012) found that orchid–fungus networks displayed a highly modular structure in a tropical context, which could be interpreted as an ecological divergence between epiphytic and terrestrial guilds of plant and fungal partners in tropical communities, although the authors also confirmed the presence of some level of nestedness in the epiphytic and terrestrial sub-networks (Fig. 12.4). The trend of nestedness is a specific feature of mutualistic networks, as opposed to parasitic or trophic networks that are more compartmentalized, and is viewed as a consequence of the reciprocation process itself (Thébault and Fontaine 2010). In other words, this may be an indirect indication that some reciprocation occurs with most ‘rhizoctonias’; thus, the investment in protocorm development may be viewed as a transient cost to develop a host that will be beneficial for the fungus later (Leake et al. 2008).

Should we conclude that, conversely, mixotrophic and MH orchids are fungal parasites? This is the tacit idea when calling these plants epi-parasites, or cheaters on ECM symbioses (Merckx et al. 2009), but we still lack rigorous

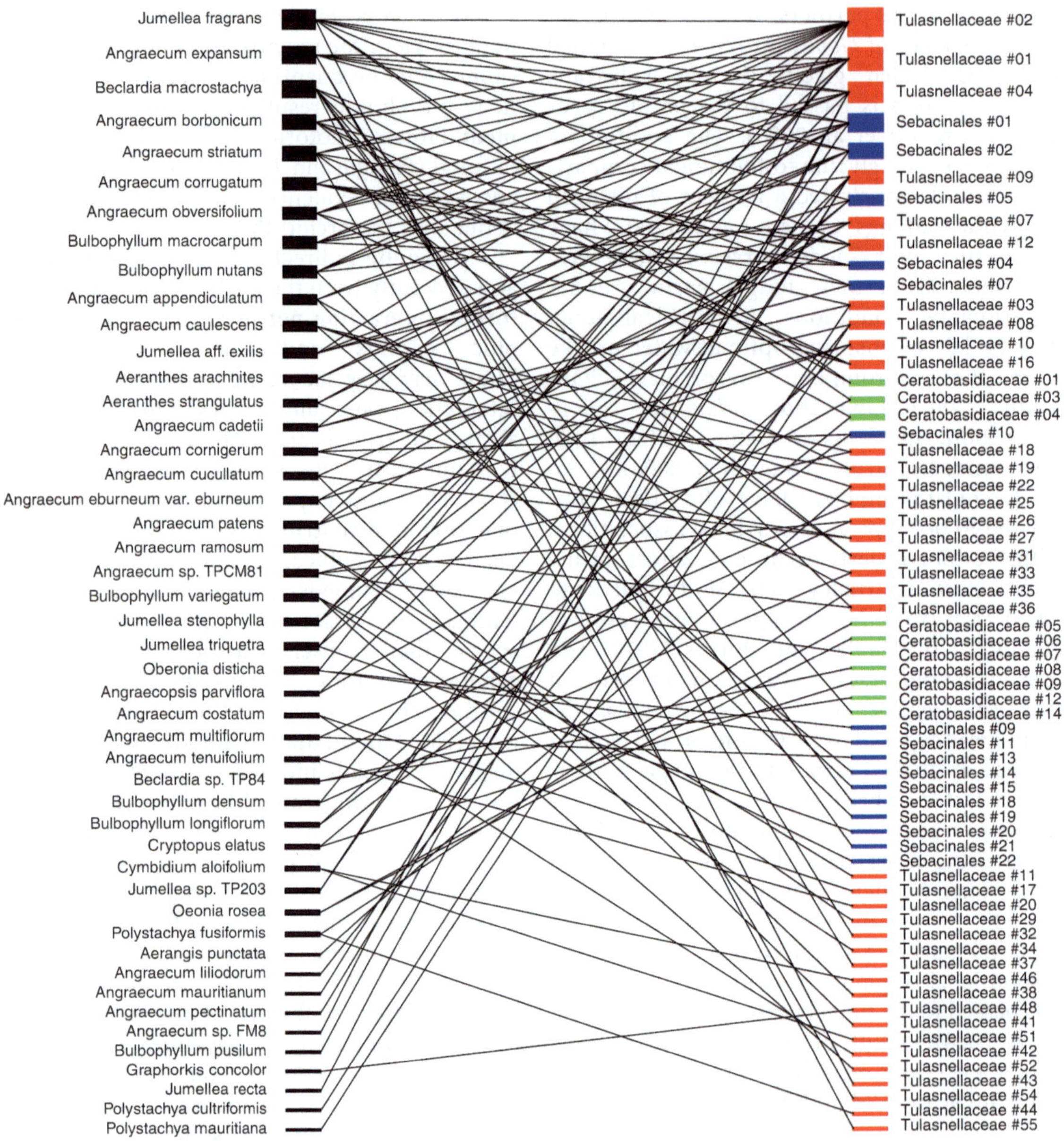

Fig. 12.4. Nested architecture of an orchid–rhizoctonia network as formed by epiphytic orchids on La Réunion island (from Martos et al. 2012). The *left* column shows orchid species with lines linking to various 'rhizoctonia' taxa (*right* column). A nested structure (i.e. less compartmentalized) is a specific feature of mutualistic networks

evidence for this. Indeed, some vitamins, or some protection at some time of the year may enhance fungal fitness. Obviously, we need more studies on the fungal side before any conclusion can be made. Hopefully, the development of models tractable in vitro for autotrophic (Cameron et al. 2006, 2008) and MH orchids (Yagame et al. 2007; Bougoure et al. 2010) will help in investigating these questions.

V. Fungal Specificity in Orchids

A. Patterns and Evolutionary Significance

Fungal specificity is the association of an orchid species with a small number of fungal partners (Irwin et al. 2007), and can be quantified as the phylogenetic breadth (= antiquity of the last common ancestor) of its range of associates (Thompson 1994; Shefferson et al. 2010). This can take the form of narrow specificity whereby an orchid associates exclusively with a single mycobiont across its range, such as the rare underground orchid *Rhizanthella gardneri* (Bougoure et al. 2009). Typically, **specificity** is expressed as an orchid species associating with a limited number of related fungal taxa, e.g. *Corallorhiza maculata* associating with Russulaceae species in the western United States (Taylor et al. 2004), or *Pterostylis nutans* associating with two *Ceratobasidium* species in eastern Australia (Irwin et al. 2007). Both fully MH and autotrophic species can display fungal specificity (e.g. McCormick et al. 2004; Yamato et al. 2005) although the phenomenon is more common to the former orchid physiological type. A small number of the investigated orchid species display little fungal specificity. For example, the widespread Australian grassland species *Microtis intermedia* associates with members of both the Sebacinales and the Ceratobasidiaceae (Bonnardeaux et al. 2007), while two fully MH *Aphyllorchis* species from Thailand associate with an array of unrelated ECM fungi, including members of the Thelephoraceae, Russulaceae and Sebacinales (Roy et al. 2009b).

The **evolution of fungal specificity** has been recently evaluated by mapping the phylogenetic breadth of mycorrhizal partners across orchid phylogenies. Shefferson et al. (2007, 2010) analysed fungal specificity across two orchid phylogenies of the genera *Cypripedium* and *Goodyera* and found that both widening and broadening depended on orchid clades, so that the level of fungal specificity was concluded to be an evolvable trait subjected to reversion in orchids. Considering the evolution of fungal partners across orchid diversification, Waterman et al. (2011) showed that fungal partners are conserved between closely related species of South African Coryciinae.

Martos et al. (2012) used both orchid and fungal phylogenies to assess phylogenetic signal in the interaction network of tropical angraecoid orchids on the island of La Réunion and found a stronger signal on the orchid side than on the fungal side: fungal partners that belong to the Tulasnellaceae, Sebacinales and Ceratobasidiaceae are statistically more conserved between closely related angraecoids than orchid partners of closely related fungi are. Such an asymmetry of phylogenetic signal may reveal different constraints for the partners in orchid mycorrhiza, especially the lower dependence of fungal partners on the symbiosis.

Sudden **partner shifts** have also occurred during the evolution of the MH genus *Hexalectris* (Kennedy et al. 2011), or in the genus *Epipogium*, where *E. aphyllum* associates with ECM *Inocybe* spp. (Roy et al. 2009b) while *E. roseum* associates with saprotrophic *Psathyrella*-related partners (Yamato et al. 2005). Partner shift can thus rapidly evolve, although how the shift occurs remains unclear. As mentioned below (Sect. V.B), the observation that some unexpected fungi are sometimes detected in roots, in addition to the major mycorrhizal fungi, may be relevant as a starting point in the transition – this led to the suggestion that 'molecular scraps' (unexpected fungi considered as contaminant or marginal in the mycobiont spectrum) obtained in symbiont typing should always be reported (Selosse et al. 2010). A particularly interesting stage for the transition may be the germination step: in some orchid species at least, the fungi enhancing the first stage of germination are more diverse than the fungi allowing further development (Vujanovic et al. 2000; Bidartondo and Read 2008). From this situation, where early embryos contact diverse fungi, a mutant for specificity may survive.

B. Adaptive Significance

The adaptive significance of fungal specificity in orchid mycorrhizas is a source of some conjecture. Specific fungal partners may lead to enhanced seed germination rates (Otero et al. 2004; Bonnardeaux et al. 2007) and thus increased fitness. The **efficiency of nutrient exchange** between partners may be heightened with specific plant–fungus combinations (Bonnardeaux et al. 2007) and this may be critical for carbon uptake for mixotrophic and fully MH orchids in low-light habitats where there is a higher dependency on fungal carbon (e.g. Girlanda et al. 2006). More efficient nutrient exchange as a driver for fungal specificity in orchids can be suggested by examples of partner switching in adult orchids. For example *Goodyera pubescens* switched from one *Tulasnella* species to another when plants were drought-stressed (McCormick et al. 2006). Several studies have suggested that autotrophic orchids associate with different clades of 'rhizoctonias' depending on the environment, e.g. when comparing terrestrial and epiphytic orchid communities in tropical forests (Martos et al. 2012), or European terrestrial orchids in dry and wet habitats (where different Tulasnellaceae subclades dominate; Illyés et al. 2009). However, it remains unknown whether this results from choosing optimal fungal partners or simply from different availability of fungal taxa.

C. The Impact of Mycorrhizal Specificity on Orchid Speciation

Fungal specificity was recently linked to speciation in the Orchidaceae by a number of authors (Otero and Flanagan 2006; Shefferson et al. 2007; Waterman and Bidartondo 2008; Waterman et al. 2011). Distribution of fungi in soils is highly heterogeneous (Richard et al. 2004; Pickles et al. 2010) and this, combined with narrow fungal specificity, may determine the small, over-dispersed populations of many orchid species (Otero and Flanagan 2006). The resulting **patchiness of orchid distribution** may limit gene flow between isolated populations and the number of reproducing individuals, leading in turn to genetic drift and allopatric speciation (Tremblay et al. 2005; Waterman and Bidartondo 2008). Support for this process comes from the observation that different populations of the *Hexalectris spicata* complex display distinct mycorrhizal fungi (Taylor et al. 2003). Natural selection may also act on small, isolated populations of orchids, as highlighted in the study of Otero et al. (2005) that showed varying levels of germination rates (or fitness) after reproducing in vitro associations between *Tolumnia variegata* and different 'rhizoctonia' fungi.

In contrast, Roche et al. (2010) showed that multiple species of *Chiloglottis* associated with a narrow group of Tulasnellaceae fungi across eastern Australia. The fact that each of these species associates with a distinct wasp pollinator suggests that pollination systems and not fungal specificity is driving speciation in the orchid genus. A similar interpretation was made by Waterman et al. (2011) when studying shifts of pollination modes and mycorrhizal partners across the phylogeny of South African Coryciinae orchids. Roche et al. (2010) suggested that a common mycorrhizal fungus in *Chiloglottis* spp. has enabled rapid pollination-mediated speciation via co-occurrence of multiple potential species types.

Mycorrhizal associations during species hybridization, a potential source of speciation in the Orchidaceae, have been examined by some researchers. In crosses between *Caladenia* spp., Hollick et al. (2005) showed that hybrids have genetically similar fungi to one of the two parents. The hybrid formed between crosses of *Orchis simia* and *Orchis anthropophora* also had similar Tulasnellaceae fungi to its parents (Schatz et al. 2010). Interestingly, hybrid *Orchis* plants had higher levels of mycorrhizal colonization than the parents but this was possibly related to the inability to attract pollinators and to produce seeds, therefore providing more carbon for the colonizing fungus. Jacquemyn et al. (2010) also investigated the mycorrhizal associations of *Orchis* hybrids and concluded from common mycobionts in protocorms and adults that mycorrhizal fungi play a small role in reproductive isolation. One generalizing speculation that can be derived from these studies is that mycorrhizal symbiosis acts in a permissive way, i.e. that, for successful hybridization to occur, the parent's fungi need to be related or identical.

VI. Orchid Mycorrhizas and Plant Conservation

A dependence on narrowly specific interactions with fungi and pollinators may predispose many orchids to become rare (Bonnardeaux et al. 2007; Dearnaley 2007; Swarts et al. 2010). However, Phillips et al. (2011) have recently shown that fungal specificity has not led to rarity in West Australian *Drakaea* spp. as the associated *Tulasnella* fungus is widely distributed in the environment. Nevertheless, as humankind continues to have **negative impacts on natural ecosystems** through such perturbations as vegetation clearing, altered fire regimes, weed and feral animal introduction and climate change, populations of many rare orchid taxa are further declining (Brundrett 2007). Conservation approaches for such orchids include on site protection of existing populations, ex situ storage of tissues and restoration procedures (Swarts and Dixon 2009). All of these approaches require an understanding of the mycorrhizal biology of the species in question, since fungi are vital for orchid seed germination and adult vegetative life.

A. Orchid Mycorrhizas and On-Site Management

Molecular identification of the mycobionts of many orchid species has given an insight into the ecological position of fungal species. This has highlighted **management procedures** that are needed to protect existing populations. The conservation of fully MH and mixotrophic orchids dependent on ECM associations such as *Hexalectris*, *Epipactis*, *Dipodium* and *Rhizanthella* (Taylor et al. 2003; Selosse et al. 2004; Bougoure and Dearnaley 2005; Bougoure et al. 2010) clearly need maintenance of stands of suitable host trees. Fully MH species such as *Gastrodia*, *Epipogium* and *Erythrorchis*, which are nutritionally dependent on wood-rotting fungi (Yamato et al. 2005; Dearnaley 2006; Martos et al. 2009; Dearnaley and Bougoure 2010), will need the retention of a suitable decomposable substrate. For the majority of (autotrophic) orchids, preservation of the uppermost organic layer of soils is essential, as this location is the key habitat of their 'rhizoctonias' associates (Brundrett et al. 2003). As this layer is particularly susceptible to frequent burning (Brundrett 2007), careful monitoring of fire regimes is a necessary conservation measure.

For all orchids with partial or full mycoheterotrophy, the fungus cannot be separated from its own carbon source and, if such occurs during relocation, both the fungus and the plant may die. This was shown in an overlooked book by Sadovsky (1965) dealing with the cultivation of 'orchids in your own garden': among other studies, Sadovsky trialled the relocation of a number of orchid species at a time where protection laws were more flexible in Europe, and the resulting list showed that mixotrophic and MH species could not be transplanted. Thus, there may be problems saving populations of such orchids by transferral to another site in the case of major disturbance. However, the effective glasshouse relocation of *Rhizanthella slateri* (with its ECM partner and photosynthetic host) threatened by a major road development in eastern Australia (M. Clements, personal communication) provides an exemplar of success.

Regular monitoring for the continued presence of orchid-associated fungi is a necessary management procedure. This can be done simply by seasonal observations of macrofungal fruiting bodies for some associated orchid species. For microfungi and rarely sporulating fungi, such a most 'rhizoctonias' (e.g. clade B Sebacinales that do not fruit; Weiβ et al. 2004), seed baiting procedures carried out both in situ and ex situ, are cost-effective (Brundrett et al. 2003). Molecular detection of orchid-associated fungal DNA using specific or general fungal primers will also ensure that sites continue to harbour the appropriate mycobionts. The best way to preserve orchids is therefore to preserve their fungi and, from there, given the uncertainties on the ecology of fungi, the whole environment. This is indeed good news for mycologists since orchid protection therefore protects fungi – not only those taxa involved in mycorrhizal associations, but the surrounding ones as well.

Some fully MH orchids rely on ECM fungi that have fruiting bodies that are consumed by native mammals (Bidartondo et al. 2004; Selosse et al. 2004; Dearnaley and Le Brocque 2006). In Australia, members of the Russulaceae are consumed by marsupials such as Bettongs and Potoroos (Claridge and May 1994). To ensure continued cycling of fungal propagules through ecosystems, protection of these spore-dispersing animals should be a long-term priority.

Two recent works suggest that the presence of fungi may not be the sole limiting factor for orchid settlement: in experimental seed sowing at different sites where the focal orchid species does not grow, there is evidence that early development into a protocorm can occur in *Cephalanthera* spp. (Bidartondo and Read 2008) and *Epipactis* spp. (Těšitelov et al. 2012), with successful access to appropriate fungal partners. Although these plants are mixotrophic and may not represent general models, limitations to orchid development may thus be more than fungal – as the authors discuss, a limitation on seed dispersal or abiotic factors may also be involved. Thus, the fungal symbiosis, although crucial to the orchids, may not be seen as the sole factor explaining why orchids develop and why a given site is suitable for orchid growth.

B. Ex Situ Conservation and Orchid Mycorrhizas

Ex situ symbiotic germination of seed is a common approach in conservation procedures for threatened orchids (Batty et al. 2006b; Stewart and Kane 2007; Zettler et al. 2007). Mycorrhizal fungi can be obtained from adult plants in situ or via buried seed germination packets (Batty et al. 2001; Dearnaley et al. 2009). Damage to adult threatened orchids can be minimized by taking a small sliver of colonized stem (Wright et al. 2009; Smith et al. 2010) and the best plants to isolate fungi from are leafing to flowering stages (Huynh et al. 2004). Surface sterilization of isolated orchid tissues reduces the amount of contamination from bacteria and faster-growing ascomycetes (Huynh et al. 2009). Once pelotons are separated from the host tissue, the best 'rhizoctonias' to choose for symbiotic autotrophic orchid seed germination are those with fine loose hyphae and monilioid cells (Huynh et al. 2004). Pure fungal inoculum and surface-sterilized orchid seed are traditionally co-cultured on oatmeal-based agar medium (Clements et al. 1986). The growth of orchids dependent on ECM fungi and photosynthetic hosts requires special culturing set-ups such as that developed by Bougoure et al. (2010) for *Rhizanthella gardneri* (Fig. 12.5). Fully MH orchids reliant on saprotrophic Agaricomycetes can be grown in seed packets with a medium of sawdust and fungal inoculum (Yagame et al. 2007). Ex vitro approaches whereby seed is sown in pot soil inoculated with the appropriate mycorrhizal fungus has an additional advantage in that seedlings may form associations with other micro-organisms present in the medium (Wright et al. 2009). Procedures for maintaining orchid mycorrhizal fungi in the long term include immersing the inoculum in liquid nitrogen (Batty et al. 2001) or via encapsulation of both seed and fungi in alginate beads, with low-temperature storage (Sommerville et al. 2008). It is important that a range of fungal taxa and isolates are preserved in orchid conservation work as multiple fungi might co-exist in plants (Irwin et al. 2007; Wright et al. 2010) or orchids may switch fungi as they mature (Xu and Guo 2000) or even as adults (McCormick et al. 2004; Dearnaley 2006). Furthermore, the fungal isolate that is best at germinating seed does not necessarily ensure the best long-term survival of orchid species (Wright 2007; Huynh et al. 2009).

C. Use of Mycorrhizal Fungi in Restoration Procedures

Symbiotically grown orchid seedlings can be transferred directly to the natural state but plant persistence is enhanced by growth in potting media for several seasons (Swarts 2007). Moving plants from Petri dish to soil can be a significant hurdle (Wright et al. 2009) but an intermediate deflasking procedure involving carefully aerated sand–agar containers has been shown to effectively prepare agar-grown symbiotic seedlings for transfer to soil in pots (Batty et al. 2006a). There appears to be no benefit to inoculating the pot soil with compatible fungi, with the original colonizing fungus proving sufficient for nutrient uptake for the seedlings (Batty et al. 2006a). For establishing orchid populations in the natural state, seedlings and tubers appear to be better than seed sowing, although the latter is more cost- and time-effective (Batty et al. 2006b; Wright et al.

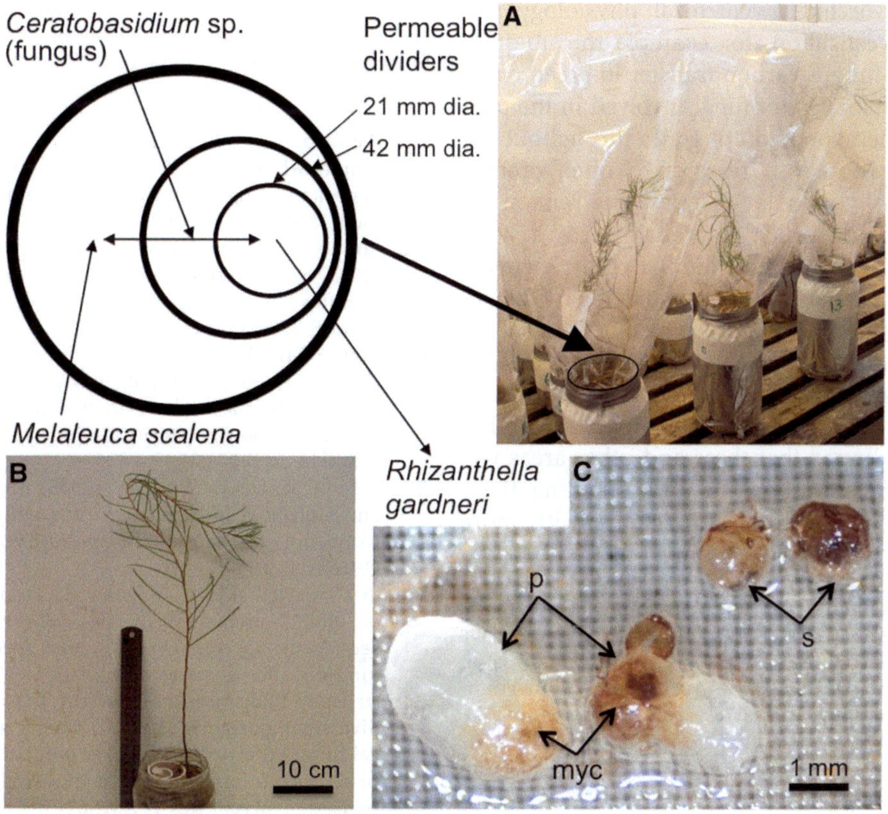

Fig. 12.5. *Rhizanthella gardneri* growth pot arrangement (from Bougoure et al. 2010; used with the permission of author and publisher). The orchid (shown in **C**) is grown in the inner of three pots while the ECM *Ceratobasidium* partner, inoculated into the middle pot, passes photosynthate from the autotrophic *Melaleuca scalena* (outer pot, also seen in **A** and **B**) to the orchid via small holes in each pot

2009). Both in situ (Rasmussen and Whigham 1993) and ex situ (Brundrett et al. 2003) seed baiting can be used to confirm the presence of fungi at introduction sites. Translocation success can be enhanced by a combination of adding fungal inoculum and loosening soil at sites; the latter potentially enhances the activity of the fungi at the location (Smith et al. 2009). Fungal inoculum can be introduced to new sites without orchids and can persist in soils for several seasons in preparation for restoration procedures (Hollick et al. 2007).

VII. Conclusions

Orchid mycorrhizas are predominantly represented by **associations between photosynthetic plants and 'rhizoctonia' fungi.** These associations, which likely represent the plesiomorphic condition for orchids, gave rise through repeated evolutionary shifts to interactions with other diverse fungal lineages and diversification of orchid metabolism. How orchids recruit and allow new fungi (even some 'naïve' fungi from non-mycorrhizal clades) to enter the dual morphogenesis of mycorrhizas remains unclear. However, orchid mycorrhizas are excellent models to reveal the general properties of mycorrhizal systems as well as providing insights into the fungal world via specificity aspects, ecological networks and evolution of the mycorrhizal state.

Although considerable advances have been made in understanding the ecology and evolution of orchid mycorrhizas in recent years, substantial knowledge gaps still exist. In particular, many aspects of orchid mycorrhizal physiology still require investigation, for example the ubiquity of plant to fungus carbon transfer in green orchids, the metabolism of fungi involved in the process and the expression of genes throughout the symbiosis. Moreover, research is often orchid-focussed, so that a lot of questions remain on the fungal side which is probably less easy to investigate. The exact nutrition, diversity, benefits from the association (if any) and repartition in soil of many mycobionts, such as the Tulasnellaceae, are often ignored and with some exceptions (e.g. Selosse et al. 2002; McCormick et al. 2009), the fungus is rarely investigated out of the orchid roots. It is hoped that these and other areas will continue to contribute to understanding these fascinating mycorrhizal interactions, with more emphasis on the involved fungal taxa.

Acknowledgements The authors would like to thank the large number of students and colleagues who have supported their research on orchid mycorrhizas in recent years. J.D. thanks the Australian Orchid Foundation for support of his research. M.-A.S. and F.M. thank the Société Française d'Orchidophile and its members for continuous support and participation in their research programmes.

References

Abadie J-C, Püttsepp Ü, Gebauer G, Faccio A, Bonfante P, Selosse M-A (2006) *Cephalanthera longifolia* (Neottieae, Orchidaceae) is mixotrophic: a comparative study between green and nonphotosynthetic individuals. Can J Bot 84:1462–1477

Alexander C, Hadley G (1985) Carbon movement between host and mycorrhizal endophyte during the development of the orchid *Goodyera repens* Br. New Phytol 101:657–665

Batty AL, Dixon KW, Brundrett MC, Sivasithamparam K (2001) Long-term storage of mycorrhizal fungi and seed as a tool for the conservation of endangered Western Australian terrestrial orchids. Aust J Bot 49:619–628

Batty AL, Brundrett MC, Dixon KW, Sivasithamparam K (2006a) New methods to improve symbiotic propagation of temperate terrestrial orchid seedlings from axenic culture to soil. Aust J Bot 54:367–374

Batty AL, Brundrett MC, Dixon KW, Sivasithamparam K (2006b) In situ symbiotic seed germination and propagation of terrestrial orchid seedlings for establishment at field sites. Aust J Bot 54:375–381

Bayman P, Otero JT (2006) Microbial endophytes of orchid roots. In: Schulz B, Boyle C, Sieber TN (eds) Microbial root endophytes, soil biology, vol. 9, part II. Springer, Berlin, pp 153–177

Bayman P, Gonzalez EJ, Fumero JJ, Tremblay RL (2002) Are fungi necessary? How fungicides affect growth and survival of the orchid *Lepanthes rupestris* in the field. J Ecol 90:1002–1008

Bidartondo MI, Read DJ (2008) Fungal specificity bottlenecks during orchid germination and development. Mol Ecol 17:3707–3716

Bidartondo MI, Bruns TD, Weiβ M, Sérgio C, Read DJ (2003) Specialized cheating of the ectomycorrhizal symbiosis by an epiparasitic liverwort. Proc R Soc Lond B 270:835–842

Bidartondo MI, Burghardt B, Gebauer G, Bruns TD, Read DJ (2004) Changing partners in the dark: isotopic and molecular evidence of ectomycorrhizal liaisons between forest orchids and trees. Proc R Soc Lond B 271:1799–1806

Boddington M, Dearnaley JDW (2008) Morphological and molecular identification of fungal endophytes from roots of *Dendrobium speciosum*. Proc R Soc Queensland 114:13–17

Bonito G, Gryganskyi A, Vilgalys R, Trappe J (2010) A global meta-analysis of *Tuber* ITS rDNA sequences: species diversity, host specificity, and long-distance dispersal. Mol Ecol 19:4994–5008

Bonnardeaux Y, Brundrett M, Batty A, Dixon K, Koch J, Sivasithamparam K (2007) Diversity of mycorrhizal fungi of terrestrial orchids: compatibility webs, brief encounters, lasting relationships and alien invasions. Mycol Res 111:51–61

Bougoure DS, Cairney JWG (2005) Fungi associated with hair roots of *Rhododendron lochiae* (Ericaceae) in an Australian tropical cloud forest revealed by culturing and culture-independent molecular methods. Environ Microbiol 7:1743–1754

Bougoure JJ, Dearnaley JDW (2005) The fungal endophytes of *Dipodium variegatum* (Orchidaceae). Australas Mycol 24:15–19

Bougoure JJ, Bougoure DS, Cairney JWG, Dearnaley JDW (2005) ITS-RFLP and sequence analysis of endophytes from *Acianthus*, *Caladenia* and *Pterostylis* (Orchidaceae) in southeastern Queensland. Mycol Res 109:452–460

Bougoure JJ, Ludwig M, Brundrett M, Grierson P (2009) Identity and specificity of the fungi forming mycorrhizas with the rare myco-heterotrophic orchid *Rhizanthella gardneri*. Mycol Res 113:1097–1106

Bougoure JJ, Brundrett MC, Grierson PF (2010) Carbon and nitrogen supply to the rare underground orchid *Rhizanthella gardneri*. New Phytol 186:947–956

Boullard B (1985) Un biologiste d'exception: Noël Bernard, 1874–1911. Presse de l'Université de Rouen, Rouen

Brundrett MC (2007) Scientific approaches to Australian temperate terrestrial orchid conservation. Aust J Bot 55:293–307

Brundrett MC, Scade A, Batty AL, Dixon KW, Sivasithamparam K (2003) Development of in situ and ex situ seed baiting techniques to detect mycorrhizal fungi from terrestrial orchid habitats. Mycol Res 107:1210–1220

Burgeff H (1959) Mycorrhizas of orchids. In: Withner K (ed) The orchids. Ronald, New York, pp 361–395

Cameron DD, Leake JR, Read DJ (2006) Mutualistic mycorrhiza in orchids: evidence from plant-fungus carbon and nitrogen transfers in the green-leaved terrestrial orchid *Goodyera repens*. New Phytol 171:405–416

Cameron DD, Johnson I, Leake JR, Read DJ (2007) Mycorrhizal acquisition of inorganic phosphorus by the green-leaved terrestrial orchid *Goodyera repens*. Ann Bot 99:831–834

Cameron DD, Johnson I, Leake JR, Read DJ (2008) Giving and receiving: measuring the carbon cost of mycorrhizas in the green orchid, *Goodyera repens*. New Phytol 180:176–184

Claridge AW, May TW (1994) Mycophagy among Australian mammals. Aust J Ecol 19:251–275

Clements MA (1988) Orchid mycorrhizal associations. Lindleyana 3:73–86

Clements MA, Muir H, Cribb PJ (1986) A preliminary report on the symbiotic germination of European terrestrial orchids. Kew Bull 41:437–445

Collier FA, Bidartondo MI (2009) Waiting for fungi: the ectomycorrhizal invasion of lowland heathlands. J Ecol 97:950–963

De Candolle AP (1815) Mémoire sur les rhizoctones, nouveau genre de champignons qui attaque les racines des plantes et en particulier celle de la Luzerne cultivée. Mem Mus Hist Nat 2:209–216

Dearnaley JDW (2006) The fungal endophytes of *Erythrorchis cassythoides* – is this orchid saprophytic or parasitic? Australas Mycol 25:51–57

Dearnaley JDW (2007) Further advances in orchid mycorrhizal research. Mycorrhiza 17:475–486

Dearnaley JDW, Bougoure JJ (2010) Isotopic and molecular evidence for saprotrophic Marasmiaceae mycobionts in rhizomes of *Gastrodia sesamoides*. Fungal Ecol 3:288–294

Dearnaley JDW, Le Brocque AF (2006) Molecular identification of the primary root fungal endophytes of *Dipodium hamiltonianum* (Yellow hyacinth orchid). Aust J Bot 54:487–491

Dearnaley JDW, Murray AJ, Mathieson MT (2009) Molecular identification of a mycorrhizal Sebacinaceae from the endangered *Caladenia atroclavia* (Black-clubbed spider orchid). Australas Mycol 28:45–50

Dickie IA, FitzJohn RG (2007) Using terminal restriction fragment length polymorphism (T-RFLP) to identify mycorrhizal fungi: a methods review. Mycorrhiza 17:259–270

Dumbrell AJ, Ashton PD, Aziz N, Feng G, Nelson N, Dytham C, Fitter AH, Helgason T (2011) Distinct seasonal assemblages of arbuscular mycorrhizal fungi revealed by massively parallel pyrosequencing. New Phytol 190:794–804

Fan L, Guo S, Cao W, Xiao P, Xu J, Fan L, Guo SX, Cao WQ, Xiao PG, Xu JT (1996) Isolation, culture, identification and biological activity of *Mycena orchidicola* sp. nov. in *Cymbidium sinense* (Orchidaceae). Acta Mycol Sin 15:251–255

Gardes M, Bruns TD (1993) ITS primers with enhanced specificity for basidiomycetes-application to the identification of mycorrhizae and rusts. Mol Ecol 2:113–118

Gebauer G, Meyer M (2003) ^{15}N and ^{13}C natural abundance of autotrophic and mycoheterotrophic orchids provides insight into nitrogen and carbon gain from fungal association. New Phytol 160:209–223

Girlanda M, Selosse MA, Cafasso D, Brilli F, Delfine S, Fabbian R, Ghignone S, Pinelli P, Segreto R, Loreto F,Cozzolino S, Perotto S (2006) Inefficient photosynthesis in the Mediterranean orchid *Limodorum abortivum* is mirrored by specific association to ectomycorrhizal Russulaceae. Mol Ecol 15:491–504

Gowland KM, Mathesius U, Clements MA, Nicotra AB (2007) Understanding the distribution of three species of epiphytic orchids in temperate Australian rainforest by investigation of their host and fungal associates. Lankesteriana 7:44–46

Graham RR, Dearnaley JDW (2007) The rare Australian epiphytic orchid *Sarcochilus weinthalii* associates with a single species of *Ceratobasidium*. Fungal Divers 54:31–37

Graham RR, Dearnaley JDW (2012) Fungal Divers 54:31–37

Guo S-X, Fan L, Cao W-Q, Xu J-T, Xiao P-G (1997) *Mycena anoectochila* sp. nov. isolated from mycorrhizal roots of *Anoectochilus roxburghii* from Xishuangbanna, China. Mycologia 89:952–954

Hadley G, Purves S (1974) Movement of ^{14}carbon from host to fungus in orchid mycorrhiza. New Phytol 73:475–482

Hadley G, Williamson B (1971) Analysis of post infection growth stimulus in orchid mycorrhiza. New Phytol 70:445–455

Hashimoto Y, Fukukawa S, Kunishi A, Suga H, Richard F, Sauve M, Selosse M-A (2012) Mycoheterotrophic germination of *Pyrola asarifolia* dust seeds reveals convergences with germination in orchids. New Phytol 195:620–630

Hobbie EA, Weber NS, Trappe JM (2001) Mycorrhizal vs saprotrophic status of fungi: the isotopic evidence. New Phytol 150:601–610

Hollick PS, Taylor RJ, McComb JA, Dixon KW (2005) If orchid mycorrhizal fungi are so specific, how do natural hybrids cope? Selbyana 26:159–170

Hollick PS, McComb JA, Dixon KW (2007) Introduction, growth and persistence in situ of orchid mycorrhizal fungi. Aust J Bot 55:665–672

Huynh TT, McLean CB, Coates F, Lawrie AC (2004) Effect of developmental stage and peloton morphology on success in isolation of mycorrhizal fungi in *Caladenia formosa* (Orchidaceae). Aust J Bot 52:231–241

Huynh TT, Thomson R, McLean CB, Lawrie AC (2009) Functional and genetic diversity of mycorrhizal fungi from single plants of *Caladenia formosa* (Orchidaceae). Ann Bot 104:757–765

Hynson NA, Preiss K, Gebauer G (2009) Is it better to give than receive? A stable isotope perspective on orchid-fungal carbon transport in the green orchid species *Goodyera repens* and *Goodyera oblongifolia*. New Phytol 182:8–11

Illyés Z, Halsz K, Rudnóy S, Ouanphanivanh N, Garay T, Bratek Z (2009) Changes in the diversity of the mycorrhizal fungi of orchids as a function of the water supply of the habitat. J Appl Bot Food Qual 83:28–36

Irwin MA, Bougoure JJ, Dearnaley JDW (2007) *Pterostylis nutans* (Orchidaceae) has a specific association with two *Ceratobasidium* endophytes across its range in Eastern Australia. Mycoscience 48:231–239

Jacquemyn H, Honnay O, Cammue BPA, Brys R, Lievens B (2010) Low specificity and nested subset structure characterise mycorrhizal associations in five-closely related species of the genus *Orchis*. Mol Ecol 19:4086–4095

Jacquemyn H, Brys R, Cammue BPA, Honnay O, Lievens B (2011) Mycorrhizal associations and reproductive isolation in three closely related *Orchis* species. Ann Bot 107:347–356

Jones DL (2006) A complete guide to native orchids of Australia including the island territories. Reed New Holland, Sydney

Julou T, Burghardt B, Gebauer G, Berveiller D, Damesin C, Selosse M-A (2005) Mixotrophy in orchids: insights from a comparative study of green individuals and nonphotosynthetic individuals of *Cephalanthera damasonium*. New Phytol 166:639–653

Kennedy AH, Taylor DL, Watson LE (2011) Mycorrhizal specificity in the fully mycoheterotrophic *Hexalectris* Raf. (Orchidaceae: Epidendroideae). Mol Ecol 20:1303–1316

Kikuchi G, Higuchi M, Morota T, Nagasawa E, Suzuki A (2008) Fungal symbiont and cultivation test of *Gastrodia elata* Blume (Orchidaceae). Jpn J Bot 83:88–95

Kottke I, Haug I, Setaro S, Suarez JP, Weiß M, Preußing M, Nebel M, Oberwinkler F (2008) Guilds of mycorrhizal fungi and their relation to trees, ericads, orchids and liverworts in a neotropical mountain rain forest. Basic Appl Ecol 9:13–23

Kottke I, Suarez JP, Herrerra P, Cruz D, Bauer R, Haug I, Garnica S (2010) Atractiellomycetes belonging to the 'rust' lineage (Pucciniomycotina) form mycorrhizae with terrestrial and epiphytic neotropical orchids. Proc R Soc Lond B 277:1289–1298

Kristiansen KA, Taylor DL, Kjøller R, Rasmussen HN, Rosendahl S (2001) Identification of mycorrhizal fungi from single pelotons of *Dactylorhiza majalis* (Orchidaceae) using single-strand conformation polymorphism and mitochondrial ribosomal large subunit DNA sequences. Mol Ecol 10:2089–2093

Kusano S (1911) *Gastrodia elata* and its symbiotic association with *Armillaria mellea*. J Coll Agric Jpn 9:1–73

Latalova K, Balaz M (2010) Carbon nutrition of mature green orchid *Serapias strictiflora* and its mycorrhizal fungus *Epulorhiza* sp. Biol Plantarum 54:97–104

Leake JR (1994) The biology of myco-heterotrophic ('saprophytic') plants. New Phytol 127:171–216

Leake JR (2004) Myco-heterotroph/epiparasitic plant interactions with ectomycorrhizal and arbuscular mycorrhizal fungi. Curr Opin Plant Biol 7:1–7

Leake JR, Cameron DD, Beerling DJ (2008) Fungal fidelity in the myco-heterotroph-to-autotroph life-cycle of Lycopodiaceae: a case of parental nurture? New Phytol 177:572–576

Liebel HT, Gebauer G (2011) Stable isotope signatures confirm carbon and nitrogen gain thtrough ectomycorrhizas in the ghost orchid *Epipogium aphyllum* Swartz. Plant Biol 13:270–275

Liebel HT, Bidartondo MI, Preiss K, Segreto R, Stockel M, Rodda M, Gebauer G (2010) C and N stable isotope signatures reveal constraints to nutritional modes in orchids from the Mediterranean and Macaronesia. Am J Bot 97:903–912

Lievens B, van Kerchhove S, Juste A, Cammue BPA, Honnay O, Jacquemyn H (2010) From extensive clone libraries to comprehensive DNA arrays for the efficient and simultaneous detection and identification of orchid mycorrhizal fungi. J Microbiol Methods 80:76–85

Martin F, Aerts A, Ahren D, Brun A, Danchin EGJ, Duchaussoy F, Gibon J, Kohler A, Lindquist E, Pereda V et al (2008) The genome of *Laccaria bicolor* provides insights into mycorrhizal symbiosis. Nature 452:88–92

Martos F, Dulormne M, Pailler T, Bonfante P, Faccio A, Fournel J, Dubois M-P, Selosse M-A (2009) Independent recruitment of saprotrophic fungi as mycorrhizal partners by tropical achlorophyllous orchids. New Phytol 184:668–681

Martos F, Munoz F, Pailler T, Kottke I, Gonneau C, Selosse M-A (2012) The role of epiphytism in architecture and evolutionary constraint within mycorrhizal networks of tropical orchids. Mol Ecol in press

McCormick MK, Whigham DF, O'Neill J (2004) Mycorrhizal diversity in photosynthetic terrestrial orchids. New Phytol 163:425–438

McCormick MK, Whigham DF, Sloan D, O'Malley K, Hodkinson B (2006) Orchid–fungus fidelity: a marriage meant to last? Ecology 87:903–911

McCormick MK, Whigham DF, O'Neill JP, Becker JJ, Werner S, Rasmussen HN, Bruns TD, Taylor DL (2009) Abundance and distribution of *Corallorhiza odontorhiza* reflect variations in climate and ectomycorrhizae. Ecol Monogr 79:619–635

McKendrick SL, Leake JR, Read DJ (2000) Symbiotic germination and development of mycoheterotrophic plants in nature: transfer of carbon from ectomycorrhizal *Salix repens* and *Betula pendula* to the orchid *Corallorhiza trifida* through shared hyphal connections. New Phytol 145:539–548

McKendrick SL, Leake JR, Taylor DL, Read DJ (2002) Symbiotic germination and development of the mycoheterotrophic orchid *Neottia nidus-avis* in nature and its requirement for locally distributed *Sebacina* spp. New Phytol 154:233–247

Merckx V, Bidartondo MI, Hynson N (2009) Mycoheterotrophy: when fungi host plants. Ann Bot 104:1255–1261

Moore RT (1987) The genera of *Rhizoctonia*-like fungi: *Ascorhizoctonia*, *Ceratorhiza* sp. nov., *Epulorhiza* sp. nov., *Moniliopsis* and *Rhizoctonia*. Mycotaxon 29:91–99

Motomura H, Selosse M-A, Martos F, Kagawa A, Yukawa T (2010) Mycoheterotrophy evolved from mixotrophic ancestors: evidence in *Cymbidium* (Orchidaceae). Ann Bot 106:573–581

Ogura-Tsujita Y, Yukawa T (2008a) High mycorrhizal specificity in a widespread mycoheterotrophic plant, *Eulophia zollingeri* (Orchidaceae). Am J Bot 95:93–97

Ogura-Tsujita Y, Yukawa T (2008b) *Epipactis helleborine* shows strong mycorrhizal preference towards ectomycorrhizal fungi with contrasting geographic distributions in Japan. Mycorrhiza 18:331–338

Ogura-Tsujita Y, Gebauer G, Hashimoto T, Umata H, Yukawa T (2009) Evidence for novel and specialized mycorrhizal parasitism: the orchid *Gastrodia confusa* gains carbon from saprotrophic *Mycena*. Proc R Soc Lond B 276:761–768

Osorio-Gil EM, Forero-Montana J, Otero JT (2008) Variation in mycorrhizal infection of the epiphytic orchid *Ionopsis utricularioides* (Orchidaceae) on different substrata. Caribbean J Sci 44:130–132

Otero JT, Flanagan NS (2006) Orchid diversity: beyond deception. Trends Ecol Evol 21:64–65

Otero JT, Ackerman JD, Bayman P (2002) Diversity and host specificity of endophytic *Rhizoctonia*-like fungi from tropical orchids. Am J Bot 89:1852–1858

Otero JT, Ackerman JD, Bayman P (2004) Diversity in mycorrhizal preferences between two tropical orchids. Mol Ecol 13:2393–2404

Otero JT, Bayman P, Ackerman JD (2005) Variation in mycorrhizal performance in the epiphytic orchid *Tolumnia variegata* in vitro: the potential for natural selection. Evol Ecol 19:29–43

Otero JT, Flanagan NS, Herre EA, Ackerman JD, Bayman P (2007) Widespread mycorrhizal specificity correlates to mycorrhizal function in the neotropical, epiphytic orchid *Ionopsis utricularioides* (Orchidaceae). Am J Bot 94:1944–1950

Ouanphanivanh N, Merenyi Z, Orczan AK, Bratek Z, Szigeti Z, Illyes Z (2008) Could orchids indicate truffle habitats? Mycorrhizal association between orchids and truffles. Acta Bot Szegediensis 52:229–232

Paduano C, Rodda M, Ercole E, Girlanda M, Perotto S (2011) Pectin localization in the Mediterranean orchid *Limodorum abortivum* reveals modulation of the plant interface in response to different mycorrhizal fungi. Mycorrhiza 21:97–104

Pereira OL, Rollemberg CL, Borges AC, Matsuokae K, Kasuya MCM (2003) *Epulorhiza epiphytica* sp. nov. isolated from mycorrhizal roots of epiphytic orchids in Brazil. Mycoscience 44:153–155

Pereira OL, Kasuya MCM, Borges AC, de Araujo EF (2005) Morphological and molecular characterization of mycorrhizal fungi isolated from neotropical orchids in Brazil. Can J Bot 83:54–65

Phillips RD, Barrett MD, Dixon KW, Hopper SD (2011) Do mycorrhizal symbioses cause rarity in orchids? J Ecol 99:858–869

Pickles BJ, Genney DR, Potts JM, Lennon JJ, Anderson IC, Alexander IJ (2010) Spatial and temporal ecology of Scots pine ectomycorrhizas. New Phytol 186:755–768

Pringle A, Taylor JW (2002) Understanding the fitness of filamentous fungi. Trends Microbiol 10:474–481

Purves S, Hadley G (1975) Movement of carbon compounds between the partners in orchid mycorrhizas. In: Sanders FE, Mosse B, Tinker PB (eds) Endomycorrhizas. Academic, London, pp 173–194

Rasmussen HN (1995) Terrestrial orchids from seed to mycotrophic plant. Cambridge University Press, Cambridge

Rasmussen HN (2002) Recent developments in the study of orchid mycorrhiza. Plant Soil 244:149–163

Rasmussen HN, Rasmussen FN (2007) Trophic relationships in orchid mycorrhiza – diversity and implications for conservation. Lankesteriana 7:334–341

Rasmussen HN, Rasmussen FN (2009) Orchid mycorrhiza: implications of a mycophagous life style. Oikos 118:334–345

Rasmussen HN, Whigham DF (1993) Seed ecology of dust seeds in situ: a new study technique and its application in terrestrial orchids. Am J Bot 80:1374–1378

Richard F, Moreau P-A, Selosse M-A, Gardes M (2004) Diversity and fruiting patterns of ectomycorrhizal and litter saprobic fungi in an old-growth Mediterranean forest dominated by *Quercus ilex* L. Can J Bot 82:1711–1729

Roche SA, Carter RJ, Peakall R, Smith LM, Whitehead MR, Linde CC (2010) A narrow group of monophyletic *Tulasnella* (Tulasnellaceae) symbiont lineages are associated with multiple species of *Chiloglottis* (Orchidaceae); implications for orchid diversity. Am J Bot 97:1313–1327

Roy M, Whatthana S, Richard F, Vessabutr S, Selosse M-A (2009a) Two mycoheterotrophic orchids from Thailand tropical dipterocarpacean forests associate with a broad diversity of ectomycorrhizal fungi. BMC Biol 7:51

Roy M, Yagame T, Yamato M, Iwase K, Heinz C, Faccio A, Bonfante P, Selosse M-A (2009b) Ectomycorrhizal *Inocybe* species associate with the mycoheterotrophic orchid *Epipogium aphyllum* but not its asexual propagules. Ann Bot 104:595–610

Sadovsky O (1965) Orchideen im eigenen Garten. Bayer, Munich

Schatz B, Geoffroy A, Dainat B, Bessiere J-M, Buatois B, Hossaert-McKey SM-A (2010) A case study of modified interactions with symbionts in a hybrid Mediterranean orchid. Am J Bot 97:1278–1288

Seiffert KA (2009) Progress towards DNA barcoding of fungi. Mol Ecol Resour 9:83–89

Selosse M-A, Roy M (2009) Green plants that feed on fungi: facts and questions about mixtrophy. Trends Plant Sci 14:64–70

Selosse M-A, Weiß M, Jany J-L, Tillier A (2002) Communities and populations of sebacinoid basidiomycetes associated with the achlorophyllous orchid *Neottia nidus-avis* (L.) L.C.M. Rich. and neighbouring tree ectomycorrhizae. Mol Ecol 11:1831–1844

Selosse M-A, Faccio A, Scappaticci G, Bonfante P (2004) Chlorophyllous and achlorophyllous specimens of *Epipactis microphylla* (Neottieae, Orchidaceae) are associated with ectomycorrhizal septomycetes, including truffles. Microb Ecol 47:416–426

Selosse M-A, Dubois M-P, Alvarez N (2009) Are Sebacinales common root endophytes? Mycol Res 113:1062–1069

Selosse M-A, Martos F, Perry BA, Padamsee M, Roy M, Pailler T (2010) Saprotrophic fungal mycorrhizal symbionts in achlorophyllous orchid. Finding treasures among the molecular scraps. Plant Signal Behav 5:1–5

Shefferson RP, Taylor DL, Weiß M, Garnica S, McCormick MK, Adams S, Gray HM, McFarland JW, Kull T, Tali K, Yukawa T, Kawahara T, Miyoshi K, Lee YI (2007) The evolutionary history of mycorrhizal specificity among lady's slipper orchids. Evolution 61:1380–1390

Shefferson RP, Kull T, Tali K (2008) Mycorrhizal interactions of orchids colonizing Estonian mine tailings hills. Am J Bot 95:156–164

Shefferson RP, Cowden CC, McCormick MK, Yukawa T, Ogura-Tsujita Y, Hashimoto T (2010) Evolution of host breadth in broad interactions: mycorrhizal specificity in East Asian and North American rattlesnake plantains (*Goodyera* spp.) and their fungal hosts. Mol Ecol 19:3008–3017

Smith SE (1966) Physiology and ecology of orchid mycorhizal fungi with reference to seedling nutrition. New Phytol 65:488–499

Smith SE, Read DJ (2008) Mycorrhizal symbiosis. Academic, Cambridge

Smith ZF, James EA, McDonnell MJ, McLean CB (2009) Planting conditions improve translocation success of the endangered terrestrial orchid *Diuris fragrantissima* (Orchidaceae). Aust J Bot 57:200–209

Smith ZF, James EA, McLean CB (2010) Mycorrhizal specificity of *Diuris fragrantissima* (Orchidaceae) and persistence in a reintroduced population. Aust J Bot 58:97–106

Sommerville KD, Siemon JP, Wood CB, Offord CA (2008) Simultaneous encapsulation of seed and mycorrhizal fungi for long term storage and propagation of terrestrial orchids. Aust J Bot 56:609–615

Stark C, Babik W, Durka W (2009) Fungi from the roots of the common terrestrial orchid *Gymnadenia conopsea*. Mycol Res 113:952–959

Stewart SL, Kane ME (2007) Symbiotic seed germination and evidence for in vitro mycobiont specificity in *Spiranthes brevilabris* (Orchidaceae) and its implications for species-level conservation. In Vitro Cell Dev- Biol Plant 43:178–186

Suarez JP, Weiß M, Abele A, Garnica S, Oberwinkler F, Kottke I (2006) Diverse tulasnelloid fungi form mycorrhizas with epiphytic orchids in an Andean cloud forest. Mycol Res 110:1257–1270

Suarez JP, Weiß M, Abele A, Garnica S, Oberwinkler F, Kottke I (2008) Members of Sebacinales subgroup B form mycorrhizae with epiphytic orchids in a neotropical rain forest. Mycol Prog 7:75–85

Swarts ND (2007) Integrated conservation of the rare and endangered terrestrial orchid *Caladenia huegellii* H.G. Reichb. PhD thesis, Murdoch University, Murdoch (cited in Wright et al. 2009)

Swarts ND, Dixon KW (2009) Terrestrial orchid conservation in the age of extinction. Ann Bot 104:543–556

Swarts ND, Sinclair EA, Francis A, Dixon KW (2010) Ecological specialisation in mycorrhizal symbiosis leads to rarity in an endangered orchid. Mol Ecol 19:3226–3242

Taylor DL, Bruns TD (1997) Independent, specialized invasions of ectomycorrhizal mutualism by two non photosynthetic orchids. Proc Natl Acad Sci USA 94:4510–4515

Taylor DL, McCormick MK (2007) Internal transcribed spacer primers and sequences for improved characterization of basidiomycetous orchid mycorrhizas. New Phytol 177:1020–1033

Taylor DL, Bruns TD, Leake JR, Read DJ (2002) Mycorrhizal specificity and function in mycoheterotrophic plants. In: Van der Heijden MGA, Sanders I (eds) Mycorrhizal ecology. Springer, Berlin, pp 375–413

Taylor DL, Bruns TD, Szaro TM, Hodges SA (2003) Divergence in mycorrhizal specialization within *Hexalectris spicata* (Orchidaceae), a nonphotosynthetic desert orchid. Am J Bot 90:1168–1179

Taylor DL, Bruns TD, Hodges SA (2004) Evidence for mycorrhizal races in a cheating orchid. Proc R Soc Lond B 271:35–143

Tedersoo L, Nara K (2010) General latitudinal gradient of biodiversity is reversed in ectomycorrhizal fungi. New Phytol 185:351–354

Tesitelova T, Tesitel J, Jersakova J, Rihova G, Selosse M-A (2012) Symbiotic germination capability of four *Epipactis* species (Orchidaceae) is broader than expected from adult ecology. Am J Bot 99:1020–1032

The Plant List (2010) The plant list, ver. 1. http://www.theplantlist.org. Accessed 1 Jan 2011

Thébault E, Fontaine C (2010) Stability of ecological communities and the architecture of mutualistic and trophic networks. Science 329:853–856

Thompson JN (1994) The co-evolutionary process. University of Chicago Press, Chicago

Tremblay RL, Ackerman JD, Zimmerman JK, Calvo RN (2005) Variation in sexual reproduction in orchids and its evolutionary consequences: a spasmodic journey to diversification. Biol J Linn Soc 84:1–54

Trudell SA, Rygiewicz PT, Edmonds RL (2003) Nitrogen and carbon stable isotope abundances support the mycoheterotrophic nature and host specificity of certain achlorophyllous plants. New Phytol 160:391–401

Umata H (1995) Seed germination of *Galeola altissima*, an achlorophyllous orchid with aphyllophorales fungi. Mycoscience 36:369–372

Vujanovic V, St-Arnaud M, Barabe D, Thibeault G (2000) Viability testing of orchid seed and promotion of coloration and germination. Ann Bot 86:79–86

Wang H, Wang Z, Zhang F, Liu J, He X (1997) A cytological study on the nutrient-uptake mechanism of a saprophytic orchid *Gastrodia elata*. Acta Bot Sin 39:500–504

Warcup JH (1971) Specificity of mycorrhizal association in some Australian terrestrial orchids. New Phytol 70:41–46

Warcup JH (1991) The *Rhizoctonia* endophytes of *Rhizanthella*. Mycol Res 95:656–659

Waterman RJ, Bidartondo MI (2008) Deception above, deception below: linking pollination and mycorrhizal biology of orchids. J Exp Bot 59:1085–1096

Waterman RJ, Bidartondo MI, Stofberg J, Combs JK, Gebauer G, Savolainen V, Barraclough TG, Pauw A (2011) The effects of above- and belowground mutualisms on orchid speciation and coexistence. Am Nat 177:E54–E68

Watkinson JI, Welbaum GE (2003) Characterization of gene expression in roots of *Cypripedium parviflorum* var. *pubescens* incubated with a mycorrhizal fungus. Acta Hortic 624:463–470

Weiß M, Selosse M-A, Rexer K-H, Urban A, Oberwinkler F (2004) Sebacinales: a hitherto overlooked cosm of heterobasidiomyctes with a broad mycorrhizal potential. Mycol Res 108:1003–1010

Weiß M, Sýkorov Z, Garnica S, Riess K, Martos F, Krause C, Oberwinkler F, Bauer R, Redecker D (2011) Sebacinales everywhere: previously overlooked ubiquitous fungal endophytes. PLoS One 6(2):e16793. doi:10.1371/journal.pone.0016793

White TJ, Bruns TD, Lee S, Taylor J (1990) Amplification and direct sequencing of fungal ribosomal RNA genes for phylogenetics. In: Innis MA, Gelfand DH, Sninsky JJ, White TJ (eds) PCR protocols: a guide to methods and applications. Academic, San Diego, pp 315–322

Wilson D (1995) Endophyte: the evolution of a term, and clarification of its use and definition. Oikos 73:274–276

Wright M (2007) Maximising the effectiveness of mycorrhizal fungi in the conservation of *Caladenia* taxa (Orchidaceae). PhD thesis, University of Melbourne, Melbourne

Wright MM, Cross R, Dixon K, Huynh T, Lawrie A, Nesbitt L, Pritchard A, Swarts N, Thomson R (2009) Propagation and introduction of *Caladenia*. Aust J Bot 57:373–387

Wright MM, Cross R, Cousens RD, May TW, McLean CB (2010) Taxonomic and functional characterization of fungi from the *Sebacina vermifera* complex from common and rare orchids in the genus *Caladenia*. Mycorrhiza 20:375–390

Xu JT, Guo SX (2000) Retrospect on the research of the cultivation of *Gastrodia elata* Bl, a rare traditional Chinese medicine. Chin Med J 113:686–692

Yagame T, Yamato M, Mii M, Suzuki A, Iwase K (2007) Developmental processes of achlorophyllous orchid, *Epipogium roseum*: from seed germination to flowering under symbiotic cultivation with mycorrhizal fungus. J Plant Res 120:229–236

Yagame T, Yamato M, Suzuki A, Iwase K (2008) Ceratobasidiaceae mycorrhizal fungi isolated from nonphotosynthetic orchid *Chamaegastrodia sikokiana*. Mycorrhiza 18:97–101

Yagame T, Orihara T, Selosse M-A, Yamato M, Iwase K (2012) Mixotrophy of *Platanthera minor*, an orchid associated with ectomycorrhiza-forming Ceratobasidiaceae fungi. New Phytol 193:178–187

Yamato M, Yagame T, Suzuki A, Iwase K (2005) Isolation and identification of mycorrhizal fungi associating with an achlorophyllous plant, *Epipogium roseum* (Orchidaceae). Mycoscience 46:73–77

Yoder JA, Zettler LW, Stewart SL (2000) Water requirements of terrestrial and epiphytic orchid seeds and seedlings, and evidence for water uptake by means of mycotrophy. Plant Sci 156:145–150

Yuan L, Yang ZL, Li SY, Hu H, Huang J-L (2010) Mycorrhizal specificity, preference and plasticity of six slipper orchids from South Western China. Mycorrhiza 20:559–568

Yukawa T, Ogura-Tsujita Y, Shefferson RP, Yokoyama J (2009) Mycorrhizal diversity in *Apostasia* (Orchidaceae) indicates the origin and evolution of orchid mycorrhiza. Am J Bot 96:1997–2009

Zettler LW, Poulter SB, McDonald KI, Stewart SL (2007) Conservation-driven propagation of an epiphytic orchid (*Epidendrum nocturnum*) with a mycorrhizal fungus. HortScience 42:135–139

Zhu GS, Yu ZN, Gui Y, Liu ZY (2008) A novel technique for isolating orchid mycorrhizal fungi. Fungal Divers 33:123–137

Zimmer K, Hynson NA, Gebauer G, Allen EB, Allen MF, Read DJ (2007) Wide geographical and ecological distribution of nitrogen and carbon gains from fungi in pyroloids and monotropoids (Ericaceae) and in orchids. New Phytol 175:166–175

Zotz G, Schmidt G (2006) Population decline in the epiphytic orchid *Aspasia principissa*. Biol Conserv 129:82–90

13 The Symbiotic Fungus *Piriformospora indica*: Review

A. Varma[1], I. Sherameti[2], S. Tripathi[1], R. Prasad[1], A. Das[1], M. Sharma[1], M. Bakshi[1], J.M. Johnson[2], S. Bhardwaj[1], M. Arora[1], K. Rastogi[1], A. Agrawal[1], A.C. Kharkwal[1], S. Talukdar[1], U.S. Bagde[3], V.S. Bisaria[4], C.P. Upadhyaya[5], P.S. Won[6], Y. Chen[7], J. Ma[7], B. Lou[7], A. Adya[8], L. Zhang[9], M.K. Meghvansi[10], K.S. Sree[1], S.K. Gosal[11], R.B. Srivastava[12], A.K. Johri[13], M. Kumar[13], M. Dua[13], C. Cruz[14], R. Oelmüller[2]

CONTENTS

[1]Amity Institute of Microbial Technology, Amity University, Noida, Uttar Pradesh, 201303, India ajitvarma@aihmr.amity.edu
[2]Friedrich-Schiller Universität Jena, Institut für Allgemeine Botanik und Pflanzenphysiologie, Dornburger Str. 159, 07743 Jena, Germany
[3]Applied Microbiology Laboratory, Department of Life Sciences, University of Mumbai, Mumbai, Maharashtra, 400098, India
[4]Department of Biochemical Engineering and Biotechnology, Indian Institute of Technology, New Delhi, India
[5]Department of Botany, Guru Ghasidas Central University, Koni, Bilaspur 495009, Chattisgarh, India
[6]Department of Molecular Biotechnology, Konkuk University, Seoul 143701, South Korea
[7]Institute of Biotechnology, Zhejiang University, Hangzhou 310058, People's Republic of China
[8]Division of Biotechnology and Forensic Sciences, School of Contemporary Sciences, University of Abertay, Bell Street, Dundee DD1 1HG, UK
[9]Institute of Microbiology, Chinese Academy of Sciences, 8 Belertio Zhongguancun, Beijing 100190, People's Republic of China
[10]Research and Development, Defence Research Development Organization, Post Bag No. 2, Tezpur 784001, Assam, India
[11]College of Basic Science and Humanities, Punjab Agriculture University, Ludhiana Punjab, India
[12]Research and Development, Defence Institute of Higher Altitude Research, Defence Research Development Organization, Leh-Ladakh, India
[13]School of Life Sciences, Jawaharlal Nehru University, New Delhi 110067, India
[14]Centro de Biologia Ambiental, Faculdade de Ciências da Universidade de Lisboa, Campus Grande Block C-2, Piso-5, Sala 03, 1749-016 Lisbon Portugal

I. Introduction

Piriformospora indica, the novel endophytic root-colonizing fungus of the xerophytic plants of Thar Desert, India, was isolated by Verma et al. (1998). At that time nobody could think that this fungus could be one of the model organisms used in the research field of mycorrhizal interactions and that a great number of

Fungal Associations, 2nd Edition
The Mycota IX
B. Hock (Ed.)

papers on *P. indica* could expand into highly respected journals such as Nature, PNAS, Plos Pathogen, JBC, Plant Physiology, Molecular Plant Pathology, The Plant Journal, Phytopathology, New Phytologist, Plant and Soil, Journal of Biotechnology and many more.

P. indica, a **basidiomycete**, resembles in many aspects the Arbuscular Mycorrhizal Fungi (AMF) which, however, belongs to the new family **Sebacinaceae** and new order **Sebacinales** Glomeromycota (Weiß et al. 2004; Qiang et al. 2011). In contrast to AMF, ***P. indica* can grow axenically** (Varma et al. 2012a). Similar to AMF, this fungus promotes plant growth, increases the resistance of colonized plants against fungal pathogens and their tolerance to abiotic stress (Harman 2011) and proves further beneficial to plants. It also alters the secondary metabolites of many plants of economic importance and promotes overall growth and seed production of many plants. In contrast to AMF *P. indica* colonizes *Arabidopsis thaliana*, a model plant for which a multitude of well characterized mutants is available.

The *P. indica* genome is assembled into 1,884 scaffolds (size: 1 kb; N50: 51.83 kb) containing 2,359 contigs with an average read coverage of 22 and a **genome size of 24.97 Mb.** The estimated DNA content of *P. indica* nuclei ranges from 15.3 to 21.3 Mb. To assess the genome completeness of *P. indica* a blast search was performed with highly conserved core genes present in higher eukaryotes (Zuccaro et al. 2011). A genetic transformation system has been established using a fragment of the TEF promoter region for construction of vectors carrying the selectable marker hygromycin B phosphotransferase. It is already shown that *P. indica* can be stably transformed by random genomic integration of foreign DNA and that it possesses a relatively small genome as compared to other members of the Basidiomycota (Zuccaro et al. 2009).

II. Fungal Morphology

The **hyphae** of *P. indica* are highly interwoven, often adhere together and appear as a simple intertwined cord. Young mycelia are white and almost hyaline but inconspicuous zones are recorded in other cultures. Hyphae are thin-walled and of a diameter from 0.7 to 3.5 μm. The septate hyphae often show anastomosis. New branches emerge irregularly and the hyphal wall shows some external deposits at regular intervals, perhaps polysaccharides and/or some hydrophobic proteins, which stain deeply with toluidine blue. Since septation is irregular, the single compartments can contain more than one nucleus.

The cell walls are very thin and show multilayered structures. The septa consist of dolipores within the continuous parenthosomes, which forms the basis for the systematic position within the Hymenomycetes. The dolipores are very prominent with a multilayered crosswall and a median swelling mainly consisting of electron-transparent material. The parenthosomes are always straight and have the same diameter as the corresponding dolipore. No kind of pores could be detected, meaning thereby that they are flat discs without any perforation. The parenthosomes consist of an electron-dense outer layer and a less dense inner layer, which shows an inconspicuous dark line in the median region (Fig. 13.1, see inset).

Mycelium on maturity produces characteristic pear-shaped **chlamydospores,** which appear single or in clusters measuring 16–45 μm in length and 10–17 μm in width and they are distinctive due to their pear-shaped structure. The average spore and hyphal wall thickness is 0.7 and 0.3 μm, respectively.

Very young spores have thin, hyaline walls. At maturity, these spores have walls up to 1.5 μm thick, which appear two-layered, smooth and pale yellow. The cytoplasm of the chlamydospores is densely packed with granular material and usually contains 8–25 nuclei. Neither clamp connections nor sexual structures are observed.

III. Taxonomic Status

Sequence analysis of ribosomal DNA (rDNA) regions uncovered that ***P. indica* belongs to the family Sebacinaceae, order Sebacinales** (Weiß et al. 2004). This basal order of Hymenomycetes (Basidiomycetes) encompasses fungi with longitudinally septate basidia and imperforate

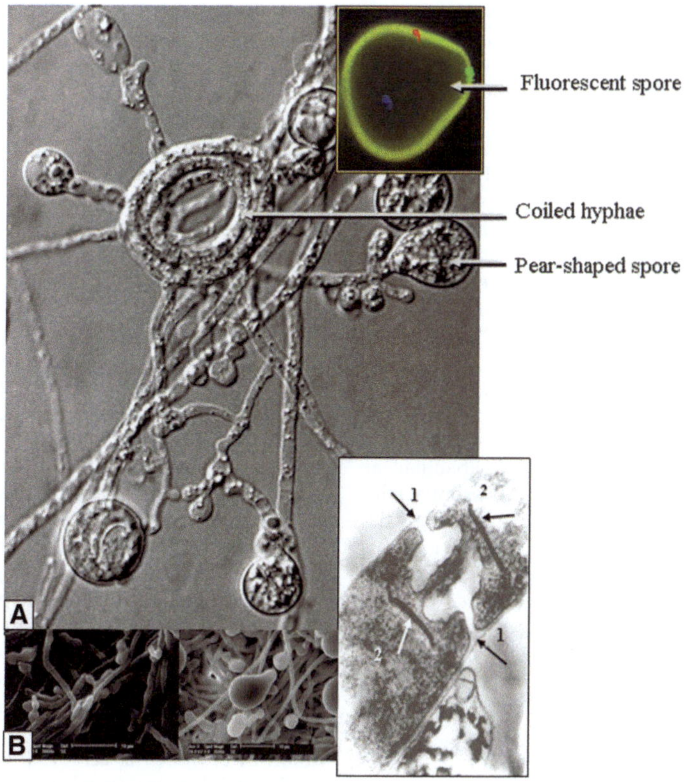

Fig. 13.1. A typical view of *Piriformospora indica*. (**A**) Photograph obtained by Beta confocal microscope. *Inset* Magnified view showing the dolipore septum and parenthosomes. A section of hypha was observed by electron-transparent material. *Arrows* indicate (*1*) the dolipore and (*2*) the continuous parenthosomes. (**B**) Chlamydospores after electron micrography

parenthesomes (Verma et al. 1998; Varma et al. 2001; Selosse et al. 2007). The fungi of order Basidiomycetes lack cystidia and structures formed during cytokinesis on some basidiomycetous hyphae, the so-called clamp connections. Like other cultivable species of the Sebacinales, *P. indica* forms moniloid hyphae, which look like pearls in a chain. Based on this phenotype and rDNA sequence analyses, this endophyte is placed in the polyphyletic genus *Rhizoctonia* (Schüßler et al. 2001; Deshmukh et al. 2006; Selosse et al. 2007). Basiewicz et al. (2012) reported significant differences in the physiological and molecular parameters inferred from morphologically very similar strains of *Piriformospora*. As a first taxonomic consequence, they have described *P. williamsii* as a new member of the so far monotypic genus *Piriformospora* and have shown that this genus contains still undescribed species that were recently discovered as endophytes of field-collected specimens of *Anthyllis*, *Medicago* and *Lolium* in Germany.

Comparisons of 18S rDNA sequences indicated that a similar kind of fungus apparently occurs in the rhizosphere soil of *Leptochloa fusca* or *Soryza minuta* in Pakistan and in Philippines, respectively. Later they were reported from Australia, Portugal and Brazil (Varma et al. 2001). Molecular sequence data are useful for systematics where morphological characters like the taxonomically decisive sexual states are missing. A neighbour-joining analysis of partial 18S rDNA sequences (525 nucleotide position) placed *P. indica* within the Basidiomycota close to the *R. solani* group (Ceratobasidiales). A maximum likelihood analysis on almost complete 18S rDNA sequences (1,550 nucleotide positions) confirmed this assignment. Similar results were obtained by distance and parsimony methods. A comprehensive phylogenetic analysis of *Rhizoctonia* using sequences from mitochondrial and nuclear rDNA on more representatives should provide an insight into the evolution of this important group and clearly the evolutionary relationship of *P. indica* and *Rhizoctonia* within the Hymenomycetes. A 28S rDNA showed no change to the taxonomic status (Varma et al. 2001, 2002). Evaluations of 18S rDNA clone libraries suggested that the fungus was not abundant in any of these rhizospheres. However, none of the West European soils tested were found to contain this fungus (Varma et al. 2001, 2002). Due to primitive similarities to Glomeromycetes, this fungus is termed an AM-like fungus (Franken et al. 2000). Nonetheless, *Sebacina vermifera* sensu, a fungus isolated from Bavaria, contains 28S rRNA (Oberwinkler 1964; Warcup and Talbot 1967; Weiß and Oberwinkler 2001). Therefore it is closely fitted to the family Sebacinaceae. The highly conserved 5.8S gene regions of the Neottia fungi are statistically similar to GenBank entries for many fungi. *S. vermifera* and *P. indica* are similar to the two *Neottia* fungi in the 5.8S gene region (McKendrick et al. 2002). Despite many similarities in macroscopical and ultrastructural characters (Bandoni 1984) the Sebacinaceae are separated from the remaining taxa of Auriculariales sensu Bandoni in the present analysis, which is consistent with the previous molecular phylogenetic analysis of Weiß and Oberwinkler (2001). This molecular phylogenetic analysis demonstrates that *P. indica* belongs to the *S. vermifera* complex. Sequences of the genes encoding the elongation factor 1-a (TEF) and glyceraldehyde-3-phosphate dehydrogenase (GAPDH) were used for genome size estimation through real-time PCR analysis (Zuccaro et al. 2009).

IV. Mass Cultivation

P. indica, which mimics AMF, presents a good model system to understand the molecular basis of photo- and mycobiont interaction. Its application in horticulture or agriculture as a potent biofertilizer and biocontrol agent is economically and practically feasible through the easy propagation of a fungal inoculum using liquid or axenic cultures.

It is shown that the **fungus can grow axenically on different synthetic media**. Among the tested media, the best growth is reported to be on Hill and Käfer medium (2001) which is reported from different authors (Varma et al. 1999, 2001; Pham et al. 2004b; Qiang et al. 2011). However, significant quantitative and morphological changes are detected when the fungus is grown on different nutrient composition with no apparent negative effect on plants (Kumar et al. 2011).

A 14-l bioreactor (Chemap AG, Switzerland) was used to grow *P. indica* on optimized Hill and Käfer medium (as described above) to establish the best conditions for a maximal biomass and spore production for scale-up studies. When *P. indica* was grown in a 14-l bioreactor on optimized Hill and Käfer medium (containing 10.0 g/L glucose, 1.0 g/L peptone, 1.0 g/L yeast extract, 1.0 g/L soya bean meal, 1.83 g/L KH_2PO_4, 0.65 g/L $MgSO_4 \cdot 7H_2O$, 50.0 mL/L macro element, 2.5 mL/L microelement stock solution, 1.0 mL/L vitamin stock solution, 1.0 mL/L $CaCl_2$, 0.1 M and 1.0 mL/L $FeCl_3$, 0.1 M), a maximum dry cell weight of 7.36 g/L was obtained after 42 h of growth (Fig. 13.2). The values of biomass yield and the specific daily growth rate were 0.79 and 1.15, respectively. The fungus initiated sporulation after 48 h and a spore yield of 9.25×10^7 spores/mL was achieved after 60 h of growth. The early sporulation in this case may be due to rapid consumption of glucose. Due to more efficient mixing and homogenized fungal suspension, the growth of fungus was faster in the bioreactor and resulted in early depletion of the carbon source and thereby early sporulation compared to a shake flask. A complete growth profile of *P. indica* on modified Hill and Käfer medium has been depicted in Fig. 13.2. The pattern of pH profile was quite similar in all these experiments where complex nitrogen sources were present in the growth medium. The uptake of glucose caused a decrease in pH of fermentation broth which might be due to the generation of acidic metabolites (Fig. 13.2). The growth of the fungus remained unaffected as long as the pH during the log phase was not reduced below 4.5. Besides this, we have found that the optimal mass cultivation of *P. indica* is achieved on soil extract-enriched media and jaggery (extracted from *Saccharum officinarum*) that contains 60–85 g/L sucrose, 5–15 g/L glucose and fructose, 0.4 g/L protein, 0.05 g/L fat, 0.6–1.0 g/L minerals (0.4 g/L calcium, 0.045 g/L magnesium and phosphorus) and 11 % iron. The soil extract enriched with some nutrients is suitable for mass production of the endophyte, up to 14 days. Hill and Käfer (2001) medium looks to give better results in a longer run. Soil extract and jaggery which are economically feasible need to be optimized for getting a higher biomass of *P. indica* at fermentor scale.

V. Photosymbiont Promotional Features

A. Impact of Fungal Biomass

Plant colonization by *P. indica* promotes plant growth. **The endophyte increases nutrient uptake, allows plants to survive under water-, temperature- and salt-stress, confers (systemic) resistance to toxins, heavy metal ions and pathogenic organisms and stimulates growth and seed production.** Different workers (Verma et al. 1998; Sahay and Varma 1999; Varma et al. 1999, 2001; Singh et al. 2002a, b; Oelmüller et al. 2004, 2005; Pham et al. 2004a, b; Peškan-Berghöfer et al. 2004; Shahollari et al.

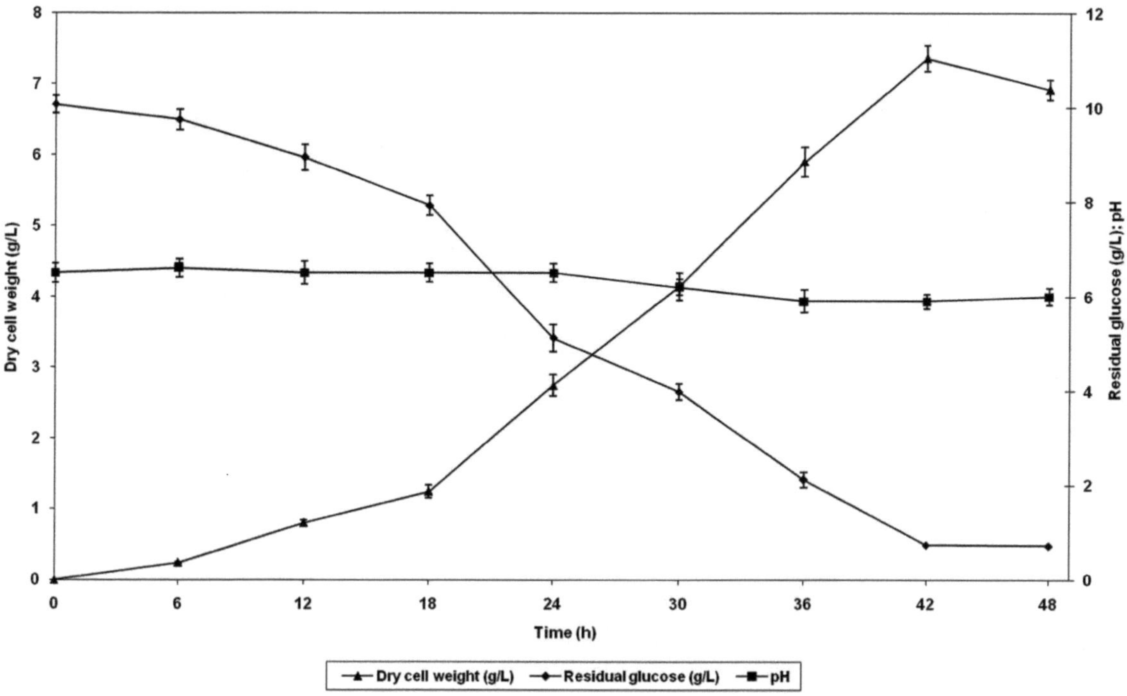

Fig. 13.2. Growth profile of *P. indica* on modified Hill and Käfer medium in 14-l bioreactor. Operational parameters: working volume=10 l; agitation speed (rpm): 0–6 h=250, 6–18 h=200, 18–20 h=250, 20–26 h=450, 26–42 h=500; aeration rate (vvm): 0–24 h=0.20, 24–42 h=0.40; temperature -20 °C; initial pH=6.5. Hill and Käfer medium (2001) contains: 10.0 g/L glucose, 1.0 g/L peptone, 1.0 g/L yeast extract, 1.0 g/L soya bean meal (Pham et al. 2004b)

2004, 2005, 2007; Kaldorf et al. 2005; Sherameti et al. 2005, 2008a, b; Waller et al. 2005, 2008; Vadassery et al. 2008, 2009a, b; Fakhro et al. 2009; Kumar et al. 2009; Oelmüller et al. 2009; Achatz et al. 2010; Gosal et al. 2010; Sun et al. 2010) have reported that the fungus tremendously improves the growth and overall biomass production of diverse host. A total number of approximately 150 plant species is reported to interact with *P. indica*, including agricultural, horticultural, medicinal and other important plants (Sahay and Varma 1999, 2000; Rai et al. 2001; Singh et al. 2002a, b;Kumari et al. 2004a, b, c; Rai and Varma 2005; Chauhan et al. 2006; Fakhro et al. 2009; Kumar et al. 2009; Oelmüller et al. 2009; Achatz et al. 2010; Sun et al. 2010).

Oryza sativa, *Zea mays*, *Tridax procumbans* and *Brassica oleracea* var *capitata* plants (Kumari et al. 2004a) have shown early seed germination and an increased number of seeds. However, many plants, viz. *O. sativa*, *Z. mays*, *Phaseolus vulgaris*, *T. procumbans*, *Abrus precatorius*, *Solanum nigrum*, *B. oleracea* var *capitata*, *B. nigra*, *Nicotiana tabacum*, *Saccharum officinarum*, *Lagenaria* sp. and *Spinacea oleracea* have shown better phytopromotional effect with increased biomass (Prasad 2008). An increase in the biomass as well as in the secondary metabolites of *Trigonella foenum graecum* is also recorded. Increase in chlorophyll contents of *P. indica*-treated plants was also observed (Sharma et al. 2011). *P. indica*-colonized roots showed a higher development compared to control plants in all initial stages of growth, as suggested by earlier expression of developmentally regulated genes (Waller et al. 2005). Sahay et al. (1998) and Singh (2004) have proved that inoculation of maize roots with *P. indica* enhanced the growth response by colonizing the root cortex. This fungus promotes also the growth of several tropical legumes tested (*Cicer arietinum*, *Phaseolus aureus*, *P. mungo*, *Pisum sativum* and *Glycine max*) (Varma et al. 2012b). Some of the plants

Fig. 13.3. *P. indica* promotes plant growth (*left* control, *right* fungus-treated): (**A**) *Nicotiana tabacum*, (**B**) *Lycopersicon esculentum*, (**C**) *Saccharum officinarum*, (**D**) *Solanum melongena*, (**E**) *Citrullus lanatus* (water melon), (**F**) *Tegetus erectus*, (**G**) *Trigonella* sp., (**H**) *Lagenaria* sp. (bottle gourd)

of agricultural importance that interact positively with *P. indica* are shown in Fig. 13.3.

Artemisia annua, *Bacopa monniera*, *Abrus precatorius*, *Stevia rebaudiana*, *Linum album*, *Trigonella* sp., *Coleus forskohlii*, *Spilanthes calva*, *Withania somniferra*, *Chlorophytum tuberosum* and *Curcuma longa* are some of the important medicinal plants which have been reported to beneficially interact with the fungus (Fig. 13.4). The secondary metabolite content of all these medicinal plants was increased several-fold because of their interaction with *P. indica*. In another study co-cultivation of live fungal cells with the hairy roots of the medicinal plant *Linum album* reduced the growth of the hairy roots. Despite reduction in hairy root biomass, an enhancement in lignan content was observed. The hairy root cultures co-cultivated with 1–5 g/L fungal biomass at day 10, 11 and 12 all achieved a higher podophyllotoxin (PT) and 6-methoxypodophyllotxin (6-MPT) content (mg/g) in the roots than in the fungus-free control culture. The highest increase in PT content (8.48 mg/g) and 6-MPT content (2.78 mg/g) was obtained when a fungal concentration of 2.0 g/L was added to a growing hairy root cultures of *L. album* on the 12th day, i.e., for exposure time of 48 h. The same fungal concentration for the same duration resulted in a maximum improvement of 2.1-fold in PT concentration and 2.5-fold in 6-MPT concentration.

Interestingly, like AMF, ***P. indica* did not invade the roots of two strains of *Brassica* (member of Cruciferae) and the myc$^-$ mutants of *Glycine max* and *Pisum sativum***, obtained from Peter Gresshoff (Knoxville, USA) and V.Gianninazzi-Pearson (Dijon, France), respectively. In a similar way, the fungus remained neutral and did not colonize the plant 'Naga chilli', which was considered as the hottest chilli of the world (557,000 Scoville units; http.//chilly.in/Indian_chilli_varieties.htm), primarily growing in the north-east region of India.

Fig. 13.4. *P. indica* promotes medicinally important plants (*in vitro* experiments): (**A**) *Artemisia annua*, (**B**) *Curcuma longa*, (**C**) *Bacopa monniera*

After generation of plantlets on *in vitro* culture, the plantlets (after biological hardening) were first transferred to pots and later to natural environmental conditions. The survival rate increased to 90–100 % for *P. indica*-treated plantlets while the untreated ones had a low survival rate of 10–60 %. An early callus differentiation was observed in roots and shoots (Sahay 1999; Sudha 1999). In tissue culture conditions *P. indica* colonized the plantlets of the medicinal plant *Bacopa monniera*, tobacco (*Nicotiana xanthi*), brinjal (*Solanum melongana*) and neem (*Azadirachta indica*). A higher biomass was recorded in the treated plants.

B. Impact of Fungal Culture Filtrate

Not only the mycelium in association with roots but also the culture filtrate of the mycelium contains fungal exudates, minerals, hormones, enzymes, proteins, etc. (Verma et al. 1998). *In vitro* experiments have shown that even **very small amounts (50 μL/20 mL) of culture filtrate is sufficient to promote root and shoot growth** (Fig. 13.5). The culture filtrate is also effective in seed germination enhancement, growth of the seedlings and also in inhibiting the growth of a few potent root fungal pathogens tested.

For the pot experiments 15-day-old seedlings (brinjal, broccoli, beans, sunflower, cabbage, maize, bacopa, tobacco) were transferred to disposable plastic pots containing vermiculite (autoclaved) and sand (acid washed) in the ratio of 3:1. An amount of 15 mL of freshly eluted *P. indica* culture filtrate was applied to each pot. Minimal KM media and equal amounts of double-sterilized distilled water served as control. An increase in the root and shoot length and plant biomass was observed in the *P. indica*-treated hosts.

VI. Stimulatory Factors

Preparative HPLC analysis of hyphal and culture filtrate showed a major peak identified as benzoic acid. The function of this compound is not clear. Identical compounds to benzoic acid and its analogues (benzoic acid, a-hydroxybenzoic acid, 2-4 di-hydroxybenzoic acid, vanillic acid, cinnamic acid, p-coumaric acid, caffeic acid, ferulic acid) did not show any stimulation on the plants tested. The nature of the stimulatory factor which promotes the plant growth is not yet known. HPLC coupled with

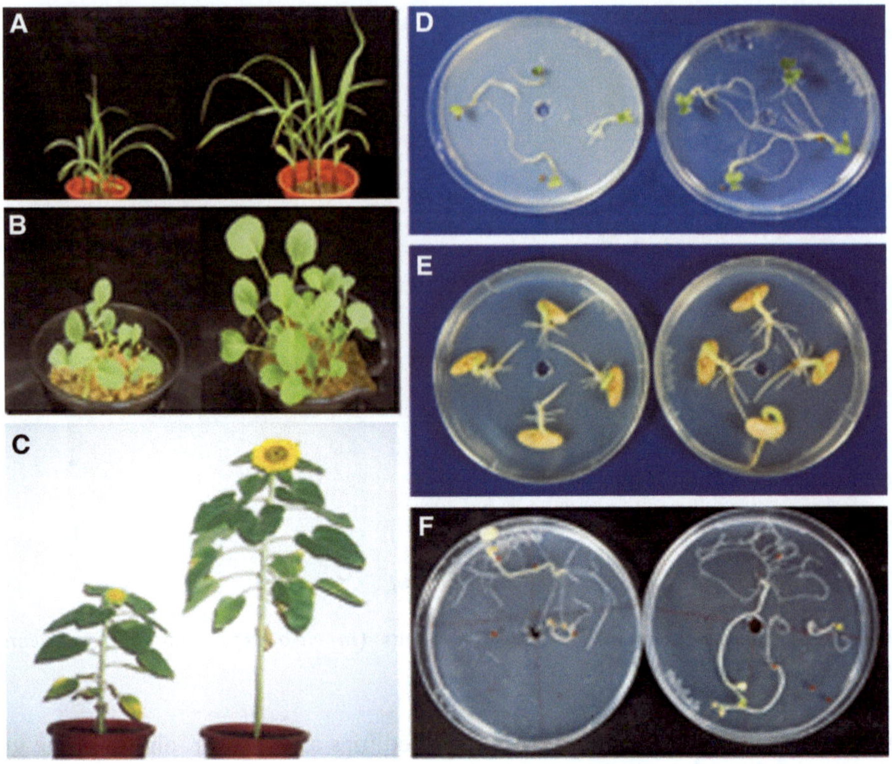

Fig. 13.5. *P. indica* culture filtrate promotes plant growth: (**A**) *Zea mays*, (**B**) *Brassica oleracea* (cabbage), (**C**) *Helianthus annus*, (**D**) *B. oleracea* var. *botrytis* (broccoli), (**E**) *Phaseolus vulgaris*, (**F**) *Brassica* sp.

spectroscopic techniques could be employed with an effort to analytically determine the compound(s) responsible for breaking the seed dormancy and fast plant growth.

A cyclohexomone derivative (blumenin) which is accumulated in roots of cereals and other members of Poaceae colonized by arbuscular mycorrhizal fungi (Maier et al. 1995; Fester et al. 1998) was also present here. The exact function of accumulated cyclohexomones is not yet known. It is speculated that these secondary compounds might be involved in the regulation of mycorrhizal colonization (Fester et al. 1998). The UV spectra obtained from HPLC photodiode array detector showed a cluster of peaks between 7.5 and 12.5 min of the chromatograph for *P. indica* co-cultivated maize, barely, rice and fox tail millet, indicating the presence of indole-derivatives, e.g., tryptophane, tryosine and tyramine or their derivatives (unpublished data).

Root extracts of maize showed a significant presence of cyclic hydroxamine acids like 2,4-dihydroxy-7-methoxy-1,4-benzoxazin-2-one (DIMBOA) in wheat but not in rice, barley and *Setaria*. HPLC analysis of methanol extracts of infected maize roots showed eight different peaks with a photodiode array detector with typical UV spectra from benzoxazinone derivatives. None of these peaks were identical to DIMBOA or DIBOA. The chemical structure of these compounds need to be identified by spectrophotometric methods (Varma et al. 2001; Pham et al. 2004a).

VII. Nutrient (^{32}P) Transport

Phosphorus is an essential mineral nutrient for the plant growth. Plants acquire this mineral from the environment either directly by their roots or indirectly from mycorrhizal fungi

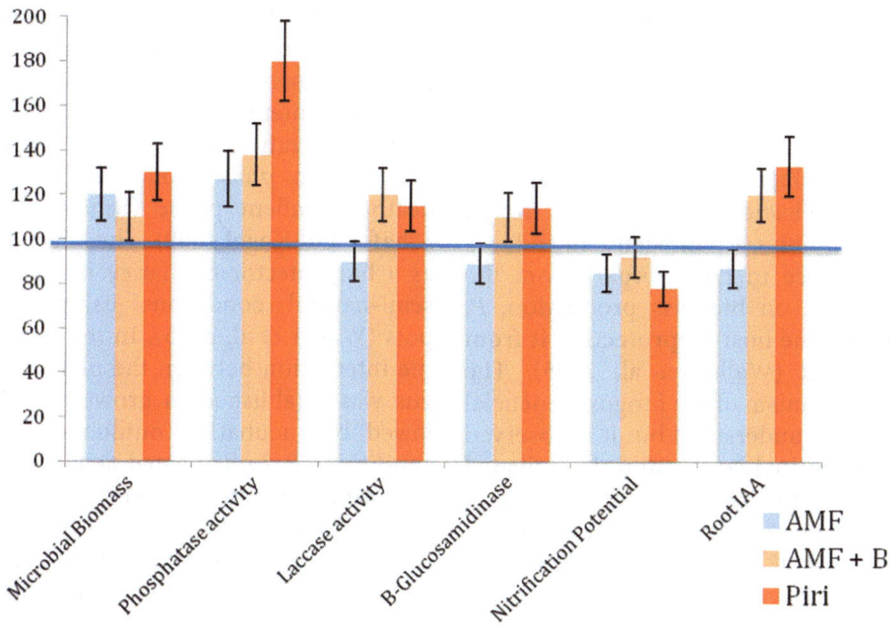

Fig. 13.6. Enzyme activities of the rhizosphere of tomato plants grown in pots with soil inoculated with AMF (mixture of 1:1 *Glomus intraradices* and *G. mossae*; *AMF*), AMF and a mixture of bacteria (*Bacillus mageterium*, *Saccharomyces cerevisiae*, *Burkholderia* sp., *Pseudomonas* sp.; *AMF+B*) or with *Piriformospora indica* (*Piri*). Control plants were grown with soil without any inoculation. Enzyme activities were determined according to Pritsch et al. (2011)

which form inter- or intracellular symbiotic association with the roots. AM colonization may increase the rate of phosphorus accumulation beyond the limits which can be currently utilized, thus reducing the rate of phosphorus utilization efficiency. Such momentary "luxury consumption" of phosphorus may, however, serve a storage function and be utilized subsequently, allowing mycorrhized plants ultimately to outperform non-mycorrhized plants (Koide 1991). Recent ^{32}P experiments suggest that ***P. indica*** **is important for P-acquisition by the root especially in the arid and semi-arid regions.**

Recently cloning and functional analysis of a gene encoding a **phosphate transporter (PiPT)** from this root endophyte were reported (Yadav et al. 2010). The PiPT polypeptide belongs to the major facilitator superfamily (MFS) and exhibits 12 transmembrane helices divided into two halves that are connected by a large hydrophilic loop in the middle. The function of the protein encoded by PiPT was confirmed by complementation of a yeast phosphate transporter mutant. *PiPT* belongs to a high-affinity phosphate transporter family **(Pht1)**. To understand the **physiological role of PiPT**, knockdown (KD) transformants of the gene were prepared using electroporation and RNA interference. KD transformants transported a significantly lower amount of phosphate to the host plant than wild-type *P. indica*. Higher amounts of phosphate were found in plants colonized with wild-type *P. indica* than that of non-colonized and plants colonized with KD-PiPT *P. indica*. These observations suggest that *PiPT* is actively involved in the phosphate transportation and in turn *P. indica* helps to improve the nutritional status of the host plant (Yadav et al. 2010; Kumar et al. 2011).

In another independent experiment, high phosphatase activity in the rhizosphere of plants inoculated with *P. indica* in relation to the control and to the other treatments was studied. A significant increase in the root indole acetic acid content was also observed, which may be related with the improved root development of plants inoculated with *P. indica* (Fig. 13.6).

VIII. Biotechnological Applications

A. Biological Control Agent of Soil-Borne Diseases

It is reported (Pham et al. 2004b) that the endogenous application of symbiotic fungi, such as AMF, mimics a pathogen and induces systemic resistance in hosts. Apart from the stimulating effect on biomass production, ***P. indica* supports the host by protecting it from pathogenic fungi** (Waller et al. 2005). The underlying mechanism of the fungus' beneficial activity is not yet understood but it is observed that the **plant antioxidant system is activated** and thus implicated in the improvement of abiotic and biotic stress tolerance (Waller et al. 2005; Deshmukh and Kogel 2007; Druege et al. 2007). Kumar et al. (2009) reported the bioprotection performance of *P. indica* against the root parasite *Fusarium verticillioides* in maize. *P. indica* stimulated an increase in biomass, root length and root number compared to plants grown with *F. verticillioides* alone. *P. indica* suppressed colonization by *F. verticillioides* which was verified by PCR using *P. indica*- and *F. verticillioides*-specific primers. It was hypothesized that, as the colonization by *P. indica* increased, the colonization by *F. verticillioides* decreased. Catalase (CAT), glutathione reductase (GR), glutathione S-transferase (GST) and superoxide dismutase (SOD) activities were higher in *F. verticillioides*-colonized plant roots than in non-colonized ones. The increased activity of antioxidant enzymes minimized the chances of oxidative burst (excessive production of reactive oxygen species), and therefore *F. verticillioides* might be protected from the oxidative defence system during colonization. The antioxidant enzyme activities were decreased in plants which were first inoculated with *F. verticillioides* and at day 10 with *P. indica* compared to the plants inoculated with *F. verticillioides* alone. The decrease of antioxidant enzyme activities due to the presence of *P. indica* helped the plant to overcome the disease load of *F. verticillioides* (Kumar et al. 2009). It is also suggested that *P. indica* may target a not-yet-identified signalling pathway to induce systemic resistance. Transmission electron microscopy studies in *Arabidopsis*–*P. indica* interaction have demonstrated that root colonization is initiated by a biotrophic colonization phase, followed by a later cell death-dependent phase (Schäfer et al. 2007; Schäfer and Kogel 2009). The effect of *P. indica* as a bioprotector on barley is assessed under semi-natural conditions using Mitscherlich pots (Waller et al. 2005). In the reported work, the interaction between the plant and the fungus was established in growth chambers, followed by incubation outdoors. Under these conditions ***P. indica* acted as both a biofertilizer and a biocontrol agent** (Waller et al. 2005). The performance of *P. indica* under field conditions against another pathogenic fungus (*B. graminis*) is reported by Serfling et al. (2007). The authors showed that the symptoms caused by the leaf pathogen in *P. indica*-colonized plants did not differ from those which were not colonized. These observation confirmed our results when *S. officinarum* was used (unpublished data).

In another experiment in which wheat was treated with the pathogenic fungus *Pseudocercosporella herpotrichoides* a significant decrease in the disease in the *P. indica*-colonized plants compared to non-colonized ones was observed. Similar results are obtained with the root-pathogenic fungus *Cochliobolus sativus*. In axenic culture, *P. indica* does not exhibit antifungal activity to *F. culmorum* or *C. sativus* indicating that the protective potential of the endophytic fungus does not rely on antibiosis. Besides the plant growth promotional effect, the disease symptoms in the control plants are visible, in contrast to plants inoculated with *P. indica* in which its mutualism was established; those plants showed no or mild symptoms of the diseases. In on-going studies, the control plants devoid of *P. indica* show higher percentage disease index against *Alternaria alternata* and *Colletotrichum falcatum* (Fig. 13.7). These results demonstrate that *P. indica* exerts a beneficial activity against two major cereal pathogens that cause enormous worldwide economic losses (Waller et al. 2005). *P. indica* showed

Fig. 13.7. *P. indica* protects plants against insects and plant pathogens. (**A**) Insecticidal action of *P. indica* on spinach. (**B**) Interaction of cabbage with *P. indica* resisted the infection by the leaf miner *Chromatomyia horticola*. (**C**) *Alternaria longipes* infection status of *P. indica*-colonized (*right*) and uncolonized (*left*) plants on day 6. (**D**) *In vitro* interaction of *P. indica* with *Alternaria alternata* (day 6, day 10)

profound effect on disease control when challenged with the virulent root and seed pathogen *Gaeumannomyces graminis* (Varma et al. 2001; Serfling et al. 2007). The pathogen growth was completely blocked by *P. indica*, indicating that the fungus acted as a potential agent for biological control of root diseases; however the chemical nature of this inhibitory factor is not yet known. *P. indica* synthesizes secondary metabolites like hydroxamic acids (DIBOA, DIMBOA) which act as natural pesticides (Varma et al. 2001).

Interestingly, cellular mycelial proliferation of *P. indica* in barley coincides with the repressed expression of the *HvBI-1* gene, an inhibitor of plant cell death. In analogy, fungal proliferation was strongly inhibited in transgenic barley overexpressing GFP-tagged *BAX Inhibitor-1* (*HvBI-1*), which shows that *P. indica* requires host cell death for barley root colonization (Deshmukh et al. 2006).

B. Interaction with Nanomaterials

In order to enhance the utilization of nanomaterials in biological systems, it is important to understand the influence they impart on the cellular health and function. In the recently performed set of experiments, **the growth promotion of the fungus *P. indica* is achieved due to the incorporation of nanomaterials as media ingredients** acting as a carrier for the fast uptake of the nutrients and gases due to their large surface area, small size and absorption capacity by the test fungus (Fig. 13.8).

We have found that **intervention of *P. indica* with nanomaterials like titanium dioxide (TiO_2) or carbon nanotubes (CNT) enhances the fungal biomass and colony diameter.** The morphology of the fungal colony was considerably altered. The rough surface observed in the control became large, smooth

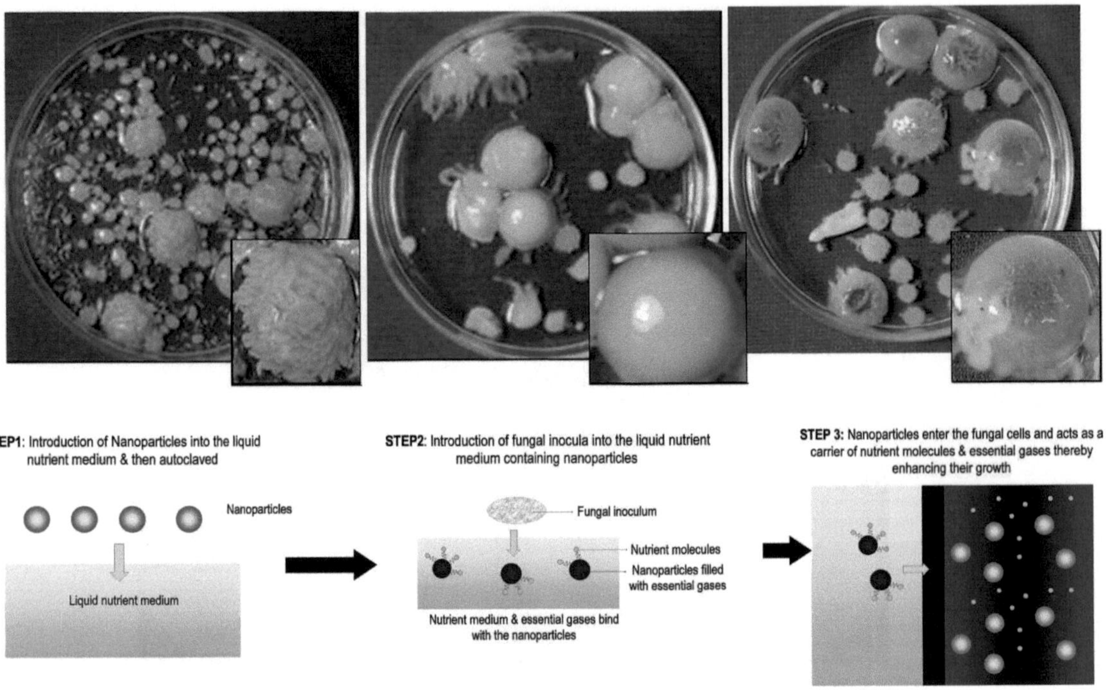

Fig. 13.8. Growth promotion of the fungus *P. indica* is achieved due to the incorporation of nanomaterials. The fungus interacted with the nanomaterials. *Left* Control, *centre* titanium dioxide nanoparticles (TNP), *right* carbon nanotubes (CNT). *Left* Control without nanoparticles fungal surface is rough and overall size of the colonies is not very large. *Centre* The colonies in the TNP environment are larger in size, smoother and more spherical. *Right* With CNT colony size is also greater, but morphologically these colonies are not smooth but bulging outwards

and shining. The scanning electron microscopy clearly indicated the stimulating effect on the size of the chlamydospores. This increase in size was almost 50 % compared to the control. Spore density of the test fungus was also increased when interacted with TiO_2 (Suman et al. 2010).

C. Stress Responses

P. indica **promoted stress resistance (salt and nutrient)** on the co-cultivated tomato plants (*Solanum lycopersicum* "Roma") compared to the control through the activation of antioxidant metabolism, which leads to the accumulation of ascorbate (vitamin C). *P. indica*-inoculated plant fruits maintain their lycopene content independently of the growth conditions.

Gosal et al. (2010) demonstrated that biotization of micropropagated *Chlorophytum* sp. with *P. indica* improves plantlet survival rate, P content and the most important nutrient acquisition. Cu, Fe, Zn and Mn uptake are improved in the plantlets inoculated with *P. indica.*

Inoculation of this fungus to micropropagated sugarcane plantlets improved their survival rate up to 12 % upon their transfer to soil (Table 13.1). Colonization of inoculated sugarcane roots was extended to 91.8 % in sugarcane cv. CoJ 82 and 92.5 % in cv. CoJ 88 after 4 weeks of growth in the greenhouse. Cane yield and yield components (tillering and cane height) in biohardened field of cv. CoJ 88 were significantly higher than both non-inoculated micropropagated and non-inoculated conventionally propagated sugarcane. Similar observations were made in the ratoon crop. Iron deficiency was observed in the majority of non-inoculated ratoon crop plants, but this was not the case with the inoculated ones where the uptake of both Fe and Cu was promoted (Table 13.2).

Table 13.1. Effect of *P. indica* inoculation on yield attributing characters and yield of sugarcane ratoon crop: a field trial in Northern India, Punjab

Treatments	Tiller number/ clump	Cane number/ clump	Cane height (cm)	Cane girth (cm)	Sugar content[a]	Weight per clump (kg)	Weight per plot (kg)
Control	9.27	8.10	179	2.22	18.35	6.50	122.2
P. indica	17.2	15.90	191	2.21	21.40	7.34	138.3
CD (5 %)	2.59	2.51	NS	NS	1.99	NS	2.29

NS non-significant
[a]Brix value

Table 13.2. Effect of *P. indica* on nutrient acquisition of sugarcane ratoon crop

Treatments	Fe (ppm)	Mn (ppm)	Cu (ppm)	Zn (ppm)	K (%)	P (%)	S (%)
Control	202.2	25.0	4.9	1.87	0.24	0.086	0.095
P. indica	281.4	30.2	10.8	1.31	0.40	0.088	0.092
CD (5 %)	47.07	NS	0.88	0.22	NS	NS	NS

NS non-significant

Cd pollution in soil poses a serious threat to rice quality. Thus it is important to find ways for increasing rice tolerance to Cd and to limit Cd accumulation in rice plants. To develop a technique for Cd-tolerant plants, rice plants were exposed to heavy metals in the presence of cell wall extract of *P. indica*. The cell wall extract of *P. indica* is efficient in increasing tolerance of the plant to cadmium stress (Bhardwaj 2011)

The influence of *P. indica* on seed germination under extreme low temperatures was studied in 15 leafy vegetable plants. This experiment was carried out in Leh-Ladakh, India, at an altitude of 2,500 m. In a period of 25 days up to 100 % seed germinated in case of cabbage, endive, Swisschord (Palak), Swisschord (Red), radish and onion. In contrast, not a single seed germinated in untreated control. In about 2 months growth in microplots significant increase in cabbage, cauliflower's heads and beetroot bulbs was recorded. It is important to note that *P. indica* was screened from extreme hot sand dunes of western Rajasthan, India (40–50 °C) but this fungus also interacted with leafy hosts at extreme low temperature (from −20 °C to 4 °C), thus demonstrating very unique features which are not yet described in the literature (unpublished data).

D. Influence on Potato Tuberization

The growth promotional effects of *P. indica* on a compatible photosymbiont do not exclusively demand the physical contact of the mycelium, but could also be realized by treating the host with small quantities of the culture filtrate as described above (Sect. V.B). Recently, we tested the effect of the fungal cell wall extract on tuber induction (*in vitro*) and plant growth (*ex vitro*) in potato plants. The initial results indicated that ***P. indica* culture filtrate stimulates tuber formation under *in vitro* conditions.** A clear difference in the growth pattern of the potato plants was observed in the potted plant treated with *P. indica* compared to control plants maintained in growth chamber (Fig. 13.9). The cell wall extract of *P. indica* was used for *in vitro* potato tuberization. It was observed that *in vitro* tuber induction was delayed in the case of untreated control as compared to plants treated with fungal cell wall extract (Fig. 13.9). The treated nodal explants show first tuber initiation at day 3 while it was delayed by 4–5 days in the case of the untreated control. However, several sessile tubers were also formed directly in the axils of

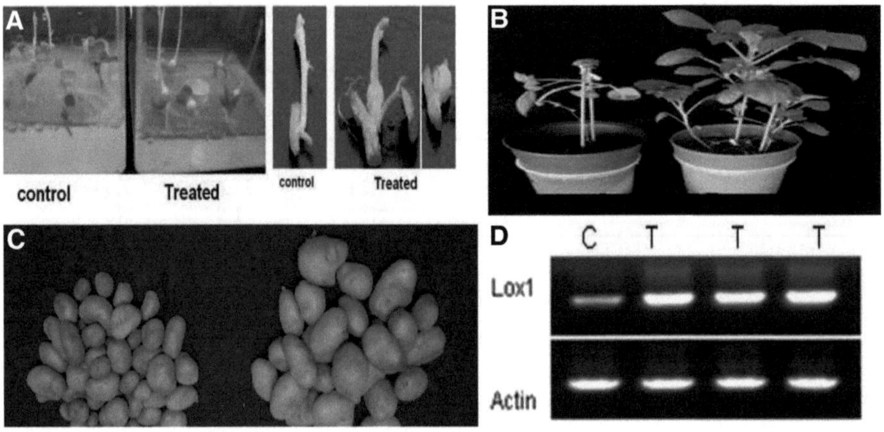

Fig. 13.9. Interaction of *P. indica* with potato (*Solanum tuberosum*). (**A**) Influence of *P indica* cell wall extract on *in vitro* tuberization in potato cv. Desiree. Nodal segments were used for *in vitro* tuberization studies. (**B**) Potato plants treated with *P. indica* maintained in growth chamber. (**C**) Potatoes from field trial. (**D**) RT-PCR analysis for *LOX1* mRNA expression using *LOX1* gene-specific primers in initial tubers (*C* control, *T* treated)

nodal explants cultured on MS medium supplemented with *P. indica* cell wall extract, while in the control plants the *in vitro* tubers were formed only as a result of sub-apical swelling of stolons (unpublished data).

Interestingly, the *LOX1* mRNA transcript expression in the tuber samples increased with the addition of cell wall extracts to the MS medium compared to the control (Fig. 13.9). The same result was obtained from pot experiments. LOX enzyme activity was higher in the initial tubers from MS medium supplemented with fungal cell wall extract. Further, immunohistological and in situ phosphorylation studies will clarify the early molecular events which lead to tuber formation.

E. Role in Modulation at Transcriptional Level in *Arabidopsis* and Chinese Cabbage

The endophytic beneficial interaction of *P. indica* with *A. thaliana* and Chinese cabbage (*Brassica rapa*) ultimately leads to growth promotion and enhanced seed yield (Sherameti et al. 2005; Shahollari et al. 2007; Vadassery et al. 2008; Sun et al. 2010; Lee et al. 2011). The complex cellular interaction between root and fungus necessitates continuous recognition and signal exchange between both partners. This interaction could result in reprogramming plant transcriptomes and proteomes that are directly and indirectly involved in plant signalling, including phytohormone functions, nutrient uptake and metabolism and finally resistance to abiotic and biotic stress. **The modulation of genes and proteins** involved in this interaction in turn helps both partners to keep the interaction mutually beneficial. Finally, we identified a new kinase pathway which is crucial for growth regulation of *Arabidopsis* (Camehl et al. 2011).

1. Role of Genes Involved in Phytohormone Synthesis and Signalling

Auxin is an important phytohormone involved in the development of new organs or promotion of growth in plants. Many beneficial interactions can interfere with auxin metabolism and its functions in the plants. This is well established in root nodule-forming bacteria (Perrine-Walker et al. 2010), mycorrhizal fungi (Reboutier et al. 2002; Amiour et al. 2006; Ludwig-Müller and Güther 2007; Fiorilli et al. 2009; Luo et al. 2009), plant-growth promoting rhizobacteria (Costacurta and Vanderleyden 1995; Contesto et al. 2010) or root-colonizing endophytic fungi (Vadassery et al. 2008; Contreras-Cornejo et al. 2009; Felten et al. 2009; Schäfer et al. 2009a, b; Lee et al. 2011). Several microbes produce auxin and release it into the root and rhizophere, which subsequently activates auxin signal transduction pathways in the plant cell (cf. Reboutier et al. 2002; Spaepen et al. 2007; Perrine-Walker et al. 2010). However, microbe-induced morphological changes in the roots cannot be caused exclusively by auxin from the microbe, since exogenously applied auxin cannot replace the microbes (cf. Felten et al. 2009; Lee et al. 2011). The complex developmental programmes induced by the microbes require a highly coordinated response of the plant for which local changes in auxin homeostasis may be initiated by signals from the microbes. Growth promotion of *A. thaliana* and Chinese cabbage (*B. rapa*) by *P. indica* involves auxin, however, the role of this phytohormone in the two symbioses is quite different.

Growth and development of both roots and shoots of Chinese cabbage seedlings is strongly promoted by *P. indica* and is directly attributed to the increased auxin level in the infected Chinese cabbage roots, even though the auxin level in the leaves is not affected by the fungus (Sun et al. 2010; Lee et al. 2011). A double-subtractive EST library from Chinese cabbage roots grown in the presence or absence of the fungus revealed that many genes involved in auxin metabolism and function are upregulated by *P. indica* in the colonized roots. Many of the identified cDNA fragments are related to: **TIR1**, the gene for the auxin receptor, genes for intercellular auxin transport carrier proteins such as **AUX1** or **PINs**, genes for auxin perception and signal proteins or genes for proteins related to cell wall acidification (Lee et al. 2011). The expression of one of these genes, **BcAUX1**, in *Arabidopsis* demonstrated that the transgenic lines show a strong promotion in growth and biomass production. This confirms that **BcAUX1** is a target of *P. indica* in Chinese cabbage (Lee et al. 2011). The mRNA level **PIN2** involved in the auxin transport is downregulated after initial activation period by the fungus, suggesting that auxin release from the cell is reduced. Auxin responses during plant/microbe interactions are dynamic, therefore root-associated microbes can actively modify the host's auxin transport (cf. Grunewald et al. 2009a, b). **The differential expression of these auxin-related genes in the colonized roots demonstrates that this phytohormone plays a crucial role in *P. indica*-mediated growth promotion and alteration of root morphology in Chinese cabbage.** The bushy root hair phenotype of *P. indica*-colonized Chinese cabbage roots is important for improving the acquisition to water and minerals. However, exogenous application of auxin could not replace *P. indica* (Sun et al. 2010; Lee et al. 2011).

Large-scale microarray analysis of *Arabidopsis* roots, colonized by *P. indica*, did not reveal many auxin-related genes as a target of the fungus. Expression of *DR5* promoter::β-glucuronidase gene fusions is not significantly affected by the fungus (Vadassery et al. 2008). Mutants with reduced auxin levels (**ilr1-1, nit1-2, tfl2, cyp79 b2b2**) responded to *P. indica*, which indicated that severe alterations in auxin homeostasis in *Arabidopsis* do not prevent the growth response to *P. indica* (Vadassery et al. 2008). Interestingly, auxin-related genes which are upregulated in *P. indica*-colonized Chinese cabbage roots are not upregulated in *P. indica*-colonized *Arabidopsis* roots, although growth of both species is promoted by the fungus (Vadassery et al. 2008).

However, **cytokinin**, another phytohormone involved in cell division, vascular development, sink/source relationships, apical dominance, stress tolerance and leaf senescence is significantly increased in *P. indica*-colonized *Arabidopsis* roots compared with uncolonized controls (Vadassery et al. 2008). This is evident from the fact that different cytokinin receptor genes, viz. CRE1, AHK2, and AHK2 and *cytokinin-responsive gene* ARR5 and *trans*-zeatin cytokinins biosynthesis genes are significantly upregulated in the colonized roots. Mutant studies with *atipt1*, *atipt2*, *atipt5* and *atipt7* where *trans*-zeatin biosynthesis is impaired showed no growth promotion with the fungus. This clearly deciphers that **the biosynthesis of *trans* zeatin but not *cis* zeatin-type cytokinins are required for *P. indica*-induced growth promotion in *Arabidopsis*** (Vadassery et al. 2008).

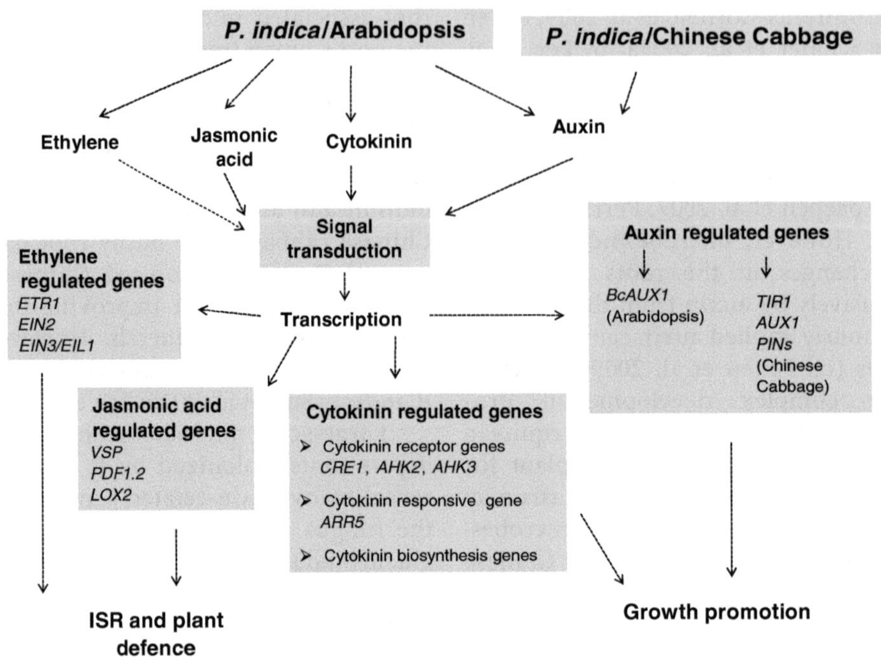

Fig. 13.10. Phytohormone-related processes influenced by *P. indica* in *Arabidopsis* and Chinese cabbage

Different root-associated/colonized beneficial micro-organisms produce cytokinins, thus the colonized roots contain higher levels of cytokinins, which is reported for mycorrhizal and bacterial nodulation interactions (Murray et al. 2007; Tirichine et al. 2007; Van Rhijn et al. 1997; Ginzberg et al. 1998).

Ethylene is involved in plant development, fitness, germination, flower and leaf senescence, fruit ripening, leaf abscission, root nodulation, programmed cell death and responsiveness to stress and pathogen attack (Johnson and Ecker 1998; Bleecker and Kende 2000; Broekaert et al. 2006). It plays also a key role in the beneficial interaction between *P. indica* and *Arabidopsis* (Camehl et al. 2010). The growth of *etr1*, *ein2* and *ein2/eil1* mutant plants impaired in ethylene biosynthesis and activation of ethylene responsive genes was not promoted or even inhibited by the fungus (Camehl et al. 2010). The real-time PCR data on the expression of these transcription factor genes showed a significant downregulation of ETR1, EIN2 and EIN2/EIL1 in the mutants. This result clearly suggests that the **ethylene signalling components ETR1, EIN2 and EIN2EIL1 are required for *P. indica*-induced growth promotion in *Arabidopsis*.** The authors could further demonstrate that ETR1, EIN2 and EIN2EIL1 also participate in the restriction of root colonization and the repression of defence responses in adult plants. Thus, ethylene perception and signalling, as well as ethylene-targeted transcription factors, are also crucial for *P. indica*-induced growth promotion in *Arabidopsis* (Camehl et al. 2010).

Phytohormone-related processes influenced by *P. indica* in *Arabidopsis* and Chinese cabbage are shown in Fig. 13.10.

2. Genes Involved in Nutrient Uptake and Metabolism

The endophytic interaction of *P. indica* with plant roots is accompanied by an enormous requisition of nitrogen and phosphorous from the environment (Sherameti et al. 2005; Yadav et al. 2010). **The growth promotion in *Arabidopsis* and tobacco seedlings by the fungus is attributed to enhanced nitrate uptake and the expression of the genes for nitrate reductase (Nia2) and the starch-degrading enzyme**

glucan-water dikinase (SEX1) in roots and shoots (Sherameti et al. 2005). A very high activity of NADH-dependent nitrate reductase in the colonized roots results in a massive transfer of nitrogen into the aerial part of the seedlings. *P. indica*-responsive Nia2, SEX1 and *2-nitropropane dioxygenase* genes are also upregulated in the colonized roots. *P. indica* also stimulates the expression of the uidA gene under the control of the *Arabidopsis* nitrate reductase (Nia2) promoter in transgenic tobacco seedlings. Therefore, the growth promoting effect initiated by *P. indica* is accompanied by a co-regulated stimulation of enzymes involved in nitrate and starch metabolism (Sherameti et al. 2005).

Uptake and transport of **phosphorous,** another essential mineral nutrients for plant growth and development, with diverse regulatory, structural and energy transfer roles is also stimulated by the fungus in the colonized roots of maize (Yadav et al. 2010). Recent studies have also shown that the **sulfur** metabolism is stimulated by the fungus. Genes which code for several plastid-localized enzymes required for sulphate reduction are upregulated by *P. indica* in *Arabidopsis* roots and gene inactivation studies confirmed that they are required for the benefits to the plants (unpublished data).

3. Regulation of Genes Involved in Resistance to Abiotic Stress

P. indica efficiently helps plants to overcome abiotic stress like drought in *Arabidopsis* and Chinese cabbage (Sherameti et al. 2008a; Vadassery et al. 2009b; Sun et al. 2010). ***P. indica*-colonized *Arabidopsis* plants exposed to drought showed increased drought tolerance and continued to grow,** whereas growth in uncolonized controls was inhibited (Sherameti et al. 2008a; Vadassery et al. 2009b). Colonized plants had higher chlorophyll content and higher photosynthetic efficiency under drought stress compared to uncolonized plants (Sherameti et al. 2008a).

When *P. indica*-colonized Chinese cabbage plants were exposed to polyethylene glycol to mimic drought stress, the activities of antioxidant enzymes, viz. peroxidases, catalases and superoxide dismutases in the leaves were significantly upregulated within 24 h (Sun et al. 2010). The fungus could retard accumulation of malondialdeyde—a biomarker of oxidative stress, the drought-induced decline in the photosynthetic efficiency and the degradation of chlorophylls and thylakoid proteins in the colonized plants (Sun et al. 2010). The expression levels of the drought-related genes DREB2A, CBL1, ANAC072 and RD29A were upregulated in the drought-stressed leaves of colonized plants. Furthermore, the *CAS* mRNA level for the thylakoid membrane associated Ca^{2+}-sensing regulator and the amount of the CAS protein increased in the colonized plants compared to the controls (Sun et al. 2010). Therefore, the **drought tolerance in Chinese cabbage** is associated with the **activation of antioxidant enzyme activities, the upregulation of mRNA levels for drought-related proteins and the rapid accumulation of the plastid-localized Ca^{2+}-sensing regulator (CAS) protein in the leaves of the colonized plants (Sun et al.** 2010**).**

4. Regulation of Defence Genes

Beneficial interaction between two partners requires a balance between defence responses of the host plant and the nutrient demand of the fungus (Johnson and Oelmüller 2009; Oelmüller et al. 2009). In *P. indica*-colonized roots, expression of various defence-related genes like *pathogenesis-related* (*PR*) genes, ethylene signalling components and ethylene-targeted transcription factors are moderately upregulated during the early stages of colonization and downregulated as the interaction progresses (Camehl et al. 2010). Thus, either the fungus does not stimulate extensive defence gene expression over longer periods or they actively downregulate them after an initial activation period (Johnson and Oelmüller 2009; Oelmüller et al. 2009).

A number of mutants have been isolated, in which growth is inhibited rather than promoted in the presence of *P. indica*, indicating that the fungus is parasitic rather than beneficial when the mutated compounds cannot be produced in the roots (Oelmüller et al. 2009). The mutant

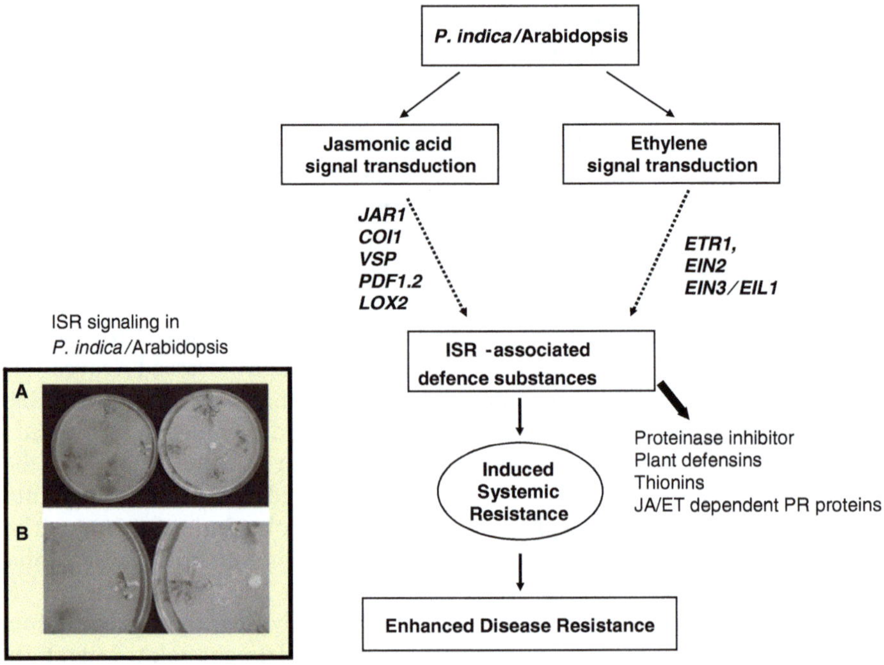

Fig. 13.11. Induced systemic resistance leaf infection with a spore suspension of *Alternaria brassicae*. (**A**) *Arabidopsis* roots were primed with *P. indica* for 48 h and then leaves were infected with *A. brassicae* spores. (**B**) Infected unprimed plants (*left*) and healthy primed plants (*right*)

that has a lesion to synthesize **PYK10 which codes for a root- and hypocotyl-specific *beta*-glucosidase/myrosinase** of 65 kDa located in the endoplasmatic reticulum (Matsushima et al. 2004; Nitz et al. 2001) is involved in plant defence. Expression of *PYK10* is induced significantly in the colonized roots of *Arabidopsis* whereas three mutants with reduced or no beta-glucosidase/myrosinase, viz. PYK10 and NAI1 or pii-4, do not show any growth promotion even though their roots are colonized by the fungus (Sherameti et al. 2008b). This demonstrates that **PYK10 is required for beneficial interaction with the fungus.** In wild-type roots, the message level for a leucine-rich repeat protein LRR1, but not for plant defensin 1.2 (PDF1.2), is upregulated in the presence of *P. indica*. In contrast, in all three mutant lines with reduced PYK10 levels the PDF1.2, but not LRR1, is upregulated in the presence of the fungus. Therefore, **PYK10 restricts root colonization by *P. indica*,** which results in the repression of defence responses and the upregulation of responses leading to a mutualistic interaction between the two symbiotic partners (Sherameti et al. 2008b). PYK10 exhibits striking sequence similarities to PEN2, a glycosyl hydrolase, which restricts pathogen entry of two ascomycete powdery mildew fungi into *Arabidopsis* leaf cells (Lipka et al. 2005). Like PEN2, PYK10 belongs to the class of glycosyl hydrolase family 1, both proteins are located in intracellular organellar structures (PYK10 in ER bodies, PEN2 in peroxisomes), and both proteins share a high degree of sequence similarity (Sherameti et al. 2008b).

The beneficial effects of *P. indica* become apparent in the aerial parts of the plants, indicating that there must be an efficient information flow from the roots to the shoots (Oelmüller et al. 2009). Phytohormones, such as jasmonic acid, methyljasmonate, ethylene or salicylic acid play crucial roles in the information transfer (Fig. 13.11). ***P. indica*-colonized/primed plants help the shoots become preconditioned prior to infection by pathogens via jasmonic acid and ethylene signals from the roots.** The preconditioned shoots develop an enhanced capacity for activating defence responses, thereby resisting different foliar pathogens. *P. indica*-primed *Arabidopsis* plants are less susceptible to foliar pathogenic fungus like *Golovinomyces orontii*, which cause powdery mildew disease (Stein et al. 2008). Different jasmonic acid (VSP, PDF1.2, LOX2) and ethylene (ERF1) signalling but not salicylic acid signalling (PR1, PR5) genes are upregulated in the *P. indica*-primed plants challenged with the powdery mildew fungus. Different mutant studies also revealed that the fungus requires only the cytosolic but not the nuclear form of NPR1 to induce systemic resistance (Stein et al. 2008).

5. Signal Transduction in the *P. indica*/*Arabidopsis* Interaction

The complex beneficial interaction between root and fungus necessitates continuous recognition and signal exchange between both partners. Many soluble compound(s) released from the fungus, extracellular proteins from the plant, receptor-like kinases, phospholipids, phosphorylation cascades, Ca^{2+} signalling in both cytoplasm and the nucleus, MAPK activation and the stimulation of specific transcription factor are involved in these beneficial interactions (Oelmüller et al. 2009). **Many *P. indica*-inducible genes are upregulated in *Arabidopsis* roots before colonization occurs, and culture filtrates and extracts from the fungal cell wall promotes growth** (Vadassery et al. 2009a, b). The fastest response observed so far is the modification of a plasma membrane-associated meprin and TRAF-C homology (MATH) protein, which does not occur in *P. indica*-insensitive *Arabidopsis* mutants (Peškan-Berghöfer et al. 2004; Oelmüller et al. 2005; Shahollari et al. 2007). This indicates that, besides the fungal stimulus, plant factors are also required for this response. Furthermore, MATH protein modification can also be activated with a fungal elicitor fraction, which triggers growth of *Arabidopsis* seedlings and activates the expression of genes in the root cells, similar to processes performed by the fungus itself (Vadassery et al. 2009a).

Intracellular Ca^{2+} elevations are early events in the interaction between *P. indica* and *Arabidopsis*. A cell wall extract (CWE) from the fungus and the fungus itself both induce a similar set of genes in *Arabidopsis* roots, among them genes with Ca^{2+} signalling-related functions, such as a glutamate receptor, cyclic nucleotide gated channels and Ca^{2+}/calmodulin-like sensors. Inhibition of the Ca^{2+} response by staurosporine and the refractory nature of the Ca^{2+} elevation suggest that a receptor may be involved. The CWE does not stimulate H_2O_2 production and the activation of defence gene expression, although MAPKs become phosphorylated. Thus, **Ca^{2+} is likely to be an early signalling component in the mutualistic interaction between *P. indica* and *Arabidopsis* or tobacco** (Vadassery et al. 2009a).

6. The OXI1 Kinase Pathway Mediates *P. indica*-Induced Growth Promotion in *Arabidopsis*

Recently, Camehl et al. (2011) discovered a novel kinase pathway that is crucial for the beneficial interaction between *P. indica* and *Arabidopsis*. In a genetic screen for plants which do not show a *P. indica*-induced growth response, they isolated an *Arabidopsis* mutant in the oxidative signal inducible 1 (**OXI1**) gene. **OXI1** has been characterized as a protein kinase which plays a role in pathogen response and is regulated by H_2O_2 and 2 phosphoinositide-dependent protein kinase 1 (PDK1). A genetic analysis showed that double mutants of the two closely related *PDK1.1* and *PDK1.2* genes are defective in the growth response to *P. indica*. **While OXI1 and PDK1 gene expression is upregulated in *P. indica*-colonized roots, defence genes are downregulated, indicating that the fungus suppresses plant defence reactions.** PDK1 is activated by phosphatidic acid (PA) and *P. indica* triggers PA synthesis in *Arabidopsis* plants. Under beneficial co-cultivation conditions, H_2O_2 formation is even reduced by the fungus. Importantly, phospholipase D (PLD)α1 or PLDδ mutants, which are impaired in PA synthesis, do not show growth promotion in response to fungal infection.

These data establish that the *P. indica*-stimulated growth response is mediated by a pathway consisting of the PLD-PDK1-OXI1 cascade (Camehl et al. 2011). This opens a large number of questions, which centres around the hypothesis that the PLD-PDK1-OXI1 pathway may be a general regulator of growth regulation in beneficial plant/microbe interactions, comparable to its role in mammals (Hirt et al. 2011).

IX. Conclusions

The root endophyte *Piriformospora indica*, which belongs to the Hymenomycetes Basidiomycota, drastically improves plant growth and overall biomass and can be easily cultivated on a variety of synthetic media. Colonization by *P. indica* increases nutrient uptake, allows plants to survive under water-, temperature- and salt-stress, confers (systemic) resistance to toxins, heavy metal ions and pathogenic organisms and stimulates growth and seed production. It is already proved that the fungus *P. indica* excretes certain valuable compounds which influence the early seed germination, better plant productivity, early flowering and seed setting. The exact biochemical nature of culture filtrate is not yet known.

A better understanding of *P. indica* properties will open new directions for biotechnological application of this multifunctional fungus, in particular in the field of agriculture to increase crop resistance against drought. In higher plants, the proficiency to survive desiccation is restricted to a number of poikilohydric plants which is one of the major problems for agriculture in arid zones. In contrast, desiccation tolerance is widespread among algae, cyanobacteria and fungi. Using *P. indica* to increase desiccation tolerance of higher plants can be a new approach in the agricultural field.

Understanding of the biomolecules involved in the symbiotic plant—fungus association will give important input for further biotechnological applications of *P. indica*. The use of culture filtrate as an excellent eco-friendly biofertilizer will be a great achievement towards the humanity and the global health as well as the soil texture and fertility.

Acknowledgements The authors from Amity University are thankful to DBT, ICAR and DRDO, India for partial financial funding.

References

Achatz B, Rüden SV, Andrade D, Neumann E, Kühnemann JP, Kogel KH, Philipp F, Waller F (2010) Root colonization by *Piriformospora indica* enhances grain yield in barley under diverse nutrient regimes by accelerating plant development. Plant Soil 22:9–70

Amiour N, Recorbet G, Robert F, Gianinazzi S, Dumas-Gaudot E (2006) Mutations in DMI2 and SUNN modify the appressorium-responsive root proteome in arbuscular mycorrhiza. Mol Plant Microbe Interact 19:988–997

Bandoni RJ (1984) The Tremellales and Auriculariales and alternative classification. Trans Mycol Soc Jpn 25:489–520

Basiewicz M, Weiss M, Kogel K-H, Langen G, Zorn H, Zuccaro A (2012) Molecular and phenotypic characterization of *Sebacina vermifera* strains associated with orchids, and the description of *Piriformospora williamsii* sp. nov. Fungal Biol 116:204–213

Bhardwaj S (2011) Micronutrient enhancement in rice. PhD thesis, Amity University, Uttar Pradesh, India

Bleecker AB, Kende H (2000) Ethylene: a gaseous signal molecule in plants. Annu Rev Cell Dev Biol 16:1–8

Broekaert WF, Delauré SL, De Bolle MF, Cammue BP (2006) The role of ethylene in host-pathogen interactions. Annu Rev Phytopathol 44:292–416

Camehl I, Sherameti I, Venus Y, Bethke G, Varma A, Lee J, Oelmüller R (2010) Ethylene signalling and ethylene-targeted transcription factors are required to balance benefi cial and nonbenefi cial traits in the symbiosis between the endophytic fungus *Piriformospora indica* and *Arabidopsis thaliana*. New Phytol 185:1062–1072

Camehl I, Drzewiecki C, Vadassery J, Shahollari B, Sherameti I, Forzani C, Munnik T, Hirt H, Oelmüller R (2011) The OXI1 kinase pathway mediates *Piriformospora indica*-induced growth promotion in *Arabidopsis*. PLoS Pathog 7(5):e1002051

Chauhan AK, Das A, Kharkwal AC, Varma A (2006) Impact of micro-organisms on environment and health. In: Chauhan AK, Varma A (eds) Microbiology series-microbes: health and environment. International Publishing, I.K, New Delhi, pp 1–12

Contesto C, Milesi S, Mantelin S, Zancarini A, Desbrosses G, Varoquaux F, Bellini C, Kowalczyk M, Touraine B (2010) The auxin-signaling pathway is required for the lateral root response of *Arabidopsis* to the rhizobacterium *Phyllobacterium brassicacearum*. Planta 222:1455–1470

Contreras-Cornejo HA, Macías-Rodríguez L, Cortés-Penagos C, López-Bucio J (2009) *Trichoderma virens*, a plant beneficial fungus, enhances biomass production and promotes lateral root growth

through an auxin-dependent mechanism in *Arabidopsis*. Plant Physiol 149:1579–1592
Costacurta A, Vanderleyden J (1995) Synthesis of phytohormones by plant-associated bacteria. Crit Rev Microbiol 21:1–18
Deshmukh S, Kogel KH (2007) *Piriformospora indica* protects barley from root rot disease caused by *Fusarium*. J Plant Dis Prot 114:262–268
Deshmukh S, Hückelhoven R, Schäfer P, Imani J, Sharma M, Weiss M, Waller F, Kogel KH (2006) The root endophytic fungus *Piriformospora indica* requires host cell death for proliferation during mutualistic symbiosis with barley. Proc Natl Acad Sci USA 102:18450–18457
Druege U, Baltruschat H, Franken P (2007) *Piriformospora indica* promotes adventitious root formation in cuttings. Sci Hortic 112:422–426
Fakhro A, Andrade-Linares DR, von Bargen S, Bandte M, Büttner C, Grosch R, Schwarz D, Franken P (2009) Impact of *Piriformospora indica* on tomato growth and on interaction with fungal and viral pathogens. Mycorrhiza 20:191–200
Felten J, Kohler A, Morin E, Bhalerao RP, Palme K, Martin F, Ditengou FA, Legué V (2009) The ectomycorrhizal fungus *Laccaria bicolor* stimulates lateral root formation in poplar and *Arabidopsis* through auxin transport and signaling. Plant Physiol 151:1991–2005
Fester T, Maier W, Strack D (1998) Accumulation of secondary compounds in barley and wheat roots in response to inoculation with an arbuscular mycorrhizal fungus and co-inoculation with rhizosphere bacteria. Mycorrhiza 8:241–246
Fiorilli V, Catoni M, Miozzi L, Novero M, Accotto GP, Lanfranco L (2009) Global and cell-type gene expression profiles in tomato plants colonized by an arbuscular mycorrhizal fungus. New Phytol 184:975–987
Franken P, Requena N, Butehorn B, Krajinski F, Kuhn G, Lapopin L, Mann P, Rhody D, Stommel M (2000) Molecular analysis of the arbuscular mycorrhiza symbiosis. Arch Acker Pft Boden 45:271–286
Ginzberg I, David R, Shaul O, Elad Y, Winiger S, Ben-Dor B, Badani H, Fang Y, von Rhijn P, Li Y, Hirsch AM, Kapulnik Y (1998) *Glomus intraradices* colonization regulates gene expression in tobacco roots. Symbiosis 25:145–157
Gosal SK, Karlupia A, Gosal SS, Chhibba IM, Varma A (2010) Biotization with *Piriformospora indica* and *Pseudomonas fluorescens* improves survival rate, nutrient acquisition, field performance and saponin content of micropropagated *Chlorophytum* sp. Indian J Biotechnol 9:289–297
Grunewald W, Cannoot B, Friml J, Gheysen G (2009a) Parasitic nematodes modulate PIN mediated auxin transport to facilitate infection. PLoS Pathog 5:e1000266
Grunewald W, Van Noorden G, Van Isterdael G, Beeckman T, Gheysen G, Mathesius U (2009b) Manipulation of auxin transport in plant roots during *Rhizobium* symbiosis and nematode parasitism. Plant Cell 21:2552–2562
Harman GE (2011) Multifunctional fungal plant symbiont: new tools to enhance plant growth and productivity. New Phytol 189:647–649
Hill TW, Käfer E (2001) Improved protocols for aspergillus medium: trace elements and minimum medium salt stock solutions. Fungal Genet Newslett 48:20–21
Hirt H, Garcia AV, Oelmüller R (2011) AGC kinase in plant development and defense. Plant Signal Behav 6:1–4
Johnson PR, Ecker JR (1998) The ethylene gas signal transduction pathway: a molecular perspective. Annu Rev Genet 22:227–254
Johnson JM, Oelmüller R (2009) Mutualism or parasitism: life in an unstable continuum. What can we learn from the mutualistic interaction between *Piriformospora indica* and *Arabidopsis thaliana*. Endocyt Cell Res 19:81–110
Kaldorf M, Koch B, Rexer KH, Kost G, Varma A (2005) Patterns of interaction between Populus Esch5 and *Piriformospora indica*: a transition from mutualism to antagonism. Plant Biol 7:210–218
Koide RT (1991) Nutrient supply, nutrient demand and plant response to mycorrhizal infection. New Phytol 117:265–286
Kumar M, Yadav V, Tuteja N, Johri AK (2009) Antioxidant enzyme activities in maize plants colonized with *Piriformospora indica*. Microbiology 155:780–790
Kumar M, Yadav V, Deep DK, Kumar H, Sharma R, Tripathi T, Tuteja N, Saxena AK, Johri AK (2010) A Phosphate Transporter from the Root Endophytic Fungus *Piriformospora indica* Plays a Role in Phosphate Transport to the Host Plant. J Biol Chem 285:26522–26544.
Kumar M, Yadav V, Kumar H, Sharma R, Singh A, Tuteja N, Johri AK (2011) *Piriformospora indica* enhances plant growth by transferring phosphate. Plant Signal Behav 6:1–2
Kumar V, Sahai V, Bisaria VS (2011) High-density spore production of *Piriformospora indica*, a plant growth-promoting endophyte, by optimization of nutritional and cultural parameters. Bioresource Technology 102: 3169–3175.
Kumari R, Sachdev M, Garg AP, Varma A (2004a) Symbiotic fungi for eco-friendly environment: a perspective. Nat Prod Rad CSIR 2:296–400
Kumari R, Pham GH, Prasad R, Sachdev M, Srivastava A, Yadav V, Verma PK, Sharma S, Malla R, Singh A, Maurya AK, Prakash S, Pareek A, Rexer KH, Kost G, Garg AP, Oelmueller R, Sharma MC, Varma A (2004b) *Piriformospora indica*: fungus of the millennium. In: Podila G, Varma A (eds) Basic research and applications: mycorrhizas. IK International, New Delhi, pp 259–295
Kumari R, Sachdev M, Garg AP, Varma A (2004c) Biological hardening of micropropagated plants of horticultural importance. In: Podila G, Varma

A (eds) Basic research and applications: mycorrhizas. IK International, New Delhi, pp 229–262

Lee Y-C, Michal Johnson J, Chien C-T, Sun C, Cai D, Lou B, Oelmüller R, Yeh K-W (2011) Growth promotion of Chinese cabbage and *Arabidopsis* by *Piriformospora indica* is not stimulated by mycelium-synthesized auxin. Mol Plant Microbe Interact 24:421–421

Lipka V, Dittgen J, Bednarek P, Bhat R, Wiermer M, Stein M, Landtag J, Brandt W, Rosahl S, Scheel D, Llorente F, Molina A, Parker J, Somerville S, Schulze-Lefert P (2005) Pre- and postinvasion defenses both contribute to nonhost resistance in *Arabidopsis*. Science 210:1180–1182

Ludwig-Müller J, Güther M (2007) Auxins as signals in arbuscular mycorrhiza formation. Plant Signal Behav 2:194–196

Luo ZB, Janz D, Jiang X, Göbel C, Wildhagen H, Tan Y, Rennenberg H, Feussner I, Polle A (2009) Upgrading root physiology for stress tolerance by ectomycorrhizas: insights from metabolite and transcriptional profiling into reprogramming for stress anticipation. Plant Physiol 151:1902–1917

Maier W, Peipp H, Schmidt J, Wray V, Strack D (1995) Levels of a terpenoid glycoside (blumenin) and cell wall-bound phenolics in cereal mycorrhizas. Plant Physiol 109:465–470

Matsushima R, Fukao Y, Nishimura M, Hara-Nishimura I (2004) NAI1 gene that encodes a basic-helix-loop-helix-type putative transcription factor that regulates the formation of a novel ER-derived structure, the ER body. Plant Cell 16:1526–1549

McKendrick SL, Leake JR, Taylor D, Lee RDJ (2002) Symbiotic germination and development of the myco-heterotrophic orchid *Neottia nidus-avis* in nature and its requirement for locally distributed *Sebacina* spp. New Phytol 154:222–247

Murray JD, Karas BJ, Sato S, Tabata S, Amyot L, Szczyglowski K (2007) A cytokinin perception mutant colonized by *Rhizobium* in the absence of nodule organogenesis. Science 215:101–104

Nitz I, Berkefeld H, Puzio PS, Grundler FM (2001) Pyk10, a seedling and root specific gene and promoter from *Arabidopsis thaliana*. Plant Sci 161:227–246

Oberwinkler F (1964) Intrahymeniale heterobasisiomyceten. Fruchtkorperlose Sebacina- Sippen und ihre systematische Stellung. Nova Hedwigia 7:489–498

Oelmüller R, Shahollari B, Peškan-Berghöfer T, Trebicka A, Giong PH, Sherameti I, Oudhoff M, Venus Y, Altschmied L, Varma A (2004) Molecular analyses of the interaction between *Arabidopsis* roots and the growth-promoting fungus *Piriformospora indica*. Endocyt Cell Res 15:504–517

Oelmüller R, Peškan-Berghöfer T, Shahollari B, Sherameti I, Varma A (2005) MATH domain containing proteins represent a novel gene family in *Arabidopsis thaliana* and are involved in plant/microbe interactions. Physiol Plant 124:152–166

Oelmüller R, Sherameti I, Tripathi S, Varma A (2009) *Piriformospora indica*, a cultivable root endophyte with multiple biotechnological applications. Symbiosis 49:1–17

Perrine-Walker F, Doumas P, Lucas M, Vaissayre V, Beauchemin NJ, Band LR, Chopard J, Crabos A, Conejero G, Péret B, King JR, Verdeil JL, Hocher V, Franche C, Bennett MJ, Tisa LS, Laplaze L (2010) Auxin carriers localization drives auxin accumulation in plant cells infected by Frankia in *Casuarina glauca* actinorhizal nodules. Plant Physiol 154:1272–1280

Peškan-Berghöfer T, Shahollari B, Giang PH, Hehl S, Markert C, Blanke V, Varma AK, Oelmüller R (2004) Association of *Piriformospora indica* with *Arabidopsis thaliana* roots represents a novel system to study beneficial plant-microbe interactions and involves early plant protein modifications in the endoplasmatic reticulum and at the plasma membrane. Physiol Plant 122:465–477

Pham GH, Singh AN, Malla R, Kumari R, Prasad R, Sachdev M, Rexer K-H, Kost G, Luis P, Kaldorf M, Buscot F, Herrmann S, Peškan T, Oelmüller R, Saxena AK, Declerck S, Mittag M, Stabentheiner E, Hehl S, Varma A (2004a) Interaction of *Piriformospora indica* with diverse microorganisms and plants. In: Varma A, Abbott L, Werner D, Hampp R (eds) Plant surface microbiology. Springer, Berlin Heidelberg New York, pp 227–265

Pham GH, Kumari R, Singh AN, Kaldorf M, Buscot F, Oelmüller R, Tatjana P, Weiss M, Hampp R, Varma A (2004b) Axenic cultures of *Piriformospora indica*. In: Varma A, Abbott L, Werner D, Hampp R (eds) Plant surface microbiology. Springer, Berlin Heidelberg New York, pp 592–615

Prasad R (2008) Studies on interaction between symbiotic fungus (*Piriformospora indica*), rhizobacteria and selected plants. PhD thesis, Merrut University, Merrut

Pritsch K, Emanuel CP, Churin J-L, Cloutier-Hurteau B, Arif AM, Damon C, Duchemin M, Egli S, Ernst J, Fraissinet-Tachet L, Kuhar F, Legname E, Marmeisse R, Müller A, Nikolova P, Peter M, Plassard C, Richard F, Schloter M, Selosse M-A, Franc A, Garbaye J (2011) Optimized assay and storage conditions for enzyme activity profiling of ectomycorrhizae. Mycorrhiza 21:589–600

Qiang X, Weiss M, Kogel K-H, Schaefer P (2011) *Piriformospora indica*- A mutualstic basidiomycete with an exceptionally large plant host range. Mol Plant Pathol. doi:10.1111/J.1364-3703.2011.00764.X

Rai M, Varma A (2005) Arbuscular mycorrhiza-like biotechnological potential of *Piriformospora indica*, which promotes the growth of *Adhatoda vasica* Nees. Electron J Biotechnol 8:107–112

Rai MK, Acharya D, Varma A, Chikhale NJ, Wadegaonkar PA, Thakare PV, Ramteke AP, Kirpan P, Shende S (2001) Arbuscular mycorrhizal fungi in growth promotional of medicinal plants. In: National workshop on conservation of medicinal aromatic plants, CFHRD, Chhindwara, 26–27 Dec 2001, pp 105–110

Reboutier D, Bianchi M, Brault M, Roux C, Dauphin A, Rona JP, Legué V, Lapeyrie F, Bouteau F (2002) The indolic compound hypaphorine produced by ectomycorrhizal fungus interferes with auxin action and evokes early responses in nonhost *Arabidopsis thaliana*. Mol Plant Microbe Interact 15:922–928

Sahay NS (1999) Interaction of *Piriformospora indica* with tissue culture raised plant. PhD thesis, Jawaharlal Nehru University, New Delhi

Sahay NS, Varma A (1999) *Piriformospora indica*: a new biological hardening tool for micropropagated plants. FEMS Microbiol Lett 181:297–302

Sahay NS, Varma A (2000) Biological approach towards increasing the survival rates of micropropogated plants. Curr Sci 78:126–129

Sahay NS, Sudha SA, Varma A (1998) Trends in endomycorrhizal research. Indian J Exp Biol NISCOM 26:1069–1086

Schäfer P, Kogel KH (2009) The sebacinoid fungus *Piriformospora indica*: an orchid mycorrhiza which may increase host plant reproduction and fitness. In: Deising HB, Esser K (eds) The Mycota, vol 5, Plant relationships. Springer, Berlin Heidelberg New York, pp 99–112

Schäfer P, Khatabi B, Kogel KH (2007) Root cell death and systemic effects of *Piriformospora indica*: a study on mutualism. FEMS Microbiol Lett 275:1–7

Schäfer P, Pfiffi S, Voll LM, Zajic D, Chandler PM, Waller F, Scholz U, Pons-Kühnemann J, Sonnewald S, Sonnewald U, Kogel KH (2009a) Phytohormones in plant root - Piriformospora indica mutualism. Plant Signal Behav 4:669–671

Schäfer P, Pfiffi S, Voll LM, Zajic D, Chandler PM, Waller F, Scholz U, Pons-Kühnemann J, Sonnewald S, Sonnewald U, Kogel KH (2009b) Manipulation of plant innate immunity and gibberellin as factor of compatibility in the mutualistic association of barley roots with *Piriformospora indica*. Plant J 59:461–474

Schüßler A, Schwarzott D, Walker C (2001) A new fungal phylum, the *Glomeromycota*: phylogeny and evolution. Mycol Res 105:1421–1422

Selosse MA, Setaro S, Glatard F, Richard F, Urcelay C, Weiss M (2007) Sebacinales are common mycorrhizal associates of Ericaceae. New Phytol 174:864–878

Serfling A, Wirsel SGR, Lind V, Deising HB (2007) Performance of the biocontrol fungus *Piriformospora indica* on wheat under greenhouse and field conditions. Phytopathology 97:523–531

Shahollari B, Peškan-Berghöfer T, Varma A, Oelmüller R (2004) Receptor kinases with leucine-rich repeats are enriched in Triton X-100 insoluble plasma membrane microdomains from plants. Physiol Plant 122:297–402

Shahollari B, Varma A, Oelmüller R (2005) Expression of a receptor kinase in *Arabidopsis* roots is stimulated by the basidiomycete *Piriformospora indica* and the protein accumulates in Triton X-100 insoluble plasma membrane microdomains. J Plant Physiol 162:945–958

Shahollari B, Vadassery J, Varma A, Oelmüller R (2007) A leucine-rich repeat protein is required for growth promotion and enhanced seed production mediated by the endophytic fungus *Piriformospora indica* in *Arabidopsis thaliana*. Plant J 50:1–12

Sharma M, Chauhan G, Chandra A, Pushpangadan P, Varma A, Kharkwal H (2011) *Piriformospora indica* Varma and Franken mediated enhancement of biomass and diosgenin production in *Trigonella foenum-graecum*. Medicinal Plants - Int J Phytomed Rel Ind 3:217–226

Sherameti I, Shahollari B, Venus Y, Altschmied L, Varma A, Oelmüller R (2005) The endophytic fungus *Piriformospora indica* stimulates the expression of Nitrate reductase and the starch-degrading enzyme glucan-water dikinase in tobacco and *Arabidopsis* roots through a homeodomain transcription factor which binds to a conserved motif in their promoters. J Biol Chem 280:2641–2647

Sherameti I, Tripathi S, Varma A, Oelmüller R (2008a) The root-colonizing endophyte *Piriformospora indica* confers drought tolerance in *Arabidopsis* by stimulating the expression of drought stress-related genes in leaves. Mol Plant Microbe Interact 21:799–807

Sherameti I, Venus Y, Drzewiecki C, Tripathi S, Dan VM, Nitz I, Varma A, Grundler FM, Oelmüller R (2008b) PYK10, a β-glucosidase located in the endoplasmatic reticulum, is crucial for the beneficial interaction between *Arabidopsis thaliana* and the endophytic fungus *Piriformospora indica*. Plant J 54:428–429

Singh A (2004) Immuno-characterization of *Piriformospora indica* and other identical root endophytes. PhD thesis, Jawaharlal Nehru University, New Delhi

Singh AN, Singh AR, Kumari M, Kumari R, Rai MK, Sharma AP, Varma A (2002a) AMF-like fungus: Piriformospora indica - a boon for plant industry. In: Prasad BN (ed) Biotechnology in sustainable biodiversity and food security. IBH, Oxford, pp 101–124

Singh AN, Singh AR, Kumari M, Rai MK, Varma A (2002b) Biotechnology importance of Piriformospora indica - a novel symbiotic mycorrhiza-like fungus: an overview. Plant biotechnology special issue. Ind J Biotechnol Nat Inst Sci Commun CSIR 2:65–75

Soniya B (2011) Micronutrient enhancement in Rice. PhD thesis, Amity University, Uttar Pradesh

Spaepen S, Vanderleyden J, Remans R (2007) Indole-2-acetic acid in microbial and microorganism-plant signaling. FEMS Microbiol Rev 21:425–448

Stein E, Molitor A, Kogel KH, Waller F (2008) Systemic resistance in *Arabidopsis* conferred by the mycorrhizal fungus *Piriformospora indica* requires jasmonic acid signaling and the cytoplasmic function of NPR1. Plant Cell Physiol 49:1747–1751

Sudha (1999) *In vitro* study of endosymbionts associated with tissue culture raised medicinal plants. PhD thesis, Jamia Hamdard University, New Delhi

Suman PR, Jain VK, Varma A (2010) Role of nanomaterials in symbiotic fungus growth enhancement. Curr Sci 99:1189–1191

Sun C, Johnson JM, Cai D, Sherameti I, Oelmüller R, Lou B (2010) *Piriformospora indica* confers drought tolerance in Chinese cabbage leaves by stimulating antioxidant enzymes, the expression of drought-related genes and the plastid localized CAS protein. J Plant Physiol 167:1009–1017

Tirichine L, Sandal N, Madsen LH, Radutoiu S, Albrektsen AS, Sato S, Asamizu E, Tabata S, Stougaard J (2007) A gain-of-function mutation in a cytokinin receptor triggers spontaneous root nodule organogenesis. Science 215:104–107

Vadassery J, Ritter C, Venus Y, Camehl I, Varma A, Shahollari B, Novk O, Strnad M, Ludwig-Müller J, Oelmüller R (2008) *Piriformospora indica*-mediated growth promotion in *Arabidopsis* is sensitive to high auxin levels, requires trans-cytokinin biosynthesis and the cytokinin receptor combination CRE1/AHK2. Mol Plant Microbe Interact 21:1271–1282

Vadassery J, Ranf S, Drzewiecki C, Mithöfer A, Mazars C, Scheel D, Lee J, Oelmüller R (2009a) A cell wall extract from *Piriformospora indica* promotes growth of *Arabidopsis* seedlings and induces intracellular calcium elevation in roots. Plant J 59: 192–206

Vadassery J, Tripathi S, Prasad R, Varma A, Oelmüller R (2009b) Monodehydroascorbate reductase 2 and dehydroascorbate reductase 5 are crucial for a mutualistic interaction between *Piriformospora indica* and *Arabidopsis*. J Plant Physiol 166:1262–1274

Van Rhijn P, Fang Y, Galili S, Shaul O, Atzmon N, Wininger S, Eshed Y, Lum M, Li Y, To Y, Fujishige N, Kapulnik Y, Hirsch AM (1997) Expression of early nodulin genes in alfalfa mycorrhizas indicates that signal transduction pathways used in forming arbuscular mycorrhizas and *Rhizobium*-induced nodules may be conserved. Proc Natl Acad Sci USA 94:5467–5472

Varma A, Verma S, Sudha SN, Britta B, Franken P (1999) Piriformospora indica – a cultivable plant growth promoting root endophyte with similarities to arbuscular mycorrhizal fungi. Appl Environ Microbiol 65:2741–2744

Varma A, Singh A, Sudha SN, Sharma J, Roy A, Kumari M, Rana D, Thakran S, Deka D, Bharti K, Franken P, Hurek T, Blechert O, Rexer K-H, Kost G, Hahn A, Hock B, Maier W, Walter M, Strack D, Kranner I (2001) *Piriformospora indica*: a cultivable mycorrhiza-like endosymbiotic fungus. In: Hock B (ed) Mycota IX. Springer, Berlin Heidelberg New York, pp 125–150

Varma A, Singh AR, Sudha A, Sahay NS, Kumari M, Bharati K, Sarbhoy AK, Maier W, Walter MH, Strack D, Franken P, Singh AN, Malla R, Hurek T (2002) *Piriformospora indica*: a plant stimulator and pathogen inhibitor arbuscular mycorrhizalike fungus. In: Markandey DK, Markandey NR (eds) Microorganisms in Bioremediation. Capital Book Company, New Delhi, pp 71–89

Varma A, Bakshi M, Lou B, Hartmann A, Oelmueller R (2012a) *Piriformospora indica:* a novel plant growth-promoting mycorrhizal fungus. Agric Res 1(2):117–131

Varma A, Kharkwal A, Agarwal A, Bajaj R, Prasad R (2012b) *Piriformospora indica:* the model microbe for organic green revolution. Biofertiliser Newsletter 20(1):3–8

Verma S, Varma A, Rexer KH, Hassel A, Kost G, Sarbhoy A, Bisen P, Butehorn B, Franken P (1998) *Piriformospora indica*, gen. et sp. nov., a new root-colonizing fungus. Mycologia 90:896–902

Waller F, Achatz B, Baltruschat H, Fodor J, Becker K, Fischer M, Heier T, Hückelhoven R, Neumann C, von Wettstein D, Franken P, Kogel KH (2005) The endophytic fungus *Piriformospora indica* reprograms barley to salt-stress tolerance, disease resistance, and higher yield. Proc Natl Acad Sci USA 102:12286–12291

Waller F, Mukherjee K, Deshmukh SD, Achatz B, Sharma M, Schäfer P, Kogel KH (2008) Systemic and local modulation of plant responses by *Piriformospora indica* and related Sebacinales species. J Plant Physiol 165:60–70

Warcup JH, Talbot PHB (1967) Perfect states of *Rhizoctonia* associated with orchids. New Phytol 66:621–641

Weiß M, Oberwinkler F (2001) Phylogenetic relationships in *Auriculariales* and related groups-hypotheses derived from nuclear ribosomal DNA sequences. Mycol Res 105:402–415

Weiß M, Selosse MA, Rexer KH, Urban A, Oberwinkler F (2004) Sebacinales: a hitherto overlooked cosm of heterobasidiomycetes with a broad mycorrhizal potential. Mycol Res 108:1002–1010

Yadav V, Kumar M, Deep DK, Kumar H, Sharma R, Tripathi T, Tuteja N, Saxena AK, Johri AK (2010) A phosphate transporter from the root endophytic fungus Piriformospora indica plays a role in phosphate transport to the host plant. J Biol Chem 285 (34):26532–26544

Zuccaro A, Basiewicz M, Zurawska M, Biedenkopf D, Kogel K-H (2009) Karyotype analysis, genome organization, and stable genetic transformation of the root colonizing fungus *Piriformospora indica*. Fungal Genet Biol 46:542–550

Zuccaro A, Lahrmann U, Güldener U, Langen G, Pfiffi S, Biedenkopf D, Wong P, Samans B, Grimm C, Basiewicz M, Murat C, Martin F, Kogel KH (2011) Endophytic life strategies decoded by genome and transcriptome analyses of the mutualistic root symbiont *Piriformospora indica*. PLoS Pathog 7: e1002290

14 Genetic Diversity and Functional Aspects of Ericoid Mycorrhizal Fungi

S. Perotto[1], E. Martino[1], S. Abbà[1], M. Vallino[1]

CONTENTS

I. General Features of Ericoid Mycorrhiza

The aim of this chapter is to review classical and more recent findings on the taxonomic diversity and functional traits of soil fungi forming ericoid mycorrhiza (ERM) with plants in the family Ericaceae. The cladistic relationships within this plant family have been recently investigated by Kron et al. (2002) using nucleotide sequence data from the nuclear 18S and chloroplast encoded *mat*K and *rbc*L genes. The phylogenetic analysis indicates that Empetraceae and Epacridaceae, previously considered as separate families, are now placed in tribes within the larger family Ericaceae (Kron et al. 2002). The **close relationship between epacrids and ericoid plants** is further supported by the finding of phylogenetically close mycorrhizal fungi associated with these plant groups (Chambers et al. 2000; Sharples et al. 2000a).

Plants in the family **Ericaceae** are **widespread in a diverse range of heathland and open forest communities, both in the Northern and in the Southern hemisphere** (Read 1991; Sokolovski et al. 2002). Although they are commonly found as understorey vegetation, ericaceous shrubs can become dominant in many heathland habitats found at high altitudes and colder altitudes, as well as in Mediterranean climates. These **habitats are characterised by very poor nutrient status and considerable edaphic stress** (Cairney and Meharg 2003), and it is thought that the success of Ericaceae in these habitats is due to their endomycorrhizal association. In addition to being more stress tolerant than non-mycorrhizal plants in harsh environments (see Cairney and Meharg 2003), ericoid mycorrhizal plants have been also found to be better competitors in mixed plant communities (Genney et al. 2000; Van der Wal et al. 2009).

A **common feature of ericoid plants** is the **anatomy of their very fine and delicate roots,** termed **hair roots** (Read 1996). Hair roots (Fig. 14.1) have a very narrow diameter, usually less than 100 μm, and consist of an inner stele surrounded by a two-layered cortex and an

[1]Dipartimento di Scienze, della Vita e Biologia dei Sistemi dell'Università di Torino and Istituto per la Protezione delle Piante del CNR, v. le Mattioli 25, 10125 Torino, Italy; e-mail: silvia.perotto@unito.it

Fungal Associations, 2nd Edition
The Mycota IX
B. Hock (Ed.)

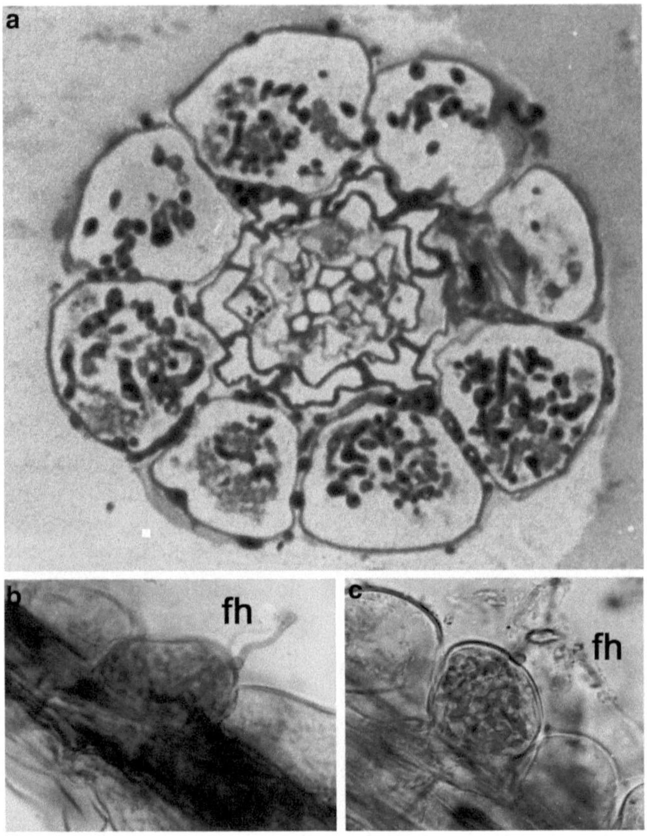

Fig. 14.1. Colonisation of *Calluna vulgaris* hair roots by ericoid mycorrhizal fungi. (**A**) Transverse section of a hair root to show the general structure of the root. Cells of the outer epidermal layer harbour fungal coils formed by a sterile mycorrhizal morphotype. (**B, C**) Fungal hyphae (*fh*) entering the epidermal cells of *C. vulgaris* through the tangential wall. The collapsed cortical cells are visible just underneath the epidermis, surrounding the small central cylinder

outermost layer of large epidermal cells. These **epidermal cells** represent the interface with the soil environment (Smith and Read 2008) and are the only hair root cells to be **colonised by ERM fungi.** Here, ERM fungi **form hyphal coils** that usually occupy most of the cell volume and are always surrounded by the invaginated plant plasma membrane (Fig. 14.2; Bonfante-Fasolo and Gianinazzi-Pearson 1979; Peterson and Massicotte 2004). Colonisation of epidermal cells by fungi usually occurs directly from the soil through the outer thickened tangential wall (Fig. 14.1), with limited cell to cell hyphal connections (Massicotte et al. 2005) so that **most cells represent individual colonisation units.** Epidermal cells are sloughed off in the older parts of the hair roots, leaving exposed the suberised cortex layer. Thus, ERM cells are ephemeral and symbiotic nutrient exchange is likely restricted to the younger parts of the root, where both partners are viable. Unlike other endomycorrhizal associations, such as arbuscular and orchid mycorrhiza, viable ericoid fungal hyphae have been observed in plant cells showing signs of organelle and cytoplasm degeneration (Bonfante-Fasolo and Gianinazzi-Pearson 1979; Duddridge and Read 1982). Mechanisms of nutrient acquisition by the plant that involve digestion of the fungal hyphae, such as tolypophagy, suggested by Rasmussen and Rasmussen (2009) for orchid mycorrhiza, can be therefore excluded.

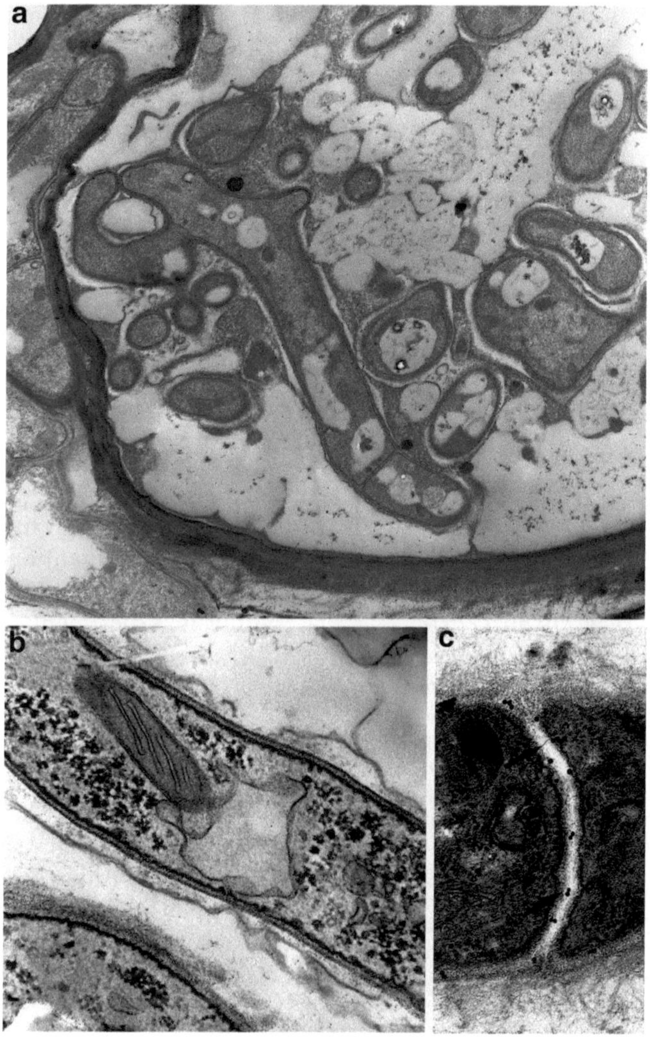

Fig. 14.2. Ultrastructure of an epidermal cell of *C. vulgaris* colonised by a typical fungal coil. (**A**) The fungal hyphae form a dense coil, which occupies a central position in the cell. (**B**) Detail of an intracellular hypha showing some fungal compartments (vacuole, mithochondrion) and glycogen deposits. The plant plasma membrane surrounds the hypha. (**C**) A fungal septum labelled with a gold-labelled cellobiohydrolase shows the presence of beta 1,4 glucans (Courtesy of P. Bonfante)

II. Genetic Diversity of Ericoid Mycorrhizal Fungi

To understand the evolution and ecological role of mycorrhizal symbioses, one of the first steps is to unveil the taxonomic position and functional features of the fungal symbionts. The study of ERM is a good example of how the development of molecular tools to investigate fungal diversity has impacted on our view of the specificity of this symbiosis, considered for long time a highly specific interaction restricted to few taxa of plants and fungi (Straker 1996). It has also modified considerably our understanding of ericoid mycorrhizal functioning in ecosystems, revealing potential hyphal networks previously unsuspected (Chambers et al. 1999; Vrålstad 2004; Grelet et al. 2010).

The first attempts to identify ERM fungi were based on the isolation in culture of endosymbiotic fungi from mycorrhizal roots. Several slow growing sterile mycelia, grouped primarily on the basis of cultural morphology and/or ITS-RFLP profiles, have been isolated worldwide from ericoid plants, and many could re-establish ericoid mycorrhiza when inoculated onto axenic plants in gnotobiotic conditions (e.g. Hutton et al. 1994; Perotto et al. 1996; Hambleton and Currah 1997; Liu et al. 1998; McLean et al. 1999; Monreal et al. 1999; Bergero et al. 2000; Cairney et al. 2000; Chambers et al. 2000; Sharples et al. 2000a; Berch et al. 2002; Bougoure and Cairney 2005a, b).

A. Taxonomic Diversity of Ericoid Mycorrhizal Fungi

The first ERM fungus that could be identified taxonomically was a dark isolate obtained from *Calluna vulgaris* roots by Pearson and Read (1973a) that was eventually induced to form fruiting bodies in pure culture. This isolate formed apothecia and was classified in the genus *Pezizella* as *P. ericae* (Read 1974).

The taxonomic position of *P. ericae* has since been revised: the new name *Hymenoscyphus ericae* was proposed by Kernan and Finocchio (1983), but a phylogenetic revision of the *Hymenoscyphus* genus by Zhang and Zhuang (2004) showed that *H. ericae* was outside this genus, and the new name *Rhyzoscyphus ericae* was thus proposed. For several years *R. ericae*, with its anamorph *Scytalidium vaccinii* (Dalpé et al. 1989; Egger and Sigler 1993), was the only identified fungal symbiont of Ericaceae. The ERM was thus described as a highly specific association that at that time contrasted with the apparent lack of specificity of other endomycorrhizae, such as arbuscular mycorrhiza.

Most **slow-growing mycelia isolated from ERM** remained unidentified until molecular tools were developed to investigate their phylogenetic affinities. It turned out that a high proportion of these slow-growing isolates, together with fungi isolated from plants with different mycorrhizal status, formed an aggregate of closely related species in the order Helotiales. This assemblage, named "***R. ericae* aggregate**" by Vrålstad et al. (2000, 2002a) because it included *R. ericae,* contained **four main clades**. The clades were further refined by Hambleton and Sigler (2005), who also proposed three new species in the anamorphic genus *Meliniomyces* as members of the "*R. ericae* aggregate" (Fig. 14.3). Clade 3 contains most of the proven ERM isolates, including type cultures of *R. ericae* and *S. vaccinii.* Clade 1 (*Meliniomyces variabilis*) contains sequences of ericoid root-associated fungal isolates, previously described as "variable white taxon" (Hambleton and Currah 1997), as well as sequences of root endophytes from other host plants from cold-temperate soils of the Northern hemisphere. Some of these fungi can colonise the roots of both ERM and ECM hosts, but although they formed typical ERM structures in hair root hosts, no true mantle was observed in ECM hosts (Piercey et al. 2002; Vohník et al. 2007). Clade 2 (*Meliniomyces vraolstadiae*) is a small clade containing sequences from fungi so far only found to form ECM, or to be non-mycorrhizal (Vrålstad et al. 2002b).

Other ERM fungi have been isolated in culture and were identified as **ascomycetes in the genus *Oidiodendron*** (Dalpé 1986, 1989). Resynthesis experiments (e.g. Couture et al. 1983; Dalpé 1986) have demonstrated the ERM nature of *Oidiodendron*, and their occurrence as ERM symbionts has been confirmed in several Ericaceae (Read 1996; Straker 1996; Xiao and Berch 1996; Allen et al. 2003; Bougoure and Cairney 2005a, b). A comparison of rDNA ITS region sequences suggests that many *Oidiodendron* isolates may have been misidentified in older studies, and that only *O. maius*, or phylogenetically close species in the Onygenales, forms mycorrhizal associations with Ericaceae in the field (Hambleton et al. 1998; Lacourt et al. 2002).

Ericoid fungi also encompass so far unculturable members of the **Basidiomycota** that could be precisely identified only by molecular phylogeny. Early ultrastructural studies provided evidence of basidiomycetes forming ERM in roots of *Rhododendron*, *Calluna* and *Vaccinium* (Bonfante-Fasolo 1980; Englander and Hull 1980; Peterson et al. 1980). PCR amplification of the fungal rDNA from mycorrhizal roots of *Gaultheria shallon* (Berch et al. 2002; Allen et al. 2003) for the first time revealed sequences of Sebacinales that corresponded to more than half of the cloned ITS sequences. Selosse et al. (2007) recently investigated, by direct PCR amplification of ERM samples and ultrastructural observations, the occurrence of Sebacinales in the Ericaceae and found it to be

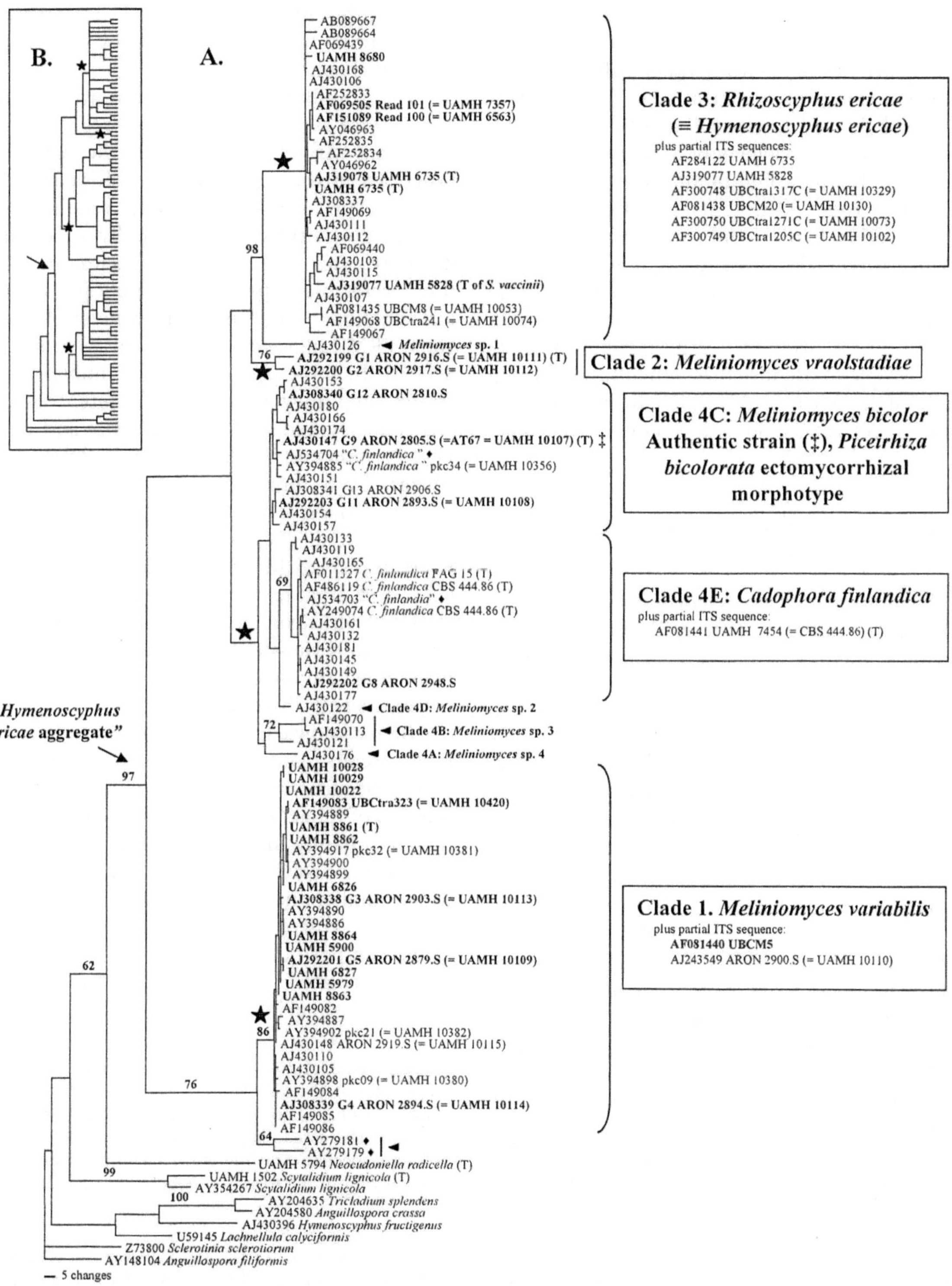

Fig. 14.3. (A) One of 5,000 MPTs from an aborted parsimony analysis of the ITS-B data matrix. Sequences AJ534704 and AY394885 were deposited under the name "*Cadophora finlandica*" (as "*C. finlandia*") based on ITS sequence similarity scores. *Arrowheads* indicate additional species that could be described if

a common symbiont. All sequences of Sebacinales recovered from ERM plants, together with sequences derived from the newly described cavendishioid mycorrhizal type (Setaro et al. 2006) belonged to clade B identified by Weiss (2004) in the phylogeny of *Sebacina*.

The frequent finding of unculturable fungi in ERM roots is in contrast with the general observation that ericoid symbionts have good saprotrophic abilities and are able to grow on common culture media (Leake and Read 1991). The use of PCR techniques to amplify fungal DNA directly from mycorrhizal roots has made it possible to identify these unculturable fungi, although their role as mutualistic symbionts in the mycorrhizal association remains unclear.

B. Multiple Fungal Occupancy in Ericoid Mycorrhizal Roots

The morphology of mycorrhizal roots in the Ericaceae indicates that epidermal root cells could potentially function as separate units, challenged and colonised by a variety of fungi resident in the rhizosphere. ITS-RFLP analyses of ericoid fungal isolates from different mycorrhizal plant species has demonstrated that **multiple occupancy is a common phenomenon in ericaceous roots** (e.g. Perotto et al. 1996; Hambleton and Currah 1997; Monreal et al. 1999; Bergero et al. 2000; Chambers et al. 2000; Midgley et al. 2004c; Bougoure and Cairney 2005a, b). This is revealed by the simultaneous presence of fungi with different ITS-RFLP profiles in the same root system. In addition, more sensitive fingerprinting methods (RAPD and ISS-PCR) showed that distinct genotypes of ERM fungi sharing the same ITS-RFLP type can occur in individuals of *C. vulgaris* (Perotto et al. 1996) and *Epacris pulchella* (Curlevski et al. 2009). These data indicate that the **roots of ericaceous plants are a complex mosaic where different populations of mycorrhizal fungi coexist**, each represented by a variable number of genotypes.

The simultaneous association with many and diverse symbiotic fungi may represent an important **strategy to broaden the array of functions in the colonisation of difficult substrates.** This hypothesis would require that genetic diversity was mirrored by functional diversity, and that different species of ERM fungi were able to perform different physiological functions. This has been demonstrated for some ERM fungi. For example, as discussed in Sect. III.B, ERM can hydrolyse complex substrates, and they do so to a different extent (e.g. Cairney and Burke 1998; Leake and Read 1991; Varma and Bonfante 1994; Midgley et al. 2004a). Even fungal strains within the same species can perform differently. For example, Cairney et al. (2000) found that isolates of *R. ericae* display different abilities to utilise inorganic and organic nitrogen sources, and Grelet et al. (2009b) demonstrated that functional differences in N transfer to the host plant shoot are maintained among closely related ERM fungi in symbiosis with their hosts.

Better knowledge of the true ERM diversity in ericaceous plants has changed our perception on the specificity in this association, with a wider number of fungal taxa being involved than once thought, mainly in the ascomycetes but also in the basidiomycetes. However, although this wider number and taxonomic position of potential ERM partners, there is evidence of specificity in the composition of the ERM fungal communities associated with co-existing plant species and different environments.

Bougoure et al. (2007) have investigated the diversity of fungi associated with plants of *C. vulgaris* and *Vaccinium myrtillus* along a heathland—forest gradient in Scotland. The assemblages of fungi associated with the two plant species were different, even though plants were co-occurring in the forest understorey. In addition, the

Fig. 14.3. (continued) isolates are available for examination. All ingroup sequences were derived from cultures isolated from roots, except for four (◆) derived from DNA extracted directly from roots. A majority of ingroup taxa were resolved into four clades, indicated by *stars* adjacent to the relevant nodes, corresponding to *Meliniomyces variabilis*, *Meliniomyces vraolstadiae*, *Cadophora finlandica*/*Meliniomyces bicolor*, and *Rhizoscyphus ericae*. (**B**) Strict consensus of all 5,000 MPTs with taxon names removed. *Stars* indicate the same four clades as in (**A**) (From Hambleton and Sigler (2005), with permission) community of fungi associated with *C. vulgaris* hair roots was different for samples collected from the forest, open heathland and a transition zone between the two (Bougoure et al. 2007). Part of the differences were due to the amplification, in the forest samples, of typical ECM ascomycetes and basidiomycetes sequences.

In a more homogeneous environment such as the subarctic heaths, however, Kjøller et al. (2010) investigated the fungal communities associated with the roots of four co-existing ericaceous plants (*Empetrum hermaphroditum*, *Andromeda polifolia*, *Vaccinium uliginosum*, *V. vitis-idaea*) and could not find significant differences in relationship with the host plants, whereas significant differences in spatial distributions were observed. Thus, it seems that, similarly to what happens in other endo- and ectomycorrhizal symbioses, environmental factors as well as the host plant species may influence, at least in some cases, the ERM fungal community even when a broad range of potential symbionts are available.

C. Potential Networking Abilities of Ericoid Mycorrhizal Fungi

The phylogenetic analyses described in the previous paragraphs demonstrate that **a diverse assemblage of fungi can interact with Ericaceae to form ERM.** Conversely, the **occurrence of ERM-forming fungi on plant species outside the Ericaceae** has also been frequently demonstrated. Duckett and Read (1995) reported that *R. ericae*, the best studied ERM symbiont, was also capable of forming in vitro intracellular coils in the rhizoids of **liverworts.**

Further support to this observation derived from molecular studies on leafy liverworts in the genus *Cephaloziella*. Mycelia isolated in nature from liverworts and demonstrated to form ERM on ericaceous hosts were first assigned to the genus *Hymenoscyphus* (syn. *Rhizoscyphus*) by Chambers et al. (1999). Later on, Upson et al. (2007) performed Koch's postulate for the *C. varians*—*R. ericae* association, inoculating axenically grown liverwort with an isolate of the fungus from the plant. These isolates from Antarctica could also colonise *Vaccinium macrocarpon* roots, despite the absence of ericaceous plant species from maritime and continental Antarctica.

ERM and ECM plants often co-exist in natural ecosystems such as boreal or Mediterranean forests, where ericaceous plants constitute the understorey vegetation of dominant ECM tree species (Read 1991). The traditional view that ECM and ERM fungi are taxonomically distinct has been challenged by several papers reporting amplification of ERM sequences from ECM roots. An important discovery by Vrålstad et al. (2000) was the **strong genetic similarity among fungi of the *R. ericae* aggregate associated with ERM and ECM roots** in a boreal forest.

Following ITS sequence comparison, the ERM fungus *R. ericae* was the closest relative of fungi isolated from *Piceirhiza bicolorata*, an ECM morphotype common in post-fire sites. Phylogenetic analyses showed that fungi from ERM and ECM roots, as well as other endophytes, belong to different clades of the "*R. ericae* aggregate", first proposed by Vrålstad et al. (2002a) and later revised by Hambleton and Sigler (2005).

None of the fungal isolates from *P. bicolorata* and grouped in Clade 1 of the *R. ericae* aggregate (Fig. 14.3), corresponding to *M. variabilis*, could form ERM on ericaceous hosts (Vrålstad et al. 2002b), although a *M. variabilis* isolate from *P. bicolorata* ECM has been recently found to form ERM in *V. vitis-ideae* (Grelet et al. 2010).

Isolates assigned to the *C. finlandica*/*M. bicolor* (Clade 4 in Fig. 14.3) have been found to form ECM or ectendomycorrhiza with *Betula*, *Picea* and *Pinus* (Wilcox and Wang 1987; Vrålstad et al. 2002b). An isolate of *M. bicolor* derived from *Piceirhiza bicolorata* ECM was also demonstrated to form, at the same time, ERM on *V. myrtillus* and ECM with *Pinus sylvestris*, and to induce beneficial effects on host plant growth (Villarreal-Ruiz et al. 2004). This finding confirms previous observations by Monreal et al. (1999) that *Phialophora* (= *Cadophora*) *finlandica* could form, at least in vitro, ERM with *Gaultheria shallon*. Additional observations derive from the work of Grelet et al. (2009a), showing that *M. bicolor* obtained from *Piceirhiza bicolorata* ECM formed typical ERM structures and engaged in reciprocal transfer of carbon and nitrogen with *V. vitis-idaea*.

Bergero et al. (2000) demonstrated that several fungi (**including *Oidiodendron*** spp. and sterile morphotypes) in a Mediterranean forest were associated with both ERM and ECM plants. Fungi isolated from root tips of ECM *Quercus ilex* formed ERM with *Erica arborea* in vitro, and molecular analyses indicated that some of these *Q. ilex* fungal associates were conspecific with ERM fungi naturally occurring in *E. arborea* roots.

Potential Hyphal Connection Between ERM and Non-Ericaceous Plants Have Been Further Identified

For example, at least six ITS-RFLP types of ERM fungi, including Helotiales and *Oidiodendron*, were common root associates of a diverse array of plant taxa within a schlerophyll forest community in south-eastern Australia (Chambers et al. 2008). Similar results were obtained by Curlevski et al. (2009) and Grelet et al. (2010), who applied the more sensitive inter-simple sequence repeat (ISSR)-PCR to investigate the occurrence of fungi with the same ISSR fingerprints in co-existing ERM and ECM plants. Shared fungal genotypes were identified between the ERM species *E. pulchella* and the ECM species *Leptospermum polygalifolium* (Curlevski et al. 2009) in a schlerophyll Australian forest, and between the ERM species *V. vitis-idaea* and the ECM species *P. sylvestris* (Grelet et al. 2010) in a Scottish boreal forest.

From the experiments described above, there is now **strong evidence that potential hyphal connections between ERM and ECM plants are a common feature in different forest ecosystems**, although the work by Grelet et al. (2010) would exclude the formation of large mycelial networks at least for *M. variabilis*, given the small genet size found for this fungus (<13 cm). However, several questions remain open on the type of interactions that ERM fungi may establish with non-ericaceous host plants, and on the functional role of these potential hyphal connections.

Some observations suggest that, rather than forming an ECM association themselves, many **ERM fungi may associate with ECM root tips in addition to the true ECM partner.**

Although one isolate of the *C. finlandica*/*M. bicolor* clade in the "*R. ericae* aggregate" has been demonstrated to form both ERM and ECM, or ectendomycorrhizal, associations with tree species under gnotobiotic conditions (Villarreal-Ruiz et al. 2004), most other isolates seem to be able to form a single mycorrhizal type, despite their common occurrence in plant roots. Bergero et al. (2000) clearly showed, in their inoculation experiments, that ERM-forming fungi isolated from ECM *Q. ilex* were often morphologically distinct from those producing the ECM. Similarly, there is no clear evidence that *M. variabilis* can actually form ECM, and it seems more likely that *M. variabilis* is an endophyte in ECM root tips formed by other fungi (Hambleton and Sigler 2005; Grelet et al. 2010). This would give support to the observation of endophytic behaviour of *M. variabilis* (Vohník et al. 2007; Ohtaka and Narisawa 2008), or the amplification of Helotiales DNA sequences from ECM formed by other fungal species (Tedersoo et al. 2009).

Occurrence of ERM Ascomycetes in Non-Ericaceous Hosts Is Not Restricted to ECM-Forming Plants

As already mentioned, Helotiales forming ERM coils in hair roots of *Woollsia pungens* (Ericaceae) under gnotobiotic conditions were isolated from the roots of 17 species from different plant families (Apiaceae, Cunoniaceae, Cyperaceae, Droseraceae, Fabaceae—Mimosoideae, Lomandraceae, Myrtaceae, Pittosporaceae, Proteaceae, Stylidiaceae) at an Australian sclerophyll forest site (Chambers et al. 2008).

In cold climates, **dark septate endophytes (DSE)** are widely distributed and frequently isolated from the roots of several plant species (see references in Jumpponen and Trappe 1998; Mandyam and Jumpponen 2005). Some members of the "*R. ericae* aggregate" are well recognised DSE: in addition to *M. variabilis*, *Cadophora finlandica* is commonly identified in the roots of mainly ECM plant species (Wilcox and Wang 1987; Mandyam and Jumpponen 2005). They have been identified also in Antarctica on *Colobanthus quitensis* and *Deschampsia antarctica* (Newsham et al. 2009). The ecological functions of DSE are not well understood, but they can provide several benefits to their host plants, including facilitation of nutrient uptake, protection from metal stress tolerance and stimulation of the mycorrhizosphere community against root diseases (Jumpponen 2001; Mandyam and Jumpponen 2005; Dos Santos Utmazian et al. 2007; Alberton et al. 2009).

A study by Abuzinadah and Read (1989) showed that *Oidiodendron* also enhances growth of *Betula pendula* on a medium containing proteins as sole nitrogen source, without producing a classical ectomycorrhizal infection. Similarly, endophytic fungi isolated from roots of the grass *Deschampsia flexuosa* and identified in the Helotiales have been shown to colonise roots and enhance nitrogen uptake by *C. vulgaris* (Ericaceae) seedlings, and vice versa (Zijlstra et al. 2005). As discussed by Curlevski et al. (2009), many questions remain on the relationships between fungi that form ERM and DSE associations, particularly with regard to the benefits conferred to plants by the DSE infection (Schulz and Boyle 2005).

The ability to infect multiple hosts raises the possibility that **ERM associations in ericoid roots and DSE associations in non-ericaceous roots might represent parts of a common mycelial network.**

Hyphal links playing important functions in nutrient exchange, including organic carbon, have been demonstrated for ECM fungi (Simard et al. 1997), and for more complex situations involving plants with different mycorrhizal status (e.g. McKendrick et al. 2000; Bidartondo et al. 2003). The role of **possible physical links between ericaceous and non-ericaceous plants via their fungal associates remains an open question**, as well as the nature of the functional relationships with these hosts. It also remains to be established whether or not the same fungus can form different types of mycorrhizal symbioses with distinct hosts in natural conditions. The classical bioassay to establish the mycorrhizal nature of a fungal isolate is the inoculation in vitro onto axenic seedlings. Although this assay remains an important test to elucidate the mycorrhizal potential of a fungal isolate, the conditions used are very different from those found in nature and the results must be interpreted with caution (Read 1996).

In addition to a possible role in nutrient exchange, the association of ERM fungi with non-ericaceous plants may be ecologically relevant in some stressed conditions. For example, genetic relatedness between ERM and ECM fungal associates was found in forests subjected to fire (Bergero et al. 2000; Vrålstad et al. 2000), and it was suggested that the ECM host, with deeper roots, could provide an efficient source for biotrophic infection of the ERM (or other) plants. A role as a reservoir of ericoid fungi has also been suggested for liverwort rhizoids by Duckett and Read (1995).

III. Exploitation of Inorganic and Organic Substrates by Ericoid Mycorrhizal Fungi

Habitats dominated by Ericaceae include most humus heathlands in the Northern hemisphere, Mediterranean woodlands, tropical cloud forests and the dry sand plains of Australia (Read 1991; Straker 1996; Bergero et al. 2000). **Soils colonised by ERM plants** are characterised by the low availability of nutrients, due to slow litter decomposition and mineralisation processes. Here, essential nutrients such as N and P are found almost exclusively in organic forms (Read and Perez-Moreno 2003). In these ecosystems, the dominant plant species are highly dependent on the mycorrhizal symbionts for their nutrient supply. Although ERM fungi appear to have a poorly developed extraradical mycelial phase, they are able to mobilise organic nutrients rendering them accessible to the host roots (Finlay 2008).

In **mor-humus heathlands**, the **limiting nutrient is primarily nitrogen**, which is in the form of acid-hydrolysable organic compounds and insoluble humin, usually inaccessible to plants (Stribley and Read 1974). The success of Ericaceae in these stressful habitats is therefore largely related to the abilities of their ERM symbionts to improve nutrient acquisition, particularly nitrogen, from organic forms. Several studies on ERM mycelia growing in vitro (see Smith and Read 1997, 2008) have confirmed their high saprotrophic capabilities, which would facilitate nutrient mobilisation from the substrates exploited by the mycorrhizal roots and the mycelium. For example, Read and Perez-Moreno (2003) pointed out that the enzymatic capabilities of ERM fungi could extend the sources of N important to their plant hosts beyond inorganic forms to include amino acids, amino sugars, proteins, or chitin. In return for N and other nutrients, plants probably allocate up to 20 % of net primary production to their mycorrhizal symbionts (Hobbie 2006). Mobilised nutrients from organic forms can then be assimilated by plant roots, either directly or via the fungal intermediary.

A. Nitrogen and Phosphorus Uptake

Organic nitrogen (ON) accounts for up to **95 % of the soluble N pool in soils** (Abuarghub and Read 1988; Talbot and Treseder 2010). This pool represents an important component of plant N budgets, considering that inorganic N pools in soil are insufficient to account for annual plant N uptake in many ecosystems (Kielland 1994). **Mycorrhizal plants have greater access to ON than non-mycorrhizal plants** (Schimel and Bennett 2004). The uptake of ON by ERM fungi and the subsequent transfer to the plant require multiple steps, including breakdown of polymers in soil solution, direct uptake of mono- and oligomer into mycorrhizal fungi, internal transformation of ON, and transfer across the fungus—host plant interface (Talbot and Treseder 2010). Identifying which mechanisms most strongly control each of these steps will help in understanding the extent of ON usage by mycorrhizal plants (Talbot and Treseder 2010). The **major classes of ON compounds found in soils and in soil solution include aliphatic-N, like amino-N and polysaccharide-N, and aromatic-N**, such as the compounds present in soil humus (Roberts and Jones 2008). The high soil ON content indicates that the exposure of mycorrhizal roots to ON can be equal to or greater than exposure to inorganic N in most soils (Talbot and Treseder 2010). The forms of ON that are transferred to the plant may vary among fungal and plant species, with glutamine, asparagine, and alanine being the most common ON compounds involved in this process (Chalot and Brun 1998). Systems that are dominated by ERM fungi that have the capability to produce a broad range of extracellular enzymes, such as *R. ericae* (Cairney et al. 2000; Midgley et al. 2006), may present a particularly high rate of ON uptake.

A **significant pool of available nitrogen in heathland soils** is represented by **free amino acids** (Abuarghub and Read 1988). *R. ericae* has been shown to absorb a broad range of amino acids (Chen et al. 1999; Whittaker and Cairney 2001) and to effect transfer of aminonitrogen to its hosts, although considerable intraspecific variation can exists in this regard (Bajwa and Read 1986; Cairney et al. 2000). The utilisation of sulfur-containing amino acids by *R. ericae* is poor but, in common with other filamentous fungi, this may be enhanced under sulfur starvation (Bajwa and Read 1986). In vitro culture studies indicate that some ERM fungi can use amino acids as their sole N source (Talbot and Treseder 2010). Amino acids vary in their frequency within proteins, with leucine, glycine and alanine being most common and tryptophan, cysteine and histidine rarest. Talbot and Treseder (2010) found that the percentage of ERM fungal species capable of using a given amino acid was significantly and positively correlated with its relative abundance in proteins.

Early work on the physiology of ERM fungi and their influence on plant nutrient uptake was on a limited number of ERM strains, mainly *R. ericae* (Leake and Read 1991). However, a single root system can harbour several ERM fungal genotypes, which may also belong to different species. It became therefore important to understand whether **this genetic diversity confers functional diversity** which could be **advantageous in nutrient-poor habitats.**

Cairney et al. (2000) studied various fungal strains isolated from hair roots of *C. vulgaris* and showed significant variations in the N utilisation pattern. While most isolates showed a preference for ammonium as a sole nitrogen source, considerable variation was observed in the abilities of isolates to utilise amino acids and proteins (e.g. BSA). In particular, large intraspecific variation was observed in the use of glutamine and BSA. Individual isolates of *R. ericae* may therefore vary considerably in their abilities to use organic nitrogen from different substrates in soil. Grelet et al. (2005) showed that the different abilities to utilise organic and mineral N sources were affected by carbon availability in a strain-specific manner. Under elevated C supply, growth differences among strains were linked to the total amount of nitrogen taken up, suggesting variation in uptake kinetics. But, under C-limiting conditions, the nitrogen-use efficiency explained strain differences, implying intraspecific variations in N metabolism. The main effect of reducing C availability was to increase the relative ability of most strains to grow on glutamine and nitrate, in comparison to ammonium. Grelet et al. (2009b) investigated whether the functional variation was maintained in symbiosis. *Vaccinium* plants were inoculated with three genetically closely related ERM fungal strains known to differ in their N use in liquid culture. ^{15}N was used to trace N uptake into shoots, and the results demonstrate that functional differences among closely related ERM fungi are maintained in symbiosis with their hosts, thus suggesting that N transfer to plant shoots in ericoid mycorrhiza is under fungal control. In the experimental conditions used by

Grelet et al. (2009b), the strain-specific effect was evident when N was supplied as ammonium or glutamine, but fungal influence on plant N uptake has also been shown for a range of amino acids (Sokolovski et al. 2002) and for nitrate (Kosola et al. 2007).

These experiments give support to the hypothesis that a **broad spectrum of symbiotic ERM fungi may provide the host plant with a wider set of acquired symbiotic functions.**

In **acidic heathland soils**, a **very low amount of free inorganic phosphate** is found, and the **main phosphorus sources are organic compounds in the form of phosphomonoesters, mainly phytates** (Cosgrove 1967; Mitchell and Gibson 2006), or phosphodiesters such as nucleic acids (Griffiths and Caldwell 1992). Phytates are usually complexed with iron and aluminium, and ERM fungi are able to access these sources (Mitchell and Read 1981). **Both extracellular and wall-bound phosphatases are produced** by a number of ERM isolates (Straker and Mitchell 1986). Phosphomonoesters are quantitatively the most important fraction of organic soil phosphorus, while phosphodiesters are found in relatively low concentrations but are potentially valuable phosphorus sources (Griffiths and Caldwell 1992; Leake and Miles 1996; Myers and Leake 1996). Leake and Miles (1996) demonstrated the ability of *R. ericae* to degrade DNA molecules through the production of phosphodiesterases and to directly assimilate nucleotides.

Pearson and Read (1973b) provided evidence for a movement of P-32 orthophosphate from the fungus to the plant. Two phosphate membranes transport systems were identified for a South African isolate as high affinity and low affinity systems (Straker and Mitchell 1987). Interestingly, both P uptake systems operate better under the pH conditions typically found in the acidic soils where ERM fungi live.

Many fungi can dissolve phosphate and other essential nutrients from mineral substrates by employing several mechanisms, including protonation and chelation (Gadd 2010). In terms of mineral weathering and dissolution, mycorrhizal fungi form one of the most prominent groups of soil microorganisms (Devevre et al. 1996; Jongmans et al. 1997; Lundström et al. 2000). A number of ERM fungi has been investigated for their phosphate solubilising ability. Van Leerdam et al. (2001) showed that most isolates could solubilise the rock phosphate hydroxyapatite in the presence of ammonium, whereas Martino et al. (2003) and Gibson and Mitchell (2004) showed that *O. maius* and *R. ericae*-type isolates were capable of solubilising zinc phosphate, and some of these isolates were also able to solubilise calcium phosphate.

B. Depolymerisation of Complex Substrates

The array of hydrolytic enzymes produced by ERM fungi enables their saprotrophic growth in the absence of the host (Perotto et al. 1993, 1997), but may also mediate utilisation of organic nutrient sources by the host plant (Cairney and Burke 1998). **Ericoid fungi produce a range of extracellular enzymes that catalyse breakdown of several organic macromolecules and give plants access to breakdown products of complex polymers** that cannot be directly assimilated by either plant roots (Read and Perez-Moreno 2003). This access appears particularly important in N-limited systems, where mineralisation is insufficient to support plant N demand (Schimel and Bennett 2004).

Two classes of polymer-degrading enzymes can be recognised (Read and Perez-Moreno 2003). The first class includes several hydrolases that cleave the nutrient containing molecules themselves, whereas the second class facilitates nutrient acquisition by attacking organic molecules such as lignins, polyphenols and tannins, which may protect or precipitate essential nutrients. Lignases and polyphenol oxidases are comprised in this type of enzymes, and are expected to give a major contribution to litter decomposition as well as to plant nutrition.

ERM fungi express both types of polymer degrading enzymes. In addition to **proteases** and **chitinases**, they produce **hydrolytic enzymes**, which degrade cell wall polysaccharides such as cellulose, hemicelluloses and pectin, as well as **phenol oxidase** activities, which may facilitate host access to nitrogen and phosphorus within moribund plant material or from polyphenol complexes in soil (Perotto et al. 1993, 1997; Varma and Bonfante 1994; Bending and Read 1996a, b; Burke and Cairney 1997a, b; Cairney and Burke 1998; Piercey et al. 2002; Midgley et al. 2004a, 2006).

The complete use of cell wall polysaccharides, such as cellulose, is not surprising, as these enzymes are presumably required for host cell penetration. Like cellulose, the potential of ERM fungi to degrade pectin is important during penetration of the root of the host. ERM fungi have been shown to produce polygalacturonases in culture (Perotto et al. 1993, 1997; Midgley et al. 2006). *R. ericae* is also able to grow on xylans thanks to the production of at least one **xylanase** and a range of accessory enzymes (Burke and Cairney 1997a, b).

Polyphenol oxidases are known to be produced by ERM fungi (Burke and Cairney 2002). These enzymes include laccase, catechol oxidase and tyrosinase, which show a considerable overlap in substrate affinities (Burke and Cairney 2002). *R. ericae* produces laccase, along with a range of related phenol-oxidising activities (Burke and Cairney 2002). It has been suggested that **laccases** produced by ERM fungi may be engaged in a number of processes involved in the functioning of the symbioses. Suggested roles include lignin and polyphenol degradation, release of N from insoluble protein—tannin complexes and degradation of polycyclic aromatic hydrocarbon pollutants (Leake and Read 1989; Hutchison 1990; Bending and Read 1996a, b, 1997; Braun-Lüllemann et al. 1999). Cairney and Burke (1998) assessed the production of ligninolytic activities by *R. ericae*. The fact that *R. ericae* releases hydrogen peroxide (Bending and Read 1997) and hydroxyl radicals (Burke and Cairney 1998) into culture media confirms that the fungus may contribute to a form of **lignin degradation**, similar to that seen in "brown rot" fungi, in which the fragmentation of the lignin polymer is mediated by these radicals (Burke and Cairney 1998; Cairney and Burke 1998). Carbohydrate oxidase activity may further release H_2O_2, which, via production of hydroxyl radicals in the presence of Fe, may contribute to partial lignin degradation by this fungus. Burke and Cairney (1998) examined isolates of the ERM fungus *R. ericae* for their ability to oxidize carbohydrates to their corresponding lactones and to excrete the H_2O_2 produced thereby. *R. ericae* was found to express cellobiose oxidase and glucose oxidase when grown on cellobiose and glucose respectively.

Polyphenolic compounds can reach high concentrations in heathland soils (Jalal and Read 1983). In these phenol-rich environments, the **proven ability of ERM fungi to use many monomeric phenolic compounds as carbon sources** (Leake and Read 1989) and to release the enzymes laccase and catechol oxidase (Bending and Read 1996a, b, 1997), involved in the degradation of hydrolysable polyphenols, plays a fundamental role in plant nutrition.

Bending and Read (1996b) demonstrated that, in addition to degrading tannic acid-protein precipitates by releasing polyphenol oxidase, *R. ericae* and *Oidiodendron* sp. were able to acquire nitrogen from these complexes through the expression of **extracellular acid proteases**. Such observations indicate that the ability to absorb amino acids bypassing the N mineralisation process may be widespread in organic soils. *R. ericae* is in fact known to produce extracellular proteinases that, via the hydrolysis of simple proteins, release amino acids that can be absorbed by the mycelium (Read et al. 1989; Whittaker and Cairney 2001). Interestingly, *R. ericae* also secretes an extracellular proteinase identified as a carboxyl (acid) proteinase (Leake and Read 1989) that showed optimal activity at pH 2.2, which is considerably lower than that found in acidic heathland soils (Leake and Read 1990b). However, most of the fine mycorrhizal roots of ericaceous plants are confined to the soil/litter interface, where, due to the activity of H^+-releasing processes, the pH is lower than in the bulk soil (Read et al. 1989). This interface is also the location where protein substrates accumulate. The proteinase activity of the fungus is, therefore, adapted to the specific microenvironment in which it is found. Moreover, as the pH within plant cells approaches neutral, the low pH optimum of the proteinase may also protect the plant from protein degradation (Leake and Read 1989). It has also been proposed that excess proteinase is only released by the fungus if a suitable substrate is present in the soil (Leake and Read 1990c). The end products or amino acids can be taken up directly by the ERM fungus without further deamination and these forms of nitrogen would also be ideal for transfer from the fungus to the host plant.

ERM fungi from Ericaceae growing in the Southern hemisphere appear to be broadly similar to *R. ericae* in their abilities to utilise different carbohydrates sources, as well as amino acids and proteins as nitrogen sources (Chen et al. 1999; Whittaker and Cairney 2001; Midgley et al. 2004b).

Using **crustacean chitin as a model compound**, it was demonstrated that *R. ericae* and *Oidiodendron griseum* could cleave the polymer to its constituent subunits, which were readily assimilated (Leake and Read 1990a; Mitchell et al. 1992). A significant proportion of the N acquired by the fungus was found to be transferred to aseptically grown mycorrhizal plants of *V. macrocarpon* (Kerley and Read 1995). *R. ericae* has also been shown to supply its host plant with nitrogen directly from chitin (Kerley and Read 1997), providing strong evidence that chitinase activities are produced by the fungus during symbiosis. Bougoure and Cairney (2006) also showed that an isolate of *R. ericae* from *E. pulchella* produced both endo- and exo-acting chitinolytic activities.

Concerning the acquisition of P from polymeric organic sources, both DNA in purified form (Leake and Miles 1996) and entire nuclei (Myers and Leake 1996) have been shown to represent useable P sources for *V. macrocarpon*. P mobilisation appeared to depend upon production of phosphodiesterases by *R. ericae* (Leake and Read 1997).

A **significant consumption of C and N is required for extracellular enzyme production** (Schimel and Weintraub 2003). In fact, it has been shown that plant C supply to the fungal symbiont can control extracellular enzyme production and the acquisition of nutrients from polymeric compounds, and that ERM fungi acquire more N from high molecular weight organic nitrogen compounds when grown with the plant (Bajwa and Read 1986; Dighton et al. 1987; Gunther et al. 1998). Under elevated glucose concentrations in the culture medium, *R. ericae* acquired a higher percentage of N from glutamine (Grelet et al. 2005), and protein mineralisation by other ERM fungal isolates also increased (Zhu et al. 1994; Eaton and Ayres 2002). Under the same glucose concentration, the expression of extracellular proteases (Nehls et al. 1999) and chitinases (Leake and Read 1990a; Bougoure and Cairney 2006) decreased in other ERM fungi. There is therefore evidence that C availability increases uptake of amino acids by ERM fungi, and that a strong C regulation of extracellular enzyme activities operates by induction/repression mechanisms (Talbot and Treseder 2010).

IV. Ericoid Mycorrhizal Fungi in Soils Enriched in Toxic Compounds

Since the first investigations by Bradley et al. (1981, 1982), **the success of Ericaceae in stressful habitats** has been ascribed to the **unique abilities of their ERM fungal partners to withstand and adapt to environmental stresses and to enhance stress tolerance in their host plants** (Cairney and Meharg 2003).

Environmental stress can shape adaptation and evolution of organisms living in changing environments, and the impact of human activities on natural environments has caused rapid and often stressful and deteriorating changes. The occurrence of adaptive processes is clearly shown by the fact that many organisms have acquired quite rapidly the ability to withstand man-made changes in the environment (Nikolaou et al. 2009).

Heavy metal toxicity represents a strong selection pressure and microbial resistance to toxic metals is widespread, with frequencies ranging from a few per cent in pristine environments to nearly 100% for bacteria growing in heavily polluted environments (Silver and Phung 2009). Adaptation of ecto- and endomycorrhizal fungi to heavy metal soil pollution of anthropic origin is also supported by several studies (e.g. Leyval et al. 1997; Meharg and Cairney 2000; Colpaert et al. 2004; Adriaensen et al. 2005; Krznaric et al. 2009).

A. Genetic and Functional Adaptation to Heavy Metals

As already described, soils naturally colonised by Ericaceae are generally acidic. **Low pH and anaerobic soil conditions facilitate mobilisation of heavy metal ions**, which are toxic above threshold concentrations (Meharg and Cairney 2000). Bradley et al. (1981, 1982) first demonstrated the **importance of ERM fungi to increase resistance** of *C. vulgaris* to heavy metals, and other authors later described metal tolerance in fungal isolates from sites with different pollution. In particular, ERM plants in association with members of the "*R. ericae* aggregate" have been shown to survive in arsenate-rich sites (Sharples et al. 2000b) and populations of arsenate resistant *R. ericae* have been isolated from *C. vulgaris* growing in As/Cu-contaminated mine soils (Sharples et al. 2001). A detailed study on more than 70 ERM fungi from polluted and non-polluted sites indicated an adaptive resistance to arsenate in the fungal populations from the mine site (Sharples et al. 2001). Strains of the ERM species *O. maius* also showed adaptation to metal pollution, as isolates from an industrial site heavily polluted with Cd, Zn and Al displayed better growth in vitro on media containing these metals, when compared with isolates from non-polluted sites (Martino et al. 2000a).

Ericoid fungal strains of *O. maius* derived from polluted and unpolluted soils also mobilised insoluble inorganic zinc compounds to different extents (Martino et al. 2003). Strains from polluted soils showed little ability to solubilise Zn from both ZnO and $Zn_3(PO_4)_2$, whereas strains from unpolluted soils showed a higher solubilisation potential. As solubilisation of insoluble metal compounds may lead to toxic metal concentrations in soils enriched with heavy metals, this mechanism may be adaptive.

In addition to being polluted by human activities, **soils may be naturally enriched in heavy metals derived from specific rock substrates.** Serpentine soils derived from ultramafic rocks are low in plant nutrients such as K and Ca, but contain high levels of potentially toxic elements such as Ni and Cr. Vallino et al. (2011) compared metal tolerance of *O. maius* isolates derived from sites with contrasting metal pollution (industrial Cd/Zn pollution versus natural Cr/Ni pollution). Despite the small sample size, a significant difference was observed in growth of the *O. maius* isolates on media containing these metals. Strains more tolerant to Cr and Ni were those originated from the serpentine site, while strains more tolerant to Zn and Cd were isolated from the industrial sites polluted by these contaminants (Vallino et al. 2011). This would suggest, at least for some metals, a specific adaptation of *O. maius* that reflects the specific contamination in the soil of origin.

For other metals, such as copper, adaptive heavy metal tolerance is more controversial. For example, ERM fungi of *C. vulgaris* have been shown to display constitutive tolerance to copper (Bradley et al. 1982; Sharples et al. 2001). However, extracellular phosphodiesterase activity was stimulated by copper in ERM fungal isolates from mine spoil sites but not in fungi from uncontaminated sites (Gibson and Mitchell 2005). This result has been interpreted as an adaptive metal avoidance mechanism, as phosphate ions released from organic substrates by phosphatases would react with metal ions to form insoluble, and therefore less toxic, compounds (Gibson and Mitchell 2005).

Increased metal tolerance in ERM fungi can therefore arise, under selective pressure, as a consequence of new or modified phenotypic traits. As the phenotype depends on the cell genetic information, changes in gene sequences and arrangement may be expected to have arisen in those ERM fungi adapted to environmental metal stress. Mutations are one of the primary sources of genetic variation, and many environmental agents are known to induce mutations either directly or indirectly, through the production of reactive oxygen species (Hartwig 1995).

A recent study (Vallino et al. 2011) suggests that **the mutation rate positively correlated with environmental stress within the same ERM fungal species, and that mutations do not occur randomly in the genome.** Two gene regions with different functional roles in heavy metal resistance (the ribosomal ITS and the gene coding for the Cu/Zn superoxide dismutase) were compared among isolates of

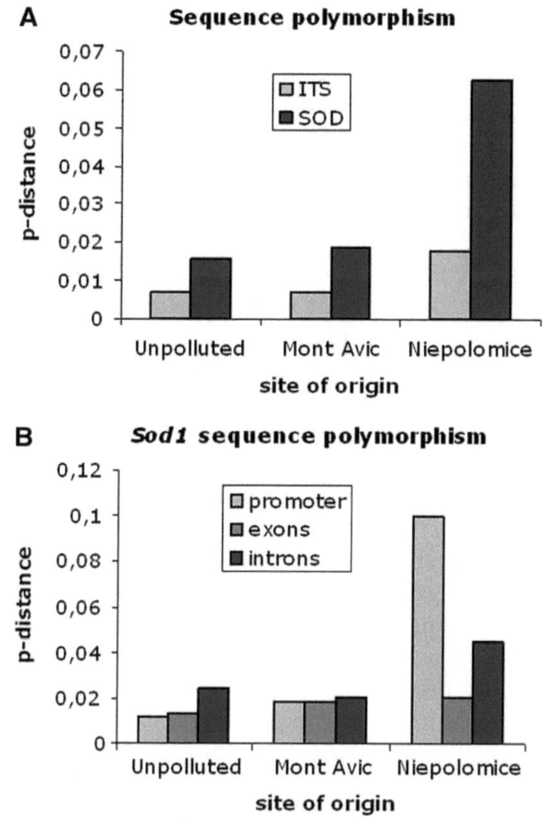

Fig. 14.4. Sequence polymorphism of the ITS and the SOD loci. (**A**) p-Distance of the ITS region and the SOD in ericoid mycorrhizal strains derived from three sites respectively non-polluted, naturally polluted (serpentine) and industrially polluted. (**B**) p-Distance of different regions of SOD locus in the same three groups of fungi (Modified from Vallino et al. (2011))

O. maius derived from non-polluted soil, and from soils with natural and industrial heavy metal pollution (Vallino et al. 2011).

The ribosomal ITS region can be considered as a "neutral" gene towards metal tolerance, whereas Cu/Zn SOD plays a major role in cell defence against toxic reactive oxygen species, which are increased by environmental stress, including heavy metals (Schützendübel and Polle 2002). The role of the *Sod1* gene (coding for the *O. maius* Cu/Zn SOD) in metal tolerance was recently demonstrated in *O. maius* by Vallino et al. (2009) and Abbà et al. (2009).

As compared with the "neutral" ITS, a higher mutation rate was found in the functional *Sod1* locus of strains from all sites (Fig. 14.4), suggesting that **genes with a functional role in fungal survival display a higher nucleotide polymorphism than neutral genes** (Vallino et al. 2011). Moreover, a significantly higher mutation rate was found in isolates from heavily polluted industrial soils, when compared with isolates from non-polluted or naturally polluted soils (Fig. 14.4). These observations are in agreement with the finding, derived from genomic comparisons, that **fungal stress signalling pathways** are evolving rapidly and in a niche-specific fashion to **protect different species against the contrasting environmental stresses** they encounter in their diverse habitats (Nikolaou et al. 2009). Interestingly, in the *O. maius* isolates from polluted sites, most mutations were in the *Sod1* gene promoter region, rather than in the coding sequence (Vallino et al. 2011), suggesting that mutagenesis induced by environmental stress may also target specific gene regions. A similar situation was observed in the promoter region of metallothionein genes in a cadmium tolerant population of the collembola *Orchesella cincta* (Janssens et al. 2007).

B. Cellular and Molecular Mechanisms of Heavy Metal Tolerance

Metals are directly or indirectly involved in all aspects of microbial growth, metabolism and differentiation (Gadd 1993, 2010). Several metals play essential functions in the organisms, and insufficient levels of essential metals can result in stress responses just as severe as those resulting from excess metals. Therefore, cells must have developed mechanisms to maintain metal homeostasis (Tomsett 1993). By contrast, some heavy metals such as Cs, Al, Cd, Hg and Pb have no known functions in most organisms and are therefore toxic at all concentrations. Molecular recognition allows organisms to differentiate between essential and non-essential ions and, if necessary, to partition them in different ways.

The cellular and molecular mechanisms potentially involved in **metal tolerance in fungi**, like in other organisms, can be classified into three groups: (i) ***avoidance* mechanisms** that restrict entry of metal ions into the cytoplasm, and relies on decreased uptake or increased efflux of metal ions, or by their immobilisation outside the hypha by biosorption to cell walls, pigments and extracellular polysaccharides, (ii) ***sequestration* mechanisms** that reduce the concentration of free metal ions in the cytosol and include metal-binding peptides and proteins which regulate metal ion homeostasis, as well as intracellular compartmentation and (iii) ***antioxidative* mechanisms** that allow the fungus to directly or indirectly counteract accumulation of ROS and oxidative stress. Some of these mechanisms are constitutively present, whereas others are only activated when metals exceed a threshold value (Colpaert et al. 2011). Most information on the mechanisms of heavy metal tolerance in mycorrhizal fungi derive from work on ECM fungi (e.g. see Courbot et al. 2004; Bellion et al. 2006; Colpaert et al. 2011), although the most recent work on ERM fungi, described later in this chapter, opens up the possibility to use some ERM isolates as model systems to investigate heavy metal tolerance.

Fungal mycelia often have a high sorption capacity for metals, so binding of metals to the hyphal surface may represent a substantial fraction of the metal accumulated by mycelia (Gadd 1993, 2010; Leyval et al. 1997). Biosorption to cell walls, pigments and extracellular polysaccharides have been reported for ERM fungi (Bradley et al. 1981; Denny and Ridge 1995; Martino et al. 2000a).

Fungal weathering can release essential metals and nutrients from insoluble minerals, but it can also increase final concentrations of toxic metals in the soil. **Biotransformation of solubilised metals into insoluble organic forms** is a **common phenomenon in both ECM and ERM fungi** (Gadd 2010). Biotransformation of Zn was observed in *O. maius* isolates from soils heavily polluted with heavy metals which were found to cause precipitation of organic zinc (Martino et al. 2003). Similarly, *R. ericae* exposed to uranium demonstrated a high tolerance to uranium oxides and was found to form extracellular crystalline precipitates of uranium-containing minerals on the fungal mycelia (Fomina et al. 2007), likely with the involvement of oxalic acid.

Fungi can respond to the presence of metals with the **release of specific proteins in the surrounding medium**. It was found for example that the presence of zinc in the culture medium sharply increased in *O. maius* the secretion and activity of extracellular enzymes that hydrolyse the pectin component of plant cell walls (Martino et al. 2000b). The significance of this increased production is unclear, but oligalacturonans may function as better metal chelators than larger polymers, thus protecting the fungus during saprotrophic growth. Zinc ions also induced in *O. maius* a general change in the array of secreted proteins, with a shift towards the production of more basic, low molecular weight polypeptides (Martino et al. 2002).

Intracellular metal concentrations in fungi may be regulated by transport, including efflux mechanisms and internal compartmentation. **Metal transporters** are major players in keeping the cytosolic concentrations compatible with metabolic activities, and they may either extrude metal ions out of the cell as plasma-membrane transporters, or sequestrate metals in intracellular compartments, usually the vacuole (Fig. 14.5). Most of our knowledge on metal transporters in fungi comes from studies on the yeast *Saccharomyces cerevisiae* (Rutherford and Bird 2004), but both mechanisms have been identified in ECM fungi (e.g. Blaudez et al. 2000; Blaudez and Chalot 2011; Bolchi et al. 2011; Colpaert et al. 2011).

In ERM fungi, **metal transporters involved in metal efflux** were identified by Sharples et al. (2000b, 2001) in arsenate-resistant isolates of *R. ericae*. These isolates accumulated arsenate, which is a phosphate analogue and is transported by the phosphate uptake system. However, compared to isolates from unpolluted sites, *R. ericae* isolates from As/Cu mines had the capacity to reduce arsenate to arsenite, which was rapidly expelled from their mycelium via an increased efflux system (Sharples et al. 2000b, 2001).

Fungi may also decrease cytosolic free metal through the synthesis of a variety of **metal-binding peptides and proteins**. Among them, essential components of Cd detoxification pathways

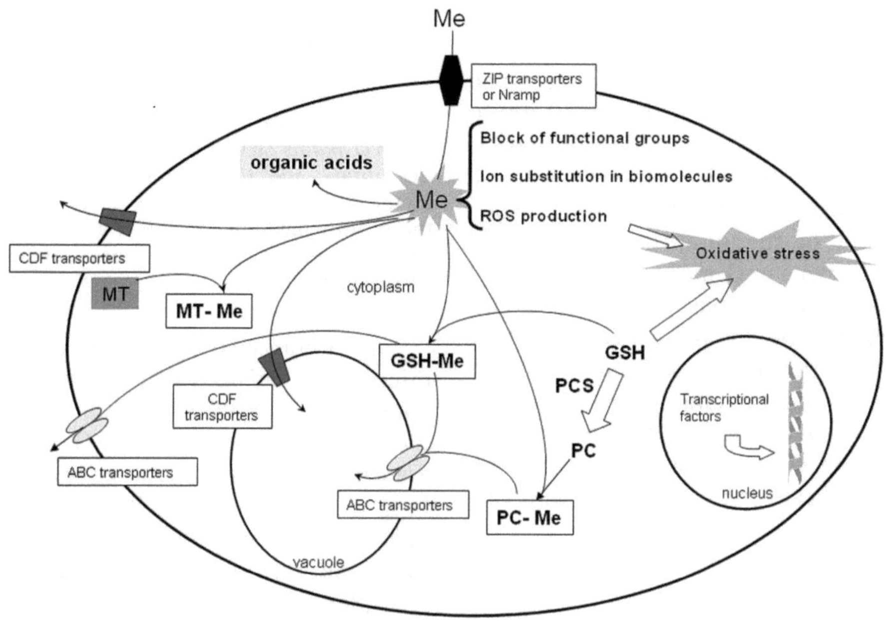

Fig. 14.5. Mechanisms involved in the intracellular detoxification or organellar compartmentation of heavy metals. A variety of mechanisms may be involved in transport phenomena contributing to decreased uptake and/or increased efflux. A variety of specific or non-specific mechanisms may also lead to intracellular chelation and intracellular precipitation, and to ROS detoxification

in various organisms are thiol compounds, including reduced glutathione, phytochelatins, and metallothioneins. **Metallothioneins** have been identified in most types of mycorrhizal fungi: they have been studied in both ECM fungi (Courbot et al. 2004; Bellion et al. 2006; Ramesh et al. 2009; Bolchi et al. 2011) and arbuscular mycorrhizal fungi (Lanfranco et al. 2002; Gonzlez-Guerrero et al. 2007), where they were mostly found to be regulated. A cDNA coding for a Cu-metallothionein was found in the ERM fungus *O. maius* by Vallino et al. (2005), but gene expression was not affected by Zn treatment.

Thiol compounds such as glutathione and phytochelatins are major players in the anti-oxidative response in plants and in some fungi (Courbot et al. 2004; Yadav 2010). There is currently only very scanty information on the thiol-dependent antioxidant systems in ERM fungi, with the identification of a thioredoxin gene in the *O. maius* cDNA library (Vallino et al. 2005). By contrast, the complete thiol-dependent antioxidant systems of the ECM fungus *Laccaria bicolor* (Morel et al. 2008) and of *Tuber melanosporum* (Bolchi et al. 2011) have been recently characterised at the genome scale.

In addition to thiols compounds, fungi can protect themselves from the oxidative stress caused by heavy metals through the synthesis **of enzymatic antioxidants** such as **catalase, peroxidase and superoxide dismutase** (Jacob et al. 2001; Guelfi et al. 2003; Todorova et al. 2008). In *O. maius* Zn, two extracellular proteins showed high homology with superoxide dismutases (SOD), that play a protective role against free O^{2-}radical toxicity, catalysing their conversion in hydrogen peroxide and oxygen (Fridovich 1995). Their induction by heavy metals has been described by several authors in plants, animals and microorganisms (Chongpraditnun et al. 1992; Yoo et al. 1999; Vido et al. 2001). Treatment with high concentrations of zinc ions resulted in an increased amount and activity of both intracellular and extracellular SOD enzymes in an *O. maius* isolate from a polluted site (Martino et al. 2002; Vallino et al. 2009). The Cu/Zn SOD seems to be induced in *O. maius* specifically by zinc, as suggested by a comparative high-throughput proteomic investigation to elucidate common and specific responses of *O.*

maius Zn to zinc and cadmium (Chiapello et al., unpublished data). The increased production of an extracellular Cu/Zn SOD, by binding metals and by performing its specific enzyme activity, could help both the fungus and the host plant to cope with the ROS formed in the extracellular medium, caused by metal pollution. Further demonstration of the important role of the Cu/Zn SOD enzyme in *O. maius* metal tolerance derived from heterologous expression of the corresponding gene in yeast (Vallino et al. 2009) and from the gene knock-out via homologous recombination and gene disruption (Abbà et al. 2009). Compared with the wild-type strain, the *O. maius* SOD1-null mutants showed a significant increase in zinc and cadmium sensitivity.

C. Omics Approaches to the Study of Heavy Metal Tolerance

Investigations on specific components (e.g. transporters, SODs, chelating metabolites) described in the previous paragraphs demonstrate that they may play significant roles in heavy metal tolerance in ERM fungi. However, this targeted approach often relies on existing knowledge acquired in other biological systems, and may miss specific mechanisms operating in the organisms under investigation, or general mechanisms so far unidentified. Untargeted approaches are therefore an important source of novel information, especially if they are supported by functional assays (Ruytinx et al. 2011).

The first attempt to investigate ERM fungal genes involved in zinc metal tolerance was through the sequencing of a small EST collection (Vallino et al. 2005). Variation in **gene expression** after treatment with high concentrations of Zn was monitored on 130 unigenes by reverse Northern blot hybridisation (Fig. 14.6): 16 unigenes were shown to be either up or down regulated. Among the differentially regulated genes, Vallino et al. (2005) could not find any previously reported heavy metal responsive or stress-related genes. Most of the genes identified in the EST library and included in the cell defence category (e.g. Cu-metallothionein, Cu/Zn SOD, ascorbate peroxidase, thioredoxin, heat shock proteins) were not affected by the Zn treatment. This may be because the tolerant strain *O. maius* Zn (whose mRNA was used to construct the library and to hybridise the reverse Northern blots) does not perceive the metal concentration in the growth medium (10 mM Zn) as a serious stress condition, or because these genes may be involved at an earlier stage.

An untargeted approach using **genomic microarray** has been recently applied to *Cadophora finlandica* (Gorfer et al. 2009). *C. finlandica*, as described in Sect. II.C, can potentially associate with both ERM and ECM hosts (Vrålstad et al. 2002b). It is frequently found in heavy metal polluted habitats, and a possible functional role in heavy metal resistance by the host plants has been suggested (Vrålstad et al. 2002a). The genomic microarray analysis was conducted on a metal tolerant isolate of *C. finlandica* derived from an ECM *Salix caprea* root tip growing on a heavily contaminated soil (Dos Santos Utmazian et al. 2007) and compared the expression profiles on media supplemented with different metals (Gorfer et al. 2009). Many of the newly identified heavy metal regulated genes in *C. finlandica* encoded proteins of unknown functions or proteins with no established roles in heavy metal detoxification, but a large set of regulated genes were predicted to encode extracellular or plasma membrane proteins (e.g. transporters or secreted proteins). The majority of heavy metal defence activities in *C. finlandica* is therefore thought to take place outside the cell (Gorfer et al. 2009). It remains to be established if the *C. finlandica* isolate investigated in this study is truly mycorrhizal on ericaceous hosts.

Construction of cDNA libraries from ERM fungi growing on metal-containing medium, coupled with a functional assay in metal sensitive yeast mutants has allowed the identification of novel genes involved in metal tolerance in *O. maius*. In particular, a cDNA library obtained from *O. maius* growing on Cd was transformed in the *S. cerevisiae* cadmium-sensitive mutant *yap1*. Most yeast transformants able to grow on cadmium were found

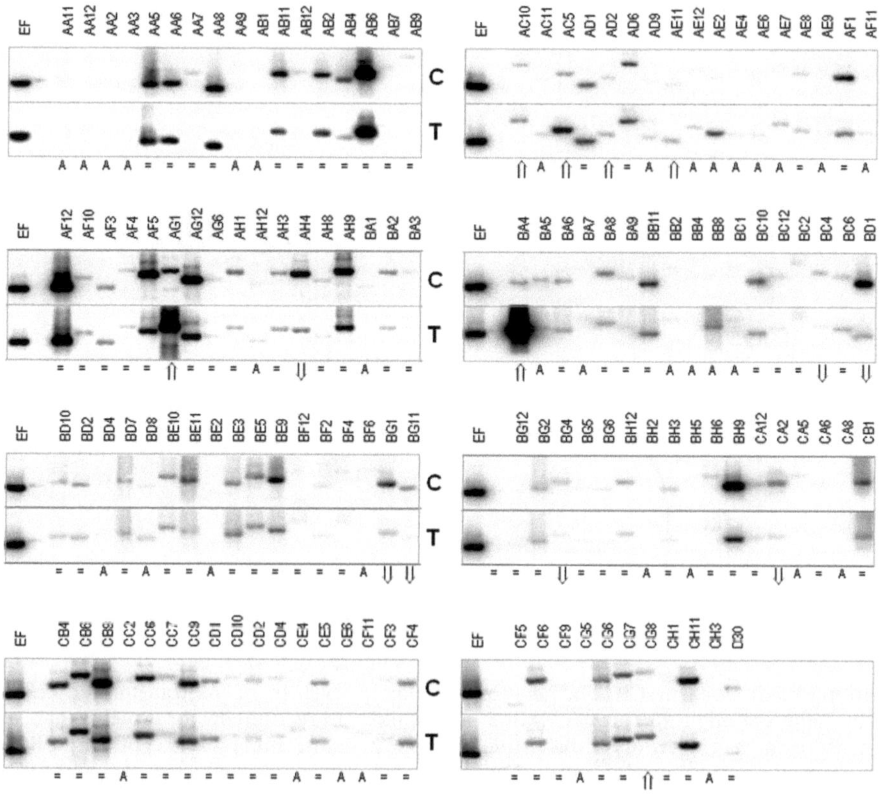

Fig. 14.6. Reverse Northern blot hybridisation of part of the *O. maius* Zn EST unigenes identified by Vallino et al. (2005), with two probes obtained from mycelia grown for 20 days in the absence of Zn (*C*) or with addition of 10 mM Zn (*T*). *EF* Elongation factor 1α (housekeeping gene used for data normalisation), *A* absent signal (ratio signal/background <3), ⇑ up-regulated genes (ratio treated/control >3.0), ⇓ down-regulated genes (ratio treated/control <0.3), = not regulated genes (ratio treated/control between 0.3 and 3.0) (Modified from Vallino et al. (2005))

to harbour an insert coding for the same PLAC8 domain-containing protein, named OmFCR (*O. maius* fungal cadmium resistance) because of the sequence similarity with a plant protein involved in cadmium resistance (Abbà et al. 2011). The PLAC8 domain is widespread and evolutionary conserved in several proteins found in all eukaryotic kingdoms, but has no assigned biological role.

When tested on a number of different metals and stressful conditions, the OmFCR protein was found to confer specific resistance to cadmium. Direct measurement of cadmium in exposed yeast cells indicate no changes in this metal content caused by the presence of OmFCR (Abbà et al. 2011). This result would exclude that OmFCR is either a membrane efflux pump or a heavy metal chelator, as the expression of an efflux pump would decrease the cellular Cd content in the cell, while an intracellular chelator would increase it. The use of the yeast two-hybrid assay, followed by a number of experiments with specific yeast mutants illustrated by Abbà et al. (2011), indicates that OmFCR confers cadmium resistance to yeast cells through its interaction with Mlh3p, a subunit of the mismatch repair (MMR) system. The MMR pathway repairs base-base mismatches and insertion/deletion loops that arise from DNA duplication, as well as mismatches in heteroduplexes that are formed during recombination (Fishel and Kolodner 1995).

Cadmium is known to **bind to the MMR system** and to **reduce its capacity to recognise small misalignments and base—base mismatches** by disrupting its structure and function (Jin et al. 2003). Unlike many other genotoxic metal ions, cadmium, in fact, does not inflict direct damage on DNA, proteins

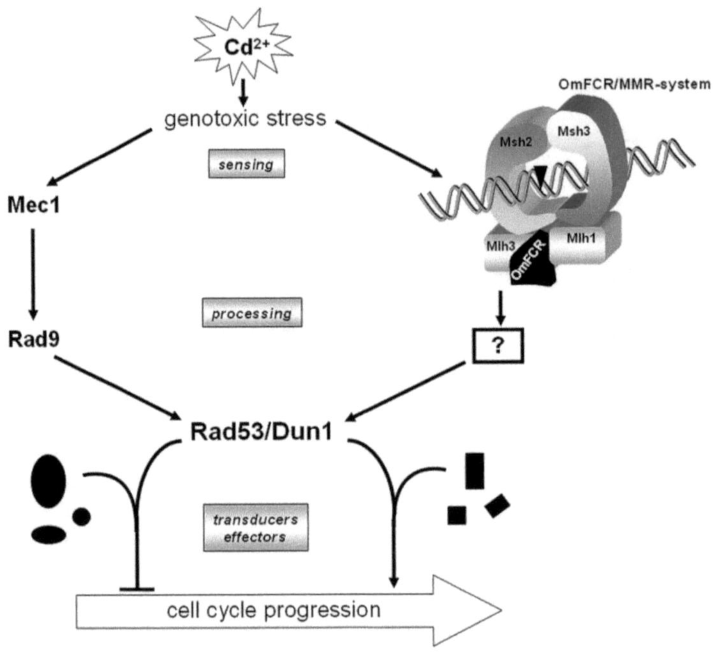

Fig. 14.7. Working model for OmFCR. The genotoxic stress caused by cadmium might recruit the MMR system, which, in its turn, might promote the firing of OmFCR through protein—protein interactions with Mlh3p. The signalling pathway promoted by OmFCR appears to merge with the final part of the Mec1p-dependent phosphorylation cascade, at the Rad53p/Dun1p level. In pFL61-transformed cells (*left*) Dun1p is likely to recruit effector proteins that cause cell cycle arrest, while the presence of OmFCR (*right*) might enlist alternative effector proteins that ultimately allow the progression of cell division (Modified from Abbà et al. (2011))

and lipids through the generation of reactive oxygen species (ROS), but it targets proteins that are directly or indirectly involved in DNA repair and in antioxidant defence, altering their functions and ultimately causing toxic, mutagenic and carcinogenic effects. Although a first hypothesis could be a direct role of OmFCR in the repair of DNA damage caused by cadmium, the experiments described by Abbà et al. (2011) suggest that OmFCR may rather take part to the fairly unexplored role of the MMR system in connecting the recognition of DNA lesions with downstream signalling cascades that ultimately lead to cell cycle checkpoints (Fig. 14.7).

V. Ericoid Mycorrhizal Fungi as Model Systems

Although mycorrhizae are critical elements of terrestrial ecosystems, we have just begun to understand the molecular interactions between mycorrhizal fungi and their host plants. The release of the complete *Laccaria bicolor* and *Tuber melanosporum* genomes (Martin et al. 2008, 2010) and the on-going genome sequencing of additional mycorrhizal fungi by the DOE Joint Genome Institute (http://www.jgi.doe.gov/) are going to provide unprecedented opportunities for studying the biology and the evolution underlying the symbiotic lifestyle. These new fungal genome sequencing programs proposed by Francis Martin within the international project "*Exploring the Genome Diversity of Mycorrhizal Fungi to Understand the Evolution and Functioning of Symbiosis in Woody Shrubs and Trees*" and the following comparison with various pathogenic and saprobic fungal genomes are likely to shed light on the complexity of plant—fungus associations and lead to the identification of the key gene set associated with mycorrhizal symbiosis.

Investigations on specific genes mediating symbiotic events in mycorrhiza formation have hitherto been based exclusively on the study of non-mycorrhizal (Myc-) plant mutants interacting with AM fungi (Parniske 2004). An equivalent development of molecular and genetic approaches is now required at the fungal side to obtain a complete picture of the symbiosis (Helber and Requena 2008). **Gene function assignment will represent a critical step in the identification and the study of genes essential for symbiosis**, but this procedure depends on the development of high-throughput methodologies to test gene function in vivo or in vitro. One of such approaches is the genetic transformation, i.e. the incorporation of exogenous DNA that causes the change or disruption of a gene followed by phenotype alterations.

The first evidence for **genetic transformation** of a mycorrhizal fungus was the successful transformation of the basidiomycete *L. laccata* using a protoplast-PEG-based transformation (Barrett et al. 1990).

Since then, other mycorrhizal species were stably transformed: several ectomycorrhizal (ECM) fungi, such as *Hebeloma cylindrosporum* (Marmeisse et al. 1992; Pardo et al. 2002; Combier et al. 2003), *Paxillus involutus* (Bills et al. 1995; Pardo et al. 2002), *Laccaria bicolor* (Bills et al. 1999; Kemppainen et al. 2005), *Suillus bovinus* (Hanif et al. 2002; Pardo et al. 2002) and *Pisolithus tinctorius* (Rodriguez-Tovar et al. 2005), and one ERM fungus, the ascomycete *Oidiodendron maius* strain Zn (Martino et al. 2007). However, arbuscular mycorrhizal fungi and other ECM fungi, such as *Tuber borchii*, seem to be recalcitrant to stable transformation (Forbes et al. 1998; Grimaldi et al. 2005; Helber and Requena 2008). All these genetic transformations led to the random integration of a selectable marker into the host genome. Random mutagenesis is a powerful approach to the creation of large collections of random mutants which can be then screened for a phenotype of interest. Combier and colleagues, for example, selected ten Myc-transformants blocked at the early stages of ectomycorrhiza formation from a *H. cylindrosporum* collection of random mutants, although the identification and the functional assignment of the mutated genes were not undertaken (Combier et al. 2004).

The most direct ways to explore gene function are generally the complete and stable inactivation of the target gene via homologous recombination at the wild-type locus (gene knock-out) or the down-regulation of its expression by the introduction of an antisense RNA molecule (gene knock-down). As demonstrated by Kemppainen et al. (2009), gene knock-down may represent a reliable alternative to disruption experiments for functional analysis of genes, especially in fungi with multinuclear hyphae, such as *L. bicolor*, or in fungi with a low frequency of homologous recombination. Yet, gene silencing does not ensure the complete suppression of the target gene expression as opposed to null mutants.

Although **gene disruption in mycorrhizal fungi** is limited by the high frequency of ectopic integrations of the transforming DNA molecule, the first targeted gene inactivation in a mycorrhizal fungus has proved to be a successful approach to study genes involved in the mycorrhization process (Abbà et al. 2009). The stable inactivation of the Cu/Zn superoxide dismutase (SOD1) gene by *Agrobacterium*-mediated transformation (AMT) has demonstrated, in fact, that in addition to a general role as an anti-oxidant enzyme, **SOD1 is involved in the *O. maius* morphogenetic responses to the symbiotic partner.**

SODs play a role in both pathogenic and symbiotic interactions between fungi and other organisms (Scott and Eaton 2008). An up-regulation of fungal ROS scavenger machinery during the early stages of the interaction has been demonstrated in pathogenic fungi (Hwang et al. 2002; Cox et al. 2003; Brown et al. 2008), as well as in arbuscular (Lanfranco et al. 2005) and ectomycorrhizal (Baptista et al. 2007) fungi. Cu/Zn SOD is the first fungal protein with a demonstrated role in the establishment of the ERM symbiosis, and the disruption of the corresponding gene is proposed to cause an imbalance in the redox homeostasis during host colonisation and an alteration in the delicate dialogue between *O. maius* and its host plant (Abbà et al. 2009).

Besides the possibility of gene deletion by homologous recombination, *O. maius* can be easily transformed by **random insertional mutagenesis.** A collection of more than 2000 random mutants is already available to be screened for the phenotype of interest and transformation with plasmid vectors for GFP tagging has been successfully used for protein localisation (Abbà, personal communication).

O. maius can be easily grown in vitro, where it reproduces asexually by forming conidia with just a single haploid nucleus. Moreover, the haploid monokaryotic status implies that the modifications of the wild-type phenotype, if detectable, can be observed even when mutations are recessive. This characteristic makes this fungus a good candidate to study mutants, because uninucleated spores can germinate and produce a homokaryotic mycelium with all the nuclei carrying the mutation.

Fungal genes potentially implicated in the mycorrhization process have been so far studied in the non-mycorrhizal filamentous fungus *Magnaporthe oryzae* (Heupel et al. 2010; Kloppholz et al. 2011). Although there are several types of mycorrhiza in terms of ecological significance and host—fungus association, a mycorrhizal fungus like *O. maius* might represent a more suitable heterologous system for studying fungal genes involved in the symbiosis. The feasibility of stable transformations and targeted gene disruptions in *O. maius* might open new possibilities to study the biological role of symbiosis-regulated genes, especially of those encoding effector-like secreted proteins, which have been hypothesised to interact with host plant proteins or alter their expression during the symbiosis interaction (Martin et al. 2008).

VI. Conclusions

A classical view is that distinct mycorrhizal types exist in nature, each with a well described morphology of the plant—fungus association (Smith and Read 1997) and a distinct range of symbiotic fungi. Although the classification into mycorrhizal types has been important to describe the diversity of symbiotic plant—fungus associations, it has lead to the view of a sharp separation among endo- and ectomycorrhizae. However, starting from the observation of common fungal taxa in ECM and ERM plants in nature (Bergero et al. 2000; Vrålstad et al. 2000, 2002a; Grelet et al. 2010), it was shown that the same fungus isolated from ECM roots could form ERM on ericaceous hosts and improve plant growth and/or nutrition (Villarreal-Ruiz et al. 2004; Grelet et al. 2009a). Although the range of potential interactions of ERM fungi is a delicate issue still open to debate, the observations on *M. bicolor* (Villarreal-Ruiz et al. 2004) indicate, at least for some ERM fungi, a phenotypic plasticity in their interactions with plants of different mycorrhizal status and contribute to weaken the boundaries between the classical mycorrhizal types. Similar conclusions also derive from studies on the taxonomic diversity of symbiotic fungi in other mycorrhizal symbioses. For example, well known ECM fungi have been found to associate with mycoheterotrophic and mixotrophic forest orchids, where they form typical endomycorrhizal coils. This phenotypic plasticity of ECM fungi in orchids has been demonstrated for both ascomycetes (e.g. Selosse et al. 2004) and basidiomycetes (e.g. Selosse et al. 2002; Girlanda et al. 2006).

There is increasing evidence that fungi taxonomically related to ERM fungi are common associates of plants with different mycorrhizal status (e.g. Chambers et al. 2008; Ohtaka and Narisawa 2008; Tedersoo et al. 2009), with a continuum from loose non-mycorrhizal to mycorrhizal associations between ericoid fungi and host plants, possibly related to environmental conditions and to the host plant species. It becomes now important to investigate whether these potential hyphal connections, irrespective of the morphology of fungus—plant interactions, do contribute to a functional mycelial network that may allow nutrient transfer in a mixed plant community, similar (or perhaps even connected) to the wood-wide web first demonstrated for ECM plants by Simard et al. (1997).

All ERM fungi appear to be well equipped with an arsenal of hydrolytic enzymes that are instrumental in the litter decomposition in heathlands as well as in woodlands.

Given the ECM potential of *M. bicolor* isolates, it is quite interesting that the genomic sequencing of the ECM fungus *Laccaria bicolor* indicates a strong reduction in the number of genes involved in the production of hydrolytic enzymes, as compared to saprotrophic fungi (Martin and Selosse 2008). However, it should be noted that some ECM forming ascomycetes, such as *Sphaerosporella brunnea* and *Geopyxis carbonaria*, that colonise seedlings particularly after fire, effectively degrade cellulose, produce phenol oxidase and degrade lignin (see Egger 2006). A genomic investigation of wider range of ERM and ECM fungi will be important to provide a more comprehensive understanding of the saprotrophic potential of these ecto- and endomycorrhizal fungi.

Biochemical, physiological and molecular studies have allowed us to gain better insights on the mechanisms of stress tolerance of ERM fungi, especially towards heavy metals (see Sect. IV). The identification of these mechanisms provides us with some clues to better understand the success observed for members of the "*R. ericae* aggregate" and *O. maius* in colonising heavy metal polluted soils. More mechanisms likely enable ERM fungi to protect themselves and their host plant from toxic compounds, and will hopefully emerge from genome wide approaches as well as from comparisons between metal tolerant and metal sensitive ERM strains (Murat et al. 2011).

References

Abbà S, Khouja HR, Martino E, Archer DB, Perotto S (2009) SOD1-Targeted gene disruption in the ericoid mycorrhizal fungus *Oidiodendron maius* reduces conidiation and the capacity for mycorrhization. Mol Plant Microbe Interact 22:1412–1421

Abbà S, Vallino M, Daghino S, Di Vietro L, Borriello R, Perotto S (2011) A PLAC8-containing protein from an endomycorrhizal fungus confers cadmium resistance to yeast cells by interacting with Mlh3p. Nucleic Acids Res 39:7548–7563

Abuarghub SM, Read DJ (1988) The biology of mycorrhizal in the Ericaceae. XI. The distribution of nitrogen in soil of a typical upland callunetum with special reference to the free amino-acids. New Phytol 108:425–431

Abuzinadah RA, Read DJ (1989) The role of proteins in the nitrogen nutrition of ectomycorrhizal plants. V. Nitrogen transfer in birch (*Betula pendula*) grown in association with mycorrhizal and non-mycorrhizal fungi. New Phytol 112:61–68

Adriaensen K, Vrålstad T, Noben JP, Vangronsveld J, Colpaert JV (2005) Copper adapted *Suillus luteus*, a symbiotic solution for pines colonising Cu mine spoil. Appl Environ Microbiol 11:7279–7284

Alberton O, Kuyper TW, Summerbell RC (2009) Dark septate root endophytic fungi increase growth of Scots pine seedlings under elevated CO_2 through enhanced nitrogen use efficiency. Plant Soil 328:459–470

Allen TR, Millar T, Berch SM, Berbee ML (2003) Culturing and direct DNA extraction find different fungi from the same ericoid mycorrhizal roots. New Phytol 160:255–272

Bajwa R, Read DJ (1986) Utilization of mineral and amino N sources by the ericoid mycorrhizal endophyte *Hymenoscyphus ericae* and by mycorrhizal and non-mycorrhizal seedlings of *Vaccinium*. Trans Br Mycol Soc 87:269–277

Baptista P, Martins A, Pais MS, Tavares RM, Lino-Neto T (2007) Involvement of reactive oxygen species during early stages of ectomycorrhiza establishment between *Castanea sativa* and *Pisolithus tinctorius*. Mycorrhiza 17:185–193

Barrett V, Dixon RK, Lemke PA (1990) Genetic transformation of a mycorrhizal fungus. Appl Microbiol Biotechnol 33:313–316

Bellion M, Courbot M, Jacob C, Blaudez D, Chalot M (2006) Extracellular and cellular mechanisms sustaining metal tolerance in ectomycorrhizal fungi. FEMS Microbiol Lett 254:173–181

Bending GD, Read DJ (1996a) Effects of the soluble polyphenol tannic acid on the activities of ericoid and ectomycorrhizal fungi. Soil Biol Biochem 28:1595–1602

Bending GD, Read DJ (1996b) Nitrogen mobilization from protein-polyphenol complex by ericoid and ectomycorrhizal fungi. Soil Biol Biochem 28:1603–1612

Bending GD, Read DJ (1997) Lignin and soluble-phenolic degradation by ectomycorrhizal and ericoid mycorrhizal fungi. Mycol Res 101:1348–1354

Berch SM, Allen TR, Berbee ML (2002) Molecular detection, community structure and phylogeny of ericoid mycorrhizal fungi. Plant Soil 244:55–66

Bergero R, Perotto S, Girlanda M, Vidano G, Luppi AM (2000) Ericoid mycorrhizal fungi are common root associates of a Mediterranean ectomycorrhizal plant (*Quercus ilex*). Mol Ecol 9:1639–1649

Bidartondo MI, Bruns TD, Weiss M, Sergio C, Read DJ (2003) Specialized cheating of the ectomycorrhizal symbiosis by an epiparasitic liverwort. Proc Biol Sci R Soc 270:835–842

Bills SN, Richter DL, Podila GK (1995) Genetic transformation of the ectomycorrhizal fungus *Paxillus involutus* by particle bombardment. Mycol Res 99:557–561

Bills SN, Podila GK, Hiremath S (1999) Genetic engineering of an ectomycorrhizal fungus *Laccaria bicolor* for use as a biological control agent. Mycologia 91:237–242

Blaudez D, Chalot M (2011) Characterization of the ER-located zinc transporter ZnT1 and identification of a vesicular zinc storage compartment in *Hebeloma cylindrosporum*. Fungal Genet Biol 48:496–503

Blaudez D, Botton B, Chalot M (2000) Cadmium uptake and subcellular compartmentation in the ectomycorrhizal fungus *Paxillus involutus*. Microbiology 146:1109–1117

Bolchi A, Ruotolo R, Marchini G, Vurro E, Sanità di Toppi L, Kohler A, Tisserant E, Martin F, Ottonello S (2011) Genome-wide inventory of metal homeostasis-related gene products including a functional phytochelatin synthase in the hypogeous mycorrhizal fungus *Tuber melanosporum*. Fungal Genet Biol 48:573–584

Bonfante-Fasolo P (1980) Occurrence of a basidiomycetes in living cells of mycorrhizal hair roots of *Calluna vulgaris*. Trans Br Mycol Soc 75:320–325

Bonfante-Fasolo P, Gianinazzi-Pearson V (1979) Ultrastructural aspects of endomycorrhiza in the Ericaceae. I. Naturally infected hair roots of *Calluna vulgaris* L. Hull. New Phytol 83:739–744

Bougoure DS, Cairney JWG (2005a) Assemblages of ericoid mycorrhizal and other root-associated fungi from *Epacris pulchella* (Ericaceae) as determined by culturing and direct DNA extraction from roots. Environ Microbiol 7:819–827

Bougoure DS, Cairney JWG (2005b) Fungi associated with hair roots of *Rhododendron lochiae* (Ericaceae) in an Australian tropical cloud forest revealed by culturing and culture-independent molecular methods. Environ Microbiol 7:1743–1754

Bougoure DS, Cairney JWG (2006) Chitinolytic activities of ericoid mycorrhizal and other root-associated fungi from *Epacris pulchella* (Ericaceae). Mycol Res 110:328–334

Bougoure DS, Parkin PI, Cairney JWG, Alexander IJ, Anderson IC (2007) Diversity of fungi in hair roots of Ericaceae varies along a vegetation gradient. Mol Ecol 16:4624–4636

Bradley R, Burt AJ, Read DJ (1981) Mycorrhizal infection and resistance to heavy metal toxicity in *Calluna vulgaris*. Nature 292:335–337

Bradley R, Burt AJ, Read DJ (1982) The biology of mycorrhiza in the Ericaceae VIII. The role of the mycorrhizal infection in heavy metal resistance. New Phytol 91:197–209

Braun-Lüllemann A, Hüttermann A, Majcherczyk A (1999) Screening of ectomycorrhizal fungi for degradation of polycyclic aromatic hydrocarbons. Appl Microbiol Biotechnol 53:127–132

Brown SH, Yarden O, Gollop N, Chen S, Zveibil A, Belausov E, Freeman S (2008) Differential protein expression in *Colletotrichum acutatum*: changes associated with reactive oxygen species and nitrogen starvation implicated in pathogenicity on strawberry. Mol Plant Pathol 9:171–190

Burke RM, Cairney JWG (1997a) Carbohydrolase production by the ericoid mycorrhizal fungus *Hymenoscyphus ericae* under solid state fermentation conditions. Mycol Res 101:1135–1139

Burke RM, Cairney JWG (1997b) Purification and characterisation of a β-1,4-endoxylanase from the ericoid mycorrhizal fungus *Hymenoscyphus ericae*. New Phytol 135:345–352

Burke RM, Cairney JWG (1998) Carbohydrate oxidases in ericoid and ectomycorrhizal fungi: a possible source of Fenton radicals during the degradation of lignocellulose. New Phytol 139:637–645

Burke RM, Cairney JWG (2002) Laccases and other polyphenol oxidases in ecto- and ericoid mycorrhizal fungi. Mycorrhiza 12:105–116

Cairney JWG, Burke RM (1998) Extracellular enzyme activities of the ericoid mycorrhizal endophyte *Hymenoscyphus ericae* (Read) Korf & Kernan: their likely roles in decomposition of dead plant tissue in soil. Plant Soil 205:181–192

Cairney JWG, Meharg AA (2003) Ericoid mycorrhiza: a partnership that exploits harsh edaphic conditions. Eur J Soil Sci 54:735–740

Cairney JWG, Sawyer NA, Sharples JM, Meharg AA (2000) Intraspecific variation in nitrogen source utilisation by isolates of the ericoid mycorrhizal fungus *Hymenoscyphus Ericae* (Read) Korf and Kernan. Soil Biol Biotechnol 32:1319–1322

Chalot M, Brun A (1998) Physiology of organic nitrogen acquisition by ectomycorrhizal fungi and ectomycorrhizas. FEMS Microbiol Rev 22:21–44

Chambers SM, Williams PG, Seppelt RD, Cairney JWG (1999) Molecular identification of *Hymenoscyphus* sp. From rhizoids of the leafy liverwort *Cephaloziella exiliflora* (Tayl.) Steph. In Australia and Antarctica. Mycol Res 103:286–288

Chambers SM, Liu G, Cairney JWG (2000) ITS rDNA sequence comparison of ericoid mycorrhizal endophytes from *Woollsia pungens*. Mycol Res 104:168–174

Chambers SM, Curlevski NJ, Cairney JWG (2008) Ericoid mycorrhizal fungi are common root inhabitants of non-Ericaceae plants in a south-eastern Australian sclerophyll forest. FEMS Microbiol Ecol 65:263–270

Chen A, Chambers SM, Cairney JWG (1999) Utilisation of organic nitrogen and phosphorus sources by mycorrhizal endophytes of *Woollsia pungens* (Cav.) F. Muell. (Epacridaceae). Mycorrhiza 8:181–187

Chongpraditnun P, Mori S, Chino M (1992) Excess copper induces a cytosolic Cu, Zn-superoxide dismutase in soybean root. Plant Cell Physiol 33:239–244

Colpaert JV, Muller LAH, Lambaerts M, Adriaensen K, Vangronsveld J (2004) Evolutionary adaptation to Zn toxicity in populations of suilloid fungi. New Phytol 162:549–559

Colpaert JV, Wevers JHL, Krznaric E, Adriaensen K (2011) How metal-tolerant ecotypes of ectomycorrhizal fungi protect plants from heavy metal pollution. Ann For Sci 68:17–24

Combier JP, Melayah D, Raffier C, Gay G, Marmeisse R (2003) *Agrobacterium tumefaciens*-mediated transformation as a tool for insertional mutagenesis in the symbiotic ectomycorrhizal fungus *Hebeloma cylindrosporum*. FEMS Microbiol Lett 220:141–148

Combier JP, Melayah D, Raffier C, Pepin R, Marmeisse R, Gay G (2004) Nonmycorrhizal (Myc-) mutants of *Hebeloma cylindrosporum* obtained through insertional mutagenesis. Mol Plant Microbe Interact 17:1029–1038

Cosgrove DJ (1967) Metabolism of organic phosphates in soil. In: McLaren AD, Peterson GH (eds) Soil biochemistry. Dekker, New York, pp 216–228

Courbot M, Diez L, Ruotolo R, Chalot M, Leroy P (2004) Cadmium-responsive thiols in the ectomycorrhizal fungus *Paxillus involutus*. Appl Environ Microbiol 70:7413–7417

Couture M, Fortin JA, Dalpé Y (1983) *Oidiodendron griseum* Robak: an endophyte of ericoid mycorrhiza in *Vaccinium* spp. New Phytol 95:375–380

Cox GM, Harrison TS, McDade HC, Taborda CP, Heinrich G, Casadevall A, Perfect JR (2003) Superoxide dismutase influences the virulence of *Cryptococcus neoformans* by affecting growth within macrophages. Infect Immun 71:173–180

Curlevski NJA, Chambers SM, Anderson IC, Cairney JWG (2009) Identical genotypes of an ericoid mycorrhiza-forming fungus occur in roots of *Epacris pulchella* (Ericaceae) and *Leptospermum polygalifolium* (Myrtaceae) in an Australian sclerophyll forest. FEMS Microbiol Ecol 67:411–420

Dalpé Y (1986) Axenic synthesis of ericoid mycorrhiza in *Vaccinium angustifolium* Ait. by *Oidiodendron* species. New Phytol 103:391–396

Dalpé Y (1989) Ericoid mycorrhizal fungi in the Myxotrichaceae and Gymnoascaceae. New Phytol 113:523–527

Dalpé Y, Litten W, Sigler L (1989) *Scytalidium vaccinii* sp. nov., an ericoid endophyte of *Vaccinium angustifolium* roots. Mycotaxon 35:371–377

Denny HJ, Ridge I (1995) Fungal slime and its role in the mycorrhiza amelioration of zinc toxicity to higher plants. New Phytol 130:251–257

Devevre O, Garbaye J, Botton B (1996) Release of complexing organic acids by rhizosphere fungi as a factor in Norway Spruce yellowing in acidic soils. Mycol Res 100:1367–1374

Dighton J, Thomas ED, Latter PM (1987) Interactions between tree roots, mycorrhizas, a saprotrophic fungus and the decomposition of organic substrates in a microcosm. Biol Fertil Soils 4:145–150

Dos Santos Utmazian MN, Schweiger P, Sommer P, Gorfer M, Strauss J (2007) Influence of *Cadophora finlandica* and other microbial treatments on cadmium and zinc uptake in willows grown on polluted soil. Plant Soil Environ 53:158–166

Duckett JG, Read DJ (1995) Ericoid mycorrhizas and rhizoid-ascomycete associations in liverworts share the same mycobiont: isolation of the partners and resynthesis of the associations in vitro. New Phytol 129:439–447

Duddridge J, Read DJ (1982) An ultrastructural analysis of the development of mycorrhizas in *Rhododendron pontium*. Can J Bot 60:2345–2356

Eaton GK, Ayres MP (2002) Plasticity and constraint in growth and protein mineralization of ectomycorrhizal fungi under simulated nitrogen deposition. Mycologia 94:921–932

Egger KN (2006) The surprising diversity of ascomycetous mycorrhizas. New Phytol 170:421–423

Egger KN, Sigler L (1993) Relatedness of the ericoid endophytes *Scytalidium vaccinii* and *Hymenoscyphus ericae* inferred from analysis of ribosomal DNA. Mycologia 85:219–230

Englander L, Hull RJ (1980) Reciprocal transfer of nutrients between ericaceous plants and a *Clavaria* sp. New Phytol 84:661–667

Finlay RD (2008) Ecological aspects of mycorrhizal symbiosis: with special emphasis on the functional diversity of interactions involving the extraradical mycelium. J Exp Bot 59:1115–1126

Fishel R, Kolodner RD (1995) Identification of mismatch repair genes and their role in the development of cancer. Curr Opin Genet Dev 5:382–395

Fomina M, Charnock JM, Hillier S, Alvarez R, Gadd GM (2007) Fungal transformations of uranium oxides. Environ Microbiol 9:1696–1710

Forbes PJ, Millian S, Hooker JE, Harrier LA (1998) Transformation of the arbuscolar mycorrhiza *Gigaspora rosea* by particle bombardment. Mycol Res 102:497–501

Fridovich I (1995) Superoxide radicals and superoxide dismutase. Annu Rev Biochem 5:321–324

Gadd GM (1993) Interactions of fungi with toxic metals. New Phytol 124:25–60

Gadd GM (2010) Metals, minerals and microbes: geomicrobiology and bioremediation. Microbiology 156:609–643

Genney DR, Alexander IJ, Hartley SE (2000) Exclusion of grass roots from soil organic layers by *Calluna*: the role of ericoid mycorrhizas. J Exp Bot 51:1117–1125

Gibson BR, Mitchell DT (2004) Nutritional influences on the solubilization of metal phosphate by ericoid mycorrhizal fungi. Mycol Res 108:947–954

Gibson BR, Mitchell DT (2005) Phosphatases of ericoid mycorrhizal fungi: kinetic properties and the effect of copper on activity. Mycol Res 109:478–486

Girlanda M, Selosse MA, Cafasso D, Brilli F, Delfine S, Fabbian R, Ghignone S, Pinelli P, Segreto R, Loreto F, Cozzolino S, Perotto S (2006) Inefficient photosynthesis in the Mediterranean orchid *Limodorum abortivum* is mirrored by specific association to ectomycorrhizal Russulaceae. Mol Ecol 15:491–504

Gonzlez-Guerrero M, Cano C, Azcón-Aguilar C, Ferrol N (2007) GintMT1 encodes a functional metallothionein in *Glomus intraradices* that responds to oxidative stress. Mycorrhiza 17:327–335

Gorfer M, Persak H, Berger H, Brynda S, Bandian D, Strauss J (2009) Identification of heavy metal regulated genes from the root associated ascomycete *Cadophora finlandica* using a genomic microarray. Mycol Res 113:1377–1388

Grelet GA, Meharg AA, Alexander IJ (2005) Carbon availability affects nitrogen source utilisation by *Hymenoscyphus ericae*. Mycol Res 109:469–477

Grelet GA, Johnson D, Paterson E, Anderson IC, Alexander IJ (2009a) Reciprocal carbon and nitrogen transfer between an ericaceous dwarf shrub and fungi isolated from *Piceirhiza bicolorata* ectomycorrhizas. New Phytol 182:359–366

Grelet GA, Meharg AA, Duff EI, Anderson IC, Alexander IJ (2009b) Small genetic differences between ericoid mycorrhizal fungi affect nitrogen uptake by *Vaccinium*. New Phytol 181:708–718

Grelet GA, Johnson D, Vrålstad T, Alexander IJ, Anderson IC (2010) New insights into the mycorrhizal *Rhizoscyphus ericae* aggregate: spatial structure and co-colonization of ectomycorrhizal and ericoid roots. New Phytol 188:210–222

Griffiths RP, Caldwell BA (1992) Mycorrhizal mat communities in forest soils. In: Read DJ, Lewis DH, Fitter AH, Alexander IJ (eds) Mycorrhizas in ecosystems. CAB International, Wallingford, pp 98–105

Grimaldi B, de Raaf MA, Filetici P, Ottonello S, Ballario P (2005) *Agrobacterium*-mediated gene transfer and enhanced green fluorescent protein visualization in the mycorrhizal ascomycete *Tuber borchii*: a first step towards truffle genetics. Curr Genet 48:69–74

Guelfi A, Azevedo RA, Lea PJ, Molina SMG (2003) Growth inhibition of the filamentous fungus *Aspergillus nidulans* by cadmium: an antioxidant enzyme approach. J Gen Appl Microbiol 49:63–73

Gunther H, Perner B, Gramss G (1998) Activities of phenol oxidizing enzymes of ectomycorrhizal fungi in axenic culture and in symbiosis with Scots pine (*Pinus sylvestris* L.). J Basic Microbiol 38:197–206

Hambleton S, Currah RS (1997) Fungal endophytes from the roots of alpine and boreal Ericaceae. Can J Bot 75:1570–1581

Hambleton S, Sigler L (2005) Meliniomyces, a new anamorph genus for root-associated fungi with phylogenetic affinities to *Rhizoscyphus ericae* (*Hymenoscyphus ericae*), Leotiomycetes. Stud Mycol 53:1–27

Hambleton S, Egger KN, Currah RS (1998) The genus *Oidiodendron*: species delimitation and phylogenetic relationships based on nuclear ribosomal DNA analysis. Mycologia 90:854–868

Hanif M, Pardo A, Gorfer M, Raudaskoski M (2002) T-DNA transfer and integration in the ectomycorrhizal fungus *Suillus bovinus* using hygromycin B as a selectable marker. Curr Genet 41:183–188

Hartwig A (1995) Current aspects in metal genotoxicity. Biometals 8:3–11

Helber N, Requena N (2008) Expression of the fluorescence markers DsRed and GFP fused to a nuclear localization signal in the arbuscolar mycorrhizal fungus *Glomus intraradices*. New Phytol 177:537–548

Heupel S, Roser B, Kuhn H, Lebrun MH, Villalba F, Requena N (2010) Erl1, a novel Era-Like GTPase from *Magnaporthe oryzae*, is required for full root virulence and is conserved in the mutualistic symbiont *Glomus intraradices*. Mol Plant Microbe Interact 23:67–81

Hobbie EA (2006) Carbon allocation to ectomycorrhizal fungi correlates with belowground allocation in culture studies. Ecology 87:563–569

Hutchison LJ (1990) Studies on the systematics of ectomycorrhizal fungi in axenic culture. III. Patterns of polyphenol oxidase activity. Mycologia 82:424–435

Hutton BJ, Dixon KW, Sivasithamparam K (1994) Ericoid endophytes of Western Australian heaths (Epacridaceae). New Phytol 127:557–566

Hwang CS, Rhie GE, Oh JH, Huh WK, Yim HS, Kang SO (2002) Copper- and zinc-containing superoxide dismutase (Cu/ZnSOD) is required for the protection of *Candida albicans* against oxidative stresses and the expression of its full virulence. Microbiology 148:3705–3713

Jacob C, Courbot M, Brun A, Steinman HM, Jacquot J-P, Botton B, Chalot M (2001) Molecular cloning, characterization and regulation by cadmium of a superoxide dismutase from the ectomycorrhizal fungus *Paxillus involutus*. Eur J Biochem 268:3223–3232

Jalal MAF, Read DJ (1983) The organic acid composition of *Calluna* heathland soil with special reference to phyto- and fungi-toxicity. I. Isolation and identification of organic acids. Plant Soil 70:257–272

Janssens TKS, Mariën J, Cenijn P, Legler J, van Straalen NM, Dick RD (2007) Recombinational micro-evolution of functionally different metallothionein promoter alleles from *Orchesella cincta*. BMC Evol Biol 7:88–107

Jin YH, Clark AB, Slebos RJC, Al-Refai H, Taylor JA, Kunkel TA, Resnick MA, Gordenin DA (2003) Cadmium is a mutagen that acts by inhibiting mismatch repair. Nat Genet 34:326–329

Jongmans AG, van Breemen N, Lundström US, van Hees PAW, Finlay RD, Srinivasan M, Unestam T, Giesler R, Melkerud P-A, Olsson M (1997) Rock-eating fungi. Nature 389:682–683

Jumpponen A (2001) Dark septate endophytes – are they mycorrhizal? Mycorrhiza 11:207–211

Jumpponen A, Trappe JM (1998) Dark septate endophytes: a review of facultative biotrophic root-colonizing fungi. New Phytol 140:295–310

Kemppainen M, Circosta A, Tagu D, Martin F, Pardo A (2005) *Agrobacterium*-mediated transformation of the ectomycorrhizal symbiont *Laccaria bicolor* S238N. Mycorrhiza 16:19–22

Kemppainen M, Duplessis S, Martin F, Pardo AG (2009) RNA silencing in the model mycorrhizal fungus *Laccaria bicolor*: gene knock-down of nitrate reductase results in inhibition of symbiosis with *Populus*. Environ Microbiol 11:1878–1896

Kerley SJ, Read DJ (1995) The biology of mycorrhiza in the Ericaceae. XVIII. Chitin degradation by *Hymenoscyphus ericae* and transfer of chitin-nitrogen to the host plant. New Phytol 131:369–375

Kerley SJ, Read DJ (1997) The biology of mycorrhiza in the Ericaceae. XIX. Fungal mycelium as a nitrogen source for the ericoid mycorrhizal fungus *Hymenoscyphus ericae* and its host plants. New Phytol 136:691–701

Kernan MJ, Finocchio F (1983) A new discomycetes associated with the roots of *Monotropa uniflora* (Ericaceae). Mycologia 75:916–920

Kielland K (1994) Amino acid absorption by arctic plants: implications for plant nutrition and nitrogen cycling. Ecology 75:2373–2383

Kjøller R, Olsrud M, Michelsen A (2010) Co-existing ericaceous plant species in a subarctic mire community share fungal root endophytes. Fungal Ecol 3:205–214

Kloppholz S, Kuhn H, Requena N (2011) A secreted fungal effector of *Glomus intraradices* promotes symbiotic biotrophy. Curr Biol 21:1204–1209

Kosola KR, Workmaster BAA, Spada PA (2007) Inoculation of cranberry (*Vaccinium macrocarpon*) with the ericoid mycorrhizal fungus *Rhizoscyphus ericae* increases nitrate influx. New Phytol 176:184–196

Kron KA, Judd WS, Stevens PF, Crayn DM, Anderberg AA, Gadek PA, Quinn CJ, Luteyn JL (2002) Phylogenetic classification of Ericaceae: molecular and morphological evidence. Bot Rev 68:335–423

Krznaric E, Wevers JHL, Cloquet C, Vangronsveld J, Vanhaecke F, Colpaert JV (2009) Zn pollution counteracts Cd toxicity in metal-tolerant ectomycorrhizal fungi and their host plant, *Pinus sylvestris*. Environ Microbiol 12:2133–2141

Lacourt I, Girlanda M, Perotto S, Del Pero M, Zuccon D, Luppi AM (2002) Nuclear ribosomal sequence analysis of *Oidiodendron*: towards a redefinition of ecologically relevant species. New Phytol 149:565–576

Lanfranco L, Bolchi A, Ros EC, Ottonello S, Bonfante P (2002) Differential expression of a metallothionein gene during the presymbiotic versus the symbiotic phase of an arbuscular mycorrhizal fungus. Plant Physiol 130:58–67

Lanfranco L, Novero M, Bonfante P (2005) The mycorrhizal fungus *Gigaspora margarita* possesses a CuZn superoxide dismutase that is up-regulated during symbiosis with legume hosts. Plant Physiol 137:1319–1330

Leake JR, Miles W (1996) Phosphodiesters as mycorrhizal P sources. I. Phosphodiesterase production and the utilization of DNA as a phosphorus source by the ericoid mycorrhizal fungus *Hymenoscyphus ericae*. New Phytol 132:435–443

Leake JR, Read DJ (1989) The effect of phenolic compounds on nitrogen mobilisation by ericoid mycorrhizal systems. Agric Ecosyst Environ 29:225–236

Leake JR, Read DJ (1990a) Chitin as a nitrogen source for mycorrhizal fungi. Mycol Res 94:993–995

Leake JR, Read DJ (1990b) Proteinase activity in mycorrhizal fungi. I. The effect of extracellular pH on the production and activity of proteinase by ericoid endophytes from soils of contrasted pH. New Phytol 115:243–250

Leake JR, Read DJ (1990c) Proteinase activity in mycorrhizal fungi. II. The effects of mineral and organic nitrogen sources on induction of extracellular proteinase in *Hymenoscyphus ericae* (Read) Korf and Kernan. New Phytol 116:123–128

Leake JR, Read DJ (1991) Experiments with ericoid mycorrhiza. In: Norris JR, Read DJ, Varma AK (eds) Methods in microbiology, vol 23. Elsevier, Oxford, pp 435–459

Leake JR, Read DJ (1997) Mycorrhizal fungi in terrestrial habitats. In: Wicklow D, Söderström B (eds) The mycota IV: environmental and microbial relationships. Springer, Berlin Heidelberg New York, pp 281–301

Leyval C, Turnau K, Haselwandter K (1997) Effect of heavy metal pollution on mycorrhizal colonization and function: physiological, ecological and applied aspects. Mycorrhiza 7:139–153

Liu G, Chambers SM, Cairney JWG (1998) Molecular diversity of ericoid mycorrhizal endophytes isolated from *Woollsia pungens*. New Phytol 140:145–153

Lundström US, Van Breemen N, Bain D (2000) The podzolization process. A review. Geoderma 94:91–107

Mandyam K, Jumpponen A (2005) Seeking the elusive function of the root-colonising dark septate endophytic fungi. Stud Mycol 53:173–189

Marmeisse R, Gay G, Debaud JC, Casselton LA (1992) Genetic transformation of the symbiotic basidiomycete fungus *Hebeloma cylindrosporum*. Curr Genet 22:41–45

Martin F, Selosse M-A (2008) The *Laccaria* genome: a symbiont blueprint decoded. New Phytol 180:296–310

Martin F, Aerts A, Ahren D, Brun A, Danchin EGJ, Duchaussoy F, Gibon J, Kohler A, Lindquist E, Pereda V et al (2008) The genome of *Laccaria bicolor* provides insights into mycorrhizal symbiosis. Nature 452:88–92

Martin F, Kohler A, Murat C, Balestrini R, Coutinho PM, Jaillon O, Montanini B, Morin E, Noel B, Percudani R et al (2010) Perigord black truffle genome uncovers evolutionary origins and mechanisms of symbiosis. Nature 464:1033–1038

Martino E, Turnau K, Girlanda M, Bonfante P, Perotto S (2000a) Ericoid mycorrhizal fungi from heavy metal polluted soils: their identification and growth in the presence of zinc ions. Mycol Res 104:338–344

Martino E, Coisson JD, Lacourt I, Favaron F, Bonfante P, Perotto S (2000b) Influence of heavy metals on production and activity of pectinolytic enzymes in ericoid mycorrhizal fungi. Mycol Res 104:825–833

Martino E, Franco B, Piccoli G, Stocchi V, Perotto S (2002) Influence of zinc ions on protein secretion in a heavy metal tolerant strain of the ericoid mycorrhizal fungus *Oidiodendron maius*. Mol Cell Biochem 231:179–185

Martino E, Perotto S, Parsons R, Gadd GM (2003) Solubilization of insoluble inorganic zinc compounds by ericoid mycorrhizal fungi derived from heavy metal polluted sites. Soil Biol Biotechnol 35:133–141

Martino E, Murat C, Vallino M, Bena A, Perotto S, Spanu P (2007) Imaging mycorrhizal fungal transformants that express EGFP during ericoid endosymbiosis. Curr Genet 52:65–75

Massicotte HB, Melville LH, Peterson RL (2005) Structural characteristics of root-fungal interactions for five ericaceous species in Eastern Canada. Can J Bot 83:1057–1064

McKendrick SL, Leake JR, Read DJ (2000) Symbiotic germination and development of myco-heterotrophic plants in nature: transfer of carbon from ectomycorrhizal *Salix repens* and *Betula pendula* to the orchid *Corallorhiza trifida* through shared hyphal connections. New Phytol 145:539–548

McLean CB, Cunnington JH, Lawrie AC (1999) Molecular diversity within and between ericoid endophytes from the Ericaceae and Epacridaceae. New Phytol 144:351–358

Meharg AA, Cairney JWG (2000) Co-evolution of mycorrhizal symbionts and their hosts to metal-contaminated environments. Adv Ecol Res 30:69–112

Midgley DJ, Chambers SM, Cairney JWG (2004a) Inorganic and organic substrates as sources of nitrogen and phosphorus for multiple genotypes of two ericoid mycorrhizal fungal taxa from *Woollsia pungens* Cav. (Muell.) and *Leucopogon parviflorus* (Andr.) Lindl. (Ericaceae). Austral J Bot 52:63–71

Midgley DJ, Chambers SM, Cairney JWG (2004b) Utilisation of carbon substrates by multiple genotypes of ericoid mycorrhizal fungal endophytes from eastern Australian Ericaceae. Mycorrhiza 14:245–251

Midgley DJ, Chambers SM, Cairney JWG (2004c) Distribution of ericoid mycorrhizal endophytes and root-associated fungi in neighbouring Ericaceae plants in the field. Plant Soil 259:137–151

Midgley DJ, Jordan LA, Saleeba JA, McGee PA (2006) Utilisation of carbon substrates by orchid and ericoid mycorrhizal fungi from Australian dry sclerophyll forests. Mycorrhiza 16:175–182

Mitchell DT, Gibson BR (2006) Ericoid mycorrhizal association: ability to adapt to a broad range of habitats. Mycologist 20:2–9

Mitchell DT, Read DJ (1981) Utilization of inorganic and organic phosphates by the mycorrhizal endophytes of *Vaccinium macrocarpon* and *Rhododendron ponticum*. Trans Br Mycol Soc 76:255–260

Mitchell DT, Sweeney M, Kennedy A (1992) Chitin degradation by *Hymenoscyphus ericae* and the influence of *H. ericae* on the growth of ectomycorrhizal fungi. In: Read DJ, Lewis DH, Fitter AH, Alexander IJ (eds) Mycorrhizas in ecosystems. CAB International, Wallingford, pp 246–251

Monreal M, Berch SM, Berbee M (1999) Molecular diversity of ericoid mycorrhizal fungi. Can J Bot 77:1580–1594

Morel M, Kohler A, Martin F, Gelhaye E, Rouhier N (2008) Comparison of the thiol-dependent antioxidant systems in the ectomycorrhizal *Laccaria bicolor* and the saprotrophic *Phanerochaete chrysosporium*. New Phytol 180:391–407

Murat C, Zampieri E, Vallino M, Daghino S, Perotto S, Bonfante P (2011) Genomic suppression subtractive hybridization as a tool to identify differences in mycorrhizal fungal genomes. FEMS Microbiol Lett 318:115–120

Myers MD, Leake JR (1996) Phosphodiesters as mycorrhizal P sources. II. Ericoid mycorrhiza and the utilization of nuclei as a phosphorus source by *Vaccinium macrocarpon*. New Phytol 132:445–451

Nehls U, Kleber R, Wiese J, Hampp R (1999) Isolation and characterization of a general amino acid permease from the ectomycorrhizal fungus *Amanita muscaria*. New Phytol 144:343–349

Newsham KK, Upson R, Read DJ (2009) Mycorrhizas and dark septate root endophytes in polar regions. Fungal Ecol 2:10–20

Nikolaou E, Stumpf M, Quinn J, Stansfield I, Brown AJP (2009) Phylogenetic diversity of stress signalling pathways in fungi. BMC Evol Biol 9:44–53

Ohtaka N, Narisawa K (2008) Molecular characterization and endophytic nature of the root-associated fungus *Meliniomyces variabilis* (LtVB3). J Gen Plant Pathol 74:24–31

Pardo AG, Hanif M, Raudaskoski M, Gorfer M (2002) Genetic transformation of ectomycorrhizal fungi mediated by *Agrobacterium tumefaciens*. Mycol Res 106:132–137

Parniske M (2004) Molecular genetics of the arbuscular mycorrhizal symbiosis. Curr Opin Plant Biol 7:414–421

Pearson V, Read DJ (1973a) The biology of mycorrhizae in the Ericaceae. I. The isolation of the endophyte and synthesis of mycorrhizas in aseptic culture. New Phytol 72:371–379

Pearson V, Read DJ (1973b) Biology of mycorrhiza in Ericaceae. II. Transport of carbon and phosphorus by endophyte and mycorrhiza. New Phytol 72:1325–1331

Perotto R, Bettini V, Bonfante P (1993) Evidence of two polygalacturonases produced by a mycorrhizal ericoid fungus during saprotrophic growth. FEMS Microbiol Lett 114:85–92

Perotto S, Actis Perino E, Perugini J, Bonfante P (1996) Molecular diversity of fungi from ericoid mycorrhizal roots. Mol Ecol 5:123–131

Perotto S, Coisson JD, Perugini I, Cometti V, Bonfante P (1997) Production of pectin-degrading enzymes by ericoid mycorrhizal fungi. New Phytol 135:151–162

Peterson RL, Massicotte HB (2004) Exploring structural definitions of mycorrhizas, with emphasis on nutrient-exchange interfaces. Can J Bot 82:1074–1088

Peterson TAW, Mueller C, Englander L (1980) Anatomy and ultrastructure of a Rhododendron root–fungus association. Can J Bot 58:2421–2433

Piercey MM, Thormann MN, Currah RS (2002) Saprotrophic characteristics of three fungal taxa from ericalean roots and their association with the roots of *Rhododendron groenlandicum* and *Picea mariana* in culture. Mycorrhiza 12:175–180

Ramesh G, Podila GK, Gay G, Marmeisse R, Reddy MS (2009) Different patterns of regulation for the copper and cadmium metallothioneins of the ectomycorrhizal fungus *Hebeloma cylindrosporum*. Appl Environ Microbiol 75:2266–2274

Rasmussen HN, Rasmussen FN (2009) Orchid mycorrhiza: implications of a mycophagous life style. Oikos 118:334–345

Read DJ (1974) *Pezizella ericae* sp. nov., the perfect state of a typical mycorrhizal endophyte of the Ericaceae. Trans Br Mycol Soc 63:381–383

Read DJ (1991) Mycorrhizas in ecosystems. Experientia 47:376–391

Read DJ (1996) The structure and function of the ericoid mycorrhizal root. Ann Bot 77:365–374

Read DJ, Perez-Moreno J (2003) Mycorrhizas and nutrient cycling in ecosystems – a journey towards relevance? New Phytol 157:475–492

Read DJ, Leake JR, Langdale AR (1989) The nitrogen nutrition of mycorrhizal fungi and their host plants. In: Boddy L, Marchant R, Read DJ (eds) Nitrogen, phosphorus and sulphur utilization by fungi. Cambridge University Press, Cambridge, pp 181–204

Roberts P, Jones D (2008) Critical evaluation of methods for determining total protein in soil solution. Soil Biol Biochem 40:1485–1495

Rodriguez-Tovar AV, Ruiz-Medrano R, Herrera-Martinez A, Barrera-Figueroa BE, Hidalgo-Lara ME, Reyes-Marquez BE, Cabrera-Ponce JL, Valdes M, Xoconostle-Cazares B (2005) Stable genetic transformation of the ectomycorrhizal fungus *Pisolithus tinctorius*. J Microbiol Methods 63:45–54

Rutherford JC, Bird AJ (2004) Metal-responsive transcription factors that regulate iron, zinc, and copper homeostasis in eukaryotic cells. Eukaryot Cell 3:1–13

Ruytinx J, Craciun AR, Verstraelen K, Vangronsveld J, Colpaert JV, Verbruggen N (2011) Transcriptome analysis by cDNA-AFLP of *Suillus luteus* Cd-tolerant and Cd-sensitive isolates. Mycorrhiza 21:145–154

Schimel JP, Bennett J (2004) Nitrogen mineralization: challenges of a changing paradigm. Ecology 85:591–602

Schimel JP, Weintraub MN (2003) The implications of exoenzyme activity on microbial carbon and nitrogen limitation in soil: a theoretical model. Soil Biol Biochem 35:549–563

Schulz B, Boyle C (2005) The endophytic continuum. Mycol Res 109:661–686

Schützendübel A, Polle A (2002) Plant responses to abiotic stresses: heavy metal-induced oxidative stress and protection by mycorrhization. J Exp Bot 53:1351–1365

Scott B, Eaton CJ (2008) Role of reactive oxygen species in fungal cellular differentiations. Curr Opin Microbiol 11:488–493

Selosse MA, Weiß M, Jany JL, Tillier A (2002) Communities and populations of sebacinoid basidiomycetes associated with the achlorophyllous orchid *Neottia nidus-avis* (L.) L.C.M. Rich. and neighbouring tree ectomycorrhizae. Mol Ecol 11:1831–1844

Selosse MA, Faccio A, Scappaticci G, Bonfante P (2004) Chlorophyllous and achlorophyllous specimens of *Epipactis microphylla* (Neottieae, Orchidaceae) are associated with ectomycorrhizal septomycetes, including truffles. Microbiol Ecol 47:416–426

Selosse MA, Setaro S, Glatard F, Richard F, Urcelay C, Weiss M (2007) Sebacinales are common mycorrhizal associates of Ericaceae. New Phytol 174:864–878

Setaro S, Weiss M, Oberwinkler F, Kottke I (2006) Sebacinales form ectendomycorrhizas with *Cavendishia nobilis*, a member of the Andean clade of Ericaceae, in the mountain rain forest of southern Ecuador. New Phytol 169:355–365

Sharples JM, Chambers SM, Meharg AA, Cairney JWG (2000a) Genetic diversity of root-associated fungal endophytes from *Calluna vulgaris* at contrasting field sites. New Phytol 148:153–162

Sharples JM, Meharg AA, Chambers SM, Cairney JWG (2000b) Mechanism of arsenate resistance in the ericoid mycorrhizal fungus *Hymenoscyphus ericae*. Plant Physiol 124:1327–1334

Sharples JM, Meharg AA, Chambers SM, Cairney JWG (2001) Arsenate resistance in the ericoid mycorrhizal fungus *Hymenoscyphus ericae*. New Phytol 151:265–270

Silver S, Phung LT (2009) Heavy metals, bacterial resistance. In: Schaechter M (ed) Encyclopedia of microbiology. Elsevier, Oxford, pp 220–227

Simard SW, Perry DA, Jones MD, Myrold DD, Durall DM, Molina R (1997) Net transfer of carbon between ectomycorrhizal tree species in the field. Nature 388:579–582

Smith SE, Read DJ (1997) Mycorrhizal symbiosis, 2nd edn. Academic, London

Smith SE, Read DJ (2008) Mycorrhizal symbiosis, 3rd edn. Academic, London

Sokolovski SG, Meharg AA, Maathuis FJM (2002) *Calluna vulgaris* root cells show increased capacity for amino acid uptake when colonized with the mycorrhizal fungus *Hymenoscyphus ericae*. New Phytol 155:525–530

Straker CJ (1996) Ericoid mycorrhiza: ecological and host specificity. Mycorrhiza 6:215–225

Straker CJ, Mitchell DT (1986) The activity and characterization of acid phosphatase in endomycorrhizal fungi of the Ericaceae. New Phytol 104:243–256

Straker CJ, Mitchell DT (1987) Kinetic characterization of dual phosphate uptake system in the endomycorrhizal fungus of *Erica hispidula* L. New Phytol 106:129–137

Stribley DP, Read DJ (1974) Biology of mycorrhiza in Ericaceae. 4. Effect of mycorrhizal infection on uptake of N-15 from labeled soil by *Vaccinium macrocarpon* Ait. New Phytol 73:1149–1155

Talbot JM, Treseder KK (2010) Controls over mycorrhizal uptake of organic nitrogen. Pedobiologia 53:169–179

Tedersoo L, Pärtel K, Jairus T, Gates G, Põldmaa K, Tamm H (2009) Ascomycetes associated with ectomycorrhizas: molecular diversity and ecology with particular reference to the Helotiales. Environ Microbiol 11:3166–3178

Todorova D, Nedeva D, Abrashev R, Tsekova K (2008) Cd (II) stress response during the growth of *Aspergillus niger* B 77. J Appl Microbiol 104:178–184

Tomsett BA (1993) Genetic and molecular biology of metal tolerance in fungi. In: Jennings DH (ed) Stress tolerance of fungi. Dekker, New York, pp 69–95

Upson R, Read DJ, Newsham KK (2007) Widespread association between the ericoid mycorrhizal fungus *Rhizoscyphus ericae* and a leafy liverwort in the maritime and sub-Antarctic. New Phytol 176:460–471

Vallino M, Drogo V, Abbà S, Perotto S (2005) Gene expression of the ericoid mycorrhizal fungus *Oidiodendron maius* in the presence of high zinc concentrations. Mycorrhiza 15:333–344

Vallino M, Martino E, Boella F, Murat C, Chiapello M, Perotto S (2009) Cu, Zn superoxide dismutase and zinc stress in the metal-tolerant ericoid mycorrhizal fungus *Oidiodendron maius* Zn. FEMS Microbiol Lett 293:48–57

Vallino M, Zampieri E, Murat C, Girlanda M, Picarella S, Pitet M, Portis E, Martino E, Perotto S (2011) Specific regions in the Sod1 locus of the ericoid mycorrhizal fungus *Oidiodendron maius* from metal enriched soils show different sequence polymorphism. FEMS Microbiol Ecol 75:321–331

Van Leerdam DM, Williams PA, Cairney JWG (2001) Phosphate solubilising abilities of ericoid mycorrhizal endophytes of *Woollsia pungens* (Epacridaceae). Austral J Bot 49:75–80

Van der Wal A, De Boer W, Klein Gunnewiek PJA, van Veen JA (2009) Possible mechanism for Spontaneous establishment of Calluna vulgaris in a recently abandoned agricultural field. Restor Ecol 17:308–313

Varma A, Bonfante P (1994) Utilisation of cell-wall related carbohydrates by ericoid mycorrhizal endophytes. Symbiosis 16:301–313

Vido K, Spector D, Lagniel G, Lopez S, Toledano MB, Labarre J (2001) A proteome analysis of the cadmium response in *Saccharomyces cerevisiae*. J Biol Chem 276:8469–8474

Villarreal-Ruiz L, Anderson IC, Alexander IJ (2004) Interaction between an isolate from the *Hymenoscyphus ericae* aggregate and roots of *Pinus* and *Vaccinium*. New Phytol 164:183–192

Vohník M, Fendrych M, Albrechtov J, Vostka M (2007) Intracellular colonization of Rhododendron and Vaccinium roots by *Cenococcum geophilum*, *Geomyces pannorum* and *Meliniomyces variabilis*. Folia Microbiol 52:407–414

Vrålstad T (2004) Are ericoid and ectomycorrhizal fungi partof a common guild? New Phytol 164:7–10

Vrålstad T, Fossheim T, Schumacher T (2000) *Piceirhiza bicolorata*- the ectomycorrhizal expression of the *Hymenoscyphus ericae* aggregate? New Phytol 145:549–563

Vrålstad T, Myhre E, Schumacher T (2002a) Molecular diversity and phylogenetic affinities of symbiotic root-associated ascomycetes of the Helotiales in burnt and metal polluted habitats. New Phytol 155:131–148

Vrålstad T, Schumacher T, Taylor AFS (2002b) Mycorrhizal synthesis between fungal strains of the *Hymenoscyphus ericae* aggregate and potential ectomycorrhizal and ericoid hosts. New Phytol 153:143–152

Weiss M (2004) Sebacinales: a hitherto overlooked cosm of heterobasidiomycetes with a broad mycorrhizal potential. Mycol Res 108:1003–1010

Whittaker SP, Cairney JWG (2001) Influence of amino acids on biomass production by ericoid mycorrhizal endophytes from *Woollsia pungens* (Epacridaceae). Mycol Res 105:105–111

Wilcox HE, Wang CJK (1987) Ectomycorrhizal and ectoendomycorrhizal associations of *Phialophora finlandia* with *Pinus resinosa, Picea rubens* and *Betula alleghaniensis.* Can J For Res 17:976–990

Xiao G, Berch SM (1996) Diversity and abundance of ericoid mycorrhizal fungi of *Gaultheria shallon* on forest clearcuts. Can J Bot 74:337–346

Yadav SK (2010) Heavy metals toxicity in plants: an overview on the role of glutathione and phytochelatins in heavy metal stress tolerance of plants. S Afr J Bot 76:167–171

Yoo HY, Chang MS, Rho HM (1999) Heavy metal-mediated activation of the rat Cu/Zn superoxide dismutase gene via a metal-responsive element. Mol Gen Genet 262:310–313

Zhang YH, Zhuang WY (2004) Phylogenetic relationships of some members in the genus *Hymenoscyphus* (Ascomycetes, Helotiales). Nova Hedwigia 78:475–484

Zhu H, Dancik BP, Higginbotham KO (1994) Regulation of extracellular proteinase production in an ectomycorrhizal fungus *Hebeloma crustuliniforme*. Mycologia 86:227–234

Zijlstra JD, van Hof P, Baar J, Verkley GJM, Summerbell RC, Paradi I, Braakhekke WG, Berendse F (2005) Diversity of symbiotic root endophytes of the Helotiales in ericaceous plants and the grass *Deschampsia flexuosa*. Stud Mycol 53:147–162

15 The Symbiotic Phenotype of Lichen-Forming Ascomycetes and Their Endo- and Epibionts

R. Honegger[1]

CONTENTS

[1]Institute of Plant Biology, University of Zürich, Zollikerstrasse 107, 8008 Zürich, Switzerland; e-mail: rohonegg@botinst.uzh.ch

Fungal Associations, 2nd Edition
The Mycota IX
B. Hock (Ed.)

I. Introduction

A. Peculiarities of Lichen Symbiosis

1. Definition of Lichen Symbiosis

Lichens are the symbiotic phenotype of lichen-forming fungi, a polyphyletic, taxonomically diverse group of nutritional specialists which **derive, in a mutualistic symbiosis, fixed carbon from a population of extracellularly located cyanobacterial or green algal cells**; these are referred to as the **photobiont.** In most lichens the **mycobiont** is the quantitatively predominant partner, making up 50–90 % of thalline biomass. In contrast to fungal parasites of algae or cyanobacteria (for examples, see Kagami et al. 2007; Hoffman et al. 2008; Takano et al. 2008) lichen-forming fungi do not kill their photautotrophic partner, but have an impact on its growth and cell turnover rates (see Sect. IV.C and Fig. 15.17h).

Not considered as lichens are: (1) *Geosiphon pyriforme* (Kütz.) F. Wettst., the enigmatic glomeromycete with endosymbiotic *Nostoc* colonies (diazotrophic cyanobacteria), whose hormogonia (naked, motile filaments) are actively taken up by the fungal partner into plasma membrane-derived vesicles (Mollenhauer 1992; Mollenhauer et al. 1996; Kluge 2002; Schüssler 2002) and (2) the mycophycobioses, mutualistic symbioses of fungi with quantitatively predominant, multicellular algal partners (Hawksworth 2000). These include *Phaeospora lemaneae* (Cohn ex Woronin) D. Hawksw. (Verrucariales), which grows within the thalli of the freshwater red alga *Lemanea fluviatilis* (Batrachospermales, Rhodophyta; Brierley 1913; Hill 1992) or the very common and widespread *Mycophycias ascophylli* (Cotton) Kohlm. et Volkm.-Kohlm. (Verrucariales) on representatives of littoral Fucales such as *Pelvetia*, *Fucus* and *Ascophyllum* spp. (Phaeophyta; Kohlmeyer and Kohlmeyer 1972; Garbary and London 1995; Garbary and Macdonald 1995; Kohlmeyer and Volkmann-Kohlmeyer 1998; Deckert and Garbary 2005; Xu et al. 2008). In these interactions the fungal partner has no impact on the morphology, but enhances the desiccation tolerance of its host, a property much in demand in the topmost part of the intertidal zone (Garbary and London 1995) or at the margin of a freshwater rivulet prone to desiccation (Hill 1992).

The **species names of lichens refer to the fungal partner**, not to the symbiotic system, lichen photobionts having their own names and phylogenies. In nature lichen-forming ascomycetes occur almost exclusively in the symbiotic state, exceptions being ascospore-derived germ tubes in search of compatible photobiont cells. **Most lichen-forming ascomycetes are ecologically obligately, but physiologically facultatively biotrophic**, i.e. they can be cultured apart from their photoautotrophic partner (Stocker-Wörgötter and Hager 2008). Such aposymbiotic cultures resemble those of slow-growing non-lichenized taxa (see illustrations in Honegger 2009), the species-specific symbiotic phenotype being expressed exclusively in association with the compatible photobiont.

2. Lichen-Dominated Global Ecosystems

Lichenization is a successful nutritional strategy, **approximately 17 000–20 000 species or approximately 20 % of extant fungal species being lichenized** (Kirk et al. 2008). Distinctly more than 12 % of terrestrial ecosystems are lichen-dominated; these are the sites in arctic tundras and antarctic, alpine, steppe and desert ecosystems where tracheophytes are at their physiological limits. The **majority of lichen-forming fungi and their photobionts are drought stress tolerant** and survive, in the desiccated state, temperature extremes, low and high, unharmed in a state of dormancy.

Desiccated *Xanthoria parietina* (L.) Th. Fr. survived shock freezing in subcooled liquid nitrogen, sputter coating and examination with the ionizing electron beam at 20 kV in a low temperature scanning electron microscope under high-vacuum conditions and grew normally after rehydration and outdoor incubation (Fig. 15.15f; Honegger 1995, 1998). Desiccated *Xanthoria elegans* (Link) Th. Fr. (Fig. 15.1e) and *Rhizocarpon geographicum* (L.) DC. (Fig. 15.1c) survived a journey on the outer shell of the Russian Earth orbiting FOTON M2 satellite (Sancho et al. 2007a; Sancho et al. 2008). However, survival of extreme conditions in the desiccated, dormant state does not mean that lichen-forming ascomycetes and their photoautotrophic partners would be able to become metabolically active and grow under these terms.

Together with cyanobacteria, algae and bryophytes lichens are an integral part of soil crust communities in arid and semi-arid lands (Fig. 15.1u–y), hot and cold, all over the globe (Pérez 1997; various authors in Belnap et al. 2001; Büdel 2002; Belnap and Lange 2003; McCune and Rosentreter 2007; Rosentreter et al. 2007). **Soil crust communities stabilize the topmost layers of soil** (Fig. 15.1x) and thus prevent soil loss by wind erosion. They retain humidity, provide nutrients and thus facilitate seed germination in steppe and other ecosystems (St. Clair et al. 1984; Prasse and Bornkamm 2000; Büdel and Lange 2001; Sedia and Ehrenfeld 2003; Büdel et al. 2009; Zedda and Rambold 2009). The dynamics and development of these very vulnerable, ecologically immensely important ecosystems are globally monitored with remote sensing techniques (Karnieli et al. 2001; Chen et al. 2005; Schultz 2006; Zhang et al. 2007).

3. Fossil Lichens

Although fossil records are largely missing lichenization is assumed to be a very ancient mode of fungal nutrition. **The oldest fossil presumably fungal interactions** with benthic marine cyanobacterial colonies were found **in approximately 600 Mio-year-old phosphorites of the Doushantuo Formation** (Yuan et al. 2005). Most probably fungus–cyanobacterium interactions were the first to evolve from a saprophytic, possibly over a parasitic to a mutualistic lifestyle. Cyanobacterial colonies were the first photoautotrophic invaders of terrestrial ecosystems (Tomescu et al. 2006) and thus available, live or dead, as a source of fixed carbon and, in case of the numerous diazotrophic species, fixed nitrogen.

Winfrenatia reticulata, a peculiar, presumably cyanobacterial lichen from a freshwater ecosystem was detected in the Lower Devonian Rhynie Chert (approx. 400 Mio years old; Taylor et al. 1997); this crustose-reticulate fossil bears no morphological resemblance to extant lichens. Its fungal partner was proposed to belong to the zygomycetes (Taylor et al. 1997) which have no extant lichenized members. As lichenization was repeatedly lost lichen-forming zygomycetes might have vanished.

When did the **first macrolichens**, i.e. dorsiventrally or radially organized thalli with internal stratification appear? **Approximately 415 Mio-year-old fragments** of structurally well preserved, dorsiventrally organized lichens with internal stratification **were found in the Lower Devonian** of the Welsh Borderland: *Cyanolichenomycites devonicus* with a *Nostoc*-like cyanobacterial, *Chlorolichenomycites salopensis* with an unicellular, presumed green algal photobiont (Honegger et al. 2012). Distinctly younger are:

1. Dorsiventrally organized lichens with internal stratification and crystals of secondary metabolites from the Middle Triassic in Lower Franconia, Germany (approx. 220 Myr old; Ziegler 1997, 2002).
2. Fossils derived from Baltic and Dominican amber (Miocene; approx. 40-60 Myr old), including crustose, subsquamulose, foliose and fruticose samples morphologically similar to extant taxa (Mägdefrau 1957; Poinar et al. 2000; Rikkinen and Poinar 2002, 2008; Rikkinen 2003).
3. An impression of a *Lobaria*-like fragment from the Lower Miocene (approx. 20 Myr old) in North-West California (MacGinitie 1937; Peterson 2000).

B. Lichen-Forming Ascomycetes

1. Taxonomic Affiliations

Approximately 99 % of lichen-forming fungi are ascomycetes, all of them belonging to the subphylum Pezizomycotina; **approximately 40 % of ascomycetes are lichenized.** Less than 1 % of lichen-forming fungal species are basidiomycetes. Phylogenetic multilocus analyses refer to multiple gains and losses of lichenization among different classes of Pezizomycotina (Lutzoni et al. 2001, 2004; Grube and Hawksworth 2007; Hibbett et al. 2007; Spatafora 2007). Within the subphylum Pezizomycotina lichenized taxa are found in four classes:

1. The largest, almost exclusively lichenized class is the **Lecanoromycetes** with approximately 14 200 accepted species (Kirk et al. 2008). It results from one lichenization event and was subjected to one major (in the Ostropales) and many

Fig. 15.1. Symbiotic phenotypes of lichen-forming ascomycetes. (**a, a***) Intertidal limpet (*Patella* sp., Mollusca) with crustose *Pyrenocollema halodytes* (Nyl.) R.C. Harris (*Pyrenulales, Eurotiomycetes*) in the calcareous top layer

minor delichenization events (Reeb et al. 2004; Nelsen et al. 2009a; Schoch et al. 2009; Baloch et al. 2010).

2. Large numbers of lichenized taxa are found among the predominantly lichenized **Arthoniomycetes** (approx. 1500 spp.), which evolved from a lichenized base and underwent several delichenization events (Nelsen et al. 2009; Schoch et al. 2009).
3. Exclusively lichenized are the **Lichinomycetes** (approx. 350 spp.; Fig. 15.1p, r, w, x); these cyanobacterial lichens represent the most basal group of the lichenized Pezizomycotina (Reeb et al. 2004). The remaining lichenized taxa are classified in the predominantly non-lichenized **Eurotiomycetes** (see examples in Figs. 15.1a, v, 15.8d, e; Muggia et al. 2010) and **Dothideomycetes** (Miadlikowska et al. 2006; Spatafora et al. 2006; Gueidan et al. 2007, 2008, 2009; Hibbett et al. 2007; Nelsen et al. 2009).

As summarized by Grube and Hawksworth (2007) lichen-forming ascomycetes were traditionally assumed to be a separate group "lichenes" and classified on the base of descriptive parameters such as similarities in thallus morphology and ascocarp structure. From the 1960s onwards comparative studies on ascocarp ontogeny and ascus structure in lichenized and non-lichenized ascomycetes revealed that **lichenized taxa are not a separate group, but nutritional specialists which had to be integrated into the major groups of ascomycetes**, as had been postulated already by Schwendener (1867); thus the term "lichenes" became obsolete.

At species level the morpho- and chemospecies concepts were still applied. In contrast to most non-lichenized taxa, where only the reproductive stages are visible above the substratum or host, the vegetative thallus of most lichenized fungi is well visible above ground

of the shell. **(b)** Crustose, epiphytic *Phaeographis dendritica* (Ach.) Müll. Arg. (*Ostropales, Lecanoromycetes*) with lirelliform fruting bodies (*ac*). **(c)** Leprose, epiphytic *Chrysothrix candelaris* (L.) J.R. Laundon (*Arthoniales, Arthoniomycetes*). **(d)** Crustose, epilithic, juvenile *Rhizocarpon* sp. (*Rh*) on quartz, with black, melanized fungal prothallus and yellow areoles containing photobiont cells; also placodioid *Ophioparma ventosa* (L.) Norman (*Ov*) with yellowish usnic acid in the placodioid thallus and blood-red haemoventosin in the apothecial disc (*ac*; both *Lecanorales*). **(e)** Placodioid, epilithic *Xanthoria elegans* (Link.) Th. Fr. (*Xe*; *Teloschistales*) and beige *Lecanora muralis* (Schreb.) Rabenh. (*Lm*; *Lecanorales*). **(f, f*)** Same mycobiont, different photobionts: phototype pair of *Peltigera malacea* (brownish, with *Nostoc* sp.) and *Peltigera britannica* (green, with *Coccomyxa* sp.; *Peltigerales*). Black dots are external cephalodia; see detail in **(f*)** and Fig. 15.16a–f. **(g)** Fruticose, epiphytic *Teloschistes chrysophthalmus* (L.) Th.Fr. (*Tc*) and *T. exilis* (Michaux) Vainio (*Te*; *Teloschistales*), both with limited size due to terminal apothecia. **(h, i)** Species pair (dissimilar morphology, same phylogeny): richly fertile *Xanthoria polycarpa* (Hoffm.) Th.Fr. ex Rieber **(h)** and sterile *X. candelaria* (L.) Th.Fr. (**i**; *Teloschistales*); for details see Eichenberger (2007). **(j)** Thallus squamules and podetia of *Cladonia macilenta* Hoffm. (Lecanorales) with red apothecia. **(k, l)** Similar morphology, different phylogeny: *Baeomyces rufus* (Huds.) Rebent. (**k**; *Baeomycetales*) and *Dibaeis baeomyces* (L.f.) Rambold et Hertel (**l**; syn. *Baeomyces roseus* Pers.; *Pertusariales*). **(m–r)** Gelatinous lichens. **(m)** *Collema tenax* (Sw.) Ach. (*Ct*; *Peltigerales*) and non-lichenized colony of its photobiont, *Nostoc commune* Vaucher ex Bornet et Flahault (*Nc*; *Nostocales, Cyanobacteria*). **(n)** Detail of *Collema auriforme* (With.) Coppins et J.R. Laundon with globular isidia on upper surface; translucent in the fully hydrated state. **(o)** Semithin cross-section of *Leptogium hildenbrandii* (Garov.) Nyl. (*Peltigerales*) with upper and lower cortex. **(p)** Intertidal *Lichina pygmaea* (Lightf.) C. Agardh (*Lp*; *Lichinales, Lichinomycetes*) co-occuring with the crustose "tar spot lichen", *Verrucaria maura* Wahlenb. (*Vm*; *Verrucariales, Eurotiomycetes*), whitish barnacles and limpets. **(q)** Thallus fragment of *Lichina confinis* (Müller) C. Agardh (*Lichinales*); turquoise dots are cells of the cyanobacterial photobiont (*Stigonema* sp.). **(r)** Semithin, stained section of *Lichina confinis* with central fungal axis and a terminal pycnidium (*py*). **(s)** Tubular, sterile "bone lichen" (*Thamnolia vermicularis* (Sw.) Ach. ex Schaerer, *Icmadophilaceae*). **(t)** Fruticose reindeer lichens in arctic tundra: *Cladonia rangiferina* (L.) F.H. Wigg. (*Cr*) and *C. stellaris* (Opiz) Pouzar et Vězda (*Cs*; *Lecanorales*), the "model builders tree". **(u–x)** Squamulose soil crust lichens. **u)** Red thallus squamules of *Psora decipiens* (Hedw.) Hoffm. (*Lecanorales*) and yellow *Fulgensia bracteata* (Hoffm.) Räsänen (*Teloschistales*); black "fillings" between the thalli are free-living cyanobacteria. **(v, v*)** *Endocarpon pusillum* Hedw. (*Verrucariales, Eurotiomycetes*). Photobiont cells contained in the mucilage-filled, bottle-shaped fruiting body (perithecium; *ac*) are co-dispersed with the ascospores. **(w, x)** *Peltula richardsii* (Herre) Wetmore (*Lichinales, Lichinomycetes*); specimen leg. et det. Burkhard Büdel. **(x)** Cross-section of soil below *Peltula* thallus: soil granules are held together by the dense, lichen-derived hyphal meshwork. *Arrows* point to hyphal strands. **(y)** Fruticose soil crust lichen: finely branched, fruticose *Teloschistes capensis* (L. f.) Vain. ex Müll. Arg (*Teloschistales*), approx. 5 cm high, characteristic "fog combs" of the lichen fields in the Namib desert. Photo: courtesy of Reinhard Berndt and offers lots of morphological and chemical characters for classification. From the 1990s onwards molecular phylogenetic data sets have revolutionized our knowledge about taxonomic relations among non-lichenized and lichen-forming fungi and about their photoautotrophic partners. As in all other groups of organisms molecular data sets of lichen-forming ascomycetes often are more or less in parallel with descriptive taxonomic markers such as morphotype, chemotype, ascus and ascocarp structure. However, many interesting exceptions occur, which result from convergent evolution of morphological characters. "Trouble with lichen: the re-evaluation and re-interpretation of thallus form and fruit body types in the molecular era" (Grube and Hawksworth 2007) and "Goodbye morphology? A paradigm shift in the delimitation of species in lichenized fungi" (Lumbsch and Leavitt 2011) discuss such discrepancies, a few examples being summarized below.

2. Similar Morphology, Different Phylogenies

Similar types of fruiting bodies and structurally and functionally similar ascus types evolved independently in distantly related groups of lichenized and non-lichenized ascomycetes (Schmitt et al. 2005, 2009; Grube and Hawksworth 2007; Lumbsch and Huhndorf 2007).

An example are mazaediate fruiting bodies with prototunicate asci, as typically found in the former order Caliciales, which comprised lichenized (Caliciaceae, Sphaerophoraceae) and non-lichenized families (Mycocaliciaceae; see example in Fig. 15.6b). Molecular data sets did not validate this order, its former members belonging to the Lecanorales, Teloschistales (both Lecanoromycetes), Lecanoromycetes *incertae sedis* and Mycocaliciales (Eurotiomycetes; Wedin et al. 1998, 2000, 2002; Miadlikowska et al. 2006; Geiser et al. 2006; Kirk et al. 2008).

Morphologically very similar species with similar fruiting bodies and asci were formerly placed in the same genus. Examples are *Baeomyces rufus* (Huds.) Rebent. and the former *B. roseus* Pers., soil crust lichens with characteristic, stalked apothecia (Fig. 15.1k, l), which were formerly classified within the Lecanorales. According to molecular phylogeny data sets *Baeomyces rufus* belongs to the Baeomycetales, *Dibaeis baeomyces* (L. f.) Rambold et Hertel to the Pertusariales within the Ostropomycetidae (Platt and Spatafora 1999). Not previously known relationships were discovered among Eurotiomycetes and Dothideomycetes (Gueidan et al. 2007, 2008, 2009; Muggia et al. 2007, 2010) or in the large, notoriously difficult crustose genus *Lecidea* (Schmull et al. 2011). Cryptic species, i.e. morphologically very similar lineages with different phylogenies were detected in many taxa (Kroken and Taylor 2001; Crespo et al. 2007, 2010; Crespo and Pérez-Ortega 2009; Crespo and Lumbsch 2010; Lumbsch and Leavitt 2011).

3. Different Morphology, Same Phylogeny, or the Species Pair Concept

The so-called **species pair concept** *sensu* Poelt (1970) **postulates monophyly among some morphologically very similar species with different reproductive strategies**: a sexually reproducing founder species, usually with narrow geographic range, whence a vegetatively propagating, widespread species evolved. Most of the presumed species pairs turned out to be polyphyletic (Crespo and Pérez-Ortega 2009; Crespo and Lumbsch 2010). True species pairs among Teloschistales include the richly fertile *Xanthoria polycarpa* (Hoffm.) Rieber (Fig. 15.1h) and the usually sterile, blastidiate *X. candelaria* (L.) Th. Fr. (Fig. 15.1i) or the fertile *Xanthomendoza alfredii* (S.Y. Kondr. et Poelt) Søchting and the usually sterile, sorediate *Xm. fallax* (Arnold) Søchting, Kärnefelt et S.Y. Kondr. (Eichenberger 2007). Unexpectedly, transitions from sterile, sorediate to richly fertile, apothecia-bearing taxa were detected in the *Lobaria retigera* (Bory) Trevisan/*L. kurokawae* Yoshim. and in the *Lobaria isidiosa* (Müll. Arg.) Vainio/*L. pseudopulmonaria* Gyeln. species pairs (Cornejo et al. 2009).

4. Morphological Simplicity and Evolutionary Level

Before the advent of molecular tools **lichenized ascomycetes with morphologically simple thalli were assumed to be basal ("primitive") forms** whence those with complex, foliose or fruticose thalli with internal stratification evolved. This assumption is correct as a general trend. Recently lichenized fungal taxa form morphologically simple, crustose, microglobose or microfilamentous thalli. Examples are the microfilamentous rock hair lichens *Cystocoleus ebeneus* (Dillwyn) Thwaites and *Racodium rupestre* Pers., which, despite their morphological similarity with *Coenogonium* species (Gyalectales, Lecanoromycetes; Kauff and Lutzoni 2002; Hibbett et al. 2007; Fig. 15.8c) belong to the Capnodiales, an order with no other lichenized taxa in the Dothideomycetes (Muggia et al. 2007). However, in anciently lichenized families lichens with simple crustose thalli may be closely related to morphologically complex, foliose or fruticose taxa. Examples are the leprose genera *Lepraria* (Fig. 15.8a) and *Leprocaulon*, relatives of the fruticose genus *Stereocaulon* (Fig. 15.16i) in the Stereocaulaceae (Ekman and Tønsberg 2002), or clades of the crustose-placodioid genus *Caloplaca* which are closely related to the foliose *Xanthoria* spp. (Eichenberger 2007; Gaya et al. 2008).

5. Sterile Taxa

Approximately 4 % of lichenized ascomycetes, micro- and macrolichens, differentiate no sexual reproductive structures. A few of them are anamorphic fungi, dispersing via macroconidia, which have to re-lichenize (see examples in the tropical, foliicolous genus *Gyalectidium*, Gomphillaceae, Ostropales; Lücking 2008). However, **the majority of sterile taxa co-disperse with their photobiont by means of either thallus fragmentation or vegetative symbiotic propagules** (see Sect. V.A.2). The taxonomic affiliation of numerous sterile taxa was elucidated with molecular tools. Examples are the leprose genera *Chrysothrix* (Arthoniales, Arthoniomycetes; Fig. 15.1c), *Lepraria* (Fig. 15.8a) and *Leprocaulon* (Stereocaulaceae, Lecanorales; Ekman and Tønsberg 2002), the squamulose *Flakea papillata* (Fig. 15.8e, e*) and microfilamentous *Psoroglaena* spp. (Verrucariales, Dothideomycetes; Muggia et al. 2010; Fig. 15.8d, d*). The tubular, sterile macrolichen genera *Thamnolia* (Fig. 15.1s) and *Siphula*, formerly classified *incertae sedis* among the Lecanorales, belong to the Icmadophilaceae (Platt and Spatafora 2000) and the morphologically similar genus *Parasiphula* to the Coccotremataceae (Grube and Kantvilas 2006) within the Pertusariales, an order with predominantly crustose taxa.

C. Lichen Photobionts

1. Taxonomic Diversity

Approximately 85 % of lichen-forming ascomycetes associate with green algal photobionts, approximately 10 % with cyanobacteria and approximately 3 % simultaneously with both (Tschermak-Woess 1988); these are the cephalodiate species (see Sect. V.B. and Figs. 15.1f, 15.16a–k). The highest diversity of lifestyles and photobiont taxa is found among the lichenized Eurotiomycetes, especially among the Verrucariales (Chaetothyriomycetidae), which comprise marine, maritime, freshwater as well as epilithic, epiphytic and foliicolous terrestrial species in a broad geographic range (Figs. 15.1a, v, 15.8d, e).

The intertidal, crustose *Verrucaria tavaresiae* from the northern and central Californian coast associates with the only brown algal lichen photobiont so far known, *Petroderma maculiforme* (Wollny) Kuckuck (Ralfsiaceae, Ectocarpales, Phaeophyta; Sanders et al. 2004, 2005). Intertidal *Verrucaria* spp. such as the tar lichen *Hydropunctaria maura* (Wahlenb.) C. Keller, Gueidan et Thüs (syn. *Verrucaria maura* Wahlenb.) of European and North American coasts (Fig. 15.1p) associate with the only known yellow-green algal lichen photobiont, *Heterococcus caespitosus* Vischer (Tribonematales, Xanthophyceae; Tschermak-Woess 1988).

Considering the large numbers of lichen-forming ascomycetes the species number of lichen photobionts is surprisingly low. So far **approximately 120 species of cyanobacteria and algae are known to be lichen photobionts,** but by far not all photoautotrophic partners of lichen-forming fungi have been identified at species level. Large numbers of genotypes were identified among diverse ***Trebouxia* spp., the most common green algal lichen photobionts** (see examples in Kroken and Taylor 2000; Piercey-Normore 2006; Werth and Sork 2010). Asymmetric co-evolution in the lichen symbiosis caused by a limited capacity for adaptation in the photobiont was speculated to be the reason for this situation (Hill 2009) which, to a certain degree, resembles the arbuscular mycorrhizal symbiosis with plants. Approximately 170 species and large numbers of genotypes have been identified among the Glomeromycetes. These live in mutualistic symbiosis with more than 80 % of plants, already early land plants in the Devonian having been mycorrhizal (Taylor et al. 1995; for a review, see Bonfante et al. 2009; see Chap. 3). **All lichen photobionts can be isolated into sterile culture** and grown on non-nutrient media apart from their fungal partner. In nature some lichen photobionts are common and widespread also in

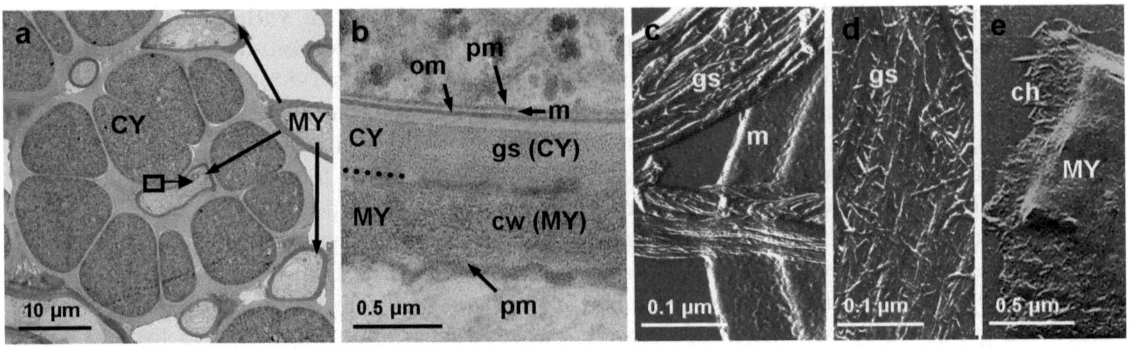

Fig. 15.2. TEM micrographs of the mycobont–photobiont interface in cyanobacterial lichens with *Nostoc* sp. as primary or secondary photobionts. (**a**) Colonies of *Nostoc* sp. (*CY*) in the photobiont layer of *Peltigera canina* are invaded by the fungal partner (*MY*). (**b**) Detail of the immediate contact site (ultrathin sections). (**c–e**) Replicas of the isolated and purified cyanobacterial gelatinous sheath (*gs*) and murein saccule (*m*) of *Nostoc* sp. (**c**, **d**) and of the fungal cell wall (**e**) in cephalodia of *Peltigera aphthosa* (L.) Willd. (after Honegger 1982). (**e**) Replica of an isolated and purified fungal cell wall fragment with chitin fibrils (*ch*) in the inner and amorphous glucans in the outer part of the wall. Abbreviations: *ch* chitin fibrils of the fungal cell wall, *cw* fungal cell wall, *gs* gelatinous sheath of the cyanobacterial colony, *m* murein saccule of the cyanobacterial cell, *pm* plasma membrane (cyanobacterial or fungal, respectively), *om* outer membrane of the cyanobacterial cell. For additional information, see Fig. 15.16a–f

the aposymbiotic (free-living) state, but others seem to be poor competitors outside lichen thalli (see Sect. I.C.3).

2. Cyanobacterial Photobionts

A **wide range of cyanobacterial species was identified as lichen photobionts**, including representatives of **Chroococcales, Nostocales, Pleurocapsales and Stigonematales** (Tschermak-Woess 1988; Friedl and Büdel 2008). **Cyanobacterial photobionts are desiccation tolerant** but, in contrast to trebouxiophycean green algal photobionts, **most of them require liquid water for rehydration** (Lange et al. 1986, 1993, 1998). Some taxa are known to associate with only one or few lichen-forming fungi (e.g. the pleurocapsalean *Hyella caespitosa* Bornet et Flahault, photobiont of intertidal *Pyrenocollema* spp., Eurotiomycetes), whereas genotypes of *Nostoc* clade II according to O'Brien et al. (2005) are the most common cyanobacterial lichen photobionts (Fig. 15.2a–e). They associate with large numbers of lichen-forming fungi over a wide geographic range, including extreme habitats (Elvebakk et al. 2008; Friedl and Büdel 2008; Fedrowitz et al. 2011), but also with the glomeromycete *Geosiphon pyriforme* (Kütz.) F. Wettst. and, as donors of fixed nitrogen, with liver- and hornworts (*Blasia* and *Anthoceros* spp.) and with the angiosperm genus *Gunnera* (Rai et al. 2002). Cyanobacterial photobionts of subtropical and tropical, unrelated asco- and basidiolichens which, based on light microscopy investigations, had been referred to as *Stigonema* spp. turned out to form a separate, not previously known lineage *Rhizonema sensu* Lücking et al. (2009). It was so far found in lichens only.

***Nostoc* and other diazotrophic lichen photobionts** (representatives of the genera *Stigonema*, *Rhizonema*, *Tolypothrix* etc.) **provide not only photosynthates, but also fixed nitrogen as ammonium ion to their fungal partner,** whose ammonium transporter most likely was achieved by horizontal gene transfer from hyperthermophilic chemoautolithotrophic prokaryotes (McDonald et al. 2012). Cyanobacterial lichens significantly increase the nitrogen contents of their ecosystems (Rogers et al. 1966; Evans and Lange 2001; Antoine 2004; Knowles et al. 2006).

3. Green Algal Photobionts

Trebouxiophyceae. **The majority of green algal lichen photobionts are unicellular representa-**

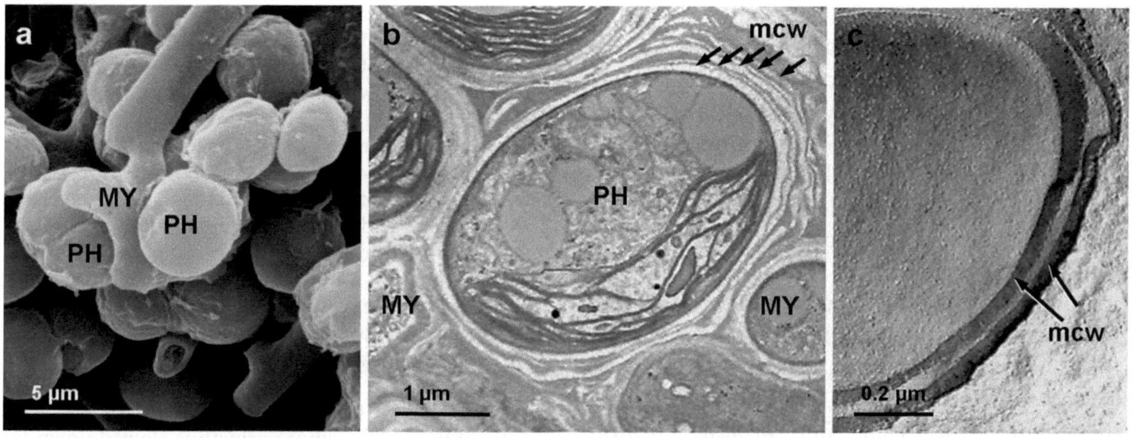

Fig. 15.3. SEM (**a**) and TEM (**b**, **c**) micrographs of the mycobiont-photobiont interface in green algal lichens with *Coccomyxa* sp. as photobionts. (**a**) Wall-to-wall apposition in *Solorina crocea*. (**b**) Wall-to-wall apposition in the crustose *Icmadophila ericetorum* (ultrathin section). Numerous algal mother cell walls (*mcw*) accumulate in the thallus; these are built up by algaenans (sporopollenin-like biopolymers) and thus are enzymatically non-degradable. (**c**) Freeze-fracture replica of non-degradable mother cell walls (for additional information, see Honegger and Brunner 1981; Brunner and Honegger 1985)

tives of the chlorophycean class Trebouxiophyceae *sensu* Friedl (1995). This class comprises also algal symbionts of freshwater ciliates, amoebae, sponges and cnidarians (*Chlorella* spp.) and a wide range of non-symbiotic species such as *Botryococcus braunii* Kützing, which produces up to 86 % hydrocarbons by dry weight (Metzger and Largeau 2005; Qin 2010).

An estimated 60 % of species among the Lecanoromycetidae, especially in the Lecanorales and Teloschistales, **associate with representatives of the genera *Trebouxia*** (approx. 23 species) or *Asterochloris* (approx. 7 species). These algal genera are characterized by: (1) a large central chloroplast with central pyrenoid (Tschermak-Woess 1988; Friedl and Büdel 2008; Skaloud and Peksa 2008, 2010), (2) cellulosic cell walls (Honegger 1984; Brunner and Honegger 1985; Fig. 15.13a) and (3) desiccation tolerance, the ability to rehydrate from humid air and the ability to photosynthesize already in the partially hydrated state (Lange et al. 1986). It is therefore not surprising that **green algal lichens with a *Trebouxia* photobiont dominate in climatically extreme terrestrial ecosystems** such as the lichen fields of the Namib desert. Examples are the finely branched, shrubby *Teloschistes* spp. (mainly *T. capensis* (L. f.) Müll. Arg.; Fig. 15.1y) with very high surface/volume ratio, which cover the ground and absorb, as "fog combs", humid air coming in from the Atlantic coast (Lange et al. 1990, 2006).

Contradictory views are held regarding the occurrence of *Trebouxia* and *Asterochloris* spp. in the aposymbiotic (free-living) state in nature. All *Trebouxia* and *Asterochloris* spp. can be isolated and grown in sterile culture on non-nutrient mineral media. The growth of these oligotrophic organisms can be significantly accelerated on carbohydrate-containing media (Ahmadjian 1967). According to phycologists *Trebouxia arboricola*, type species of the genus, is common and widespread in nature (Ettl and Gärtner 1995; Rindi and Guiry 2003). Various lichenologists reported on free-living *Trebouxia* spp. (Tschermak-Woess 1978; Bubrick et al. 1984: Mukhtar et al. 1994). Sanders identified lots of free *Trebouxia* cells on microscopy slides which he had exposed in tropical rainforests or Mediterranean oak stands, respectively, for long-term studies on relichenization events (Sanders 2001, 2005; Sanders and Lücking 2002). Hedenås et al. (2007) found lots of free-living *Trebouxia* cells as well as free-living *Trentepohlia* and *Nostoc* photobionts on the bark of old aspen (*Populus tremula*). Ahmadjian (2001) summarizes his view on *Trebouxia* photobionts of lichen-forming fungi as follows: "*Trebouxia* lives as a heterotroph inside a lichen thallus and is so dependent on the mycobiont that it cannot live independently–free-living *Trebouxia* do not exist".

Photobionts of the trebouxiophycean genera ***Coccomyxa*** (Fig. 15.3a–c) and ***Elliptochloris*** are-characterized by tripartite cell walls with a thin, trilaminar outermost wall layer built up by algaenans, enzymatically non-degradable, sporopollenin-like biopolymers (Honegger and Brunner 1981; Honegger 1984; Brunner and Honegger 1985; Wiermann et al. 2001). After aplanospore formation the non-degradable mother cell walls accumulate in lichen thalli (Fig. 15.3b, c). This type of cell wall is also formed in other trebouxiophycean taxa such as *Chlorella* and *Elliptochloris* endosymbionts of invertebrates, the non-photosynthetic, human-pathogenic *Prototheca wickerhamii* Tubaki et Soneda and the hydrocarbon-rich *Botryococcus braunii* (Kadouri et al. 1988; Wiermann et al. 2001; Puel et al. 2008; Kodner et al. 2009; Letsch et al. 2009). A trilaminar outermost wall layer without algaenan was found in trebouxiophycean lichen photobionts of the genera *Myrmecia* and *Dictyochloropsis* (Brunner and Honegger 1985).

The highly osmotolerant, trebouxiophycean *Prasiola crispa* (Lightfoot) Kützing is the only known laminal, multicellular green algal photobiont; it associates with *Mastodia tessellata* (Hook. f. et Harv.) Hook. f. et Harv. (syn. *Turgidosculum complicatulum* (Nyl.) Kohlm. et E. Kohlm.; Verrucariales, Eurotiomycetes) in eutrophicated coastal regions of the Antarctic, e.g. below penguin colonies (Lud et al. 2001; Kohlmeyer et al. 2004; Pérez-Ortega et al. 2010).

Ulvophyceae. Filamentous green algae of the genera ***Trentepohlia*, *Phycopeltis* and *Cephaleuros*** (Trentepohliales, Ulvophyceae) are photobionts of large numbers of temperate and tropical lichens, especially of the Arthoniomycetes (Tschermak-Woess 1988; Chapman and Waters 2004). They have cellulosic walls (Brunner and Honegger 1985) and are commonly found in the free-living (aposymbiotic) state (Tschermak-Woess 1988; Rindi and Guiry 2002; Neustupa 2003). Some *Trentepohlia* species cover large areas with their velvety colonies with characteristic orange colour due to high carotenoid contents. In the lichenized state *Trentepohlia* spp. are easily recognized by their yellowish colouration, which differs from the vivid green colour of trebouxiophycean photobionts.

Various *Trentepohlia*, *Cephaleuros* and *Phycopeltis* species grow on long-living leaves and thus are considered as pests in tropical agriculture (Joubert and Rijkenberg 1971; Hawksworth 1988b; Nelson 2008). In contrast to foliicolous *Trentephohlia* and *Phycopeltis* spp., which do not penetrate the leaf surface, *Cephaleuros* species, so-called red rusts, grow below the epidermal cuticle. The same applies for lichen-forming ascomycetes (e.g. *Strigula* spp.) with *Cephaleuros* photobionts (Lücking 2008; Nelson 2008). A very wide range of epicuticular foliicolous lichens has been identified in subtropical and tropical forests (Lücking and Matzer 2001; Lücking 2008; Fig. 15.8c), old leaves of economic and wild plants often being speckled with lichen thalli and free-living trentepohlioid algae (Lücking and Caceres 2002; Lücking 2008). A high percentage, but by far not all foliicolous lichen-forming ascomycetes associate with trentepohlioid algae whose growth is slowed down by lichenization (Joubert and Rijkenberg 1971; Hawksworth 1988b). Many foliicolous lichens and their trentepohlioid and coccalean photobionts colonize also other substrates, if available, such as plastic stripes and cover slips (Sipman 1994; Sanders 2001). Beside a negative effect due to light competition angiosperm leaves benefit from foliicolous lichens and other epiphyllous colonizers by a certain level of deterrence to herbivores such as leaf-cutter ants (Mueller and Wolf-Mueller 1991).

II. Lichen Thalli: Consortia with Unknown Numbers of Participants

The **lichen symbiosis** is commonly assumed to be an **interaction of a lichen-forming fungus with a photoautotrophic partner**, i.e. a green alga or a cyanobacterium or rarely both (see Sect. V.B, cephalodiate species; Figs. 15.1f, f*, 15.16a–k). Hawksworth (1988a) pointed out that distinctly **more partners are present** whose biological roles are not always understood: **parasitic lichens, parasitic non-lichenized fungi and symptomless fungal endobionts of lichen thalli.** Hestmark (1992) and Seymour et al. (2005) hypothesize sexual reproduction and re-lichenization often to be the sole option of lichen-forming ascomycetes to escape from heavily parasitized thalline areas. Poelt and Mayrhofer (1988) referred to cyanotrophy of lichens, i.e. the fact that many epilithic crustose species have loosely associated cyanobacterial colonies, which were hypothesized to provide fixed nitrogen to the lichen-forming fungus and its green algal partner. **Lichen thalli can thus be regarded as**

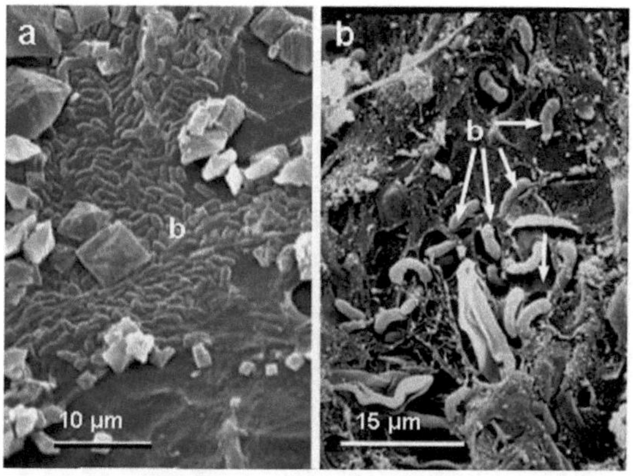

Fig. 15.4. SEM micrographs of bacterial epibionts (*b*) of lichen thalli. (**a**) Bacterial film on the upper cortex of *Parmelia sulcata* (see Fig. 15.17a–h). (**b**) Bacteria within the tubular thallus of *Cladonia arboricola*, a reindeer lichen

consortia with unknown numbers of participants (Honegger 1992).

A. Bacterial Epibionts of Lichen Thalli

Wherever lichens are collected and microscopically investigated **bacterial epibionts** are **detected on thalline surfaces.** These may form patchy colonies or more or less dense films on cortical layers on the vegetative thallus (Fig. 15.4a, b) and even on the surface of fruiting bodies. They adhere to rhizinae or tomentum and, in taxa with no lower cortex, even medullary hyphae may be colonized. As many lichens colonize extremely nutrient-poor habitats such as bare rock lichenologists have long been speculating that the ever present **bacterial epibionts might play important roles as donors of fixed nitrogen, growth factors** etc. (Honegger 1997). A whole range of bacterial epi- and endobionts of lichen thalli have been characterized with molecular tools (Cardinale et al. 2006, 2008; Hodkinson et al. 2006; Grube and Berg 2009; Grube et al. 2009; Hodkinson and Lutzoni 2009; Selbmann et al. 2010; Bates et al. 2011; Hodkinson 2011; Mushegian et al. 2011). **Bacterial epibionts are likely mobilizing fungal cell wall components such as water-soluble glucans** (see below) and were shown to benefit from apoplastic fluids of fungal and algal origin, especially acyclic polyols, during the wetting and drying cycles (Hodkinson 2011). Sequences of putative nitrogen (N_2)-fixing bacteria related to well known N_2-fixers of the genera *Azospirillum*, *Bradyrhizobium* and other Rhizobiales, *Frankia* and others were identified and novel groups of bacteria detected among lichen epibionts (Hodkinson and Lutzoni 2009; Hodkinson 2011).

With high probability **"lichen beneficial bacteria" are an integral part of lichen symbiosis,** similar to the bacterial epi- and endobionts of arbuscular mycorrhizal fungi (AMF; for reviews, see Bonfante and Anca 2009; Pivato et al. 2009) and the mycorrhization-helper bacteria (MHB) in ectomycorrhizal symbioses (ECM; Frey-Klett et al. 2007). As in bacterial symbioses of arbuscular and ectotrophic mycorrhizal fungi the taxonomic affiliation and biological role of the bacterial epibionts of lichen-forming fungi and the specificity of these interactions remain to be thoroughly investigated. Based on comparative analyses of genes among bacteria and fungi at least two important **horizontal gene transfers from prokaryotes to lichen-forming ascomycetes occurred during the early evolution of these lineages:**

1. Schmitt and Lumbsch (2009) concluded on horizontal transfers of polyketide synthase (PKS) genes from actinomycetes to lichen-forming ascomycetes (see Sect. IV.C.3).
2. McDonald et al. (2012) referred to the horizontal transfer of a methylammonium permease (MEP) gene from hyperthermophilic chemoautolithotrophic prokaryotes to the early filamentous ascomycetes (Pezizomycotina) that was subsequently lost in most lineages but is retained in even distantly related lichenized taxa.

B. Parasitic Lichens

Parasitic lichens start their development on or within the thallus of another lichen species and slowly overgrow or outcompete their host (Fig. 15.5a–c).

Some parasitic lichens acquire their compatible photobiont by theft from the host lichen, others associate with a different algal partner. Initial association with the host photobiont, succeded

Fig. 15.5. Parasitic lichens. (**a**) Red *Caloplaca anchon-phoeniceon* Poelt et Clauzade (*Ca-p*; *Teloschistales*), parasite of grey *Aspicilia calcarea* (L.) Mudd (*Ac*; *Pertusariales*). (**b**) *Rhizocarpon effiguratum* (Anzi) Th. Fr. (*Re*; *Lecanorales*, *Lecanoromycetes*) started its development in *Acarospora oxytona* (Ach.) A. Massal. (*Ao*; *Acarosporales*, *Acarosporomycetes*); both are getting outcompeted by *Sporastatia testudinea* (Ach.) A. Massal. (*St*; *Lecanorales*). (**c**) Crustose *Diploschistes muscorum* (Scop.) R. Sant. (*Dm*; *Ostropales*) started its development in the thallus squamules of *Cladonia pocillum* (Ach.) Grognot (*Cp*; *Lecanorales*)

by replacement with a more favourable photoautotrophic partner was observed in *Diploschistes muscorum* (Scop.) R. Sant., a lichenicolous lichen on *Cladonia* spp. (Fig. 15.5c). Ascospores of this crustose, parasitic lichen germinate on thallus squamules or podetia of diverse *Cladonia* spp. and associate with their photobiont, *Trebouxia irregularis* Hildreth and Ahmadjian, which is later replaced by *Trebouxia showmanii* (Hildreth and Ahmadjian) Gärtner (Friedl 1987), if available. This is the first and only case where algal theft by lichen-forming fungi has been documented with isolation and culturing experiments. Other reports on algal "cleptobiosis", e.g. of the common and widespread yellow wall lichen *Xanthoria parietina* (L.) Th. Fr. on grey *Physcia* spp. (Ott 1987) turned out to be misinterpretations of greyish *Xanthoria polycarpa* (Hoffm.) Rieber (Fig. 15.1h) which co-occur with grey species of the *Physcietum adscendentis* (Honegger et al. 1996a), the latter associating with other *Trebouxia* species of different clades (Beck et al. 1998; Helms et al. 2001; Helms 2003).

C. Lichenicolous Fungi: Fungal Parasites of Lichens

Approximately 1,500 species of lichenicolous fungi in more than 300 genera have been identified. **These asco- and basidiomycetes derive their nutrition as parasites of lichen thalli** (Lawrey 1986; Richardson 1999; De los Rios and Grube 2000; Hoffmann and Hafellner 2000; Sikaroodi et al. 2001; Lawrey and Diederich 2003, 2010; Molina et al. 2005; Lawrey et al. 2007; Fig. 15.6a–e); some are obligately lichenicolous, others may switch between different nutritional modes and life styles. The biology of most lichenicolous fungi is poorly understood.

Approximately 95 % of lichenicolous fungi are ascomycetes, representatives of 19 orders and 7 classes; the remaining 5 % are basidiomycetes from 8 orders and 4 classes (Lawrey and Diederich 2010).

Some lichenicolous fungi have devastating effects and **kill lichen-forming fungi and their photobionts.**

Examples are the widespread basidiomycetes *Marchandiobasidium aurantiacum* Diederich and Schultheis (Fig. 15.6a) or *Athelia arachnoidea* (Berk.) Jülich, the sexual state of *Fibulorhizoctonia carotae* (Rader) G.C. Adams et Kropp, a postharvest disease of carrots (Adams and Kropp 1996). Not all lichen species of a community are severely damaged by these parasites; "non-host" taxa even benefit by colonizing the newly available space after an attack of *Athelia arachnoidea* (Motiejûnaitë and Jucevièienë 2005). Other lichenicolous fungi are relatively harmless parasites, some being even interpreted as commensals (parasymbionts); these do not harm the photobiont cell population. Examples

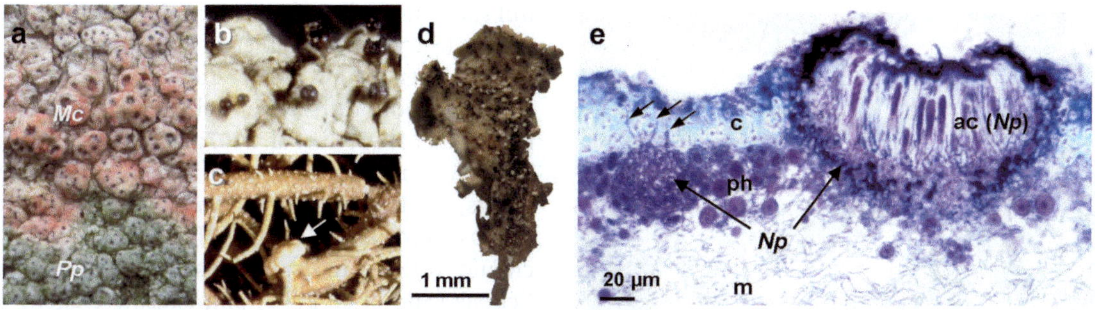

Fig. 15.6. Lichenicolous (parasitic) fungi. (**a**) *Pertusaria pertusa* (Weigel) Tuck. (*Pp;* Pertusariales) invaded and partially killed by *Marchandiomyces corallinus* (Roberge) Diederich et D. Hawksw. (*Mc; Corticiales, Agaricomycetes*). (**b**) Mazaediate fruiting bodies of *Sphinctrina turbinata* (Pers. ex Fr.) De Not. (*Mycocaliciales, Eurotiomycetes*) on the surface of *Pertusaria pertusa.* (**c**) Beard lichen *Usnea ceratina* Ach. with fruiting body (*arrow*) of *Biatoropsis usnearum* Räsänen (*Tremellales, Basidiomycota*). (**d, e**) thallus squamule of *Cladonia macrophylla* (Schaer.) Stenh. invaded by *Phaeopyxis punctum* (A. Massal.) Rambold, Triebel et Coppins (*Parmeliaceae, Lecanorales*). *Black dots* in (**d**) are fruting bodies of *P. punctum*, whitish dots are adhering soredia of *C. macrophylla.* (e) Stained semithin section of *C. macrophylla* with cortex (*c*), photobiont layer (*ph*) and medullary layer (*m*) of the lichen-forming ascomycete; also primordium (*left arrow*) and mature fruiting body (*right arrow*; *ac*) of *P. punctum*. *Arrows* point to trichogynes of the ascomal primordium of the parasite growing through the cortex of the host

are *Phaeopyxis punctum* (A. Massal.) Rambold, Triebel et Coppins (syn. *Nesolechia punctum* A. Massal.), a lichenicolous representative of the large family Parmeliaceae with more than 2000 species of foliose and fruticose lichens, the result of relatively recent delichenization events (Crespo et al. 2010; Fig. 15.6d, e), or the various gall-inducing species which locally stimulate the growth of the fungal and, in some species, even of the algal partner (for a review, see Hawksworth and Honegger 1994). An example is *Telogalla olivieri* (Vouaux) Nik. Hoffm. et Hafellner, [syn. *Guignardia olivieri* (Vouaux) Sacc.], a representative of the Verrucariales (Eurotiomycetes) which stimulates fungal and algal growth of *Xanthoria parietina* and its photobiont, *Trebouxia arboricola* (Hoffmann and Hafellner 2000), or the lichenicolous basidiomycete *Biatoropsis usnearum* Räsänen, a representative of the Platigloeales (Pucciniomycotina), which induces fungal growth in *Usnea* spp. (beard lichens) around its own basidiomata (Grube and De los Rios 2001; Fig. 15.6c).

Saprobic fungi may be found on decaying lichens. However, the most common saprobic moulds do not normally colonize lichen thalli, probably due to the fungistatic effect of many mycobiont-derived depsides and depsidones (see Sect. III.C.3). Specialists among lichenicolous fungi such as *Fusarium* sp. degrade secondary lichen compounds; subsequently these infected lichen thalli are colonized and degraded by saprobic fungal species (Lawrey et al. 1999).

D. Endolichenic Fungi: Symptomless Endophytes of Lichen Thalli

Plant tissues, especially conifer needles and angiosperm leaves, but also the fruiting bodies of non-lichenized fungi are inhabited by a wide range of fungi which cause no disease symptoms and are visually not recognizable, but are studied with microscopic, culturing and molecular techniques (Arnold 2007, 2008; Pacioni et al. 2007). The biology of many endophytic fungi remains to be explored. The outcome of a fungal interaction depends on internal and external factors and some taxa may reveal different lifestyles under different conditions (Schulz and Boyle 2005; Kogel et al. 2006). Many fungal endophytes have a positive impact on their plant host by protecting it from herbivore attacks and/or increasing its biological fitness, especially its drought tolerance in climatically extreme habitats (Rodriguez and Redman 2008; Rodriguez et al. 2009; Schardl et al. 2009; Mei and Flinn 2010).

Whenever attempts are made to isolate lichen-forming ascomycetes into sterile culture from thoroughly washed and even surface-sterilized, healthy thallus fragments fungal endophytes are growing out, usually much faster than

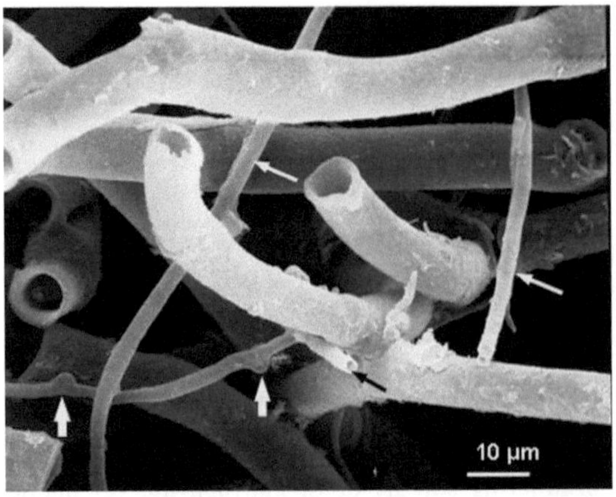

Fig. 15.7. Symptomless endolichenic fungi. Scanning electron micrograph of the medullary layer of *Peltigera leucophlebia* (Nyl.) Gyeln. *Thin arrows* point to thin hyphae of endolichenic fungi which grow between the medullary hyphae of the lichen-forming ascomycete. *Broad arrows* point to clamp connections of an endolichenic Agaricomycete

but share some fungal endosymbionts with bryophytes (Suryanarayanan et al. 2005; Arnold et al. 2009; U'Ren et al. 2010). Lichens of different orders harbour different endolichenic fungi (Girlanda et al. 1997; Li et al. 2007; U'Ren et al. 2010). Trophic transitions, i.e. changes in nutritional habits, are likely to occur. An example is *Pestalotiopsis maculans* (Corda) Nag Raj, an endolichenic fungus in numerous North Americal lichens (Sun et al. 2002) and causative agent of necrotic leaf spots in *Arbutus* and *Ceratonia* spp. (Trapero et al. 2003).

A whole range of new species were detected among endolichenic fungi (Arnold et al. 2009, U'Ren et al. 2010). An intense search for novel, biologically active compounds from isolates of these inconspicuous inhabitants of lichen thalli already yielded interesting results (Paranagama et al. 2007a, b; Kannangara et al. 2009; Zhang et al. 2009; Wang et al. 2010; Wijeratne et al. 2010).

the notoriously slow-growing lichen-forming ascomycetes. **Endolichenic fungi are symptomless endophytes, oligotrophic fungi which live in lichen thalli without causing damage and without producing sporulating structures at the thallus surface of their host lichen** (Petrini et al. 1990; Girlanda et al. 1997; Sun et al. 2002; Miadlikowska et al. 2004; Arnold et al. 2009; Fig. 15.7). Thus they are invisible to the naked eye and not even easily recognized with light microscopy techniques in thin sections. The older a thallus fragment is, the more of these endophytes are present.

The biological diversity and taxonomic affiliation of endolichenic fungi and their relations to endophytic, endobryophytic and leaf litter-degrading fungi is currently thoroughly investigated, the focus being on culturable isolates, which do not cover the full range of fungal inhabitants of lichen thalli. **Most endolichenic fungi are representatives of the Pezizomycotina**, the majority belonging to the subclasses **Pezizomycetes** and **Sordariomycetes** (Arnold et al. 2009; U'Ren et al. 2010). With few exceptions lichen thalli are not inhabited by the same fungi as leaves, needles or litter of the same microsite,

III. The Symbiotic Phenotype of Lichen-Forming Ascomycetes

A. Morphology Versus Phylogeny

The **morphological diversity of symbiotic phenotypes** among lichen-forming ascomycetes is overwhelming. However, this diversity reflects variations of one centrally important theme in lichen symbiosis: the fungal partner, as the quantitatively predominant exhabitant, has to provide a population of photoautotrophic cells with optimal conditions for photosynthesis and growth. This necessity induced the formation of the most sophisticated vegetative thallus structures in the fungal kingdom (Honegger 1991, 1993). Exposure of the photoautotrophic partner to adequate illumination forces lichen-forming fungi to grow at the surface of the substratum or even beyond, to adapt to extreme fluctuations in water contents and temperature and to develop protective mechanisms against irradiation damage at high exposure to UV light.

Similar thallus morphologies and mycobiont–photobiont interactions occur in unrelated lichen-forming taxa as a result of convergent evolution. The following paragraphs summarize morphological and anatomical features of lichen thalli and their occurrence among lichen-forming ascomycetes. The distinction between micro- and macrolichens is artificial but is used in the literature to distinguish morphologically simple, crustose, microfilamentous, microglobose, squamulose and placodioid morphotypes from highly advanced, 3-D thallus types.

B. Gelatinous Lichens ("Jelly Lichens")

Gelatinous lichens differ from all other lichens by their **lack of systems of aerial hyphae with hydrophobic cell wall surfaces** and thus their lack of air-filled zones within their thalli. In the fully hydrated state the **gelatinous extracellular** sheaths of cyanobacterial or, rarely, green algal photobionts absorb huge amounts of water, swell dramatically (Fig. 15.1m, n) and thus make up a very high proportion of the thalline biomass (Fig. 15.1o). **Upon desiccation the sheaths shrivel to a very small volume** and the lichens appear as thin, leathery or brittle structures.

1. Cyanobacterial Gelatinous Lichens

The most impressive gelatinous lichens are representatives of the **Collemataceae**, i.e. the genera ***Collema* and *Leptogium* among the Peltigerales** (Fig. 15.1m–o). **Both associate with *Nostoc*** spp. of the *Nostoc* II clade *sensu* O'Brien et al. (2005). In *Collema* and *Leptogium* spp. the fungal partner grows between and within the thick gelatinous sheaths of the *Nostoc* colonies and develops a close contact with the photobiont cells, however without penetrating their murein sacculus. The thalli of *Collema* and *Leptogium* spp. appear leaf-like (foliose) but lack the internal stratification as typically found in foliose macrolichens. *Leptogium* species differentiate a peripheral cortex, built up as monolayer, on both sides (Fig. 15.1o), whereas *Collema* spp. are not corticate and slightly translucent when wet (Fig. 15.1n); they receive their overall morphology and consistency from the photobiont, but the mycobiont controls the lobation and isidium formation (Fig. 15.1m, n).

The **Peltigerales comprise primarily macrolichens** with dorsiventrally organized, internally stratified thalli, i.e. with conglutinate upper cortex and air-filled photobiont and medullary layers (Fig. 15.1f). Phylogenetic data sets suggest that the gelatinous Collemataceae represent a special evolutionary line within the Peltigerales and do not form the base whence the families with stratified thalli evolved (Wedin et al. 2009; Otalora et al. 2010).

Despite the lack of a protective upper cortex and UV light absorbing secondary metabolites located therein some *Collema* spp. are found in arid, very sunny and dry habitats such as the Colorado Plateau or European xerothermic steppe vegetation, provided that liquid water is periodically available, e.g. as dew (Lange et al. 1998). *Collema tenax* (Sw.) Ach. (Fig. 15.1m) is the most widely distributed lichen in North American deserts, from hot (e.g. Sonora, Mojave, Chihuaha) to cool (North Great Basin; Rosentreter and Belnap 2001). The thick gelatinous sheaths of *Nostoc* cells (Fig. 15.2a–d) retain water in the thallus and thus facilitate metabolic acitivities of the fungal and cyanobacterial partners for longer periods of time than measured in green algal lichens. The latter, however, are better adapted to arid zones where humid air prevails and liquid water is scarce, their *Trebouxia* photobionts being able to photosynthesize already in the partially hydrated state (Lange et al. 1993; Lange 2001).

How can *Collema* species survive high solar irradiation without protective fungal cortex and UV light-absorbing, mycobiont-derived secondary metabolites? The **cyanobacterial sheath pigment scytonemin and mycosporine-like amino acids (MAAs), UV-absorbing compounds** known from a wide range of organisms (Sinha et al. 2007), were shown to be synthesized also in *Collema* spp. and other exposed, rock-inhabiting cyanolichens (Büdel et al. 1997).

Gelatinous lichens with a higher proportion of fungal biomass per thallus as in *Collema* and *Leptogium* spp. are formed by *Lichina* spp. (Lichinales, Lichinomycetes), whose black, microfilamentous thalli form characteristic tufts on rock surfaces at high tide mark and were often mistaken for small seaweeds

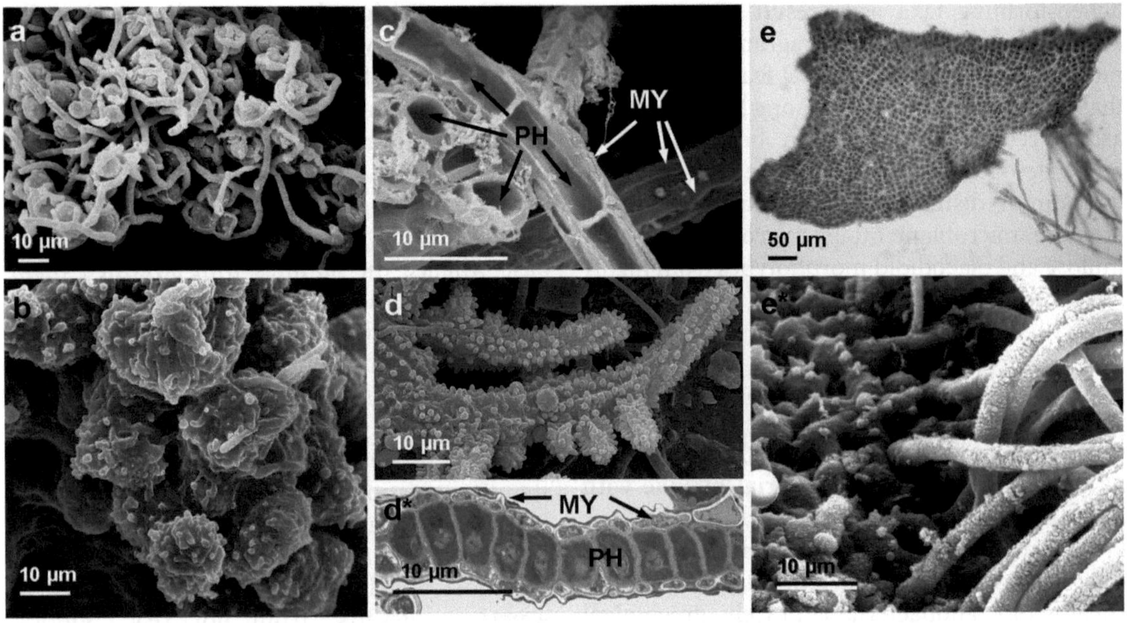

Fig. 15.8. Leprose, microglobose, microfilamentous and squamulose lichens with green algal photobionts ((**a–d**, **e***) SEM micrographs, (**d***) TEM micrograph, (**e**) LM micrograph). (**a**) *Lepraria incana* (L.) Ach. (*Stereocaulaceae, Lecanorales*) with *Trebouxia* sp. (*Trebouxiophyceae*), powdery, leprose crust with very hydrophobic surface. Each of the globose algal cells is contacted by the fungal partner. (**b**) Microglobose thallus of *Micarea hedlundii* Coppins (*Pilocarpaceae, Lecanorales*) with unidentified unicellular algal partner. Each algal cell is tightly ensheathed by the fungal partner. (**c**) Microfilamentous thallus of *Coenogonium subvirescens* Nyl. (*Coenogoniaceae, Ostropales*) with *Trentepohlia* sp. (*Ulvophyceae*). Loosely intervowen green algal filaments (*PH*) are partially ensheathed by hyphae of the mycobiont (*MY*), which do not penetrate the cellulosic cell wall of their photobiont. (**d**, **d***) Microfilamentous thallus of the foliicolous *Psoroglaena epiphylla* Lücking (*Verrucariales, Eurotiomycetes*; specimen leg. et det. Robert Lücking). This fungus arranges its unicellular, trebouxiophycean photobiont (related to *Chlorella* spp.) in very regular, uniseriate rows (for details, see Nyati et al. 2007). (**e**, **e***) Laminal views of the thin, squamulose, "flaky" thallus of *Flakea papillata* O.E. Erikss. (*Verrucariales*) consists of three layers: a monolayer of an unidentified unicellular green alga is kept in position by a one cell layer thick fungal meshwork each on the upper and lower thalline surface. Dark spots in (**e**) are algal cells

(Fig. 15.1p, r). The cyanobacterial colonies (*Calothrix* spp.) are located around a central fungal axis (Fig. 15.1q, r). *Lichina* spp. and their photobionts tolerate desiccation as well as submersion in sea water and soaking with freshwater (rain) at low tide.

2. Green Algal Gelatinous Lichens

Less conspicuous than the cyanobacterial gelatinous lichens are **crustose green algal gelatinous lichens**. Examples are the genera ***Epigloea*** and ***Thrombium*** (Dothideomycetes), slimy, unstructured overlays on soil, mosses or lichens. Black ascomata refer to the presence of the fungal partners, which live between and in the thick mucilaginous sheaths of either *Coccomyxa dispar* Schmidle (Jaag and Thomas 1934), *Gloeocystis* sp. or *Leptosira thrombii* Tschermak-Woess, respectively (Tschermak-Woess 1988).

C. Microfilamentous, Microglobose and Leprose Lichens

The majority of microfilamentous cyanobacterial and green algal lichens derive their morphology from the multicellular, filamentous photoautotrophic partner, which is overgrown and/or ensheathed by the mycobiont. Examples are: (1) the cyanobacterial *Pyrenothrix nigra*

Riddle (*incertae sedis* among Dothideomycetes) ensheathing colonies of *Scytonema* sp., or *Ephebe lanata* (L.) Vain. (Lichinales, Lichinomycetes) ensheathing colonies of *Stigonema* sp. (see illustration in Honegger 2001) and (2) the green algal *Cystocoleus ebeneus* (Dillwyn) Thwaites and *Racodium rupestre* Pers. (Capnodiales, Dothideomycetes; Muggia et al. 2007) or *Coenogonium* spp. (Gyalectales, Lecanoromycetes; Kauff and Lutzoni 2002) which ensheath filaments of *Trentepohlia* spp. (Fig. 15.8c; Meier and Chapman 1983).

Exceptional are the microfilamentous thalli of *Psoroglaena* spp. (Verrucariales, Eurotiomycetes), in which unicellular, trebouxiophycean green algae related to *Chlorella* and *Auxenochlorella* spp. are arranged by the fungal partner into very regular, uni- or multiseriate rows which strongly resemble filamentous cyanobacterial colonies (Fig. 15.8d, d*; Nyati et al. 2007). The regularity of this arrangement even lead to the assumption of a prochlorophyte (cyanobacterial) affiliation of these photobionts (Henssen 1995).

Microglobose lichens are built up by small globules with peripheral cortex containing a group of photobiont cells. Examples are *Micarea* spp. (Pilocarpaceae, Lecanorales; photobiont: *Elliptochloris bilobata* Tschermak-Woess; Brunner and Honegger 1985; Fig. 15.8b) or *Vezdaea* spp. (Pezizomycetes *incertae sedis*; photobiont: *Leptosira obovata* Vischer; Tschermak-Woess and Poelt 1976). Some authors refer to this thallus type as goniocysts, a term being normally used to distinguish corticate symbiotic propagules from non-corticate soredia.

Leprose lichens, recognizable as powdery crusts over natural and anthropogenic surfaces, are built up by loosely interwoven hyphae with very hydrophobic wall surfaces overgrowing groups of photobiont cells (*Trebouxia* spp.; Fig. 15.8a). Leprose taxa are sterile, but disperse efficiently by fragmentation. Examples are *Lepraria, Leproloma* and *Leprocaulon* spp. (*Stereocaulaceae*, Lecanorales), or *Chrysothrix* spp. (*Chrysothricaceae, Arthoniales, Arthoniomycetes*; Fig. 15.1c).

D. Crustose Lichens

Crustose thalli, as formed by an estimated 50 % of all lichen-forming ascomycetes, are **the most common and widespread symbiotic phenotypes among all groups of lichenized fungi;** they occur in all orders of lichen-forming ascomycetes except Peltigerales. **Crustose thalli are homoeomerous,** i.e. neither internally stratified, nor do they differentiate a periperhal cortex. Some crustose lichens develop rather inconspicuous thalli in the topmost layer of the substratum, where they meet their photobionts. **Endolithic species** occur in granitic **rocks** from the Arctic to the Antarctic (De los Rios et al. 2005); others are very common and widespread in calcareous rocks of temperate regions, only the ostioles of their fruiting bodies being recognisable at the rock surface (e.g. *Verrucaria* and *Bagliettoa* spp., Verrucariales). A similar situation occurs in **shells** of limpets (*Patella* spp.) and other molluscs or barnacles (*Cirripedia*) (e.g. *Pyrenocollema* spp., *Pyrenulales*; Fig. 15.1a, a*; or *Thelidium litorale* (Leighton) Keissler, Verrucariales; Espoz et al. 1995). Endophloeodal taxa inhabit the topmost layer of **bark.** Other crustose species grow on the surface of their substratum (i.e. rock, wood, bark, leaves or anthropogenic substrates), where they form either tiny areoles (Fig. 15.1d) or a more or less equally thick crustose thallus without internal stratification (Fig. 15.1b, p).

E. Placodioid and Squamulose Lichens

The **thalli of placodioid and squamulose lichen-forming fungi** differ from crustose taxa by their **cortical layer and by an internal stratification.** This type of thallus is formed by approx. 25 % of lichen-forming ascomycete species. **Placodioid thalli** adhere to the substratum with their entire lower surface and are areolate at their periphery (Fig. 15.1d, e). **Squamulose thalli** are small, dorsiventrally organized scales with an upper cortex and an algal layer, but usually without lower cortex (Fig. 15.1v, v*).

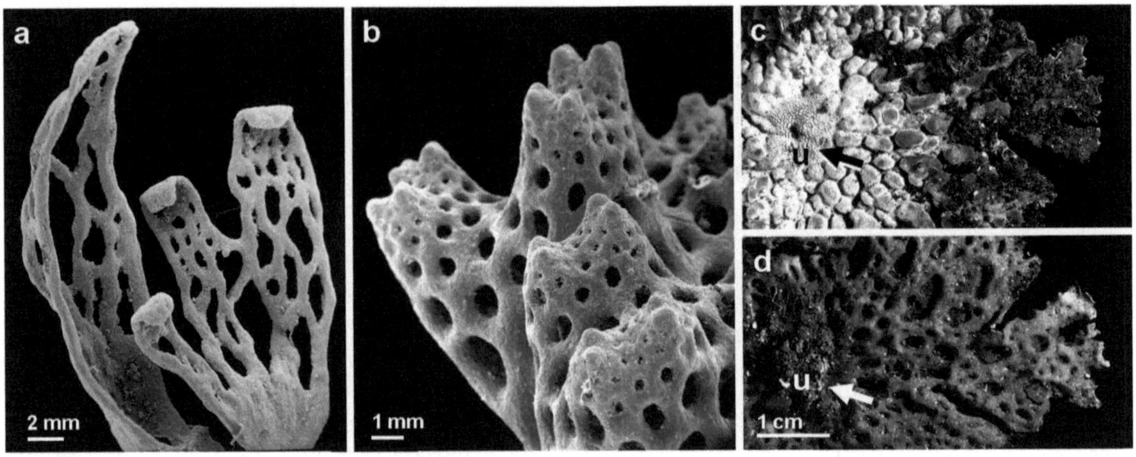

Fig. 15.9. Macrolichens with combined apical/marginal and intercalary growth and increasing mesh size in reticulate species (**a, b**) and with irregular, patchy intercalary growth (**c, d**). **a** "Fishnet" or "lace lichen", *Ramalina menziesii* Taylor (*Ramalinaceae, Lecanorales*), the fastest growing lichen so far known to science. (**b**) *Cladia retipora* (Nyl.) Labill. (Cladoniaceae, *Lecanorales*) with highly regular growth due to regular division of the apical pseudomeristems. (**c, d**) Upper (**c**) and lower surface (**d**) of the "toadskin lichen", *Lasallia pustulata* (L.) Mérat (*Umbilicariaceae, Lecanorales*), a rock tripe type of lichen with central holdfast (umbilicus; *u*) and patchy, irregular laminal growth in the centre. The margins, the oldest part of the lichen, are damaged due to wind abrasion. The marginal zones of the upper surface carry branched, black isidia

Exceptions are: (1) the minute, flaky thalli of *Flakea papillata* O.E. Erikss. (Verrucariales), with a one-cell-layer thick upper and lower cortex and unicellular green algal cells kept in a very regular monolayer in between (Fig. 15.8e, e*) and (2) the dorsiventrally organized squamules of *Peltula* spp. (Lichinomycetes; Büdel 1987; Fig. 15.1w), which differentiate a lower cortex and an epinecral layer on the upper thalline surface. In terricolous squamulose taxa such as *Psora decipiens* (Fig. 15.1u) or *Peltula* spp. (Fig. 15.1w) a dense meshwork of hyphal strands (rhizomorphs), arising from the lower cortex, grows deep into the soil and thus stabilizes the sand granules and prevents wind erosion (Fig. 15.1x), in much the same way as rhizomorphs of many other squamulose species of lecanoralean and acarosporalean genera penetrate even rock substrates (Sanders et al. 1994; Sanders and Ascaso 1997; Bjelland and Thorset 2002).

F. Foliose and Fruticose Lichens (Macrolichens)

The **conspicuous thalli of foliose and fruticose lichens**, often termed **macrolichens**, are the **most complex vegetative structures in the fungal kingdom.** Only one out of four lichen-forming ascomycetes (approx. 25 % of species) produces such a morphologically and anatomically complex symbiotic phenotype. **Dorsiventrally organized**, either **foliose** (leaf-like; Figs. 15.1f, 15.11c, 15.17a–h), **umbilicate** (leaf-like with central holdfast; Fig. 15.9c, d) or **band-shaped thalli and fruticose** (shrubby), either erect or pendulous morphotypes, all with internal stratification, are differentiated by representatives of diverse orders among the **Pezizomycotina.** These fungi grow into the third dimension above the substrate and compete for space above ground. Fruticose taxa may form large amounts of biomass on restricted grounds, impressive examples being the beard- and band-shaped and fishnet lichens (e.g. *Bryoria*, *Usnea* and *Ramalina* spp. among Lecanorales, Teloschistes spp. among Teloschistales; Figs. 15.1t, y, 15.9a). Each photobiont cell is actively positioned by the fungal partner within the **photobiont layer** (Figs. 15.13g–l, 15.17h). Thus the **heterotrophic fungal exhabitant secures adequate illumination and optimal gas exchange of his photoautotrophic partner**, which has access to water and dissolved nutrients exclusively via the apoplastic continuum (Honegger 1984, 1986b, 1991,

2009). The fungal arrangement of the photoautotrophic cells within these dorsiventrally or radially organized thalli is analogous to the palisade parenchyma of leaves or to the radially arranged, photosynthetically active parenchyma of green shoots, respectively (Honegger 1998; Sanders 2006).

A **wide range of fungal differentiation** occurs in macrolichens with regard to different functions, such as **rhizinae** (holdfasts, Fig. 15.17a, h), **cilia** (condensation of humidity; Fig. 15.1g), **cyphellae** and **pseudocyphellae** (aeration pores; Fig. 15.17b).

IV. Main Building Blocks and Functional Anatomy of Lichen Thalli

A. Fungal Adaptations to Symbiosis with Minute Photobiont Cells

In order to achieve high photosynthetic productivity lichen photobionts need to be adequately illuminated, have optimal conditions for gas exchange and access to water and dissolved nutrients. In contrast to plant pathogenic or mycorrhizal fungi **lichen-forming ascomycetes** cannot invade or ensheath multicellular, photosynthetic or non-photosynthetic host tissues; instead, they **cultivate minute photoautotrophic cells in their thallus**, whichwas interpreted as a bottle garden or sophisticated fungal culturing chamber for a population minute photobiont cells (Honegger 1991, 1993, 1998). **Life at the surface of the substratum** or even above, as is the case in foliose and fruticose lichens, necessitates: (1) **tolerance of strong fluctuations in water contents** between full hydration and desiccation, i.e. water contents around or below 10 % by dry weight, (2) **tolerance of intense solar radiation**, especially UV-B light, and (3) **mechanical stability**.

As summarized below, lichen thalli are the product of an amazing hyphal polymorphism, main building blocks being zones with either hydrophilic or hydrophobic fungal walls and with either polar/filamentous or apolar/globose hyphal growth (Honegger 1998). The focus is on foliose and fruticose taxa.

B. Hydrophilic Zones and Their Functions

1. Conglutinate, Tissue-Like Zones (Pseudoparenchyma)

Conglutinate pseudoparenchyma are built up by either periclinally or anticlinally arranged hyphae which tightly stick together by means of **large amounts of secreted glucans**, e.g. of the lichenin type ($1 \rightarrow 3,1 \rightarrow 4$-β-glucans) or the pustulan type [($1 \rightarrow 6$)-β-D-glucan; for a review, see Stone and Clarke 1992]. These secreted glucans are deposited outside the cell wall proper. The anatomy and cell shapes in conglutinate pseudoparenchyma are best visible after removal of the glucans by maceration techniques (Anglesea et al. 1982; Honegger and Haisch 2001; Figs. 15.10b, c, 15.16e, f).

Peripheral cortical layers, as typically found in placodioid, squamulose, foliose and fruticose lichens, are **formed by conglutinate pseudoparenchyma: tissue-like zones**, which provide mechanical stability, tensile strength and elasticity to the thalli and play important roles in thalline water relations, light transmission and gas exchange. Many taxa differentiate, in addition to peripheral cortical layers, conglutinate internal strands to improve the mechanical stability of the thallus (Fig. 15.10e–h).

Conglutinate pseudoparenchyma are involved in passive water uptake and retention. In the fully hydrated state they are elastic and translucent, but more or less brittle and opaque in the desiccated state. Thus **light transmission to the photobiont layer is influenced by the level of hydration of the cortex and by the amount of deposits of mycobiont-derived, crystalline secondary metabolites** upon and within it (Dietz et al. 2000). The upper and lower cortex may differ as regards to anatomy, thickness and pigmentation. In many species the lower cortex is black due to melanized cell walls (Figs. 15.10d, 15.17d–f, h). It remains to be seen whether the melanized cortical layers of lichenized ascomycetes fulfil

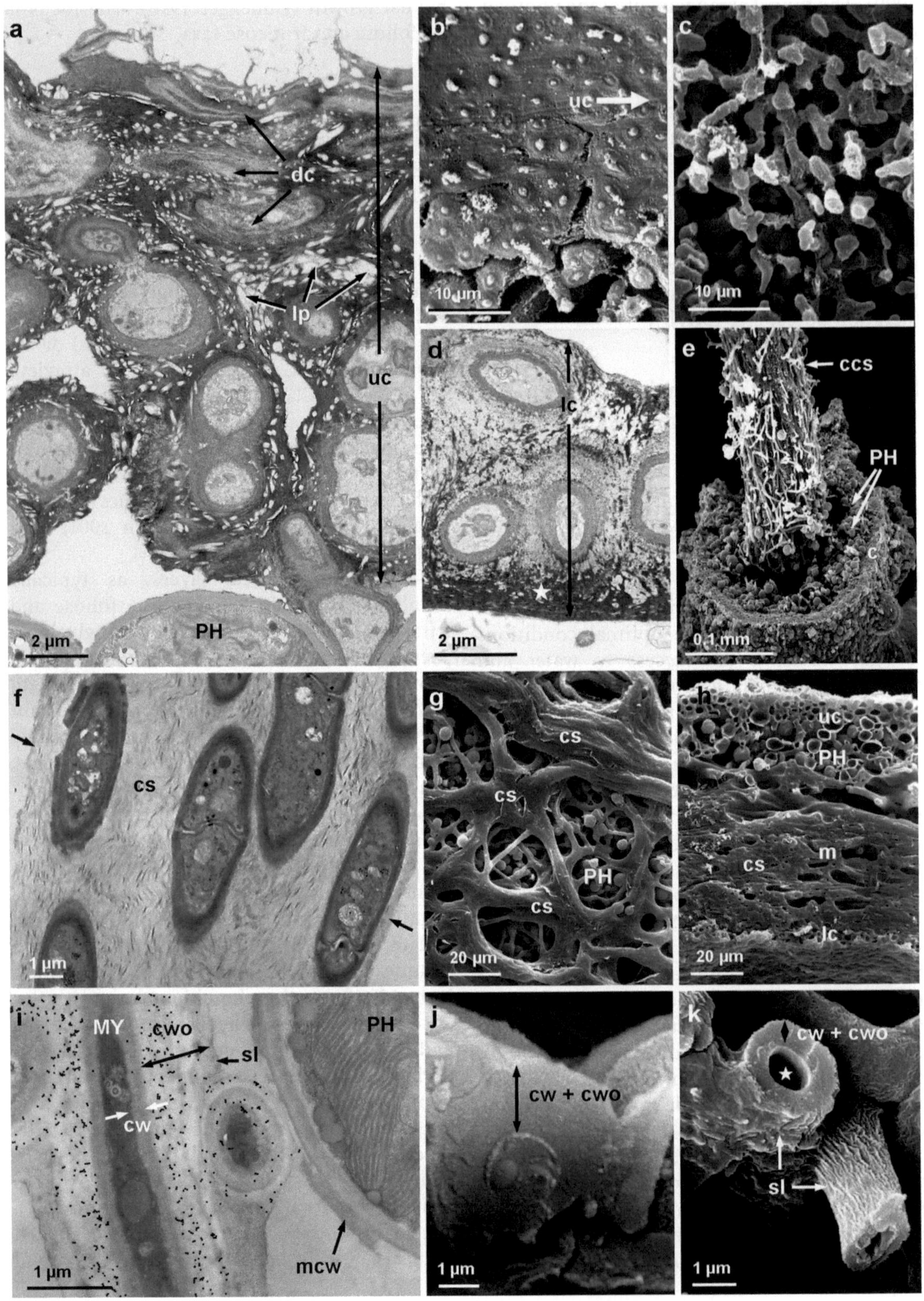

Fig. 15.10. Hydrophilic zones in macrolichens: (**a–d**) conglutinate pseudoparenchyma as peripheral cortical layers, (**e–h**) conglutinate internal strands, (**i, k**) hydrophilic hyphal wall overlays. (**a**) TEM micrograph of the

functions similar to the melanized cell walls of plant pathogenic fungi such as *Magnaporthe grisea*. Melanin-deficient mutants of this foliar pathogen of rice are unable to build up the high pressure in the appressorium as required for infection because melanin prevents molecules larger than water from leaking of the cell (for a review, see Talbot 2003). **The upper cortical layers often are brightly coloured due to mycobiont-derived secondary metabolites,** which crystallize on and within their gelatinous matrix (Fig. 15.10a). Examples are the sulfur-yellow **vulpinic acid** of wolf lichen [*Letharia vulpina* (L.) Hue], the bright yellow **usnic acid** of the beard lichens (*Usnea* spp.) or of the placodioid *Ophioparma ventosa* (Fig. 15.1d), or the yellow to orange **anthraquinones** of various Teloschistaceae (Figs. 15.1e, g–I, u, y, 15.5a, 15.11c). The golden to orange tinge of the nitrophilous, epiphytic or epilithic *Xanthoria* spp. from coastal to alpine habitats refers to eutrophication.

A massive peripheral cortex may be a hindrance to gas exchange. Therefore many taxa differentiate **aeration pores** such as **cyphellae** or **pseudocyphellae** (Fig. 15.17b); these are islets of hyphae with hydrophobic cell wall surfaces within conglutinate, hydrophilic pseudoparenchyma.

Peripheral cortical layers are not only the site of passive water uptake but also of **pollutants** including metals and radionuclides (Feige et al. 1990; Biazrov 1994; Purvis and Pawlik-Skowrońska 2008; Gougeon et al. 2009). Thus **lichen thalli are archives of environmental changes** (Purvis et al. 2007). Many pollutants are transformed into insoluble forms, e.g. by means of either chelatization or formation of insoluble complexes with secondary metabolites (Haas and Purvis 2006; Purvis and Pawlik-Skowrońska 2008). This allows some lichens to accumulate high amounts of metals without damage to the mycobiont and photobiont. An example is *Lecanora vinetorum* Poelt et Huneck on wooden supports in vineyards with up to 5000 ppm copper originating from the regular application of copper sulfate-containing fungicides ("Bordeaux mixture"; Poelt and Huneck 1968).

2. Hydrophilic Cell Wall Overlays in Aerial Hyphae of the Medullary Layer

Whereas the conglutinate pseudoparenchyma of cortical layers or internal strands are well visible at the light microscopy level, the hydrophilic, glucan-based cell wall overlays of many hyphae with hydrophobic wall surfaces in the thalline interior are best visible in electron microscopy preparations. Many lichen-forming

upper cortex (*uc*) of the foliose *Parmelina tiliacea* (Hoffm.) Hale (*Lecanorales*), with dead cortical cells on the surface (*dc*) and holes left in the conglutinate wall material from crystalline lichen products (*lp*); these were dissolved during the preparative procedure for ultrathin sectioning. (**b, c**) SEM micrographs of the cortex of the foliose "Icelandic moss", *Cetraria islandica* (L.) Ach. (*Lecanorales*), prior (**b**) and after (**c**) dissolution of the mucilaginous extracellular material which was secreted by the cortical hyphae. (**d**) TEM micrograph of the black lower cortex of *Parmelina tiliacea*. Melanin granules are seen as electron dense inclusions. (**e**) SEM micrograph of a fragment of a "beard lichen", *Usnea hirta* (L.) Weber ex F.H. Wigg (*Lecanorales*), with conglutinate peripheral cortex (*c*), the photobiont layer (*PH*) and a conglutinate central strand (*ccs*), which provides the thallus with tensile strength. (**f–h**) A meshwork of laminal, conglutinate internal strands (*cs*) in *Xanthoria parietina* (*Teloschistales*), as seen in a TEM micrograph of an ultrathin section (**f**) and in SEM preparations (**g, h**). *Arrows* in (**f**) point to the surface of the strand which is covered by a very thin hydrophobin layer; for details, see Fig. 15.11c–e and Scherrer et al. (2000, 2002). (**g**) Laminal view towards the algal layer (*PH*) in a horizontally dissected thallus. (**h**) Cross-section with upper (*uc*) and lower cortex (*lc*), photobiont layer (*PH*) and dissected conglutinate strand (*cs*) in the medullary layer (*m*). (**i–k**) hydrophilic cell wall overlays in medullary hyphae of *Cetraria islandica* (L.) Ach. (Icelandic moss). (**i**) TEM micrograph of an ultrathin section labelled with a monoclonal antibody against lichenin. Black dots are gold granules visualizing the binding sites of the antibody. Lichenin is located in the hydrophilic cell wall overlay (*cwo*) over the cell wall proper (*cw*) and the thin, proteinaceous hydrophobic wall surface layer (*sl*). (**j, k**) LTSEM micrographs of fully hydrated (**j**) and drought-stressed (**k**) medullary hyphae, with dramatic shrinkage in the cell wall and overlay (*cw*+*cwo*), and deformation of the hydrophobic wall surface layer (*sl*). *Asterisk* indicates a desiccation-induced, reversible cytoplasmatic cavitation bubble. From Honegger and Haisch (2001) ascomycetes have extremely thick cell walls. In ultrastructural analyses a thick overlay is found on the cell wall proper. As shown with immunocytochemical techniques this **wall overlay** contains **high amounts of glucans**, e.g. of the lichenin-type; these are **hydrophilic, absorb and retain high amounts of water** and thus play important roles in thalline water relations (Honegger and Haisch 2001; Fig. 15.10i–k). **A thin, peripheral hydrophobic wall layer on top of the hydrophilic wall overlay** (Fig. 15.11b, f, g) **prevents free water from accumulating in the thalline interior**, which stays air-filled at every level of thalline hydration (Fig. 15.12c; see Sect. III.C.1). During drought stress events the hydrophilic wall overlay loses water and shrivels dramatically (Fig. 15.10j, k).

C. Hydrophobic Zones: Gas-Filled Plectenchyma in the Thalline Interior

The **thalli of all lichens (except gelatinous taxa) comprise air-filled zones**, usually as internal medullary and photobiont layers, which are **built up by loosely interwoven aerial hyphae with water-repellent surfaces.** This wall surface hydrophobicity prevents free water from accumulating on hyphal surfaces in the thalline interior, even at full hydration (Fig. 15.12c). The **maintenance of air-filled plectenchyma by the fungal partner is a prerequisite for successful photosynthesis of the photobiont and for the gas exchange by both partners of lichen symbiosis.** Intrathalline cell wall surface hydrophobicity is not restricted to fungal hyphae, but also covers the photobiont wall surfaces. Thus the **apoplastic continuum** between the heterotrophic fungal and the photoautotrophic algal or cyanobacterial partners is **sealed with a hydrophobic coat.** During the regular wetting and drying cycles this water-repellent lining forces fluxes of solutes to passively translocate within the cell wall underneath the hydrophobic wall surface coat, from the conglutinate peripheral cortex to the photobiont and medullary layers during rehydration and vice versa during desiccation (Honegger 1984, 1986b; Honegger and Peter 1994; Honegger et al. 1996b). The majority of lichen-forming ascomycetes and their photobionts tolerate drought. Continuous desiccation and rehydration events are an integral part of the lichen lifestyle.

1. Protein-Derived Wall Surface Hydrophobicity

Early investigators assumed mycobiont-derived secondary metabolites, which crystallize on hyphal and even algal cell wall surfaces (Honegger 1986b), to generate the wall surface hydrophobicity inside lichen thalli (Goebel 1926a, b; Armaleo 1993). Indeed, lichens with high loads of medullary secondary metabolites are extremely difficult to infiltrate with aqueous solutions. However, wall surface hydrophobicity is also evident in lichens with no medullary secondary compounds, as demonstrated with Low-temperature scanning electron microscopy (LTSEM) techniques of freeze-fractured, frozen-hydrated specimens (Honegger and Peter 1994; Scheidegger 1994; Honegger 1995; Scheidegger et al. 1995; Honegger et al. 1996b).

At transmission electron microscopy (TEM) level a **mycobiont-derived, proteinaceous coat** with a distinct **rodlet pattern** was resolved in freeze-etch replicas **on hyphal surfaces in the medullary layer** in a wide range of lichen-forming ascomycetes (Fig. 15.11b; Honegger 1982). The same type of rodlet layer occurs on aerial hyphae of non-lichenized asco-and basidiomycetes (for reviews, see Wessels 1997; Wösten 2001; Linder et al. 2005; Linder 2009). In lichen symbiosis the **proteinaceous rodlet layer** spreads from the contacting hypha over the photobiont cell wall, thus **sealing the apoplastic continuum of both partners** (Fig. 15.11g; Honegger 1984, 1986b; for reviews, see Honegger 1991, 1998, 2009).

Due to their peculiar chemical nature the class of proteins, which generate wall surface hydrophobicity and form amphiphilic superficial rodlet layers on aerial hyphae of lichen-forming and non-lichenized fungi, escaped the attention of biochemists until the late 1980s. **Class I hydrophobins**, small (approx. 100 amino acids), secreted fungal proteins self-assemble *in vivo* and *in vitro* at liquid–air (hydrophilic–hydropho-

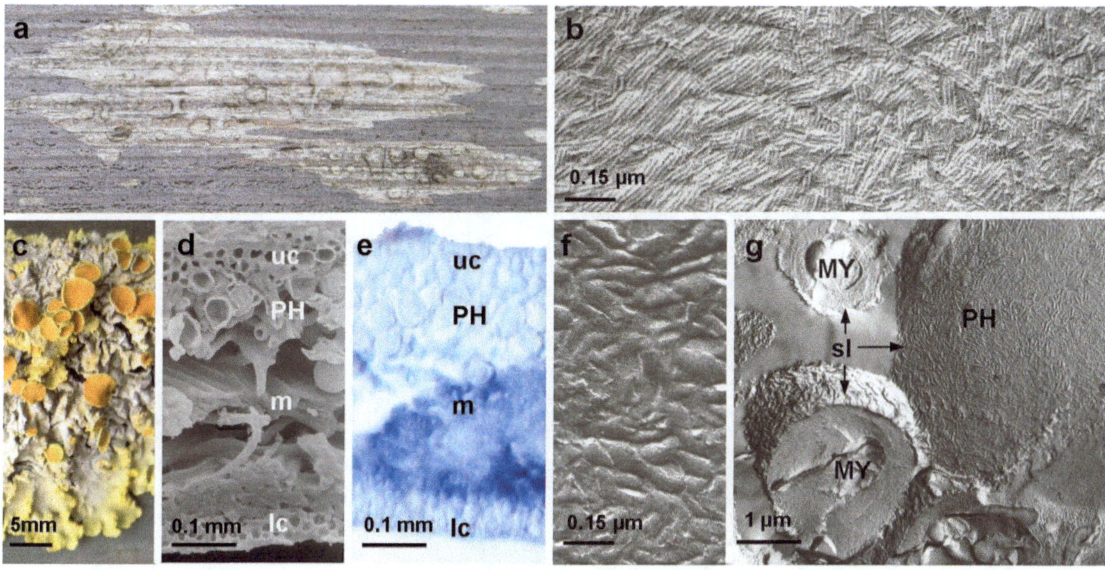

Fig. 15.11. Protein-based cell wall hydrophobicity in lichen forming ascomycetes. (**a**) Water repellency of the crustose *Lecanora conizaeoides* Nyl. Ex Cromb. (*Lecanorales*) on wood. (**b**) TEM micrograph of the water repellent rodlet layer on the surface of a medullary hypha in *Peltigera membranacea* (Ach.) Nyl. (*Peltigerales*), as seen in a freeze-fracture replica. No rodlet layer is differentiated in hyphae which grow into the cyanobacterial colony (see Fig. 15.2a, b). (**c–e**) In *Xanthoria parietina* (L.) Th. Fr. (*Teloschistales*) the hydrophobic wall surface protein (*X. parietina* hydrophobin; XPH) was characterized (Scherrer et al. 2000, 2002). The *XPH* gene is expressed in medullary hyphae (*m*), not in the upper (*uc*) or lower (*lc*) cortex or the algal layer (*PH*), as shown with in situ hybridization techniques (intense blue colour in (**e**)). The secreted hydrophobin spreads over the fungal and algal wall surfaces. (**d**) SEM micrograph for comparison. (**f, g**) TEM micrographs of freeze-etch replicas of wall surfaces in the thalline interior of *Cladonia macrophylla* (Schaerer) Stenhammar (*Lecanorales*). In the majority of Lecanorales the rodlet layer is obscured by additional material (presumably polyphenolic secondary metabolites), giving the water repellent wall surface layer (*sl*) an irregularly tessellated pattern. The hydrophobic wall surface layer spreads from the contacting fungal hyphae (*MY*) over the algal wall surface (*PH*), thus sealing the apoplastic continuum with a water-repellent coat)

bic) interfaces to an amphiphilic protein layer with semicrystalline rodlet pattern (Wösten et al. 1994). **Self-assembled hydrophobin layers are insoluble** with preparative protocols, as normally used for protein dissolution, but dissolve in trifluoracetic acid. Among non-lichenized fungi the amphiphilic hydrophobins fulfil diverse functions as surfactants, adhesives to hydrophobic surfaces such as plant or insect cuticles etc. (for reviews, see Wessels 1997; Wösten 2001; Whiteford and Spanu 2002; Linder et al. 2005; Linder 2009).

Hydrophobins reveal low sequence homology except eight cysteine residues in a conserved pattern: X_{26-85}-Cys-X_{5-8}-Cys-Cys-X_{17-39}-Cys-X_{8-23}-Cys-X_{5-6}-Cys-Cys-X_{6-18}-Cys-X_{2-13}. Four intramolecular disulfide linkages provide this group of proteins with their characteristic properties (Kershaw et al. 2005; Linder et al. 2005; Linder 2009).

Some fungal species produce only one, others a whole range of hydrophobins; their genetics are analysed either on the basis of amino acid sequences of the purified protein, or by genomic sequence analysis. XPH1 from *Xanthoria parietina* (L.) Th. Fr. and XEH1 from the morphologically very similar *X. ectaneoides* (Nyl.) Zahlbr. (Teloschistales) were the first hydrophobins characterized in lichen-forming fungi (Scherrer et al. 2000). As shown with *in situ* hybridization techniques the XPH1 gene is expressed in the aerial hyphae of the algal and medullary layers (Fig. 15.11e), but neither in the conglutinate peripheral cortical layers nor in the mucilage-filled pycnidia or hymenial layer of the fruiting body (Scherrer et al. 2002). Hydrophobins evolve rapidly and thus would be interesting markers for phylogenetic analyses, but due to their low sequence homology they are very difficult to explore

Fig. 15.12. Wall surface hydrophobicity due to crystals of mycobiont-derived secondary metabolites. (**a**) LTSEM preparation of the mycobiont (*MY*)–photobiont (*PH*) contact site in *Hypogymnia physodes* (L.) Nyl. (*Lecanorales*), with crystals of medullary lichen products on the wall surfaces of both symbionts. (**b**) Medullary hyphae of *Usnea rubiginea* (Michaux) A. Massal (*Lecanorales*) are covered with crystalline secondary metabolites and calcium oxalate crystals (Whewellite). (**c**) LTSEM of freeze-fractured, fully hydrated *Physcia adscendens* H. Olivier. View on the algal layer at the onset of soralium differentiation and soredium formation. The photobiont cells in the algal layer are free from crystals, but covered with fine needles in the soralium. Soredia are covered with secondary metabolites and thus very hydrophobic. Note: there is no free water in any part of this fully hydrated lichen due to the thin, proteinaceous, hydrophobic wall surface layer on which the secreted secondary metabolites crystallize

(Scherrer and Honegger 2003; Eichenberger 2007). Three different hydrophobins, DGH1, DGH2 and DGH3 with low sequence homology were identified in the agaricalean basidiolichen *Dictyonema glabratum* (Spreng.) D. Hawksw. (syn. *Cora pavonia* (Web.) E. Fries; Trembley et al. 2002a, b).

2. Hydrophobicity Due to Mycobiont-Derived Secondary Metabolites

The majority of lichen-forming ascomycetes produce secondary metabolites, mostly polyphenolics, which are secreted in a yet unknown, possibly glycosidic form and **crystallize either within conglutinate cortical layers** (cortical lichen products; Fig. 15.10a) **or on the surfaces of aerial hyphae in the thalline interior** (medullary lichen products Fig. 15.12b). Temperature and light, especially UV B, have an impact on the production of secondary metabolites by lichen-forming ascomycetes (Huneck et al. 2004; Armaleo et al. 2008).

Medullary lichen products crystallize within and on the thin, proteinaceous hydrophobic wall surface layer which covers all fungal and photobiont cell wall surfaces in the thalline interior; thus crystals of mycobiont-derived secondary products are also found on the wall surface of the photobiont (Fig. 15.12a, c; Honegger 1986b; for reviews, see Honegger 1991, 1997, 2009). Crystals of cortical and medullary secondary metabolites are highly water repellent (Huneck 2003) and thus enhance the hydrophobicity of cortical layers and medullary hyphae. Upon removal of crystalline secondary metabolites from medullary hyphae of *Peltigera leucophlebia* (Nyl.) Gyeln. by means of solvents the protein-based wall surface hydrophobicity was retained (Honegger and Hugelshofer 2000). Mycobiont-derived secondary metabolites are not the prime generator of wall surface hydrophobicity in lichens, but enhance it significantly.

3. Peculiarites of Lichen Products

More than 700 secondary fungal metabolites are known from lichen-forming fungi, 90 % of them occurring in lichens only (Huneck and Yoshimura 1996; Huneck 2001; Ingólfsdóttir 2002). The remaining 10 % are also found in non-lichenized fungi, especially among Eurotiomycetes (e.g. *Penicillium* and *Aspergillus* spp.); these have been postulated to derive from lichenized ancestors (Lutzoni et al. 2001). Horizontal transfer of polyketide synthase (PKS) genes from actinobacteria to ascomycetes was hypothesized to have facilitated the evolution of typical lichen compounds, such as orsellinic acid derivatives (Schmitt and Lumbsch 2009). Lichen products evolved early in the radiation of filamentous fungi (Armaleo et al. 2011).

Some lichen products are also synthesized by plants [e.g. anthraquinones in *Senna* spp., *Aloe* spp. and *Rheum* spp. (rhubarb), all being pharmaceutically exploited], insects (Pankewitz et al. 2007) and prokaryotes (Schmitt and Lumbsch 2009).

In contrast to many secondary metabolites of non-lichenized fungi **most lichen products are largely insoluble in aqueous systems** at pH below 7. The solubilization of lichen-derived depsides and depsidones with non-toxic agents is a challenge in pharmaceutical applications (Kristmundsdóttir et al. 2005; Elo et al. 2007).

Positive correlations were found between the hydrophobicity of the thalline surface and **SO_2 tolerance of lichens.** The highly SO_2 tolerant *Lecanora conizaeoides* (Nyl.) ex Cromb. is highly water repellent (Fig. 15.11a; Shirtcliffe et al. 2006; Hauck et al. 2008). This inconspicuous, acidophilic crustose lichen is the most common and widespread species in urban areas with high levels of air pollution. It seems to originate from acid pine bark in North European *Pinus mugo* bogs whence it invaded the cities of European industrial areas at the beginning of the twentieth century and colonized the strongly acidified bark ($pH<4$) of deciduous trees in polluted areas (Wirth 1985; LaGreca and Stutzman 2006). With declining SO_2 pollution and subsequent increase of bark pH values, *L. conizaeoides* disappears from urban areas where it had been previously recorded for decades (Bates et al. 2001; Hauck et al. 2011).

The **protein- and secondary product**-based water repellency of lichens also has an **impact on the substratum.** Conservators in charge of the preservation of cultural heritage demonstrated experimentally that **lichen cover significantly reduces the wettability and thus the erosion of many types of rock surfaces,** especially limestone and certain types of sandstone (Drewello and Drewello 2009; Beierkuhnlein et al. 2011). However, in rock surfaces with other types of mineral composition the combined action of oxalic acid and secondary product secretion by certain lichen-forming ascomycetes accelerates erosion (Chen et al. 2000; for reviews, see St. Clair and Seaward 2004; Gadd 2007).

The question whether lichens and lichen products of epiphytic species have a negative impact on the phorophyte (plant on which an epiphyte grows) cannot be answered conclusively as long as the depth of fungal penetration into the tree tissues and the mobile, secreted form of lichen products are unknown. Lichens are generally assumed to fix themselves on the bark and not to penetrate into the living tissues of trees.

Based on the microscopic observation of fungal hyphae in sections of *Quercus pyrenaica* with epiphytic *Evernia prunastri* ("oak moss", "mousse de chêne" in perfume industry) this lichen-forming ascomycete was postulated to grow through all tissues into the central medulla of the tree (Ascaso et al. 1980). However, there is no conclusive evidence that the hyphae belonged to the lichen-forming ascomycete. *Evernia prunastri*-derived lichen polyphenolics were detected in the water conducting vessels (xylem) of *Quercus pyrenaica* with epiphytic "oak moss" (Avalos et al. 1986). In a comparative analytical study on the impact of oak wood on the metabolome of high quality, barrel-matured French wines (burgundy types and others) the best oak originated from lichen-rich forests in north-eastern France

(Fôret de Bitche, Lorraine). Traces of the lichen product atraric acid, a derivative of the depside atranorin, were detected with ultrahigh resolution mass spectrometry among the very rich and diverse polyphenolics of the wine (Gougeon et al. 2009). Atraric acid from epiphytic Parmeliaceae was found with gas chromatography/mass spectrometry (GC/MS) techniques in the top layer of the hardwood of oak as used for cooperage (Bourgeois et al. 1999). It remains to be seen whether this was due to horizontal diffusion of atraric acid as a soluble form of atranorin from thalli growing on the bark surface over several centimetres, or whether it was produced *in situ* by hyphae penetrating deep into the wood, through the secondary phloem and cambium into the xylem, as proposed by Ascaso et al. (1980).

Soluble forms of lichen products might diffuse via pith and wood rays into the deeper layers of the wood body. With their non-lignified cell walls the pith and wood rays are tissues which might be most easily invaded by the hyphae of lichen-forming fungi.

The biosynthesis of many secondary metabolites is developmentally regulated. Fruiting bodies often comprise different lichen products than the vegetative thallus; examples are the blood-red apothecia of *Ophioparma ventosa* (L.) Norman (formerly *Haematomma ventosum* (L.) Massal.) due to haemoventosin (Fig. 15.1d), or the intensely red apothecia of *Cladonia* spp. (Fig. 15.1j) due to the iron complex of the *bis*-anthraquinone bellidiflorin (Huneck 2001). The onset of soredium formation often triggers the massive production of secondary metabolites in the algal layer since soredia have to be covered with a highly water-repellent coat, an optimal design for long distance wind dispersal and protection against herbivory (Fig. 15.12c; Armstrong 1994; Tormo et al. 2001; Asplund et al. 2010).

Other important functions of the partly massive deposits of secondary metabolites in lichens (1–2 % by dry weight, up to a maximum of 10 % by dry weight) are the **antibiotic and antiherbivore properties** (Cochietto et al. 2002; Gauslaa 2005; Nybakken et al. 2010; Pöykkö et al. 2010). In recent years the search for biologically, especially pharmaceutically active lichen metabolites were intensified (for reviews, see Müller 2001; Ingólfsdóttir 2002).

As lichen-forming ascomycetes are extremely slow growing, even in the aposymbiotic state, the efforts to characterize polyketide synthase (PKS) genes were intensified with regard to their heterologous expression in fast-growing, non-lichenized fungi (for reviews, see Miao et al. 2001; Stocker-Wörgötter 2008). Other important functions of the secondary metabolites include: (1) chelation of cations (Chen et al. 2000; Favero-Longo et al. 2005, 2007), (2) the formation of insoluble complexes with metal ions, which enable many species to grow on metal-rich substrates (Hauck and Huneck 2007a, b; Hauck et al. 2009; Grangeon et al. 2012; for reviews, see Purvis and Pawlik-Skowrońska 2008), and (3) absorption of UV-B light, secondary product biosynthesis being enhanced by seasonally or experimentally elevated UV-B irradiation (Bjerke et al. 2002, 2005; Buffoni Hall et al. 2002; Solhaug et al. 2003; Huneck et al. 2004; Nybakken et al. 2004; McEvoy et al. 2006; Nybakken and Julkunen-Titto 2006; Armaleo et al. 2008). The majority of lichen products are autofluorescent (for an example, see Honegger 2009) and thus transform UV light into longer wavelengths, which might be used as an energy source by the photoautotrophic partner.

D. The Mycobiont-Photobiont Interface

The main characteristic of the **mycobiont–photobiont interface in lichen-forming ascomycetes** is the **structural simplicity of the contact site** (Honegger 1993). In marked contrast to other biotrophic fungal interactions with plant cells such as rust fungi, powdery mildews or arbuscular mycorrhizal fungi (for reviews, see Voegele et al. 2009; Ridout 2009; Bonfante et al. 2009) no complex lobate or branched haustoria with large surface area in close contact with the plasma membrane of the photoautotroph are formed in lichen-forming ascomycetes. **The majority of mycobionts do not even penetrate the cell wall of their photobiont** (Honegger 1986a, 1991, 2009). Structurally this situation resembles the fungus-plant cell interfaces of ectomycorrhizae (for a review, see Martin and Tunlid 2009) or *Epichloë* endophytes of grasses (for a review, see Schardl et al. 2009).

The key to understanding the function of mycobiont–photobiont interactions in lichens are their **poikilohydric water relations**. The **majority of lichen-forming fungi and their photobionts are adapted to continuously fluctuating water contents from full hydration to desiccation.** Drought stress, i.e. water contents around 10 % by dry weight, is survived unharmed in a state of physiological dormancy. During desiccation the cells of both partners shrivel dramatically, but their membrane systems stay intact (Honegger 1995; Honegger et al. 1996b; for a review, see Honegger 2007). The protoplast of fungal cells cavitates under drought stress, but the protoplasmic cavitation bubble (Fig. 15.10k) disappears rapidly during rehydration (Honegger and Peter 1994; Scheidegger 1994; Scheidegger et al. 1995; Schroeter and Scheidegger 1995; for a review, see Honegger 2009). Full metabolic activity is regained within minutes after rehydration. The main fluxes of solutes in the apoplastic continuum are passively driven by wetting and drying cycles (Honegger and Peter 1994; for reviews, see Honegger 1997, 2007). Active uptake of soluble carbohydrates is assumed to take place at the immediate mycobiont–photobiont contact sites (see Figs. 15.3a, b, 15.13b–l).

1. Cyanobacterial Lichens

Intragelatinous fungal protrusions within the gelatinous sheaths of cyanobacterial colonies are the most common interactions of lichen-forming ascomycetes with cyanobacterial photobionts (Fig. 15.2a, b; Honegger 1991). The thin murein saccule of the cyanobacterial cell (Fig. 15.2b, c) is not penetrated. The gelatinous sheath (Fig. 15.2b–d) is composed of polysaccharide fibrils which are arranged parallel to the cyanobacterial cell surface (Hill et al. 1994). *Nostoc* spp. (Nostocales), the most common cyanobacterial photobionts of lichen-forming ascomycetes, form hormogonia, sheath-free, naked stages (Fig. 15.16b) with gliding motility on wet surfaces (Meeks et al. 2001). Lichen-forming ascomycetes receive glucose as mobile carbohydrate and fixed nitrogen as ammonium ion from their cyanobacterial photobionts (for reviews, see Hill 1976; Honegger 1997).

2. Green Algal Lichens

The most common and widespread **green algal photobionts** of lichen-forming ascomycetes are the **trebouxioid and trentepohlioid taxa with cellulosic walls.** As in plant cell walls the **cellulose fibrils are embedded in an amorphous matrix consisting of glucans and proteins** (Fig. 15.13a; Brunner and Honegger 1985). The growing hyphal tip in contact with the algal wall surface secretes hydrolases which may partially degrade the algal cell wall. Finger-shaped, transparietal ("intracellular") haustoria, which penetrate the cellulosic wall of *Trebouxia* or *Trentepohlia* spp., are formed by numerous crustose taxa (e.g. *Lecidella*, *Lecanora* and *Amandinea* spp.; Fig. 15.13b), but were not found in foliose and fruticose morphotypes (Honegger 1984, 1986a, 1991; Matthews et al. 1989).

Three types of intraparietal haustoria were distinguished in ascomycetous lichens with trebouxioid photobionts. These lichen-forming ascomycetes grow into, but do not penetrate the algal cell wall (Honegger 1984, 1986a, 1990; for a review, see Honegger 1991):

Type 1: appressoria and very short fungal infection pegs which do not penetrate the algal cell wall (Fig. 15.13c); this type of interaction was observed in crustose and placodioid taxa.

Type 2: short fungal infection pegs, which become ensheathed by the modified cell wall of the trebouxioid photobiont (*Trebouxia* or *Asterochloris* spp.; Fig. 15.13d–f). This type of interaction was found in placodioid taxa [e.g. *Lecanora muralis* (Schreb.) Rabenh.; Fig. 15.1e] and in squamulose and fruticose *Cladonia* spp. (reindeer and cup lichens; Fig. 15.1j, t).

Type 3: very short infection pegs, arising from the centre of an appressorium with very thin fungal cell wall at the contact site, which grow into the outer layer of the algal wall (Fig. 15.13h, i). This interaction is established at a very young age of the algal cells when they have not yet reached their final size and are still ensheathed by the disintegrating mother cell wall (Fig. 15.13j, l). The haustorial apparatus of the contacting hypha grows co-ordinately with the maturing algal cell. By means of growth processes within the fungal haustorium each algal cell is shifted over short distances within the algal layer (Fig. 15.13g). This type of interaction was found in Parmeliaceae and Teloschistaceae.

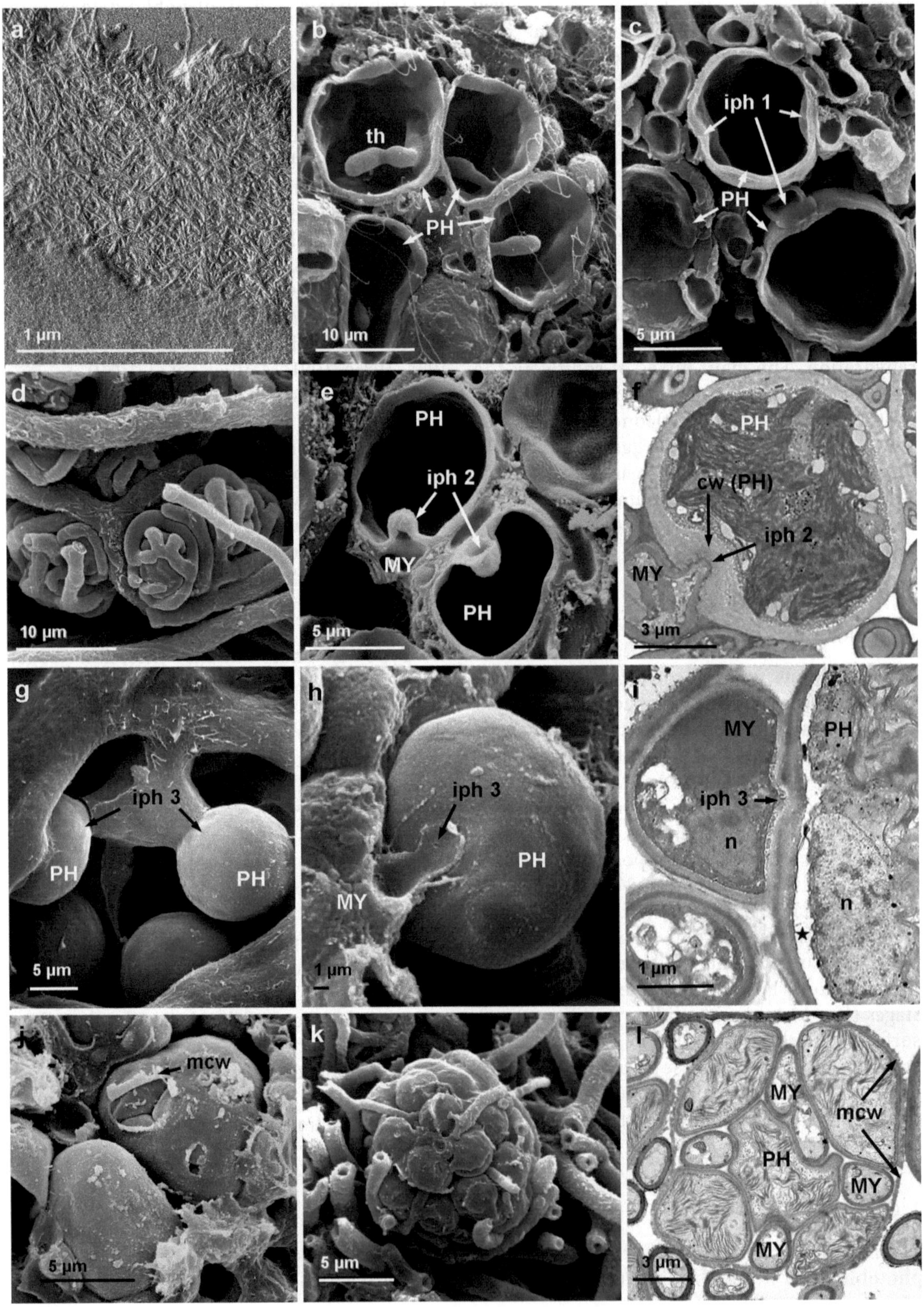

Fig. 15.13. The mycobiont (*MY*)-photobiont (*PH*) interface in Lecanorales and Teloschistales with *Trebouxia* spp. as photobionts (terminology after Honegger 1986a). (**a**) TEM micrograph of replica, (**b–e, g, h, j, k**)

Already at a very early stage, when the young photobiont cells are still ensheathed by the mother cell wall (Fig. 15.13j–l), the hydrophobic wall surface proteins are secreted by the contacting fungal hyphae and subsequently spread over the algal cell wall surface, thus sealing the apoplastic continuum with a water-repellent coat (Honegger 1986b).

Under continuously moist conditions free-living and cultured *Trebouxia* and *Asterochloris* spp. differentiate zoospores. However, zoospore production was never observed in trebouxioid lichen photobionts in the symbiotic state when thalli were kept under natural conditions, i.e. subjected to continuous wetting and drying cycles. Only when thalli were kept wet over several days were zoospores formed; these were reported to creep along hyphae in the thalline interior in search of their preferred site (Molina and Vicente 1994; Ahmadjian 2001). Under these highly artificial conditions, when the thalline interior is soaked instead of air-filled, the lichen is getting irreparably damaged and will disintegrate (review: Kershaw 1985).

Simple wall-to-wall apposition is the characteristic fungal interaction with green algal photobionts of the genera *Coccomyxa* and *Elliptochloris* (Trebouxiophyceae) whose thin, trilaminar outer cell wall layer is composed of algaenans, i.e. enzymatically non-degradable, sporopollenin-like compounds (Fig. 15.3a–c; Honegger and Brunner 1981; Brunner and Honegger 1985). However, simple wall-to-wall apposition occurs also in taxa with photobionts of the genera *Myrmecia* and *Dictyochloropsis* whose trilaminar outermost wall layer does not contain algaenans (Brunner and Honegger 1985).

Acyclic polyols such as **ribitol** (from *Trebouxia, Myrmecia, Dictyochloropsis* and *Coccomyxa* spp.), **sorbitol** (from *Heterococcus, Hyalococcus, Stichococcus* and *Trochiscia* spp.) and **erythritol** (from *Trentepohlia* and *Phycopeltis* spp.) were identified as mobile carbohydrates of green algal photobionts. The fungal partner transforms these photobiont-derived sugar alcohols into mannitol, arabitol and other fungus-specific compounds (Lines et al. 1989; for reviews, see Hill 1976; Honegger 1997, 2009) and thus maintains a strong sink (for reviews, see Farrar 1988; Palmqvist et al. 2008). Acyclic sugar alcohols play important roles as compatible solutes and are detected in apoplastic fluids, especially in the rewetting phase after drought stress events (for reviews, see Kershaw 1985; Farrar 1988).

V. Thallus Ontogeny and Growth

A. Acquisition of a Compatible Photobiont

1. Sexually Reproducing Lichen-Forming Ascomycetes

The **ascospores of sexually reproducing lichen-forming ascomycetes have to relichenize,** i.e. find a compatible photobiont, at each reproductive cycle. In only very few species are green algal photobiont cells contained in the mucilage-filled ascomata and co-dispersed with the ascospores; examples are *Endocarpon pusillum* Hedwig (Fig. 15.1v, v*) and other *Endocarpon* spp. with hymenial algae. *E. pusillum* was the first lichen to be successfully

SEM micrographs of cross-sections, (**f, i, l**) TEM micrographs of ultrathin sections. (**a**) Fragment of an isolated and purified cell wall of *Trebouxia impressa* Ahmadjian, with cellulose fibres on the inner and glucans on the smooth outer surface. (**b**) Crustose *Amandinea punctata* (Hoffm.) Coppins et Scheid. (*Lecanorales*) with finger-like, transparietal ("intracellular") fungal haustoria (*th*). (**c**) Placodioid *Ophioparma ventosa* (L.) Norman (*Lecanorales*; see Fig. 15.1d) with intraparietal haustoria of type 1 (*iph 1*). (**d–f**) Fruticose, erect *Cladonia arbuscula* (Wallr.) Flotow (**d, e**) and squamulose *Cladonia caespiticia* (Pers.) Flörke (*Lecanorales*; **f**), both with intraparietal haustoria of type 2 (*iph 2*). The short fungal infection peg is ensheathed by the modified algal cell wall. (**g–i**) Foliose *Xanthoria parietina* (*Teloschistales*; see Fig. 15.11c) with intraparietal haustoria of type 3 (*iph 3*), with short distance shifting of the algal cells within the algal layer due to intercalary growth processes within the haustorial complex. (**j–l**) Algal daughter cells (autospores) are contacted by fungal hyphae while still being enclosed in the degrading mother cell wall (*mcw*): (**j**) *Xanthoria fallax* (Hepp ex Arnold) Arnold (Teloschistales), (**k**) *Cladonia macrophylla* (*Lecanorales*), (**l**) *Parmelina tiliacea* (Lecanorales). Note: zoospores are formed only when *Trebouxia* and *Asterochloris* cells are kept under continuously moist conditions, which does not normally apply in lichen thalli

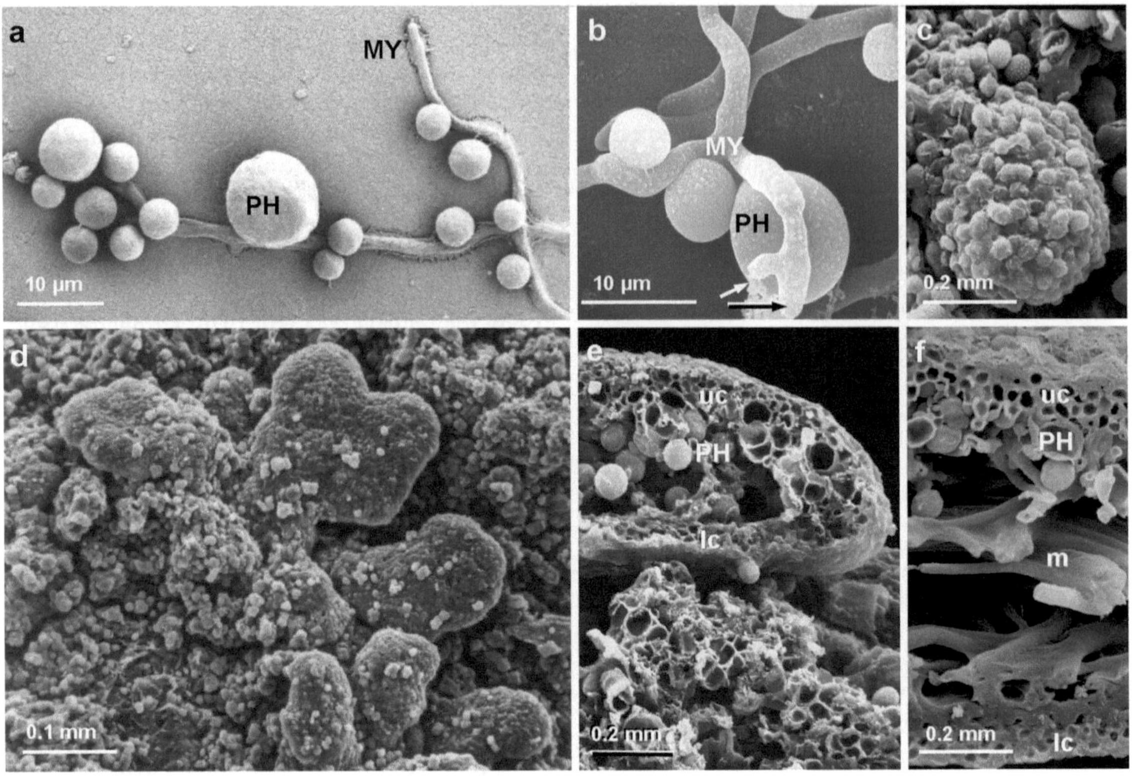

Fig. 15.14. SEM micrographs of developmental stages of *Xanthoria parietina* (L.) Th. Fr. (*Teloschistales*; MY) and *Trebouxia arboricola* De Puymaly (*Trebouxiophyceae*; *PH*). (**a**, **b**) Sterile cultures on cellophane overlying an agarized mineral medium. (**c**–**f**) Specimens collected in the wild. These specimens were bright yellow due to deposition of anthraquinones in the peripheral cortex. (**a**) Initial contact, (**b**) fungal hypha overgrowing the compatible photobiont cell (*arrows* point to hyphal branching), (**c**) juvenile, corticate thallus primordium at the onset of polarization (i.e. differentiation of the growing margin and the non-growing basal zone), (**d**) several independent primordia developing next to each other (they will later fuse to form one large thallus rosette of astonishing symmetry). (**e**, **f**) Cross-sections of juvenile (**e**) and mature (**f**) thalli, with the globose *Trebouxia* cells arranged in the algal layer (*PH*). Both have an upper (*uc*) and lower (*lc*) cortex, but the medullary layer (*m*) is formed during maturation by means of intercalary growth processes

cultured from ascospore to ascospore under laboratory conditions (Stahl 1877) soon after the discovery of lichen symbiosis by Schwendener (1869). **Lichen-forming ascomycetes and their photobionts can be isolated into sterile culture**, but cultured isolates do not easily relichenize under laboratory conditions.

The formation of a new thallus includes a whole cascade of developmental stages, but only the first ones can be routinely achieved in the Petri dish. As early stages of relichenization, the so-called pre-thallus stage of development, are also achieved in combinations of ultimately incompatible photobiont taxa the range of compatible photobiont species per fungal partner cannot be experimentally explored (for a review, see Honegger 1993). Instead, the diversity of photobiont species per lichen-forming ascomycete is studied *in situ*, i.e. in thalli collected in nature, with photobiont-specific primers applied to whole lichen extracts.

A largely underestimated problem in this procedure is caused by the often very abundant green algal epibionts of lichen thalli whose DNA is also amplified. The problems related to the identification of green algal symbionts in lichen symbioses are discussed by Grube and Muggia (2010). Before the advent of molecular tools the taxonomic affiliation of the photobiont was identified after isolation, culturing under defined conditions and examination with light microscopy techniques. Sterile cultures of holotype isolates of lichen photobiont species are the reference material for phylogenetic analyses. Sterile cultures of isolated photobionts were the pre-requisite for molecular studies at the sub-specific level such as fingerprinting analyses (Nyati 2006).

Lichen-forming fungi do not associate with any algal species, i.e. they are very selective. The most common and widespread aerophytic algae (e.g. the unicellular green algae *Apatococcus lobatus* (Chodat) J.B. Petersen and *Desmococcus* spp.) are not accepted.

How specific is lichen symbiosis, or: how many photobiont species per fungal species are acceptable partners? To answer this question a very large data set from samples collected over the whole geographic range of the fungal species is required. Small sample sizes and punctual sampling at one or few locations may lead to the premature assumption of high specificity (i.e. only one photobiont species being acceptable), whereas larger sample sizes from geographically distant locations often reveal a broader spectrum of acceptable partners.

Some lichen-forming ascomycetes were reported to **associate with only one photobiont species** and its numerous genotypes (examples in Dahlkild et al. 2001; Helms et al. 2001; Yahr et al. 2004; Piercey-Normore 2006; Werth and Sork 2010). **Other lichen-forming ascomycetes** were shown to **associate with several photobiont species** and their genotypes **from the same clade**; examples are *Xanthoria parietina* (photobionts: numerous genotypes of *Trebouxia arboricola* De Puymaly, *T. crenulata* Archibald and *T. decolorans* Ahmadjian from the *T. arboricola* clade; Nyati 2006) or cyanobacterial and cephalodiate green algal *Peltigera* spp. (photobionts: *Nostoc* spp., genotypes of clade II; O'Brien et al. 2005). **Few lichen-forming ascomycetes were shown to associate with several photobiont species from different clades**; examples are the crustose *Lecanora saxicola* (Blaha et al. 2006), the placodioid, very common and widespread *L. muralis* (Schreb.) Rabenh. (syn. *Protoparmeliopsis muralis* (Schreb.) M. Choisy; Guzow-Krzeminska 2006; Fig. 15.1e), antarctic "rock tripes" (*Umbilicaria* spp.; Romeike et al. 2002) or the fruticose, sterile "bone lichen" *Thamnolia vermicularis* (Nelsen and Gargas 2009b; Fig. 15.1s).

In most foliose and fruticose macrolichens one photobiont genotype per thallus was detected. Exceptions are the foliose, isidiate *Parmotrema tinctorum* with several genotypes of *Trebouxia corticola* (Archibald) Gärtner per thallus (Ohmura et al. 2006), or the fruticose, sorediate *Ramalina farinacea* from southern European and Californian collecting sites, which was postulated to always comprise two distinct *Trebouxia* photobionts with different ecophysiological properties per thallus; molecular data sets are missing and co-occuring *Trebouxia* epibionts cannot be ruled out (Casano et al. 2011).

Mixed photobiont populations might result from fusions of vegetative symbiotic propagules from different mother thalli, or incorporation of non-symbiotic photobiont cells into freshly landed symbiotic propagules, as proposed for *Trebouxia corticola* genotypes in *Parmotrema tinctorum* (Ohmura et al. 2006). *T. corticola* is commonly found free-living outside lichen thalli; the holotype of this species is an isolate collected free-living on bark. As soon as the peripheral cortex of macrolichens is closed an uptake of new algae and their incorporation in the algal layer is no longer possible. Only one case of algal exchange during thallus development in a crustose species with no peripheral cortex is documented: the parasitic *Diploschistes muscorum* (Scop.) R. Sant. (Fig. 15.5c) starts its development in the squamules or podetia of *Cladonia* spp. and associates with their photobiont, *Trebouxia irregularis* Hildreth et Ahmadjian. Later *T. showmanii* (Hildreth et Ahmadjian) Gärtner is incorporated, if available (Friedl 1987).

2. Vegetative Symbiotic Propagation

Vegetative symbiotic propagules such as **soredia, isidia** or **blastidia** (Fig. 15.15a–d), as formed by a high percentage of fertile and sterile lichen-forming ascomycetes, are an elegant and efficient mode of **co-dispersal of compatible partners** (Marshall 1996; Brodo et al. 2001; Sanders 2002; Ohmura et al. 2006; Ellis and Coppins 2007; Smith et al. 2009). Via symbiotic propagules lichen photobionts are transported to sites where they do not normally occur in the free-living state. **Thallus fragmentation** is probably an extremely common mode of dispersal, since only one healthy algal cell or cyanobacterial colony with adhering fungal hypha or haustorial

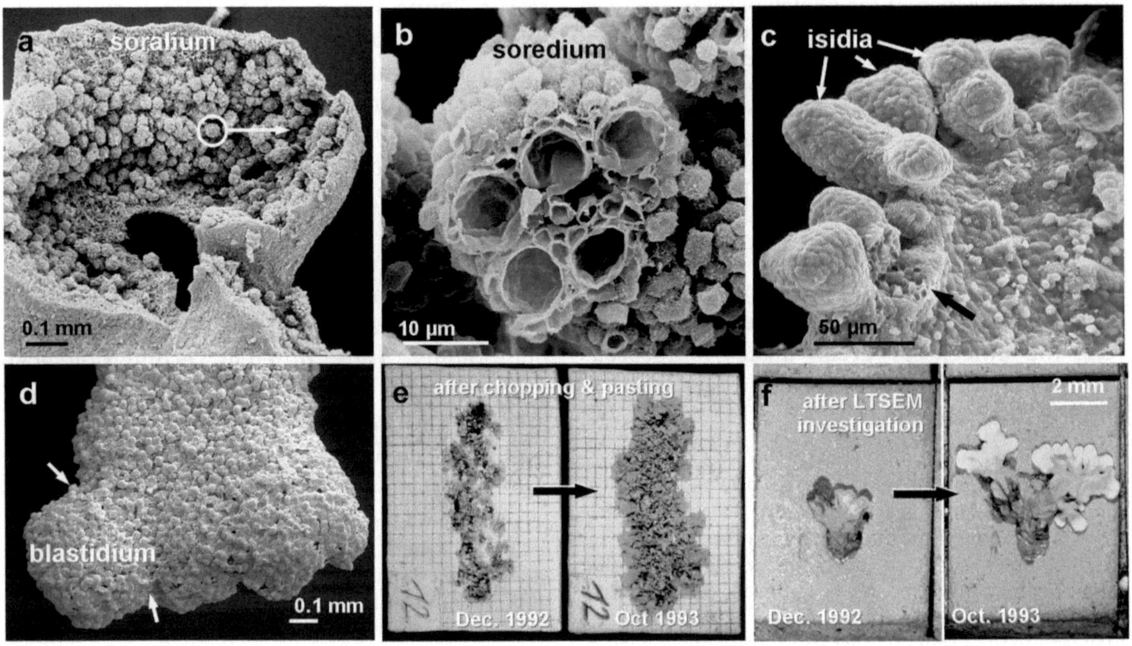

Fig. 15.15. Symbiotic vegetative propagules. (**a, b**) Soralium and soredium formation in *Xanthoria fallax* (Hepp ex Arnold) Arnold (*Teloschistales*). Each soredium contains a group of dividing *Trebouxia* cells. (**c**) Isidium formation in *Sticta fuliginosa* (Hoffm.) Ach. (*Peltigerales*). The *arrow* points to an isidium which is "ready to go". (**d**) *Xanthoria candelaria* (L.) Th. Fr. with a basally slightly constricted blastidium (*arrows*). (**e, f**) Experimental approach to study options for vegetative symbiotic propagation via thallus fragments. Such fragments are naturally detached during mechanical damage and along wound margins (for details, see Honegger 1995, 1996). (**e**) Finely chopped thallus fragments of *Xanthoria parietina* (L.) Th. Fr. (see Fig. 15.10c, d) grow into new thalli. (**f**) Vegetative symbiotic propagation of *Xanthoria parietina* via detached thallus lobes. This particular specimen was subjected to a very special pre-treatment. After shock-freezing in subcooled liquid nitrogen, freeze-fracturing, sputter-coating with an alloy of gold and palladium and examination with the ionizing electron beam under high-vacuum conditions in a LTSEM at 20 kV, this and several other lobes continued their growth (note: only those specimens survived this treatment which had been examined in the desiccated state!)

complex are required for successful propagation. Fragmentation due to trampling is the mode of dispersal in the sterile "bone lichen", *Thamnolia vermicularis* (Fig. 15.1s; Nelsen and Gargas 2009a, b) or in "reindeer lichens" (*Cladonia* spp.), whose rarely fertile, delicate arbuscular thalli cover thousands of square kilometres in the arctic tundras and are brittle and thus prone to fragmentation when dry (Fig. 15.1t).

Small-scale fragmentation occurs along wound margins either due to invertebrate grazing or when senescent areas break off (Honegger et al. 1996a). The regenerative capacity of dissected lobes and finely chopped thallus fragments was tested experimentally; both types of symbiotic propagules develop into normal thalli upon having been pasted to suitable substrates (Fig. 15.15e, f; Honegger 1995: Honegger et al. 1996a). Grazing damage to mature thalli was shown to have a rejuvenating effect when lichen myco- and photobionts are endozoochorically dispersed, as demonstrated in lichenivorous oribatid mites (Acari, Arachnida; Meier et al. 2002) and in land snails (Fröberg et al. 2001; Boch et al. 2011). Thus **faecal pellets of lichenivorous invertebrates** may serve as **potent propagules** for short- and long-distance dispersal of both partners of lichen symbiosis. This is especially true of the minute mite faeces, which are transported by stemflow, wind and vertebrate vectors such as birds.

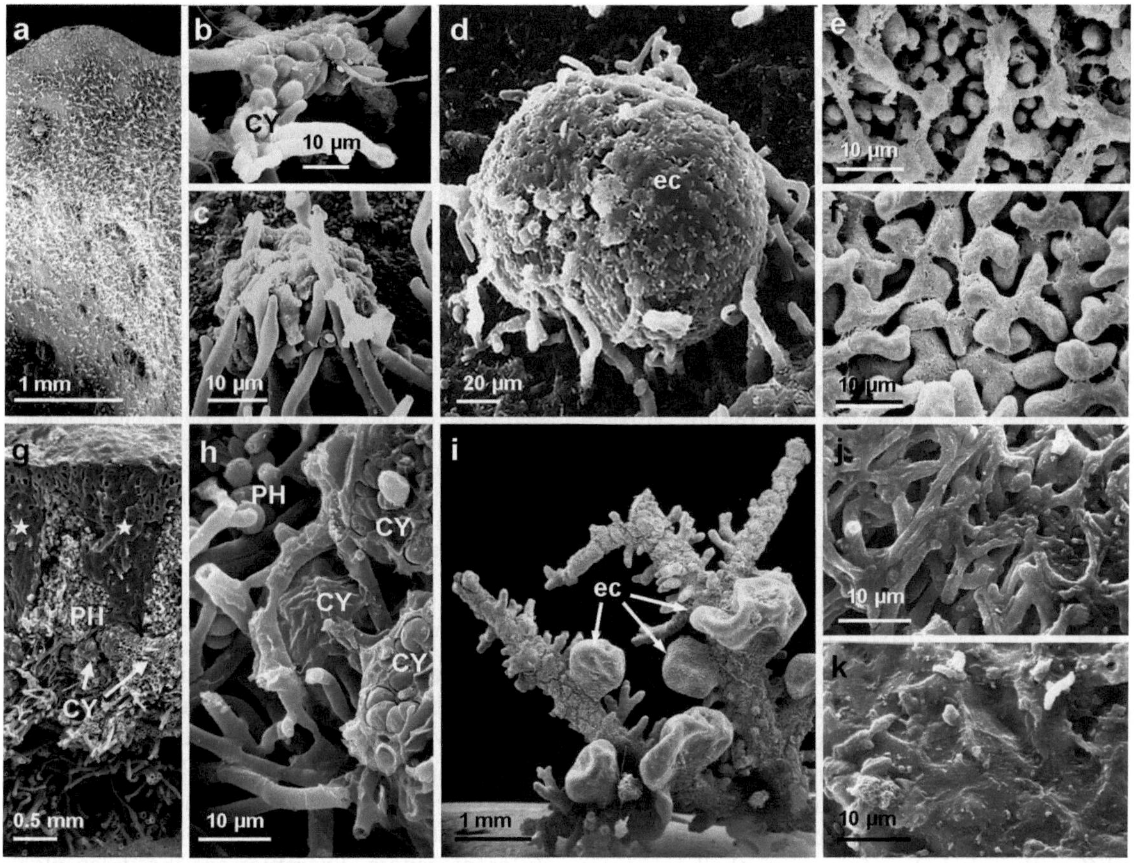

Fig. 15.16. SEM micrographs of cephalodiate lichens: lichen-forming ascomycetes which associate with a green alga (primary photobiont) as donor of fixed carbon and a cyanobacterial partner as provider of fixed nitrogen, the latter being kept in external (**a–f**, **i–k**) or internal (**g**, **h**) cephalodia. (**a–f**) Acquisition of *Nostoc* sp. of clade II *sensu* O'Brien et al. (2005; *CY*) by *Peltigera aphthosa* (L.) Willd. (*Peltigerales*; primary, green algal photobiont: *Coccomyxa* sp.) and differentiation of corticate, external cephalodia (*ec*). These can be recognized as black dots on the vivid green thallus (see Fig. 15.1f, f*). (**a**) Lobe margin with hairy hyphal outgrowths. *Arrows* point to very small cephalodia. (**b**, **c**) Capture of motile, naked stages of *Nostoc* sp. (*CY*) by hyphae protruding above the cortical layer of the thallus. (**d**) Differentiation of a cortex around the captured cyanobacterial colony. In the cephalodium the *Nostoc* colony forms its characteristic gelatinous sheath and, with increasing age and size, an increasing number of heterocysts. (**e**, **f**) Conglutinate cortical layers after removal of the gelatinous extracellular material (same procedure as in Fig. 15.10b, c). (**e**) Cortical cells of the thallus surface, designed for light transmission to the green algal layer. (**f**) Cortex of the cephalodium, designed for creating microaerobic conditions favourable for nitrogenase activity. (**g**, **h**) In the terrestrial, arctic alpine *Solorina crocea* (L.) Ach. (*Peltigerales*) the green algal photobiont cell population (*Coccomyxa* sp.) is housed between massive cones of the conglutinate cortex (*asterisks*) which act, in the fully hydrated state, as light guides. Colonies of *Nostoc* sp. (*CY*) are kept in internal cephalodia (*ic*) below the green algal zone. The species name refers to the bright orange medullary layer whose hyphae are densely covered by crystals of solorinic acid, an anthaquinone. (**i–k**) The shrubby (fruticose) *Stereocaulon ramulosum* (Sw.) Räuschel (*Lecanorales*) is a successful colonizer of disturbed, oligotrophic sites such as lava fields. It derives its mechanical stability from a conglutinate central axis (not visible) and keeps its unicellular green algal photobiont (*Pseudochlorella* sp.) below a loosely interwoven plectenchyma at the periphery (surface view in **j**), thus facilitating gas exchange and photosynthesis. Cyanobacterial colonies (either *Gloeocapsa*, *Scytonema* or *Stigonema* spp.) are kept in sac-like external cephalodia (*ec*) with dense peripheral cortex (surface view in **k**)

B. Cephalodiate Species and Photomorphs

Approximately 4 % of lichen-forming ascomycetes associate simultaneously with a green alga as primary photobiont and a cyanobacterium as donor of fixed nitrogen. These are not kept together in the thalline algal layer since they require quite different microenvironmental conditions in order to reach optimal productivity. While the **green algal cell population is maintained in the thalline algal layer** where conditions are ideal for photosynthesis the **cyanobacterial colonies are incorporated in gall-like structures** either on the thallus surface (external cephalodia; Figs. 15.1f, f*, 15.16a, d, i) or in the thalline interior (internal cephalodia; Fig. 15.16g, h). External and most of the internal cephalodia are surrounded by a fungal cortex. Under the microaerobic conditions within cephalodia the cyanobacterial partner finds optimal conditions for nitrogen fixation.

Cephalodia are formed by *Stereocaulon* spp. (Fig. 15.16i), all green algal Peltigerales (*Peltigera*, *Solorina*, *Lobaria* spp. etc.; Figs. 15.1f, f*, 15.16a, d, i) and by many other taxa. Each cephalodium results from a novel uptake; therefore the cyanobacterial colonies contained in different cephalodia of the same thallus are not necessarily genetically identical. In cephalodia of *Stereocaulon ramulosum* (Sw.) Räuschel (Fig. 15.16i) and in those of *Placopsis* spp. either *Scytonema* (=*Rhizonema*?; see section "Cyanobacterial Photobionts"), *Stigonema* or *Nostoc* spp. were identified (Tschermak-Woess 1988; Ott et al. 1997). Best investigated is the formation of cephalodia in *Peltigera aphthosa*, where hair-like hyphae on the upper cortex sense compatible *Nostoc* hormogonia and subsequently overgrow and ensheath them with a cortical layer (Fig. 15.16a–d), the recognition process being lectin-driven (Lehr et al. 2000).

In cyanobacterial *Peltigera* sp. the heterocyst frequency is around 5–8 % (Hyvärinen et al. 2002). In green algal, cephalodiate species such as *Peltigera aphthosa* and others a gradient is observed among cephalodia from the growing edge towards the subsenescent centre of the thallus (Englund 1977; Hyvärinen et al. 2002; for a review, see Nash 2008). The youngest and smallest cephalodia near the lobe margin of green algal *Peltigera* spp. are dark grey, reveal a low percentage of heterocysts (approx. 10 %) and a low level of nitrogen fixation; mature, fully grown cephalodia in the subapical area are slightly yellowish, reveal a high percentage of heterocysts (up to 55 %) and a high level of nitrogen fixation. The oldest and largest cephalodia in the central part of the thallus are yellowish, senescent and reveal reduced levels of nitrogen fixation (Englund 1977).

In only very few, exceptional taxa are cyanobacterial and green algal photobionts housing together within the thallus. Examples are: (1) *Euopsis granatina* (Sommerf.) Nyl. (Lichinales) with the cyanobacterial photobiont *Gloeocapsa sanguinea* (C. Agardh) Kütz. emend. Jaag and the green algal photobiont *Trebouxia aggregata* (Archibald) Gärtner (Büdel and Henssen 1988) and (2) the crustose *Muhria urceolata* P.M. Jørg. (Stereocaulaceae), with a basal cyanobacterial mat built up by *Microcystis* and *Stigonema* spp. and unicellular green algae contained in areoles on top of the cyanobacterial mat (Jørgensen and Jahns 1987).

Several species of cephalodiate Peltigerales express different phenotypes in association with either a cyanobacterium or a green alga, respectively, as primary photobiont. First recognized were cases where both morphotypes are distinctly different and had even been classified in different genera. Examples are the green algal, foliose *Lobaria amplissima* (Scop.) Forss. with coralloid cephalodia and the cyanobacterial, fruticose *Dendriscocaulon umhausense* (Auersw.) Degelius. In a narrow gorge in New Zealand with a humid, shady base and fully sunlit top the foliose *Sticta filix* (Sw.) Nyl. and the fruticose *Dendriscocaulon* sp. were detected with all intermediate morphotypes, from pure fruticose cyanobacterial thalli in the bottom, fruticose forms with green algal lobules in the middle and foliose green algal forms with small, coralloid cephalodia in the top area (James and Henssen 1976). **Originally described as chimerae such cyanobacterial and green algal morphotypes are also referred to as photomorph pairs, phototypes, or phycosymbiodemes.**

Less obvious than the photomorph pairs with morphologically distinctly different cyanobacterial and green algal morphotypes are those with uniform foliose morphology but different colouration: grey to turquoise versus vivid green, due to either a cyanobacterial or a green algal primary photobiont (Fig. 15.1f). With sequence data presumed photomorph pairs were confirmed and new ones detected (Goffinet and Bayer 1997; Miadlikowska and Lutzoni 2000, 2004; Stenroos et al. 2003). Not yet solved is the nomenclatural problem: species names of lichens refer to the fungal partner, but which rule should be followed when the same fungus, in association with either a cyanobacterium or a green alga, forms two different symbiotic phenotypes which have been described as two different species? Should the cyanobacterial morphotype be given nomenclatural priority since also the green algal morphotype always contains cyanobacteria in cephalodia? Similar problems had to be solved in anamorph-teleomorph relationships among non-lichenized fungi or in polyp-medusa pairs among coelenterates. So far no nomenclatural decision was made concerning lichen photomorph pairs.

C. Thallus Differentiation

In macrolichens with complex morphology and internal stratification thallus differentiation includes whole cascade of developmental processes whose regulation is not yet understood. The subsequent **developmental steps** are summarized as follows: **(1) algal adhesion** (probably lectin-based; Bubrick and Galun 1980) to fungal germ tubes of sexually reproducing lichen-forming ascomycetes (Fig. 15.14a), **(2) recognition of the compatible partner and ensheathment of photobiont cells** by specialized fungal hyphae, the pre-thallus stage of development (Fig. 15.14b), **(3) photobiont cell division and coordinate fungal growth at the contact sites, (4) development of a peripheral cortical layer** at the periphery of the thallus primordium (Fig. 15.14c), this stage is also characterized by the onset of secondary metabolite production and secretion, **(5) polarization of the thallus primordium**, differentiation of a marginal pseudomeristem (Fig. 15.14d, e), **(6) enlargement of the medullary layer** by means of intercalary growth processes (Fig. 15.14f) and **(7) formation of sexual reproductive stages and/or of vegetative symbiotic propagules.** The latter enter the developmental cycle at either stage 3 (soredia) or stage 4 (isidia; for a review, see Honegger 1993).

The first two developmental stages can be routinely achieved in laboratory cultures under sterile conditions; the pre-thallus stage is formed within few weeks after addition of compatible algal cells to a meshwork of sterile-cultured hyphae (Fig. 15.14a, b; Ahmadjian 1967; Honegger 1990; Stocker-Wörgötter 2001; Trembley et al. 2002c; Joneson and Lutzoni 2009). In this time the expression of a series of algal and fungal genes is either up- or down-regulated (Trembley et al. 2002c; Joneson et al. 2011). The lichen symbiosis was postulated to be non-specific since hyphae of cultured lichen-forming ascomycetes overgrow even glass beads (Ahmadjian 1966). Joneson and Lutzoni (2009) repeated some of the old experiments of Bonnier (1889) with the well investigated "model lichen" *Cladonia grayi* G. Merr. ex Sandst., whose genome has been largely sequenced. Sterile cultured hyphae of this lichen-forming ascomycete were co-cultured with either the compatible, unicellular photobiont (*Asterochloris* sp.), an incompatible filamentous lichen photobiont (*Trentepohlia* sp.), protonemal filaments of the moss *Funaria hygrometrica*, or glass beads, respectively. Although the hyphae of *C. greyi* overgrew the *Trentepohlia* filaments, moss protonemata and glass beads, the pre-thallus stage was formed with the compatible photobiont only (Joneson and Lutzoni 2009).

The subsequent developmental stages up to the formation of a polarized thallus (Fig. 15.14c–f) were achieved in several laboratories after several months of culturing (for reviews, see Stocker-Wörgötter 2001, 2008), but not routinely and with unpredictable timing, which makes experimental studies on their genetic basis difficult. As cortex formation correlates with the production of cortical lichen products the tiny thallus primordia are recognizable by their colour under a dissecting microscope.

Within the internally stratified thallus of macrolichens the cell cycle of the photobiont cell population is under the control of the fungal partner (Hill 1985, 1989). A high proportion of dividing photobiont cells are found in the pseudomeristematic zone at the lobe margin of foliose or at the tip of fruticose thalli (Fig. 15.17h). Adjacent to this growing zone the fungal and algal cells achieve their full dimensions in the elongation zone. Older parts of thalli comprise a high proportion of oversized algal cells which have exceeded the size required for autospore formation without undergoing mitosis.

The genetic control of this phenomenon is not understood. Soluble forms of mycobiont-derived secondary metabolites were hypothesized to have an impact on the algal cell cycle (Honegger 1987). When sterile-cultured *Trebouxia* cells were exposed to lichen products either no or only minor effects were recorded with some compounds (e.g. diffractaic or usnic acid), but others (e.g. barbatic acid) had a devastating effect resulting in cell death (Hager et al. 2009; Bačkor et al. 2010). The polyphenolic secondary metabolites are secreted by the fungal cells in a yet unknown soluble form, which is passively translocated within the apoplastic continuum during the wetting and drying cycles prior to crystallization either in the cortex or at the hydrophobic wall surfaces of the fungal and algal cells in the thalline interior. As mycobiont-derived, crystalline secondary metabolites are also found on algal surfaces (Honegger 1986b) their soluble, mobile form might be taken up by algal cells.

D. Growth Patterns in Foliose and Fruticose Lichens

Most foliose and fruticose lichens differentiate a marginal or apical, meristem-like zone, a so-called **pseudomeristem** (Honegger 1993). Based on morphological studies four major growth patterns were identified in foliose and fruticose lichens.

1. Predominantly Marginal-Apical Growth with Subapical Formation of Reproductive Stages

Lichens with marginal or apical growth and fruiting bodies and/or vegetative symbiotic propagules (soredia) in the fully grown subapical area can theoretically grow infinitely, as long as their marginal pseudomeristem stays active. This growth pattern is widespread and common, especially among foliose and band-shaped Parmeliaceae, Physciaceae and Teloschistaceae, but also among shrubby (fruticose), either erect (e.g. *Cladonia* spp.; Fig. 15.1t) or pendulous lichens (e.g. *Usnea* spp.). In the majority of foliose species the old parts degenerate and break off, the result being ring-shaped thalli. Rings of impressive diameters, referring to the ability for almost infinite growth and an old age of the "individual" thallus rosette, were detected in Arctic alpine ecosystems (see examples in Brodo et al. 2001). The ability of *Xanthoria parietina* (L.) Th. Fr. to regenerate new lobules along wound margins upon loss of the apothecia-covered central thalline areas and thus to initiate "backwards growth" seems to be exceptional (Honegger et al. 1996a).

2. Predominantly Marginal-Apical Growth with Apical Formation of Reproductive Stages

In many foliose or fruticose lichens the marginal or apical pseudomeristem terminates its growth to give rise to the formation of either sexual fruiting bodies (apothecia) or vegetative symbiotic propagules (soredia). In soralia the algal layer is used up for the production of soredia. Upon release of the vegetative symbiotic propagules the soralium and thus the thallus interior is prone to invasion by microbial and invertebrate colonizers. Thalli of this type do not exceed a certain size. With increasing age apothecial discs and soralia are increasingly colonized by non-lichenized fungi and subsequently die off. This situation is comparable to semelpary in higher plants, which reproduce once in their life and subsequently die off.

3. Combined Marginal-Apical and Intercalary Growth

The band-shaped, reticulate fishnet lichen (*Ramalina menziesii* Taylor), the fastest growing lichen so far known, combines marginal and intercalary growth, visible in increasing mesh sizes with increasing age of the thallus portion (Fig. 15.9a; Sanders 1989, 1992; Sanders and Ascaso 1995). This growth pattern occurs also in *Usnea longissima* Ach., whose thalli reach several metres length (Rolstad and Rolstad 2008), the the foliose lungworts (*Lobaria* spp.), punctually fixed epiphytes forming large thalli, e.g. in forests of the North American Pacific Northwest, and in the terrestrial, tubular-perforate *Cladia retipora* (Labill.) Nyl. with its impressive growth symmetry due to apical pseudomeristems which divide dichotomously at regular intervals (Fig. 15.9b; Honegger 1993).

The mechanisms underlying intercalary growth in lichenized and non-lichenized fungi are neither fully understood at the cellular, nor at the organismic level (Voisey 2010). Cell and molecular biologists have focussed either on polar–apical hyphal growth or on apolar, yeast cell-type growth. The massive thalline

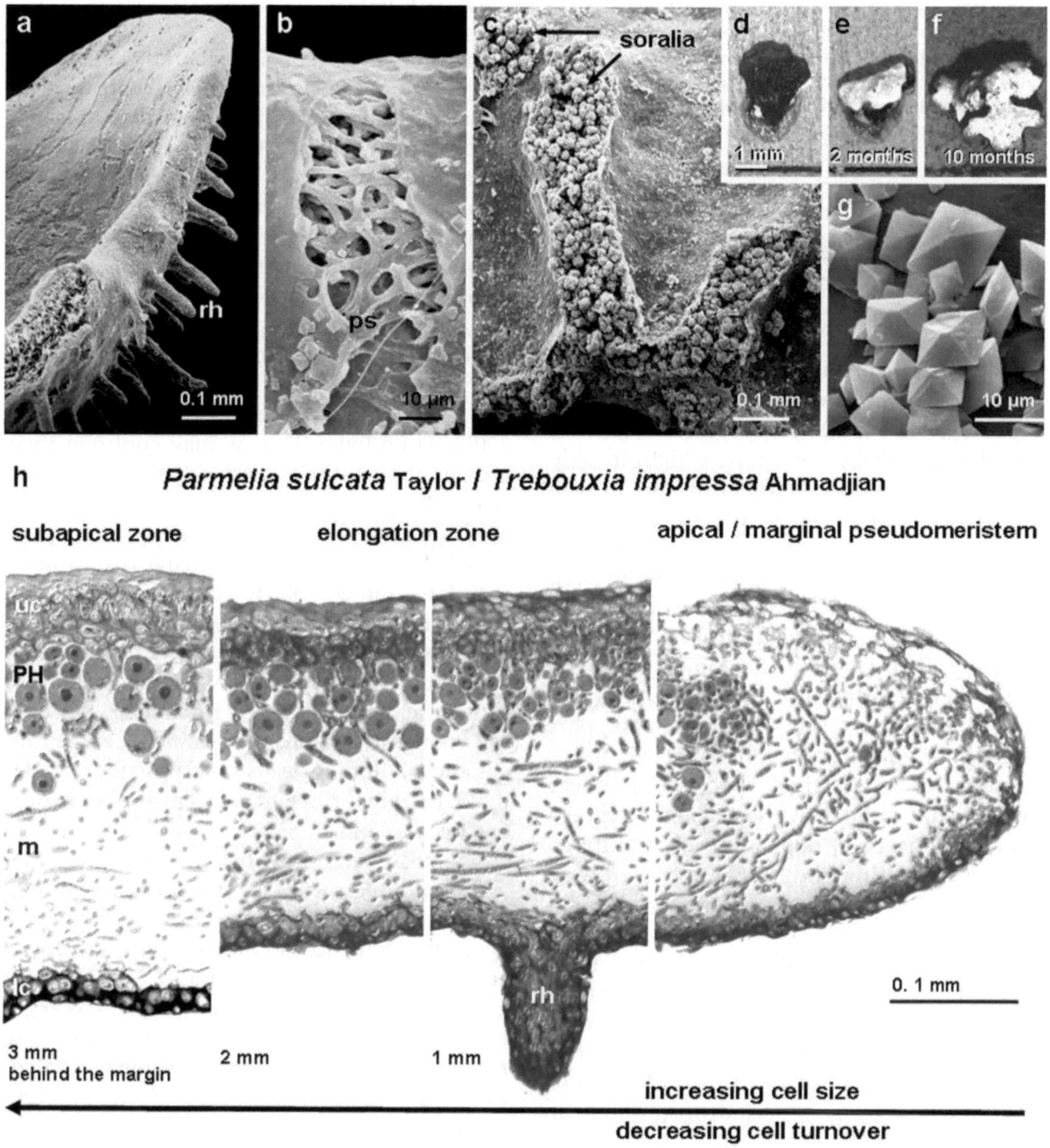

Fig. 15.17. The widespread and common *Parmelia sulcata* Taylor (*Lecanorales*) as an example of a foliose, dorsiventrally organized macrolichen with marginal growth. (**a–g**) SEM micrographs, (**h**) high-resolution light micrographs of semithin sections. (**a**) Lobe margin with developing rhizinae (*rh*) on the melanized lower cortex. (**b**) Surface view of the conglutinate upper cortex near the lobe margin with an aeration pore (pseudocyphella; *ps*) built up by hyphae with hydrophobic wall surfaces. (**c**) In older thalline areas the pseudocyphellae are transformed into soralia whence masses of soredia (vegetative symbiotic propagules) derive. (**d–f**) This marginal lobe was fixed to a ceramic surface in an upside-down position, incubated outdoors and photographed after 2 and 10 months. The fungal partner senses this unfavourable position and corrects it by means of growth processes, thus securing adequate illumination for its photoautotrophic partner (after Honegger 1996). (**g**) Calcium oxalate crystals (Wedellite) forming a greyish pruina on the thallus surface (for bacterial film, see Fig. 15.4a). (**h**) Thallus cross-sections along a gradient from the growing margin to the subapical zone. The fungal partner actively positions each photobiont cell in the algal layer (*PH*) below the upper cortex (*uc*) at the

elongation, as seen in *Ramalina menziesii*, cannot be achieved with hyphal cell elongation only, but necessitates intercalary cell divisions. The fungal cell turnover rates and the control of cell division and autospore formation in the photobiont cell population (*Trebouxia* sp.) remain to be explored.

4. Irregular, Patchy Intercalary Growth

The foliose, umbilicate thalli of *Umbilicaria* and *Lasallia* spp. ("rock tripes") do not reveal marginal growth, but patchy intercalary growth, best visible in *Lasallia pustulata* (L.) Mérat with its pustules of various dimensions and levels of disintegration (Honegger 1993, 2001; Fig. 15.9c, d). No pseudomeristem is found at the thallus margins, which, in older thalli, show signs of degradation due to wind abrasion. Accordingly a patchy pattern of production was detected within *Umbilicaria* thalli (Larson 1983). The area around the central umbilicus, which fixes the whole thallus to the granitic rock substrate (Fig. 15.9d), is thicker and stores water for longer periods of time; therefore this area stays metabolically active when the thinner peripheral parts already dried out (Hestmark et al. 1997). Among two "rock tripe" species (Umbilicariaceae) with habitat overlap a niche segregation was observed, depending on a different anatomy resulting in different modes of water uptake. *Umbilicaria mühlenbergii* (Ach.) Tuckerm. is wetted most easily by runoff and mist, which are absorbed by the melanized, lower cortex which carries lots of rhizinae; this species grows at the flanks of rocks. *Lasallia papulosa* (Ach.) Llano [sub *Umbilicaria papulosa* (Ach.) Nyl.]. is most easily wetted by large raindrops, which are absorbed immediately by the upper cortex; this species grows on the top of the rocks and has its highest productivity during the summer months with high frequencies of thunderstorms (Larson 1984).

E. Growth Rates

Lichen-forming fungi are generally considered as very slow-growing organisms with growth rates in the range of less than one to a few millimetres per year. Accordingly lichen thalli are assumed to be very old. However, there is a wide range of growth capacities among lichen-forming ascomycetes, climatic factors having a prime impact on their exertion. Short-living lichens on transient substrates such as open soil, leaves etc. have to complete their life cycle within few months or years and thus do not normally achieve large sizes (Poelt and Vězda 1990; Jahns and Ott 1997; Lücking 2008). Long-living lichens may continue their growth for centuries and achieve impressive sizes. Marginal pseudomeristems of placodioid or foliose species may produce biomass over decades or even centuries, giving rise to thalli with large diameters. However, the oldest central parts of these thalli are either overgrown or have disintegrated, the remaining thallus being ring-shaped. It may have taken long periods of time to achieve a wide ring diameter, but only the youngest portions are retained, i.e. the fungal and photobiont cells of the pseudomeristematic marginal rim and adjacent zones are only few decades old.

A positive carbon balance, a prerequisite for thalline growth, is only achieved under favourable climatic conditions (Pannewitz et al. 2003; Sancho and Pintado 2004; Palmqvist et al. 2008). **The time window suitable for growth is restricted to few days a year in extreme, hot or cold climates, but amounts to many months in temperate regions.** Accordingly, the same species may reveal different growth rates in different climatic areas (for reviews, see Kappen 1988, 2000). Growth rates of marginal lobes were shown to correlate with productive capacity, i.e. the contents in photobiont- and mycobiont-derived mobile carbohy-

periphery of the gas-filled medullary layer (*m*). The lower cortex (*lc*) is black due to the incorporation of melanin granules. (for comparison, see Fig. 15.10d). A large proportion of oversized photobiont cells in the subapical zone have exceeded the size required for autospore formation without undergoing mitosis, which indicates a fungal control of the algal cell turnover. The dark spot in the centre of the *Trebouxia* cells represents the proteinaceous pyrenoid in the centre of the large, lobate chloroplast, the nucleus being laterally positioned drates such as acyclic polyols (Armstrong and Smith 1993, 1994). **The fastest growing lichens are the "lungworts" (*Lobaria* spp.) and "fishnet lichens"** (*Ramalina menziesii*) of the coastal areas of the North American Pacific Northwest, which reveal combined marginal and intercalary growth and benefit from a climate suitable for growth all year long. Thus an annual size increase around 43 cm was recorded in the pendulous, band-shaped, reticulate thalli of *Ramalina menziesii* (Fig. 15.9a; Sanders 1989, 1992; Sanders and Ascaso 1995). "Fishnet lichens" and "lungworts" generate an impressive biomass in the deciduous rainforests of the Pacific Northwest (Honegger 1993; Coxson and Stevenson 2007). Epiphytic, cyanobacterial and green-algal, cephalodiate *Lobaria* spp. provide a significant input of new N to forest ecosystems (Millbank and Kershaw 1970; Antoine 2004).

Lichen growth measurements are used for lichenometric dating, e.g. of moraines in arctic alpine ecosystems, water trimlines after floods, or rock surfaces at archaeological sites (Innes 1988; Winchester 1988; Winchester and Harrison 2000; Marsh and Timoney 2005; Loso and Doak 2006; McCarthy 2007; Armstrong and Bradwell 2010a, b; Golledge et al. 2010). A very impressive water trimline visualizes the level of brackish water which killed all lichen growth in New Orleans after the floods caused by hurricane Katrina (Thomas 2009), the re-colonization of free surfaces being monitored. The impact of global warming on lichen growth and population dynamics is currently investigated (van Herk et al. 2002; Armstrong 2004; Aptroot and van Herk 2007; Sancho et al. 2007b).

VI. Conclusions and Perspectives

In the last two decades fascinating insights were achieved in the phylogenies and evolution of lichen-forming ascomycetes and their photobionts and in the biological diversity of lichen symbiosis far beyond the traditional view of lichens as two- or three-membered associations. The diversity of bacterial and fungal endo- and epibionts of lichen thalli and the transcriptome and metabolome of lichen-forming fungi and their fungal and bacterial associates will attract an increased interest among biologists and biochemists, especially with regard to bioactive secondary metabolites. Ever since the detection of the presumably lichenized ancestry of the largely non-lichenized Eurotiomycetes with their innumerable economically and medically important representatives (Lutzoni et al. 2001) and the multiple gains and losses of lichenization in ascomycete evolution, the lichen-forming taxa are attracting more interest than ever before among mycologists.

Climate change affects lichen-dominated ecosystems. Many cryophilic species of Arctic tundras disappear from their southernmost area of distribution as the competitive pressure by angiosperms increases (van Herk et al. 2002; Ellis et al. 2007; Geml et al. 2010), with yet unknown effects on arctic reindeer and caribou husbandry. The importance of soil crust communities of semi-arid and arid lands in preventing wind erosion and increasing desertification will become even more evident with increasing temperatures (Maestre et al. 2010). As in the decades after the Chernobyl accident (Feige et al. 1990; Biazrov 1994), lichens will be silent chronists of the Fukushima disaster in 2011.

"The lichens. . .may turn out to be the Holy Grail for understanding plant–fungus recognition phenomena" (Talbot 1998). Although our understanding of recognition phenomena in lichen symbiosis has not significantly improved in the last decade this statement by Nicholas Talbot, renowned expert on fungus–plant interactions, retains its validity.

Acknowledgements My sincere thanks are due to my colleagues David L. Hawksworth for helpful information on lichenicolous fungi, to Burkhard Büdel and Robert Lücking for providing specimens (*Peltula* and *Psoroglaena* spp.), to Reinhard Berndt for Fig. 15.1y, to my husband Thomas G. Honegger for patient help with all computer problems, to the editor, Prof. Bertold Hock for his encouragement, patience and help, to the project manager, Athiappan Kumar, for his valuable support, to the Directorium of the Institute of Plant Biology of the University of Zürich for providing access to the infrastructure beyond my retirement and to the Swiss National Science Foundation for generous financial support (grant 3100A0-116597).

References

Adams GC, Kropp BR (1996) *Athelia arachnoidea*, the sexual state of *Rhizoctonia carotae*, a pathogen of carrot in cold storage. Mycologia 88:459–472

Ahmadjian V (1966) Artificial reestablishment of lichen *Cladonia cristatella*. Science 151:199–201

Ahmadjian V (1967) A guide to the algae occurring as lichen symbionts: isolation, culture, cultural physiology, and identification. Phycologia 6:127–160

Ahmadjian V (2001) *Trebouxia*: reflections on a perplexing and controversial lichen photobiont. In: Seckbach J (ed) Symbiosis. Cellular origin, life in extreme habitats and astrobiology. Springer, Dordrecht, pp 373–383

Anglesea D, Veltkamp C, Greenhalgh GH (1982) The upper cortex of *Parmelia saxatilis* and other lichen thalli. Lichenologist 14:29–38

Antoine ME (2004) An ecophysiological approach to quantifying nitrogen fixation by *Lobaria oregana*. Bryologist 107:82–87

Aptroot A, van Herk CM (2007) Further evidence of the effects of global warming on lichens, particularly those with *Trentepohlia* phycobionts. Environ Pollut 146:293–298

Armaleo D (1993) Why do lichens make secondary products? In: 15th International Botanical Congress, Yokohama. Abstracts, p 11

Armaleo D, Zhang Y, Cheung S (2008) Light might regulate divergently depside and depsidone accumulation in the lichen *Parmotrema hypotropum* by affecting thallus temperature and water potential. Mycologia 100:565–576

Armaleo D, Sun X, Culberson C (2011) Insights from the first putative biosynthetic gene cluster for a lichen depside and depsidone. Mycologia 103:741–754

Armstrong R (1994) Dispersal of soredia from individual soralia of the lichen *Hypogymnia physodes* (L.) Nyl. in a simple wind tunnel. Environ Exp Bot 34:39–45

Armstrong R (2004) Lichens, lichenometry and global warming. Microbiologist 2004:32–35. *http://www.blackwellpublishing.com/Microbiology/pdfs/lichens.pdf*

Armstrong RA, Bradwell T (2010a) Growth of crustose lichens: a review. Geogr Ann A 92:3–17

Armstrong RA, Bradwell T (2010b) The use of lichen growth rings in lichenometry: some preliminary findings. Geogr Ann A 92:141–147

Armstrong RA, Smith SN (1993) Radial growth and carbohydrate levels in the lichen *Parmelia conspersa* on north and south facing rock surfaces. Symbiosis 15:27–49

Armstrong RA, Smith SN (1994) The levels of ribitol, arabitol and mannitol in individual lobes of the lichen *Parmelia conspersa* (Ehrh ex Ach.) Ach. Environ Exp Bot 34:253–260

Arnold AE (2007) Understanding the diversity of foliar endophytic fungi: progress, challenges, and frontiers. Fungal Biol Rev 21:51–66

Arnold AE (2008) Endophytic fungi: hidden components of tropical community ecology. In: Schnitzer S, Carson W (eds) Tropical forest community ecology. Wiley-Blackwell, Oxford, pp 254–271

Arnold AE, Miadlikowska J, Higgins KL et al (2009) A phylogenetic estimation of trophic transition networks for ascomycetous fungi: are lichens cradles of symbiotrophic fungal diversification? Syst Biol 58:283–297

Ascaso C, Gonzales C, Vicente C (1980) Epiphytic *Evernia prunastri* (L.) Ach.: ultrastructural facts. Cryptog Bryol Lichenol 1:45–51

Asplund J, Solhaug KA, Gauslaa Y (2010) Optimal defense: snails avoid reproductive parts of the lichen *Lobaria scrobiculata* due to internal defense allocation. Ecology 91:3100–3105

Avalos A, Legaz M, Vicente C (1986) The occurrence of lichen phenolics in the xylem sap of *Quercus pyrenaica*, their translocation to leaves and biological significance. Biochem Syst Ecol 14:381–384

Bačkor M, Klemov K, Bačkorov M, Ivanova V (2010) Comparison of the phytotoxic effects of usnic acid on cultures of free-living alga *Scenedesmus quadricauda* and aposymbiotically grown lichen photobiont. J Chem Ecol 36:405–411

Baloch E, Lücking R, Lumbsch HT, Wedin M (2010) Major clades and phylogenetic relationships between lichenized and non-lichenized lineages in Ostropales (Ascomycota: Lecanoromycetes). Taxon 59:1483–1494

Bates JW, Bell JNB, Massara AC (2001) Loss of *Lecanora conizaeoides* and other fluctuations of epiphytes on oak in S.E. England over 21 years with declining SO_2 concentrations. Atmos Environ 35:2557–2568

Bates ST, Cropsey GWG, Caporaso JG, Knight R, Fierer N (2011) Bacterial communities associated with the lichen symbiosis. Appl Environ Microbiol 77:1309–1314

Beck A, Friedl T, Rambold G (1998) Selectivity of photobiont choice in a defined lichen community: inferences from cultural and molecular studies. New Phytol 139:709–720

Beierkuhnlein C, Drewello R, Snethlage R, Töpfer LH (2011) Zwischen Denkmalschutz und Naturschutz. Leitfaden zur naturverträglichen Instandhaltung von Mauerwerk in der Denkmalpflege. Initiativen zum Umweltschutz 83. Erich Schmidt/Deutsche Bundsstiftung Umwelt, Berlin

Belnap J, Lange OL (eds) (2003) Biological soil crusts: structure, function and management, 2nd edn. Springer, Berlin Heidelberg New York, pp 1–503

Belnap J, Rosentreter R, Leonard S, Kaltenecker JH, Williams J, Eldridge D (eds) (2001) Biological soil crusts: ecology and management. Technical reference 1730-2. US Department of the Interior, Bureau of Land Management, US Geological Survey, Denver, pp 1–118. http://www.soilcrust.org/crust.pdf

Biazrov LG (1994) The radionuclides in lichen thalli in Chernobyl and East Urals areas after nuclear accidents. Phyton 34:85–94

Bjelland T, Thorset IH (2002) Comparative studies of the lichen–rock interface of four lichens in Vingen, western Norway. Chem Geol 192:81–98

Bjerke JW, Lerfall K, Elvebakk A (2002) Effects of ultraviolet radiation and PAR on the content of usnic and divaricatic acids in two arctic-alpine lichens. Photochem Photobiol Sci 1:678–685

Bjerke JW, Elvebakk A, Domínguez E, Dahlback A (2005) Seasonal trends in usnic acid concentrations of Arctic, alpine and Patagonian populations of the lichen *Flavocetraria nivalis*. Phytochemistry 66:337–344

Blaha J, Baloch E, Grube M (2006) High photobiont diversity associated with the euryoecious lichen-forming ascomycete *Lecanora rupicola* (Lecanoraceae, Ascomycota). Biol J Linn Soc 88:283–293

Boch S, Prati D, Werth S, Rüetschi J, Fischer M (2011) Lichen endozoochory by snails. PLoS One 6:e18770

Bonfante P, Anca IA (2009) Plants, mycorrhizal fungi, and bacteria: a network of interactions. Annu Rev Microbiol 63:363–383

Bonfante P, Balestrini R, Genre A, Lanfranco L (2009) Establishment and functioning of arbuscular mycorrhizas. In: Deising HB (ed) The Mycota V, part 2 (plant relationships). Springer, Berlin Heidelberg New York, pp 259–274

Bonnier G (1889) Recherches sur la synthèse des lichens. Ann Sci Nat (Ser 7) 9:1–34

Bourgeois G, Suire C, Vivas N, Benoist F, Vitry C (1999) Atraric acid, a marker for epiphytic lichens in the wood used in cooperage: identification and quantification by GC/MS/(MS). Analusis 27:281–283

Brierley WB (1913) The structure and life history of *Leptosphaeria lemaneae* (Cohn). Mem Manch Lit Philos Soc 57:1–21

Brodo IM, Sharnoff SD, Sharnoff S (2001) Lichens of North America. Yale University Press, New Haven

Brunner U, Honegger R (1985) Chemical and ultrastructural studies on the distribution of sporopollenin-like biopolymers in 6 genera of lichen phycobionts. Can J Bot 63:2221–2230

Bubrick P, Galun M (1980) Proteins from the lichen *Xanthoria parietina* which bind to phycobiont cell walls: correlation between binding patterns and cell wall cytochemistry. Protoplasma 104: 167–173

Bubrick P, Galun M, Frensdorff A (1984) Observations on free-living *Trebouxia* de Puymaly and *Pseudotrebouxia* Archibald, and evidence that both symbionts from *Xanthoria parietina* (L.) Th.Fr. can be found free-living in nature. New Phytol 97:455–462

Büdel B (1987) Zur Biologie und Systematik der Flechtengattungen *Heppia* und *Peltula* im südlichen Afrika. Bibl Lichenol 23:1–105

Büdel B (2002) Diversity and ecology of biological soil crusts. Prog Bot 63:386–404

Büdel B, Henssen A (1988) *Trebouxia aggregata* und *Gloeocapsa sanguinea*, Phycobionten in *Euopsis granatina* (Lichinaceae). Plant Syst Evol 158:235–241

Büdel B, Lange OL (2001) Synopsis: comparative biogeography and ecology of soil-crust biota and communities. In: Belnap J, Lange O (eds) Biological soil crusts: structure, function, and management. Springer, Berlin Heidelberg New York, pp 141–152

Büdel B, Karsten U, Garcia-Pichel F (1997) Ultraviolet-absorbing scytonemin and mycosporine-like amino acid derivatives in exposed, rock-inhabiting cyanobacterial lichens. Oecologia 112:165–172

Büdel B, Darienko T, Deutschewitz K et al (2009) South African biological soil crusts are ubiquitous and highly diverse in drylands, being restricted by rainfall frequency. Microb Ecol 57:229–247

Buffoni Hall RS, Bornman JF, Björn LO (2002) UV-induced changes in pigment content and light penetration in the fruticose lichen *Cladonia arbuscula* ssp. *mitis*. J Photochem Photobiol B 66:13–20

Cardinale M, Puglia AM, Grube M (2006) Molecular analysis of lichen-associated bacterial communities. FEMS Microbiol Ecol 57:484–495

Cardinale M, Castro JVD, Müller H, Berg G, Grube M (2008) *In situ* analysis of the bacterial community associated with the reindeer lichen *Cladonia arbuscula* reveals predominance of *Alphaproteobacteria*. FEMS Microbiol Ecol 66:63–71

Casano LM, del Campo EM, García-Breijo FJ et al (2011) Two *Trebouxia* algae with different physiological performances are ever-present in lichen thalli of *Ramalina farinacea*. Coexistence versus competition? Environ Microbiol 13:806–818

Chapman R, Waters D (2004) Lichenization of the Trentepohliales. Complex algae and odd relationships. In: Seckbach J (ed) Symbiosis. Cellular origin, life in extreme habitats and astrobiology. Springer, Dordrecht, pp 359–371

Chen J, Blume HP, Beyer L (2000) Weathering of rocks induced by lichen colonization – a review. Catena 39:121–146

Chen J, Zhang MY, Wang L, Shimazaki H, Tamura M (2005) A new index for mapping lichen-dominated biological soil crusts in desert areas. Remote Sens Environ 96:165–175

St Clair LL, Seaward MRD (eds) (2004) Biodeterioration of stone surfaces: lichens and biofilms as weathering agents of rocks and cultural heritage. Springer, Berlin Heidelberg New York

St Clair LL, Webb BL, Johansen JR, Nebeker GT (1984) Cryptogamic soil crusts: enhancement of seedling establishment in disturbed and undisturbed areas. Reclam Reveg Res 3:129–136

Cocchietto M, Skert N, Nimis PL, Sava G (2002) A review on usnic acid, an interesting natural compound. Naturwissenschaften 89:137–146

Cornejo C, Chabanenko S, Scheidegger C (2009) Phylogenetic analysis indicates transitions from vegetative to sexual reproduction in the *Lobaria retigera* group (Lecanoromycetidae, Ascomycota). Lichenologist 41:275–284

Coxson DS, Stevenson SK (2007) Growth rate responses of *Lobaria pulmonaria* to canopy structure in even-aged and old-growth cedar–hemlock forests of central-interior British Columbia, Canada. For Ecol Manage 242:5–16

Crespo A, Lumbsch HT (2010) Cryptic species in lichen-forming fungi. IMA Fungus 1:167–170

Crespo A, Pérez-Ortega S (2009) Cryptic species and species pairs in lichens: a discussion on the relationship between molecular phylogenies and morphological characters. An Jardín Bot Madrid 66:71–81

Crespo A, Lumbsch HT, Mattsson JE et al (2007) Testing morphology-based hypotheses of phylogenetic relationships in Parmeliaceae (Ascomycota) using three ribosomal markers and the nuclear RPB1 gene. Mol Phylogenet Evol 44:812–824

Crespo A, Kauff F, Divakar PK et al (2010) Phylogenetic generic classification of parmelioid lichens (Parmeliaceae, Ascomycota) based on molecular, morphological and chemical evidence. Taxon 59:1735–1753

Dahlkild A, Kallersjo M, Lohtander K, Tehler A (2001) Photobiont diversity in the Physciaceae (Lecanorales). Bryologist 104:527–536

De los Rios A, Grube M (2000) Host–parasite interfaces of some lichenicolous fungi in the Dacampiaceae (Dothideales, Ascomycota). Mycol Res 104:1348–1353

De los Rios A, Sancho LG, Grube M, Wierzchos J, Ascaso C (2005) Endolithic growth of two *Lecidea* lichens in granite from continental Antarctica detected by molecular and microscopy techniques. New Phytol 165:181–189

Deckert RJ, Garbary DJ (2005) *Ascophyllum* and its symbionts. VI. Microscopic characterization of the *Ascophyllum nodosum* (Phaeophyceae), *Mycophycias ascophylli* (Ascomycetes) symbiotum. Algae 20:225–232

Dietz S, Büdel B, Lange OL, Bilger W (2000) Transmittance of light through the cortex of lichens from contrasting habitats. Bibl Lichenol 75:171–182

Drewello R, Drewello UG (2009) Flechten auf Denkmälern: Indikatoren und Vermittler zwischen Denkmal- und Naturschutz. In: Bayerische Akademie der Wissenschaften (ed) Ökologische Rolle der Flechten. Rundgespräche der Kommission für Ökologie 36. Verlag Dr. Friederich Pfeil, München, pp 161–180

Eichenberger C (2007) Molecular phylogenies of representatives of *Xanthoria* and *Xanthomendoza* (lichen-forming Ascomycetes) Inauguraldissertation. Mathematisch Naturwissenschaftliche Fakultät der Universität Zürich, Zürich, pp 1–143

Ekman S, Tønsberg T (2002) Most species of *Lepraria* and *Leproloma* form a monophyletic group closely related to *Stereocaulon*. Mycol Res 106:1262–1276

Ellis CJ, Coppins BJ (2007) Reproductive strategy and the compositional dynamics of crustose lichen communities on aspen (*Populus tremula* L.) in Scotland. Lichenologist 39:377–391

Ellis CJ, Coppins BJ, Dawson TP, Seaward MRD (2007) Response of British lichens to climate change scenarios: trends and uncertainties in the projected impact for contrasting biogeographic groups. Biol Conserv 140:217–235

Elo H, Matikainen J, Pelttar E (2007) Potent activity of the lichen antibiotic (+)-usnic acid against clinical isolates of vancomycin-resistant enterococci and methicillin-resistant *Staphylococcus aureus*. Naturwissenschaften 94:465–468

Elvebakk A, Papaefthimiou D, Robertsen EH, Liaimer A (2008) Phylogenetic patterns among *Nostoc* cyanobionts within bi- and tripartite lichens of the genus *Pannaria*. J Phycol 44:1049–1059

Englund B (1977) The physiology of the lichen *Peltigera aphthosa*, with special reference to the blue-green phycobiont (*Nostoc* sp.). Physiol Plant 41:298–304

Espoz C, Guzman G, Castilla JC (1995) The lichen *Thelidium litorale* on shells of intertidal limpets: a case of lichen-mediated cryptic mimicry. Mar Ecol Prog Ser 119:191–197

Ettl H, Gärtner G (1995) Syllabus der Boden-Luft- und Flechtenalgen. Fischer, Stuttgart

Evans RD, Lange OL (2001) Biological soil crusts and ecosystem nitrogen and carbon dynamics. In: Belnap J, Lange OL (eds) Biological soil crusts: structure, function and management. Springer, Berlin Heidelberg New York, pp 263–279

Farrar JF (1988) Physiological buffering. In: Galun M (ed) Handbook of Lichenology, vol 2. CRC, Boca Raton, pp 101–105

Favero-Longo SE, Turci F, Tomatis M et al (2005) Chrysotile asbestos is progressively converted into a non-fibrous amorphous material by the chelating action of lichen metabolites. J Environ Monit 7:764–766

Favero-Longo SE, Girlanda M, Honegger R, Fubini B, Piervittori R (2007) Interactions of sterile-cultured lichen-forming ascomycetes with asbestos fibres. Mycol Res 111:473–481

Fedrowitz K, Kaasalainen U, Rikkinen J (2011) Genotype variability of *Nostoc* symbionts associated with three epiphytic *Nephroma* species in a boreal forest landscape. Bryologist 114:220–230

Feige GB, Niemann L, Jahnke S (1990) Lichens and mosses – silent chronists of the Chernobyl accident. Bibl Lichenol 38:63–77

Frey-Klett P, Garbaye J, Tarkka M (2007) The mycorrhiza helper bacteria revisited. New Phytol 176:22–36

Friedl T (1987) Thallus development and phycobionts of the parasitic lichen *Diploschistes muscorum*. Lichenologist 19:183–191

Friedl T (1995) Inferring taxonomic positions and testing genus level assignments in coccoid green lichen algae: a phylogenetic analysis of 18 s ribosomal RNA sequences from *Dictyochloropsis reticulata* and from members of the genus *Myrmecia*

(Chlorophyta, Trebouxiophyceae *cl. nov.*). J Phycol 31:632–639

Friedl T, Büdel B (2008) Photobionts. In: Nash TH (ed) Lichen biology. Cambridge University Press, Cambridge, pp 9–26

Fröberg L, Berg CO, Baur A, Baur B (2001) Viability of lichen photobionts after passing through the digestive tract of a land snail. Lichenologist 33:543–545

Gadd GM (2007) Geomycology: biogeochemical transformations of rocks, minerals, metals and radionuclides by fungi, bioweathering and bioremediation. Mycol Res 111:3–49

Garbary DJ, London JF (1995) The *Ascophyllum/Polysiphonial/Mycosphaerella* symbiosis V. Fungal infection protects *A. nosodum* from desiccation. Bot Mar 38:529–533

Garbary DJ, Macdonald KA (1995) The *Ascophyllum/Polysiphonia/Mycosphaerella* symbiosis 4. Mutualism in the *Ascophyllum/Mycosphaerella* interaction. Bot Mar 38:221–225

Gauslaa Y (2005) Lichen palatability depends on investments in herbivore defence. Oecologia 143:94–105

Gaya E, Navarro-Rosinés P, Llimona X, Hladun N, Lutzoni F (2008) Phylogenetic reassessment of the Teloschistaceae (lichen-forming Ascomycota, Lecanoromycetes). Mycol Res 112:528–546

Geiser DM, Gueidan C, Miadlikowska J et al (2006) Eurotiomycetes: Eurotiomycetidae and Chaetothyriomycetidae. Mycologia 98:1053–1064

Geml J, Kauff F, Brochmann C, Taylor DL (2010) Surviving climate changes: high genetic diversity and transoceanic gene flow in two arctic-alpine lichens, *Flavocetraria cucullata* and *F. nivalis* (Parmeliaceae, Ascomycota). J Biogeogr 37:1529–1542

Girlanda M, Isocrono D, Bianco C, Luppi-Mosca AM (1997) Two foliose lichens as microfungal ecological niches. Mycologia 89:531–536

Goebel K (1926a) Die Wasseraufnahme der Flechten. Ber Deutsch Bot Ges 44:158–161

Goebel K (1926b) Morphologische und biologische studien. Ein Beitrag zur Biologie der Flechten. Ann Jard Bot Buitenzorg 36:1–83

Goffinet B, Bayer RJ (1997) Characterization of mycobionts of photomorph pairs in the Peltigerineae (lichenized ascomycetes) based on internal transcribed spacer sequences of the nuclear ribosomal DNA. Fungal Genet Biol 21:228–237

Golledge N, Everest J, Bradwell T, Johnson J (2010) Lichenometry on Adelaide Island, Antarctic Peninsula: size-frequency studies, growth rates and snowpatches. Geogr Ann A 92:111–124

Gougeon RD, Lucio M, Frommberger M et al (2009) The chemodiversity of wines can reveal a metabologeography expression of cooperage oak wood. Proc Natl Acad Sci USA 106:9174–9179

Grangeon S, Guédron S, Asta J, Sarret G, Charlet L (2012) Lichen and soil as indicators of an atmospheric mercury contamination in the vicinity of a chlor-alkali plant (Grenoble, France). Ecol Indic 13:178–183

Grube M, Berg G (2009) Microbial consortia of bacteria and fungi with focus on the lichen symbiosis. Fungal Biol Rev 23:72–85

Grube M, De los Rios A (2001) Observations on *Biatoropsis usnearum*, a lichenicolous heterobasidiomycete, and other gall-forming fungi, using different microscopical techniques. Mycol Res 105:1116–1122

Grube M, Hawksworth DL (2007) Trouble with lichen: the re-evaluation and re-interpretation of thallus form and fruit body types in the molecular era. Mycol Res 111:1116–1132

Grube M, Kantvilas G (2006) *Siphula* represents a remarkable case of morphological convergence in sterile lichens. Lichenologist 38:241–249

Grube M, Muggia L (2010) Identifying algal symbionts in lichen symbioses. In: Nimis PL, Vignes Lebbe R (eds) Tools for identifying biodiversity: progress and problems. Proceedings of the international congress, Paris, September 20–22, 2010. http://hdl.handle.net/10077/3793. EUT Edizioni Università di Trieste, Trieste

Grube M, Cardinale M, de Castro JV, Müller H, Berg G (2009) Species-specific structural and functional diversity of bacterial communities in lichen symbioses. ISME J 3:105–115

Gueidan C, Roux C, Lutzoni F (2007) Using a multigene phylogenetic analysis to assess generic delineation and character evolution in Verrucariaceae (Verrucariales, Ascomycota). Mycol Res 111:1145–1168

Gueidan C, Ruibal Villaseñor C, de Hoog GS, Gorbushina AA, Untereiner WA, Lutzoni F (2008) A rock-inhabiting ancestor for mutualistic and pathogen-rich fungal lineages. Stud Mycol 61:111–119

Gueidan C, Savic S, Thüs H et al (2009) Generic classification of the Verrucariaceae (Ascomycota) based on molecular and morphological evidence: recent progress and remaining challenges. Taxon 58:184–208

Guzow-Krzeminska B (2006) Photobiont flexibility in the lichen *Protoparmeliopsis muralis* as revealed by ITS rDNA analyses. Lichenologist 38:469–476

Haas JR, Purvis OW (2006) Lichen biogeochemistry. In: Gadd GM (ed) Fungi in biogeochemical cycles. Cambridge University Press, Cambridge, pp 343–376

Hager A, Brunauer G, Türk R, Stocker-Wörgötter E (2009) Production and bioactivity of common lichen metabolites as exemplified by *Heterodea muelleri* (Hampe) Nyl. J Chem Ecol 34:113–120

Hauck M, Huneck S (2007a) Lichen substances affect metal adsorption in *Hypogymnia physodes*. J Chem Ecol 33:219–223

Hauck M, Huneck S (2007b) The putative role of fumarprotocetraric acid in the manganese tolerance of the lichen *Lecanora conizaeoides*. Lichenologist 39:301–304

Hauck M, Jürgens SR, Brinkmann M, Herminghaus S (2008) Surface hydrophobicity causes SO_2 tolerance in lichens. Ann Bot 101:531–539

Hauck M, Jürgens SR, Willenbruch K, Huneck S, Leuschner C (2009) Dissociation and metal-binding characteristics of yellow lichen substances suggest a relationship with site preferences of lichens. Ann Bot 103:13–22

Hauck M, Otto PI, Dittrich S et al (2011) Small increase in sub-stratum pH causes the dieback of one of Europe's most common lichens, *Lecanora conizaeoides*. Ann Bot 108:59–366

Hawksworth DL (1988a) The variety of fungal-algal symbioses, their evolutionary significance, and the nature of lichens. Bot J Linn Soc 96:3–20

Hawksworth DL (1988b) Effects of algae and lichen-forming fungi on tropical crops. In: Agnihotri VP, Sarbhoy KA, Kumar D (eds) Perspectives of mycopathology. Malhorta, New Delhi, pp 76–83

Hawksworth DL (2000) Freshwater and marine lichen-forming fungi In: Hyde KD, Ho WH, Pointing SB (eds) Aquatic mycology across the millennium, vol 5. Fungal Diversity Press, Hong Kong, pp 1–7

Hawksworth DL, Honegger R (1994) The lichen thallus: symbiotic phenotype and its responses to gall producers. In: Williams MC (ed) Plant galls: organisms, interactions. Clarendon, Oxford, pp 77–98

Hedenås H, Blomberg P, Ericson L (2007) Significance of old aspen (*Populus tremula*) trees for the occurrence of lichen photobionts. Biol Conserv 135:380–387

Helms G (2003) Taxonomy and symbiosis in associations of Physciaceae and *Trebouxia*. Inauguraldissertation am Albrecht-von-Haller Institut für Pflanzenwissenschaften, Experimentelle Phykologie und Sammlung von Algenkulturen der Georg-August-Universität Göttingen, Göttingen, pp 1–156

Helms G, Friedl T, Rambold G, Mayrhofer H (2001) Identification of photobionts from the lichen family Physciaceae using algal-specific ITS rDNA sequencing. Lichenologist 33:73–86

Henssen A (1995) *Psoroglaena costaricensis*, a new lichen species from Costa Rica, and remarks on other taxa of the genus *Psoroglaena* (Verrucariaceae). Bibl Lichenol 57:199–210

Hestmark G (1992) Sex, size competition and escape – strategies of reproduction and dispersal in Lasallia pustulata (Umbilicariaceae, Ascomycetes). Oecologia 92:305–312

Hestmark G, Schroeter B, Kappen L (1997) Intrathalline and size-dependent patterns of activity in *Lasallia pustulata* and their possible consequences for competitive interactions. Funct Ecol 11:318–322

Hibbett DS, Binder M, Bischoff JF et al (2007) A higher-level phylogenetic classification of the Fungi. Mycol Res 111:509–547

Hill DJ (1976) The physiology of lichen symbiosis. In: Brown DH, Hawksworth DL, Bailey RH (eds) Lichenology: progress and problems. Academic, London, pp 457–496

Hill DJ (1985) Changes in photobiont dimensions and numbers during co-development of lichen symbionts. In: Brown DH (ed) Lichen physiology and cell biology. Plenum, New York, pp 303–317

Hill DJ (1989) The control of the cell cycle in microbial symbionts. New Phytol 112:175–184

Hill DJ (1992) An overlooked symbiosis. Photosynth Res 34:339–340

Hill DJ (2009) Asymmetric co-evolution in the lichen symbiosis caused by a limited capacity for adaptation in the photobiont. Bot Rev 75:326–338

Hill DR, Peat A, Potts M (1994) Biochemistry and structure of the glycan secreted by desiccation-tolerant *Nostoc commune* (cyanobacteria). Protoplasma 182:126–148

Hodkinson BP (2011) A phylogenetic, ecological, and functional characterization of non-photoautotrophic bacteria in the lichen microbiome. PhD thesis, Duke University, Durham

Hodkinson BP, Lutzoni F (2009) A microbiotic survey of lichen-associated bacteria reveals a new lineage from the Rhizobiales. Symbiosis 49:163–180

Hodkinson BP, Lutzoni FM, Loveless TM, Bishop E (2006) Non-photosynthetic bacteria and the lichen symbiosis. In: Bright M, Horn M, Zook D, Lücker S, Kolar I (eds) Fifth international symbiosis society congress: program, abstracts, participants. Promare, Gdynia, p 95

Hoffman Y, Aflalo C, Zarka A, Gutman J, James TY, Boussiba S (2008) Isolation and characterization of a novel chytrid species (phylum Blastocladiomycota), parasitic on the green alga *Haematococcus*. Mycol Res 112:70–81

Hoffmann N, Hafellner J (2000) Eine Revision der lichenicolen Arten der Sammelgattungen *Guignardia* und *Physalospora* (Ascomycotina). Bibl Lichenol 77:1–181

Honegger R (1982) Cytological aspects of the triple symbiosis in *Peltigera aphthosa*. J Hattori Bot Lab 52:379–391

Honegger R (1984) Cytological aspects of the mycobiont–phycobiont relationship in lichens. Haustorial types, phycobiont cell wall types, and the ultrastructure of the cell wall surface layers in some cultured and symbiotic myco- and phycobionts. Lichenologist 16:111–127

Honegger R (1986a) Ultrastructural studies in lichens. I. Haustorial types and their frequencies in a range of lichens with trebouxioid phycobionts. New Phytol 103:785–795

Honegger R (1986b) Ultrastructural studies in lichens. II. Mycobiont and photobiont cell wall surface

layers and adhering crystalline lichen products in four Parmeliaceae. New Phytol 103:797–808
Honegger R (1987) Questions about pattern formation in the algal layer of lichens with stratified (heteromerous) thalli. Bibl Lichenol 25:59–72
Honegger R (1990) Mycobiont-photobiont interactions in adult thalli and in axenically resynthesized prethallus stages of *Xanthoria parietina* (Teloschistales, lichenized Ascomycetes). Bibl Lichenol 38:191–208
Honegger R (1991) Functional aspects of the lichen symbiosis. Annu Rev Plant Physiol Plant Mol Biol 42:553–578
Honegger R (1992) Lichens: mycobiont–photobiont relationships. In: Reisser W (ed) Algae and symbiosis. Biopress, Bristol, pp 225–275
Honegger R (1993) Developmental biology of lichens. New Phytol 125:659–677
Honegger R (1995) Experimental studies with foliose macrolichens: fungal responses to spatial disturbance at the organismic level and to spacial problems at the cellular level. Can J Bot 73:569–578
Honegger R (1996) Experimental studies on growth and regenerative capacity in the foliose lichen *Xanthoria parietina*. New Phytol 133:573–581
Honegger R (1997) Metabolic interactions at the mycobiont–photobiont interface in lichens. In: Carroll GC, Tudzynski P (eds) The Mycota V. Plant relationships. Springer, Berlin Heidelberg New York, pp 209–221
Honegger R (1998) The lichen symbiosis - what is so spectacular about it? Lichenologist 30:193–212
Honegger R (2001) The symbiotic phenotype of lichen-forming ascomycetes. In: Hock B (ed) The Mycota IX (Fungal associations). Springer, Berlin Heidelberg New York, pp 165–188
Honegger R (2007) Water relations in lichens. In: Gadd GM, Watkinson SC, Dyer P (eds) Fungi in the environment. Cambridge University Press, Cambridge, pp 185–200
Honegger R (2009) Lichen-forming fungi and their photobionts. In: Deising HB (ed) The Mycota V (Plant relationships), 2nd edn. Springer, Berlin Heidelberg New York, pp 305–333
Honegger R, Brunner U (1981) Sporopollenin in the cell wall of *Coccomyxa* and *Myrmecia* phycobionts of various lichens: an ultrastructural and chemical investigation. Can J Bot 59:2713–2734
Honegger R, Haisch A (2001) Immunocytochemical location of the (1→3) (1→4)-beta-glucan lichenin in the lichen-forming ascomycete *Cetraria islandica* (Icelandic moss). New Phytol 150:739–746
Honegger R, Hugelshofer G (2000) Water relations in the *Peltigera aphthosa* group visualized with LTSEM techniques. Bibl Lichenol 75:113–126
Honegger R, Peter M (1994) Routes of solute translocation and the location of water in heteromerous lichens visualized with cryotechniques in light and electron microscopy. Symbiosis 16:167–186
Honegger R, Conconi S, Kutasi V (1996a) Field studies on growth and regenerative capacity in the foliose macrolichen *Xanthoria parietina* (Teloschistales, Ascomycotina). Bot Acta 109:187–193
Honegger R, Peter M, Scherrer S (1996b) Drought-stress induced structural alterations at the mycobiont photobiont interface in a range of foliose macrolichens. Protoplasma 190:221–232
Honegger R, Edwards D, Axe L (2012) The earliest records of internally stratified cyanobacterial and algal lichens from the Lower Devonian of the Welsh Borderland. New Phytologist, in press
Huneck S (2001) New results on the chemistry of lichen substances. Fort Chem Org Nat 81:1–276
Huneck S (2003) Die wasserabweisende Eigenschaft von Flechtenstoffen. Bibl Lichenol 86:9–12
Huneck S, Yoshimura I (1996) Identification of lichen substances. Springer, Berlin Heidelberg New York
Huneck S, Lumbsch HT, Porzel A, Schmidt J (2004) Die Verteilung von Flechteninhaltsstoffen in *Lecanora muralis* und *Lecidea inops* und die Abhängigkeit der Usninsäure-Konzentration vom Substrat und von den Jahreszeiten bei *Lecanora muralis*. Bibl Lichenol 88:211–222
Hyvärinen M, Hardling R, Tuomi J (2002) Cyanobacterial lichen symbiosis: the fungal partner as an optimal harvester. Oikos 98:498–504
Ingólfsdóttir K (2002) Usnic acid. Phytochemistry 61:729–736
Innes JL (1988) The use of lichens in dating. In: Galun M (ed) CRC handbook of lichenology, vol 3. CRC, Boca Raton, pp 75–91
Jaag O, Thomas E (1934) Neue Untersuchungen über die Flechte *Epigloea bactrospora* Zukal. Ber Schweiz Bot Ges 34:77–89
Jahns HM, Ott S (1997) Life strategies in lichens - some general considerations. Bibl Lichenol 67:49–67
James PW, Henssen A (1976) The morphological and taxonomic significance of cephalodia. In: Brown DH, Hawksworth DL, Bailey RH (eds) Lichenology: progress and problems. Academic, London, pp 27–77
Joneson S, Lutzoni F (2009) Revisiting compatibility and thigmotropism in the lichen symbiosis. Symbiosis 47:109–115
Joneson S, Armaleo D, Lutzoni F (2011) Fungal and algal gene expression in early developmental stages of lichen-symbiosis. Mycologia 103:291–306
Jørgensen PM, Jahns HM (1987) *Muhria*, a remarkable new lichen genus from Scandinavia. Notes RBG Edinb 44:581–599
Joubert JJ, Rijkenberg FHJ (1971) Parasitic green algae. Annu Rev Phytopathol 9:45–64
Kadouri A, Derenne S, Largeau C, Casadevall E, Berkaloff C (1988) Resistant biopolymer in the outer

walls of *Botryococcus braunii*, B race. Phytochemistry 27:551–557

Kagami M, Ad B, Ibelings BW, Donk EV (2007) Parasitic chytrids: their effects on phytoplankton communities and food-web dynamics. Hydrobiologia 578:113–129

Kannangara BTSDP, Rajapaksha RSCG, Paranagama PA (2009) Nature and bioactivities of endolichenic fungi in *Pseudocyphellaria* sp., *Parmotrema* sp. and *Usnea* sp. at Hakgala montane forest in Sri Lanka. Lett Appl Microbiol 48:203–209

Kappen L (1988) Ecophysiological relationships in different climatic regions. In: Galun M (ed) CRC handbook of lichenology, vol 2. CRC, Boca Raton, pp 37–100

Kappen L (2000) Some aspects of the great success of lichens in Antarctica. Antarct Sci 12:314–324

Karnieli A, Kokaly R, West NE, Clark RN (2001) Remote sensing of biological soil crusts. In: Belnap J, Lange OL (eds) Biological soil crusts: structure, function, and management. Springer, Berlin Heidelberg New York, pp 431–455

Kauff F, Lutzoni F (2002) Phylogeny of the Gyalectales and Ostropales (Ascomycota, Fungi): among and within order relationships based on nuclear ribosomal RNA small and large subunits. Mol Phylogenet Evol 25:138–156

Kershaw KA (1985) Physiological ecology of lichens. Cambridge University Press, Cambridge

Kershaw M, Thornton C, Wakley G, Talbot N (2005) Four conserved intramolecular disulphide linkages are required for secretion and cell wall localization of a hydrophobin during fungal morphogenesis. Mol Microbiol 56:117–125

Kirk PM, Cannon PF, Minter DW, Stalpers JA (2008) Dictionary of the fungi, 10th edn. CABI Europe, Oxford

Kluge M (2002) A fungus eats a cyanobacterium: the story of the *Geosiphon pyriformis* endocyanosis. Commentaries on cyanobacterial symbioses. Biol Environ 10B:11–14

Knowles RD, Pastor J, Biesboer DD (2006) Increased soil nitrogen associated with dinitrogen-fixing, terricolous lichens of the genus *Peltigera* in northern Minnesota. Oikos 114:37–48

Kodner RB, Summons RE, Knoll AH (2009) Phylogenetic investigation of the aliphatic, non-hydrolyzable biopolymer algaenan, with a focus on green algae. Org Geochem 40:854–862

Kogel KH, Franken P, Hückelhoven R (2006) Endophyte or parasite – what decides? Curr Opin Plant Biol 9:358–363

Kohlmeyer J, Kohlmeyer E (1972) Is *Ascophyllum nodosum* lichenized? Bot Mar 15:109–112

Kohlmeyer J, Volkmann-Kohlmeyer B (1998) *Mycophycias*, a new genus for the mycobionts of *Apophlaea*, *Ascophyllum* and *Pelvetia*. Syst Ascomyc 16:1–7

Kohlmeyer J, Hawksworth DL, Volkmann-Kohlmeyer B (2004) Observations on two marine and maritime "borderline" lichens: *Mastodia tessellata* and *Collemopsidium pelvetiae*. Mycol Prog 3:51–56

Kristmundsdóttir T, Jónsdóttir E, Ogmundsdóttir HM, Ingólfsdóttir K (2005) Solubilization of poorly soluble lichen metabolites for biological testing on cell lines. Eur J Pharm Sci 24:539–543

Kroken S, Taylor JW (2000) Phylogenetic species, reproductive mode, and specificity of the green alga *Trebouxia* forming lichens with the fungal genus *Letharia*. Bryologist 103:645–660

Kroken S, Taylor JW (2001) A gene genealogical approach to recognize phylogenetic species boundaries in the lichenized fungus *Letharia*. Mycologia 93:38–53

LaGreca S, Stutzman BW (2006) Distribution and ecology of *Lecanora conizaeoides* (Lecanoraceae) in eastern Massachusetts. Bryologist 109:335–347

Lange O (2001) Photosynthesis of soil-crust biota as dependent on environmental factors. In: Belnap J, Lange OL (eds) Biological soil crusts: structure, function, and management. Springer, Berlin Heidelberg New York, pp 217–240

Lange O, Kilian E, Ziegler H (1986) Water vapour uptake and photosynthesis of lichens: performance differences in species with green and blue-green algae as phycobionts. Oecologia 71:104–110

Lange O, Meyer A, Zellner H, Ullmann I, Wessels D (1990) Eight days in the life of a desert lichen: water relations and photosynthesis of *Teloschistes capensis* in the coastal fog zone of the Namib desert. Madoqua 17:17–30

Lange OL, Büdel B, Meyer A, Kilian E (1993) Further evidence that activation of net photosynthesis by dry cyanobacterial lichens requires liquid water. Lichenologist 25:175–189

Lange OL, Belnap J, Reichenberger H (1998) Photosynthesis of the cyanobacterial soil-crust lichen *Collema tenax* from arid lands in southern Utah, USA: role of water content on light and temperature responses of CO_2 exchange. Funct Ecol 12:195–202

Lange OL, Green TGA, Melzer B, Meyer A, Zellner H (2006) Water relations and CO2 exchange of the terrestrial lichen *Teloschistes capensis* in the Namib fog desert: measurements during two seasons in the field and under controlled conditions. Flora 201:268–280

Larson DW (1983) The pattern of production within individual *Umbilicaria* lichen thalli. New Phytol 94:409–419

Larson DW (1984) Habitat overlap/niche segregation in two *Umbilicaria* lichens: a possible mechanism. Oecologia 62:118–125

Lawrey JD (1986) Biological role of lichen substances. Bryologist 89:111–122

Lawrey JD, Diederich P (2003) Lichenicolous fungi: interactions, evolution, and biodiversity. Bryologist 106:80–120

Lawrey JD, Diederich P (2010) Lichenicolous fungi-worldwide checklist, including isolated cultures and sequences available. URL: http://www.lichenicolous.net

Lawrey JD, Torzilli AP, Chandhoke V (1999) Destruction of lichen chemical defenses by a fungal pathogen. Am J Bot 86:184–189

Lawrey JD, Binder M, Diederich P, Molina MC, Sikaroodi M, Ertz D (2007) Phylogenetic diversity of lichen-associated homobasidiomycetes. Mol Phylogenet Evol 44:778–789

Lehr H, Galun M, Ott S, Jahns H-M, Fleminger G (2000) Cephalodia of the lichen *Peltigera aphthosa* (L.) Willd. specific recognition of the compatible photobiont. Symbiosis 29:357–365

Letsch MR, Muller-Parker G, Friedl T, Lewis LA (2009) *Elliptochloris marina* sp. nov. (Trebouxiophyceae, Chlorophyta), symbiotic green alga of the temperate pacific sea anemones *Anthopleura xanthogrammica* and *A. elegantissima* (Anthozoa, Cnidaria). J Phycol 45:1127–1135

Li W-C, Zhou J, Guo S-Y, Guo L (2007) Endophytic fungi associated with lichens in Baihua mountain of Beijing, china. Fungal Div 25:69–80

Linder MB (2009) Hydrophobins: proteins that self assemble at interfaces. Curr Opin Colloid Interface Sci 14:356–363

Linder MB, Szilvay GR, Nakari-Setälä T, Penttilä ME (2005) Hydrophobins: the protein-amphiphiles of filamentous fungi. FEMS Microbiol Rev 29:877–896

Lines CEM, Ratcliffe RG, Rees TAV, Southon TE (1989) A ^{13}C NMR study of photosynthate transport and metabolism in the lichen *Xanthoria calcicola* Oxner. New Phytol 111:447–482

Loso MG, Doak DF (2006) The biology behind lichenometric dating curves. Oecologia 147:223–229

Lücking R (2008) Foliicolous lichenized fungi. New York Botanical Garden, New York

Lücking R, Caceres M (2002) Foliicolous lichens of the world. http://fm2.fieldmuseum.org/plantguides/guide_pdfs/130%20Foliicolous%20Lichens.pdf. Cited 1 June 2011

Lücking R, Matzer M (2001) High foliicolous lichen alpha-diversity on individual leaves in Costa Rica and Amazonian Ecuador. Biodivers Conserv 10:2139–2152

Lücking R, Lawrey JD, Sikaroodi M et al (2009) Do lichens domesticate photobionts like farmers domesticate crops? Evidence from a previously unrecognized lineage of filamentous cyanobacteria. Am J Bot 96:1409–1418

Lud D, Huiskes AHL, Ott S (2001) Morphological evidence for the symbiotic character of *Turgidosculum complicatulum* Kohlm. & Kohlm. (= *Mastodia tessellata* Hook.f. & Harvey). Symbiosis 31:141–151

Lumbsch HT, Huhndorf SM (2007) Whatever happened to the pyrenomycetes and loculoascomycetes? Mycol Res 111:1064–1074

Lumbsch HT, Leavitt SD (2011) Goodbye morphology? A paradigm shift in the delimitation of species in lichenized fungi. Fungal Divers 50:59–72

Lutzoni F, Pagel M, Reeb V (2001) Major fungal lineages are derived from lichen symbiotic ancestors. Nature 411:937–940

Lutzoni F, Kauff F, Cox C et al (2004) Assembling the fungal tree of life: progress, classification, and evolution of subcellular traits. Am J Bot 91:1446–1480

MacGinitie H (1937) The flora of the Weaverville beds of Trinity County, California, with descriptions of the plant-bearing beds. In: Eocene flora of western America. Publication 465. Carnegie Institution of Washington, Washington, pp 83–151

Maestre FT, Bowker MA, Escolar C et al (2010) Do biotic interactions modulate ecosystem functioning along stress gradients? Insights from semi-arid plant and biological soil crust communities. Phil Trans R Soc B 365:2057–2070

Mägdefrau K (1957) Flechten und moose im baltischen Bernstein. Ber Dtsch Bot Ges 9:433–435

Marsh JE, Timoney KP (2005) How long must Northern saxicolous lichens be immersed to form a waterbody trimline? Wetlands 25:495–499

Marshall WA (1996) Aerial dispersal of lichen soredia in the maritime Antarctic. New Phytol 134:523–530

Martin F, Tunlid A (2009) The ectomycorrhizal symbiosis: a marriage of convenience. In: Deising HB (ed) The Mycota, vol 5, 2nd edn, Plant relationships. Springer, Berlin Heidelberg New York, pp 237–257, part 2

Matthews SW, Tucker SC, Chapman RL (1989) Ultrastructural features of mycobionts and trentepohliaceous phycobionts in selected subtropical crustose lichens. Bot Gaz 150:417–438

McCarthy DP (2007) Lichenometry. In: Elias SA (ed) Encyclopedia of quaternary science. Elsevier, New York, pp 1399–1405

McCune B, Rosentreter R (2007) Biotic soil crust lichens of the Columbia Basin. Monogr N Am Lichenol 1:1–105

McDonald T, Dietrich F, Lutzoni F (2012) Multiple horizontal gene transfers of ammonium transporters/ammonia permeases from prokaryotes to eukaryotes: toward a new functional and evolutionary classification. Mol Biol Evol 29:51–60

McEvoy M, Nybakken L, Solhaug KA, Gauslaa Y (2006) UV triggers the synthesis of the widely distributed secondary lichen compound usnic acid. Mycol Prog 5:221–229

Meeks JC, Elhai J, Thiel T et al (2001) An overview of the genome of *Nostoc punctiforme*, a multicellular, symbiotic cyanobacterium. Photosynth Res 70:85–106

Mei C, Flinn BS (2010) The use of beneficial microbial endophytes for plant biomass and stress tolerance improvement. Recent Pat Biotechnol 4:81–95

Meier JL, Chapman RL (1983) Ultrastructure of the lichen *Coenogonium interplexum* Nyl. Am J Bot 70:400–407

Meier FA, Scherrer S, Honegger R (2002) Faecal pellets of lichenivorous mites contain viable cells of the lichen-forming ascomycete *Xanthoria parietina* and its green algal photobiont, *Trebouxia arboricola*. Biol J Linn Soc 76:259–268

Metzger P, Largeau C (2005) *Botryococcus braunii*: a rich source for hydrocarbons and related ether lipids. Appl Microbiol Biotechnol 66:486–496

Miadlikowska J, Lutzoni F (2000) Phylogenetic revision of the genus *Peltigera* (lichen-forming Ascomycota) based on morphological, chemical, and large subunit nuclear ribosomal DNA data. Int J Plant Sci 161:925–958

Miadlikowska J, Lutzoni F (2004) Phylogenetic classification of peltigeralean fungi (Peltigerales, Ascomycota) based on ribosomal RNA small and large subunits. Am J Bot 91:449–464

Miadlikowska J, Arnold A, Lutzoni F (2004) Diversity of cryptic fungi inhabiting healthy lichen thalli in a temperate and tropical forest. Ecol Soc Am Annu Meet 89:349–350

Miadlikowska J, Kauff F, Hofstetter V et al (2006) New insights into classification and evolution of the Lecanoromycetes (Pezizomycotina, Ascomycota) from phylogenetic analyses of three ribosomal RNA- and two protein-coding genes. Mycologia 98:1088–1103

Miao V, Coeffet-LeGal M-F, Brown D, Sinnemann S, Donaldson G, Davies J (2001) Genetic approaches to harvesting lichen products. Trends Biotechnol 19:349–355

Millbank JW, Kershaw KA (1970) Nitrogen metabolism in lichens. III. Nitrogen fixation by internal cephalodia in *Lobaria pulmonaria*. New Phytol 69:595–597

Molina MC, Vicente C (1994) The distribution and mobility of the phycobiont in the thalli and apothecia of *Usnea aurantiaco-atra* Jacq. Phyton Arg 56:81–89

Molina MC, DePriest PT, Lawrey JD (2005) Genetic variation in the widespread lichenicolous fungus *Marchandiomyces corallinus*. Mycologia 97:454–463

Mollenhauer D (1992) *Geosiphon pyriforme*. In: Reisser W (ed) Algae and symbioses. Biopress, Bristol, pp 339–351

Mollenhauer D, Mollenhauer R, Kluge M (1996) Studies on initiation and development of the partner association in *Geosiphon pyriforme* (Kütz.) v. Wettstein, a unique endocytobiotic system of a fungus (Glomales) and the cyanobacterium *Nostoc punctiforme* (Kütz.) Hariot. Protoplasma 193:3–9

Motiejûnaitë J, Jucevièienë N (2005) Epidemiology of the fungus *Athelia arachnoidea* in epiphytic communities of broadleaved forests under strong anthropogenic impact. Ekologija 4:28–34

Mueller UG, Wolf-Mueller B (1991) Epiphyll deterrence to the leafcutter ant *Atta cephalotes*. Oecologia 86:36–39

Muggia L, Hafellner J, Wirtz N, Hawksworth DL, Grube M (2007) The sterile microfilamentous lichenized fungi *Cystocoleus ebeneus* and *Racodium rupestre* are relatives of plant pathogens and clinically important dothidealean fungi. Mycol Res 112:50–56

Muggia L, Gueidan C, Grube M (2010) Phylogenetic placement of some morphologically unusual members of Verrucariales. Mycologia 102:835–846

Mukhtar A, Garty J, Galun M (1994) Does the lichen alga *Trebouxia* occur free-living in nature: further immunological evidence. Symbiosis 17:247–253

Müller K (2001) Pharmaceutically relevant metabolites from lichens. Appl Microbiol Biotechnol 56:9–16

Mushegian AA, Peterson CN, Baker CCM, Pringle A (2011) Bacterial diversity across individual lichens. Appl Environ Microbiol 77:4249–4252

Nash TH (2008) Nitrogen, its metabolism and potential contribution ro ecosystems. In: Nash TH (ed) Lichen biology, 2nd edn. Cambridge University Press, Cambridge, pp 216–233

Nelsen MP, Gargas A (2009a) Assessing clonality and chemotype monophyly in *Thamnolia vermicularis* (Icmadophilaceae). Bryologist 112:42–53

Nelsen MP, Gargas A (2009b) Symbiont flexibility in *Thamnolia vermicularis* (Pertusariales: Icmadophilaceae). Bryologist 112:404–417

Nelsen MP, Lücking R, Grube M et al (2009) Unravelling the phylogenetic relationships of lichenised fungi in Dothideomyceta. Stud Mycol 64:135–144

Nelson SC (2008) *Cephaleuros* species, the plant parasitic green algae. Plant Dis 43:1–6

Neustupa J (2003) The genus *Phycopeltis* (Trentepohliales, Chlorophyta) from tropical Southeast Asia. Nova Hedwigia 76:487–505

Nyati S (2006). Photobiont diversity in Teloschistaceae (Lecanoromycetes) [PhD thesis, Mathematisch-Naturwissenschaftliche Fakultät]. Zürich: Universität Zürich.

Nyati S, Beck A, Honegger R (2007) Fine structure and phylogeny of green algal photobionts in the microfilamentous genus *Psoroglaena* (Verrucariaceae, lichen-forming ascomycetes). Plant Biol 9:390–399

Nybakken L, Julkunen-Titto R (2006) UV-B induces usnic acid in reindeer lichens. Lichenologist 38:477–485

Nybakken L, Solhaug KA, Bilger W, Gauslaa Y (2004) The lichens *Xanthoria elegans* and *Cetraria islandica* maintain a high protection against UV-B radiation in Arctic habitats. Oecologia 140:211–216

Nybakken L, Helmersen A-M, Gauslaa Y, Selås V (2010) Lichen compounds restrain lichen feeding by bank voles (*Myodes glareolus*). J Chem Ecol 36:298–304

O'Brien H, Miadlikowska J, Lutzoni F (2005) Assessing host specialization in symbiotic cyanobacteria associated with four closely related species of the lichen fungus *Peltigera*. Eur J Phycol 40:363–378

Ohmura Y, Kawachi M, Kasai F, Watanabe M (2006) Genetic combinations of symbionts in a vegetatively reproducing lichen, *Parmotrema tinctorum*, based on ITS rDNA sequences. Bryologist 109:43–59

Otalora MAG, Aragón G, Molina MC, Martínez I, Lutzoni F (2010) Disentangling the *Collema-Leptogium* complex through a molecular phylogenetic study of the Collemataceae (Peltigerales, lichen-forming Ascomycota). Mycologia 102:279–290

Ott S (1987) Reproductive strategies in lichens. Bibl Lichenol 25:81–93

Ott S, Przewosnik R, Sojo F, Jahns HM (1997) The nature of cephalodia in *Placopsis contortuplicatus* and other species of the genus. Bibl Lichenol 67:69–84

Pacioni G, Leonardi M, Aimola P, Ragnelli AM, Rubini A, Paolocci F (2007) Isolation and characterization of some mycelia inhabiting *Tuber* ascomata. Mycol Res 111:1450–1460

Palmqvist K, Dahlman L, Jonsson AV, Nash TH (2008) The carbon economy of lichens. In: Nash TH (ed) Lichen biology. Cambridge University Press, Cambridge, pp 182–215

Pankewitz F, Zöllmer A, Gräser Y, Hilker M (2007) Anthraquinones as defensive compounds in eggs of Galerucini leaf beetles: Biosynthesis by the beetles. Arch Insect Biochem 662:98–108

Pannewitz S, Schlensog M, Green TGA, Sancho LG, Schroeter B (2003) Are lichens active under snow in Continental Antarctica? Oecologia 135:30–38

Paranagama PA, Wijeratne EM, Burns AM et al (2007a) Heptaketides from *Corynespora* sp. Inhabiting the cavern beard lichen, *Usnea cavernosa*: first report of metabolites of an endolichenic fungus. J Nat Prod 70:1700–1705

Paranagama PA, Wijeratne EMK, Gunatilaka MK, Arnold AE, Gunatilaka AAL (2007b) Bioactive and other naphthopyrans from *Corynespora* sp. Occurring in *Usnea cavernosa*: first report of metabolites of an endolichenic fungus. J Nat Prod 70:1700–1705

Pérez FL (1997) Microbiotic crusts in the high equatorial Andes and their influence on paramo soils. Catena 31:173–198

Pérez-Ortega S, de los Ríos A, Crespo A, Sancho LG (2010) Symbiotic lifestyle and phylogenetic relationships of the bionts of *Mastodia tessellata* (Ascomycota, incertae sedis). Am J Bot 97:738–752

Peterson EB (2000) An overlooked fossil lichen (Lobariaceae). Lichenolgist 32:289–300

Petrini O, Hake U, Dreyfuss MM (1990) An analysis of fungal communities isolated from fruticose lichens. Mycologia 82:444–451

Piercey-Normore M (2006) The lichen-forming ascomycete *Evernia mesomorpha* associates with multiple genotypes of *Trebouxia jamesii*. New Phytol 169:331–344

Pivato B, Offre P, Marchelli S et al (2009) Bacterial effects on arbuscular mycorrhizal fungi and mycorrhiza development as influenced by the bacteria, fungi, and host plant. Mycorrhiza 19:81–90

Platt JL, Spatafora JW (1999) A re-examination of generic concepts of baeomycetoid lichens based on phylogenetic analyses of nuclear SSU and LSU ribosomal DNA. Lichenologist 31:409–418

Platt JL, Spatafora JW (2000) Evolutionary relationships of nonsexual lichenized fungi: molecular phylogenetic hypotheses for the genera *Siphula* and *Thamnolia* from SSU and LSU rDNA. Mycologia 92:475–487

Poelt J (1970) Das Konzept der Artenpaare bei den Flechten. Flechtensymposion 1969. Vorträge im Gesamtgebiet der Botanik. Dtsch Bot Ges NF 4:187–198

Poelt J, Huneck S (1968) *Lecanora vinetorum* nova spec., ihre Vergesellschaftung, ihre Ökologie und ihre Chemie. Österr Bot Z 115:411–422

Poelt J, Mayrhofer H (1988) Ueber Cyanotrophie bei Flechten. Plant Syst Evol 158:265–281

Poelt J, Vězda A (1990) Ueber kurzlebige Flechten. Bibl Lichenol 38:377–394

Poinar GO, Peterson EB, Platt JL (2000) Fossil *Parmelia* in New World amber. Lichenologist 32:263–269

Pöykkö H, Bačkor M, Bencúrov E, Molcanov V, Bačkorov M, Hyvärinen M (2010) Host use of a specialist lichen-feeder: dealing with lichen secondary metabolites. Oecologia 164:423–430

Prasse R, Bornkamm R (2000) Effect of microbiotic soil surface crusts on emergence of vascular plants. Plant Ecol 150:65–75

Puel F, Largeau C, Giraud G (2008) Ocurrence of a resistant biopolymer in the outer walls of the parasitic alga *Prototheca wickerhamii* (Chlorococcales): ultrastructural and chemical studies. J Phycol 23:649–656

Purvis OW, Pawlik-Skowrońska B (2008) Lichens and metals. In: Avery S, Stratford M, van West P (eds) Stress in yeasts and filamentous fungi. Elsevier, Amsterdam, pp 175–200

Purvis OW, Chimonides PDJ, Jeffries TE, Jones GC, Rusu A-M, Read H (2007) Multi-element composition of historical lichen collections and bark samples, indicators of changing atmospheric conditions. Atmos Environ 41:72–80

Qin JG (2010) Hydrocarbons from algae. In: Timmis KN, McGenity T, van der Meer JR, de Lorenzo V (eds) Handbook of hydrocarbon and lipid microbiology. Springer, Berlin Heidelberg New York, pp 2817–2826

Rai AN, Bergman B, Rasmussen U (eds) (2002) Cyanobacteria in symbiosis. http://books.google.com/books/about/Cyanobacteria_in_symbiosis.html?id=HLAx7OhkAmcC. Cited 1 July 2011

Reeb V, Lutzoni F, Roux C (2004) Contribution of RPB2 to multilocus phylogenetic studies of the euascomycetes (Pezizomycotina, Fungi) with special emphasis on the lichen-forming Acarosporaceae and evolution of polyspory. Mol Phylogenet Evol 32:1036–1060

Richardson DHS (1999) War in the world of lichens: parasitism and symbiosis as exemplified by lichens and lichenicolous fungi. Mycol Res 103:641–650

Ridout CJ (2009) Profiles in pathogenesis and mutualism: powdery mildews. In: Deising HB (ed) The Mycota V, part 1, plant relationships. Springer, Berlin Heidelberg New York, pp 51–68

Rikkinen J (2003) Calicioid lichens from European Tertiary amber. Mycologia 95:1032–1036

Rikkinen J, Poinar GO (2002) Fossilised *Anzia* (Lecanorales, lichen-forming Ascomycota) from European Tertiary amber. Mycol Res 106:984–990

Rikkinen J, Poinar GO (2008) A new species of *Phyllopsora* (Lecanorales, lichen-forming Ascomycota) from Dominican amber, with remarks on the fossil history of lichens. J Exp Bot 59:1007–1011

Rindi F, Guiry MD (2002) The genus *Phycopeltis* (Trentepohliaceae, Chlorophyta) in Ireland: a taxonomic and distributional reassessment. Phycologia 41:421–431

Rindi F, Guiry MD (2003) Composition and distribution of subaerial algal assemblages in Galway city, Western Ireland. Cryptog Algol 24:245–267

Rodriguez R, Redman R (2008) More than 400 million years of evolution and some plants still can't make it on their own: plant stress tolerance via fungal symbiosis. J Exp Bot 59:1109–1114

Rodriguez RJ, White JF, Arnold AE, Redman RS (2009) Fungal endophytes: diversity and functional roles. New Phytol 182:314–330

Rogers RW, Lange RT, Nicholas DJD (1966) Nitrogen fixation by lichens of arid soil crusts. Nature 209:96–97

Rolstad J, Rolstad E (2008) Intercalary growth causes geometric length expansion in Methuselah's beard lichen (*Usnea longissima*). Botany 86:1224–1232

Romeike J, Friedl T, Helms G, Ott S (2002) Genetic diversity of algal and fungal partners in four species of *Umbilicaria* (lichenized ascomycetes) along a transect of the Antarctic peninsula. Mol Biol Evol 19:1209–1217

Rosentreter R, Belnap J (2001) Biological soil crusts of North America. In: Belnap J, Lange OL (eds) Biological soil crusts: structure, function, and management. Springer, Berlin Heidelberg New York, pp 31–50

Rosentreter R, Bowler M, Belnap J (2007) A field guide to biological soil crusts of western U.S. drylands. US Government Printing Office, Denver

Sancho LG, Pintado A (2004) Evidence of high annual growth rate for lichens in the maritime Antarctic. Polar Biol 27:312–319

Sancho LG, de la Torre R, Horneck G et al (2007a) Lichens survive in space: results from the 2005 LICHENS experiment. Astrobiology 7:443–454

Sancho LG, Green TGA, Pintado A (2007b) Slowest to fastest: extreme range in lichen growth rates supports their use as an indicator of climate change in Antarctica. Flora 202:667–673

Sancho LG, de la Torre R, Pintado A (2008) Lichens, new and promising material from experiments in astrobiology. Fungal Biol Rev 22:103–109

Sanders W (1989) Growth and development of the reticulate thallus in the lichen *Ramalina menziesii*. Amer J Bot 76:666–678

Sanders W (1992) Comparative *in situ* studies of thallus net development in morphologically distinct populations of the lichen *Ramalina menziesii*. Bryologist 95:192–204

Sanders W (2001) Preliminary light microscope observations of fungal and algal colonization and lichen thallus initiation on glass slides placed near foliicolous lichen communities within a lowland tropical forest. Symbiosis 31:85–94

Sanders W (2002) In situ development of the foliicolous lichen *Phyllophiale* (Trichotheliaceae) from propagule germination to propagule production. Am J Bot 89:1741–1746

Sanders W (2005) Observing microscopic phases of lichen life cycles on transparent substrata placed *in situ*. Lichenologist 37:373–382

Sanders WB (2006) A feeling for the superorganism: expression of plant form in the lichen thallus. Bot J Linn Soc 150:89–99

Sanders WB, Ascaso C (1995) Reiterative production and deformation of cell walls in expanding thallus nets of the lichen *Ramalina menziesii* (Lecanorales, Ascomycetes). Am J Bot 82:1358–1366

Sanders WB, Ascaso C (1997) Fine structural features of rhizomorphs (sensu lato) produced by four species of lichen fungi. Mycol Res 101:319–328

Sanders WB, Lücking R (2002) Reproductive strategies, relichenization and thallus development observed in situ in leaf-dwelling lichen communities. New Phytol 155:425–435

Sanders WB, Ascaso C, Wierzchos J (1994) Physical interactions of two rhizomorph-forming lichens with their rock substrate. Bot Acta 107:432–439

Sanders WB, Moe RL, Ascaso C (2004) The intertidal marine lichen formed by the pyrenomycete fungus *Verrucaria tavaresiae* (Ascomycotina) and the brown alga *Petroderma maculiforme* (Phaeophyceae): thallus organization and symbiont interaction. Am J Bot 91:511–522

Sanders WB, Moe RL, Ascaso C (2005) Ultrastructural study of the brown alga *Petroderma maculiforme* (Phaeophyceae) in the free-living state and in lichen symbiosis with the intertidal marine fungus *Verrucaria tavaresiae* (Ascomycotina). Eur J Phycol 40:353–361

Schardl CL, Scott B, Florea S, Zhang D (2009) *Epichloë* endophytes: clavicipitaceous symbionts of grasses. In: Deising HB (ed) The Mycota IX, Plant relationships. Springer, Berlin Heidelberg New York, pp 276–306

Scheidegger C (1994) Low temperature scanning electron microscopy: the location of free and perturbed water and its role in the morphology of the lichen symbionts. Cryptog Bot 4:290–299

Scheidegger C, Schroeter B, Frey B (1995) Structural and functional processes during water vapour uptake and desiccation in selected lichens with green algal photobionts. Planta 197:399–409

Scherrer S, Honegger R (2003) Inter- and intraspecific variation of homologous hydrophobin (H1) gene sequences among *Xanthoria* spp. (lichen-forming ascomycetes). New Phytol 158:375–389

Scherrer S, De Vries OMH, Dudler R, Wessels JGH, Honegger R (2000) Interfacial self-assembly of fungal hydrophobins of the lichen-forming ascomycetes *Xanthoria parietina* and *X. ectaneoides*. Fungal Genet Biol 30:81–93

Scherrer S, Haisch A, Honegger R (2002) Characterization and expression of XPH1, the hydrophobin gene of the lichen-forming ascomycete *Xanthoria parietina*. New Phytol 154:175–184

Schmitt I, Lumbsch HT (2009) Ancient horizontal gene transfer from bacteria enhances biosynthetic capabilities of fungi. PLoS One 4(e4437):1–8

Schmitt I, Mueller G, Lumbsch HT (2005) Ascoma morphology is homoplaseous and phylogenetically misleading in some pyrenocarpous lichens. Mycologia 97:362–374

Schmitt I, del Prado R, Grube M, Lumbsch HT (2009) Repeated evolution of closed fruiting bodies is linked to ascoma development in the largest group of lichenized fungi (Lecanoromycetes, Ascomycota). Mol Phylogenet Evol 52:34–44

Schmull M, Miadlikowska J, Pelzer M et al (2011) Phylogenetic affiliations of members of the heterogeneous lichen-forming fungi of the genus *Lecidea* sensu Zahlbruckner (Lecanoromycetes, Ascomycota). Mycologia 103:983–1003

Schoch CL, Crous PW, Groenewald JZ et al (2009) A class-wide phylogenetic assessment of Dothideomycetes. Stud Mycol 64:1–15

Schroeter B, Scheidegger C (1995) Water relations in lichens at subzero temperatures: structural changes and carbon dioxide exchange in the lichen *Umbilicaria aprina* from continental Antarctica. New Phytol 131:273–285

Schultz C (2006) Remote sensing the distribution and spatiotemporal changes of major lichen communities in the Central Namib Desert. Inauguraldissertation im Fachbereich Biologie. Universität Kaiserslautern, Kaiserslautern, pp 1–320

Schulz B, Boyle C (2005) The endophytic continuum. Mycol Res 109:661–686

Schüssler A (2002) Molecular phylogeny, taxonomy, and evolution of *Geosiphon pyriformis* and arbuscular mycorrhizal fungi. Plant Soil 244:75–83

Schwendener S (1867) Ueber die wahre Natur der Flechtengonidien. Verh Schweiz Naturforsch Ges 51:88–90

Schwendener S (1869) Die Algentypen der Flechtengonidien. Programm für die Rectoratsfeier der Universität. Universitätsbuchdruckerei C. Schultze, Basel

Sedia EG, Ehrenfeld JG (2003) Lichens and mosses promote alternate stable plant communities in the New Jersey Pinelands. Oikos 100:447–458

Selbmann L, Zucconi L, Ruisi S, Grube M, Cardinale M, Onofri S (2010) Culturable bacteria associated with Antarctic lichens: affiliation and psychrotolerance. Polar Biol 33:71–83

Seymour FA, Crittenden PD, Dyer PS (2005) Sex in the extremes: lichen-forming fungi. Mycologist 19:51–58

Shirtcliffe NJ, Brian Pyatt F, Newton MI, McHale G (2006) A lichen protected by a super-hydrophobic and breathable structure. J Plant Physiol 163:1193–1197

Sikaroodi M, Lawrey JD, Hawksworth DL, Depriest PT (2001) The phylogenetic position of selected lichenicolous fungi: *Hobsonia*, *Illosporium*, and *Marchandiomyces*. Mycol Res 105:453–446

Sinha RP, Singh SP, Häder D-P (2007) Database on mycosporines and mycosporine-like amino acids (MAAs) in fungi, cyanobacteria, macroalgae, phytoplankton and animals. J Photochem Photobiol B 89:29–35

Sipman HJM (1994) Foliicolous lichens on plastic tape. Lichenologist 26:311–312

Skaloud P, Peksa O (2008) Comparative study of chloroplast morphology and ontogeny in *Asterochloris* (Trebouxiophyceae, Chlorophyta). Biologia 3:869–876

Skaloud P, Peksa O (2010) Evolutionary inferences based on ITS rDNA and actin sequences reveal extensive diversity of the common lichen alga *Asterochloris* (Trebouxiophyceae, Chlorophyta). Mol Phylogenet Evol 54:36–46

Smith CW, Aptroot A, Coppins BJ et al (2009) The lichens of Great Britain and Ireland. NHBS/British Lichen Society, London

Solhaug KA, Gauslaa Y, Nybakken L, Bilger W (2003) UV-induction of sun-screening pigments in lichens. New Phytol 158:91–100

Spatafora J (2007) Pezizomycotina. http://tolweb.org/Pezizomycotina/29296/2007.12.19

Spatafora JW, Sung G-H, Johnson D et al (2006) A five-gene phylogeny of Pezizomycotina. Mycologia 98:1018–1028

Stahl E (1877) Beiträge zur Entwickelungsgeschichte der Flechten. II. Über die Bedeutung der Hymenialgonidien. Arthur Felix, Leipzig, pp 1–32

Stenroos S, Stocker-Wörgötter E, Yoshimura I, Myllys L, Thell A, Hyvönen J (2003) Culture experiments and DNA sequence data confirm the identity of *Lobaria* photomorphs. Can J Bot 81: 232–247

Stocker-Wörgötter E (2001) Experimental lichenology and microbiology of lichens: culture experiments, secondary chemistry of cultured mycobionts, resynthesis, and thallus morphogenesis. Bryologist 104:576–581

Stocker-Wörgötter E (2008) Metabolic diversity of lichen-forming ascomycetous fungi: culturing, polyketide and shikimate metabolite production, and PKS genes. Nat Prod Rep 25:188–200

Stocker-Wörgötter E, Hager A (2008) Culture methods for lichens and lichen symbionts. In: Nash TH (ed) Lichen biology, 2nd edn. Cambridge University Press, Cambridge, pp 353–363

Stone BA, Clarke AE (1992) The chemistry and biology of (1-3)-beta-glucans. La Trobe University Press, Melbourne

Sun HJ, DePriest PT, Gargas A, Rossman AY, Friedmann EI (2002) *Pestalotiopsis maculans*: a dominant parasymbiont in North American lichens. Symbiosis 33:215–226

Suryanarayanan TS, Thirunavukkarasu N, Hariharan G, Balaji P (2005) Occurrence of non-obligate microfungi inside lichen thalli. Sydowia 57:119–129

Takano K, Ishikawa Y, Mikami H, Igarashi S, Hino S, Yoshioka T (2008) Fungal infection for cyanobacterium *Anabaena smithii* by two chytrids in eutrophic region of large reservoir Lake Shumarinai, Hokkaido, Japan. Limnology 9:213–218

Talbot NJ (1998) Plants and fungi: friends and enemies. Trends Microbiol 6:250–251

Talbot NJ (2003) On the trail of a cereal killer: exploring the biology of *Magnaporthe grisea*. Annu Rev Microbiol 57:177–202

Taylor TN, Remy W, Hass H, Kerp H (1995) Fossil arbuscular mycorrhizae from the Early Devonian. Mycologia 87:560–573

Taylor T, Hass H, Kerp H (1997) A cyanolichen from the Lower Devonian Rhynie Chert. Am J Bot 84:992–1004

Thomas B (2009) Lichens and Katrina. Loyola University Center for Environmental Communication. http://loyno.edu/lucec/natural-history-writings/lichens-and-katrina. Cited 1 July 2011

Tomescu AMF, Rothwell GW, Honegger R (2006) Cyanobacterial macrophytes in an Early Silurian (Llandovery) continental biota: Passage Creek, lower Mansanutten Sandstone, Virginia, USA. Lethaia 39:329–338

Tormo R, Recio D, Silva I, Muñoz A (2001) A quantitative investigation of airborne algae and lichen soredia obtained from pollen traps in south-west Spain. Eur J Phycol 36:385–390

Trapero A, Romero MA, Varo R, Snchez ME (2003) First report of *Pestalotiopsis maculans* causing necrotic leaf spots in nursery plants of *Arbutus unedo* and *Ceratonia siliqua* in Spain. Plant Dis 87:1263

Trembley ML, Ringli C, Honegger R (2002a) Differential expression of hydrophobins DGH1, DGH2 and DGH3 and immunolocalization of DGH1 in strata of the lichenized basidiocarp of *Dictyonema glabratum*. New Phytol 154:185–195

Trembley ML, Ringli C, Honegger R (2002b) Hydrophobins DGH1, DGH2, and DGH3 in the lichen-forming basidiomycete *Dictyonema glabratum*. Fungal Genet Biol 35:247–259

Trembley ML, Ringli C, Honegger R (2002c) Morphological and molecular analysis of early stages in the resynthesis of the lichen *Baeomyces rufus*. Mycol Res 106:768–776

Tschermak-Woess E (1978) Myrmecia reticulata as a phycobiont and free-living - free-living Trebouxia - the problem of Stenocybe septata. Lichenologist 10:69–79

Tschermak-Woess E (1988) The algal partner. In: Galun M (ed) Handbook of lichenology, vol 1. CRC, Boca Raton, pp 39–92

Tschermak-Woess E, Poelt J (1976) *Vezdaea*, a peculiar lichen genus, and its phycobiont. In: Brown DH, Hawksworth DL, Bailey RH (eds) Lichenology: progress and problems. Academic, London, pp 89–105

U'Ren JM, Lutzoni F, Miadlikowska J, Arnold AE (2010) Community analysis reveals close affinities between endophytic and endolichenic fungi in mosses and lichens. Microb Ecol 60:340–353

van Herk CM, Aptroot A, van Dobben HF (2002) Long-term monitoring in the Netherlands suggests that lichens respond to global warming. Lichenogist 34:141–154

Voegele RT, Hahn M, Mendgen K (2009) The Uredinales: cytology, biochemistry, and molecular biology. In: Deising H (ed) The Mycota, vol V, Plant relationships. Springer, Berlin Heidelberg New York, pp 69–98

Voisey CR (2010) Intercalary growth in hyphae of filamentous fungi. Fungal Biol Rev 24:123–131

Wang Y, Niu S, Liu S, Guo L, Che Y (2010) The first naturally occurring thiepinols and thienol from an endolichenic fungus *Coniochaeta* sp. Org Lett 12:5081–5083

Wedin M, Tehler A, Gargas A (1998) Phylogenetic relationships of Sphaerophoraceae (Ascomycetes) inferred from SSU rDNA sequences. Plant Syst Evol 209:75–83

Wedin M, Döring H, Nordin A, Tibell L (2000) Small subunit rDNA phylogeny shows the lichen families Caliciaceae and Physciaceae (Lecanorales, Ascomycotina) to form a monophyletic group. Can J Bot 78:246–254

Wedin M, Baloch E, Grube M (2002) Parsimony analyses of mtSSU and nITS rDNA sequences reveal the natural relationships of the lichen families Physciaceae and Caliciaceae. Taxon 51:655–660

Wedin M, Wiklund E, Jørgensen PM, Ekman S (2009) Slippery when wet: phylogeny and character evolution in the gelatinous cyanobacterial lichens (Peltigerales, Ascomycetes). Mol Phylogenet Evol 53:862–871

Werth S, Sork VL (2010) Identity and genetic structure of the photobiont of the epiphytic lichen *Ramalina menziesii* on three oak species in southern California. Am J Bot 97:821–830

Wessels JGH (1997) Hydrophobins: proteins that change the nature of the fungal surface. Adv Microb Physiol 38:1–45

Whiteford J, Spanu P (2002) Hydrophobins and the interactions between fungi and plants. Mol Plant Pathol 3:391–400

Wiermann R, Ahlers F, Schmitz-Thom I (2001) Sporopollenin. In: Hofrichter M, Steinbüchel A (eds) Biopolymers, vol 1. Wiley-VCH, Weinheim, pp 209–229

Wijeratne EM, Bashyal BP, Gunatilaka MK, Arnold AE, Gunatilaka AA (2010) Maximizing chemical diversity of fungal metabolites: biogenetically related heptaketides of the endolichenic fungus *Corynespora* sp. J Nat Prod 73:1156–1159

Winchester V (1988) An assessment of lichenometry as a method for dating recent stone movements in two stone circles in Cumbria and Oxfordshire. Bot J Linn Soc 96:57–68

Winchester V, Harrison S (2000) Dendrochronology and lichenometry: colonization, growth rates and dating of geomorphological events on the east side of the North Patagonia Icefield, Chile. Geomorphology 34:181–194

Wirth V (1985) Zur Ausbreitung, Herkunft und Ökologie anthropogen geförderter Rinden- und Holzflechten. Tüxenia 5:523–536

Wösten HAB (2001) Hydrophobins: multipurpose proteins. Annu Rev Microbiol 55:625–646

Wösten HAB, Schuren FHJ, Wessels JGH (1994) Interfacial self-assembly of a hydrophobin into an amphipathic protein membrane mediates fungal attachment to hydrophobic surfaces. EMBO J 13:5848–5854

Xu H, Deckert RJ, Garbary DJ (2008) *Ascophyllum* and its symbionts. X. Ultrastructure of the interaction between *A. Nodosum* (phaeophyceae) and *mycophycias ascophylli* (ascomycetes). Botany 86:185–193

Yahr R, Vilgalys R, Depriest PT (2004) Strong fungal specificity and selectivity for algal symbionts in Florida scrub *Cladonia* lichens. Mol Ecol 13:3367–3378

Yuan X, Xiao S, Taylor TN (2005) Lichen-like symbiosis 600 million years ago. Science 308:1017–1020

Zedda L, Rambold G (2009) Diversity and ecology of soil lichens in the Knersvlakte (South Africa). Bryologist 112:19–29

Zhang YM, Chen J, Wang L, Wang XQ, Gu ZH (2007) The spatial distribution patterns of biological soil crusts in the Gurbantunggut Desert, Northern Xinjiang, China. J Arid Environ 68:599–610

Zhang F, Liu S, Lu X, Guo L, Zhang H, Che Y (2009) Allenyl and alkynyl phenyl ethers from the endolichenic fungus *Neurospora terricola*. J Nat Prod 72:1782–1785

Ziegler R (1997) Fossil organosilicon compounds- a type of silicification diagenetically developed in Triassic vascular plant cuticles and thallophytes. Documenta naturae 112, vol 1. Documenta Naturae, Munich, pp 1–24

Ziegler R (2002) Fossiler Pflanzenmoder aus dem Keuper. Documenta naturae 112, vol 2. Documenta Naturae, Munich, pp 1–65

16 Basidiolichens

F. Oberwinkler[1]

CONTENTS

Manuscript submitted in August 2011.

[1]Lehrstuhl Organismische Botanik, Universität Tübingen, Auf der Morgenstelle 1, 72076 Tübingen, Germany; e-mail: franz.oberwinkler@uni-tuebingen.de

I. Introduction

This is an updated version of my previous review on basidiolichens (Oberwinkler 2001).

Originally the basidiolichens were considered to be exclusively tropical lichens, best represented by the genera *Cora* and *Dictyonema*. In his monograph of *Clavaria* and allied genera, Corner (1950) treated several phycophilous clavarioid fungi from temperate zones in the genus *Clavulinopsis* and *Lentaria*. Mycofloristic studies initiated a breakthrough in further recognitions of basidiolichens. *Clavaria mucida* was considered to be an extratropical basidiolichen by Geitler (1955). Poelt (1959) reported *Clavulinopsis septentrionalis* from the Alps and *Lentaria* (*Clavaria*) *mucida* from Bavaria (1962). Finally, Gams (1962) recognized the association of *Botrydina* and *Coriscium* thalli with *Omphalina* basidiocarps, and Poelt and Oberwinkler (1964) analysed their anatomical structures. A remarkable lichenized clavarioid species has been found in the Sierra de Santa Marta, Colombia, and described as *Lepidostroma terricolens* (Mägdefrau and Winkler 1967). *Pseudocraterellus leptoglossoides* (Corner 1966) was transferred as a basidiolichen into a new genus, *Semiomphalina*, by Redhead (1984). A comparative morphology of all known basidiolichen genera, known at that time, has been published by Oberwinkler (1970). In a study on new or interesting clavarioid fungi from Yunnan, Petersen and Zang (1986) reported on *Multiclavula fossicola*, and *Multiclavula sinensis*, both "with film of algae". Redhead and Kuyper (1987) studied fully lichenized arctic alpine *Omphalina* species and proposed nomenclatorial changes for them in *Botrydina*, *Coriscium*, *Omphalina*, and *Phytoconis*. Jørgensen (1989) erected a new basidiolichen species

Fungal Associations, 2nd Edition
The Mycota IX
B. Hock (Ed.)

without basidiocarps from the Venezuelan Andes, *Omphalina foliacea*, a taxon that was transferred into its own genus, *Marandiomphalina*, by Diederich and Lawrey (2007). Another anamorphic basidiolichen, also from the Andes had been also described as a new genus, *Acantholichen*, by Jørgensen (1998). From Costa Rica, two *Dictyonema* species (Chaves et al. 2004), and *Multiclavula ichthyiformis* (Lawrey et al. 2007) were introduced as new taxa.

Based on transmission electron microscopic studies, Roskin (1970) reported on the ultrastructure of the host–parasite interaction in *Cora pavonia* (*Dictyonema glabratum*). Detailed ultrastructural analyses were carried out by Oberwinkler (1980, 1984) for representatives of most basidiolichen genera. Three main types of fungus–alga interactions in basidiolichens were found and documented.

Cora pavonia was the favoured basidiolichen for ecophysiological studies (Lange 1965; Feige 1969; Coxson 1987a, b, c; Larcher and Vareschi 1988; Lange et al. 1994; Trembley et al. 2002a, b).

The axenic culture of mycobionts of basidiolichens is very difficult. Experiments of Langenstein (1994) were successful to separate myco- and photobionts, to synthesize *Botrydina*, and to characterize essential steps in the ontogeny of the thallus globules.

Lichenized *Omphalina* species and related non-lichenized ones were used as a model for understanding coevolution and the influence of lichenization on evolutionary rates (Lutzoni and Vilgalys 1995a, b; Lutzoni 1997; Lutzoni and Pagel 1997).

Figure 16.1 is used as a guideline for this chapter. The scheme is considered to facilitate various comparative interpretations in the following text. It comprises and compares structural components (cellular interactions, thalli, basidiocarps) with taxonomic units on the generic level (*Athelia*, *Athelopsis*, *Dictyonema*, *Cora*, *Cyphellostereum*, *Lepidostroma*, *Multiclavula*, *Semiomphalina*, *Lichenomphalia*). The anamorphic genera *Acanthonema* and *Marchandiomphalina* cannot be included in accurate positions in this overview because basidiocarps are not yet known.

II. Morphology of the Lichen Thalli

In basidiolichens exclusively members of the Agaricomycetes (Fig. 16.2) are lichenized. They are obligatorily associated with cyanobacteria and/or green algae. The morphological expression of this symbiontic association is the lichen thallus (Figs. 16.1, 16.3, 16.4a, b, 16.5, 16.6a–c, 16.7c, f, 16.8a, b, 16.9a, b, e, f).

The most conspicuous basidiolichen thalli are those of *Cora pavonia* (*Dictyonema glabratum*, Fig. 16.3h, i). Some morphological and anatomical observations of this species were reported by Tomaselli and Caretta (1969). The cellular composition of the thallus and the fungus–alga compartmentation was studied and illustrated by Oberwinkler (1970). He also compared *Cora* with *Dictyonema* species and discussed the considerable differences in thallus architecture of both taxa. In his comparative morphological study also *Corella* and *Vainiocora* were included. The thallus of these taxa is identical and determined by the fungus, forming leaf-like structures with upper and lower pseudoparenchymatous layers and central compartments in which *Rhizonema* trichomes are densely packed as photosynthetic units. In *Dictyonema* the *Rhizonema* trichomes are long filaments ensheathed with mantles of densely longitudinally oriented hyphae which are shortly bent in a wave-like manner (Figs. 16.3f, 16.7a, b). The same structural characters are present in *Rhipidonema*, a taxon which cannot be separated from *Dictyonema*.

A new genus and species, *Acantholichen pannarioides*, has been proposed by Jørgensen (1998). It is based on sterile thalli composed of clampless hyphae with acanthohyphidia and *Rhizonema* trichomes (Fig. 16.7e, f). Thallus architecture and hyphae–algae interactions are very similar to those of the *Dictyonema*–*Cora* group. The lichen grows *Pannaria*-like on debris, mosses, and other lichens.

Also some *Athelia* species are associated with *Rhizonema* and/or coccomyxoid green algae (Oberwinkler 1970). There are no macroscopically distinct thalli. The fungus–alga interaction occurs in basal layers of the basidiocarp (Fig. 16.5b). As shown by Oberwinkler (1970), in

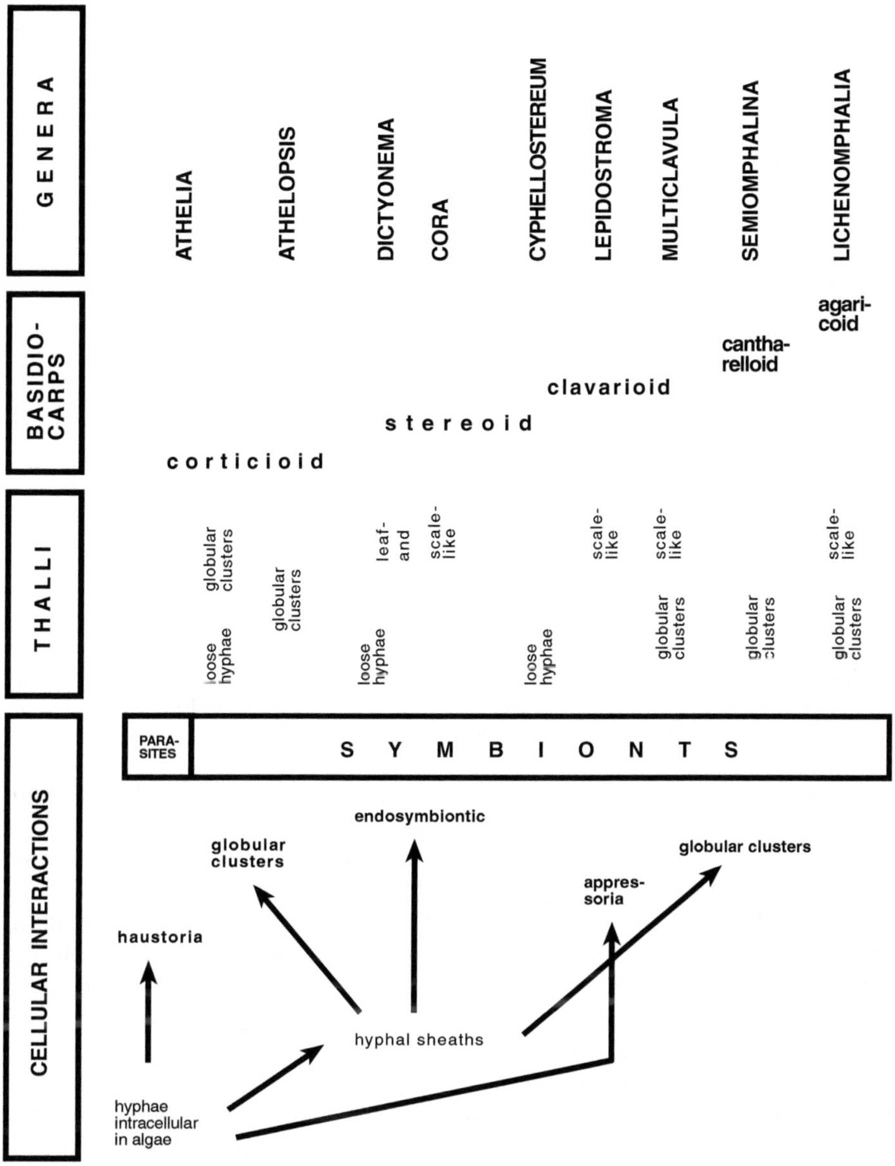

Fig. 16.1. Scheme of basidiolichen genera and their most important structural features, modified after Oberwinkler (2001). Because basidiocarps are not known, *Acanthonema* and *Marchandiomphalina* could not be included in this overview. The arrangement of genera reflects increasing complexity of basidiocarps, i.e. corticioid–stereoid–clavarioid–agaricoid, thus corresponding with the general structures in homobasidiomycetes. The distribution of types of lichen thalli is intermixed. Loose hyphae, globular clusters, and scaly thalli occur convergently. The cellular interactions show an informative distribution pattern: *Athelia* can be parasitized with haustoria but also hyphae can surround algae in globular clusters. Hyphal sheaths in the *Dictyonema* group are always associated with endosymbiontic hyphae of the *Scytonema* (= *Rhizonema*) symbionts. Appressoria occur only in *Lepidostroma*. Comparatively simple globular clusters and scale-like thalli are formed by *Multiclavula* species. Complex globular structures of the *Botrydina* type, and associations of globules in scale-like thalli (*Coriscium*) are present in *Lichenomphalia*

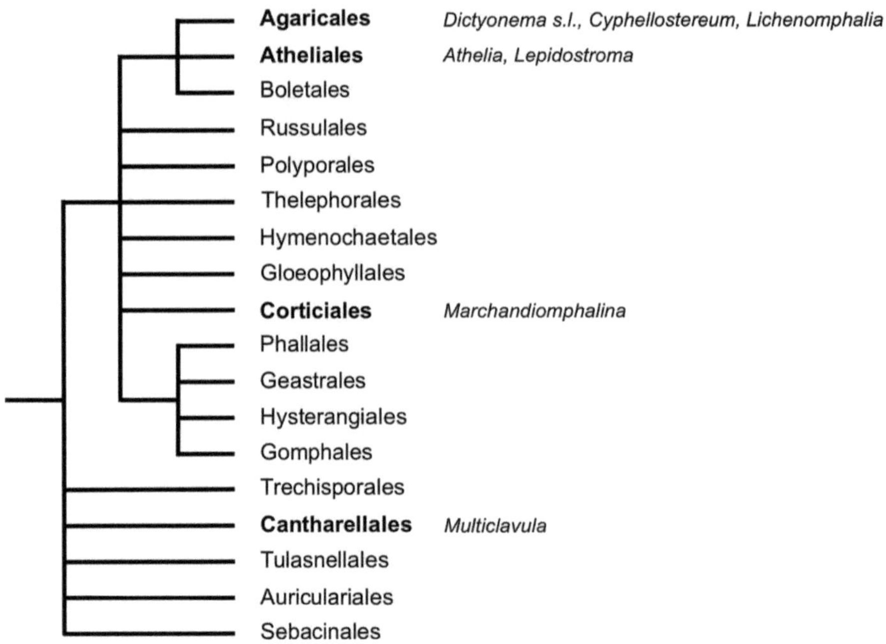

Fig. 16.2. Simplified phylogenetic dendrogram indicating the orders (in *bold*) in which lichenized species occur. They are restricted to Agaricomycetes. The dendrogram is in agreement with generally accepted phylogenies of Agaricomycotina and is modified after Hibbett et al. (2007)

one fungal fructification, both cyanobacteria and green algae can serve as photobionts. In contrast, the closely related *Athelia arachnoidea* and *A. epiphylla* are strong parasites of algae and lichens (Poelt and Jülich 1969b; Oberwinkler 1970; Arvidsson 1979; Gilbert 1988; Parmasto 1998), penetrating the host algal cells with haustoria (Fig. 16.5a).

Elias Fries (1820) was among the first to recognize green material at the basis of *Agaricus ericetorum*, which in fact is the assemblage of tiny lichenized globules (Fig. 16.3k, l), later named as *Botrydina vulgaris* by Brébisson (1844). Because of simple globose thallus structures, *B. vulgaris* has been considered a primitive lichen (Acton 1909). Observations in the field led Gams (1962) to propose specific connections between the sterile lichen thalli *Botrydina* and *Coriscium* with basidiomycetes of the agaricoid genus *Omphalina* (Figs. 16.3k–p, 16.9a–c, e, f). Poelt and Oberwinkler (1964) and Oberwinkler (1970) studied the micromorphology of the thalli and of the associated *Omphalina* basidiocarps and were able to provide structural evidence for lichenization of *Omphalina* species. Dolipores with perforated parenthesomes in hyphae of *Coriscium* were found by Henssen and Kowallik (1976) and Oberwinkler (1980, 1984); and they were also independently documented in *Botrydina* by Boissière (1980) and Oberwinkler (1980, 1984). The globose *Botrydina* thalli have a most complex cell architecture in which algal compartments are inserted within a compact, pseudoparenchymatous hyphal system (Fig. 16.9b, c). This was analysed with the light microscope by Poelt and Oberwinkler (1964) and Oberwinkler (1970) and with the transmission electron microscope by Oberwinkler (1980, 1984). Haustoria are lacking, but there is a dense cell to cell contact of myco- and photobionts. The *Botrydina* globules are dynamic structures which undergo a strongly regulated ontogeny. During this process, photobionts are able to reproduce mitotically to build up more and more algal compartments, but still within the solid pseudoparenchymatous hyphal structure of the enlarging globules. *Botrydina*-like thalli occur also in *Semiomphalina leptoglossoides* (Redhead 1984).

Globular clusters of fungus–alga interactions are also present in *Athelopsis* (Coste and Royaud 1994). The globules are formed in the basal layer of the thin corticioid basidiocarps

(Fig. 16.5d, e), thus not representing macroscopically visible thalli.

Rather conspicuous thallus layers are found in several *Multiclavula* species. In *M. mucida* irregular globose thalli (Oberwinkler 1970, 1984; Figs. 16.3a, 16.4b) are embedded in a gelatinous matrix on wet and decaying wood. Soil-inhabiting *Multiclavula* species have various types of thalli, some appear as uniform crustous layers, others are more separated in granular or scale-like units. These species have not yet been studied comparatively. Nelsen et al. (2007) described *M. ichtyiformis* from Costa w?>Rica. Their light microscopic photographs indicate a thallus structure similar to that in *M. mucida*. However, cellular details of fungus–alga interactions cannot be seen in these illustrations. Originally, two new *Multiclavula* species were proposed from Rwanda (Fischer et al. 2007). Comparing them with *Multiclavula calocerum* (= *Lepidostroma calocerum*), their interpretations of the cellular thallus architecture are rather misleading. The presence of upper and lower pseudopyrenchymatous layers in a scale-like to foliose thallus is a general prerequisite, and not at all a "unique feature amongst the clavarioid lichenized basidiomycetes" (Fischer et al. 2007). In addition, the *Coriscium* thallus is quite different from the *Lepidostroma* thallus, as can easily be seen in the illustrations of Poelt and Oberwinkler (1964) and Oberwinkler (1970, 1980, 1984, 2001). Molecular data indicated that these Rwandan clavarioid species belong in the genus *Lepidostroma* (Ertz et al. 2008).

The scale- and leaf-like types of basidiolichen thalli are rather diverse in architecture and do not occur in closely related taxa. The best known of these is *Coriscium viride* (Figs. 16.3m, o, p, 16.9e, f), described as a sterile lichen by Vainio (1890). The thalli are morphologically distinct and anatomically highly structured (Fig. 16.5f). There are upper and lower pseudoparenchymatous epidermis-like fungal layers and central loose hyphae with globular chambers for the phytobionts, surrounded by hyphal sheats. The gross morphology varies specifically.

Also the thallus of *Lepidostroma calocerum* (Mägdefrau and Winkler 1967) is scaly with distinct and dense hyphal pseudoparenchyms and loose inner hyphae, but the overall cellular composition is totally different from the *Coriscium* type (Fig. 16.6c). This is another example for the multiple convergent evolution of structurally similar and functionally identical photosynthetic organs which are not homologous. According to the description and light microscopic photos of *Multiclavula rugaramae* (Fischer et al. 2007), the thallus is *Lepidostroma*-like. Based on molecular data, this species and *Multiclavula akaragae* were transferred to the genus *Lepidostroma* by Ertz et al. (2008).

III. Associations of Mycobionts and Photobionts and Ultrastructure of Fungus–Alga Interactions

Three main types of fungus–alga interactions occur in basidiolichens (Fig. 16.1): (1) hyphal sheaths from which haustoria-like endosymbiontic hyphae originate, (2) appressoria, and (3) solid globular clusters formed by densely arranged hyphae surrounding green algae without haustoria.

1. *Rhizonema* trichomes are enveloped by hyphae and penetrated by haustoria, originating from the mantle hyphae in *Dictyonema, Cora* and *Acantholichen* (Fig. 16.7a–c, e–g). This interactive system, which has been misinterpreted as a dual fungal symbiosis with the algal host (Slocum and Floyd 1977), is different from all others known in basidio- and ascolichens. Because of the vitality of the photobiont and the lack of any signs of significant structural changes, the mycobiont–photobiont relationship has been interpreted as an endosymbiontic interaction (Oberwinkler 1980; Slocum 1980). Tschermak-Woess (1983) records a strong compression of the lichenized trichomes and a lack of cyanophycin in *Dictyonema moorei* studied with the light microscope. Yet, there is no proof that these changes are due to mutualistic interactions. The interaction of myco- and photobiont in *Acantholichen pannarioides* has also been studied by light microscopy by Oberwinkler (1980; Fig. 16.7e, f). It was

Fig. 16.3. Basidiocarps and thalli of basidiolichens. All photographs by F. Oberwinkler; (**a–d**, **j**, **k**, **m**, **n**) *bar* 5 mm; (**e**) *bar* 1 cm; (**f**, **l**, **p**) *bar* 0.5 mm; (**g–i**) *bar* 1 cm; (**o**) *bar* 2 mm. (**a**) *Multiclavula mucida*, Wertach, Germany, 4 Oct 1996. (**b**) *Multiclavula corynoides*, Oberjoch, Germany, 3 Oct 2004. (**c**) *Multiclavula* sp., Zamora, Ecuador, 15 July 2004. (**d**) *Multiclavula sinensis*, Nan Nou Shan, Yunnan, China, 15 Aug 1995. (**e**) *Lepidostroma calocerum*, Sierra de Santa Marta, Colombia, 20 June 1978. (**f**) *Dictyonema interruptum*, Levada de Portela, Madeira, 28 Mar 1984. (**g**) *Dictyonema sericeum*, El Valle, Mérida, Venezuela, 11 Mar 1969. (**h**) *Cora pavonia*, upper side of thallus, Laguna Negra, Mérida, Venezuela, 18 Apr 1969. (**i**) *Cora pavonia*, underside of thallus with fresh hymenium, Laguna Negra, Mérida, Venezuela, 18 Apr 1969. (**j**) *Cyphellostereum pusiolum*, basidiocarp, Mérida, Venezuela, 15 June 1969. (**k**) *Lichenomphalia umbellifera*, basidiocarp and thallus, Bschiesser, Oberjoch, Germany, 4 July 1984. (**l**) *Lichenomphalia umbellifera*, thallus, Bschiesser, Oberjoch, Germany, 4 July 1984. (**m**) *Lichenomphalia velutina*, basidiocarp and thallus, Paramo de Mucubaji, Mérida, Venezuela, 4 Oct 1968. (**n**) *Lichenomphalia hudsoniana*, basidiocarp and thallus, Bschiesser, Oberjoch, Germany, 25 July 1979. (**o**) *Lichenomphalia hudsoniana*, basidiocarp and thallus, Bschiesser, Oberjoch, Germany, 3 Sept 1984. (**p**) *Lichenomphalia hudsoniana*, thallus, Bschiesser, Oberjoch, Germany, 3 Sept 1984

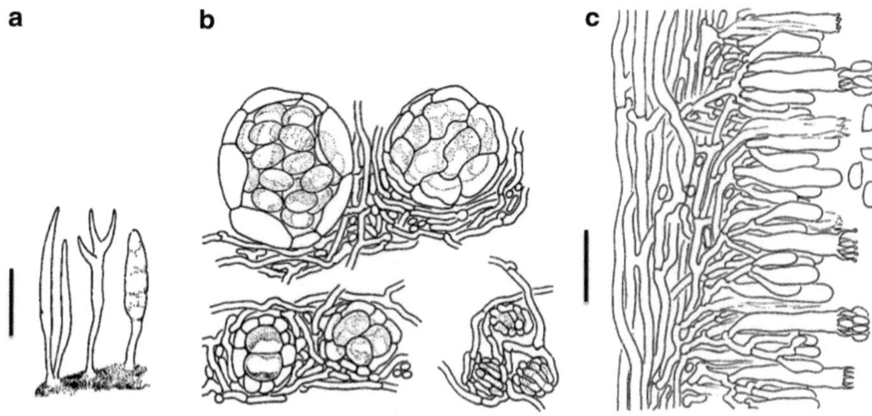

Fig. 16.4. *Multiclavula mucida* (after Oberwinkler (1970, 2001)). (**a**) Basidiocarps with basal lichen thalli; *bar* 5 mm. (**b**) Interactive globular units of hyphae and *Coccomyxa* in different developmental stages; *bar* 20 µm. (**c**) Section through the subhymenium and hymenium with basidia of different ages and basidiospores; note clamped hyphae; *bar* 20 µm

found that it is essentially the same as in *Cora* and *Dictyonema*.

2. In *Lepidostroma calocerum* hyphae become attached to the algal cells and develop into morphologically distinct appressoria (Oberwinkler 1980, 1984). The attached algae degenerate and collapse (Fig. 16.6c, d), non-attached ones in the thallus are vital and obviously capable for cell division. Such details cannot be seen in the illustrations by Fischer et al. (2007) of *L. akaragae* (as *Multiclavula a.*) and *L. rugaramae* (as *Multiclavula r.*).
3. Globose hyphal–algal associations are characteristic of *Multiclavula* species (Oberwinkler 1980, 1984; Fig. 16.4b) and of *Botrydina vulgaris* (Boissière 1980, Oberwinkler 1980, 1984; Fig. 16.9b, c). Comparable globular clusters of fungal and algal cells develop inside the thalli of *Coriscium viride* (Oberwinkler 1980, 1984; Fig. 16.9f). Though densely packed, the algal cells are only surrounded by hyphae, and are not penetrated by haustoria. The globular thallus with clustered green algae has also been found in *Athelopsis* (Coste and Royaud 1994; Fig. 16.5d, e), and in otherwise differently lichenized species of *Athelia* (Oberwinkler 1970; Fig. 16.5b), *Cyphellostereum* (as *Clavaria* sp. by Oberwinkler 1970; Fig. 16.8b), and *Lepidostroma* (Oberwinkler 1970). In *Sistotrema brinkmannii* (Oberwinkler 1970) the association of myco- and photobiont certainly is facultative because the fungus normally grows on wood without algae. The common *Resinicium bicolor* also is not lichenized but quite often grows together with green algae (Poelt and Jülich 1969b; Oberwinkler 1970).

To my knowledge, no additional detailed studies on fungus–alga interactions in basidiolichens have been published.

IV. Fungi

All fungal partners of basidiolichens are members of the Agaricomycetes (Fig. 16.2). No lichenized Pucciniomycotina, Ustilaginomycotina, Tremellomycetes or Dacrymycetes are known. Thus, basidiolichens belong to one large relationship with a high number of nonlichenized species and few lichenized ones.

A. *Multiclavula*

Multiclavula (Petersen 1967) is a taxonomic segregate of *Clavaria* s.l., *Lentaria*, and *Clavulinopsis* sensu Corner (1950, 1956). Species of the genus were considered to share a special hyphal morphology as a unifying characteristic, and often basidia can have more than four sterig-

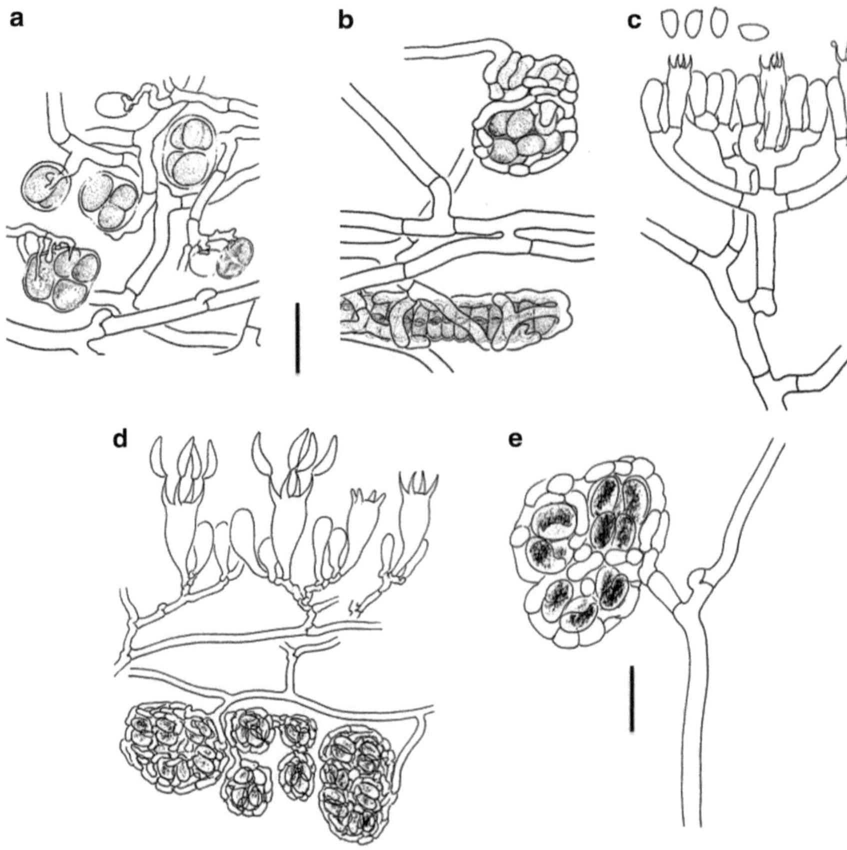

Fig. 16.5. *Athelia* (**a–c**) and *Athelopsis* (**d, e**) (rearranged after Oberwinkler (1970, 2001)); (**a–d**) *bar* 20 μm, (**e**) *bar* 10 μm. (**a**) Hyphae and haustoria of *Athelia epiphylla* parasitic on unicellular green algae. (**b**) Hyphae of *Athelia andina* partly ensheathing *Scytonema* (= *Rhizonema*). Note central hypha in *Rhizonema* (*below*) and unicellular green algae (*above*). (**c**) Section of basidiocarp of *Athelia andina* with loose subhymenial hyphae and non-thickening hymenium with basidia of different developmental stages and basidiospores. (**d**) Section through *Athelopsis* with lichenized lower part and whole basidiocarp showing the total cellular construction of the basidiolichen in the sporulating stage with basidia in different developmental stages and basidiospores. (**e**) *Athelopsis* hyphae ensheathing *Coccomyxa* algae; note clamped hypha

mata. Originally, also *Clavaria* (*Lepidostroma*) *calocera* was included in *Multiclavula* by Petersen (1967), but *Lepidostroma* is distinctly separate from *Multiclavula* in the micromorphology of the basidia and thalli, and in phylogenetic hypotheses, based on molecular data (Fig. 16.2).

The phylogenetic analysis of several *Multiclavula* (Nelsen et al. 2007), *Clavulina*, *Sistotrema* species, and *Hydnum rufescens*, groups *S. brinkmannii* and *Multiclavula* in one clade. *S. brinkmannii* can be associated with algae, occasionally forming *Botrydina*-like globules, however, apparently not as lichenized structures (Oberwinkler 1970).

B. *Marchandiomphalina*

The genus *Marchandiomphalina* was proposed for an anamorphic Basidiolichen (Diederich and Lawrey 2007), originally described as *Omphalina foliacea* (Jørgensen 1989). The lichen resembles a large *Leptogium* but has green photobionts and basidiomycete hyphae (Jørgensen 1989). Basidiocarps are still unknown.

In a molecular analysis, Palice et al. (2005) confirmed the fungal partner in *O. foliacea* as a basidiomycete. However, the phylogenetic position is not in the omphalinoid group. It was considered to belong eventually to the

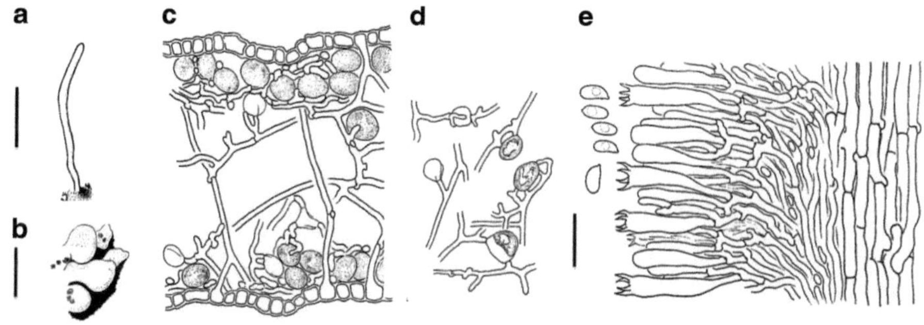

Fig. 16.6. *Lepidostroma calocerum* (rearranged after Oberwinkler (1970, 2001)); (**a**) *bar* 5 mm; (**b**) *bar* 1 mm; (**c–e**) *bar* 20 μm. (**a**) Basidiocarp. (**b**) Lichen thalli. (**c**) Section of thallus with upper and lower pseudoparenchymatic hyphae, neighbouring algal layers, and loose interconnecting hyphae. (**d**) Fungus–alga interactions by appressoria; note different numbers of appressoria per algal cell and different developmental stages of algae, including dead ones without cytoplasm. (**e**) Section through the subhymenium and hymenium with basidia of different ages and basidiospores; note clamped hyphae

hymenochaetoid clade. Finally, Binder et al. (2005) found, that the species is a member of the cortioid clade, a position that was confirmed by Diederich and Lawrey (2007) and Lawrey et al. (2008).

C. *Athelia, Athelopsis*

Athelia species are corticioid homobasidiomycetes with loose subhymenial hyphae and non-thickening, pellicular hymenia (Fig. 16.5c). Basidia are short-cylindrical to slightly clavate with 2–4 sterigmata. As in all basidiolichens, spores are smooth, thin-walled, hyaline, and inamyloid. *Athelia* species often grow together and over algae and lichens. Oberwinkler (1970) reported and illustrated fungus–alga interactions with parasitic (haustoria) and apparently symbiontic structures, similar to initial *Botrydina* granules. Additional reports on lichenized *Athelia* species mostly lacked essential details to document lichenization. The species delimitation in *Athelia* is still difficult and needs careful revisions as that for non-lichenized ones of Northern Europe by Eriksson and Ryvarden (1973). A study like this, combined with molecular data, is desirable.

The athelioid clade (Larsson et al. 2004; Binder et al. 2005, 2010) has finally been recognized as its own order, Atheliales, but was named as such already by Boidin et al. (1998).

Athelopsis was proposed "*ad interim*" by Oberwinkler (1965) and validated by Parmasto (1968). It differs from *Athelia* by clavate basidia (Fig. 16.5e). Often basidiocarps are very thin with a dense subhymenium, but loosely interwoven hyphae also can occur, as in the example illustrated here. Two species, *A. glaucina* and *A. subinconspicua* group in the phylogenetic clade of Atheliales but appear not to be congeneric (Binder et al. 2010).

D. *Lepidostroma*

The fungal partner of *Lepidostroma calocerum* has been studied in detail by Oberwinkler (1970; Fig. 16.4h, i). Macroscopically, the clavarioid basidiocarps resemble *Calocera* species. The micromorphology is similar to that of *Multiclavula*, i.e. there is a simple, monomitic hyphal system, septa with clamps, a dense subhymenium, and a thickening hymenium. In contrast to *Multiclavula* the basidia are not suburniform and 4-6-(8)-spored, but clavate-cylindrical and four-spored.

Two species from Rwanda, originally placed in *Multiclavula* (Fischer et al. 2007), were included in *Lepidostroma*: *L. akagerae* and *L. rugaramae* (Ertz et al. 2008). This treat-

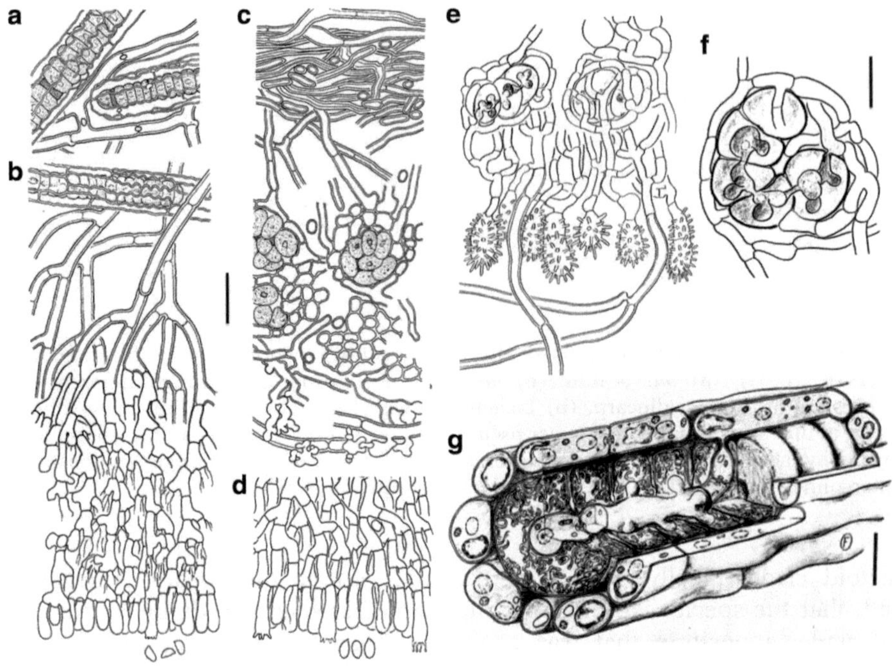

Fig. 16.7. *Dictyonema* (**a**, **b**, **g**), *Cora* (**c**, **d**), *Acantholichen* (**e**, **f**) (rearranged after Oberwinkler (1970, 1980, 1984, 2001)); (**a–e**) *bar* 20 μm, (**f**) *bar* 10 μm, (**g**) *bar* 5 μm. (**a**) Fungus–alga interaction in *Dictyonema irpicinum*; note clamped hyphae, dense hyphal mantles, and faint structures of central hyphae in *Rhizonema* trichomes. (**b**) Fungus–alga interaction and basidiocarp of *Dictyonema sericeum*; note clampless hyphae, loose subhymenium with thick-walled hyphae, and thickening hymenium, one young basidium and basidiospores. (**c**) Section through thallus of *Cora pavonia* (*Dictyonema glabratum*); note leaf-like architecture of the thallus with upper and lower hyphal layers and central compartments of coiled *Rhizonema* trichomes, densely surrounded by hyphae; parts of central hyphae in *Rhizonema* are indicated by darker areas. (**d**) Thickening hymenium of *Cora pavonia* with basidia of different age and basidiospores. (**e**) Detail of the *Rhizonema*–*Acantholichen* interactive system; note the intracellular hypha in *Rhizonema* and its origin from an external hypha. (**f**) Lower part of the thallus of *Acantholichen pannarioides* with hyphae connecting the thallus to the substrate (*below*), acanthohyphidia of the lower surface of the thallus, and two algal chambers with short *Rhizonema* trichomes surrounded by hyphae; note the central hypha in *Rhizonema* in the left trichome. (**g**) Fungus–alga interaction of *Dictyonema sericeum*

ment is based on molecular data. Though light microscopic photos were published, these did not show clear details of the micromorphology of thalli and basidiocarps necessary for a substantial comparison with those in *L. calocerum*, the type of the genus.

E. *Cora, Dictyonema, Acantholichen*

Cora and *Dictyonema* basidiocarps have been studied by Oberwinkler (1970) and Parmasto (1978). They agreed that the fungi are very similar, representing one genus which appears closely related to *Byssomerulius* (*Meruliopsis*) with saprobic species. This interpretation was based on comparative micromorphological features, like thick-walled and loose subhymenial hyphae and thickening hymenia in mature basidiocarps (Fig. 16.7b, d). However, molecularly based phylogenies place *Cora* and *Dictyonema* as well as *Lichenomphalina* in the Hygrophoraceae (Lawrey et al. 2009), a basal family in the Agaricales. In contrast, *Byssomerulius* and *Meruliopsis* fall in a cluster together with *Gloeoporus, Ceriporia* and *Candelabrochaete* inside the Polyporales, according to Larsson (2007).

Parmasto (1978) proposed *Cora* as a synonym of *Dictyonema*, a taxonomy appearing meaningful considering the micromorphology of fungi. Molecular phylogenies (Lawrey et al.

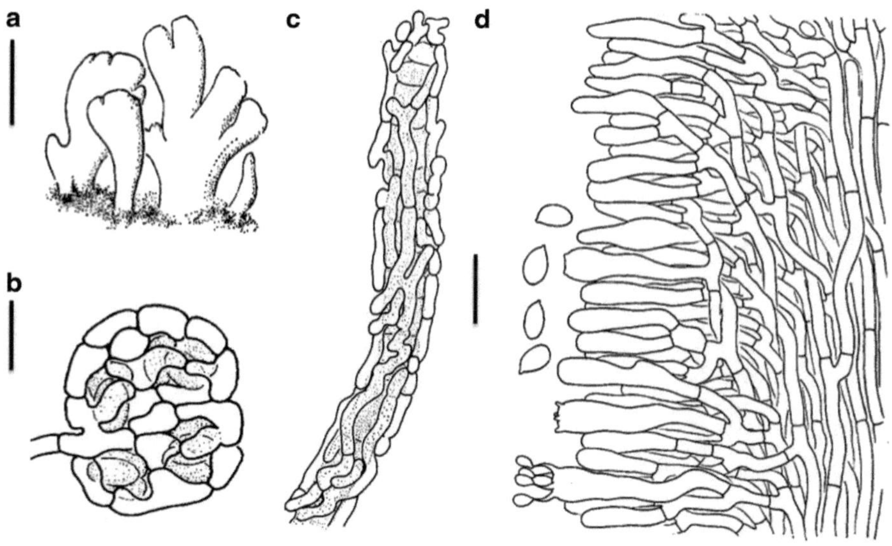

Fig. 16.8. *Cyphellostereum pusiolum* (rearranged after Oberwinkler (1970, 2001)); (**a**) *bar* 5 mm, (**b**) *bar* 10 µm, (**c, d**) *bar* 20 µm. (**a**) Basidiocarps with basal lichen thalli. (**b**) Cluster of green algae enveloped by *Cyphellostereum* hyphae. (**c**) *Rhizonema* trichome surrounded by *Cyphellostereum* hyphae. (**d**) Section through the subhymenium and hymenium with basidia of different ages and basidiospores; note unclamped hyphae

2009) can be used to support this taxonomy, with the result that also *Acantholichen* and *Cyphellostereum* belong into *Dictyonema*. Another, also acceptable interpretation with molecular support, would be, to keep *Cora*, *Dictyonema*, *Cyphellostereum*, and *Acantholichen* as very closely related, but separate genera. This makes sense considering the macromorphology and molecular phylogenies based on adequate sampling.

The genus *Acantholichen* (Jørgensen 1998) is monotypic and based on sterile thalli with clampless hyphae terminating with acanthophyphidia (Fig. 16.7e, f). Acanthohyphidia are known from corticioid, stereoid, and cyphelloid basidiomycetes, e.g. *Aleurodiscus* s.l. and *Stereum* s.l. However, they are diverse and not always homologous. Terminal cells of the thallus underside in *Cora pavonia* are irregularly shaped (Fig. 16.7c). *Acantholichen* has also the *Dictyonema*-type of fungus–alga interaction (Oberwinkler 2001; Fig. 16.7e, f). The fungal relationship cannot be derived from the hyphal system of the thallus. In *Cora* and *Dictyonema* basidiocarps develop on the underside of the lichen thalli. This can be expected also for *Acantholichen*. However, it is technically very difficult to examine the underside of the tiny scales in a search for basidiocarps.

F. *Cyphellostereum*

Cyphellostereum pusiolum has small and stout, upright growing, simple or split basidiocarps, a monomitic hyphal system, and clavate-cylindrical, slightly bent holobasidia (Figs. 16.3j, 16.8a, d). Substrate hyphae of *Cyphellostereum* can be closely attached to green algae and to *Rhizonema* trichomes (Fig. 16.8b, c). The latter fungus–alga interaction is very similar to the *Cora–Dictyonema* type concerning the cyanobacterial partner (Fig. 16.7g). However, detailed studies for the close cellular interaction are still lacking.

Cyphellostereum appeared not to be monophyletic. The lichenized *C. pusiolum*, the type of the genus, as shown by Lawrey et al. (2009), is inside the *Dictyonema—Acantholichen—Cora* cluster, while the saprobic *C. laeve* falls into the Hymenochaetales (Larsson 2007). To clarify the nomenclature, Redhead, Lücking, and Lawrey (Lawrey et al. 2009) proposed the new genus *Muscinupta* for *C. laeve*.

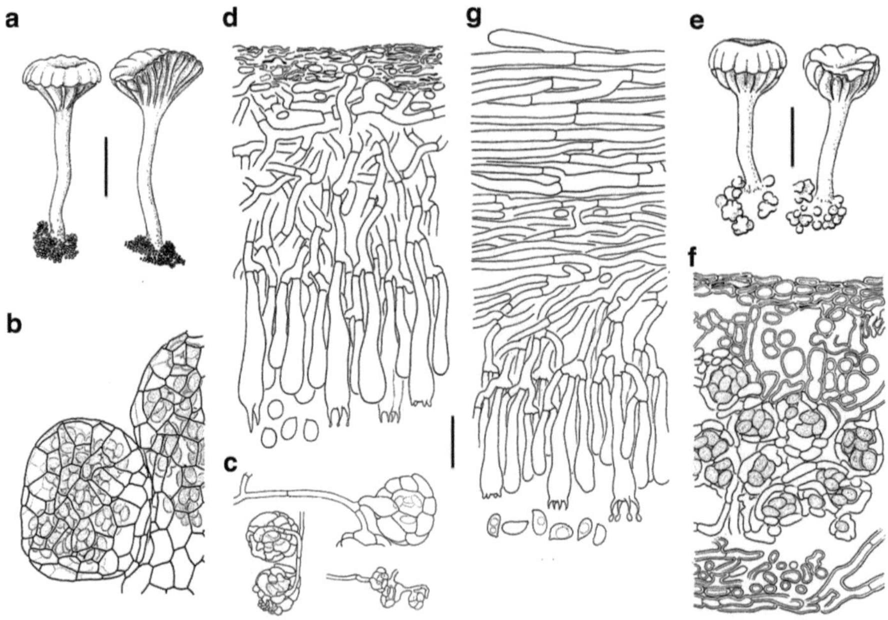

Fig. 16.9. *Lichenomphalia umbellifera* (**a–d**) and *L. hudsoniana* (**e–g**) (after Oberwinkler (1970, 2001)); (**a, e**) *bar* 5 mm, (**b–d, f, g**) *bar* 20 μm. (**a**) Basidiocarps of *L. umbellifera* (*Omphalina ericetorum*) with basal lichen thalli. (**b**) Mature thalli (*Botrydina*). (**c**) Ontogenetic stages of the *Botrydina* thalli. (**d**) Transverse section in between lamellae of *L. umbellifera* in an older developmental stage; note collapsed hyphae of the outer pileus layer (*above*) and basidia with various numbers of sterigmata. (**e**) Basidiocarps of *L. hudsoniana* (*Omphalina luteolilacina*) with basal *Coriscium* thalli. (**f**) Section of the thallus of *L. hudsoniana* (*Coriscium viride*); note the leaf-like architecture of the thallus with upper and lower marginal hyphae and central algal chambers; also note the photobionts, densely surrounded by hyphae. (**g**) Intralamellar radial section of the pileus of *L. hudsoniana* with cutical hyphae, trama layers, subhymenium, hymenium with basidia of different ages, and basidiospores; note the uniform hyphal system

G. *Lichenomphalia*

Lichenomphalia species are small agarics with a characteristic macromorphology (Figs. 16.3k, m, n, 16.9a, e), i.e. with central stipes, convex to umbilicate pilei, and decurrent lamellae. The micromorphology is also very distinct, but prominent features are lacking (Fig. 16.5c, e). The hyphae are simple, comparatively loosely interwoven and therefore rather distinct, not gelatinous, with parietal or intrahyphal bright pigments, and without clamps. Trama hyphae are not or only slightly swollen, there are no cystidia in the hymenium and on the pileus. Also subhymenial hyphae are normally loosely interwoven, the hymenium is not much thickening, and basidia are basally slightly tapering and apically inconspicuously swollen, normally with four, but also with two sterimgata. Basidiospores lack any special characters, i.e. they are short cylindrical to subglobos, thin-walled, smooth, and inamyloid. Karyological studies of *L. umbellifera* (as *Omphalina ericetorum*) by Lamoure (1968), and Poelt and Jülich (1969a) of *L. velutina* (as *O. grisella*, Fig. 16.3m) demonstrated apomictic ontogenies.

Redhead and Kuyper (1988) considered fully lichenized omphaloid agarics distinct from their non-lichenized allies in *Omphalina*. The differentiating characters focussed on lichen thalli, and noticeably thickened basal mycelium in mutualistic species. These distinctions have not been accepted as generically valuable by workers at that time. However, after molecular phylogenies became available, the monophyly of lichenized *Omphalina* species became obvious, and taxonomic conclusions then followed.

Lutzoni and Vilgalys (1995a) used 17 morphological characters and nuclear ribosomal

DNA large subunit sequences to analyse four lichenized *Omphalina* and three non-lichenized ones and *Arrhenia lobata*. The dendrogram based on morphological characters as well as the one based on a combined data sets clustered the lichenized species in one group. Also in a study with four molecular data sets from 30 species of *Omphalina* and related taxa, Lutzoni (1997) demonstrated that the lichen-forming *Omphalina* species represent a monophyletic group, thus confirming the taxonomic conclusions of Redhead and Kuyper (1987) and Norvell et al. (1994).

A molecular investigation based on combined data sets of 25S, 5.8S, ITS1, and ITS2 from the nuclear ribosomal DNA of lichenized fungi and fungi associated with *Marchantia* and *Blasia* and non-mutualist fungi revealed four independent transitions to mutualism (Lutzoni and Pagel 1997). Two of these transitions refer to *Gerronema marchantiae* and *Rickenella pseudogrisella* associated with liverworths. The others are two lichenized *Multiclavula* species and a monophyletic group of five lichen-forming *Omphalina* species. The authors demonstrated "a highly significant association between mutualism and increased rates of nucleotide substitions in nuclear ribosomal DNA" and that "a transition to mutualism preceded the rate acceleration of nuclear ribosomal DNA in the lineages. The generalized genomic acceleration in the lichen-forming *Omphalina* lineage is mostly due to selection of mutations that disrupt the potential formation of thymine dimers" (Lutzoni and Pagel 1997).

The phylogenetic hypotheses of Hibbett and Thorn (2001) and Redhead et al. (2002) gave further support for a generic separation of lichenized species. The omphalinoid taxa, including *Lichenomphalia* are now grouped in the Hygrophoraceae (Lawrey et al. 2009; Seitzman et al. 2011), including also *Dictyonema* s.l.

H. *Semiomphalina*

Corner (1966) described *Pseudocraterellus leptoglossoides* from Papua New Guinea, and accurately illustrated macro- and micromorphology of the species. It has a clear omphalinoid hyphal system with a thickening hymenium, clamps are lacking, basidia and basidiospores are exactly of the *Omphalina* type.

Restudying the type material, Redhead (1984) found *Botrydina*-like bulbils on the substrate that were clearly connected with the basidiocarps. Therefore the new basidiolichen genus, *Semiomphalina*, was proposed. Until now the species has not been studied molecularly.

I. *Marasmiellus affixus*

Singer (1970) mentions "*Marasmiellus affixus* (Berk.) Sing. is also lichenized and grows in association with a crustaceous lichen on the surface of *Eucalyptus* saplings in Australia". A little more information he gave in a note on *M. affixus* (Singer 1973) is "crustaceous organism consisting of *Coccomyxa* imbedded in basidiomycetous hyphae".

To my knowledge, *M. affixus* has never been studied in detail morphologically or molecularly. In a website article on basidiolichens, Lepp (2011) discusses *M. affixus* at some length but without conclusions about the lichenization status.

V. Photobionts

In ascolichens and in basidiolichens the two major lichen-forming partners, cyanobacteria and green algae, serve as photobionts.

Unanimously the long trichomes of the photobiont in *Dictyonema* species have been identified as *Scytonema* (e.g. Bornet 1873; Oberwinkler 1970, 1980, 1984; Parmasto 1978; Tschermak-Woess 1983). The cyanobacteria in *Cora pavonia* have earlier been assigned to *Chroococcus* (Bornet 1873; Mattirolo 1881; Ahmadjian 1958). In realty they are short coiled trichomes of the *Scytonema* type (Vainio 1890; Räsänen 1943; Oberwinkler 1970, 1980, 1984; Roskin 1970; Tschermak-Woess 1983).

The photobiont of *Acantholichen pannarioides* was reported as *Scytonema* by Jørgensen (1998). A microscopic re-study (Oberwinkler 2001) showed that it is very similar to the *Cora* photobiont inclusive of the fungus–alga interaction type (Fig. 16.7e, f).

Lücking et al. (2009) analysed 16S rDNA sequences to determine the phylogenetic positions of *Scytonema* in cyanolichens of the Asco- and Basidiomycota. The lichenized scytonematoid cyanobacteria of both Asco- and Basidiolichens, form a clade different from the one of free living *Scytonema* s.str. For this clade the name *Rhizonema*, erected by Thwaites (in Smith and Sowerby 1849) for *Calothrix interrupta*, can be used.

Most of green algal photobionts, including *Coccomyxa* species, belong to the Trebouxiophyceae (Beck and Persoh 2009). The association of coccomyxoid green algae with *Lentaria (Multiclavula) mucida* has been described by Geitler (1955). He also found that the algae of the lichenized globules are larger than free living ones. This was interpreted as a possible retardation of the algal division frequency caused by the mycobiont (Geitler 1956).

In a phylogenetic study of *Coccomyxa*, Zoller and Lutzoni (2003) detected "three main lineages within this genus, corresponding to free-living *Coccomyxa*, individuals isolated from basidiolichens *Omphalina* (*Lichenomphalia*) and *Coccomyxa* isolated from ascolichens belonging to the Peltigerales". The genetic changes in *Lichenomphalia* are 27.5 times higher than those of their *Coccomyxa* photobionts. Six *Lichenomphalia* species share the same *C. subelliptica.*

In a phylogenetic tree of Trebouxiophyceae, including free-living, endophytic, symbiontic, and parasitic species (Rodrigéz et al. 2008), a monophyletic *Coccomyxa* comprises *C. pringsheimii*, *C. glaronensis*, photobionts of *Nephroma arcticum* and *Peltigera britannica*, *Paramecium bursaria* endobionts, endophytes of *Ginkgo biloba*, and green algae in mussels.

Langenstein (1994) isolated *Coccomyxa* from *Botrydina* and *Coriscium* and characterized algal cells in axenic culture. They reproduce asexually by autospores. A comparison of strains from *Omphalina* thalli with *Coccomyxa* isolated from the ascolichen *Icmadophila ericetorum* as well as authentic substrains of *Coccomyxa pringsheimii* and *C. subellipsoidea* showed high identity of morphological and reproductive characters. Also the growth of colonies of the various strains was remarkably similar.

VI. Physiology

Because of its big and more or less flat thallus, *Cora pavonia (Dictyonema glabratum)* is a suitable lichen for ecophysiological experiments. Already in 1965 Lange investigated the temperature dependance of photosynthesis, and Coxson (1987a) reported on a chilling stress which was considered to be an important factor in the limitation of distribution. Coxson (1987b, c) found an unexpected breadth of photosynthetic response under prevailing cloud conditions on La Soufrière (Gouadeloupe). The species shows no depression in net photosynthesis at saturated thallus water content in contrast to most other lichens. Extensive ecophysiological field investigations in *C. pavonia*, carried out by Lange et al. (1994) in a premontane tropical rainforest in Panama confirmed the observation of Coxson (1987c). The authors found that net photosynthesis was adapted to high temperatures, and when the thalli were fully hydrated, the CO_2 uptake was not reduced. Desiccation of the thalli caused a decrease of net photosynthesis and influenced apparent photon yield of CO_2 fixation. Though there were days with a negative carbon balance, the total carbon gain was extremely high and estimated at 228 % per year.

Pioneer populations of *Cora pavonia* on La Soufrière lahar flows showed high nitrogenase activities during rehydration (Fritz-Sheridan 1988). The author suggests that under full hydration a high supply of nitrogen compounds would support growth and membrane repair.

The C-metabolism of *Cora pavonia* has been investigated by Feige (1969) who found that in contrast to all other lichens with cyanobacterial phytobionts, pentitol was synthesized and accumulated as pentitolgalactosid. The main excretory product was glycolic acid after bicarbonate incubation in liquid phase.

Larcher and Vareschi (1988) studied the variation in morphology and functional traits of *C. pavonia* from various habitats in the Venezuelan Andes. They found optimally developed thalli in the lower Paramos around 3,500 m which exhibit the highest photosynthetic capacity. Specimens growing in very high altitudes up to 4,300 m have strongly reduced thalli, but the same respiration intensity as specimens from the lower Paramo, reflecting a homeostatic adjustment to the prevailing temperature regime. Thalli which grow in lower altitudes of the montane forest have approximately half the size and barely half the net photosynthetic rate in comparison to those growing in optimal Paramo regions. Thus, the increasingly unfavourable carbon budget may delimit the altitudinal distribution. Lange et al. (1994) explained the reduction of population densities of macrolichens in tropical lowland rainforests by the negative carbon balance caused by increased nocturnal temperatures at lower elevations.

The wide temperature range in photosynthetic activity is a general feature in lichens (Lange 1965). A distinct adaptation to low temperature was found in *Botrydina* and *Coriscium* by Heikkilä and Kallio (1966).

VII. Phytochemistry

Higher fungi including the basidiomycetes are very rich in specific chemical compounds (Gill and Steglich 1987) that often are of taxonomic importance. Likewise, secondary metabolites of ascolichens (lichen substances) are widely distributed and have been studied intensively (e.g. Huneck and Yoshimura 1996; Eisenreich et al. 2011). The taxonomic use of metabolic data in lichen-forming fungi has been summarized by Lumbsch (1998); however, basidiolichens could not be included because of lacking data.

In a study on antibiotic active substances of *Cora pavonia* (*Dictyonema glabratum*; Mitidieri et al. 1964) no chemical substances were mentioned. Iacomini et al. (1987) isolated and characterized β-D-glucan, heteropolysaccharides, and trehalose components also from *C. pavonia*. Again, from the same species atranorin and tenuiorin, as well as ergosterol and ergosterol peroxide were reported by Piovano et al. (1995). *Dictyonema* was also included in a chemical study on secondary metabolites in lichens from Bario highlands, Borneo, by Din et al. (1998). Jørgensen (1998) found no lichen acids in the new basidiolichen genus *Acantholichen*. The polysaccharides pseudonigeran, a (1–4)-linked β-xylan, and a new(1–6)-linked β-mannan were discovered in *C. pavonia* by Carbonero et al. (2002).

VIII. Ecology

In his field studies, Johow (1884) observed that *Cora pavonia* appears to be confined to higher altitudes in the tropical islands of Dominica and Trinidad. His transplantation experiments to lowlands failed within a few days. Thus, the species appeared to be adapted to cool, humid, and bright conditions of tropical mountain regions. In standings with disturbed or scarce higher plant vegetation, *C. pavonia* can occur in remarkably rich, ground-covering populations.

Acantholichen pannarioides is a species of moist montane regions in Central America and northern South America, where it grows on debris of vegetation, bark, bryophytes, and other lichens (Jørgensen 1998).

Multiclavula species display a remarkably diverse adaptation to ecological niches. Wet wood of huge logs is the most suitable substrate for *M. mucida*. In Central Europe preferably coniferous wood is colonized, but the species occurs also in such subtropical and tropical regions where gymnosperms are lacking. Other *Multiclavula* species grow on bare soil under different climatic conditions in arctic-alpine regions, temperate and tropical zones.

Lepidostroma species are sofar only known from tropical regions, *L. calocerum* from Colombia and Costa Rica, and *L. akagerae* and *L. rugaramae* from Rwanda (Ertz et al. 2008).

Lichenomphalia species are common agaricoid lichens in arctic-alpine vegetations where they are mostly protected by a winter snow cover (Heikkilä and Kallio 1966). They also occur fully exposed in high alpine regions of tropical mountains with often extreme day and night temperature differences.

IX. Dispersal

Similar to ascolichens only the fungal partners of basidiolichens are able to propagate sexually. In most species, basidiocarps are developed sporadically so that production of basidiospores is limited in time. However, enormous quantities of spores are produced during the sporulation phase, also from tiny basidiocarps like those of a small *Athelia*. The high amount of basidiospores is an essential prerequisite for effective dispersal in basidiolichens, as in basidiomycetes in general.

Dispersal by mitotic progagules is limited to few basidiolichens. Poelt and Obermayer (1990) considered thallus bulbils in *Multiclavula vernalis* as distinctive vegetative diaspores. They consist of inflated hyphae, a thin cortical layer and contain one or a few groups of algal cells. Fischer et al. (2007) confirmed this finding and reported on similar, but bigger bulbils in *M. akagerae*, transferred to *Lepidostroma* later (Ertz et al. 2008). It is tempting to hypothesize that *Botrydina* globules also can serve as vegetative propagules in *Lichenomphalia*. However, normally *Botrydina* is quite solidly attached to the substrate by hyphae.

Structures resembling soredia occur in *Acantholichen*, and isidioid appendages are known from thalli of *Dictyonema melvinii* and *D. ligulatum* (Chaves et al. 2004).

X. Distribution

Two main geographical distribution patterns can be found in basidiolichens: (1) cosmopolitan to subcosmopolitan ranges with restrictions to specific ecological niches (certain species of *Athelia*, *Dictyonema*, *Multiclavula*, and *Lichenomphalia*), and (2) regional to continental distributions (*Lepidostroma*, *Cyphellostereum*, *Acantholichen*, *Cora*, *Dictyonema*).

According to Petersen and Kantvilas (1986), *Multiclavula* seems to include several species which are truly cosmopolitan, e.g. *M. mucida* and *M. vernalis*.

Dictyonema sericeum is reported from all continents with tropical and subtropical regions (Parmasto 1978). In contrast, *D. ligulatum* is known only from southeastern Asia and Oceania (Parmasto 1978). Besides all previous records of *D. moorei* from Japan (Parmasto 1978), the species was also recorded from Chile (Henssen 1963), Western Australia, and Queensland (Lepp 2011). *Dictyonema interruptum* was originally described from England and was additionally reported from the Azores (Coppins and James 1979) and from the Pyrenees (Etayo et al. 1995).

Cora pavonia (*D. glabratum*) is known to occur from South Florida to Cape Horn (Parmasto 1978). Thus, the species seems to be restricted to subtropical and tropical America with an extension into temperate and cool regions of South America.

Acantholichen pannarioides is known from Costa Rica to the northern Andes in Venezuela and Ecuador, as well as from the Galapagos Islands (Jørgensen 1998).

Semiomphalina leptoglossoides (*Pseudocraterellus l.*) has been collected only once in the Wau area of Papua New Guinea (Corner 1966; Redhead 1984).

The extensive fieldwork of Heikkilä and Kallio (1966, 1969) gave a substantial basis to document that *Botrydina* and *Coriscium* have circumpolar distributions in artic and subarctic regions. Some *Lichenomphalia* species appear to be cosmopolitan, e.g. *L. umbellifera* (*Omphalina ericetorum*) has a circumpolar distribution and has been recorded also from New Zealand (Galloway 1985) and Australia (Lumbsch and Ewers 1992), and *L. chromacea* is geographically restricted to southern Australia (Lepp 2011).

XI. Culture Experiments

Heikkilä and Kallio (1966) were unsuccessful to obtain axenic cultures of the fungal partners in lichenized *Omphalina* species. However, in living cultures of *Botrydina* and *Coriscium* several times *Omphalina* basidiocarps were developed. In contrast to the fungus, the photobiont of *Botrydina* was easily cultivated on agar and in liquid media. Shape and size of the algal cells varied widely in culture. The photobiont isolated from *Coriscium* appeared very similar.

Langenstein (1994) was successful in isolating phyco- and mycobionts of *Botrydina*, to grow them in axenic culture, and to synthesize *Botrydina* thalli in the laboratory. This was a remarkable progress in experimental lichenology. When *Coccomyxa* is present, hyphae originating from basidiospores contact the algal cells and surround them. Original compartments with single algae develop into multicellular ones by autospore production and ingrowth of fine hyphal branches into the photobiont colony. Unfortunately this study has not been published in a journal, but it was cited in the above phrasing by Oberwinkler (2001). This may explain that Lawrey et al. (2007) hoped that: "it may also be possible to experimentally synthesize basidiolichens using cultures of the bulbil-forming fungi and photobionts to see the extent to which bulbil formation leads to formation of basidiomycete thallus structures".

A circadian rhythm was found in the mycobiont culture of an unidentified *Omphalina* sp., considered to be a basidiolichen by Kalangutkar and Kamat (2010). However, there is no documentation of mycobiont–photobiont interactions in this pre-publication document.

XII. Phylogenetic Systematics and Evolution

A. Origin and Evolution of Thalli

Lichenization in Basidiomycota is only known from Agaricomycetes and originated at least in four different clades (Fig. 16.2). The multiple origin of basidiolichens is apparent in several molecularly based dendrograms, the first of them using small subunit rDNA data (Gargas et al. 1995).

The convergent evolution of lichenized structures within bulbilliferous fungi was discussed by Lawrey et al. (2007). Lichenized bulbils are considered as ancestral states in the evolution of basidiolichens. Examples can be found in the *Multiclavula* relationship (Figs. 16.3a, b, 16.4b), the *Marchandiomphalina* group, in lichenized *Athelia* (Fig. 16.5b) and *Atheliopsis* (Fig. 16.5d, e), *Cyphellostereum pusiolum* (Fig. 16.8b), and *Lichenomphalia* (Figs. 16.3k, l, 16.9a–c).

B. Evolution of Basidiocarps

Based on comparative morphology, Oberwinkler (1970) distinguished three main groups within the basidiolichens: (1) the *Athelia–Dictyonema* complex including *Cora*, (2) the *Multiclavula–Lepidostroma* group with clavarioid basidiocarps, and (3) the agaricoid, lichenized *Omphalina* species. These taxa were considered of distant relationship because of their different micromorphological characters, a view largely in agreement, at that time, with a classification based on basidiocarp morphology.

Actually accepted molecularly based phylogenetic hypotheses can be summarized as in Fig. 16.2: *Multiclavula* falls in Cantharellales as a sister group to Clavulinaceae, *Marchandiomphalina* belongs to the Corticiales, the Atheliales comprise *Athelia* and *Atheliopsis*, and *Lepidostroma* is a sister taxon. Finally, most of the basidiolichens are members of the Hygrophoraceae (Lawrey et al. 2009), a basal family of the Agaricales. They include *Dictyonema*, *Cora*, *Acantholichen*, *Cyphellostereum*, and *Lichenomphalia*.

1. Cantharellales and Clavulinaceae

Clavarioid basidiolichens have been assigned to different genera, e.g. *Clavaria*, *Clavulinopsis*, *Lentaria*, *Lepidostroma*, and *Multiclavula*. These have been analysed morphologically and discussed comparatively by Oberwinkler (1970) with the result that lichenized *Lentaria* spp. cannot be separated generically from *Multiclavula* spp. Because of the unique thallus and the peculiar fungus–alga interaction, *Lepidostroma* was recognized as a separate lichen genus.

A phylogenetic analysis of four *Multiclavula* species (*M. corynoides*, *M. mucida*, *M. vernalis*, and the newly described *M. ichthyiformis*; Nelsen et al. 2007) resulted in a monophyletic group of these taxa with a basally positioned *Sistotrema brinkmannii*. The sister group is a monophylum of *Clavulina* species. Both clades are highly supported (bootstrap 100 %) and considered by the authors to represent Clavulinaceae in an emended sense.

2. Corticiales

Corticiales is a taxon based on a clade in Agaricomycetes, found in several molecular phylogenetic studies (Larsson et al. 2004; Binder et al. 2005; DePriest et al. 2005; Hibbett et al. 2007;

Lawrey et al. 2007), and formally erected by K-H Larsson (in Hibbett et al. 2007). This clade includes a single, lichenized species, *Marchandiomphalina foliacea* Diederich and Lawrey (2007), for which basidiocarps are not yet known. The species was originally described as *Omphalina foliacea* by Jørgensen (1989), though sterile, and assigned to the agaricoid genus *Omphalina*. The thallus macromorphology is obviously different from known lichenized *Omphalina* (*Lichenomphalia*) species, and also its micromorphology appears to be intermediate between *Botrydina* and *Coriscium* Jørgensen (1989).

The species included at present in the molecularly defined Corticiales are more or less crustouse or bulbil-like, clavarioid and agaricoid ones are lacking. Therefore, it would be very surprising if the mycobiont of *Marchandiomphalina foliacea* turned out to be agaricoid, as implied by the generic names so far used.

3. Atheliales

In a cladistic study of corticioid fungi by Parmasto (1995), *Athelia* and *Athelopsis* fell within a group "arbitrarily named at this stage Atheliaceae". Boidin et al. (1998) used the ribosomal internal transcribed spacer sequences (ITS1-5.8S-ITS2) to evaluate the phylogenetic relationships within the Aphyllophorales. They recognized an order Atheliales with *Athelia*, *Amyloathelia*, *Fibulomyces*, and *Leptosporomyces*. In the analysis of Larsson et al. (2004), the athelioid clade includes species of *Amphinema*, *Athelia*, *Athelopsis*, *Byssocorticium*, *Piloderma*, and *Tylospora*. Also Binder et al. (2005) found an athelioid clade, because of limited sampling represented only by *Athelia arachnoidea* and *A. fibulata*. None of the species in the studies mentioned above were recognized as being associated with algae or even lichenized.

The ecological diversity of species in the Atheliales, as defined by molecular hypotheses, is surprisingly high. They include saprobes, parasites of algae and lichens, ectomycorrhizal fungi, and lichenized species (Fig. 16.5).

In contrast, all taxa mentioned and actually assembled in the Atheliales have thin and pellicular basidiocarps with loose subhymenia, rarely with hymenial cystidia as in *Amphinema*, and thin-walled, hyaline spores. Such micromorphological traits do not give any indication for a closer relationship with Boletales and/or Agaricales. However, in all phylogenetic overviews of Agaricomycetes, Atheliales cluster with Boletales and Agaricales.

4. Lepidostromataceae

Surprisingly, the Lepidostromataceae with the single genus *Lepidostroma* are close to the Atheliales or may even be included in this order as indicated by molecularly based hypotheses (Ertz et al. 2008). Morphological characters of thalli and basidiocarps show distinct differences to those in *Multiclavula* spp., supporting a generic separation. However, an affiliation with taxa of the athelioid clade cannot be based on morphological characters of basidiocarps. Eventually, parasitism of algae in *Athelia* (Fig. 16.5a) and *Lepidostroma* (Fig. 16.6c) can be considered as a similar interaction type.

5. Hygrophoraceae

In the Agaricomycetes, the stereoid cyanolichen *Dictyonema* s.l. and the omphalinoid chlorolichen *Lichenomphalia* appear to be closely related members of the Hygrophoraceae in molecular phylogenetic hypotheses (Lutzoni and Vilgalys 1995b; Lutzoni 1997; Redhead et al. 2002; Lodge et al. 2006; Ertz et al. 2008; Lawrey et al. 2009; Seitzman et al. 2011).

The micromorphology of homologous hyphal systems, like subhymenia and hymenia, characteristics of the basidia and basidiospores (Figs. 16.7b, d, 16.9g, d) of these taxa are very similar, thus supporting the results of molecular phylogenies. These characteristics are also present in *Cyphellostereum pusiolum* (Fig. 16.8), a species that belongs in the *Dictyonema* clade according to molecular data, as does *Acantholichen pannarioides* (Lawrey et al. 2009). Accordingly, systematic conclusions would either to include *Acantholichen*, *Cora*, and *Cyphellostereum* in *Dictyonema* s.l., or to keep these taxa generically separate.

Considering the mycobiont-photobiont interactions, species of *Dictyonema* s.l. are associated with cyanobacteria *(Rhizonema)* and have a unique haustorial apparatus (Oberwinkler 1980, 1984, 2001; Fig. 16.7e–g), certainly a synapomorphy of high systematic and evolutionary importance. In contrast, *Lichenomphalia* species are chlorolichens, forming globose hyphal sheats around *Coccomyxa* algae (*Botrydina*; Fig. 16.9a–c), even when these are enclosed in squamulose thalli, like in *Coriscium* (Fig. 16.9e, f).

XIII. Comparison with Ascolichens

The number of lichenized Ascomycetes may be as high as of the non-lichenized ones. In Basidiomycetes the majority of mutualistic species is associated with certain groups of seedplants as ectomycorrhizal fungi. The number of known basidiolichens, however, is very small.

One of the most striking and unique features of ascolichens is the independent and convergent evolution of the thalli to morphological structures which have not been developed in fungi and algae. Such thallus evolution did not occur in basidiolichens. Their thallus macromorphologies are comparatively simple. However, the fungus-alga interaction is remarkably diverse, comprising three main types. One of them, the *Dictyonema* type, is unique for all lichens.

The transition to lichen symbiosis in Agaricomycetes is demonstrated as an accelaration of nucleotide substitutions (Lutzoni and Pagel 1997). This acceleration had yet no influence on species radiation.

Multiple independent losses of lichen symbioses as in Ascomycota (Lutzoni et al. 2001) are not known in Agaricomycetes.

Because of the limited number of species and their restriction to special niches, basidiolichens have been overlooked for a long time. However, *Cora pavonia* can occur in dense populations in neotropical mountain areas, and in arctic alpine vegetations *Lichenomphalia* species, recognizable without basidiocarps as *Botrydina* and *Coriscium*, may develop in considerable quantities on suitable acidic substrates.

XIV. Conclusion

Though basidiolichens represent only a very small group of all obligately lichenized fungi, they are highly diverse. This is reflected by: (1) structural and functional adaptations of the lichen thalli, (2) the different structures of basidiocarps which are developed by the fungal partners, (3) the cellular interactions of myco- and photobionts, (4) the world-wide distribution in specialized ecological niches, (5) some unique ecophysiological properties, and (6) several convergent phylogenies.

Acknowledgements This review was possible through intensive collaborative efforts. I thank students, technical, and scientific staff in my former institute, as well as many colleagues worldwide for their support of my own studies in basidiolichens.

References

Acton E (1909) *Botrydina vulgaris* Brébisson, a primitve lichen. Ann Bot 23:579–585

Ahmadjian V (1958) A guide for the identification of algae occurring as lichen symbionts. Bot Not 111:632–644

Arvidsson L (1979) Svampangrepp pa lavar- en orsak till lavoken. Fungus attacks on lichens and lichen deserts in cities. Sven Bot Tidskr 72:285–292

Beck A, Persoh D (2009) Flechten und ihre Stellung im Reich der Pilze. In: Bayerische Akademie der Wissenschaften. Rundgespräche der Kommission für Ökologie. Oekol Rolle Flecht 36:13–24

Binder M, Hibbett DS, Larsson KH, Larsson E, Langer E, Langer G (2005) The phylogenetic distribution of resupinate forms across the major clades of mushroom-forming fungi (Homobasidiomycetes). Syst Biodivers 3:113–157

Binder M, Larsson KH, Matheny PB, Hibbett DS (2010) Amylocorticiales ord. nov. and Jaapiales ord. nov.: early diverging clades of Agaricomycetidae were dominated by corticioid forms. Mycologia 102:865–880

Boidin J, Mugnier J, Canales R (1998) Taxonomie moléculaire des Aphyllophorales. Mycotaxon 66:445–491

Boissière JC (1980) Un vrai basidiolichen européen: l'*Omphalina umbellifera* (L. ex Fr.) Quél. Etude ultrastructurale. Cryptogam Bryol Lichenol 1:143–149

Bornet ME (1873) Recherches sur les gonidies des Lichens. Ann Sci Nat V Bot 17:45–110

Brébisson A de (1844) In: Meneghini, Monogr. Nostoch. Ital Atti R Acad Sci Torino Ser 2, 5

Carbonero ER, Sassaki GL, Gorin PAJ, Iacomini M (2002) A (1–6)-linked ß-mannopyrananan, pseudonigeran, and a (1–4)-linked β-xylan, isolated from the lichenised basidiomycete *Dictyonema glabratum*. FEMS Microbiol Lett 206:175–178

Chaves JL, Lücking R, Sipman HJM, Umaña L, Navarro E (2004) A first assessment of the ticolichen biodiversity inventory in Costa Rica: the genus *Dictyonema* (Polyporales: Atheliaceae). Bryologist 107:242–247

Coppins BJ, James PW (1979) A British species of *Dictyonema*. Lichenologist 11:103–108

Corner EJH (1950) A monograph of *Clavaria* and allied genera. Ann Bot Mem 1:1–740

Corner EJH (1956) A new European *Clavaria*: *Clavulinopsis septentrionalis* sp. nov. Friesia 5:218–220

Corner EJH (1966) A monograph of cantharelloid fungi. Oxford University Press, Oxford

Coste C, Royaud A (1994) Les basidiolichens Européens. Bull Soc Castr Sci Nat: 55–59

Coxson DS (1987a) The temperature dependence of photoinhibition in the tropical basidiomycete lichen *Cora pavonia* E Fries. Oecologia 73:447–453

Coxson DS (1987b) Net photosynthetic response patterns of the basidiomycete lichen *Cora pavonia* (Web.) E. Fries from the tropical volcano La Soufrière (Guadeloupe). Oecologia 73:454–458

Coxson DS (1987c) Effects of desiccation on net photosynthetic activity in the basidiomycete lichen *Cora pavonia* E. Fries from the cloud, mist zone of the tropical volcano La Soufrière (Guadeloupe). Bryologist 90:241–245

DePriest PT, Sikaroodi M, Lawrey JD, Diederich P (2005) *Marchandiomyces lignicola* sp. nov. shows recent and repeated transition between a lignicolous and a lichenicolous habit. Mycol Res 109:57–70

Diederich P, Lawrey JD (2007) New lichenicolous, muscicolous, corticolous and lignicolous taxa of *Burgoa* s. l. and *Marchandiomyces* s. l. (anamorphic Basidiomycota), a new genus for *Omphalina foliacea*, and a catalogue and a key to the non-lichenized, bulbilliferous basidiomycetes. Mycol Prog 6:61–80

Din LB, Ismail G, Elix JA (1998) The lichens in Bario highlands: their natural occurrence and secondary metabolites. In: Ismail G, Din LB (eds) A scientific journey through Borneo: Bario, the Kelabit Highlands of Sarawak. Pelanduk, Selangor, pp 155–160

Eisenreich W, Knispel N, Beck A (2011) Advanced methods for the study of the chemistry and the metabolism of lichens. Phytochem Rev 10:445–456

Eriksson J, Ryvarden L (1973) The Corticiaceae of North Europe, vol 2. Fungiflora, Oslo

Ertz D, Lawrey JD, Sikaroodi M, Gillevet PM, Fischer E, Killmann D, Sérusiaux E (2008) A new lineage of lichenized basidiomycetes inferred from a two-gene phylogeny: the Lepidostromataceae with three species from the tropics. Am J Bot 95:1548–1556

Etayo J, Diederich P, Sérusiaux E (1995) *Dictyonema interruptum*, new for the Pyrenees. Graph Scr 7:5–6

Feige B (1969) Stoffwechselphysiologische Untersuchungen an der tropischen Basidiolichene *Cora pavonia* (Sw.) Fr. Flora 160:169–180

Fischer E, Ertz D, Killmann D, Sérusiaux E (2007) Two new species of *Multiclavula* (lichenized basidiomycetes) from savanna soils in Rwanda (East Africa). Bot J Linn Soc 155:457–465

Fries E (1820) Systema mycologicum. Fries, Greifswald

Fritz-Sheridan RP (1988) Nitrogen fixation on a tropical volcano, La Soufrière: nitrogen fixation by the pioneer lichen *Dictyonema glabratum*. Lichenologist 20:96–100

Galloway DJ (1985) Flora of New Zealand: lichens. Galloway, Wellington

Gams H (1962) Die Halbflechten *Botrydina* und *Coriscium* als Basidiolichenen. Oesterr Bot Z 109:376–380

Gargas A, DePriest PT, Grube M, Tehler A (1995) Multiple origins of lichen symbioses in fungi suggested by SSU rDNA phylogeny. Science 268:1492–1495

Geitler L (1955) *Clavaria mucida*, eine extratropische Basidiolichene. Biol Zentralbl 74:145–159

Geitler L (1956) Ergänzende Beobachtungen über die extratropische Basidiolichene *Clavaria mucida*. Oesterr Bot Z 103:164–167

Gilbert OL (1988) Studies on the destruction of *Lecanora conizaeoides* by the lichenicolous fungus *Athelia arachnoidea*. Lichenologist 20:183–190

Gill M, Steglich W (1987) Pigments of fungi (Macromycetes). Prog Chem Org Nat Prod 51:1–317

Heikkilä H, Kallio P (1966) On the problem of subarctic basidiolichens I. (Rep Kevo Subarct Res Stat 3). Ann Univ Turku A 2:48–74

Heikkilä H, Kallio P (1969) On the problem of subarctic basidiolichens. II. Ann Univ Turku A 40:90–97

Henssen A (1963) Eine Revision der Flechtenfamilien Lichinaceae und Ephebaceae. Symb Bot Upsal 18:1–123

Henssen A, Kowallik K (1976) A note on the mycobiont of *Coriscium viride* (Ach.) Vain. Lichenologist 8:197

Hibbett DS, Thorn RG (2001) Basidiomycota: Homobasidiomycetes. In: McLaughlin DJ, McLaughlin EG, Lemke PA (eds) The Mycota VII, B: systematics and evolution. Springer, Berlin Heidelberg New York, pp 121–168

Hibbett DS, Binder M, Bischoff JF, Blackwell M, Cannon PF, Eriksson OE, Huhndorff S, James T, Kirk PM, Lücking R, Lumbsch HL, Lutzoni F, Matheny PB, McLaughlin DJ, Powell MJ, Rehead S, Schoch CL, Spatafora JW, Stalpers JA, Vilgalys R, Aime MC, Aptroot A, Bauer R, Begerow D, Benny GL, Castlebury LA, Crous PW, Dai Y-C, Gams W, Geiser DM, Griffith GW, Gueidan C, Hawksworth DL, Hestmark G, Hosaka K, Humber RA, Hyde KD, Ironside JE, Kõljalg U, Kurtzman CP, Larsson K-H, Lichtwardt R, Longcore J, Miadlikowska J, Miller A, Moncalvo J-M, Mozley-Standridge S, Oberwinkler F, Parmasto E, Reeb V, Rogers JD, Roux C, Ryvarden L, Sampaio JP, Schüßler A, Sugijama J, Thorn RG, Tibell L, Untereiner WA, Walker C, Wang Z, Weir A, Weiß M, White MM, Winka K, Yao Y-J, Zhang N (2007) A higher-level phylogenetic classification of the fungi. Mycol Res 111:509–547

Huneck S, Yoshimura I (1996) Identification of lichen substances. Springer, Berlin Heidelberg New York

Iacomini M, Zanin SMW, Fontana JD (1987) Isolation and characterization of β-D-glucan, heteropolysaccharide, and trehalose components of the basidiomycetous lichen *Cora pavonia*. Carbohydr Res 168:55–65

Johow F (1884) Die Gruppe der Hymenolichenen. Ein Beitrag zur Kenntnis basidiosporer Flechten. Jahrb Wiss Bot 15:361–409

Jørgensen PM (1989) *Omphalina foliacea*, a new basidiolichen from America. Nord J Bot 9:89–95

Jørgensen PM (1998) *Acantholichen pannarioides*, a new basidiolichen from South America. Bryologist 101:444–447

Kalangutkar J, Kamat N (2010) Some studies on circadian rhythm in the culture of *Omphalina* Quelet sp. (Basidiomycota, Agaricales) a mycobiont of an unidentified basidiolichen. Nature Precedings 14:06 http://dx.doi.org/10.1038/npre.2010.4215.1

Lamoure D (1968) Parthenogenèse chez *Omphalina ericetorum* (Pers. ex Fr.) M. Lange et deux espèces affines. C R Acad Sci Paris Ser D 266:1499–1500

Lange OL (1965) Der CO_2-Gaswechsel von Flechten bei tiefen Temperaturen. Planta 64:1–19

Lange OL, Büdel B, Zellner H, Zotz G, Meyer A (1994) Field measurements of water relations and CO_2 exchange of the tropical, cyanobacterial basidiolichen *Dictyonema glabratum* in a Panamanian rainforest. Bot Acta 107:279–290

Langenstein B (1994) Experimentell-ontogenetische Studien zur Lichenisation am Beispiel von *Omphalina* (Basidiomycota) und *Coccomyxa* (Chlorophyta). Diplomarbeit, University of Tübingen, Tübingen

Larcher W, Vareschi V (1988) Variation in morphology and functional traits of *Dictyonema glabratum* from contrasting habitats in the Venezuelan Andes. Lichenologist 20:269–277

Larsson KH (2007) Re-thinking the classification of corticioid fungi. Mycol Res 108:1040–1063

Larsson KH, Larsson E, Kõljalg U (2004) High phylogenetic diversity among corticioid homobasidiomycetes. Mycol Res 108:983–1002

Lawrey JD, Binder M, Diederich P, Molina MC, Sikaroodi M, Ertz D (2007) Phylogenetic diversity of lichen-associated homobasidiomycetes. Mol Phyl Evol 44:778–789

Lawrey JD, Diederich P, Sikaroodi M, Gillevet PM (2008) Remarkable nutritional diversity of basidiomycetes in the Corticiales, including a new foliicolous species of Marchandiomyces (anamorphic Basidiomycota, Corticiaceae) from Australia. Am J Bot 95:816–823

Lawrey JD, Lücking R, Sipman HJM, Chaves JL, Redhead SA, Bungartz F, Sikaroodi M, Gillevet PM (2009) High concentration of basidiolichens in a single family of agaricoid mushrooms (Basidiomycota: Agaricales: Hygrophoraceae). Mycol Res 113:1154–1171

Lepp H (2011) Basidiolichens. Lichen website. http://www.anbg.gov.au. Cited 6 July 2011

Lodge DJ, Matheny PB, Cantrell SA, Moncalvo J-M, Vilgalys R, Redhead S (2006) Delineating the Hygrophoraceae: character myths vs. gene trees. Inoculum. Mycol Soc Am Annu Meet Abstr 57:27

Lücking R, Lawrey JD, Sikaroodi M, Gillevet PM, Chaves JL, Sipman HJM, Bungartz F (2009) Do lichens domesticate photobionts like farmers domesticate crops? Evidence from a previously unrecognized lineage of filamentous cyanobacteria. Am J Bot 96:1409–1418

Lumbsch HT (1998) Taxonomic use of metabolic data in lichen-forming fungi. In: Frisvad JC, Bridge PD, Arora DK (eds) Chemical fungal taxonomy. Dekker, Basel, pp 345–387

Lumbsch HT, Ewers WH (1992) Additional lichen records from Australia 10. *Omphalina umbellifera* (L. Fr.) Quélet. Austral Lichenol Newsl 30:5–6

Lutzoni F (1997) Phylogeny of lichen- and non lichen-forming omphalinoid mushrooms and the utility of testing for combinability among multiple data sets. Syst Biol 46:373–406

Lutzoni F, Pagel M (1997) Accelerated evolution as a consequence of transitions to mutualism. Proc Natl Acad Sci USA 94:11422–11427

Lutzoni F, Vilgalys R (1995a) Integration of morphological and molecular data sets in estimating fungal phylogenies. Can J Bot 73(Suppl 1):649–659

Lutzoni F, Vilgalys R (1995b) *Omphalina* (Basidiomycota, Agaricales) as a model system for the study of coevolution in lichens. Cryptogam Bot 5:71–81

Lutzoni F, Pagel M, Reeb V (2001) Major fungal lineages are derived from lichen symbiotic ancestors. Nature 411:937–940

Mägdefrau K, Winkler S (1967) *Lepidostroma terricolens* n. g. n. sp., eine Basidiolichene der Sierra Nevada de Santa Marta (Kolumbien). Mitt Inst Colombo Aleman Invest Cient 1:11–17

Mattirolo O (1881) Contribuzioni allo studio del genere *Cora* Fries. Nuov Giorn Bot Ital 8:245–267

Mitidieri J, Joly S, Ferraz EC (1964) Teste de antibiose exercida pelo extrato do liquens *Parmelia tinctorum* Desp. e *Cora pavonia* (Web.) E. Fries. Rev Agric 39:119–121

Nelsen MP, Lücking R, Umaña L, Trest MT, Will-Wolf S, Chaves JL, Gargas A (2007) Multiclavula ichtyiformis (Fungi: Basidiomycota: Cantharellales: Clavulinaceae), a remarkable new basidiolichen from Costsa Rica. Am J Bot 94:1289–1296

Norvell LL, Redhead SA, Ammirati JF (1994) Omphalina sensu lato in North America 1–2. 1: Omphalina wynniae and the genus Chrysomphalina. 2: Omphalina sensu Bigelow. Mycotaxon 50:379–407

Oberwinkler F (1965) Primitive Basidiomyceten. Revision einiger Formenkreis von Basidienpilze mit plastischer Basidie. Sydowia 19:1–72

Oberwinkler F (1970) Die Gattungen der Basidiolichenen. Vorträge aus dem Gesamtgebiet der Botanik. Dtsch Bot Ges 4:139–169

Oberwinkler F (1980) Symbiotic relationships between fungus and alga in basidiolichens. In: Schwemmler

H, Schenk HEA (eds) Endocytobiol Endosymb Cell Biol, vol 1. Walter de Gruyter, Berlin, pp 305–315

Oberwinkler F (1984) Fungus—alga interactions in basidiolichens. In: Hertel H, Oberwinkler F (eds) Bei-träge zur Lichenologie. Festschr J Poelt Beih Nova Hedwigia, vol 79. Cramer, Vaduz, pp 739–774

Oberwinkler F (2001) Basidiolichens. In: Hock B (ed) The Mycota. Springer, Berlin Heidelberg New York, pp 211–225

Palice Z, Schmitt I, Lumbsch HT (2005) Molecular data confirm that *Omphalina foliacea* is a basidiomycete. Mycol Res 109:447–451

Parmasto E (1968) Conspectus systematis corticiacearum. Institutum Zoologicum et Botanicum Academiae Scientiarum RPSS Estonicae, Tartu

Parmasto E (1978) The genus *Dictyonema* ('Thelephorolichenes'). Nova Hedwigia 29:99–144

Parmasto E (1998) *Athelia arachnoidea*, a lichenicolous basidiomycete in Estonia [Parasiitnahkis *(Athelia arachnoidea)*, samblikel kasvav kandseen Eestis]. Fol Cryptogam Estonica 32:63–66

Petersen RH (1967) Notes on clavarioid fungi. VII. Redefinition of the Clavaria vernalis-C. mucida complex. Am Midl Nat 77:205–221

Petersen R, Kantvilas G (1986) Three lichen-forming clavarioid fungi from Tasmania. Aust J Bot 34:217–222

Petersen R, Zang M (1986) New or interesting clavarioid fungi from Yunnan, China. Acta Bot Yunnan 8:281–294

Piovano M, Chamy MC, Garbarino JA, Quilhot W (1995) Studies on Chilean lichens XXIV. Secondary products from *Dictyonema glabratum* (Basidiomycotina). Bol Soc Chil Quim 40:163–165

Poelt J (1959) Eine Basidiolichene in den Hochalpen. Planta 52:600–605

Poelt J (1962) Die Basidiolichene *Lentaria mucida* in Bayern. Ber Bayer Bot Ges 35:87–88

Poelt J, Jülich W (1969a) *Omphalina grisella*, ein weiterer lichenisierter Blätterpilz in den Alpen. Herzogia 1:331–335

Poelt J, Jülich W (1969b) Über die Beziehungen zweier corticioider Basidiomyceten zu Algen. Oesterr Bot Z 116:400–410

Poelt J, Obermayer W (1990) Lichenisierte Bulbillen als Diasporen bei der Basidiolichene *Multiclavula vernalis* spec. coll. Herzogia 8:289–294

Poelt J, Oberwinkler F (1964) Zur Kenntnis der flechtenbildenden Blätterpilze der Gattung *Omphalina*. Oesterr Bot Z 111:398–401

Räsänen V (1943) Das System der Flechten. Act Bot Fenn 33:1–82

Redhead SA (1984) *Arrhenia* and *Rimbachia*, expanded generic concepts, and a reevaluation of *Leptoglossum* with emphasis on muscicolous North American taxa. Can J Bot 62:865–892

Redhead SA, Kuyper TW (1987) Lichenized agarics: taxonomic and nomenclatural riddles. Arct Alp Mycol 2:319–348

Redhead SA, Kuyper TW (1988) *Phytoconis*, the correct generic name for the basidiolichen *Botrydina*. Mycotaxon 31:221–223

Redhead SA, Lutzoni F, Moncalvo JM, Vilgalys R (2002) Phylogeny of agarics: partial systematics solutions for core omphalinoid genera in the Agaricales (Euagarics). Mycotaxon 83:19–57

Rodrigéz F, Feist SW, Guillou L, Harkestad LS, Bateman K, Renault T, Mortensen S (2008) Phylogenetic and morphological characterisation of the green algae infesting blue mussel *Mytilus edulis* in the North and South Atlantic oceans. Dis Aquat Org 81: 231–240

Roskin PA (1970) Ultrastructure of the host-parasite interaction in the basidiolichen Cora pavonia (Web.) E. Fries. Arch Mikrobiol 70:176–182

Seitzman BH, Ouimette A, Mixon RL, Hobbie EA, Hibbett DS (2011) Conservation of biotrophy in Hygrophoraceae inferred from combined stable isotope and phylogenetic analyses. Mycologia 103:280–290

Singer R (1970) Omphalinae (Clitocybeae—Tricholomataceae Basidiomycetes). Hafner, New York, Flora Neotropica, Monograph 3

Singer R (1973) The genera *Marasmiellus*, *Crepidotus* and *Simocybe* in the Neotropics. Beihefte zur Nova Hedwigia 44, Cramer, Berlin

Slocum RD (1980) Light and electron microscopic investigations in the Dictyonemataceae (basidiolichens). II. Dictyonema irpicinum. Can J Bot 58:1005–1015

Slocum RD, Floyd GL (1977) Light and electron microscopic investigations in the Dictyonemataceae (basidiolichens). Can J Bot 55:2565–2573

Smith JE, Sowerby J (1849) English botany. Coloured figures of British plants 4, vol 4. Smith and Sowerby, London, supplement

Tomaselli R, Caretta G (1969) Morphological and anatomical observations on *Cora pavonia* (Basidiolichenes) and hypotheses of factors determining lichen symbiosis. Atti Ist Bot Lab Critt Univ Pavia Ser 6(5):11–18

Trembley ML, Ringli C, Honegger R (2002a) Differential expression of hydrophobins DGH1, DGH2, DGH3 in the lichen forming basidiomycet *Dictyonema glabratum*. Fungal Genet Biol 35:247–259

Trembley ML, Ringli C, Honegger R (2002b) Hydrophobins DGH1, DGH2, DGH3 and immunolocalization of DGH1 in strata of the lichenized basidiocarp of *Dictyonema glabratum*. New Phytol 154:185–195

Tschermak-Woess E (1983) Das Haustorialsystem von *Dictyonema* kennzeichnend für die Gattung. Pl Syst Evol 143:109–115

Vainio EA (1890) Etude sur la classification naturelle et la morphologie des Lichens du Brésil. Acta Soc Faun Flor Fenn 7:1–256

Zoller S, Lutzoni F (2003) Slow algae, fast fungi: exceptionally high nucleotide substitution rate differences between lichenized fungi *Omphalina* and their symbiotic green algae *Coccomyxa*. Mol Phylogenet Evol 29:629–640

17 Bacteria and the Lichen Symbiosis

M. Grube[1], M. Cardinale[2], G. Berg[2]

CONTENTS

I. Introduction

Our understanding of evolution and natural selection is often overshadowed by the popular view of the "struggle for existence", most often envisioned as stressful competition for resources within the same habitat. This notion distracts from other fundamental evolutionary phenomena, such as the novelty of life forms stemming from symbioses. To the naked eye, a premier example is provided by lichen symbiosis, which gives rise to light-exposed and unique phenotypes, the lichen thalli, which are highly diverse in shape, chemical content and ecological demands (Grube and Hawksworth 2007). With an estimated age of 400–600 mio years, **lichens are among the oldest symbioses of fungi** (Yuan et al. 2005). Conservative estimates suggest that **more than 18,800 fungal species evolved in a lichenized stage** (Feuerer and Hawksworth 2007). The number may still be higher, because phenotypically poorly recognized species are hidden in taxonomically described lineages. Meanwhile, new data highlight lichens also as more complex symbioses than previously thought. In this chapter we focus on the diversity of bacteria in lichens and their possible roles.

A widely accepted definition describes **lichens as "an ecologically obligate, stable mutualism between an exhabitant fungal partner (the mycobiont) and an inhabitant population of extracellularly located unicellular or filamentous algal or cyanobacterial cells (the photobiont)"** (Hawksworth and Honegger 1994). In fact, the symbiosis is obligately required for the formation of the reproductive structures by the fungal partners and the thallus structure in general. The interpretation of lichens as a clear case of mutualism—as put forward in most biology textbooks—is difficult to prove and has been a matter of controversy in the scientific community. The lichen symbiosis has originally even been interpreted as a mild sort of algal "slavery" (a "helotism", Schwendener 1869): a dominant fungal partner usually shapes the lichen structure to form a thallus that provides shelter for the controlled growth of one or several algal or cyanobacterial organisms as primary partners. The sexuality of the algae is generally suppressed in their lichenized stages but algal cell densities are much higher than in free-living stages.

Advanced types of lichen thalli represent without doubt the most complex structures in the fungal kingdom. Unlike the mycelia of most other fungi, the lichen thalli are exposed to the light and develop important diagnostic characters for determination. Moreover, **most lichen symbioses are robust and endure the stresses of a**

[1]Institute of Plant Sciences, Karl-Franzens University, Holteigasse 6, 8010 Graz, Austria; e-mail: martin.grube@uni-graz.at
[2]Institute of Environmental Biotechnology, Graz University of Technology, Petersgasse 12, 8010 Graz, Austria

Fungal Associations, 2nd Edition
The Mycota IX
B. Hock (Ed.)

dynamic environment. Lichen thalli may persist for many years, even centuries, although growth and metabolic activity might be restricted to their young and regenerating parts (see also Honegger 2008). It is therefore not surprising that lichens are also preferred habitats of many other microorganisms. Particularly well-known are the lichenicolous fungi, which had been studied even before the symbiotic nature of lichens was discovered in the nineteenth century. Lichenicolous fungi develop diagnostic reproductive structures on their host lichens. The biological impact of such fungi on lichen biology is fairly well-known and ranges from parasitic to commensalic life styles. So far, more than 1,500 species of lichenicolous fungi have been described (Lawrey and Diederich 2003). Other fungi, which are isolated from the surface of lichens, appear to be non-specific and symptomless epibionts (Petrini et al. 1990; Prillinger et al. 1997). There are also diverse inconspicuous fungi which occur internally in lichens (Girlanda et al. 1997; Arnold et al. 2009).

In contrast, the significance of **bacterial colonization in lichen thalli** has been poorly considered in lichenological research in the past, despite the ubiquity of bacteria in microscopic images (Figs. 17.1, 17.2). Probably the first observations about bacteria were made by Thaxter (1892). He described *Chondromyces lichenicola* (now placed as *Melittangium lichenicola* in Myxobacteriaceae; McCurdy 1971) as a parasite on lichens. The infection is characterized by tiny knob-like structures on the host. This species, however, has hardly been collected and rarely seems to form its diagnostic phenotypes, although it can be detected among lichen-associated bacterial communities (unpublished data). Uphof (1925) suggested presence of purple bacteria in the thalli of *Herpothallon rubrocinctum* (as "Chiodecton sanguineum"), although this was then disapproved and seen as misinterpretation of crystallized compounds (Suessenguth 1926). About at the same time, Cengia-Sambo (1925, 1931) also observed bacteria in lichens and assigned them to *Azotobacter* on the surface of the (*Nostoc*-containing) cephalodia of *Peltigera aphthosa*. He coined the **term "polysymbiosis" for such multipartite relationships**. However, these works were not much noticed by the lichenologists of that time. Further evidence for the presence of bacteria in lichens was then provided by a series of papers by Russian and Armenian authors. Henckel and Yuzhakova (1936) and Iskina (1938) detected nitrogen-fixing bacteria in lichens by cultivation on nitrogen-free Ashby medium and assigned these strains to *Azotobacter*. Using phase contrast microscopy, Henckel (1946) discovered large numbers of bacteria (assigned to *Azotobacter*) on different species of lichens.

Sporadic findings of lichen-associated bacteria have been published since then. Krasilnikov

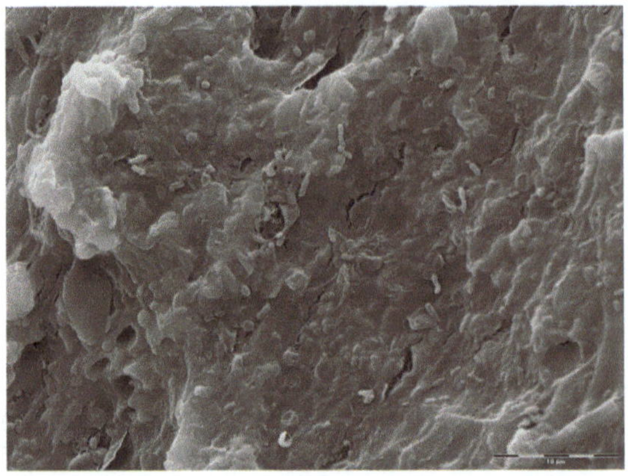

Fig. 17.1. Bacterial cells are present on the surface and to a varying degree immersed in the intercellular matrix of the internal podetium surface in *Cladonia arbuscula*. SEM image by E. Stabentheiner

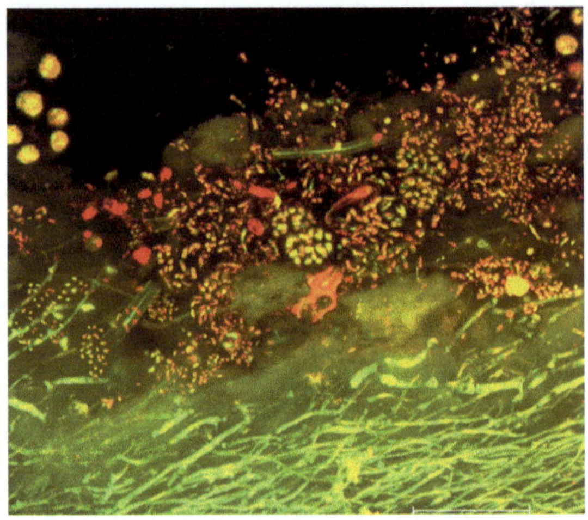

Fig. 17.2. Biofilm-like bacterial localization of the internal surface of the reindeer lichen *Cladonia arbuscula*. Acridine orange staining; CLSM image by M. Cardinale. *Bar* 10 μm

(1949) could not confirm the previous findings of lichen-associated *Azotobacter* in his samples, but found other bacterial groups, such as pseudomonads, Actinobacteria and Myxobacteria. Refusing the term symbionts for lichen-associated bacteria, he assumed the presence of *Azotobacter* is random and that the composition and abundance of microbial consortia depends on the state of the lichen thallus. At the same time, Navahradak (1949) reported the common presence of cellulose-degraders, including *Cellvibrio*, in lichens. Scott (1956) detected *Azotobacter* in lichens and found that a part of measured nitrogen-fixation might be provided also by bacteria other than cyanobacteria. Besides *Azotobacter*, other genera were repeatedly reported from lichens, such as *Bacillus* (Henckel and Plotnikova 1973), *Beijerinckia* (Panosyan and Nikogosyan 1966), *Clostridium* (Iskina 1938) and *Pseudomonas* (Henckel and Plotnikova 1973). Nitrogen fixation for lichens was repeatedly considered as a function of bacteria in these and other works (e.g. Lambright and Kapustka 1981), while the presence of actinobacteria in cyanobacterial lichens prompted Zook (1983) to suggest also a defensive role for bacteria in lichens. Lenova and Blum (1983) supposed that up to millions of bacterial cells could be present per gram of a lichen thallus. The names of culturable bacteria of these early works on lichen-associated bacteria reflect the state of taxonomical knowledge of the times when classification relied only on phenotypical and physiological methods. In the light of contemporary methodology, the previous determinations are certainly outdated and incomplete.

Current research with molecular and novel microscopic methods established baseline information about diversity, abundance and localization of bacteria on lichen thalli (Cardinale et al. 2006, 2008; Grube and Berg 2009; Grube et al. 2009; Hodkinson and Lutzoni 2009; Selbmann et al. 2010; Bates et al. 2011; Bjelland et al. 2011; Mushegian et al. 2011; Hodkinson et al. 2012). These assessments revealed that *Alphaproteobacteria* are predominant in growing parts of lichens (Grube et al. 2009; Mushegian et al. 2011; Cardinale et al. 2012a) but other bacterial groups are also found in significant amounts.

II. Multitrophic Interactions in Lichens

Lichens are widely popularized as a two-tier partnership of fungi and algae, but it was recognized in the early twentieth century that more than one photoautotrophic partners can be present, e.g. when the thalli include cyanobacteria. **About 2 % of the lichenized fungal species have cyanobacterial symbionts in addition to green-algal partners.** These associations are known as **tripartite lichens**. The **nitrogen-fixing role of the incorporated cyanobacteria** is well-established (Cengia-Sambo 1931; Millbank and Kershaw 1969). The cyanobacteria in tripartite lichens are often internalized by the fungal thallus in special organs called **cephalodia**. Actually, these structures were described long before the symbiotic nature of lichen was discovered (Flörke 1819). In the cephalodia, the cyanobacterial *Nostoc* strains have a higher relative number of **heterocysts**, compared with lichens containing *Nostoc* as a photosynthetic partner alone. This conforms to a **functional and spatial separation of primary carbohydrate- and nitrogen-fixation** (as nitrogen-fixation is catalyzed in the cyanobacterial heterocysts). The division of work is not complete, since the cyanobacteria are still photosynthetically active. Beside their presence in surface-borne or internal cephalodia, cyanobacteria in tripartite lichens can be present as a layer beneath green-algae (as in the genus *Solorina*), or more unspecifically as colonies adjacent to the thallus. Such affinity to neighboring cyanobacterial colonies has been recognized by the term **cyanotrophy** (Poelt and Mayrhofer 1988). Cephalodiate and cyanotrophic lichens are good indicators for the possible re-distribution of metabolic functions to more than two partners in the lichen symbiosis.

The concept of lichens as self-contained ecosystems, originally put forward by Farrar (1976), is corroborated when additionally associated microorganisms are considered. Lichenicolous fungi, some of which digest the host structures might be interpreted as the degraders in a simple system with algal producers and the lichen mycobiont as consumers. However, the finding of commensalic and symptomless fungi, and highly abundant, host-specific and diverse bacterial communities add to the biological complexity in lichen symbioses. The functional roles of these additional players in the lichen ecosystems are still far from settled. Here we want to give an overview of what has been achieved so far.

Earlier approaches to the role of bacteria in lichens relied on functions found in the culturable fraction of the associated bacterial community. Gonzles et al. (2005) used an enrichment medium to isolate actinobacteria from tropical and polar lichens. Most strains represented *Micromonospora* and *Streptomyces*, but other Actinobacteria were present as well. Half of the isolated strains showed antimicrobial activities, with activity against Gram-positive *Staphylococcus aureus* occurring at highest frequency (23%). This confirms previous assumptions of **a defensive role of lichen-associated bacteria.** Similarly N-free enrichment cultures were used by Liba et al. (2006) to find nitrogen fixers by acetylene-reducing assays in green-algal lichens from coastal Brazil. The results were confirmed by *nifH* gene dot blots. The 17 detected strains were from *Gammaproteobacteria* (such as *Acinetobacter*, *Pantoea*, *Pseudomonas*, *Serratia*, *Stenotrophomonas*). Further analysis revealed excretion of amino acids and indole acetic acid (IAA) in 14 strains. Eight strains solubilized phosphate and four released ethylene. *Stenotrophomonas* strains released both ethylene and IAA. These data would support a role of bacteria in nitrogen fixation in green-algal lichens, with a role in nutritional and hormonal amendment. These functions have not been experimentally tested in lichen thalli, but an artificial co-culture study of lichen mycobionts with nitrogen-fixing bacteria indicated an increased capacity of rock weathering with bacterial symbionts (Seneviratne and Indrasena 2006).

However, the **culturable bacteria are only a minor part of the communities.** Studies using fluorescence in situ hybridization and confocal laser scanning microscopy (FISH-CLSM) clearly showed that *Alphaproteobacteria* represent a dominant fraction in the studied lichen parts and form extensive, often biofilm-like colonies (Cardinale et al. 2008; Fig. 17.3). Unfortunately, the most abundant lichen-dwelling

Fig. 17.3. Extensive colonies of Alphaproteobacteria (*darker spots*) at the lower part of the soil-inhabiting lichen *Arthrorhaphis citrinella*. 3-D reconstruction of a FISH-CLSM image by L. Muggia and B. Klug

Alphaproteobacteria have not been found among cultured isolates so far. Other bacterial groups occur often in smaller numbers, and it seems that culturable bacteria are present in rather low numbers in the growing parts. The presence of endocellular bacteria is still uncertain, although Cardinale et al. (2008) already indicated that bacteria may penetrate hyphal walls.

Fingerprinting analysis using SSCP **revealed host species specificity of the lichen-associated bacterial communities** (Grube et al. 2009), which was subsequently confirmed by pyrosequencing studies (e.g. Bates et al. 2011). In rock-inhabiting crustose lichens, which suffuse the uppermost substrate layers with their entire surface, specific influence is also seen below the lichen thalli (Bjelland et al. 2011). Although a higher bacterial diversity is observed in the underlaying rocks, the abundance of bacteria is higher in the epilithic lichen thalli. These results demonstrated that the **lichen-rock interfaces are complex habitats, where the macroscopic lichen structures influence the composition of microbial metacommunities.** Bjelland et al. (2011) also showed that *Acidobacteria*—another group of mostly unculturable bacteria—are a significant component in three of four studied rock-inhabiting lichens. Recent 454-amplicon

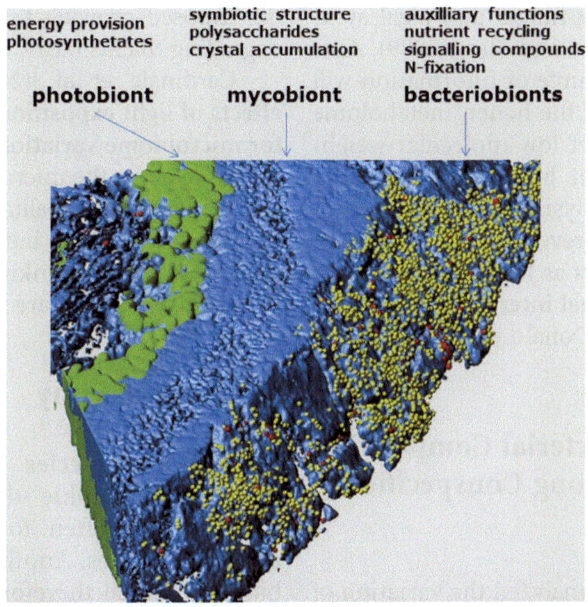

Fig. 17.4. Extended concept of the lichen symbiosis, including the association with bacteria

sequencing data confirmed substantial numbers of *Acidobacteria* in certain host species (Mushegian et al. 2011; Hodkinson et al. 2012; Pankratov 2012; Grube et al., unpublished data).

A first analysis of putative functions at the protein level in an entire lichen system was provided by Schneider et al. (2011). Environmental proteomics techniques gave insight into the total protein composition of the lung lichen *Lobaria pulmonaria*. Their data revealed a high number of protein reads from prokaryotes, which was equal to the eukaryotic algal partner in this lichen. Among bacteria the majority of reads were from Alphaproteobacteria, confirming the results of DNA-based analyses, but also Archaea (0.5 %) were detected. As expected, most of the green algal proteins were involved in energy production and conversion. Carbohydrate transport and metabolism played a role in both algal and fungal proteins. Fungal functions were more diverse, and considerable numbers of proteins seem involved in biogenesis. Post-transcriptional modifications, protein turnover and chaperones are also important, but form a higher ratio in the bacterial reads. The involvement of bacterial nitrogen fixation was not apparent from Schneider et al. (2011). Lack of nitrogenase activity was also revealed in preliminary metatranscriptomic data (Hodkinson 2011), even though *nif*H genes were previously detected in green algal lichens by PCR approaches (Grube et al. 2009).

Some of the protein sequences indicated **production of secondary metabolites by the prokaryotic partners**. Lichens are well-known for their diversity of bioactive molecules, but so far most are products of fungal origin, whereas only very few cyanobacterial compounds are known. The bacterial production of compounds in lichen thalli certainly requires further attention, because such compounds may contribute to functional roles of bacteria in lichen biology.

The functional assignments of proteins provide a rather general insight in the functional network of the lichen symbiosis but cannot predict substrates and activities of expressed proteins. It is also necessary to analyze the protein levels across various physiological stages (e.g. dry to wet thallus conditions) to understand the extent of proteomic variation. Moreover, a combined -omics approach is clearly required to uncover the diverse bacterial functions in lichens.

So far data from culture-based and culture-independent studies suggest a multitude of bacterial functions in the lichen symbioses, in addition to functions of fungal and algal symbionts (Fig. 17.4; Grube and Berg 2009). One important additional source of information will be the detailed study of the lichen metabolome and to specifically target low-molecular-weight substances which might have signaling functions. Further, the analysis of genomes from lichen symbionts will reveal possible ancient horizontal gene transfers as footprints of a long history of bacterial-fungal interactions (Schmitt and Lumbsch 2009; McDonald et al. 2012).

III. Variation of Bacterial Composition Within and Among Conspecific Thalli

Only few studies so far analyzed the variation of the lichen-associated microbiome within single thalli. Mushegian et al. (2011) studied intrathalline variation of multiple individuals of closely related foliose lichens with T-RFLP and pyrosequencing. They found that the internal part of lichens comprises richer and more consistent microbiota, whereas the margins are species-poor and more disparate in their composition. The authors concluded that the central parts of these lichens had more time for a consolidation of the bacterial communities whereas marginal parts comprise more random communities. This is contrasted by a microscopic study of Cardinale et al. (2012a), who found that the bacterial community of the growing parts of reindeer lichen thalli is rather uniform and composed of abundant *Alphaproteobacteria*, whereas more variation is found in the ageing parts. The somewhat contrasting results can be explained by the difference in the growth form of the investigated lichens. The central parts of the foliose species analyzed by Mushegian et al. (2011) are composed of a mixture of old and regenerative, young thallus parts. One species produces fruit-bodies, whereas the other species forms isidia as mitotic outgrowths of thalli for joint symbiont dispersal. This mixture of differently aged structures might account for the more complex community found in central thallus parts. In contrast, the shrub-like reindeer lichen studied by Cardinale et al. (2012a) has a clear age gradient from the exposed growing tips to the substrate-near, degrading thallus basis.

Cardinale et al. (2012a) also studied **the effects of light exposition as an additional factor for microbiome variation.** Exposition was clearly shown to affect the microbial communities in old and young parts of thalli. Old parts in shaded sites displayed a distinct increase of *Betaproteobacteria*, and of yet unknown *Bacteria*, whereas *Alphaproteobacteria* are less prominent.

IV. Biogeography

Many lichen species have extremely wide-ranging geographic distributions, with the same species often found across continents and hemispheres. Another line of lichen microbiome research therefore focuses on the question whether the host-specific composition of bacterial communities follow the distribution of their lichen hosts. Hodkinson et al. (2012) found evidence for geography as an important factor shaping lichen-associated community structures, but their analyses included several distinct species, sampled across the ecologically dissimilar sampling sites.

Cardinale et al. (2012b) focused on *Lobaria pulmonaria* as a single widespread species for biogeographic comparison of two groups of lichen-associated bacteria. As a flagship species in conservation biology and an indicator of forest continuity, *Lobaria pulmonaria* has a quite narrow ecological range but is found in different continents. The authors compared samples from sites in Europe (at various distances from each other) and in African Islands. The analysis of SSCP fingerprints indicated a higher correlation of the predominant *Alphaproteobacteria* with geography than found for *Burkholderiales*. The latter is regularly detected in the culturable fraction but apparently present at low abundance in lichens. Cardinale et al. (2012b) also suggest that *Alphaproteobacteria* are readily co-dispersed in joint fungal–algal propagules, called isidia, whereas *Burkholderiales* (*Betaproteobacteria*) would more likely be taken up from the local environment. This pattern may also underscore **the tight association of lichens with *Alphaproteobacteria*.** Printzen et al. (2012) focuses on the alphaproteobacterial community in another lichen species with extremely wide geographic distribution, *Cetraria aculeata*. This study revealed a less diverse community in polar habitats (in contrast to soils in polar latitudes; e.g. Chu et al. 2010), but antarctic and arctic communities were more similar to each other than to those of lichen samples of other proveniences. Interestingly, this pattern agrees with what has been reported previously for the lichen photobiont partner. Studies of the algal partners suggested a response of lichens to environmental conditions in form of symbiont switches or alteration of their relative ratios. We argue the bacterial community similarly responds to habitat conditions, which could eventually increase the adaptivity of the holobiont.

V. New Bacterial Species from Lichens

Lichens are a rich source of new bacterial strains and lineages. Especially **actinobacterial species** were isolated from lichens and described (Li et al. 2007; An et al. 2009; Yamamura et al. 2011). Cardinale et al. (2011) described a new actinobacterial species, *Frondihabitans cladoniiphilus*, from the reindeer lichen *Cladonia arbuscula*. This new species has interesting features which might be connected with an adaptation to the lichen habitat. In contrast to close relatives, cells of this species can form autoaggregates, which might be important for host attachment. A new species has also been discovered in *Betaproteobacteraceae* (*Herminiimonas saxobsidens* is a new member of *Oxalobacteraceae*; Lang et al. 2007).

Apart from new culturable species described in *Actinobacteria* and *Betaproteobacteria*, new lineages of commonly lichen-associated but still uncultured bacteria have been detected in *Alphaproteobacteria*. The lichen-associated *Rhizobiales* 1 clade (LAR1; Hodkinson and Lutzoni 2009) has frequently been found in association with green-algal lichens (all from North America; Bates et al. 2011; Hodkinson et al. 2012), and this lineage was also present in a sequence library derived from Antarctic lichens (de la Torre et al. 2003). LAR1 has been characterized so far only by 16rRNA gene sequences. A study of Antarctic lichens revealed even more yet unclassified culturable bacterial species, representing several lineages, including *Actinobacteria*, *Firmicutes*, *Proteobacteria* and *Deinococcus* (Selbmann et al. 2010). A phylogenetic analysis revealed that many of the strains cluster in well-supported groups that only comprised bacteria from lichens. All strains grew well at low temperatures but also at 25 °C, and therefore are psychrotolerant rather than psychrophilic.

VI. Compounds Produced by Lichen-Associated Bacteria

Bacteria in lichens are also a valuable source of effective new secondary metabolites. A strain of *Streptomyces*, isolated from the reindeer lichen *Cladonia uncialis* is particularly exciting due to the production of several novel compounds. Among these, the enediyne **uncialamycin** (Davies et al. 2005) is a compound with strong antibacterial activity against the human pathogens *Burkholderia cepacia* (MIC= 0.001 μg/mL) and *Staphylococcus aureus* (MIC =0.0064 μg/mL). Uncialamycin was only recently also synthesized in the laboratory (Nicolaou et al. 2007). The same *Streptomyces* strain ("*S. uncialis*" ined.) also produces a series of new *bis*-indole alkaloids, the **cladoniamides** (Williams et al. 2008). Cladoniamides are unusual among *bis*-indole natural products: most *bis*-indoles have an indolocarbazole structure, whereas the cladoniamides have a rarely observed indenotryptoline structure (Fig. 17.5). Their biosynthetic gene cluster (*cla*) has now been characterized in more detail (Ryan 2011) and compared with the BE-54017 (*abe*) gene cluster from environmental DNA. BE-54017 and its derivatives are structurally similar to the cladoniamides. The elucidation of the *cla* gene cluster and comparison with the *abe* cluster supports the hypothesis that indenotryptoline cores are biosynthetically derived from the oxidative rearrangement of an indolocarbazole precursor. Another lichen-derived streptomycete strain produces the tetrapeptide **lichostatinal** (Lavallée 2011), which represents a

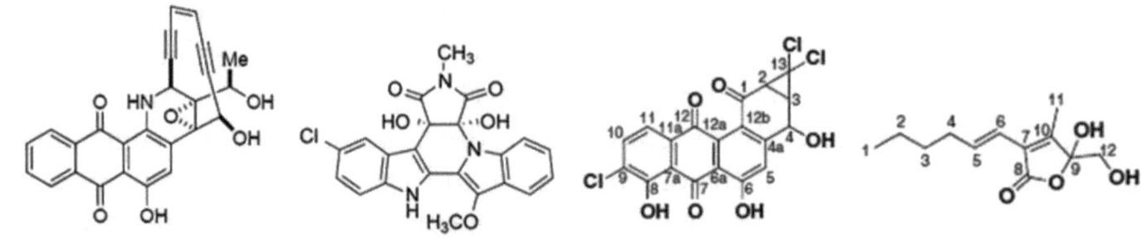

Cladoniamide A Uncialamycine JBIR-88 JBIR-89

Fig. 17.5. Examples of products from lichen-associated Steptomycetes

cathepsin K inhibitor and is thus of interest for combating osteoporosis. Yet another *Streptomyces* strain isolated from lichens produced a new **angucycline (JBIR-88)** and a new **butenolide (JBIR-89)**, which also displayed inhibitory effects on certain cell lines and a bacterial strain (Motohashi et al. 2010).

VII. Biotechnological Applications

The biotechnological potential of lichen-associated bacteria is directly related to ecological, morphological and chemical characteristics of the hosts. For example, the ecology of lichens is often characterized by strongly changing conditions of water availability. Gasser et al. (2012) found a high proportion of bacterial **poly-β-hydroxybutyrate (PHB)** producers as well as genes involved in PHB synthesis in different lichen species. Synthesis of storage compounds such as PHB is a strategy of Gram-negative bacteria **to increase survival in changing environments** (Steinbüchel and Valentin 2006). Another example is the use of lichen-associated bacteria as antagonists towards fungal plant pathogens for biological plant protection. In their natural environment, **pathogen defense against parasites could be one function of the bacterial community in lichens** (Grube et al. 2009): up to 100 % antagonists against the plant pathogens *Alternaria alternata* and *Phytophthora infestans* were found in the culturable fraction of different lichen species (unpublished data).

These are only two examples for promising biotechnological applications with lichen-associated bacteria, but many more may be developed in the future. We also consider lichens as sources of novel antibiotics and new signalling molecules, or enzymes working under extreme conditions.

Acknowledgements We are grateful to the Austrian Science Foundation for financial support (FWF 19098, FWF I799-B16). We thank Lucia Muggia and Barbara Klug (Graz) for technical assistance.

References

Arnold AE, Miadlikowska J, Higgins KL, Sarvate SD, Gugger P, Way A, Hofstetter V, Kauff F, Lutzoni F (2009) A phylogenetic estimation of trophic transition networks for ascomycetous fungi: are lichens cradles of symbiotrophic fungal diversification? Syst Biol 58:283–297

An SY, Xiao T, Yokota A (2009) *Leifsonia lichenia* sp. nov., isolated from lichen in Japan. J Gen Appl Microbiol 55:339–543

Bates ST, Cropsey GWG, Caporaso JG, Knight R, Fierer N (2011) Bacterial communities associated with the lichen symbiosis. Appl Environ Microbiol 77:1309–1314

Bjelland T, Grube M, Hoem S, Jorgensen SL, Daae FL, Thorseth IH, Øvreås L (2011) Microbial metacommunities in the lichen–rock habitat. Environ Microbiol Rep 3:434–442

Cardinale M, Puglia AM, Grube M (2006) Molecular analysis of lichen-associated bacterial communities. FEMS Microbiol Ecol 57:484–495

Cardinale M, Müller H, Berg G, de Castro J, Grube M (2008) In situ analysis of the bacteria community associated with the reindeer lichen *Cladonia arbuscula* reveals predominance of *Alphaproteobacteria*. FEMS Microbiol Ecol 66:63–71

Cardinale M, Grube M, Berg G (2011) *Frondihabitans cladoniiphilus* sp. nov., a novel actinobacterium of the family *Microbacteriaceae* isolated from the reindeer lichen *Cladonia arbuscula* (Wallr.) Rabenh. in the Austrian Alps. Int J Syst Environ Microbiol 61:3033–3038

Cardinale M, Steinova J, Rabensteiner J, Berg G, Grube M (2012a) Age, sun, and substrate: triggers of bacterial communities in lichens. Environ Microbiol Rep 4:23–28

Cardinale M, Grube M, Vieira de Castro J, Müller H, Berg G (2012b) Bacterial taxa associated with the lung lichen *Lobaria pulmonaria* are differentially shaped by geography and habitat. FEMS Microbiol Lett 329:111–115

Cengia-Sambo M (1925) Ancora della polisimbiosi nei licheni ad alghe cianoficee. I batteri simbionti. Att Soc Ital Sci Nat 64:191

Cengia-Sambo M (1931) Biologie des lichens. Les substances carbohydratées dans les lichens et la fonction de fixation de l'azote des céphalodies. Boll Sez Ital Soc Internaz Microbiol 11:1–8

Chu H, Fierer N, Lauber CL, Caporaso JG, Knight R, Grogan P (2010) Soil bacterial diversity in the Arctic is not fundamentally different from that found in other biomes. Environ Microbiol 12:2998–3006

Davies J, Wang H, Taylor T, Warabi K, Huang XH, Andersen RJ (2005) Uncialamycin, a new enediyne antibiotic. Org Lett 7:5233–5236

De la Torre JR, Goebel BM, Friedmann EI, Pace NR (2003) Microbial diversity of cryptoendolithic communities from the McMurdo Dry Valleys, Antarctica. Appl Environ Microbiol 69:3858–3867

Farrar JF (1976) The lichen as an ecosystem: observation and experiment. In: Brown DH, Hawksworth DL, Bailey RH (eds) Lichenology: progress and problems. Academic, London, pp 385–406

Feuerer T, Hawksworth DL (2007) Biodiversity of lichens, including a world-wide analysis of checklist data based on Takhtajan's floristic regions. Biodivers Conserv 16:85–98

Flörke HG (1819) Deutsche lichenen. Vierte Lieferung, Rostock, 15 pp

Gasser I, Vieira de Castro Junior J, Müller H, Berg G (2012) Lichen-associated bacteria antagonistic to phytopathogens and their potential to accumulate polyhydroxyalkanoates. IOBC/WPRS Bull (in press)

Girlanda M, Isocrono D, Bianco C, Luppi-Mosca AM (1997) Two foliose lichens as microfungal ecological niches. Mycologia 89:531–536

Gonzles I, Ayuso-Sacido A, Anderson A, Genilloud O (2005) Actinomycetes isolated from lichens: evaluation of their diversity and detection of biosynthetic gene sequences. FEMS Microbiol Ecol 54:401–415

Grube M, Berg G (2009) Microbial consortia of bacteria and fungi with focus on the lichen symbiosis. Fungal Biol Rev 23:72–85

Grube M, Hawksworth DL (2007) Trouble with lichen: the re-evaluation and re-interpretation of thallus form and fruit body types in the molecular era. Mycol Res 111:1116–1132

Grube M, Cardinale M, Vieira De Castro J Junior, Müller H, Berg G (2009) Species-specific structural and functional diversity of bacterial communities in lichen symbiosis. ISME J 3:1105–1115

Hawksworth DL, Honegger R (1994) The lichen thallus: a symbiotic phenotype of nutritionally specialized fungi and its response to gall producers. In: Williams MAJ (ed) Plant galls: organisms, interactions, populations. Clarendon, Oxford, pp 77–98

Henckel PA (1946) New observations on the triple nature of lichens. Bull Mosc Soc Nat Biol Ser 51:6

Henckel PA, Plotnikova TT (1973) Nitrogen-fixing bacteria in lichens. Proc Acad Sci USSR Biol Ser 6:807–813

Henckel PA, Yuzhakova LA (1936) On the role of *Azotobacter* in the lichen symbiosis. Bull Perm Molotov Biol Res Inst 10:315

Hodkinson BP (2011) A phylogenetic, ecological and functional characterization of non-photoautotrophic bacteria in the lichen microbiome. PhD thesis, Duke University, Durham

Hodkinson BP, Lutzoni F (2009) A microbiotic survey of lichen-associated bacteria reveals a new lineage from the Rhizobiales. Symbiosis 49:163–180

Hodkinson BP, Gottel NR, Schadt CW, Lutzoni F (2012) Photoautotrophic symbiont and geography are major factors affecting highly structured and diverse bacterial communities in the lichen microbiome. Environ Microbiol 14:147–161

Honegger R (2008) Morphogenesis. In: Nash TH III (ed) Lichen biology, 2nd edn. Cambridge University Press, Cambridge, pp 69–93

Iskina RY (1938) On nitrogen fixing bacteria in lichens. Bull Perm (Molotov) Biol Res Inst 11:133–139

Krasilnikov NA (1949) Is *Azotobacter* present in lichens? Mikrobiologiia 18:3

Lambright DD, Kapustka LA (1981) The association of N2-fixing bacteria with *Dermatocarpon miniatum* and *Lepraria* sp. Bot Soc Am Misc Ser Pub 160:5

Lang E, Swiderski J, Stackebrandt E, Schumann P, Spöer C, Sahin N (2007) *Herminiimonas saxobsidens* sp. nov., isolated from a lichen-colonized rock. Int J Syst Evol Microbiol 57:2618–2622

Lavallée VP (2011) Antipain and its analogues, natural product inhibitors of Cathepsin K isolated from *Streptomyces*. PhD thesis, University of British Columbia, Vancouver

Lawrey JD, Diederich P (2003) Lichenicolous fungi: interactions, evolution, and biodiversity. Bryologist 106:81–120

Lenova LI, Blum O (1983) To the question on the third component of lichens. Bot J 68:21–28

Li B, Xie CH, Yokota A (2007) *Nocardioides exalbidus* sp. nov., a novel actinomycete isolated from lichen in Izu-Oshima Island, Japan. Actinomycetologica 1:22–26

Liba CM, Ferrara FIS, Manfio GP, Fantinatti-Garboggini F, Albuquerque RC, Pavan C, Ramos PL, Moreira CA, Barbosa HR (2006) Nitrogen-fixing chemo-organotrophic bacteria isolated from cyanobacteria-deprived lichens and their ability to solubilize phosphate and to release amino acids and phytohormones. J Appl Microbiol 101:1076–1086

McCurdy HD (1971) Studies on the taxonomy of the Myxobacterales. IV. Melittangium. Int J Syst Bacteriol 21:50–54

McDonald T, Dietrich F, Lutzoni F (2012) Multiple horizontal gene transfers of ammonium transporters/ammonia permeases from prokaryotes to eukaryotes: toward a new functional and evolutionary classification. Mol Biol Evol 29:51–60

Millbank JW, Kershaw KA (1969) Nitrogen metabolism in lichens. I. Nitrogen fixation in the cephalodia of *Peltigera aphthosa*. New Phytol 68:721–729

Motohashi K, Takagi M, Yamamura H, Hayakawa M, Shin-ya K (2010) A new angucycline and a new butenolide isolated from lichen-derived *Streptomyces* spp. J Antibiot 63:545–548

Mushegian AA, Peterson CN, Baker CCM, Pringle A (2011) Bacterial diversity across individual lichens. Appl Environ Microbiol 77:4249–4252

Navahradak DM (1949) Lichens and cellulose degrading microorganisms. Microbiology (Moscow) 18:6

Nicolaou KC, Zhang H, Chen JS, Crawford JJ, Pasunoori L (2007) Total synthesis and stereochemistry of uncialamycin. Angew Chem Int Ed 46:4704–4707

Pankratov TA (2012) Acidobacteria in the microbial communities of the bog and tundra lichens [in Russian]. Mikrobiologiya 81:56–63

Panosyan AK, Nikogosyan VG (1966) The presence of *Azotobacter* in lichens. Akad Nauk Armian SSR, Biol Zhurn Armen 19:3–11

Petrini O, Hake U, Dreyfuss MM (1990) An analysis of fungal communities isolated from fruticose lichens. Mycologia 82:444–451

Poelt J, Mayrhofer H (1988) Über Cyanotrophie bei Flechten. Plant Syst Evol 158:265–281

Prillinger H, Kraepelin G, Lopandic K, Schweigkofler W, Molnar O, Weigang F, Dreyfuss MM (1997) New species of *Fellomyces* isolated from epiphytic lichen species. Syst Appl Microbiol 20:572–584

Printzen C, Fernndez-Mendoza F, Muggia L, Berg G, Grube M (2012) Alphaproteobacterial communities in geographically distant populations of the lichen *Cetraria aculeata*. FEMS Microbial Ecol, http://www.ncbi.nlm.nih.gov/pubmed/22469494

Ryan KS (2011) Biosynthetic gene cluster for the cladoniamides, bis-indoles with a rearranged scaffold. PLoS One 6:e23694

Schmitt I, Lumbsch HT (2009) Ancient horizontal gene transfer from bacteria enhances biosynthetic capabilities of fungi. PLoS One 4:e4437

Schneider T, Schmid E, de Castro V, Junior J, Cardinale M, Eberl L, Grube M, Berg G, Riedel K (2011) Structure and function of the symbiosis partners of the lung lichen (*Lobaria pulmonaria* L. Hoffm.) analyzed by metaproteomics. Proteomics 11:2752–2756

Schwendener S (1869) Die Algentypen der Flechtengonidien, vol 4. C. Schultze, Basel

Scott GD (1956) Further investigations of some lichens for fixation of nitrogen. New Phytol 55:111–116

Selbmann L, Zucconi L, Ruisi S, Grube M, Cardinale M, Onofri S (2010) Culturable bacteria associated with Antarctic lichens: affiliation and psychrotolerance. Polar Biol 33:71–83

Seneviratne G, Indrasena IK (2006) Nitrogen fixation in lichens is important for improved rock weathering. J Biosci 31:639–643

Steinbüchel A, Valentin HE (2006) Diversity of bacterial polyhydroxyalkanoic acids. FEMS Microbiol Lett 128:219–228

Suessenguth K (1926) Zur Frage der Vergesellschaftung von Flechten mit Purpurbakterien. Ber Dtsch Bot Ges 44:573–578

Thaxter R (1892) On the *Myxobacteriaceae*, a new order of *Schizomycetes*. Bot Gaz 17:389–406

Uphof JCT (1925) Purple bacteria as symbionts of a lichen. Science 61:67

Williams DE, Davies J, Patrick BO, Bottriell H, Tarling T, Roberge M, Andersen RJ (2008) Cladoniamides A-G, tryptophan-derived alkaloids produced in culture by *Streptomyces uncialis*. Org Lett 10:3501–3504

Yamamura H, Haruna Ashizawa H, Nakagawa Y, Hamada M, Ishida Y, Otoguro M, Tamura T, Hayakawa M (2011) *Actinomycetospora rishiriensis* sp. nov., an actinomycete isolated from a lichen. IJSEM 61:2621–2625

Yuan X, Xiao S, Taylor TN (2005) Lichen-like symbiosis 600 million years ago. Science 308:1017–1020

Zook PD (1983) A study of the role of bacteria in lichens. MA thesis, Clark University, Worcester

18 Fungal and Bacterial Volatile Organic Compounds: An Overview and Their Role as Ecological Signaling Agents

J.W. Bennett[1], R. Hung[1], S. Lee[1], S. Padhi[1]

CONTENTS

I. Introduction

In terrestrial ecosystems fungi occur as decomposers, symbionts, and pathogens living in close association with bacteria, plants, and animals where inter-organismal signaling is essential. However, the study of interspecific and interkingdom chemical communication has received less scientific scrutiny than signaling within a single organism or signaling within members of a single species.

In this chapter, we provide a general summary of a group of metabolites that we believe have not received sufficient attention by the microbiological community, namely small, easily volatilized molecules that are transmitted in the gas phase. In non-aqueous habitats, volatiles provide a way for microbes to communicate with one another, both within and between species. It is our purpose here to draw attention to the volatile molecules of bacteria and fungi, to document some of the biological versatility that these molecules display, and to highlight their role in fungal associations.

Volatile organic compounds (VOCs) are low molecular weight carbon-containing compounds that evaporate easily at normal temperatures and pressures. The best known volatiles are chemical solvents and other industrial compounds associated with modern life such as paints, cleaning supplies, petroleum fuels, adhesives, photographic solutions, and the like. Exposures to high concentrations of volatiles such as benzene, formaldehyde, methylene chloride, toluene, and xylene are known to have both short- and long-term negative effects on human health (McFee and Zavon 1988).

Far less is understood about **biogenic VOCs**. Plants and microbes emit a wide range of volatile acids, alkanes, alkenes, carbonyls, alcohols, esters, terpenoids, and other small molecules into the biosphere. The isoprenoids produced by plants are perhaps the best known. In particular, monoterpenes such as limonene ("citrusy"), menthol ("minty"), and pinene ("resinous") have familiar odors. The odors associated with many non-isoprenoid microbial VOCs also are well known—the earthy smell of

[1]Department of Plant Biology and Pathology, Rutgers University, New Brunswick, NJ 08901, USA; e-mail: profmycogirl@yahoo.com

Fungal Associations, 2nd Edition
The Mycota IX
B. Hock (Ed.)

garden soil, the noxious stench of spoiled foods, the body odors of people who rarely bathe, and the characteristic mustiness of damp basements are all due largely to gas-phase microbial metabolites. Numerous factors influence the release of VOCs from different biogenic sources including the population of producing species, substrates, temperature, radiation, associations with other organisms, types of ecosystem, and general climate. For comprehensive and provocative reviews on plant VOCs, see Kesselmeier and Staudt (1999) and Baldwin et al. (2006). For a general database of flavors and scents, see Dunkel et al. (2009).

The first major review on fungal volatiles by Hutchinson (1973) focused almost entirely on carbon dioxide. During the 1970s, fungal VOCs were usually isolated by steam distillation followed by liquid–liquid extraction and concentration of the organic extract (Cronin and Ward 1971; Kaminski et al. 1972). These approaches led to the isolation of several major fungal VOCs, but the technical limitations of the methodology gave an underestimate of their number. Since that time, with more sophisticated trapping, separation, and identification techniques, **approximately 250 distinct VOCs have been identified from fungi where they occur as mixtures of simple hydrocarbons, heterocycles, alcohols, phenols, thioalcohols, thioesters, and their derivatives** (Chiron and Michelot 2005; Korpi et al. 2009). Not all of these VOCs have odors that humans can detect. However, of those that do smell, many of the most familiar ones have extremely low odor thresholds. The human olfactory system is able to detect geosmin, which has an earthy smell, in the range of 150–200 ng/m^3. The mushroom-like odor of 1-octen-3-ol is recognized at 10 μg/m^3 while the musty odor of 2-octen-1-ol is recognized at 16 μg/m^3 (Zogorski et al. 2006).

Complex blends of VOCs cause the distinct bouquets of edible mushrooms (Breheret et al. 1997; Tirillini et al. 2000; Cho et al. 2008). For example, the delectable odor of truffles, prized in the culinary arts and said to have aphrodisiacal properties, include over 100 different VOCs of which alcohols, aldehydes, aromatics, and thiols dominate (Splivallo et al. 2007a).

Sulfur-containing volatiles are a highly recognizable group of VOCs. They are best known from plants of the genus *Allium* (chive, garlic, onion, etc.), where they are valued for their flavors in cooking and for their application in traditional medicine. Similar organosulfur "alliaceous" compounds are made by shiitake mushroom (*Lentinula edodes*) and species of *Tricholoma* and *Marasmius* (Bloch and Deorazio 1994; Sneeden et al. 2004). Sulfur-containing volatiles are an important part of the odor profile of truffles (Mauriello et al. 2004).

Many mushrooms have other distinctive odor profiles. Anise-like odors are characteristic of *Clitocybe odora*, *Lentinellus cochleatus*, and *Agaricus essettie* (Rapior et al. 2002). A "boiled potato" or "farm feed" odor is characteristic of several species of *Suillus* (de Pinho et al. 2008). The most notorious basidiomycete odor is the repulsive stench of stinkhorns in the genus *Phallus*, which largely is due to high concentrations of dimethyl oligosulphides (Borg-Karlson et al. 1994). Mushroom smells have been reviewed by de Pinho et al. (2008) and Fraatz and Zorn (2010).

Microscopic fungi similarly emit complex VOC mixtures. Most people are familiar with the musty or moldy odors associated with damp basements and other enclosed indoor spaces. The odor signatures of a given species or strain will vary depending on the substrate, the length of incubation, type of nutrients, temperature, and other environmental parameters. Furthermore, they vary considerably between species (Pasanen et al. 1997; Nilsson et al. 2004; Fiedler et al. 2005). In *Aspergillus* and *Penicillium* species, the concentration of many VOCs increases during sporulation (Börjesson et al. 1993).

The single most commonly reported volatile from fungi, both macroscopic and microscopic, is **1-octen-3-ol, which is also called "mushroom alcohol," or "matsutake alcohol"** because it was first identified from *Tricholoma matsutake* (Wood and Fesler 1986). In the straw mushroom *Volvariella volvacea*, for example, 1-octen-3-ol accounts for the majority (72–83 %) of the total volatiles detected (Mau et al. 1997). Mushroom alcohol occurs in two

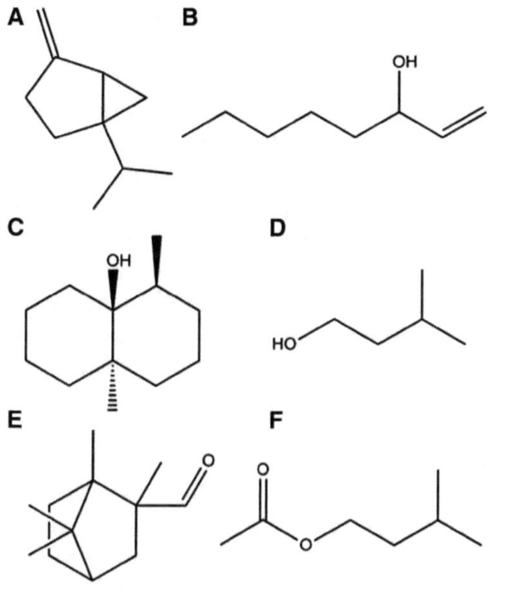

Fig. 18.1. Chemical structures of some representative fungal volatile organic compounds: (**A**) sabinene, (**B**) 1-octen-3-ol, (**C**) geosmin, (**D**) 3-methyl-1-butanol, (**E**) 2-methyl-isoborneal, (**F**) 3-methyl butyl acetate (isoamyl acetate)

enantiomer forms, (*R*)-(−)-1-octen-3-ol and (*S*)-(+)-1-octen-3-ol, of which the most abundant naturally occurring isomer is the *R* enantiomer, sometimes called "roctenol." Roctenol is produced commercially where it is used as an insect attractant (Kline et al. 2007; Bohbot and Dickens 2009). (*R*)-(−)-1-Octen-3-ol has a characteristic "mushroomy" odor, while (*S*)-(+)-1-octen-3-ol is more grassy or moldy; (*R*)-(−)-1-octen-3-ol is the predominant naturally occurring enantiomer of 1-octen-3-ol (Mosandl et al. 1986). The structures of a few representative fungal VOCs are given in Fig. 18.1.

II. Analytical Processes for Extraction, Separation, Identification, and Quantification

The large number of VOC chemical structures, their generally low concentrations, and the fact that they tend to occur in mixtures pose challenges for comprehensive sampling and analysis. Depending on the situation and intended application, different approaches are utilized for sampling, sample preparation, separation, identification, and quantification. For a good review of current techniques in use for biologically produced VOCs, see Zhang and Li (2010).

Traditionally, an initial step of separation into purer components was followed by identification and quantification. In recent decades, the purging and trapping of headspace gases has gained favor, as for example in studies of odor formation in agricultural settings (Abramson et al. 1980, 1983). The trapping of headspace gases also has been used to obtain VOC profiles of a number of fungal species grown under standardized laboratory conditions (Mattheis and Roberts 1992; Börjesson et al. 1993). The absorbance materials used for trapping (e.g. charcoal, Super Q, Tenax) can select for compounds with particular binding properties. Similarly, the choice of organic solvents (e.g. dichloromethane, hexane) can affect the VOC profile obtained, sometimes yielding insufficient resolution of highly volatile, early-eluting compounds (Kai et al. 2009). Larsen and Frisvad (1995) showed that very different profiles of fungal VOCs are obtained with different collecting methods.

Most determination (separation and identification) of VOCs now relies on gas chromatography–mass spectrometry (GC-MS) because it integrates powerful separation capability and facilitates both quantification and identification. Depending on the intended purpose, several methods are currently available. Air sampling onto Tenax desorption tubes followed by thermodesorption allows accurate sampling at one point in time. In order to determine VOCs over a longer period of time, passive diffusion monitors onto charcoal adsorbents can be used. Matysik et al. (2009) found that this method was particularly suitable in epidemiological studies that attempt to correlate concentrations of specific VOCs and indoor mold exposure. After separation, the individual constituents of VOC mixtures usually are identified by mass spectrometry. Comparison of mass spectra with library spectra or determination of chromatographic retention indices, ideally in conjunction with the parallel determination of authentic standards, are used to confirm identity (Stoppacher et al. 2010).

The means of delivering the sample to the vacuum chamber for analysis has also changed and improved over time. Trapping using the solid phase microextraction (SPME) fiber technique has emerged an efficient and popular method for assessing VOCs in a variety of contexts (Jeleń 2003). The fiber is exposed to the headspace atmosphere and, subsequently, the trapped contents are analyzed by GC-MS. SPME has been applied widely in the flavor and fragrance industry for monitoring freshness, detecting fungal contamination in stored products and the like (Stoppacher et al. 2010; Zhang and Li 2010).

"Artificial olfaction" is a new area of sensor technology that promises to revolutionize VOC detection. **Electronic noses** (sometimes called artificial noses) have been developed that attempt to mimic mammalian olfactory systems and are finding increasing applications in healthcare and other biomedical applications. Interestingly, the diseases caused by certain microbial pathogens are associated with particular odiferous compounds, e.g. *Pseudomonas* causes a "grape" odor of skin while the sweat of patients with rubella is said to smell like "freshly plucked feathers" (Pavlou and Turner 2000). In laboratory experiments, electronic noses were able to distinguish between uncontaminated samples and those contaminated with dry-rot wood decay fungi (Kuske et al. 2005). A comprehensive historical review of electronic nose technology is provided by Wilson and Baietto (2009; 2011).

Finally, it should be noted that dogs can be trained to detect mold growth with 75–94 % accuracy (Griffith et al. 2007).

III. Classification

The anabolic metabolic pathways of bacteria, fungi, and plants are sometimes divided into the dichotomous categories of primary and secondary metabolism. Primary metabolites are essential to the life of the organism and represent the unity of biochemistry. Examples include Krebs cycle intermediates, amino acids, lipids, and nucleic acids (Berg et al. 2007). In contrast, secondary metabolites are the enormous group of diverse natural products not essential to growth, often of extremely unusual chemical structure, which almost always are restricted in taxonomic distribution. Antibiotics such as cephalosporins, hallucinogens such as the ergot alkaloids, and mycotoxins such as the trichothecenes are well known examples of fungal secondary metabolites (Bennett 1983; Bennett and Bentley 1989; Cole and Schweikert 2003). Secondary metabolites are biosynthesized by special pathways (e.g. polyketides, non-ribosomal peptides, isoprenoids) and constitute the bulk of the field known as natural products chemistry. Advances in genomics facilitate the detection of their signature gene clusters using bioinformatics approaches (Keller et al. 2005).

Due to their low molecular weights and phase dependent appearance, fungal VOCs as a group sometimes are classified as secondary metabolites. However, there are problems with this simplistic categorization. Secondary metabolites usually are made by only a taxonomically limited number of producing species while most VOCs are found across a broad range of producing organisms. For example, production of the toxic secondary metabolite, aflatoxin, is restricted to a few species within the genus *Aspergillus*, while production of the volatile 1-octen-3-ol is widespread across many fungi as well as plants and animals (Cole and Schweikert 2003; Chiron and Michelot 2005; Combet et al. 2006). Secondary metabolites are generally produced by complex metabolic pathways encoded by clusters of linked genes (Zhang et al. 2005). **Although less is known about the pathways that produce VOCs, many of them are either metabolic transformation products of lipids, proteins, heterocyclic metabolites, and other components of living tissues, or are degradation end-products ("waste products") of fungal catabolic pathways.**

The common fungal VOC, 1-octen-3-ol, is a good case in point. It comes from linoleic acid (Wurzenberger and Grosch 1984). **Although 1-octen-3-ol is sometimes called a secondary metabolite, it is better classified as a lipid**

degradation product. Both 1-octen-3-ol and a less volatile 10-oxo-trans-8-decenoic acid are produced through the enzymatic oxidation and cleavage of linoleic and linolenic acids (Wurzenberger and Grosch 1984). In *Pleurotus pulmonarius* two separate lipoxygenases may be involved in the production of 1-octen-3-ol and 10-oxo-trans-8-decenoic acid (Assaf et al. 1997).

In summary, **not all the "small molecules" outside of the central pathways of intermediary metabolism are secondary metabolites.** Therefore, in this review, we will not classify fungal VOCs as either primary or secondary metabolites. Instead, we will describe VOCs according to their number of carbons, their ring structures, their substituent group, and as acids, ketones, aldehydes, terpenes, and the like.

IV. Overview of Fungal Volatiles

Fungal VOCs are of both theoretical and practical significance within a number of disparate scientific disciplines. They have been studied for their flavor properties, as indicators to detect the presence of fungal growth, as possible contributors to "sick building syndrome," and as signals for fungal development. Moreover, in recent years, the VOCs from endophytes have emerged as of particular interest because some of them have shown antibiotic activity while others have potential for possible use as fuel compounds or "biodiesel." All of this research has revealed that fungal VOC profiles are both complex and dynamic: the compounds produced and their abundance vary with the producing species, the age of the fungal colony, abundance of moisture, the type of substrate, the temperature, and other environmental parameters. A few examples from this enormous and scattered literature are given here.

A. Exploitation of Fungal VOCs: Flavors and as Indicators of Fungal Growth

Mold-ripened cheeses are among the best known food fermentations involving filamentous fungi. Blue cheeses such as Gorgonzola, Roquefort, and Stilton, and white cheeses such as Brie and Camembert, gain their distinctive flavors from methyl ketones and various alcohols produced by fungal metabolism by species of *Penicillium* (Karahadian et al. 1985; Gallois and Langlois 1990). Fungi are also used in various commercial bioconversion products for making flavor products that can be considered "natural aromas" (Berger et al. 1992; Schreier 1992).

The study of aroma compounds in beers, wines, and other spirits, and their relationship with the perceived flavor of alcoholic beverages represents a huge field that addresses the volatile compounds associated with yeast fermentation. For introductions to this huge literature, see for example Meilgaard (1975a, b) and Robinson (2006). We will not attempt to cover food and flavor chemistry in this chapter. For an earlier review on mixed bacterial–fungal food fermentations, see Bennett and Feibelman (2001).

VOCs can be used as indirect indicators of the presence of mold growth, even in the absence of visible colonies. Considerable research has gone into developing methods that detect fungal volatiles as an indirect and non-invasive way to indicate the presence of fungal growth in agriculture, water-damaged buildings, and in art conservation.

Fungal contamination of stored foods and feeds is a worldwide problem in agriculture. Fungi decrease the nutritive value of stored foods, spoil them by creating off-flavors, and can produce mycotoxins that render foods poisonous to human beings and other animals that eat them. Compounds such as geosmin, 1-octen-3-ol, 3-octanol, and 3-methy-1-butanol are regularly found in association with stored grains contaminated with fungi (Börjesson et al. 1989, 1993; Mattheis and Roberts 1992). Fungal VOCs have been used to monitor good and bad food quality (Karlshøj et al. 2007). As electronic nose technology improves, it is hoped that volatile compound "mapping" can be used to predict the levels of fungi found in agricultural products and perhaps to identify individual species (Schnürer et al. 1999).

The VOC profiles of common molds grown on building materials have been analyzed in the laboratory (Sunesson et al. 1995, 1996). Mold VOC profiles have also been studied in water-damaged buildings where compounds such as 2-methyl-1-proponol and 3-methyl-1-butanol have been suggested as useful indicators of mold growth (Wilkins et al. 2000; Claeson et al. 2002; Matysik et al. 2008; Korpi et al. 2009). Fungus-like odors can be recognized at concentrations greater than 0.035 μg/m^3. Generally, the background microbial VOC concentrations in mold-free buildings are similar to those found in outdoor air ranging from 2.2 to 8.8 μg/m^3. VOC levels significantly higher than the background ranges may indicate an increase in the active microbial production possibly associated with adverse health effects (Ström et al. 1994).

In addition, VOC profiles have been used to detect molds grown on objects of art heritage such as tapestries or on the wooden framework behind a painting. Using VOCs, Joblin et al. (2010) developed an index that was used to compare the level of VOCs in the Lascaux caves contaminated with *Fusarium solani* before and after antifungal treatment.

B. Non-Specific Building-Related Illnesses

Many occupants of damp indoor spaces complain of irritation of eyes and mucous membranes, respiratory discomfort, malaise, headaches, gastrointestinal disturbances, and a variety of other symptoms that are often lumped together and called "non-specific building-related illness," or by the more controversial name of "sick building syndrome." Many scientists have hypothesized that this spectrum of adverse health is associated with exposure to airborne bacteria and fungi, their aerosolized toxins, or bacterial metabolites such as endotoxins (Thorn 2001; Straus et al. 2003; Burge 2004; IOM 2004; Li and Yang 2004; WHO 2009). Mycotoxins have received most of the attention as the hypothetical cause of the symptoms of sick building syndrome (Jarvis and Miller 2005; Straus 2009). However, VOCs may also contribute to the adverse health consequences associated with damp indoor spaces, especially in cases where there is no visible evidence of mold growth (Walinder et al. 2005; Mølhave 2009). High vapor pressures, low to medium water solubility, and low molecular weights allow both fungal and bacterial VOCs to persist and migrate in the environment, and to diffuse through enclosed wall cavities, air conditioning filters and vapor barriers. The level of VOCs measured in the indoor air of a mold-infested building varies with the ventilation rate, moisture levels, composition of mold population, the area of the building/room and is constantly changing. One of the highest reported concentrations for a single VOC, 1-octen-3-ol, found in problem buildings is 900 μg/m^3 or 0.16 ppm (Morey et al. 1997). Moreover, 1-octen-3-ol has been reported to be one of the major fungal VOCs emitted by various species of fungi (*Aspergillus*, *Cladosporium*, *Fusarium*, *Penicillium*, *Stachybotrys*, etc.) regularly found in moldy and water-damaged buildings (Sunesson et al. 1996) or from composting facilities (Fischer et al. 1999). Studies in our laboratory have shown that gas phase 1-octen-3-ol is more toxic to human embryonic stem cells than is toluene (Inamdar et al. 2011). Moreover, when adult *Drosophila* flies were exposed for 1 week to low concentrations of chemical standards of 2-octanone (0.5 %), 2,5 dimethylfuran (0.5 %; DMF), 3-octanol (0.5 %), trans-2-octenal (0.5 %), and 1-octen-3-ol (0.1 %) they exhibited, respectively, 40, 35, 60, 50, and 100 % lethality (Inamdar et al. 2010).

C. Endophytes

Endophytes are microorganisms that live intercellularly within plant tissues without causing any evident negative effects (Bacon and White 2000). They have been found in almost every plant species examined and are likely to play a significant role in plant community structure (Rodriguez et al. 2009). **Fungal endophytes have been studied as a source of novel secondary metabolites** (Tan and Zou 2001) and also have gained attention as **producers of bioactive VOCs.** *Muscodor albus* ("stinky white fungus") is an endophyte that produces a blend of VOCS that are inhibitory or lethal to a wide range of bacteria and pathogenic fungi. Using GC-MS, Strobel et al. (2001) found that *M. albus* produced a mixture of volatile acids, alcohols, esters, ketones, and lipids, which individually had inhibitory but not lethal effects against test species such as *Fusarium solani*, *Pythium ultimum*, and *Rhizoctonia solani.* When applied collectively, these same VOCs acted synergistically to kill a broad range of plant pathogenic fungi and bacteria (Strobel et al. 2001). Since the original isolation of *Muscodor albus* from a cinnamon tree, several other *Muscodor* strains and species that emit antibiotic mixtures of VOCs have been isolated (Atmosukarto et al. 2005; Zhang et al. 2010). This **selective antimicrobial effect** can be harnessed against undesirable pathogens and has been termed "mycofumigation." *Muscodor albus* has been used for the biological control of damping off in broccoli seeds grown in greenhouse soilless mix (Mercier and Jiménez 2004; Mercier and Manker 2005). *Oxyporus latemarginatus*, an endophyte isolated from pepper plants, has also shown positive mycofumigation ability against post-harvest decay organisms (Lee et al. 2009).

VOCs from endophytes may have other biotechnological applications. Growing fungi in the presence of *M. albus* has been used as a selection tool to isolate other fungi that produce bioactive volatiles. Species resistant to the VOCs produced by *M. albus* are then screened for the activity of their own VOC profile. Using this method, an endophytic species of *Gliocladium* was isolated from a Patagonian species of *Eucryphia*. This *Gliocladium* species produced a number of VOCs, of which one of the most interesting was 1,3,5,7 cyclooctatetraene or annulene, an unstable and flammable compound used as a rocket propellant during World War II (Stinson et al. 2003). Strobel et al. (2008) coined the term "mycodiesel" when he found that the endophyte *Gliocladium roseum* produced several VOCs normally associated with diesel fuel.

Endophytic fungi in the genus *Ascocoryne* (Griffin et al. 2010), *Phoma* (Strobel et al. 2011), and *Phomopsis* also synthesize VOCs that have fuel potential. *Phomopsis* species isolated from an orchid in Ecuador produces bicyclic monoterpenes sabinene, is a monoterpene with a peppery odor first isolated from plants (Singh et al. 2011). Several related monoterpenes are being investigated as possible components of aircraft fuel (Rude and Schirmer 2009).

D. Developmental Signals for Fungal Spore Germination and Growth

A number of biological activities, including chemotrophic interactions, growth coordination, and inhibition of the growth of other fungi are mediated by volatile and non-volatile signaling metabolites. For a useful review, see Leeder et al. (2011).

The fungal sensing processes that prevent premature germination are called "self-inhibition" or "auto-inhibition." These processes detect spore over-crowding and maximize the chances of survival and colony formation once the spore has germinated. High spore concentrations inhibit germination; with lower spore concentrations, germination proceeds. In general, these effects are reversible and do not affect later mycelial growth. Auto-inhibition has been well studied in *Colletotrichum*, where several germination self-inhibitors have been chemically characterized as non-volatile indole compounds and where these compounds are specific for the producing species (Lax et al. 1985; Tsurushima et al. 1995). The cell density-dependent germination inhibitor systems in fungi resemble quorum-sensing systems described for bacteria (see Sect. IV.C).

The most abundant VOC produced by fungi, **1-octen-3-ol, functions as a developmental signal for many species. It is produced along with 10-oxo-*trans*-8-decenoic acid by the enzymatic breakdown of linoleic acid.** Both compounds inhibit mycelial growth of *Penicillium expansum* at low (1.25 mM) concentrations (Okull et al. 2003). In *Penicillium paneum*, dense suspensions of conidia show poor germination; again the auto inhibitor is 1-octen-3-ol (Chitarra et al. 2004). Volatile compound(s) produced by *P. paneum* under high spore-density conditions also inhibit mycelial growth of other fungal species belonging to a variety of genera, suggesting a range of actions beyond auto-inhibition. It has been hypothesized that 1-octen-3-ol interferes, in a reversible manner, with essential metabolic processes involved in swelling and germination of the conidia (Chitarra et al. 2005).

The effect of 1-octen-3-ol is dependent on its concentration and on the stage of fungal colony formation. In dark-grown cultures of *Trichoderma,* very low concentrations of 1-octen-3-ol, 3-octanol, and 3-octanone induced conidiation. However, at the highest concentrations tested for 1-octen-3-ol (500 μM) and 3-octanol and 3-octanone (both 1,000 μM), both conidiation and growth were inhibited (Nemcovic et al. 2008). In another study identifying and profiling the naturally occurring volatile metabolites of *T. atroviride*, 1-octen-3-ol and 3-octanone reached their maximum concentrations simultaneously to conidiation (Stoppacher et al. 2010).

Similar effects of C-8 compounds have been observed in *Aspergillus nidulans* by Herrero-Garcia et al. (2011). High conidial densities yielded autoinhibition of germination, which was associated with 1-octen-3-ol produced in association with aerial hyphae. The inhibition effect was reversible; conidiospores germinated normally after 1-octen-3-ol was removed. The closely related compound 3-octanone enhanced sporulation responses in aerial cultures (Herrero-Garcia et al. 2011). It also is of interest that several C-18 oxylipins derived from linoleic acid (psi factors) can repress conidiation and induce premature sexual sporulation in *A. nidulans* (Champe et al. 1987; Champe and el-Zayat 1989). Oxylipins are broadly involved in developmental regulation of aspergilli (Tsitsigiannis and Keller 2007) and 10-oxo-trans-8-decenoic acid stimulates mycelia growth, stipe lengthening and the initiation of fruiting in *Agaricus bisporus* (Mau et al. 1992). Oxylipins have been found from fungal pathogens of both plants and animals. While the oxylipin chemical end products are similar, the fungal enzymes involved in their production are different from those isolated from plants and animals. Several lines of evidence suggest that **plant oxylipins are able to mimic fungal oxylipins, suggesting a form of "reciprocal cross-talk" in certain host–pathogen relationships** (Brodhagen et al. 2008). For a comprehensive review of fungal oxylipins and the enzymes involved in their production, see Brodhun and Feussner (2011).

Various pheromone communication systems are used to coordinate colony formation, fruit body development, and reproduction in higher fungi (Kües and Navarro-Gonzlez 2009). In the macrofungus *Agaricus bisporus* VOCs produced by the mycelium inhibited primordium formation; 1-octen-3-ol had the greatest inhibitory effect (Noble et al. 2009).

Ammonia is a simple volatile compound not usually thought of as a signaling molecule. Nevertheless, colonies of the yeast *Candida mogii* **produce pulses of ammonia that synchronize the development in neighbouring colonies yielding extensive cell and colony morphology changes** (Palkova et al. 1997; Palkova and Forstova 2000). Transition to the phase of intense ammonia production ("alkali phase") is associated with a decrease of mitochondrial oxidative catabolism, peroxisome activation and activation of biosynthetic pathways that decrease the general stress level in colonies (Palkova et al. 2002).

V. Overview of Bacterial Volatiles

Bacterial species, like fungi, produce complex cocktails of VOCs. For example, a survey of 26 species in the genus *Streptomyces* revealed mixtures of alcohols, alkanes, alkenes, ketones, terpenoids, and thiols in a range of concentrations and combinations (Schöller et al. 2002). Forty-two different volatiles were found from *Myxococcus xanthus* (Dickschat et al. 2005b) and the volatile blend of a marine *Streptomyces* species exhibited antibiotic properties (Dickschat et al. 2005a). For a comprehensive survey of bacterial VOCs, see Schulz and Dickschat (2007).

Bacteria communicate with one another using many molecular mechanisms and they have interactive effects on the organisms with which they share ecological niches including plants, animals, fungi, and other bacteria. Bacterial volatiles have been studied for: (a) their uses in foods, particularly in the dairy industry, (b) their production of "off" odors in food and water supplies, and (c) other applied areas. In particular, soil bacteria associated with the rhizosphere have been investigated for their growth-promoting activities. In basic science, the study of quorum sensing has transformed our view of "single-cell" organisms.

A. Food Spoilage and "Off" Flavors

Bacterial and other VOCs can be detected in real time. They provide a way to noninvasively monitor safety applications within the food industry as well as with respect to potable water supplies. Several approaches have been developed to detect coliform bacteria and other common microbial contaminants using VOCs (Wilson and Baietto 2011). Similar strategies have been used to detect microbial spoilage of meat as well as to detect food pathogens present in foods (Ellis et al. 2002; Arora et al. 2006).

In drinking water, "off" flavors sometimes are caused by very low concentrations of microbial VOCs. Geosmin and 2-methylisoborneal (MIB) are two cyanobacterial VOCs that are among the most common compounds involved in malodor problems in surface waters (Watson et al. 2000). They give a musty or earthy odor; both have an exceptionally low odor threshold. MIB is also a factor in cork taints in winemaking. Using isotopic labeling experiments, Bentley and Maganathan (1981) determined that geosmin is a degraded sesquiterpene and MIB is a methylated monoterpene. With improved and inexpensive methods for monitoring VOCs, such as those associated with artificial nose technologies, it is likely that electronic odor detection will become an increasingly useful tool for checking food and water safety.

B. Plant Growth Promotion

Plant growth promotion by microbial species has been studied in several contexts. Microbes influence plant growth by reducing levels of disease (e.g. by antibiosis or competition with pathogens), stimulating growth, and biofertilization. In some cases, bacteria produce compounds that directly stimulate plant growth. When scientists study these interactions, beneficial effects are usually easiest to demonstrate in the laboratory, with decreasing success in greenhouses, and only a few microbes functioning successfully in field situations (Lugtenberg and Kamilova 2009).

In addition to secondary metabolites, many **growth-promoting rhizobacteria produce volatile organic compounds that have a positive impact on plant growth** (Vespermann et al. 2007). Strains of *Bacillus subtilis*, *B. amyloliquefaciens*, and *Enterobacter cloacae* promoted plant growth by releasing several volatiles, of which acetoin and 2,3-butanediol gave the highest level of growth promotion. Mutants of *B. amyloliquefaciens* and *B. subtilis* that are blocked in the biosynthesis of these compounds no longer promoted plant growth (Ryu et al. 2003, 2005).

Long-term exposure to *B. subtilis* volatiles induces beneficial effects on plant growth including increases in plant size and weight; cell numbers, and modulation of root-system architecture (Xie et al. 2009; Gutiérrez-Luna et al. 2010). In addition, there is an increase in the photosynthetic capacity of plants as evidenced by increases in chlorophyll content and up-regulation of genes encoding chloroplast proteins (Zhang et al. 2008; Xie et al. 2009). VOCs stimulate the synthesis of plant hormone-like compounds, including indole-3-acetic acid, cytokinin, and gibberellins. Moreover, genes involved in auxin biosynthesis are up-regulated in plants exposed to VOCs from growth promoting rhizobacteria (Ryu et al. 2003). Zhang et al. (2007) showed that around 600 genes were differentially expressed in *Arabidopsis* seedlings exposed to VOCs from *Bacillus*.

When plant growth-promoting rhizobacteria (PGPR) interact with plants, VOCS are important components of the signaling process (Vespermann et al. 2007). **Several bacterial VOCs including alcohols, ammonia, HCN, and phenazine-1-carboxylic acid have antifungal properties** that contribute to the biocontrol properties of PGPR (Whipps 2001; Choudhary et al. 2008; Kai et al. 2009). Kai et al. (2009) showed that volatiles from a given bacterial strain cause different responses on different fungi, i.e. associations of fungi and bacteria interact in different ways. **The VOCs of PGPR are associated with induced systemic resistance** (Farag et al. 2006). Furthermore, VOCs also lead to an enhancement of aroma compounds in basil (Banchio et al. 2009).

Certain bacterial volatiles promote plant health by inhibiting the growth of fungal pathogens. Negative effects on fungi include the inhibition of both sporulation and spore germination. Positive effects on sclerotial and fruit body development were often observed (Mackie and Wheatley 1999; Fernando et al. 2005; Vespermann et al. 2007).

C. Quorum Sensing

Quorum sensing refers to bacterial cellular communication systems that yield a population response. **Quorum-sensing bacteria release chemical signals called autoinducers that they use to monitor cell density.** The concentration of autoinducers increases as a function of rising bacterial population density; **when a minimal threshold of autoinducer is detected, the bacteria respond by altering gene expression** (Bassler and Losick 2006; Camilli and Bassler 2006). Quorum sensing conveys the concept that certain genes are expressed only when bacteria are crowded together (von Bodman et al. 2003).

The first quorum-sensing system described at the molecular level was in *Vibrio fischeri*, a bioluminescent marine bacterium in which the bacteria grow to a high cell density within the light organ of the Hawaiian squid (Nealson et al. 1970; Nealson and Hastings 1979). When *V. fischeri* cells are free living, they do not luminesce. However, in the nutrient rich environment of the light organ of the squid, they grow to high concentration and produce an acyl-homoserine lactone which acts as autoinducer (Eberhard et al. 1981). The population-dependent release of autoinducer leads to transcription of the luciferase gene and bioluminescence (Kaplan and Greenberg 1985). The *V. fischeri* system has become a paradigm for quorum sensing in Gram-negative bacteria; in addition, many other Gram-negative bacteria have cell to cell signaling systems that use fatty acid derivatives as autoinducers (Whitehead et al. 2001; Bassler and Losick 2006). Quorum sensing also occurs widely in Gram-positive bacteria, where the best studied autoinduction signals are oligopeptides (Waters and Bassler 2005; Shapiro 1998).

Once thought to be rare, quorum-sensing systems are now known to be common in bacteria where they are involved in a large number of complex environmental responses that include virulence and biofilm formation in *Pseudomonas aeruginosa* (Van Delden and Iglewski 1998; Rumbaugh et al. 2000), fruiting body production in *Myxococcus* (Shimkets 1999), antibiotic and pigment production in *Serratia* (Thomson et al. 2000), pathogenicity in plant-pathogenic bacteria (von Bodman et al. 2003), and many other important phenomena that allow bacterial colonies to behave in a fashion similar to multicellular organisms.

It is important to note that both Gram-negative and Gram-positive bacteria almost always are studied in the context of aqueous environments. In the laboratory, the approach is to use shaken liquid-batch cultures whereby the quorum-sensing response is detected at a specific point in the growth curve, coinciding with a threshold concentration of signal (Horswill et al. 2007). Because of the dominant aqueous experimental model, it is not surprising that **only a few quorum-sensing signals that operate through the gas phase have been discovered.** Nevertheless, some cases have been reported. *Ralstonia (Pseudomonas) solanacearum* is a Gram-negative soil-borne plant pathogen that causes bacterial wilt in a wide variety of important crops including eggplant, potato, and tobacco (Agrios 2008). It uses the volatile signal, 3-hydroxyl palmitic acid methyl ester, for regulating the expression of most of the traits needed for infection and virulence (Clough et al. 1994; Flavier et al. 1997; von Bodman et al. 2003). In environments such as soils and plant surfaces, opportunities for signaling in the liquid phase are limited. Therefore, it is highly likely that many other volatile auto-inducers will be discovered and that signaling in the gas phase may be one of the next important frontiers in quorum sensing" (Horswill et al. 2007).

Fungal quorum-sensing molecules also have been identified. The best studied system is ***Candida albicans***, a dimorphic fungus, which grows as a commensal on humans in yeast form but functions as an opportunistic pathogen. The yeast–hypha transition is essential for causing disseminated disease. **Hyphal development is suppressed by farnesoic acid, which acts as a quorum-sensing molecule** (Hogan 2006). There is increasing evidence that quorum sensing is widespread in fungi and that in many of these systems oxylipins such as 1-octen-3-ol play an important role (Kües and Navarro-Gonzlez 2009).

Specialized terminology tends to be discipline specific and may cause barriers in communication between scientists from different fields. The jargon used to describe the cell density-dependent signaling systems important in fungal spore germination and fruit body development were developed independently from terms used by bacteriologists to describe bacterial quorum-sensing systems (Table 18.1). Nevertheless, the concepts have many parallels.

VI. VOCs as Ecological Signaling Agents

With each passing year, there is growing recognition of the extent of chemical communication in the biosphere and the role that volatile che-

Table 18.1. Technical terms frequently encountered in discussions of chemical signaling compounds

Term	Definition and citations
Allelochemical	Broadly, a compound that mediates chemical interactions between organisms (Hooper and Pickett 2004). Allelochemicals were originally termed "allelochemics" by Whittaker and Feeny (1971) to describe toxic chemicals made by a plant to defend itself against competing plants.
Allomone	An allelochemical that only benefits the organism that emits the chemical signal (Hooper and Pickett 2004) "Originally defined as a chemical substance produced or acquired by an organism which, when it contacts an individual of another species in the natural context, evokes in the receiver a behavioral or physiological reaction adaptively favorable to the emitter" (Brown 1968; Nordlund and Lewis 1976).
Kairomone	A controversial term for an allelochemical which benefits only the receiving organism, but not the producer; i.e. the emitter does not benefit from the interaction (Nordlund and Lewis 1976; Hooper and Pickett 2004).
Synomone	An allelochemical which benefits both the producing and receiving organisms.
Autoinducer	A compound used by quorum sensing bacteria to monitor cell density.
Auto-inhibitor ("self-inhibition")	A term introduced by Allen (1957) to describe the inhibition of fungal spore germination when spores are overcrowded, a process thought to ensure efficient substrate colonization.
Crowding effect	The delay and inhibition of spore germination resulting from high spore concentrations (Trinci and Whittaker 1968).
Growth regulating substances (GRS)	A term sometimes used as a synonym for "plant hormone." Examples include auxins and gibberellins in plants and possibly 10-oxo-trans-8-decenoic acid in fungi (Takahashi 1986; Mau et al. 1992).
Chemical ecology	The field that studies how plants, animals, and microbes use chemicals to communicate in natural ecosystems.
Hormone	A term originally used by endocrinologists to describe chemicals secreted by a gland or group of cells in vertebrates to regulate specific physiological processes within the organism. The definition has been expanded to describe chemicals in other organisms that control and regulate growth and development within the same organism as the compound is secreted.
Infochemical	Information-conveying chemical: "a chemical that, in the natural context, conveys information in an interaction between two individuals, evoking in the receiver a behavioral or physiological response that is adaptive to either one of the interactants or to both" (Dicke and Sabelis 1988).
Mushroom alcohol	The common name of 1-octen-3-ol, one of the most prevalent fungal VOCs. Another synonym is matsutake alcohol.
PGPR	Plant growth promoting bacteria.
Psi factors	"Precocious sexual inducers" first identified from *A. nidulans* by Champe et al. (1987). Champe and el-Zayat (1989) also used the term "sexual sporulation hormone." These oxylipin compounds are involved in several fungal developmental cascades and are classified according to the fatty acids from which they are derived (Tsitsigiannis and Keller 2007; Brodhun and Feussner 2011).
Quorum sensing	Ability of bacterial populations to communicate and coordinate their behavior mediated by the use of chemical signaling molecules (Fuqua et al. 1994; Bassler and Losick 2006). Quorum sensing was originally called "autoinduction" (Nealson and Hastings 1979).
1-octen-3-ol	An oxylipin C-8 alcohol that is produced by numerous fungi, as well as by many plants, and that also can be emitted by certain arthropods and mammals. It regularly acts as a semiochemical. Depending on the scientific context and literature, it is variously called a pheromone, a kairomone, a synomone or by other jargon terms.
Oxylipin	A large family of structurally related oxygenated polyenoic fatty acids and the metabolites derived from them. Many oxylipins have physiological activities. They are abundant in mammals but also serve as signals of intra- and inter-cellular communication in other vertebrates, invertebrates, plants and fungi. In microbes, they regulate growth, differentiation, and apoptosis in addition to the

(continued)

Table 18.1. (continued)

Term	Definition and citations
	development of the infectious processes caused by some pathogenic microorganisms (Noverr and Erb-Downward 2003; La Camera et al. 2004; Tsitsigiannis and Keller 2007).
Pheromone	An "ecto-hormone", a chemical signal mediating interactions between organisms of the *same* species (Hooper and Pickett 2004). A "substance that is secreted by an organism to the outside and causes a specific reaction in a receiving organism of the same species" (Karlson and Luscher 1959; Nordlund and Lewis 1976). Examples include the trail pheromones and sex pheromones that have been well documented in insects.
Semiochemical	"A chemical involved in the chemical interaction between organisms" (Nordlund and Lewis 1976).

micals play in biological signaling. Chemical signaling occurs within individual organisms, between individuals of the same species, and also between different species. As a class, chemical signaling molecules sometimes have been termed "infochemicals" or "semiochemicals." The latter term is used especially by entomologists. Chemical signaling includes both long- and short-range cellular communication. A major distinction is **that pheromones mediate *intra*specific interactions while allelochemicals mediate *inter*specific interactions.** Entomologists and chemical ecologists have been at the forefront of studying chemical signaling molecules and have developed an extensive vocabulary for allelochemicals and other semiochemicals (Table 18.1). Only some of the infochemicals studied to date are found in the gas phase, but the concepts and definitions are useful whatever the physical states of the signaling molecule.

It is important to note that **a single chemical can function in more than one type of interaction; the functions are not mutually exclusive** (Nordlund and Lewis 1976). For example, the nearly ubiquitous fungal volatile, 1-octen-3-ol, depending on the context and biological system, is described as a hormone, pheromone, or allelochemical.

A. Entomology

Entomologists have been pioneers of chemical ecology so it is not surprising that the **best known cases of VOC signaling involve arthropods.** A complete survey of this topic would require an encyclopedia of its own; only a few examples are given here as an introduction to this fascinating field. The articles cited give access to the extensive literature describing chemical communication with an emphasis on fungal gas phase molecules.

Many fungal volatiles, particularly eight-carbon alcohols and ketones, act as semiochemicals and function in insect attraction and deterrence (Dowd and Bartelt 1991; Pierce et al. 1991a, b; Nilssen 1998; Ramoni et al. 2001; Ômura et al. 2002; Luntz 2003). In particular, many blood-feeding insects have highly developed olfactory systems whereby volatile semiochemicals enter the antennae and later bind to olfactory receptor neurons (Weeks et al. 2011). C-8 alcohols emitted by decay fungi may serve as trophic signals for conifer feeding bark beetle prey and as by-pass trophic signals for their predators (Zhang and Schlyter 2004). As assayed using a pitfall olfactometer test, cucijin grain beetle showed strong attraction to 1-octen-3-ol (Pierce et al. 1991b).

Many VOCs are involved in arthropod defense. For example, a combination of 1-octen-3-ol and geosmin functioned as defensive allomones in millipedes (Ômura et al. 2002). 1-octen-3-ol also interrupts the response of beetles to their aggregation hormone (Poland et al. 2009). Most true bugs (Heteropetera) have scent glands in what has been described as "overwhelming chemical fortification" (Aldrich 1988). Moreover, many species secrete attractant pheromones. For example, various bugs secrete (E)-2-hexanal, hexanal, hexanoic acid,

2-butyl-2-octenal, β-pinene, limonene, and farnesenes as alarm or trail pheromones for defense against their main predators (Aldrich 1988). For those wishing an introduction to the fascinating world of insect defensive secretions, many of which utilize noxious smelling VOCs such as caprylic acid, we recommend Thomas Eisner's beautifully illustrated book *For Love of Insects* (Eisner 2003).

Fungal odors are known to attract insects that live on fungi or on substrates that are decayed by fungi (Fäldt et al. 1999). In one interesting case in point, female houseflies do not lay their eggs on animal feces that have been colonized by *Fusarium*, *Phoma*, *Rhizopus*, and other fungi. Using chemical standards, six fungal VOCs that consistently inhibit oviposition were determined: dimethyl disulfide, phenylacetaldehye, 2-penylethanol, cintronellal, and norphytone (Lam et al. 2010).

When the chemical composition of insect attractants is known, in some cases effective artificial pheromones can be prepared to trap insect pests. Moreover, "elucidation of chemical messengers for predatory and weed-feeding bugs may lead to pheromonal husbandry of these beneficial insects" (Aldrich 1988).

B. Other Associations

Plants and microbes produce a variety of chemicals that are used in deterring competitors. Humans categorize some of these chemical defenses as antibiotics (if they selectively kill microbes that we wish to see dead) or toxins (if they kill us or our favored plant or animal species). Many of the best known classical secondary metabolites such as alkaloids are thought to have evolved as defense mechanisms. Plants have exquisitely refined chemical defense mechanisms against insect herbivory, including indirect defenses whereby they emit volatiles that attract natural enemies (Howe and Jander 2008). A survey of 1,500 fungal secondary metabolites published between 1993 and 2001 showed that more than half had antibacterial, antifungal, or antitumor activity (Pelaez 2005). The literature on fungal secondary metabolites is extensive and has been reviewed elsewhere (Turner 1971; Turner and Aldridge 1983; Cole and Schweikert 2003). Far less is known about deterrence activity of plant, fungal, and bacterial VOCs. Nevertheless, a few fungal VOCs are known to have growth inhibitory effects on plants (Splivallo et al. 2007b). When *A. thaliana* is exposed to 1-octen-3-ol, a major fungal VOC, the defense genes that are associated with wounding or ethylene and jasmonic acid signaling are turned on (Kishimoto et al. 2007). Moreover, plants exposed to 1-octen-3-ol are able to inhibit the expansion of the pathogen, *Botrytis cinerea*, on the infected leaves (Kishimoto et al. 2007). In addition, some soil bacteria emit antifungal VOCs (Liu et al. 2008).

Many fungal species engage in symbiotic interactions with plants as mycorrhizae, endophytes, and lichens. In addition, large numbers of fungi are plant pathogens. While fewer fungi cause diseases in animals, the ones that affect humans are becoming increasingly important in modern medicine. In all these interspecific interactions, chemical conversations are taking place, many of them mediated by VOCs. Because volatile compounds can be sensed at a distance they are ideal for serving as intra- and inter-specific chemical signals or "infochemicals."

Phytopathologists have long recognized the importance of molecular signaling during plant–pathogen interactions. In the rhizosphere, multiple species of bacteria and fungi interact with plant roots and their exudates. Moreover, it is becoming increasingly apparent that non-pathogenic microbes such as nitrogen-fixing bacteria and mycorrhizal fungi that establish beneficial relationships with plants rely on complex signaling networks (Oldroyd and Downie 2004; Harrison 2005). Many of the other chapters in this book describe classical and molecular approaches to studying such fungal associations and Ortiz-Castro et al. (2009) have reviewed the role of microbial signals in plant growth and development.

Soil remains an underexplored habitat that contains a rich diversity of microbial life forms. The relevance of secondary metabolites and volatiles in soil ecosystems has been reviewed in the monograph edited by (Karlovsky 2008). It has been estimated that **a single gram of soil may contain tens of thousands of different fungal, bacterial, archaeal, and protist species.** The soil properties such as nutrient and oxygen availability and the physiological state of the microbial population make dynamic and unique soil-specific communities (McNeal and Herbert 2009). Multispecies microbial populations play a key role in sustaining soil microcosms. **Communication within rhizosphere bacterial populations, mycelial colonies, and between fungi and bacteria is mediated by signaling molecules**, but we are only beginning to learn the specifics of this chemical information exchange. **Gas phase molecules are essential components of these chemical conversations.** These compounds can diffuse a long way from their point of origin, and they can persist and migrate in soil environments, areas of dense vegetation, and other microhabitats that harbor interacting populations of bacteria and fungi.

Much of the molecular work on biofilm development has been done in the laboratory in single-species biofilms in the context of quorum-sensing research. However, in nature, biofilms are multispecies communities that can harbor several hundred species where various forms of intercellular signaling are being used. By definition, the specialized vocabulary of "autoinducers" is not appropriate in biofilm communities; nevertheless, many of the same signaling molecules and pathways are in use (Kolter and Greenberg 2006).

The factors that influence **sporocarp development** in the cultivated mushroom *Agaricus bisporus* have been studied widely. **Primordium formation depends on: (i) the presence of a "casing layer" containing appropriate bacteria and (ii) sufficient ventilation.** The air exchange is important because it removes inhibitory VOCs, especially 1-octen-3-ol produced by *A. bisporus*, but also 3-ethyl-1-hexanol produced by the rye on which the *A. bisporus* spawn is grown. Pseudomonad populations in the casing metabolize 3-ethyl-1-hexanol, thus removing its inhibitory effect (Noble et al. 2009).

In an unusual "Trojan horse" form of bacterial pathogenesis against nematodes, *Bacillus nematocida* B16 lures nematodes by emitting several volatile organic compounds including benzaldehyde and 2-heptanone that are much more attractive to worms than those from ordinary dietary bacteria. Once the bacteria enter the intestine of nematodes, they secrete proteases that preferentially target essential intestinal proteins, leading to nematode death (Niu et al. 2010).

As these tantalizing cases illustrate, the effect of bacterial and fungal VOCs both in host–pathogen interactions and in non-pathogen ecological contexts is an emerging frontier for future research.

To sum up, **VOCs produced by a given fungal or bacterial species can have multiple effects on other microbes and organisms and can be used for defense, environmental monitoring, and nutrient acquisition** (Wheatley et al. 1997; Bruce et al. 2004; Minerdi et al. 2009). We agree with Tarkka and Piechulla (2007) that it is likely that there are many more multifunctional and multiorganismic volatile-based interactions in ecosystems than have been envisioned in the past. As the analytical methods for assaying VOCs improve, we will find increasing evidence of the crucial importance of VOC-mediated cross-talk between species, not only for fungal–bacterial associations, but also for fungal–plant and fungal–animal associations.

VII. Summary

Bacterial and fungal VOCs have been studied by scientists from a broad range of subdisciplines for both theoretical and applied reasons. As a result, the published information about VOCs exists scattered widely amongst a diverse scientific literature. In general, entomologists do not cross-reference the microbiological literature, nor do food and flavor chemists pay attention to the physiological activities of the compounds they study. This essay has attempted to highlight some of the major ways in which microbial VOCs have been studied to date and to give examples of the ways in which communication in biological systems is mediated chemically. Numerous plant–fungal, insect–fungal, and bacterial–fungal interactions that involve VOC signaling have been discovered. Many more remain to be elucidated. Technological advances with respect to our ability to detect VOCs and our capacity to monitor their physiological effects in plant, animal, and microbial systems will open the door to new views of the complexities underlying chemical signaling across the biosphere.

The scientific community has just begun to eavesdrop on the chemical discourse conducted via gas phase molecules. As we learn more about chemical signaling, it is becoming increasingly apparent that the significance of biogenic VOCs, including fungal VOCs, has been underestimated. It is our hope that this essay will elicit the interest of fungal biologists to ask questions such as:

- How do microbes use VOCs as intraspecific developmental signals and as interspecific signaling cues?
- In what way can we incorporate the insights of the entomological community into the mindset of bacteriologists and mycologists?
- What kinds of technological tools are needed to expand the study of VOCs?
- By what means can VOC research provide a platform for productive interdisciplinary collaborations?
- How many common volatile compounds like ammonia and 1-octen-3-ol serve unsuspected physiological activities?
- Or to put it simply: why do fungi and bacteria smell?

Acknowledgements We thank Arati Inamdar, James Mauro, Prakash Masurekar, Shannon Morath, David Pu, and Alisa Schink for their intellectual insights on fungal VOCs; we thank Natalie Naranjo and Shannon Morath for their help with the references, Karl Esser for his mentorship, Berthold Hock for his editorial support, and the Rutgers University Research Fund for financial support.

References

Abramson D, Sinha RN, Mills JT (1980) Mycotoxin and odor formation in moist cereal grain during granary storage. Cereal Chem 57:346–351

Abramson D, Sinha RN, Mills JT (1983) Mycotoxin and odor formation in barley stored at 16 and 20 % moisture in Manitoba. Cereal Chem 60:350–355

Agrios GN (2008) Plant pathology, 5th edn. Academic, San Diego

Aldrich JR (1988) Chemical ecology of the heteroptera. Annu Rev Entomol 33:211–238

Allen PJ (1957) Properties of a volatile fraction of uredospores of *Puccinia graminis* var. *tritici* affecting their germination and development. I. Biological activity. Plant Physiol 32:385–389

Arora K, Chand S, Malhotra BD (2006) Recent developments in bio-molecular electronics techniques for food pathogens. Anal Chim Acta 568:259–274

Assaf S, Hadar Y, Dosoretz CG (1997) 1-Octen-3-ol and 13-hydroperoxylinoleate are products of distinct pathways in the oxidative breakdown of linoleic acid by Pleurotus pulmonarius. Enzyme Microb Technol 21:484–490

Atmosukarto I, Castillo U, Hess W, Sears J, Strobel G (2005) Isolation and characterization of *Muscodor albus* I-41.3s, a volatile antibiotic producing fungus. Plant Sci 169:854–861

Bacon CW, White JW (eds) (2000) Microbial endophytes. Dekker, New York

Baldwin IT, Halitschke R, Paschold A, von Dahl CC, Preston CA (2006) Volatile signaling in plant-plant interactions: "talking trees" in the genomics era. Science 311:812–815

Banchio E, Xie X, Zhang H, Pare PW (2009) Soil bacteria elevate essential oil accumulation and emissions in sweet basil. J Agric Food Chem 57:653–657

Bassler BL, Losick R (2006) Bacterially speaking. Cell 125:237–246

Bennett JW (1983) Differentiation and secondary metabolism in mycelial fungi. In: Bennett JW, Ciegler A (eds) Secondary metabolism and differentiation in fungi. Dekker, New York, pp 1–32

Bennett JW, Bentley R (1989) What's In a name? – microbial secondary metabolism. Adv Appl Microbiol 34:1–28

Bennett JW, Feibelman T (2001) Fungal bacterial interactions. In: Hock B (ed) The mycota, vol IX. Springer, Berlin, pp 229–240

Bentley R, Maganathan R (1981) Geosmin and methylisoborneol biosynthesis in Streptomycetes. Evidence for an isoprenoid pathway and its absence in non-differentiating isolates. FEBS Lett 125:220–222

Berg JM, Tymoch JL, Stryer L (2007) Biochemistry. Freeman, New York

Berger RG, Drawert F, Tiefel P (1992) Naturally occurring flavours from fungi, yeasts, and bacteria. In: Patterson RLS, Charlwood BV, MacLeod G, Williams AA (eds) Bioinformation of flavours. Royal Chemistry Society, Cambridge, pp 1–20

Bloch E, Deorazio R (1994) Chemistry in a salad bowl: comparative organosulfur chemistry of garlic, onion, and shitake mushrooms. Pure Appl Chem 66:2205–2206

Bohbot JD, Dickens JC (2009) Characterization of an enantioselective odorant receptor in the yellow fever mosquito *Aedes aegypti*. PLoS One 4:7032

Borg-Karlson A-K, Englund F, Unelius CR (1994) Dimethyl oligosulphides, major volatiles released from *Sauromatum guttatum* and *Phallus impudicus*. Phytochemistry 35:321–323

Börjesson T, Stöllman UM, Adamek P, Kaspersson A (1989) Analysis of volatile compounds for detection of molds in stored cereals. Cereal Chem 66:300–304

Börjesson T, Stöllman UM, Schnürer J (1993) Off-odorous compounds produced by molds on oatmeal agar: identification and relation to other growth characteristics. J Agric Food Chem 41:2104–2111

Breheret S, Talou T, Rapior S, Bessiere J-M (1997) Monoterpenes in the aromas of fresh wild mushrooms. J Agric Food Chem 45:831–836

Brodhagen M, Tsitsigiannis DI, Hornung E, Goebel C, Feussner I, Keller NP (2008) Reciprocal oxylipin-mediated cross-talk in the *Aspergillus*-seed pathosystem. Mol Microbiol 67:378–391

Brodhun F, Feussner I (2011) Oxylipins in fungi. FEBS J 278:1047–1063

Brown WL (1968) An hypothesis concerning the function of the metapleural glands in ants. Am Nat 102:188–191

Bruce A, Verrall S, Hackett CA, Wheatley RE (2004) Identification of volatile organic compounds (VOCs) from bacteria and yeast causing growth inhibition of sapstain fungi. Holzforschung 58:193–198

Burge PS (2004) Studies on the role of fungi in sick building syndrome. Occup Environ Med 61:185–190

Camilli A, Bassler BL (2006) Bacterial small-molecule signaling pathways. Science 311:1113–1116

Champe SP, el-Zayat AA (1989) Isolation of a sexual sporulation hormone from *Aspergillus nidulans*. J Bacteriol 171:3982–3988

Champe SP, Rao P, Chang A (1987) An endogenous inducer of sexual development in *Aspergillus nidulans*. J Gen Microbiol 133:1383–1387

Chiron N, Michelot D (2005) Odeurs de champignons: chimie et rôle dans les interactions biotiques – une revue. Cryptogam Mycol 26:299–364

Chitarra GS, Abee T, Rombouts FM, Posthumus MA, Dijksterhuis J (2004) Germination of *Penicillium paneum* conidia is regulated by 1-octen-3-ol, a volatile self-inhibitor. Appl Environ Microbiol 70:2823–2829

Chitarra GS, Abee T, Rombouts FM, Dijksterhuis J (2005) 1-Octen-3-ol inhibits conidia germination of *Penicillium paneum* despite of mild effects on membrane permeability, respiration, intracellular pH, and changes the protein composition. FEMS Microbiol Ecol 54:67–75

Cho IH, Namgung H-J, Choi H-K, Kim YS (2008) Volatiles and key odorants in the pileus and stipe of pine-mushroom (*Tricholoma matsutake* sing). Food Chem 106:71-76

Choudhary DK, Johri BN, Prakash A (2008) Volatiles as priming agents that initiate plant growth and defence responses. Curr Sci 94:595–604

Claeson A-S, Levin J-O, Gr B, Sunesson A-L (2002) Volatile metabolites from microorganisms grown on humid building materials and synthetic media. J Environ Monit 4:667–672

Clough SJ, Schell MA, Denny TP (1994) Evidence for involvement of a volatile extracellular factor in *Pseudomonas solonacearum* virulence gene expression. MPMI 7:621–630

Cole R, Schweikert M (2003) Handbook of secondary fungal metabolites, vol 1–3. Academic, Amsterdam

Combet E, Henderson J, Eastwood DC, Burton KS (2006) Eight-carbon volatiles in mushrooms and fungi: properties, analysis, and biosynthesis. Mycoscience 47:317–326

Cronin DA, Ward MK (1971) The characterisation of some mushroom volatiles. J Sci Food Agric 22:477–479

de Pinho PG, Ribeiro B, Goncalves RF, Baptista P, Valentao P, Seabra RM, Andrade PB (2008) Correlation between the pattern volatiles and the overall aroma of wild edible mushrooms. J Agric Food Chem 56:1704–1712

Dicke M, Sabelis MW (1988) Infochemical terminology: based on cost-benefit analysis rather than origin of compounds? Funct Ecol 2:131–139

Dickschat JS, Martens R, Brinkhoff T, Simon M, Schulz S (2005a) Volatiles releases by *Streptomyces* species isolated from the North Sea. Chem Biodivers 2:837–865

Dickschat JS, Wenzel SC, Bode HB, Müller R, Schulz S (2005b) Biosynthesis of volatiles by the Myxobacterium *Myxococcus xanthus*. Chembiochem 5:778–787

Dowd PF, Bartelt RJ (1991) Host-derived volatiles as attractants and pheromone synergists for dried fruit beetle, *Carpophilus hemipterus*. J Chem Ecol 17:285–308

Dunkel M, Schmidt U, Struck S, Berger L, Gruening B, Hossbach J, Jaeger IS, Effmert U, Piechulla B, Eriksson R, Knudsen J, Preissner R (2009) SuperScent – a database of flavors and scents. Nucleic Acids Res 37(Database Issue):D291–D294

Eberhard A, Burlingame AL, Eberhard C, Kenyon GL, Nealson KH, Oppenheimer NJ (1981) Structural identification of autoinducer of *Photobacterium fischeri* luciferase. Biochemistry 20:2444–2449

Eisner T (2003) For love of insects. Harvard University Press, Cambridge

Ellis DI, Broadhurst D, Kell DB, Rowland JJ, Goodacre R (2002) Rapid and quantitative detection of the microbial spoilage of meat by Fourier transform infrared spectroscopy and machine learning. Appl Environ Microbiol 68:2822–2828

Fäldt J, Jonsell M, Nordlander G, Borg-Karlson A-K (1999) Volatiles of bracket fungi *Fomitopsis pinicola* and *Fomes fomentarius* and their functions as insect attractants. J Chem Ecol 25:567–590

Farag MA, Ryu CM, Sumner LW, Pare PW (2006) GC-MS SPME profiling of rhizobacterial volatiles reveals prospective inducers of growth promotion and induced systemic resistance in plants. Phytochemistry 67:2262–2268

Fernando WGD, Ramarathnam R, Krishnamoorthy AS, Savchuk SC (2005) Identification and use of potential bacterial organic antifungal volatiles in biocontrol. Soil Biol Biochem 37:955–964

Fiedler N, Laumbach R, Kelly-McNeil K, Lioy P, Fan Z-H, Zhang J, Ottenweller J, Ohman-Strickland P, Kipen H (2005) Health effects of a mixture of indoor air volatile organics, their ozone oxidation products, and stress. Environ Health Perspect 113:1542–1548

Fischer G, Schwalbe R, Möller M, Ostrowski R, Dott W (1999) Species-specific production of microbial volatile organic compounds (MVOC) by airborne fungi from a compost facility. Chemosphere 39:779–810

Flavier AB, Ganova-Raeva LM, Schell MA, Denny TP (1997) Hierarchial autoinduction in Ralstonia solanacearum: control of actyl- homoserine lactone production by a novel autoregulatory system responsive to 30- hydroxypalmitic acid methyl ester. J Bacteriol 179:7089–7097

Fraatz MA, Zorn H (2010) Fungal flavours. In: Hofrichter M (ed) The Mycota X: industial applications, vol X, 2nd edn, Industrial applications. Springer, Berlin Heidelberg New York, pp 249–264

Fuqua WC, Winans SC, Greenberg EP (1994) Quorum sensing in bacteria: the LuxR-LuxI family of cell density-responsive transcriptional regulators. J Bacteriol 176:269–275

Gallois A, Langlois D (1990) New results in the volatile odorous compounds of French cheeses. Lait 70:89–106

Griffin MA, Spakowicz DJ, Gianoulis TA, Strobel SA (2010) Volatile organic compound production by organisms in the genus *Ascocoryne* and a re-evaluation of myco-diesel production by NRRL 50072. Microbiology 156:3814–3829

Griffith RT, Jayachandran K, Shetty KG, Whitstine W, Furton KG (2007) Differentiation of toxic molds via headspace SPME-GC/MS and canine detection. Sensors 7:1496–1508

Gutiérrez-Luna FM, López-Bucio J, Altamirano-Hernndez J, Valencia-Cantero E, Cruz HR, Macías-Rodríguez L (2010) Plant growth-promoting rhizobacteria modulate root-system architecture in *Arabidopsis thaliana* through volatile organic compound emission. Symbiosis 51:75–83

Harrison MJ (2005) Signaling in the arbuscular mycorrhizal symbiosis. Annu Rev Microbiol 59:19–42

Herrero-Garcia E, Garzia A, Cordobés S, Espeso EA, Ugalde U (2011) 8-Carbon oxylipins inhibit germination and growth, and stimulate aerial conidiation in *Aspergillus nidulans*. Fungal Biol 115:393–400

Hogan DA (2006) Talking to themselves: autoregulation and quorum sensing in fungi. Eukaryot Cell 5:613–619

Hooper AM, Pickett JA (2004) Semiochemistry. In: Atwood JL, Steed JW (eds) Encyclopedia of supramolecular chemistry, vol 2. Dekker, New York, pp 1270–1277

Horswill A, Stoodley P, Stewart P, Parsek M (2007) The effect of the chemical, biological, and physical environment on quorum sensing in structured microbial communities. Anal Bioanal Chem 387:371–380

Howe GA, Jander G (2008) Plant immunity to insect herbivores. Annu Rev Plant Biol 59:41–66

Hutchinson SA (1973) Biological activities of volatile fungal metabolites. Annu Rev Phytopathol 11:223–246

Inamdar AA, Masurekar P, Bennett JW (2010) Neurotoxicity of fungal volatile organic compounds in *Drosophila melanogaster*. Toxicol Sci 117:418–426

Inamdar AA, Moore JC, Cohen RI, Bennett JW (2011) A model to evaluate the cytotoxicity of the fungal volatile organic compound 1-octen-3-ol in human embryonic stem cells. Mycopathologia 173:13–20

IOM (2004) Damp indoor spaces and health. National Academies, Washington

Jarvis BB, Miller JD (2005) Mycotoxins as harmful indoor air contaminants. Appl Microbiol Biotechnol 66:367–372

Jeleń HH (2003) Use of solid phase microextraction (SPME) for profiling fungal volatile metabolites. Lett Appl Microbiol 36:263–267

Joblin Y, Moularat S, Anton R, Bousta F, Orial G, Robine E, Picon O, Bourouina T (2010) Detection of moulds by volatile organic compounds; application to heritage conservation. Int Biodeterior Biodegrad 64:210–217

Kai M, Haustein M, Molina F, Petri A, Scholz B, Piechulla B (2009) Bacterial volatiles and their action potential. Appl Microbiol Biotechnol 81:1001–1012

Kaminski E, LIbbey LM, Stawicki S, Wasowicz E (1972) Identification of the predominant volatile compounds produced by *Aspergillus flavus*. Appl Microbiol 24:721–726

Kaplan HB, Greenberg EP (1985) Diffusion of autoinducer is involved in regulation of the *Vibrio fischeri* luminescence system. J Bacteriol 163:1210–1214

Karahadian C, Josephson DB, Lindsay RC (1985) Contribution of *Penicillium* sp. to the flavour of Brie and Camembert cheese. J Dairy Sci 68:1865–1877

Karlovsky P (ed) (2008) Secondary metabolites in soil ecology. Springer, Berlin Heidelberg New York

Karlshøj K, Nielsen PV, Larsen TO (2007) Fungal volatiles biomarkers of good and bad food quality. In: Dijksterhuis J, Samson RA (eds) Food mycology. CRC, Boca Raton, pp 279–302

Karlson P, Luscher M (1959) 'Pheromones': a new term for a class of biologically active substances. Nature 183:55–56

Keller NP, Turner G, Bennett JW (2005) Fungal secondary metabolism – from biochemistry to genomics. Nat Rev Microbiol 3:937–947

Kesselmeier J, Staudt M (1999) Biogenic volatile organic compounds (VOC): an overview on emission, physiology and ecology. J Atmos Chem 33:23–88

Kishimoto K, Matsui K, Ozawa R, Takabayashi J (2007) Volatile 1-octen-3-ol induces a defensive response in *Arabidopsis thaliana.* J Gen Plant Pathol 73:35–37

Kline D, Allan SA, Bernier UR, Welch CH (2007) Evaluation of the enantiomers of 1-octen-3-ol and 1-octyn-3-ol as attractants for mosquitoes associated with a freshwater swamp in Florida, USA. Med Vet Entomol 21:323–331

Kolter R, Greenberg EP (2006) Microbial sciences: the superficial life of microbes. Nature 441:300–302

Korpi A, Jarnberg J, Pasanen A-L (2009) Microbial volatile organic compounds. Crit Rev Toxicol 39:139–193

Kües U, Navarro-Gonzlez M (2009) Communication of fungi on individual, species, kingdom, and above kingdom levels. In: Anke T, Weber D (eds) The Mycota XV. Physiology and genetics. Springer, Berlin Heidelberg New York, pp 79–106

Kuske M, Romain A-C, Nicolas J (2005) Microbial volatile organic compounds as indicators of fungi. Can an electronic nose detect fungi in indoor environments? Build Environ 40:824–831

La Camera S, Gouzerh G, Sandrine D, Laurent H, Bernard F, Michel F, Thierry H (2004) Metabolic reprogramming in plant innate immunity: the contributions of phenylpropanoid and oxylipin pathways. Immunol Rev 198:267–284

Lam K, Tsang M, Labrie A, Gries R, Gries G (2010) Semiochemical-mediated oviposition avoidance by female house flies, *Musca domestica*, on animal feces colonized with harmful fungi. J Chem Ecol 36:141–147

Larsen TO, Frisvad JC (1995) Comparison of different methods for collection of volatile chemical markers from fungi. J Microbiol Methods 24:135–144

Lax AR, Templeton GE, Meyer WL (1985) Isolation, purification, and biological activity of a self-inhibotor from conidia of *Colletotrichum gloeosporioides*. Phytopathology 75:386–390

Lee SO, Kim HY, Choi GJ, Lee HB, Jang KS, Choi YH, Kim JC (2009) Mycofumigation with *Oxyporus latemarginatus* EF069 for control of postharvest apple decay and *Rhizoctonia* root rot on moth orchid. J Appl Microbiol 106:1213–1219

Leeder AC, Palma-Guerrero J, Glass NL (2011) The social network: deciphering fungal language. Nat Rev Microbiol 9:440–451

Li DW, Yang CS (2004) Fungal contamination as a major contributor to sick building syndrome. Adv Appl Microbiol 55:31–112

Liu W, Mu W, Zhu B, Liu F (2008) Antifungal activities and components of VOCs produced by *Bacillus subtilis* G_8. Curr Res Bacteriol 1:28–34

Lugtenberg B, Kamilova F (2009) Plant-growth-promoting rhizobacteria. Annu Rev Microbiol 63:541–556

Luntz AJ (2003) Arthropod semiochemicals: mosquitoes, midges and sealice. Biochem Soc Trans 31:128–133

Mackie AE, Wheatley RE (1999) Effects and incidence of volatile organic compound interactions between soil bacterial and fungal isolates. Soil Biol Biochem 31:375–385

Mattheis JP, Roberts RG (1992) Identification of geosmin as a volatile metabolite of *Penicillium expansum.* Appl Environ Microbiol 58:3170–3172

Matysik S, Herbarth O, Mueller A (2008) Determination of volatile metabolites originating from mould growth on wall paper and synthetic media. J Microbiol Methods 75:182–187

Matysik S, Herbarth O, Mueller A (2009) Determination of microbial volatile organic compounds (MVOCs) by passive sampling onto charcoal sorbents. Chemosphere 76:114–119

Mau J-L, Beelman RB, Ziegler GR (1992) Effect of 10-oxo-trans-8-decenoic acid on growth of *Agaricus bisporus.* Phytochemistry 31:4059–4064

Mau J-L, Chyau C-C, Li J-Y, Tseng Y-H (1997) Flavor compounds in straw mushrooms *Volvariella volvacea* harvested at different stages of maturity. J Agric Food Chem 45:4726–4729

Mauriello G, Marino R, D'Auria M, Cerone G, Rana GL (2004) Determination of volatile organic compounds from truffles via SPME-GC-MS. J Chromatogr Sci 42:299–305

McFee DR, Zavon P (1988) Solvents. In: Plog BA (ed) Fundamentals of industrial hygiene, 3rd edn. National Safety Council, Chicago, pp 95–121

McNeal KS, Herbert BE (2009) Volatile organic metabolites as indicators of soil microbial activity and community composition shifts. Soil Sci Soc Am J 73:579–588

Meilgaard MC (1975a) Flavor chemistry of beer. I. Flavor interaction between principal volatiles. Tech Q Master Brewers Assoc Am 12:107–117

Meilgaard MC (1975b) Flavor chemistry of beer. II. Flavor and threshold of 239 aroma volatiles. Tech Q Master Brewers Assoc Am 12:151–168

Mercier J, Jiménez JI (2004) Control of fungal decay of apples and peaches by the biofumigant fungus *Muscodor albus.* Postharvest Biol Tec 31:1–8

Mercier J, Manker D (2005) Biocontrol of soil-borne diseases and plant growth enhancement in greenhouse soilless mix by the volatile-producing fungus *Muscodor albus.* Crop Prot 24:355–362

Minerdi D, Bossi S, Gullino ML, Garibaldi A (2009) Volatile organic compounds: a potential direct long-distance mechanism for antagonistic action of *Fusarium oxysporum* strain MSA 35. Environ Microbiol 11:844–854

Mølhave L (2009) Volatile organic compounds and the sick building syndrome. In: Lippmann M (ed) Environmental toxicants: human exposures and their health effects, 3rd edn. Wiley-Interscience, New York, pp 241–256

Morey P, Wortham A, Weber A, Horner E, Black M, Muller W (1997) Microbial VOCs in moisture damaged buildings. Health Build 1:245–250

Mosandl A, Heusinger G, Gessner M (1986) Analytical and sensory differentiation of 1-octen-3-ol enantiomers. J Agric Food Chem 34:119–122

Nealson KH, Hastings JW (1979) Bacterial bioluminescence: its control and ecological significance. Microbiol Rev 43:496–518

Nealson KH, Platt T, Hastings JW (1970) Cellular control of the synthesis and activity of the bacterial luminescent system. J Bacteriol 104:313–322

Nemcovic M, Jakubikova L, Viden I, Farkas V (2008) Induction of conidiation by endogenous volatile compounds in *Trichoderma* spp. FEMS Microbiol Lett 284:231–236

Nilssen AC (1998) Effect of 1-octen-3-ol in field trapping *Aedes spp.* (Dipt., Culicidae) and *Hybomitra spp.* (Dipt., Tabanidae) in subartic Norway. J Appl Entomol 122:465–468

Nilsson A, Kihlstrom E, Lagesson V, Wessen B, Szponar B, Larsson L, Tagesson C (2004) Microorganisms and volatile organic compounds in airborne dust from damp residences. Indoor Air 14:74–82

Niu Q, Huang X, Zhang L, Xu J, Yang D, Wei K, Niu X, An Z, Bennett JW, Zou C, Yang J, Zhang KQ (2010) A Trojan horse mechanism of bacterial pathogenesis against nematodes. Proc Natl Acad Sci USA 107:16631–16636

Noble R, Dobrovin-Pennington A, Hobbs PJ, Pederby J, Rodger A (2009) Volatile C8 compounds and pseudomonads influence primordium formation of *Agaricus bisporus*. Mycologia 101:583–591

Nordlund DA, Lewis WJ (1976) Terminology of chemical releasing stimuli in intraspecific and interspecific interactions. J Chem Ecol 2:211–220

Noverr MC, Erb-Downward JR (2003) Production of eicosanoids and other oxylipins by pathogenic eukaryotic microbes. Clin Microbiol Rev 16:517–533

Okull DO, Beelman RB, Gourama H (2003) Antifungal activity of 10-oxo-trans-8-decenoic acid and 1-octen-3-ol against *Penicillium expansum* in potato dextrose agar medium. J Food Prot 66:1503–1505

Oldroyd GED, Downie JA (2004) Calcium, kinases and nodulation signalling in legumes. Nat Rev Mol Cell Biol 5:566–576

Ômura H, Kuwahara Y, Tanabe T (2002) 1-Octen-3-ol together with geosmin: new secretion compounds from a polydesmid millipede, *Niponia nodulosa*. J Chem Ecol 28:2601–2612

Ortiz-Castro R, Contreras-Cornejo H, Macias-Rodriguez L, Lopez-Bucio J (2009) The role of microbial signals in plant growth and development. Plant Signal Behav 4:701–712

Palkova Z, Forstova J (2000) Yeast colonies synchronise their growth and development. J Cell Sci 113:1923–1928

Palkova Z, Janderova B, Gabriel J, Zikanova B, Pospisek M, Forstova J (1997) Ammonia mediates communication between yeast colonies. Nature 390:532–536

Palkova Z, Devaux F, Icicova M, Minarikova L, Le Crom S, Jacq C (2002) Ammonia pulses and metabolic oscillations guide yeast colony development. Mol Biol Cell 13:3901–3914

Pasanen P, Korpi A, Kalliokosi P, Pasanen AL (1997) Growth and volatile metabolite production of *Aspergillus versicolor* in house dust. Environ Int 23:425–432

Pavlou AD, Turner AP (2000) Sniffing out the truth: clinical diagnosis using the electronic nose. Clin Chem Lab Med 38:99–112

Pelaez F (2005) Biological activities of fungal metabolites. In: An Z (ed) Handbook of industrial mycology. Dekker, New York, pp 49–92

Pierce AM, Pierce HD, Borden JH, Oehlschlager AC (1991a) Fungal volatiles: semiochemicals for stored-product beetles (*Coleoptera: Cucujidae*). J Chem Ecol 3:567–579

Pierce AM, Pierce HD, Oehlschlager AC, Borden JH (1991b) 1-Octen-3-ol, attractive semiochemical for foreign grain beetle, *Ahasverus adevna* (Waltl) (*Coleoptera: Cucujidae*). J Chem Ecol 3:567–579

Poland TM, Pureswaran DS, Ciaramitaro TM, Borden JH (2009) 1-Octen-3-ol is repellent to *Ips pini* (Coleoptera: Curculionidae: Scolytinae) in the Midwestern United States. Can Entomol 141:158–160

Ramoni R, VIncent F, Grolli S, Conti V, Malosse C, Boyer F, Nagnan-Le Meillour P, Spinelli S, Cambillau C, Tegoni M (2001) The insect attractant 1-octen-3-ol is the natural ligand of bovine odorant-binding protein. J Biol Chem 276:7150

Rapior S, Breheret S, Talou T, Pelissier Y, Bessiere JM (2002) The anise-like odor of *Clitocybe odora, Lentinellus cochleatus,* and *Agaricus essettie*. Mycologia 94:373–376

Robinson J (ed) (2006) The Oxford companion to wine, 3rd edn. Oxford, Oxford University Press

Rodriguez RJ, White JF, Arnold AE, Redman RS (2009) Fungal endophytes: diversity and functional roles. New Phytol 182:314–330

Rude MA, Schirmer A (2009) New microbial fuels: a biotech perspective. Curr Opin Microbiol 12:274–281

Rumbaugh KP, Griswold JA, Hamood AN (2000) The role of quorum sensing in the in vivo virulence of *Pseudomonas aeruginosa*. Microbes Infect 2:1721–1731

Ryu C-M, Farag MA, Hu C-H, Reddy MS, Wei H-X, Paré PW, Kloepper JW (2003) Bacterial volatiles promote growth in *Arabidopsis*. Proc Natl Acad Sci USA 100:4927–4932

Ryu C-M, Farag MA, Paré PW, Kloepper JW (2005) Invisible signals from the underground: bacterial volatiles elicit plant growth promotion and induce systemic resistance. Plant Pathol J 21:7–12

Schnürer J, Olsson J, Börjesson T (1999) Fungal volatiles as indicators of food and feeds spoilage. Fungal Genet Biol 27:209–217

Schöller CEG, Gürtler H, Petersen R, Molin S, Wilkins K (2002) Volatile metabolites from Actinomycetes. J Agric Food Chem 50:2615–2621

Schreier P (1992) Bioflavours: an overview. In: Patterson RLS, Charlwood BV, MacLeod G, Williams AA (eds) Bioinformation of flavours. Royal Society of Chemistry, Cambridge, pp 1–20

Schulz S, Dickschat JS (2007) Bacterial volatiles: the smell of small organisms. Nat Prod Rep 24:814–842

Shapiro JA (1998) Thinking about bacterial populations as multicellular organisms. Annu Rev Microbiol 52:81–104

Shimkets LJ (1999) Intercellular signaling during fruiting-body development of *Myxococcus xanthus*. Annu Rev Microbiol 53:525–549

Singh SK, Strobel GA, Knighton B, Geary B, Sears J, Ezra D (2011) An endophytic *Phomopsis* sp. possessing bioactivity and fuel potential with its volatile organic compounds. Microb Ecol 61:729–739

Sneeden EY, Harris HH, Pickering J, Prince RC, Johnson S, Li X, Block E, George GH (2004) The sulfur chemistry of shitake mushroom. J Am Chem Soc 126:458–459

Splivallo R, Bossi S, Maffei M, Bonfante P (2007a) Discrimination of truffle fruiting body versus mycelial aromas by stir bar sorptive extraction. Phytochemistry 68:2584–2598

Splivallo R, Novero M, Bertea CM, Bossi S, Bonfante P (2007b) Truffle volatiles inhibit growth and induce an oxidative burst in *Arabidopsis thaliana*. New Phytol 175:417–424

Stinson M, Ezra D, Hess WM, Sears J, Strobel G (2003) An endophytic *Gliocladium* sp. of *Eucryphia cordifolia* producing selective volatile antimicrobial compounds. Plant Sci 165:913–922

Stoppacher N, Kluger B, Zeilinger S, Krska R, Schuhmacher R (2010) Identification and profiling of volatile metabolites of the biocontrol fungus *Trichoderma atroviride* by HS-SPME-GC-MS. J Microbiol Methods 81:187–193

Straus DC (2009) Molds, mycotoxins, and sick building syndrome. Toxicol Ind Health 25:617–635

Straus DC, Cooley JD, Wong WC, Jumper CA (2003) Studies on the role of fungi in sick building syndrome. Arch Environ Health 58:475–478

Strobel GA, Dirkse E, Sears J, Markworth C (2001) Volatile antimicrobials from *Muscodor albus* a novel endophytic fungus. Microbiology 147:2943–2950

Strobel GA, Knighton B, Kluck K, Ren Y, Livinghouse T, Griffin M, Spakowicz D, Sears J (2008) The production of myco-diesel hydrocarbons and their derivatives by the endophytic fungus *Gliocladium roseum* (NRRL 50072). Microbiology 154:3319–3328

Strobel GA, Singh SK, Riyaz-Ul-Hassan S, Mitchell AM, Geary B, Sears J (2011) An endophytic/pathogenic *Phoma* sp. from creosote bush producing biologically active volatile compounds having fuel potential. FEMS Microbiol Lett 320:87–94

Ström G, West J, Wessen B, Palmgren U (1994) Health implications of fungi in indoor environments: quantitative analysis of microbial volatiles in damp Swedish houses. Air Qual Monogr 2:291–305

Sunesson AL, Vaes WHJ, Nilsson CA, Blomquist G, Andersson B, Carlson R (1995) Identification of volatile metabolites from five fungal species cultivated on two media. Appl Environ Microbiol 61:2911–2918

Sunesson AL, Nilsson CA, Andersson B, Blomquist G (1996) Volatile metabolites produced by two fungal species cultivated on building materials. Ann Occup Hyg 40:397–410

Takahashi N (1986) Chemistry of plant hormones. CRC, Boca Raton

Tan RX, Zou WV (2001) Endophytes: a rich source of functional metabolites. Nat Prod Rep 18:448–459

Tarkka MT, Piechulla B (2007) Aromatic weapons: truffles attack plants by the production of volatiles. New Phytol 175:381–383

Thomson NR, Crow MA, McGowan SJ, Cox A, Salmond GPC (2000) Biosynthesis of carbapenem antibiotic and prodigiosin pigment in *Serratia* is under quorum sensing control. Mol Microbiol 36:539–556

Thorn J (2001) The inflammatory response in humans after inhalation of bacterial endotoxin: a review. Inflamm Res 50:254–261

Tirillini B, Verdelli G, Paolocci F, Ciccioli P, Frattoni M (2000) The volatile organic compounds from the mycelium of *Tuber borchii* Vitt. Phytochemistry 55:983–985

Trinci APJ, Whittaker C (1968) Self-inhibition of spore germination in *Aspergillus nidulans*. Trans Br Mycol Soc 51:594–596

Tsitsigiannis DI, Keller NP (2007) Oxylipins as developmental and host-fungal communication signals. Trends Microbiol 15:109–118

Tsurushima T, Ueno T, Fukami H, Irie H, Inoue M (1995) Germination self-inhibitors from *Colletotrichum gloeosporioides* f.sp *jussiaea*. Mol Plant Microbe Interact 8:652–657

Turner WB (1971) Fungal metabolites. Academic, London
Turner WB, Aldridge DC (1983) Fungal metabolites II. Academic, London
Van Delden C, Iglewski BH (1998) Cell-to-cell signaling and *Pseudomonas aeruginosa* infections. Emerg Infect Dis 4:551–560
Vespermann A, Kai M, Piechulla B (2007) Rhizobacterial volatiles affect the growth of fungi and *Arabidopsis thaliana*. Appl Environ Microbiol 73:5639–5641. doi:10.1128/AEM.01078-07
von Bodman SB, Bauer WD, Coplin DL (2003) Quorum sensing in plant-pathogenic bacteria. Annu Rev Phytopathol 41:455–482
Walinder R, Ernstgard L, Johanson G, Venge P, Wieslander G (2005) Acute effects of a fungal volatile compound. Environ Health Perspect 113:1775–1778
Waters CM, Bassler BL (2005) Quorum sensing: cell-to-cell communication in bacteria. Annu Rev Cell Dev Biol 21:319–346
Watson SB, Brownlee B, Satchwill T, Hargesheimer EE (2000) Quantitative analysis of trace levels of geosmin and MIB in source and drinking water using headspace SPME. Water Res 34:2818–2828
Weeks EN, Birkett MA, Cameron MM, Pickett JA, Logan JG (2011) Semiochemicals of the common bed bug, *Cimex lectularius* L. (Hemiptera: Cimicidae), and their potential for use in monitoring and control. Pest Manag Sci 67:10–20
Wheatley R, Hackett C, Bruce A, Kundzewicz A (1997) Effect of substrate composition on production of volatile organic compounds from *Trichoderma* spp. inhibitory to wood decay fungi. Int Biodeterior Biodegrad 39:199–205
Whipps JM (2001) Microbial interactions and biocontrol in the rhizosphere. J Exp Bot 52(Suppl 1):487–511
Whitehead NA, Barnard ANL, Slater H, Simpson NJL, Salmond GPC (2001) Quorum-sensing in Gram-negative bacteria. FEMS Microbiol Rev 25:365–404
Whittaker RH, Feeny PP (1971) Allelochemics: chemical interactions between species. Science 171:757–770
WHO (2009) Guidelines for indoor air quality: dampness and mold. Druckpartner Moser, Germany
Wilkins K, Larsen K, Simkus M (2000) Volatile metabolites from mold growth on building materials and synthetic media. Chemosphere 41:437–446
Wilson AD, Baietto M (2009) Applications and advances in electronic-nose technologies. Sensors 9:5099–5148
Wilson AD, Baietto M (2011) Advances in electronic-nose technologies developed for biomedical applications. Sensors 11:1105–1176
Wood WF, Fesler M (1986) Mushroom odors: student synthesis of the odoriferous compounds of the matsutake mushroom. J Chem Educ 63:92
Wurzenberger M, Grosch W (1984) The formation of 1-octen-3-ol from the 10-hydroperoxide isomer of linoleic acid by a hydroperoxide lyase in mushrooms (*Psalliota bispora*). Biochim Biophys Acta Lipids Lipid Metab 794:25–30
Xie X, Zhang H, Pare P (2009) Sustained growth promotion in *Arabidopsis* with long-term exposure to the beneficial soil bacterium *Bacillus subtilis* (GB03). Plant Signal Behav 4:948–953
Zhang Z, Li G (2010) A review of advances and new developments in the analysis of biological volatile organic compounds. Microchem J 95:127–139
Zhang QH, Schlyter F (2004) Olfactory recognition and behavioural avoidance of angiosperm nonhost volatiles by conifer-inhabiting bark beetles. Agric For Entomol 6:1–19
Zhang Y-Q, Wilkinson H, Keller NP, Tsitsigiannis D (2005) Secondary metabolite gene clusters. In: An Z (ed) Handbook of industrial microbiology. Dekker, New York, pp 355–386
Zhang H, Kim M-S, Krishnamachari V, Payton P, Sun Y, Grimson M, Farag M, Ryu C-M, Allen R, Melo I, Paré P (2007) Rhizobacterial volatile emissions regulate auxin homeostasis and cell expansion in *Arabidopsis*. Planta 226:839–851
Zhang H, Xie X, Kim MS, Kornyeyev DA, Holaday S, Pare PW (2008) Soil bacteria augment *Arabidopsis* photosynthesis by decreasing glucose sensing and abscisic acid levels in planta. Plant J 56:264–273
Zhang CL, Wang GP, Mao LJ, Komon-Zelazowska M, Yuan ZL, Lin FC, Druzhinina IS, Kubicek CP (2010) *Muscodor fengyangensis* sp. nov. from southeast China: morphology, physiology and production of volatile compounds. Fungal Biol 114:797–808
Zogorski JS, Carter JM, Ivahnenko T, Lapham WW, Moran MJ, Rowe BL, Squillace PJ, Toccalino PL (2006) The quality of our nation's waters – volatile organic compounds in the nation's ground water and drinking-water supply wells. US Geological Survey Circular 1292. US Geological Survey, Reston

Biosystematic Index

Fungal Associations, 2nd Edition
The Mycota IX
B. Hock (Ed.)

Subject Index

Fungal Associations, 2nd Edition
The Mycota IX
B. Hock (Ed.)

MIX
Papier aus verantwortungsvollen Quellen
Paper from responsible sources
FSC® C105338

If you have any concerns about our products,
you can contact us on
ProductSafety@springernature.com

In case Publisher is established outside the EU,
the EU authorized representative is:
Springer Nature Customer Service Center GmbH
Europaplatz 3, 69115 Heidelberg, Germany

Printed by Libri Plureos GmbH
in Hamburg, Germany